PYTHAGOREAN THEOREM

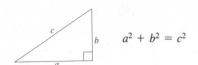

$$a^2 + b^2 = c^2$$

FORMULAS FROM SOLID GEOMETRY

h = height, V = volume, K = lateral surface area, S = total surface area.

Right Circular Cylinder

$$V = \pi r^2 h \qquad K = 2\pi rh$$
$$S = 2\pi(r^2 + rh)$$

Right Circular Cone

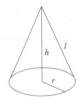

$$V = \frac{\pi}{3}r^2 h \qquad K = \pi rl$$
$$S = \pi rl + \pi r^2$$

Frustum of a Cone

$$V = \frac{\pi}{3}h(r^2 + rR + R^2)$$
$$K = \pi(r + R)l \qquad S = \pi(r + R)l + \pi(r^2 + R^2)$$

Sphere

$$V = \frac{4}{3}\pi r^3, \; S = 4\pi r^2$$

Parallelpiped

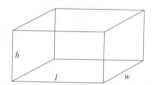

$$V = wlh, \; S = 2(hl + hw + lw)$$

LAW OF SINES

$$\frac{\sin \alpha}{a} = \frac{\sin \beta}{b} = \frac{\sin \gamma}{c}$$

LAW OF COSINES

$$a^2 = b^2 + c^2 - 2bc \cos \alpha$$
$$b^2 = a^2 + c^2 - 2ac \cos \beta$$
$$c^2 = a^2 + b^2 - 2ab \cos \gamma$$

AREA FORMULAS

$$A = \tfrac{1}{2}bc \sin \alpha = \tfrac{1}{2}ac \sin \beta = \tfrac{1}{2}ab \sin \gamma$$
$$A = \sqrt{s(s - a)(s - b)(s - c)}, \text{ where } s = \tfrac{1}{2}(a + b + c)$$
$$A = \frac{a^2 \sin \beta \sin \gamma}{2 \sin \alpha} = \frac{b^2 \sin \alpha \sin \gamma}{2 \sin \beta} = \frac{c^2 \sin \alpha \sin \beta}{2 \sin \gamma}$$

Prelude to Calculus

PRELUDE TO
CALCULUS

Warren L. Ruud
Santa Rosa Junior College

Terry L. Shell
Santa Rosa Junior College

Wadsworth Publishing Company
Belmont, California
A Division of Wadsworth, Inc.

Mathematics Editor: Barbara Holland
Editorial Assistant: Leslie With
Cover Design: Harry Voigt
Text and Managing Designer: James Chadwick
Production Editor: Harold Humphrey
Print Buyer: Karen Hunt
Copy Editor: Yvonne Howell
Technical Illustrators: Alexander Teshin Associates
Compositor: Syntax International
Signing Representative: Robin Levy

Printed in the United States of America   48

              4 5 6 7 8 9 10——94 93 92 91

Library of Congress Cataloging-in-Publication Data

Ruud, Warren L., 1953–
    Prelude to calculus / Warren L. Ruud, Terry L. Shell.
        p.   cm.
    Includes index.
    ISBN 0-534-10290-5
    1. Mathematics.   I. Shell, Terry L., 1953–.   II. Title.
QA39.2.R88   1990
510—dc20   89-9102
                                                CIP

Students
A *Solutions Manual* containing complete step-
by-step solutions to one-half of the odd-
numbered problems in this book is available
from your bookstore.

TO FANNY AND LEIF

TO POAD

CONTENTS

PREFACE

Prelude to Calculus provides a solid foundation for the successful study of calculus; it covers all of the standard topics of a one semester or one quarter precalculus course. At the same time, it takes an approach not found in existing precalculus books, presenting the material as a unified body of knowledge rather than a miscellaneous list of skills.

The text is designed to fit many different audiences, since most chapters beyond the core of the text are independent of each other. However, it is primarily intended for students who will be serious users of mathematics, taking at least a year of calculus as well as courses in physical science and engineering. The prerequisite is a good background in intermediate algebra.

Prelude to Calculus is intended to:

1) Prepare students for the entire calculus sequence, including complete discussion of the topics required for the second, third, and fourth semesters of calculus (vectors, parametric equations, coordinate geometry in space, etc.), as well as mathematics students will encounter in physics and engineering.

2) Discuss precalculus topics as they relate to calculus specifically. For example, the discussion of composition of functions in Section 2.6 emphasizes the decomposition of functions, a crucial concept for learning the chain rule in calculus. In the algebra review of Chapter 1, the student encounters manipulations that arise in calculus (such as simplification of difference quotients or the simplification of product and quotient rule expressions).

3) Offer a complete development of graphing, and to use graphing to motivate and explain. The value of drawing a picture to

understand and even extend an idea should never be underestimated, so the entire book takes a visual approach whenever appropriate.

Prelude to Calculus provides the extensive toolbox for graphing that students need for success in calculus. Some specific examples include: graphing concepts such as translation, expansions and compression, and reflection are introduced in Chapter 2 using familiar curves. A catalog of common graphs is also introduced in Chapter 2, and these concepts are a unifying thread throughout the text, as seen in the discussion of the graphs of exponential and log functions in Chapter 4 and the graphs of trigonometric functions in Chapters 5, 6, and 7. Graphing is used to motivate an intuitive approach to the limit of a function in Chapters 3, 4, and 5 and to the limit of a sequence in Chapter 11.

4) Develop the intuition, problem-solving ability, and mathematical maturity required for success in calculus. Students are encouraged to apply mathematics and to incorporate several mathematical concepts to solve a problem. For example, Section 2.2 emphasizes expressing a relation between quantities in a real-world situation as a function, in the same way that students are expected to when working max-min or related rates problems in calculus.

Section 1.6 discusses methods of problem-solving which students can use throughout the course. Also, an intuitive discussion of proof introduces students to this new concept.

Other Features

Exercises and Examples The exercise sets and examples in this text are its backbone. They reinforce, expand, and synthesize the concepts of each section and incorporate ideas from previous sections. They are designed to stimulate the students' curiosity and to prod them to think about the "big picture."

Each exercise set is graded into A, B, and C categories. The A exercises are for drill of the basic ideas and skills. The B exercises are more stimulating and require synthesis of what has been discussed. The C exercises are designed to challenge even the most capable problem-solvers. At the end of each chapter is a set of ungraded miscellaneous exercises.

Classroom Testing The text has been tested in whole and in part by us and by others over a period of three years.

Calculators Exercises and examples assume the availability of a calculator. In many instances, such as in the evaluation of logarithmic and exponential expressions in Chapter 4, the student is shown in detail how to use the calculator.

Historical Vignettes At the end of each chapter is a historical vignette conveying the personal qualities and remarkable genius of some of the giants of mathematics.

Text Organization

The text divides naturally (with some overlap) into four parts. Parts I and II form a core, while Parts III and IV allow the instructor to tailor the course to the requirements of the syllabus, the ability of the class, and his or her personal interest.

Part I Chapter 1 reviews algebra and basic analytic geometry in a unified way and at a relatively brisk pace. Emphasis is placed on those concepts specifically prerequisite to calculus. Chapter 2 offers extensive discussion of a function and its graph. A catalog of graphs and graphing techniques are developed in Chapters 2 and 3.

Part II Chapters 3, 4, and 5 apply the concepts developed in Part I to those algebraic and transcendental functions encountered in calculus. Chapter 3 addresses polynomials, rational functions, and the theory of equations. Chapter 4 deals with exponential and logarithmic functions, emphasizing logarithms as functions rather than as computational tools. Chapter 5 discusses trigonometric functions and their graphs.

Part III Chapters 5, 6, and 7 give a complete treatment of trigonometry for calculus. Chapter 5 introduces the trigonometric functions in terms of a right triangle. The instructor, however, may reorder the sections in Chapter 5 to initially define the trigonometric functions using the unit circle. Chapter 6 exercise sets emphasize the use of identities in situations encountered in calculus. Inverse trigonometric functions are covered fully in Section 6.4. For those students with prior experience with trigonometry, many sections of Part III may be skimmed or omitted entirely.

Part IV Chapters 8 through 12 cover four independent sets of topics important to the later parts of calculus as well as physical science and engineering courses. Chapters 8 and 12 cover plane and solid analytic geometry. Parametric representations and polar coordinates are treated fully in Chapter 9. Chapter 10 introduces concepts of basic linear algebra and their applications to calculus and science. Chapter 11 solidly introduces the ideas of sequences and series in calculus.

Supplementary Materials

Instructor's Manual The instructor's manual contains a full set of answers to the exercises and suggestions on organizing and teaching the course.

Student Solutions Manual A student solutions manual features detailed solutions to one-half of the odd-numbered exercises.

Test Booklet A printed and bound test bank provides sample tests and a battery of test questions for constructing exams.

EXP Test This computerized test generation system is available for IBM PCs and compatibles and allows fast and easy creation of tests, with all technical symbols and fonts appearing on the screen as they will appear when printed.

Acknowledgements

Many valuable contributions were made by those who reviewed the text as it moved through its various stages. In particular we thank Edith Ainsworth, University of Alabama; Pat Arpaia, St. John Fisher College; Karen Barker, Indiana University–South Bend; Bob Collings, North Harris Community College; Antonella Cupillari, Penn State Erie–the Behrend College; Louis Hoelzle, Bucks County Community College; Laurence Maher, North Texas State University; Gary Phillips, Oakton Community College; Alicia Sevilla, Moravian College; George Szoke, University of Akron; Lynn Tooley, Bellevue Community College; and Sandy Wager, University of North Carolina at Wilmington.

We also express our appreciation to our students who used *Prelude to Calculus* in its manuscript form. We would like to acknowledge five of them who meticulously checked every problem in the book: Nancy Leong, Erik Oehm, Jay Dawes, Bryce Emunson, and Rand Van Dyke. Our appreciation also goes to Ken Seydel of Skyline College for using the manuscript in class and offering many valuable suggestions for improvement, and to Karl Seydel for his careful checking of the manuscript. Finally, our special thanks go to Barbara Holland and the staff at Wadsworth Publishing Company for their professionalism and enthusiasm throughout this project.

Warren L. Ruud
Terry L. Shell
Santa Rosa, California

Prelude to Calculus

CHAPTER 1

TOPICS FROM ALGEBRA

REAL NUMBERS AND THE COORDINATE LINE

The set of **real numbers** (represented by the symbol ℝ) plays a fundamental role in calculus. Understanding the properties and structure of these numbers is paramount to the successful study of calculus. One way to visualize a set of numbers is with a **coordinate line** (Figure 1).

FIGURE 1

There is a one-to-one correspondence between the set of real numbers ℝ and the set of points on the coordinate line. What this means is that each real number is represented by a point on the coordinate line and that each point represents a real number. The point on the line that corresponds to zero is called the **origin**. A set of real numbers can be represented by a **graph** on the coordinate line (Figure 2).

FIGURE 2 *Graph of the integers*

We start with the set of **integers:**

$$\ldots -3, -2, -1, 0, 1, 2, 3, 4 \ldots.$$

The positive integers 1, 2, 3, 4 . . . are called the **natural numbers.** The non-negative integers 0, 1, 2, 3, 4 . . . are called the **whole numbers.**

A rational number is a number that can be written as a ratio p/q, where p and q are integers and $q \neq 0$ (the set of rational numbers is conventionally represented by the symbol ℚ). Examples of rational numbers are

$$\frac{1}{4}, \frac{12}{5}, -\frac{3}{5}, \frac{125}{239}, \quad \text{and} \quad -\frac{9}{2}.$$

Of course, any integer n is also a rational number, since $n = n/1$.

One of the interesting facts about rational numbers is that an infinite number of other rational numbers exist between any two of them (see Problem 63). A direct result of this fact is that no matter how many rational numbers we plot on a number line, infinitely many more can still be plotted. This would seem to imply that any point on the coordinate line must correspond to a rational number.

This is not, however, the case. Consider a right triangle with legs both of length 1 unit. The hypotenuse, by the Pythagorean theorem (see Section

1

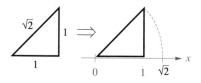

FIGURE 3 *Locating $\sqrt{2}$ on the number line*

1.6), is $\sqrt{2}$. We can locate the point on the coordinate line that corresponds to the number $\sqrt{2}$ by the scheme suggested by Figure 3. We can show that this number cannot be written as the ratio of two integers.

Suppose we assume that the number $\sqrt{2}$ is a rational number, that is,

$$\sqrt{2} = \frac{a}{b}$$

for some positive integers a and b. Let's also assume that a and b are not both even (if this were not true, the ratio could be reduced to make it true). Multiplying through by b, we obtain

$$b\sqrt{2} = a.$$

Squaring both sides yields

$$2b^2 = a^2.$$

This implies that a^2 is an even integer, which in turn implies that a itself is an even integer (the only integers whose squares are even are themselves even). Thus a is twice another integer, say, $a = 2c$. So,

$$2b^2 = a^2$$
$$2b^2 = (2c)^2$$
$$2b^2 = 4c^2.$$

Dividing each side by 2 gives

$$b^2 = 2c^2.$$

This last statement implies that b is also an even integer, which is in obvious contradiction to our original assumption that $\sqrt{2}$ is the ratio of two integers, neither of which is even. This forces us to the inescapable conclusion that our assumption is wrong; $\sqrt{2}$ cannot be expressed as the ratio of two integers. (This elegant proof that $\sqrt{2}$ is not rational is usually attributed to the Greek geometer Hippasus who lived about 470 B.C.).

A real number such as $\sqrt{2}$ that cannot be written as the ratio of two integers is called an **irrational number.** There are many irrational numbers; specifically, any real number $\sqrt{n}$ (where n is not the square of a rational number) or $\sqrt[3]{n}$ (where n is not the cube of a rational number) is an example of an irrational number. One of the more celebrated irrational numbers is π, the ratio of the circumference to the diameter of a circle. For hundreds of years, π was suspected to be irrational, but it was not proven so until 1760, by the Swiss mathematician Johann Lambert (his proof is too difficult to be presented here).

Figure 4 shows the relationship of the four sets of numbers.

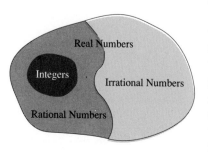

FIGURE 4

EXAMPLE 1   Determine which of these numbers,

$$-\tfrac{13}{5}, \sqrt{4}, \sqrt{13}, \pi/2,$$

are **a)** integers, **b)** rational numbers, **c)** irrational numbers, **d)** real numbers. Write them in increasing order.

SOLUTION Because $-\tfrac{13}{5}$ is the ratio of the integers -13 and 5, it is a rational number. Its decimal representation is -2.6.

Since $\sqrt{4} = 2$, this number is an integer (and therefore also a rational number).

On the other hand, $\sqrt{13}$ is irrational since 13 is not the perfect square of a rational number. It follows from $3^2 = 9$ and $4^2 = 16$ that $3 < \sqrt{13} < 4$ (you could also use your calculator to determine that $\sqrt{13} \approx 3.61$).

The symbol $\approx$ is read "approximately equal to."

Now consider the real number $\pi/2$. Suppose for a moment that this number is rational; that is, there exist integers a and b such that $\pi/2 = a/b$. Think about what this would imply about the irrational number π:

$$\pi = 2\left(\frac{\pi}{2}\right) = 2\left(\frac{a}{b}\right) = \frac{2a}{b} \Rightarrow \pi \text{ is the ratio of two integers, } 2a \text{ and } b.$$

We know that this cannot be the case since π is an irrational number. No such a and b exist; $\pi/2$ is an irrational number. Since $\pi \approx 3.14$, $\pi/2 \approx 1.57$.

In summary:

Integers	$\sqrt{4}$
Rational Numbers	$-\tfrac{13}{5}, \sqrt{4}$
Irrational Numbers	$\sqrt{13}, \pi/2$
Real Numbers	$-\tfrac{13}{5}, \sqrt{4}, \sqrt{13}, \pi/2$

We can use the decimal representations or approximations to order this set of the numbers:

$$-\tfrac{13}{5}, \pi/2, \sqrt{4}, \sqrt{13}$$

The decimal representation of a rational number either terminates (such as $\tfrac{7}{4} = 1.75$) or eventually repeats (such as $\tfrac{3}{22} = 0.1363636\ldots$). The decimal representations of irrational numbers neither terminate nor repeat.

The notation $a \le x \le b$ is called a **compound inequality.** It describes the set of real numbers x such that both

$$a \le x \quad \text{and} \quad x \le b.$$

For example, $-1 \le x \le 3$ describes the set shown in Figure 5.

FIGURE 5

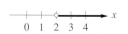

FIGURE 6

The filled-in circles at the endpoints are used to indicate that the numbers represented by these endpoints are part of the set. This is a **closed** interval of real numbers since the endpoints of the graph are part of the set.

The set of all real numbers that are greater than 2 can be described as $x > 2$. On the coordinate-line graph of Figure 6, the number represented by the endpoint of the graph, $x = 2$ (the open circle), is not part of the set. This is an **open** interval.

Sets of real numbers can also be described using **interval notation.** For example, the set $-1 \leq x \leq 3$ can be described as $[-1, 3]$. A bracket indicates that the corresponding endpoint is part of the set. The set $x > 2$ can be shown as $(2, \infty)$. A parenthesis shows that the corresponding endpoint is not part of the set. Also, the symbol ∞ (read as "infinity") is used to indicate that the interval is unbounded on the right. The classification and notation for other intervals are shown in the table below.

*Interval and Inequality
Notation*

Description and Coordinate Line Graph	Inequality Notation	Interval Notation
Open Interval ○———○ x a b	$a < x < b$	(a, b)
Closed Interval ●———● x a b	$a \leq x \leq b$	$[a, b]$
Half-Open Intervals ●———○ x a b	$a \leq x < b$	$[a, b)$
○———● x a b	$a < x \leq b$	$(a, b]$
Open Infinite Intervals ○———— x a	$x > a$	(a, ∞)
————○ x a	$x < a$	$(-\infty, a)$
Closed Infinite Intervals ●———— x a	$x \geq a$	$[a, \infty)$
————● x a	$x \leq a$	$(-\infty, a]$

EXAMPLE 2   Sketch the graph of the set described by the inequality notation. Describe the set using interval notation.

a) $-4 \leq x < 2$ **b)** $x \leq 3$ or $x \geq 5$

SOLUTION

FIGURE 7

a)   The half-open, bounded interval described by this compound inequality has endpoints -4 and 2 (Figure 7). In interval notation, this is $[-4, 2)$.

b) This is the union of two unbounded intervals (Figure 8). In interval notation, this is $(-\infty, 3]$ or $[5, \infty)$.

FIGURE 8

EXAMPLE 3 Sketch the graph of the set described by the interval notation. Describe the set using inequality notation.

a) $(-1, 3)$ **b)** $(-3, 2)$ or $[5, \infty)$

SOLUTION

a)  $-1 < x < 3$

b) $-3 < x < 2$ or $x \geq 5$

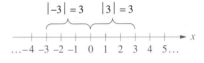

FIGURE 9

The **absolute value** of a real number a, denoted by $|a|$, is the distance on the coordinate line between the origin and the point corresponding to a (Figure 9). For example, $|3| = 3$ and $|-3| = 3$ since both are 3 units from the origin (Figure 10).

$|-3| = 3$ $|3| = 3$

...−4 −3 −2 −1   0   1   2   3   4   5...

FIGURE 10

Because the absolute value of a real number a represents a distance, $|a| \geq 0$ with $|a| = 0$ if and only if $a = 0$.

The **absolute value** or **magnitude** of a real number a is

$$|a| = \begin{cases} a & \text{if} \quad a \geq 0 \\ -a & \text{if} \quad a < 0 \end{cases}.$$

Absolute Value

In essence, the absolute value of a nonnegative real number is the same as the number, while the absolute value of a negative real number is the opposite of the number (notice that if $a < 0$, then $-a$ is a positive quantity).

EXAMPLE 4 Write the expression without absolute value symbols.

a) $\left| \sqrt{10} - 3 \right|$ **b)** $\left| \sqrt{10} - 4 \right|$

c) $\left| x^2 + 2 \right|$ **d)** $\left| \dfrac{t^3 - 2}{3t^4} \right|$ for $t < 0$

SOLUTION

a) The value of $\sqrt{10}$ is larger than 3. This implies that $\sqrt{10} - 3 > 0$. By the definition above then,

$$\left| \sqrt{10} - 3 \right| = \sqrt{10} - 3.$$

b) The value of $\sqrt{10}$ is smaller than 4. This implies that $\sqrt{10} - 4 < 0$. By the definition above then,

$$|\sqrt{10} - 4| = -(\sqrt{10} - 4) = 4 - \sqrt{10}.$$

c) Consider the terms of the sum $x^2 + 2$: The first term x^2 is nonnegative (why?), and the second term 2 is positive. This implies that the sum is positive. Hence,

$$|x^2 + 2| = x^2 + 2.$$

d) Because $t < 0$, the numerator $t^3 - 2$ is negative. The denominator $3t^4$ is positive. This implies that

$$\frac{t^3 - 2}{3t^4} < 0.$$

It follows that

$$\left| \frac{t^3 - 2}{3t^4} \right| = -\left(\frac{t^3 - 2}{3t^4} \right) = \frac{2 - t^3}{3t^4}.$$

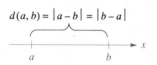

$$d(a, b) = |a - b| = |b - a|$$

FIGURE 11

For two real numbers a and b, the quantity $|a - b|$ has an important geometric meaning on the coordinate line (Figure 11). This is the distance, denoted $d(a, b)$, from a to b. Of course, $d(a, b) = d(b, a)$.

Distance between Two Real Numbers

The **distance** between two real numbers is

$$d(a, b) = |a - b| = |b - a|.$$

EXAMPLE 5 Sketch the graph of the set of real numbers described on the coordinate line.

a) $|x| < 5$

b) $|x| \geq 2$

c) $|x - 2| < 0.3$

d) $|x + 3| = 4$

FIGURE 12

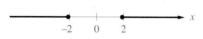

FIGURE 13

SOLUTION

a) From the definition of absolute value, this is the set of all points that are within 5 units of the origin. This interval can be described using interval notation as $(-5, 5)$ (Figure 12).

b) This is the set of all points that are 2 or more units from the origin. It can be described using interval notation as $(-\infty, -2]$ or $[2, \infty)$ (Figure 13).

c) This inequality describes the set of real numbers within 0.3 units of $x = 2$. This set can be described as (1.7, 2.3) (Figure 14).

FIGURE 14

d) This equation can be rewritten as $|x - (-3)| = 4$. It is the set of points that are exactly 4 units from -3. They are $-3 + 4 = 1$ and $-3 - 4 = -7$ (Figure 15).

FIGURE 15

One of the concerns in calculus is to describe a set of points that are either less than, equal to, or greater than a specified distance δ units from a real number a (the symbol δ is the Greek lowercase letter delta). The table below describes each of these situations in detail.

Description		Set of Points	Graph
$\lvert x - a \rvert < \delta$	All points within δ units of a	$(a - \delta, a + \delta)$	
$\lvert x - a \rvert = \delta$	All points exactly δ units from a	$x = a \pm \delta$	
$\lvert x - a \rvert > \delta$	All points farther than δ units from a	$(-\infty, a - \delta)$ or $(a + \delta, \infty)$	

Using Your Calculator

Scientific hand-held calculators can be used to approximate the values of irrational numbers. For example, suppose that we wish to approximate the value of

$$\frac{\sqrt{30} - \sqrt{6}}{2}$$

to the nearest 0.001:

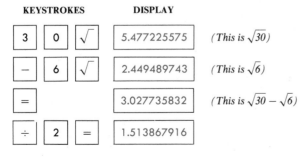

KEYSTROKES			**DISPLAY**	
3	0	√	5.477225575	*(This is $\sqrt{30}$)*
−	6	√	2.449489743	*(This is $\sqrt{6}$)*
=			3.027735832	*(This is $\sqrt{30} - \sqrt{6}$)*
÷	2	=	1.513867916	

To the nearest 0.001, $\dfrac{\sqrt{30} - \sqrt{6}}{2} \approx 1.514$.

EXERCISE SET 1.1

A

In Problems 1 through 6, determine which of the numbers in the given list are integers, which are rational numbers, and which are irrational numbers. Write the numbers in increasing order.

1. $\frac{5}{3}, -\frac{12}{4}, \sqrt{7}, \sqrt{19}$

2. $\frac{8}{4}, -\frac{11}{3}, \sqrt{9}, \sqrt{17}$

3. $\pi, 3.14, 0, \sqrt{\frac{4}{9}}$

4. $-\frac{5}{12}, -\frac{12}{4}, \sqrt{1}, \sqrt{36}$

5. $\pi/2, -\frac{3}{2}, \sqrt{4}, \sqrt[3]{-8}$

6. $-2/\pi, \frac{10}{4}, \sqrt{4}, \sqrt[3]{27}$

In Problems 7 through 18, a set of real numbers is described using inequality notation. Describe this set using interval notation.

7. $x > 4$

8. $x \leq 4$

9. $-2 < x < 4$

10. $-3 \leq x \leq 10$

11. $-4 < x \leq 2$

12. $0 \leq x < 5$

13. $x \leq -2$ or $x > 2$

14. $x < -4$ or $x > 0$

15. $x > -5$ and $x \leq 2$

16. $x \geq -1$ and $x \leq 6$

17. $-3 < x \leq 1$ or $x > 4$

18. $-4 \leq x < 2$ or $x > 2$

In Problems 19 through 30, a set of real numbers is described using interval notation. Describe this set using inequality notation.

19. $(-\infty, -2)$

20. $[-2, \infty)$

21. $[4, 9]$

22. $(-1, 6)$

23. $[-4, 0)$

24. $(-5, 1]$

25. $(-\infty, -2)$ or $(2, \infty)$

26. $(-\infty, 0]$ or $[3, \infty)$

27. $[-1, \infty)$ and $(-\infty, 6]$

28. $[7, \infty)$ and $(-\infty, 2]$

29. $(-5, -2)$ or $(0, \infty)$

30. $(-\infty, 1]$ or $[3, 7)$

B

In Problems 31 through 42, write the expression without absolute value symbols.

31. $|4 - \sqrt{10}|$

32. $|3 - \sqrt{10}|$

33. $|\sqrt{13} - 6|$

34. $|\sqrt{13} + 6|$

35. $|\sqrt{10} - \pi|$

36. $|2\pi - 6|$

37. $|x^2|$

38. $|2 + t^4|$

39. $|5 - n^3|$, for $n > 2$

40. $|2 + n^3|$, for $n > 0$

41. $\dfrac{x}{|x|}$, for $x > 0$

42. $\dfrac{x}{|x|}$, for $x < 0$

In Problems 43 through 54, sketch the graph of the set of real numbers described on the coordinate line.

43. $|x| > 3$

44. $|x| < \frac{3}{2}$

45. $|x| = 6$

46. $|x - 7| < 4$

47. $|x - 2| = 5$

48. $|x + 3| > 1$

49. $|x - 3| < 0.5$

50. $|x - \sqrt{2}| < \frac{1}{6}$

51. $|x - \pi| \geq \pi$

52. $|x + \sqrt{2}| \leq \sqrt{2}$

53. $|x + 1| < 0.2$

54. $|x + 4| < 0.1$

In Problems 55 through 60, use your calculator to approximate the real numbers given to the nearest 0.0001. Write the numbers in increasing order.

55. $2 + \sqrt{7}, 2 - \sqrt{7}, \sqrt{5} + \sqrt{3}$

56. $\sqrt{2} - 10, \sqrt{7} + 5, \pi + \frac{12}{7}$

57. $\frac{1}{2}(\sqrt{2} + \sqrt{6}), \sqrt{2 + \sqrt{3}}$

58. $-\frac{\pi}{3}, 0.84 \left(\dfrac{17 + 15\sqrt{5}}{7 + 15\sqrt{5}} \right)$

59. $\dfrac{\sqrt{5} + \sqrt{3}}{\sqrt{5} - \sqrt{3}}, 4 + \sqrt{15}, \dfrac{1}{4 - \sqrt{15}}$

60. $\dfrac{(\sqrt{5} + 1)}{2}, \dfrac{2}{(\sqrt{5} + 1)}, \left[\dfrac{(\sqrt{5} + 1)}{2} \right]^2$

61. Determine an inequality of the form $|x - a| < \delta$ that is equivalent to the compound inequality $-3 < x < 7$.

62. Determine an inequality of the form $|x - a| < \delta$ that is equivalent to the compound inequality $-2.1 < x < 2.5$.

63. **a)** Give five rational numbers between $\frac{9}{4}$ and $\frac{17}{7}$. (Hint: First, show that the interval $[2.3, 2.4]$ is contained in the interval $(\frac{9}{4}, \frac{17}{7})$; next, find five rational numbers in this closed interval.)

b) Give ten rational numbers between $\frac{9}{4}$ and $\frac{17}{7}$.

64. a) Give ten rational numbers between $\sqrt{2}$ and $\sqrt{3}$. (Hint: First, show that the interval $[1.5, 1.6]$ is contained in the interval $(\sqrt{2}, \sqrt{3})$; next, find five rational numbers in this closed interval.)

b) Describe how you would determine 100 rational numbers between $\sqrt{2}$ and $\sqrt{3}$.

C

65. Show that $\sqrt{3}$ is irrational. (Hint: Assume that $\sqrt{3} = a/b$, and show that this leads to a contradiction.)

66. Show that $\sqrt[3]{2}$ is irrational. (Hint: Assume that $\sqrt[3]{2} = a/b$, and show that this leads to a contradiction.)

67. Give five irrational numbers between $\frac{9}{4}$ and $\frac{17}{7}$.

ALGEBRA FOR CALCULUS: ALGEBRAIC EXPRESSIONS

S E C T I O N   1.2

Manipulating algebraic expressions is an essential skill for the study of mathematics and science. **Algebraic expressions** can be described as mathematically meaningful combinations of constants and variables involving the operations of addition, subtraction, multiplication, and division, as well as raising numbers to powers and taking their roots. The purpose of this section is to review these manipulations in the context in which they arise in the study of calculus.

You are, no doubt, familiar with the concept of integer exponents. We state them here for the purpose of review.

Exponents

Given that b is a real number and that n is an integer, then

for $n > 0$, $\qquad b^n = \underbrace{b \cdot b \cdot b \cdot b \cdot \;\cdots\; \cdot b \cdot b}_{n \text{ factors of } b}$;

for $n = 0$, $\qquad b^n = 1 \quad (b \neq 0)$;

for $n < 0$, $\qquad b^n = \dfrac{1}{b^{-n}} = \dfrac{1}{\underbrace{b \cdot b \cdot b \cdot b \cdot \;\cdots\; \cdot b \cdot b}_{-n \text{ factors of } b}}, \quad (b \neq 0)$.

You should also be familiar with the properties of exponents. These properties follow directly from the definitions of integer exponents.

Properties of Exponents

> Given that a and b are real numbers and that n and m are integers, then
>
> $$a^m \cdot a^n = a^{m+n}, \qquad \frac{a^m}{a^n} = a^{m-n} \quad (a \neq 0),$$
>
> $$(ab)^n = a^n b^n, \qquad \left(\frac{a}{b}\right)^n = \frac{a^n}{b^n} \quad (b \neq 0),$$
>
> $$(a^m)^n = a^{mn}.$$

EXAMPLE 1   Use the properties of exponents to simplify the expression.

a) $(2x^2 y^{-1})^{-2}$ **b)** $\dfrac{a^{-2} b^3}{2a^{-3} b^0}$ **c)** $\dfrac{x^{n+2}}{x^3} \cdot x^{2n}$

SOLUTION

a) $(2x^2 y^{-1})^{-2} = 2^{-2}(x^2)^{-2}(y^{-1})^{-2} = \dfrac{1}{4} x^{-4} y^2 = \dfrac{y^2}{4x^4}.$

b) $\dfrac{a^{-2} b^3}{2a^{-3} b^0} = \dfrac{\dfrac{1}{a^2} \cdot b^3}{2 \cdot \dfrac{1}{a^3} \cdot 1} = \dfrac{\left(\dfrac{b^3}{a^2}\right)}{\left(\dfrac{2}{a^3}\right)} = \dfrac{b^3}{a^2} \div \dfrac{2}{a^3} = \dfrac{b^3}{a^2} \cdot \dfrac{a^3}{2} = \dfrac{ab^3}{2}.$

c) $\dfrac{x^{n+2}}{x^3} \cdot x^{2n} = x^{(n+2)-3} \cdot x^{2n} = x^{(n+2)-3+2n} = x^{3n-1}.$

The discussion of exponents leads to a question. Given a real number a and a positive integer n, does there exist a number (or numbers) x such that $x^n = a$? Such a number is called an **nth root** of the number a. For example, the 4th roots of the real number 16 are 2 and -2 since $2^4 = 16$ and $(-2)^4 = 16$, and the 3rd root of the real number 8 is 2 since $2^3 = 8$. In general, for any real positive number a and even positive integer n, it should be apparent that a has two nth roots, one positive and one negative. We call the positive root the **principal** nth root of a, $\sqrt[n]{a}$. If n is odd, then there is only one nth root of a real number a. In this case, this root is also called the principal nth root of a. These cases are summarized next.

Suppose that a is a real number.   *Definition of $\sqrt[n]{a}$*

1) If n is an even, positive integer and $a > 0$, then $\sqrt[n]{a}$ is the positive nth root of a.

2) If n is an even, positive integer and $a < 0$, then $\sqrt[n]{a}$ is not a real number.

3) If n is an odd, positive integer, then $\sqrt[n]{a}$ is the nth root of a.

The expression $\sqrt[n]{a}$ is called a **radical.** The real number a is the **radicand,** and the positive integer n is the **index** of the radical (the index 2 of a square root is conventionally omitted).

EXAMPLE 2   Evaluate the given root.

a) $\sqrt[3]{64}$ **b)** $\sqrt[3]{-64}$ **c)** $\sqrt[6]{64}$ **d)** $-\sqrt[6]{64}$ **e)** $\sqrt[6]{-64}$

SOLUTION

a) We seek a real number x such that $x^3 = 64$. Since $4^3 = 64$, it follows that $\sqrt[3]{64} = 4$.

b) $(-4)^3 = -64 \Rightarrow \sqrt[3]{-64} = -4$.

c) Since $(-2)^6 = 64$ and $2^6 = 64$, the two sixth roots of 64 are -2 and 2. It follows that $\sqrt[6]{64} = 2$, since 2 is positive.

d) From part c), $-\sqrt[6]{64} = -2$.

e) We seek a real number x such that $x^6 = -64$. However, since any real number raised to an even power is always nonnegative, no such real number x exists. Thus, $\sqrt[6]{-64}$ is not a real number.

As with exponents, there are properties that permit the manipulation and simplification of algebraic expressions involving radicals.

Given that a and b are nonnegative real numbers and that n and   *Properties of Radicals*
m are positive integers, then

1) $\sqrt[mn]{a^m} = \sqrt[n]{a};$ 2) $\sqrt[n]{ab} = \sqrt[n]{a}\,\sqrt[n]{b};$

3) $\sqrt[n]{\dfrac{a}{b}} = \dfrac{\sqrt[n]{a}}{\sqrt[n]{b}}, \quad b \neq 0;$ 4) $\sqrt[m]{\sqrt[n]{a}} = \sqrt[mn]{a}.$

5) If n is an odd integer, then $\sqrt[n]{-a} = -\sqrt[n]{a}.$

We will offer a proof for the first property. The others are left to you (see Exercises 92 and 93).

Suppose that we let $x = \sqrt[n]{a}$. If we can show that $x = \sqrt[mn]{a^m}$ also, then our proof is complete. Now,

$$x = \sqrt[n]{a} \Rightarrow x^n = a \Rightarrow (x^n)^m = (a)^m \Rightarrow x^{mn} = a^m \Rightarrow x = \sqrt[mn]{a^m}.$$

This completes the proof.

EXAMPLE 3   Use the properties of radicals to simplify the expressions (assume that all variables are positive).

a) $\sqrt[3]{8a^6\sqrt{c}}$

b) $\sqrt[4]{\dfrac{16x^6y^4}{z^8}}$

SOLUTION

a) $\sqrt[3]{8a^6\sqrt{c}} = \sqrt[3]{8}\sqrt[3]{a^6}\sqrt[3]{\sqrt{c}} = \sqrt[3]{2^3}\sqrt[3]{(a^2)^3}\sqrt[6]{c}$

$\quad\quad = 2a^2\sqrt[6]{c}.$

b) $\sqrt[4]{\dfrac{16x^6y^4}{z^8}} = \dfrac{\sqrt[4]{16x^6y^4}}{\sqrt[4]{z^8}} = \dfrac{\sqrt[4]{16}\sqrt[4]{x^6}\sqrt[4]{y^4}}{\sqrt[4]{(z^2)^4}} = \dfrac{\sqrt[4]{2^4}\sqrt[4]{x^4}\sqrt[4]{x^2}\sqrt[4]{y^4}}{\sqrt[4]{(z^2)^4}}$

$\quad\quad = \dfrac{2xy\sqrt{x}}{z^2}.$

Rational Exponents

The concept of an nth root of a real number gives meaning to the use of a rational number as an exponent. For example, suppose that we try to attach some significance to the expression $7^{1/5}$. Assuming that the usual properties of exponents apply, notice that

$$(7^{1/5})^5 = 7^{(1/5)(5)} = 7^1 = 7.$$

That is, to preserve the properties of exponents, $7^{1/5}$ must be the 5th root of 7, $\sqrt[5]{7}$.

In a similar fashion, $7^{2/5}$ can be rewritten as

$$7^{2/5} = (7^{1/5})^2 = (\sqrt[5]{7})^2 \quad \text{or} \quad 7^{2/5} = (7^2)^{1/5} = \sqrt[5]{7^2}.$$

These definitions are generalized in the following box.

Definition of $a^{1/n}$ and $a^{m/n}$

Given that n is a positive integer and a is a real number, then

$$a^{1/n} = \sqrt[n]{a}.$$

Also, if m and n are integers ($n > 0$) and a is a real number, then

$$a^{m/n} = \sqrt[n]{a^m} = (\sqrt[n]{a})^m.$$

EXAMPLE 4   Evaluate the expression, if possible.

    a) $16^{1/4}$ **b)** $(-16)^{1/4}$ **c)** $27^{2/3}$

    d) $(-27)^{2/3}$ **e)** $27^{-2/3}$

SOLUTION

    a) $16^{1/4} = \sqrt[4]{16} = 2.$

    b) $(-16)^{1/4} = \sqrt[4]{-16}$ is not a real number.

    c) $27^{2/3} = (\sqrt[3]{27})^2 = (3)^2 = 9.$

    d) $(-27)^{2/3} = (\sqrt[3]{-27})^2 = (-3)^2 = 9.$

    e) $27^{-2/3} = \dfrac{1}{27^{2/3}} = \dfrac{1}{9}.$

Quite often, a radical expression can be simplified by rewriting it in terms of rational exponents.

EXAMPLE 5   Rewrite the expression using rational exponents and simplify it using the properties of exponents.

    a) $\dfrac{\sqrt{x\sqrt[3]{y}}}{\sqrt[6]{xy}}$ **b)** $\sqrt[6]{\dfrac{8x^{2/5}}{y^{1/4}}}$

SOLUTION

    a) $\dfrac{\sqrt{x\sqrt[3]{y}}}{\sqrt[6]{xy}} = \dfrac{(xy^{1/3})^{1/2}}{(xy)^{1/6}} = \dfrac{x^{1/2}y^{1/6}}{x^{1/6}y^{1/6}} = \dfrac{x^{1/2}}{x^{1/6}} = x^{1/2-1/6} = x^{1/3}.$

    b) $\sqrt[6]{\dfrac{8x^{2/5}}{y^{1/4}}} = \left(\dfrac{8x^{2/5}}{y^{1/4}}\right)^{1/6} = \dfrac{(8x^{2/5})^{1/6}}{(y^{1/4})^{1/6}} = \dfrac{8^{1/6}(x^{2/5})^{1/6}}{(y^{1/4})^{1/6}} = \dfrac{2^{1/2}x^{1/15}}{y^{1/24}}.$

A **monomial** in the variable x is an algebraic expression of the form ax^n,   Polynomials
where a is a real number and n is a nonnegative integer. A monomial, or a
sum of monomials, is called a polynomial.

> A **polynomial** in the variable x is an algebraic expression that can
> be written in the form
> $$a_nx^n + a_{n-1}x^{n-1} + a_{n-2}x^{n-2} + \cdots + a_1x^1 + a_0,$$
> where n is a nonnegative integer and $a_n, \ldots, a_1, a_0$ are real numbers
> $(a_n \neq 0)$.

In the term $a_k x^k$, the real number a_k is called the **coefficient** of the term. For the sake of our discussion, a polynomial is **simplified** if it is expressed in the form

$$a_n x^n + a_{n-1} x^{n-1} + a_{n-2} x^{n-2} + \cdots + a_1 x^1 + a_0$$

where any term with coefficient of zero is deleted. Notice that the terms are arranged in descending powers of x. If the term $a_n x^n$ is the term with the highest power of x, then the real number a_n is called the **leading coefficient** of the polynomial. Also, if the term $a_n x^n$ is the term with highest power of x, then the polynomial is said to have **degree n.** A polynomial in which all coefficients are zero is called the **zero polynomial;** this polynomial is not assigned a degree. A polynomial with exactly two terms is a **binomial;** one with exactly three terms is a **trinomial.**

According to these definitions, a nonzero real number c can be considered a polynomial of degree zero since for $x \neq 0$, $c = cx^0$.

In certain contexts, we will want to consider polynomials in more than one variable. For example, a polynomial with terms of the form $ax^n y^m$ is a polynomial in two variables.

Depending on the situation, a polynomial in the form of a product may need to be simplified, or a polynomial in the form of a sum may need to be factored (that is, written as a product). These transformations from products to sums or from sums to products are accomplished by using product-factoring formulas (these should be mostly familiar to you). They are listed here for convenience.

Product-Factoring Formulas

$$a(x + y) = ax + ay$$
$$(ax + by)(cx + dy) = acx^2 + (ad + bc)xy + bdy^2$$
$$(x + y)(x - y) = x^2 - y^2$$
$$(x + y)^2 = x^2 + 2xy + y^2$$
$$(x - y)^2 = x^2 - 2xy + y^2$$
$$(x + y)^3 = x^3 + 3x^2 y + 3xy^2 + y^3$$
$$(x - y)^3 = x^3 - 3x^2 y + 3xy^2 - y^3$$
$$(x + y)(x^2 - xy + y^2) = x^3 + y^3$$
$$(x - y)(x^2 + xy + y^2) = x^3 - y^3$$

Each of these formulas can be verified directly by multiplying the factors on the left side.

EXAMPLE 6   Simplify the following polynomials.

a) $2(x^2 + 2x - 7) + (2x^3 - x) - (x^2 - 5x + 2)$

b) $(x^2 - 2x)(4x^2 + 3x - 1)$

c) $(x + 3y)^2 - (x - 2y)(x + 3y)$

SOLUTION

a) $2(x^2 + 2x - 7) + (2x^3 - x) - (x^2 - 5x + 2)$

$$= 2x^2 + 4x - 14 + 2x^3 - x - x^2 + 5x - 2$$

$$= 2x^3 + x^2 + 8x - 16.$$

b) $(x^2 - 2x)(4x^2 + 3x - 1)$

$$= (x^2 - 2x)4x^2 + (x^2 - 2x)3x - (x^2 - 2x)1$$

$$= 4x^4 - 8x^3 + 3x^3 - 6x^2 - x^2 + 2x$$

$$= 4x^4 - 5x^3 - 7x^2 + 2x.$$

c) $(x + 3y)^2 - (x - 2y)(x + 3y)$

$$= (x^2 + 6xy + 9y^2) - (x^2 + xy - 6y^2)$$

$$= x^2 + 6xy + 9y^2 - x^2 - xy + 6y^2$$

$$= 5xy + 15y^2.$$

There is another way to simplify the expression in part c). The expression can be factored as

$$(x + 3y)^2 - (x - 2y)(x + 3y) = (x + 3y)[(x + 3y) - (x - 2y)]$$

$$= (x + 3y)[x + 3y - x + 2y]$$

$$= (x + 3y)[5y]$$

$$= 5xy + 15y^2.$$

Notice that regardless of the method of solution the answer is the same.

If a polynomial with integer coefficients can be written as a product of polynomials of positive degree and with integer coefficients, it is said to be **factorable** over the set of integers. Otherwise, it is said to be **irreducible** over the set of integers. If a polynomial is written as the product of factors that cannot be factored further, then the polynomial is said to be completely factored.

EXAMPLE 7   Factor completely over the integers.

    a) $x^4 - 16$ **b)** $2y^3 + 128z^3$ **c)** $u^2 - 2$

SOLUTION

    a) Since this expression can be written as a difference of squares, $x^4 - 16 = (x^2)^2 - (4)^2$, it factors as

$$x^4 - 16 = (x^2 - 4)(x^2 + 4).$$

    The factor $(x^2 - 4)$ is itself a difference of squares, so

$$x^4 - 16 = (x - 2)(x + 2)(x^2 + 4).$$

    Each of these factors is irreducible over the set of integers.

    b) First, note that

$$2y^3 + 128z^3 = 2[y^3 + (4z)^3].$$

    So, using the sum of cubes formula, we get

$$2y^3 + 128z^3 = 2(y + 4z)(y^2 - 4yz + 16z^2).$$

    Pause here and convince yourself that this is the complete factorization.

    c) The polynomial $u^2 - 2$ is not factorable over the set of integers.

If we were to relax the restriction of factoring over the integers in part c) of Example 7, we could factor $u^2 - 2$ as

$$u^2 - 2 = (u - \sqrt{2})(u + \sqrt{2}).$$

(You should verify this directly by simplifying the expression on the right.) We say that this expression is factorable over the real numbers.

EXAMPLE 8   Factor completely over the integers.

    a) $x^3 - 4x^2 - x + 4$ **b)** $8x^3 - 36x^2 + 54x - 27$

SOLUTION

a) At first glance, this polynomial does not appear to yield to any of the factoring formulas. However, if we group the terms of the polynomial into pairs as

$$x^3 - 4x^2 - x + 4 = (x^3 - 4x^2) - (x - 4),$$

then

$$x^3 - 4x^2 - x + 4 = x^2(x - 4) - (x - 4)$$
$$= (x - 4)(x^2 - 1)$$
$$= (x - 4)(x - 1)(x + 1).$$

b) Proceeding as we did in part a), we obtain

$$8x^3 - 36x^2 + 54x - 27 = 4x^2(2x - 9) + 27(2x - 1).$$

This seems to lead us to a dead end since there is no common factor in the two terms. Let's rearrange the terms and start over. Suppose that we rewrite the polynomial as $8x^3 - 27 - 36x^2 + 54x$. Then

$$8x^3 - 36x^2 + 54x - 27 = 8x^3 - 27 - 36x^2 + 54x$$
$$= (8x^3 - 27) - (36x^2 - 54x)$$
$$= (2x - 3)(4x^2 + 6x + 9) - 18x(2x - 3)$$
$$= (2x - 3)[(4x^2 + 6x + 9) - 18x]$$
$$= (2x - 3)[4x^2 - 12x + 9]$$
$$= (2x - 3)[(2x - 3)^2]$$
$$= (2x - 3)^3$$

EXERCISE SET 1.2

A

1. Express using nonnegative exponents.

  a) $x^{-2/3}$ **b)** $2y^{-1}$ **c)** $(2y)^{-1}$ **d)** x^5y^{-2}

2. Express using nonnegative exponents.

  a) $a^{-4/3}$ **b)** $7x^{-4}$ **c)** $(7x)^{-4}$ **d)** $a^{-3}b^3$

3. Express the radical expression in the form x^r.

  a) $\sqrt[5]{x^2}$ **b)** $\sqrt{x^{-3}}$ **c)** $\dfrac{1}{\sqrt[7]{x^4}}$ **d)** $\dfrac{1}{\sqrt{x^{-5}}}$

4. Express the radical expression in the form x^r.

  a) $\sqrt[4]{x^7}$ **b)** $\sqrt[3]{x^{-2}}$ **c)** $\dfrac{1}{\sqrt{x^5}}$ **d)** $\dfrac{1}{\sqrt[8]{x^{-9}}}$

5. Express in the form $\sqrt[q]{x^p}$ (where p and q are integers).

    a) $x^{2/5}$ **b)** $x^{-3/7}$ **c)** $\dfrac{1}{x^{3/2}}$ **d)** $\dfrac{1}{x^{-2/9}}$

6. Express in the form $\sqrt[q]{x^p}$ (where p and q are integers).

    a) $x^{4/3}$ **b)** $x^{-3/5}$ **c)** $\dfrac{1}{x^{4/9}}$ **d)** $\dfrac{1}{x^{-7/3}}$

In Problems 7 through 18, use the properties of exponents to simplify the expression. Write your answers using positive exponents only. Assume that all variables are positive.

7. $x^{-3}(2x^7)$ **8.** $3x^{-2}(x^3)^4$

9. $(-5x^3)^2(4x)^{-1}$ **10.** $(4x^{-3}y^2)^2$

11. $(6x^4y^{-6})^0(3x^6y^4)$ **12.** $(6z^4)^{-2}(-2z)^0$

13. $(25x^4)^{-1/2}$ **14.** $(8x^6)^{-1/3}$

15. $\left(\dfrac{x^{12}}{y^8}\right)^{3/4}$ **16.** $\left(\dfrac{27z^3}{8x^3y^9}\right)^{-2/3}$

17. $(x^4y^2z^{-6})^{5/2}$

18. $(xy^2)^{1/3}(x^5y^{-2}z^{-6})^{-2/3}$

In Problems 19 through 24, use the properties of radicals to simplify the expression. Your answer should have at most one radical. Assume that all variables are positive.

19. $\sqrt[3]{x^2y} \cdot \sqrt[3]{xy^2} \cdot \sqrt[3]{x^2y^2}$ **20.** $\sqrt[3]{16\sqrt{16}}$

21. $\sqrt[3]{8x^{12}}$ **22.** $\sqrt[3]{t\sqrt{t^7}}$

23. $\sqrt[3]{\sqrt{x^{1/2}}}$ **24.** $\sqrt[3]{\dfrac{64x^7y^2}{(x^2)^2y^5}}$

In Problems 25 through 44, simplify the polynomial expression.

25. $(2x^2 + 4x - 7) - 2x(x - 8)$

26. $(3x^3 + 1) - 2x(x^2 - 5x + 2)$

27. $x^2y(2xy - 5x + y^2) + xy^2(x^2 - 2xy + 2)$

28. $xy(2x^2 + xy + 5y^2) - 2y^2(x^2 - xy + 3y^2)$

29. $(2x - 1)(3x + 1)$ **30.** $(4x + 9)(x + 3)$

31. $(2x + y)^3$ **32.** $(x - 3)^3$

33. $(x^2 - x)(x + 1)$

34. $(x^2 - 2x + 4)(x^2 + 2x + 4)$

35. $(x + 5)(x^2 - 5x + 25)$

36. $(x - 2z)(x^2 + 2xz + 4z^2)$

37. $(x - 2)(x + 4) + (x + 4)^2$

38. $(x - 3y)^2 - (x + 4y)^2$ **39.** $3(2x^2 + 1)^2(4x)$

40. $3(1 - 3x^2)^2(-6x)$ **41.** $(x - 2)(x^2 + 4)(x + 2)$

42. $(2x - 1)(4x^2 - 1)(2x + 1)$

43. $x^5[2(4x - 1)(4)] + 5x^4(4x - 1)^2$

44. $[2(x^2 - 3)2x](3x + 1)^2 + (x^2 - 3)^2 2(3x + 1)(3)$

In Problems 45 through 56, factor the polynomial over the integers.

45. $3x^2 - 2x - 1$ **46.** $8x^2 - 10x - 12$

47. $x^3 - 3x^2 - 10x$ **48.** $x^4 - 5x^3 + 6x^2$

49. $25x^2 - 36$ **50.** $8x^2 - 50$

51. $2x^3 - 16$ **52.** $8x^3 + y^3$

53. $x^4 - 27x$ **54.** $x^2 - 64x^4$

55. $x^3 - x^2 + 4x - 4$ **56.** $9x^3 - x^2 + 18x - 2$

B

In Problems 57 through 74, simplify the expressions (assume that all variables are positive).

57. $(\sqrt{x} + 2)^2$

58. $(\sqrt{x} + \sqrt{6})(\sqrt{x} - \sqrt{6})$

59. $(\sqrt{x + y} + \sqrt{x})(\sqrt{x + y} - \sqrt{x})$

60. $(x\sqrt{x} + \sqrt{x})^2$

61. $x^{1/2}(x^{3/2} + 2x^{1/2} - x^{-1/2})$

62. $3x^{1/3}(x^{1/6} - 2x^{2/3}) - x^{1/2}$

63. $(x^{1/2} + x^{3/2})^2$

64. $(x^{-2} - x^2)^2$

65. $(x^2 + y^2)(x^4 - x^2y^2 + y^4)$

66. $(x^{1/3} - 2)(x^{2/3} + 2x^{1/3} + 4)$

67. $(x + 2)^3 - (x - 2)^3$

68. $(1 + 2x)^3 + (1 - 2x)^3$

69. $2x^{4/3} - x^{2/3}(x^{2/3} - 3)$

70. $x^{1/2}(x^2 - 3x^{3/2}) + x^{5/2}$

71. $(x^{4/3} - x^{1/3})^2$

72. $(x^{3/2} + x^{1/2})^3$

73. $(x - 1)(x^6 + x^5 + x^4 + x^3 + x^2 + x + 1)$

74. $(x + 1)(x^5 - x^4 + x^3 - x^2 + x - 1)$

In Problems 75 through 86, factor the given polynomial over the integers.

75. $x^6 - 18x^4y^2 + 81x^2y^4$ **76.** $x^7 - 8x^4y^3$

77. $2x^3 + x^2 + 8x + 4$

78. $x^3 + 3x^2 - x - 3$

79. $x^4 - 2x^2 - 3$

80. $x^6 - 9x^3 + 8$

81. $x^4 - 16y^8$

82. $2x^3 + 54$

83. $x^3 - 12x^2 + 48x - 64$

84. $8x^3 + 12x^2 + 6x + 1$

85. $(x + 2)^2 + 3(x + 2) - 4$

86. $(2x + 3)^3 - 2(2x + 3)^2 - 3(2x + 3)$

87. Write $2x^{8/3} - 5x^{5/3} + 2x^{2/3}$ as a product of a second-degree polynomial and a factor of the form x^r (r is a rational number).

88. Write $x^{13/5} + 4x^{3/5} - 3x^{-2/5}$ as a product of a third-degree polynomial and a factor of the form x^r (r is a rational number).

89. Write as a sum in which each term is of the form ax^r (r is a rational number):

$$\frac{x^{5/2} - 2x^{8/3} + 4x - 2}{x^2}.$$

90. Write as a sum in which each term is of the form ax^r (r is a rational number):

$$\frac{x^3 - 2x^2 + 5x - x^{-1}}{x^{1/3}}.$$

91. Factor over the real numbers.

   a) $x^2 - 6$ **b)** $2x^2 - 9$ **c)** $3x^2 - 6$ **d)** $x^2 + 7$

C

92. **a)** Prove that $\sqrt[n]{ab} = \sqrt[n]{a}\sqrt[n]{b}$.

   b) Prove that $\sqrt[n]{\dfrac{a}{b}} = \dfrac{\sqrt[n]{a}}{\sqrt[n]{b}}$, $b \neq 0$.

93. **a)** Prove that $\sqrt[m]{\sqrt[n]{a}} = \sqrt[mn]{a}$.

   b) Prove that if n is an odd integer, then $\sqrt[n]{-a} = -\sqrt[n]{a}$.

94. **a)** Show that $x^6 - y^6$ can be factored as $(x^2 - y^2)(x^4 + x^2y^2 + y^4)$.

   b) Show that $x^6 - y^6$ can be factored as $(x^3 - y^3)(x^3 + y^3)$. Complete the factorization of $x^6 - y^6$ by factoring each of these factors.

   c) Use part a) and part b) to determine the factorization of $x^4 + x^2y^2 + y^4$.

95. **a)** Show that $x^4 + x^2y^2 + y^4$ can be expressed as

$$x^4 + x^2y^2 + y^4 = (x^4 + 2x^2y^2 + y^4) - x^2y^2.$$

   Write this expression as a difference of squares.

   b) Use your result in part a) to factor $x^4 + x^2y^2 + y^4$ over the integers.

96. Use the method of Problem 95 to factor $4x^4 + 3x^2y^2 + y^4$ over the integers.

97. Use the method of Problem 95 to factor $x^4 + 2x^2y^2 + 4y^4$ over the real numbers.

ALGEBRA FOR CALCULUS: ALGEBRAIC FRACTIONS

S E C T I O N   1.3

A quotient of algebraic expressions is called an **algebraic fraction.** In particular, we are often concerned with those algebraic fractions in which both the numerator and denominator are polynomials.

A rational expression is an algebraic expression that can be written as the quotient of two polynomials.

Rational Expression

The relation between the set of rational expressions and the set of polynomials is very much the same as the relation between the set of rational numbers and the set of integers. Rational expressions can be reduced, added, subtracted, multiplied, and divided in much the same way as rational numbers.

In reducing a rational expression, it is absolutely essential that the numerator and denominator are written in factored form.

EXAMPLE 1   Simplify these rational expressions by reducing them to lowest terms.

a) $\dfrac{2x^2 - x - 10}{2x^2 + 7x + 6}$

b) $\dfrac{4x - x^2}{3x^2 - 7x - 20}$

SOLUTION The plan of attack is to factor both the numerator and denominator and to eliminate common factors.

a) $\dfrac{2x^2 - x - 10}{2x^2 + 7x + 6} = \dfrac{(x + 2)(2x - 5)}{(x + 2)(2x + 3)} = \dfrac{2x - 5}{2x + 3}.$

b) $\dfrac{4x - x^2}{3x^2 - 7x - 20} = \dfrac{x(4 - x)}{(3x + 5)(x - 4)}$

$\qquad\qquad = \dfrac{x}{(3x + 5)} \cdot \dfrac{(4 - x)}{(x - 4)}$

$\qquad\qquad = \dfrac{x}{(3x + 5)} \cdot (-1) \qquad \text{(since } (4 - x) = -(x - 4))$

$\qquad\qquad = -\dfrac{x}{(3x + 5)}.$

EXAMPLE 2   Perform the operations and simplify.

a) $\dfrac{x^2 - 2x}{x^2 - 9} \cdot \dfrac{x^2 + 3x}{x^4 - 4x^3 + 4x^2}$

b) $(x^3 - 8) \div \dfrac{3x^2 + 6x + 12}{3x - 7}$

SOLUTION

a) $\dfrac{x^2 - 2x}{x^2 - 9} \cdot \dfrac{x^2 + 3x}{x^4 - 4x^3 + 4x^2} = \dfrac{x(x - 2)}{(x - 3)(x + 3)} \cdot \dfrac{x(x + 3)}{x^2(x - 2)^2}$

$\qquad\qquad = \dfrac{x^2(x - 2)(x + 3)}{x^2(x - 3)(x + 3)(x - 2)^2}$

$\qquad\qquad = \dfrac{1}{(x - 3)(x - 2)} \quad \text{or} \quad \dfrac{1}{x^2 - 5x + 6}.$

b) $(x^3 - 8) \div \dfrac{3x^2 + 6x + 12}{3x - 7} = \dfrac{x^3 - 8}{1} \cdot \dfrac{3x - 7}{3x^2 + 6x + 12}$

$$= \frac{(x - 2)(x^2 + 2x + 4)}{1} \cdot \frac{3x - 7}{3(x^2 + 2x + 4)}$$

$$= \frac{(x - 2)(3x - 7)}{3} \quad \text{or} \quad x^2 - \frac{13}{3}x + \frac{14}{3}.$$

As you recall, adding or subtracting arithmetic fractions involves expressing each of the terms with a common denominator and adding or subtracting the numerators. A similar process allows us to add or subtract rational expressions.

EXAMPLE 3   Perform the operations and simplify.

a) $\dfrac{2}{2x - 5} + \dfrac{3}{x}$

b) $\dfrac{x + 3}{2x^2 - x - 1} + \dfrac{3}{4x^2 + 4x + 1} - \dfrac{2}{x - 1}$

SOLUTION

a) The least common denominator is the product $x(2x - 5)$.

$$\frac{2}{(2x - 5)} + \frac{3}{x} = \frac{2x}{x(2x - 5)} + \frac{3(2x - 5)}{x(2x - 5)}$$

$$= \frac{2x + 3(2x - 5)}{x(2x - 5)}$$

$$= \frac{8x - 15}{x(2x - 5)} \quad \text{or} \quad \frac{8x - 15}{2x^2 - 5x}.$$

b) We start by factoring the denominator of each of the terms.

$$\frac{x + 3}{2x^2 - x - 1} + \frac{3}{4x^2 + 4x + 1} - \frac{2}{x - 1}$$

$$= \frac{x + 3}{(2x + 1)(x - 1)} + \frac{3}{(2x + 1)^2} - \frac{2}{(x - 1)}.$$

The common denominator is the product $(2x + 1)^2(x - 1)$. Next, we multiply the numerator and denominator of each term by an expression that gives each term this common denominator, and then we simplify.

$$\frac{x + 3}{2x^2 - x - 1} + \frac{3}{4x^2 + 4x + 1} - \frac{2}{x - 1}$$

$$= \frac{(x + 3)(2x + 1)}{(2x + 1)^2(x - 1)} + \frac{3(x - 1)}{(2x + 1)^2(x - 1)} - \frac{2(2x + 1)^2}{(2x + 1)^2(x - 1)}$$

$$= \frac{(x + 3)(2x + 1) + 3(x - 1) - 2(2x + 1)^2}{(2x + 1)^2(x - 1)}$$

$$= \frac{(2x^2 + 7x + 3) + (3x - 3) - 2(4x^2 + 4x + 1)}{(2x + 1)^2(x - 1)}$$

$$= \frac{2x^2 + 7x + 3 + 3x - 3 - 8x^2 - 8x - 2}{(2x + 1)^2(x - 1)}$$

$$= \frac{-6x^2 + 2x - 2}{(2x + 1)^2(x - 1)}.$$

A rational expression such that the degree of the numerator is greater than or equal to the degree of the denominator is called an **improper** rational expression. For example, the expressions

$$\frac{x^5 + x^3 - 13}{x^2}, \quad \frac{6x^4 + 5x^3 - 4x^2 - 5x + 7}{2x + 3}, \quad \text{and} \quad \frac{2x^4 + x^2 - 3}{x^4 - 2x}$$

are improper rational expressions. As you might suspect, a rational expression such that the degree of the numerator is less than the degree of the denominator is called a **proper** rational expression.

Just as an improper fraction such as $\frac{13}{4}$ can be expressed as $3\frac{1}{4}$ (the sum of an integer and a proper fraction), an improper rational expression can be expressed as the sum of a polynomial and a proper rational expression. For example,

$$\frac{x^5 + x^3 - 13}{x^2} = \frac{x^5}{x^2} + \frac{x^3}{x^2} - \frac{13}{x^2} = \underbrace{x^3 + x}_{\text{polynomial}} + \underbrace{-\frac{13}{x^2}}_{\text{proper rational expression}}.$$

Often, this transformation involves **polynomial long division,** which is reviewed in the next two examples.

EXAMPLE 4   Write the rational expression as the sum of a polynomial and a proper rational expression:

$$\frac{6x^4 + 5x^3 - 4x^2 - 5x + 7}{2x + 3}.$$

SOLUTION The first step is to place the denominator and numerator in their appropriate places in the division tableau (notice that they are written in descending powers):

$$2x + 3 \overline{\smash{\big)}\, 6x^4 + 5x^3 - 4x^2 - 5x + 7}.$$

This process is much like the division of integers. We divide the first term of the dividend, $6x^4$, by the first term of the divisor, $2x$, and place the result, $3x^3$, in the tableau as shown:

$$\begin{array}{r} 3x^3 \\ 2x + 3 \overline{\smash{\big)}\, 6x^4 + 5x^3 - 4x^2 - 5x + 7}. \end{array}$$

Next, we multiply the divisor by this first term of the quotient to get $6x^4 + 9x^3$, which we place in the following tableau as shown. After subtracting, we bring down the next term:

$$\begin{array}{r} 3x^3 \\ 2x + 3 \overline{\smash{\big)}\, 6x^4 + 5x^3 - 4x^2 - 5x + 7} \\ \underline{6x^4 + 9x^3 } \\ -4x^3 - 4x^2. \end{array}$$

These steps are repeated until the remainder is zero or has degree less than the divisor:

$$\begin{array}{r} 3x^3 - 2x^2 + x - 4 \\ 2x + 3 \overline{\smash{\big)}\, 6x^4 + 5x^3 - 4x^2 - 5x + 7} \\ \underline{6x^4 + 9x^3 } \\ -4x^3 - 4x^2 \\ \underline{-4x^3 - 6x^2 } \\ 2x^2 - 5x \\ \underline{2x^2 + 3x } \\ -8x + 7 \\ \underline{-8x - 12} \\ 19. \end{array}$$

Interpreting this result, we get

$$\frac{6x^4 + 5x^3 - 4x^2 - 5x + 7}{2x + 3} = \underbrace{3x^3 - 2x^2 + x - 4}_{\text{polynomial}} + \underbrace{\frac{19}{2x + 3}}_{\text{proper rational expression}}.$$

EXAMPLE 5 Write the rational expression as the sum of a polynomial and a proper rational expression:

$$\frac{2x^4 + x^2 - 3}{x^4 - 2x}.$$

SOLUTION Notice that the terms $0x^3$ and $0x$ are placed in the tableau to help keep the proper alignment of terms:

$$
\begin{array}{r}
2 \\
x^4 - 2x \overline{\smash{\big)}\ 2x^4 + 0x^3 + x^2 + 0x - 3} \\
\underline{2x^4 \qquad\qquad\ - 4x} \\
x^2 + 4x - 3.
\end{array}
$$

Thus

$$
\frac{2x^4 + x^2 - 3}{x^4 - 2x} = \underbrace{2}_{\text{polynomial}} + \underbrace{\frac{x^2 + 4x - 3}{x^4 - 2x}}_{\text{proper rational expression}}.
$$

A compound rational expression is one in which either the numerator or the denominator has a term that is itself a rational expression. There are two general methods by which a compound rational expression can be simplified. They are both illustrated in the next example.

*A compound rational expression is also called a **compound fraction** or a **complex fraction**.*

EXAMPLE 6 Simplify $\dfrac{1 - \dfrac{x}{x + 2}}{1 + \dfrac{1}{x}}$.

SOLUTION In the first method, we simplify the numerator and the denominator as expressions in their own right, and then treat the expression as a quotient:

$$
\frac{1 - \dfrac{x}{x + 2}}{1 + \dfrac{1}{x}} = \frac{\dfrac{x + 2}{x + 2} - \dfrac{x}{x + 2}}{\dfrac{x}{x} + \dfrac{1}{x}} = \frac{\dfrac{2}{x + 2}}{\dfrac{x + 1}{x}}
$$

$$
= \frac{2}{x + 2} \div \frac{x + 1}{x} = \frac{2}{x + 2} \cdot \frac{x}{x + 1}
$$

$$
= \frac{2x}{(x + 2)(x + 1)} \quad \text{or} \quad \frac{2x}{x^2 + 3x + 2}.
$$

In the second method, we begin by determining a common denominator of the rational expressions in both the numerator and denominator, and then we multiply both the numerator and denominator by this product (this eliminates the smaller fractions). In this case our common denominator is $x(x + 2)$, so

$$\frac{1 - \dfrac{x}{x+2}}{1 + \dfrac{1}{x}} = \frac{1 - \dfrac{x}{x+2}}{1 + \dfrac{1}{x}} \cdot \frac{x(x+2)}{x(x+2)}$$

$$= \frac{x(x+2) - \dfrac{x}{x+2} \cdot x(x+2)}{x(x+2) + \dfrac{1}{x} \cdot x(x+2)} = \frac{x(x+2) - x^2}{x(x+2) + (x+2)}$$

$$= \frac{x^2 + 2x - x^2}{x^2 + 2x + x + 2} = \frac{2x}{x^2 + 3x + 2}.$$

Of course, the answer is the same regardless of the method of solution.

Many of the techniques used to simplify rational expressions can also be used to simplify algebraic fractions in general. The expression in the next example is typical of those you will encounter in calculus.

EXAMPLE 7   Simplify the algebraic expression:

$$\frac{x^{1/3} \cdot 2(x-4) - \frac{1}{3}x^{-2/3}(x-4)^2}{[x^{1/3}]^2}.$$

SOLUTION

$$\frac{x^{1/3} \cdot 2(x-4) - \frac{1}{3}x^{-2/3}(x-4)^2}{[x^{1/3}]^2} = \frac{x^{1/3} \cdot 2(x-4) - \left[\dfrac{(x-4)^2}{3x^{2/3}}\right]}{x^{2/3}} \cdot \frac{3x^{2/3}}{3x^{2/3}}$$

$$= \frac{x^{1/3} \cdot 2(x-4) \cdot 3x^{2/3} - \left[\dfrac{(x-4)^2}{3x^{2/3}}\right]3x^{2/3}}{x^{2/3} \cdot 3x^{2/3}}$$

$$= \frac{6x(x-4) - (x-4)^2}{3x^{4/3}}$$

$$= \frac{6x^2 - 24x - (x^2 - 8x + 16)}{3x^{4/3}}$$

$$= \frac{5x^2 - 16x - 16}{3x^{4/3}}$$

Quite often, the numerator or denominator of an algebraic fraction involves radicals. It is often to your advantage to manipulate such an algebraic fraction to eliminate the radicals from either the numerator or denominator. The **conjugate** of a sum $a + b$ is defined to be $a - b$. For example, the conjugate of $x + 4$ is $x - 4$, and the conjugate of $\sqrt{2} - \sqrt{3}$ is $\sqrt{2} + \sqrt{3}$. The next example shows how a conjugate can be used to rationalize a denominator.

EXAMPLE 8　Simplify this algebraic expression by rationalizing the denominator:

$$\frac{x - 1}{\sqrt{x + 1} - \sqrt{2}}.$$

SOLUTION　The conjugate of the denominator $\sqrt{x + 1} - \sqrt{2}$ is the sum $\sqrt{x + 1} + \sqrt{2}$. Multiplying both the numerator and denominator by this sum, we obtain

$$\frac{x - 1}{\sqrt{x + 1} - \sqrt{2}} = \frac{x - 1}{\sqrt{x + 1} - \sqrt{2}} \cdot \frac{\sqrt{x + 1} + \sqrt{2}}{\sqrt{x + 1} + \sqrt{2}}$$

$$= \frac{(x - 1)(\sqrt{x + 1} + \sqrt{2})}{(\sqrt{x + 1})^2 - (\sqrt{2})^2}$$

$$= \frac{(x - 1)(\sqrt{x + 1} + \sqrt{2})}{(x + 1) - 2}$$

$$= \frac{(x - 1)(\sqrt{x + 1} + \sqrt{2})}{x - 1}$$

$$= \sqrt{x + 1} + \sqrt{2}.$$

EXERCISE SET 1.3

A

In Problems 1 through 12, reduce each rational expression to lowest terms.

1. $\dfrac{2x^4 + x^2}{1 + 2x^2}$

2. $\dfrac{x^3 + 3x^2}{3 + x}$

3. $\dfrac{x^2 - 7x + 12}{x^2 - 16}$

4. $\dfrac{x^2 - 2x - 3}{x^2 - 9}$

5. $\dfrac{(x^3 + 3x^2 + 4x)(2x + 4)}{(x^2 + 2)(x^2 + 3x + 4)}$

6. $\dfrac{(x^2 + 3x - 10)(x^2 + 6x + 8)}{(x^2 - 4)(x^2 - x - 20)}$

7. $\dfrac{9 - x^2}{6x^2 - 11x - 21}$

8. $\dfrac{3x - 2x^2}{4x^2 - 12x + 9}$

9. $\dfrac{x^2 - y^2}{(x - y)^2}$

10. $\dfrac{x^3 + y^3}{(x + y)^3}$

11. $\dfrac{(x + 3)^2 - 16}{(x - 4)^2 - 9}$

12. $\dfrac{4 - (x - 1)^2}{25 - (x + 2)^2}$

In Problems 13 through 24, perform the operation and simplify. The answer should be left in factored form.

13. $\dfrac{4x^2 + 4x + 1}{4x^2 - 1} \cdot \dfrac{2x^2 + x - 1}{2x^2 - x - 1}$

14. $\dfrac{2x^2 + 13x + 21}{3x^2 - 2x} \cdot \dfrac{3x^2 + x - 2}{x^2 + 4x + 3}$

15. $\dfrac{x^2 - 2x - 3}{x^2 - 4} \div \dfrac{x - 3}{x - 2}$

16. $\dfrac{x^4 - 16}{x^2 - 2x} \div \dfrac{5x^2 + 12x + 4}{25x^2 + 20x + 4}$

17. $\dfrac{3}{x + 1} + \dfrac{4}{x - 2}$

18. $\dfrac{3x}{x - 4} + \dfrac{2x}{x + 3}$

19. $\dfrac{2}{x + 5} - \dfrac{3}{x - 5}$

20. $\dfrac{x - 2}{x + 2} - \dfrac{x + 2}{x - 2}$

21. $\dfrac{x}{x^2 - 7x + 6} - \dfrac{x}{x^2 - 2x - 24}$

22. $\dfrac{3}{(x + 1)(x - 1)^2} + \dfrac{2}{(x + 1)^2(x - 1)}$

23. $\dfrac{x - 1}{x + 1} + \dfrac{2x}{3x + 3} - \dfrac{4}{5 + 2x}$

24. $\dfrac{x}{x + 1} - \dfrac{x + 1}{x + 2} + \dfrac{1}{x - 1}$

B

In Problems 25 through 36, simplify the compound fraction.

25. $\dfrac{\dfrac{1}{x + 2} - \dfrac{1}{x}}{2}$

26. $\dfrac{\dfrac{1}{x - 1} + \dfrac{2}{x}}{2}$

27. $\dfrac{1 - \dfrac{1}{x}}{\dfrac{1}{x} - x}$

28. $\dfrac{\dfrac{1}{x} - \dfrac{1}{y}}{\dfrac{1}{x^2} - \dfrac{1}{y^2}}$

29. $\dfrac{\dfrac{x + h - 1}{x + h + 1} - \dfrac{x - 1}{x + 1}}{h}$

30. $\dfrac{\dfrac{x + h}{x + h - 2} - \dfrac{x}{x - 2}}{h}$

31. $\dfrac{\dfrac{1}{(x - h)^2} - \dfrac{1}{x^2}}{h}$

32. $\dfrac{\dfrac{1}{(x + h)^3} - \dfrac{1}{x^3}}{h}$

33. $\dfrac{\dfrac{1}{\sqrt{x}} - \sqrt{x}}{\sqrt{x}}$

34. $\dfrac{\dfrac{1}{x^{1/3}} - x^{2/3}}{x^{4/3}}$

35. $\dfrac{\dfrac{x^2}{y^2} - \dfrac{y^2}{x^2}}{\dfrac{x^4}{y^4} - 2 + \dfrac{y^4}{x^4}}$

36. $\dfrac{x^2 + \dfrac{x}{y} - \dfrac{6}{y^2}}{x^2 + \dfrac{4x}{y} + \dfrac{3}{y^2}}$

In Problems 37 through 60, perform the operation and simplify.

37. $(x^2 - 4)\left(\dfrac{x^2 - 2x - 8}{x - 2}\right)$

38. $(x^2 + 3x - 10)\left(\dfrac{x^2 - 4x + 4}{x + 5}\right)$

39. $x^2\left(\dfrac{1}{\sqrt{x}} + \sqrt{x}\right)\left(\dfrac{1}{\sqrt{x}} - \sqrt{x}\right)$

40. $x\left(\dfrac{2}{\sqrt{x}} + x\sqrt{x}\right)\left(\dfrac{2}{\sqrt{x}} - 3x\sqrt{x}\right)$

41. $\left(\dfrac{\sqrt{x} + 2}{x^2 - 4x - 12}\right)(\sqrt{x} - 2)$

42. $\left(\dfrac{\sqrt{x} - 3}{x^3 - 9x}\right)(x\sqrt{x} + 3x)$

43. $x\sqrt{x} + \dfrac{1}{\sqrt{x}}$

44. $\dfrac{1}{\sqrt{x + 2}} + \sqrt{x + 2}$

45. $\dfrac{x^2}{\sqrt{1 - x^2}} + \sqrt{1 - x^2}$

46. $\sqrt{1 - x^2} - \dfrac{1}{\sqrt{1 - x^2}}$

47. $\dfrac{x - x^{-1}}{x + x^{-1}}$

48. $\dfrac{x^2 - x^{-2}}{x - x^{-1}}$

49. $\sqrt{3x + 5} \cdot \dfrac{1}{2\sqrt{x - 4}} + \sqrt{x - 4} \cdot \dfrac{3}{2\sqrt{3x + 5}}$

50. $\sqrt{x - 3} \cdot \dfrac{5}{2\sqrt{5x + 2}} + \sqrt{5x + 2} \cdot \dfrac{1}{2\sqrt{x - 3}}$

51. $\dfrac{\sqrt{x^2 + 2} - x \cdot \dfrac{x}{\sqrt{x^2 + 2}}}{x^2 + 2}$

52. $\dfrac{\sqrt{4 + x} - x \cdot \dfrac{1}{2\sqrt{4 + x}}}{4 + x}$

53. $\dfrac{3x^2(1 - x^2)^{1/3} - x^3 \cdot \frac{1}{3}(1 - x^2)^{-2/3}(-2x)}{(1 - x^2)^{2/3}}$

54. $\dfrac{2x(x^2 + 5)^{1/2} - x^2(x^2 + 5)^{-1/2} \cdot 2x}{x^2 + 5}$

55. $\dfrac{\dfrac{2 + x^2}{2\sqrt{x}} - 2x \cdot \sqrt{x}}{(2 + x^2)^2}$

56. $\dfrac{(x + 5) \cdot \frac{4}{3}x^{1/3} - x^{4/3}}{(x + 5)^2}$

57. $\sqrt{\left(\dfrac{3}{x^4} - 2\right)^2 + \dfrac{24}{x^4}}$

58. $\sqrt{\left(\dfrac{1}{x^2} + 1\right)^2 - \dfrac{4}{x^2}}$ $x \geq 1$

59. $\dfrac{1}{\sqrt{1 - \dfrac{x^2}{1 + x^2}}}$

60. $\dfrac{1}{\sqrt{x^2 + \dfrac{x^4}{1 - x^2}}}$

In Problems 61 through 66, simplify the algebraic expression by rationalizing the denominator.

61. $\dfrac{1}{\sqrt{x} + 1}$

62. $\dfrac{1}{\sqrt{x+4} + 2}$

63. $\dfrac{x - 6}{\sqrt{x} - \sqrt{6}}$

64. $\dfrac{x + 3}{\sqrt{x+5} - \sqrt{2}}$

65. $\dfrac{x + \sqrt{1 - x^2}}{x - \sqrt{1 - x^2}}$

66. $\dfrac{\sqrt{x^2 + 1} - x}{\sqrt{x^2 + 1} + x}$

In Problems 67 through 78, use polynomial long division to express the improper rational expression as a sum of a polynomial and a proper rational expression.

67. $\dfrac{x - 4}{x + 2}$

68. $\dfrac{3x + 5}{x - 3}$

69. $\dfrac{x^2 + 3x + 7}{x + 4}$

70. $\dfrac{2x^2 - 3x + 10}{x - 2}$

71. $\dfrac{3x^4 - 2x^2 - 5}{x + 1}$

72. $\dfrac{x^5 - 3x^3 + 4x - 1}{x - 3}$

73. $\dfrac{x^2 - 6x - 3}{x^2 - 2}$

74. $\dfrac{x^2 + 5x + 2}{x^2 + 3}$

75. $\dfrac{4x^3 + 2x + 4}{2x^2 + 1}$

76. $\dfrac{2x^3 - 10x - 4}{x^2 + 3}$

77. $\dfrac{2x^4 + x^3 - 5x^2 - x - 18}{x^2 - 4}$

78. $\dfrac{x^5 - 2x^4 - 3x^3 - 5x^2 + 12x + 9}{x^2 - 2x - 3}$

C

79. a) Evaluate the compound fractions.

$$1 + \cfrac{2}{1 + \cfrac{2}{1 + 2}} \qquad 1 + \cfrac{2}{1 + \cfrac{2}{1 + \cfrac{2}{1 + 2}}}$$

$$1 + \cfrac{2}{1 + \cfrac{2}{1 + \cfrac{2}{1 + \cfrac{2}{1 + 2}}}}$$

$$1 + \cfrac{2}{1 + \cfrac{2}{1 + \cfrac{2}{1 + \cfrac{2}{1 + \cfrac{2}{1 + 2}}}}}$$

b) Suppose that the pattern initiated in part **a)** is continued indefinitely. The expression below is used to indicate this pattern; it is called a **continued fraction.** Make a conjecture about the value of this continued fraction.

$$1 + \cfrac{2}{1 + \cfrac{2}{1 + \cfrac{2}{1 + \cfrac{2}{1 + \cfrac{2}{1 + \ddots}}}}}$$

In Problems 80 and 81, make a conjecture about the value of the continued fraction (see Problem 79).

80. $2 + \cfrac{3}{2 + \cfrac{3}{2 + \cfrac{3}{2 + \cfrac{3}{2 + \ddots}}}}$

81. $2 - \cfrac{2}{3 - \cfrac{2 \cdot 3}{1 - \cfrac{1 \cdot 2}{3 - \cfrac{4 \cdot 5}{1 - \cfrac{3 \cdot 4}{3 - \cfrac{6 \cdot 7}{1 - \cfrac{5 \cdot 6}{3 - \ddots}}}}}}}$

ALGEBRA FOR CALCULUS: EQUATIONS
S E C T I O N 1.4

A **solution** or **root** of an equation in one variable is a real number that makes the equation true. For example, $x = 4$ and $x = -3$ are solutions to the equation $x^2 - x = 12$ since

$$(4)^2 - (4) = 12 \quad \text{and} \quad (-3)^2 - (-3) = 12.$$

An equation can be placed generally into one of three categories, depending on the nature of its solutions. The **replacement set** for a variable of an equation is the set of all real numbers for which the equation makes sense. An equation that is true for all members of its replacement set is called an **identity.** For example,

$$x + x = 2x, \qquad x^3 - 27 = (x - 3)(x^2 + 3x + 9), \qquad \frac{x^2 - 4}{x + 2} = x - 2,$$

are all identities since they are true for any meaningful substitution of x. On the other hand, an equation that is false for all members of its replacement set is called a **contradiction.** The equations

$$x = x + 2, \qquad \frac{10}{x} = 0, \qquad \frac{(x + 2)^2}{x + 2} = x + 4,$$

are examples of contradictions (pause here and determine why none of these is true for any of its replacements).

Our interest will mainly lie with the set of equations that are true for only some of its replacements. These are called **conditional equations.** Solving conditional equations is the topic of our review in this section.

Two equations are **equivalent** if they have the same set of solutions. There are two basic operations that we can perform on an equation to arrive at another that is equivalent to the original. They are the following:

Adding (or subtracting) the same quantity from both sides of an equation produces an equivalent equation.

Multiplying (or dividing) both sides of an equation by the same nonzero quantity produces an equivalent equation.

Solving an equation is more or less a process in which the equation is manipulated so that we may discover its roots. Much of this manipulation uses the preceding two properties to arrive at a simpler, equivalent equation from which we can ascertain the roots.

Linear Equations

A **linear equation** in one unknown can be written as $ax + b = 0$, where x is the variable and a and b are real numbers. The next three examples involve linear equations.

EXAMPLE 1   Solve the equation

$$(x - 2)(2x + 3) = x^2 + (x - 3)^2.$$

SOLUTION We start by simplifying each side of the equation:

$$(x - 2)(2x + 3) = x^2 + (x - 3)^2$$
$$2x^2 - x - 6 = x^2 + (x^2 - 6x + 9)$$
$$2x^2 - x - 6 = 2x^2 - 6x + 9.$$

Then

$$-x - 6 = -6x + 9 \qquad \text{subtracting } 2x^2 \text{ from both sides}$$
$$5x - 6 = 9 \qquad \text{adding } 6x \text{ to both sides}$$
$$5x = 15 \qquad \text{adding 6 to both sides}$$
$$x = 3 \qquad \text{multiplying each side by } \tfrac{1}{5}$$

The root of the equation is apparently $x = 3$. To verify this solution we substitute $x = 3$ into the original equation and check to see if it makes the equation true:

$$(3 - 2)(2(3) + 3) \overset{?}{=} 3^2 + (3 - 3)^2$$
$$(1)(9) \overset{?}{=} 9 + 0^2$$
$$9 = 9.$$

The only root of the equation is $x = 3$.

The next example involves equations that contain rational expressions, which are called **rational equations.** In general, a rational equation can be simplified by multiplying each side by the least common denominator of the rational expressions in the equation. This procedure can, however, introduce roots into the solution that are not actually solutions to the original equation. These are called **extraneous roots.** Our only recourse is to check any potential root in the original equation to make sure that it is an actual root.

EXAMPLE 2   Solve the equation

$$\frac{6}{x} - 1 = \frac{9}{2x}.$$

SOLUTION To clear the fractions from the equation, we first multiply each side of the equation by $2x$, the least common denominator of the fractions $6/x$ and $9/2x$.

$$2x\left(\frac{6}{x} - 1\right) = 2x\left(\frac{9}{2x}\right)$$

$$2x\left(\frac{6}{x}\right) - 2x(1) = 2x\left(\frac{9}{2x}\right)$$

It bears repeating that this step is valid only if we multiply each side by a nonzero quantity; this means that x cannot be zero. We then simplify each side and proceed as usual:

$$12 - 2x = 9$$

$$-2x = -3$$

$$x = \tfrac{3}{2}.$$

This solution, $x = \tfrac{3}{2}$, can be verified directly by substituting this value for x into the original equation:

$$\frac{6}{\left(\frac{3}{2}\right)} - 1 \overset{?}{=} \frac{9}{2\left(\frac{3}{2}\right)}$$

$$4 - 1 = 3.$$

EXAMPLE 3   A body is released with an initial velocity of v_0 and is allowed to fall. The velocity v of this body (neglecting air resistance) after t seconds is given by

$$v = -gt + v_0.$$

Solve this equation for t.

SOLUTION Since we are solving for t, we treat v, g, and v_0 as constants. So,

$$v = -gt + v_0$$

$$v - v_0 = -gt$$

$$\frac{v - v_0}{-g} = t \quad \Rightarrow \quad t = \frac{v_0 - v}{g}.$$

To check: $v \overset{?}{=} -g\left[\dfrac{v_0 - v}{g}\right] + v_0$

$$v \overset{?}{=} -(v_0 - v) + v_0$$

$$v \overset{?}{=} -v_0 + v + v_0$$

$$v = v.$$

Solving Nonlinear Equations

As you are probably well aware, a variety of methods are used to solve non-linear equations. The remainder of this section reviews some of the more common of these techniques.

An equation that can be written in the form

$$ax^2 + bx + c = 0$$

is a **quadratic** or **second-degree equation.** You recall, no doubt, that many of these can be solved by factoring.

The method of **factoring** relies on the fact that if a product of real numbers is zero, then at least one of the factors is itself zero. We state this property of real numbers, the **zero factor property,** more formally in the following box.

Zero Factor Property

> $AB = 0$, if and only if either $A = 0$ or $B = 0$.

It is important to remember that this property holds only for products that are equal to zero.

EXAMPLE 4   Solve the equation.

a) $2x^2 - 3x - 2 = 0$ **b)** $x^3 + 4 = 4x^2 + x$

SOLUTION

a)   This quadratic equation yields to factoring:

$$2x^2 - 3x - 2 = 0$$

$$(2x + 1)(x - 2) = 0.$$

Setting each of the factors equal to zero gives us two linear equations to solve:

$$2x + 1 = 0 \Rightarrow x = -\tfrac{1}{2}$$
$$x - 2 = 0 \Rightarrow x = 2.$$

The roots are $-\tfrac{1}{2}$ and 2.

b) First, we write the equation so that the left side is zero:

$$x^3 - 4x^2 - x + 4 = 0.$$

Although this is not a quadratic equation, it can be factored (see Example 8 of Section 1.2):

$$x^3 - 4x^2 - x + 4 = 0$$
$$(x - 4)(x - 1)(x + 1) = 0$$
$$x - 4 = 0 \Rightarrow x = 4$$
$$x - 1 = 0 \Rightarrow x = 1$$
$$x + 1 = 0 \Rightarrow x = -1.$$

The roots are 4, -1, and 1.

The obvious shortcoming of the factoring method is that many expressions that arise in equations defy even the most dogged attempts at factoring.

Many equations can also be solved by the method of **extracting roots.** The next example illustrates this technique.

EXAMPLE 5   Solve the equation.

a) $4(x + 1)^2 - 5 = 0$ **b)** $(3 - x)^3 = 13$

SOLUTION

a) We first solve for the quantity $(x + 1)^2$:

$$4(x + 1)^2 - 5 = 0$$
$$4(x + 1)^2 = 5$$
$$(x + 1)^2 = \tfrac{5}{4}.$$

In the next step, we take the square root of each side of the equation, taking care to account for both the positive and negative roots (this is because we are taking even roots).

$$x + 1 = \pm\frac{\sqrt{5}}{2}$$

$$x = -1 \pm \frac{\sqrt{5}}{2} \quad \text{or} \quad \frac{-2 \pm \sqrt{5}}{2}$$

b) We proceed as we did before except that we need concern ourselves with only one root:

$$(3 - x)^3 = 13$$

$$3 - x = \sqrt[3]{13}$$

$$x = 3 - \sqrt[3]{13}$$

We can use this technique to solve a quadratic equation in the form $ax^2 + bx + c = 0$. The trick is first to manipulate this equation to a form $(x - h)^2 = k$, as we had in part a) of Example 5. This process, called **completing the square,** has three steps. We will use $x^2 - 6x + 7 = 0$ for an example.

1)   Manipulate the equation to the form

$$x^2 - 6x = -7.$$

2)   Add the square of half of the coefficient of the x-term to each side and write as a perfect square:

$$x^2 - 6x + (-3)^2 = -7 + (-3)^2$$

$$x^2 - 6x + 9 = 2$$

$$(x - 3)^2 = 2.$$

3)   Extract roots:

$$x - 3 = \pm\sqrt{2}.$$

The roots are $x = 3 + \sqrt{2}, 3 - \sqrt{2}.$

EXAMPLE 6   Solve the equation by completing the square:

$$2x^2 - 3x - 2 = 0.$$

SOLUTION Multiplying each side by $\frac{1}{2}$ gives us

$$x^2 - \tfrac{3}{2}x - 1 = 0.$$

Adding 1 to both sides gives

$$x^2 - \tfrac{3}{2}x = 1.$$

Proceeding as before,

$$x^2 - \frac{3}{2}x + \frac{9}{16} = 1 + \frac{9}{16}$$

$$\left(x - \frac{3}{4}\right)^2 = \frac{25}{16}$$

$$x - \frac{3}{4} = \pm\frac{5}{4}$$

$$x = \frac{3}{4} \pm \frac{5}{4} \quad \text{or} \quad x = 2, -\frac{1}{2}.$$

You may recognize this as the equation that was solved in part a) of Example 4. The roots, of course, are the same as we discovered in that example.

Rather than repeat these steps for completing the square over and over again for each equation, we can solve the general equation $ax^2 + bx + c = 0$ by using this process and arrive at the following well-known formula.

The solutions of the quadratic equation $ax^2 + bx + c = 0$ are

$$x = \frac{-b + \sqrt{b^2 - 4ac}}{2a} \quad \text{and} \quad x = \frac{-b - \sqrt{b^2 - 4ac}}{2a}.$$

Quadratic Formula

The proof of this is left as an exercise (see Problem 73).

EXAMPLE 7   Solve the equation by using the quadratic formula.

   a) $x^2 - 2x - 4 = 0$ **b)** $3x^2 = 2\sqrt{3}x - 1$ **c)** $2x - \frac{1}{2}x^2 = \frac{7}{3}$

SOLUTION

   a) In this particular case, we have $a = 1$, $b = -2$, and $c = -4$. Substituting these values into the quadratic formula gives

$$x = \frac{-(-2) \pm \sqrt{(-2)^2 - 4(1)(-4)}}{2(1)}$$

$$= \frac{2 \pm \sqrt{20}}{2} = 1 \pm \sqrt{5}.$$

   The roots are $1 + \sqrt{5}$ and $1 - \sqrt{5}$.

   b) The first step is to write the equation in the form $ax^2 + bx + c = 0$:

$$3x^2 - 2\sqrt{3}x + 1 = 0.$$

Using $a = 3$, $b = -2\sqrt{3}$, and $c = 1$ in the quadratic formula, we get

$$x = \frac{-(-2\sqrt{3}) \pm \sqrt{(-2\sqrt{3})^2 - 4(3)(1)}}{2(3)}$$

$$= \frac{2\sqrt{3} \pm \sqrt{0}}{6} = \frac{\sqrt{3}}{3}.$$

The one real root is $\sqrt{3}/3$.

c) Writing this in the form $ax^2 + bx + c = 0$, we get

$$-\tfrac{1}{2}x^2 + 2x - \tfrac{7}{3} = 0.$$

We could substitute these coefficients into the quadratic formula to determine the roots. However, multiplying each side of the equation by -6 eliminates the fractions and makes simplifying the roots much easier:

$$-6(-\tfrac{1}{2}x^2 + 2x - \tfrac{7}{3}) = -6(0)$$

$$3x^2 - 12x + 14 = 0.$$

So,

$$x = \frac{-(-12) \pm \sqrt{(-12)^2 - 4(3)(14)}}{2(3)} = \frac{12 \pm \sqrt{-24}}{6}.$$

The expressions

$$\frac{-12 + \sqrt{-24}}{6} \quad \text{and} \quad \frac{-12 - \sqrt{-24}}{6}$$

are not real numbers. There are no real roots to this equation.

The number of solutions of a given quadratic equation of course depends on the value of the radicand $b^2 - 4ac$. This particular expression is called the **discriminant** of the quadratic equation.

The Discriminant

> The **discriminant** of the quadratic equation $ax^2 + bx + c = 0$ is the expression $b^2 - 4ac$.
>
> 1) If $b^2 - 4ac > 0$, then the equation has two distinct real roots.
>
> 2) If $b^2 - 4ac = 0$, then the equation has exactly one real root, namely $x = -b/2a$.
>
> 3) If $b^2 - 4ac < 0$, then the equation has no real roots.

For example, if we check the discriminant of $-\frac{1}{2}x^2 + 2x - \frac{7}{3} = 0$, the equation that we tried to solve in part c) of Example 7, we get

$$b^2 - 4ac = (2)^2 - 4(-\tfrac{1}{2})(-\tfrac{7}{3}) = 4 - \tfrac{14}{3} = -\tfrac{2}{3} < 0.$$

This implies that there are no real roots for this equation, as we found.

EXAMPLE 8   For what values of k does the equation

$$4x^2 + kx + 9 = 0$$

have exactly one real root?

SOLUTION   If the equation has exactly one real root, then the discriminant must be zero. The discriminant of this equation is

$$k^2 - 4(4)(9) = k^2 - 144.$$

Setting this equation to zero and solving yields

$$k^2 - 144 = 0$$
$$k^2 = 144$$
$$k = \pm 12.$$

Thus, each of the equations

$$4x^2 - 12x + 9 = 0 \quad \text{and} \quad 4x^2 + 12x + 9 = 0$$

has exactly one root (pause here and verify this assertion).

Equations of the form $|u| = a \ (a \geq 0)$ were discussed in Section 1.1. As you recall, this equation has two roots, $u = a$ and $u = -a$.

Equations Involving Absolute Values

EXAMPLE 9   Solve the equation.

    a) $|2x + 1| = 7$ **b)** $|2x^2 - x + 1| = 7$

SOLUTION

    a)   As before, $|2x + 1| = 7$ means that either

$$2x + 1 = 7 \quad \text{or} \quad 2x + 1 = -7.$$

So,

$$2x = 6 \qquad 2x = -8$$
$$x = 3. \qquad x = -4.$$

The two roots of this equation are $x = 3, -4$ (you should pause here and verify this).

b) In a similar fashion $|2x^2 - x + 1| = 7$ implies that either

$$2x^2 - x + 1 = 7 \quad \text{or} \quad 2x^2 - x + 1 = -7.$$

$2x^2 - x + 1 = 7$	$2x^2 - x + 1 = -7$
$2x^2 - x - 6 = 0$	$2x^2 - x + 8 = 0$
$(2x + 3)(x - 2) = 0$	$2x^2 - x + 8 = 0.$
$2x + 3 = 0 \Rightarrow x = -\frac{3}{2}$	Since the discriminant is
$x - 2 = 0 \Rightarrow x = 2.$	$(-1)^2 - 4(2)(8) < 0,$
	there are no real roots.

The roots are $x = -\frac{3}{2}, 2$.

Equations Containing Radicals

An equation that involves radical expressions can usually be reduced to one without radicals by raising each side of the equation to the appropriate power. Care must be taken, however, since this process does not necessarily yield an equivalent equation.

There are basically two different cases. Raising both sides of an equation to an odd power produces an equivalent equation. Raising both sides of an equation to an even power does not in general produce an equivalent equation; there may be extraneous roots.

For example, the equation $x - 1 = 3$ has only the root $x = 4$. Squaring both sides of this equation yields a new equation $(x - 1)^2 = 9$. It has both $x = 4$ and $x = -2$ as roots; it is not equivalent to the original equation.

EXAMPLE 10   Solve the equation.

   a) $\sqrt{2x + 1} = x - 1$ **b)** $2\sqrt{x + 2} - \sqrt{2x + 3} = 3$

SOLUTION

   a) We start by squaring both sides of the equation:

$$(\sqrt{2x + 1})^2 = (x - 1)^2$$
$$2x + 1 = x^2 - 2x + 1.$$

This reduces the problem to that of solving a quadratic equation. Simplifying, we get

$$0 = x^2 - 4x$$

$$0 = x(x - 4)$$

$$x = 0 \quad \text{or} \quad x = 4.$$

These are roots to the quadratic equation, but they may not be roots to the radical equation with which we started. We need to test each one in the original equation:

$$\sqrt{2(0) + 1} \stackrel{?}{=} 0 - 1 \qquad \sqrt{2(4) + 1} \stackrel{?}{=} 4 - 1$$

$$\sqrt{1} \neq -1 \qquad\qquad \sqrt{9} = 3.$$

Of the two potential roots 0 and 4, only 4 proves to be an actual root of the equation. Since the root 0 does not check in the original equation, it is an extraneous root.

b) It is to our advantage to add $\sqrt{2x + 3}$ to each side of the equation before squaring:

$$2\sqrt{x + 2} = 3 + \sqrt{2x + 3}$$

$$(2\sqrt{x + 2})^2 = (3 + \sqrt{2x + 3})^2$$

$$2^2(\sqrt{x + 2})^2 = 3^2 + 2(3)(\sqrt{2x + 3}) + (\sqrt{2x + 3})^2$$

$$4(x + 2) = 9 + 6\sqrt{2x + 3} + (2x + 3)$$

$$4x + 8 = 2x + 12 + 6\sqrt{2x + 3}$$

$$2x - 4 = 6\sqrt{2x + 3}$$

$$x - 2 = 3\sqrt{2x + 3}$$

$$x^2 - 4x + 4 = 9(2x + 3)$$

$$x^2 - 22x - 23 = 0$$

$$(x - 23)(x + 1) = 0$$

$$x - 23 = 0 \quad \text{or} \quad x + 1 = 0$$

$$x = 23 \quad | \quad x = -1.$$

Checking these potential roots,

$$2\sqrt{(23) + 2} - \sqrt{2(23) + 3} \stackrel{?}{=} 3 \qquad\qquad 2\sqrt{(-1) + 2} - \sqrt{2(-1) + 3} \stackrel{?}{=} 3$$

$$2\sqrt{25} - \sqrt{49} \stackrel{?}{=} 3 \qquad\qquad 2\sqrt{1} - \sqrt{1} \stackrel{?}{=} 3$$

$$10 - 7 = 3. \qquad\qquad 2 - 1 \neq 3.$$

The only root is $x = 23$. The other potential root, $x = -1$, is extraneous; it fails to check in the original equation.

We finish the section with two examples that are typical of some equations that you will encounter in calculus.

EXAMPLE 11   Solve the equation $x^{2/3}(2x^{2/3} - 7) = 4$.

SOLUTION The first step is to simplify the left side and arrange to have the right side of the equation equal to zero:

$$2x^{4/3} - 7x^{2/3} = 4$$

$$2x^{4/3} - 7x^{2/3} - 4 = 0.$$

Even though $2x^{4/3} - 7x^{2/3} - 4$ is not a second-degree polynomial, this particular equation can be thought of as a quadratic equation in $x^{2/3}$:

$$2(x^{2/3})^2 - 7x^{2/3} - 4 = 0.$$

Solving this particular equation for $x^{2/3}$ is equivalent to solving $2u^2 - 7u - 4 = 0$ for u (pause here and solve this equation in u by factoring). So,

$$(x^{2/3} - 4)(2x^{2/3} + 1) = 0$$

$$x^{2/3} - 4 = 0 \quad \text{or} \quad 2x^{2/3} + 1 = 0$$

$$
\begin{array}{c|c}
x^{2/3} = 4 & x^{2/3} = -\tfrac{1}{2} \\
x^2 = 4^3 & x^2 = (-\tfrac{1}{2})^3 \\
x^2 = 64 & x^2 = -\tfrac{1}{8} \\
x = \pm 8. & \text{no real roots.}
\end{array}
$$

The two real roots to this equation are 8 and -8.

EXAMPLE 12   Solve the equation

$$\frac{x^{2/3} \cdot 2(x^2 - 3) \cdot 2x - \tfrac{2}{3}x^{-1/3} \cdot (x^2 - 3)^2}{(x^{2/3})^2} = 0.$$

SOLUTION Multiplying each side of the equation by $(x^{2/3})^2$ gives us

$$x^{2/3} \cdot 2(x^2 - 3) \cdot 2x - \tfrac{2}{3}x^{-1/3} \cdot (x^2 - 3)^2 = 0.$$

(It should seem reasonable that an algebraic fraction is equal to zero only if the numerator of the fraction is zero.) Writing this with positive exponents, we get

$$4x^{5/3}(x^2 - 3) - \frac{2}{3x^{1/3}} \cdot (x^2 - 3)^2 = 0.$$

Next, we multiply each side by $3x^{1/3}$ to clear the fractions from the equation:

$$3x^{1/3}\left[4x^{5/3}(x^2-3)-\frac{2}{3x^{1/3}}\cdot(x^2-3)^2\right]=3x^{1/3}(0)$$

$$12x^2(x^2-3)-2(x^2-3)^2=0$$

$$2(x^2-3)[6x^2-(x^2-3)]=0$$

$$2(x^2-3)(5x^2+3)=0$$

$$x^2-3=0 \quad \text{or} \quad 5x^2+3=0$$

$$x^2=3 \qquad\qquad x^2=-\tfrac{3}{5}$$

$$x=\pm\sqrt{3}. \qquad \text{no real roots.}$$

The potential roots of this equation are $-\sqrt{3}$ and $\sqrt{3}$. Checking the potential root $x=\sqrt{3}$ is easier if we express $\sqrt{3}$ as $3^{1/2}$:

$$\frac{(3^{1/2})^{2/3}\cdot2((3^{1/2})^2-3)\cdot2(3^{1/2})-\tfrac{2}{3}(3^{1/2})^{-1/3}\cdot((3^{1/2})^2-3)^2}{((3^{1/2})^{2/3})^2}\overset{?}{=}0$$

$$\frac{3^{1/3}\cdot2(3-3)\cdot2(3^{1/2})-\tfrac{2}{3}\cdot3^{-1/6}\cdot(3-3)^2}{3^{2/3}}\overset{?}{=}0$$

$$\frac{3^{1/3}\cdot2(0)\cdot2(3^{1/2})-\tfrac{2}{3}\cdot3^{-1/6}\cdot(0)^2}{3^{2/3}}\overset{?}{=}0$$

$$\frac{0}{3^{2/3}}=0$$

This verifies that $x=\sqrt{3}$ is an actual root of the equation. We leave it to you to verify that $x=-\sqrt{3}$ is also an actual root.

A

In Problems 1 through 6, solve the equation.

1. $\dfrac{x}{2}-\dfrac{x}{3}=4$

2. $\dfrac{2x}{3}=5-\dfrac{x}{6}$

3. $(x-2)(x+4)=x^2-3$

4. $(2x+1)(x-3)=x^2+(x-1)^2$

5. $\dfrac{2}{x}-3=\dfrac{5}{x}-2$

6. $\dfrac{5}{x+1}=2-\dfrac{3}{x+1}$

In Problems 7 through 12, solve the equation for the specified variable.

7. $A=\tfrac{1}{2}h(a+b)$ for b

8. $PV=nRT$ for n

9. $FR=mv^2$ for m

10. $A=P+Prt$ for P

11. $\dfrac{1}{R}=\dfrac{1}{R_1}+\dfrac{1}{R_2}$ for R

12. $G=\dfrac{a}{1-r}$ for r

In Problems 13 through 18, solve the equation by factoring.

13. $12x^2 + 7x = 12$ **14.** $3x^2 = x + 4$

15. $10x^2 + 13x = 3$ **16.** $-5x = 6 - 4x^2$

17. $2x^3 - x = x^2$ **18.** $x^2(x + 2) = 143x$

In Problems 19 through 24, solve the equation by extracting roots.

19. $x^2 + 15 = 20$ **20.** $2x^2 - 5 = 11$

21. $(x - 4)^2 = 12$ **22.** $4(x + 1)^2 = 5$

23. $12(x + 4)^2 - 5 = 1$ **24.** $3(2x - 1)^2 - 7 = 5$

In Problems 25 through 30, use the quadratic formula to find all real roots of the equation.

25. $2x^2 - 5x - 42 = 0$ **26.** $x^2 - 2x = 2$

27. $2x^2 - 5x = 8$ **28.** $x^2 - 2x - 6 = 0$

29. $2x - x^2 = 5$ **30.** $5x^2 + 1 = 2x$

In Problems 31 through 36, complete the square to express the equation in the form $(x - h)^2 = k$; then solve the equation by extracting roots.

31. $x^2 - 6x = 0$ **32.** $x^2 + 8x = 4$

33. $x^2 = 2x - 4$ **34.** $x^2 + 5x = 2$

35. $4x^2 + 6x = 9$ **36.** $3x^2 + 5x = 1$

In Problems 37 through 42, solve the absolute value equation.

37. $|2x - 1| = 7$ **38.** $\left|\frac{2}{3}x + 4\right| = 3$

39. $\left|-\frac{1}{2}x + 6\right| = 4$ **40.** $|2x^2 - 5x| = 3$

41. $|x^3 + 14| = 13$ **42.** $|x^3 - x| = 0$

B

In Problems 43 through 72, solve the equation.

43. $x^3 - 25 = 2$ **44.** $x^4 + 14 = 30$

45. $(x + 2)^3 = 24$ **46.** $(x - 6)^{1/3} = 3$

47. $(x - 1)^{1/4} = 1$ **48.** $x^4 = (2x - 1)^2$

49. $\dfrac{1}{x + 1} + \dfrac{1}{x} = 1$ **50.** $\dfrac{3}{x - 2} - \dfrac{5}{x + 2} = 0$

51. $\dfrac{2x}{x - 2} = \dfrac{4}{x - 2} + 1$

52. $1 + \dfrac{7}{x - 3} = \dfrac{2}{x} + \dfrac{21}{x(x - 3)}$

53. $|x|^2 - 8|x| = 0$ **54.** $|x|^2 - 2 = |x|$

55. $\left|\dfrac{1}{x - 2}\right| = 3$ **56.** $\left|\dfrac{2}{x + 1}\right| = x$

57. $x^4 - 2x^2 = 3$ **58.** $x^4 + x^2 = 2$

59. $x^{1/3} - 9x^{1/6} + 8 = 0$ **60.** $2x^{-2} - 5x^{-1} - 3 = 0$

61. $\left(\dfrac{1}{x - 3}\right)^2 - \left(\dfrac{1}{x - 3}\right) = 12$

62. $(3x + 1) - 3\sqrt{3x + 1} = 4$

63. $\sqrt{x^2 + 3x} = x + 1$

64. $\sqrt{x^2 - 3x + 7} = x + 1$

65. $\sqrt{x + 12} = 2 + \sqrt{x}$

66. $2 + \sqrt{x} = \sqrt{x + 16}$

67. $\sqrt{10 + 3\sqrt{x}} = \sqrt{x} + 2$

68. $\sqrt{x + \sqrt{x + 4}} = 4$

69. $\sqrt{3x + 1} \cdot \dfrac{1}{2\sqrt{x - 2}} + \sqrt{x - 2} \cdot \dfrac{3}{2\sqrt{3x + 1}} = 0$

70. $5(x^2 - 3x + 1)^4(2x - 3) = 0$

71. $\dfrac{2x(x^2 + 5)^{1/2} - x^2(x^2 + 5)^{-1/2}}{x^2 + 5} = 0$

72. $\dfrac{\dfrac{12 + x^2}{2\sqrt{x}} - 2x \cdot \sqrt{x}}{(12 + x^2)^2} = 0$

73. The quadratic formula can be derived by solving the general quadratic equation $ax^2 + bx + c = 0$ by completing the square and extracting roots. The steps that follow outline this solution.

a) Show that $ax^2 + bx + c = 0$ can be written

$$x^2 + \frac{b}{a}x = -\frac{c}{a}.$$

b) Complete the square to arrive at

$$\left(x + \frac{b}{2a}\right)^2 = \frac{b^2 - 4ac}{4a^2}.$$

c) Solve this equation for x by extracting roots.

74. Without solving, explain why each of the following has no real roots.

a) $\sqrt{x + 2} = -8$ **b)** $|x^2 - 10| = -1$

c) $\dfrac{10}{2 - 5x} = 0$ **d)** $\sqrt{x - 2} - \sqrt{x} = 13$

75. **a)** Find a positive number that is one less than its square.

b) Find a positive number that is one greater than its reciprocal.

C

76. Explain what is wrong with the following "solution."

$$x = -4$$

$$x + (5x + 8) = -4 + (5x + 8)$$

$$6x + 8 = 5x + 4$$

$$x^2 + 6x + 8 = x^2 + 5x + 4$$

$$(x + 4)(x + 2) = (x + 4)(x + 1)$$

$$x + 2 = x + 1$$

$$2 = 1?$$

Just as there is a formula to solve any quadratic equation, there is also a formula to solve a cubic equation. We look at a special case of it here. Suppose we wish to find a root for the third-degree equation $x^3 + ax + b = 0$. One of the roots is given by

$$x = \sqrt[3]{-\frac{b}{2} + \sqrt{\frac{b^2}{4} + \frac{a^3}{27}}} + \sqrt[3]{-\frac{b}{2} - \sqrt{\frac{b^2}{4} + \frac{a^3}{27}}}.$$

77. Use the preceding cubic formula and your calculator to approximate (to the nearest 0.0001) a root of $x^3 - 2x - 2 = 0$.

78. Use the cubic formula and your calculator to approximate (to the nearest 0.0001) a root of $x^3 - 3x - 4 = 0$.

79. Use the cubic formula and your calculator to approximate (to the nearest 0.0001) a root of $x^3 - 2x - 8 = 0$.

ALGEBRA FOR CALCULUS: INEQUALITIES   S E C T I O N   1.5

Just as with an equation, the **solution** to an inequality in one variable is a real number that makes the inequality true. For example, consider the inequality $2x + 1 < 7$. The real number $x = 2$ is a solution because $2(2) + 1 < 7$ is a true statement. As you might suspect, two inequalities are **equivalent** if they have the same set of solutions.

Solutions to inequalities tend to be intervals of real numbers. These sets of solutions are described by using inequality notation, interval notation, or a graph on a coordinate line.

It should seem reasonable that $a > b$ if and only if $a - b$ is a positive number. Similar statements can be made for the other types of inequalities.

$a > b \Leftrightarrow a - b$ is a positive number.
$a < b \Leftrightarrow a - b$ is a negative number.
$a \geq b \Leftrightarrow a - b$ is a nonnegative number.
$a \leq b \Leftrightarrow a - b$ is a nonpositive number.

Inequalities

We can use the preceding statements to prove the familiar properties that enable us to solve inequalities. We state these properties in terms of inequalities of the form $a > b$. There are similar properties for inequalities of the type $a < b$, $a \geq b$, and $a \leq b$.

Properties of Inequalities

> 1) If $a > b$, then $a + c > b + c$ (adding the same quantity to each side yields an equivalent inequality).
>
> 2) If $a > b$ and $c > 0$, then $ac > bc$ (multiplying each side by same positive quantity yields an equivalent inequality).
>
> 3) If $a > b$ and $c < 0$, then $ac < bc$ (multiplying each side by same negative quantity and reversing the inequality sign yields an equivalent inequality).
>
> 4) If $a > b$ and $b > c$, then $a > c$.

We will prove parts 1) and 4); the others are left as exercises (see Problems 56 and 57).

To prove the first property, notice that, by definition, $a > b$ implies that there is some positive real number p such that $a - b = p$. We need to show that $(a + c) - (b + c)$ is positive. So,

$$(a + c) - (b + c) = a + c - b - c = a - b = p > 0.$$

This completes the proof of 1).

To prove the fourth property, we start by defining two positive numbers p and q such that $a - b = p$ (since $a > b$) and $b - c = q$ (since $b > c$). We need to show that $a - c$ is positive. So,

$$a - c = a - \boldsymbol{b} + \boldsymbol{b} - c = (a - b) + (b - c) = p + q > 0.$$

This completes the proof of the fourth property.

EXAMPLE 1   Solve $2x + 1 > 7$. Sketch the solution on a coordinate line.

SOLUTION The solution of this linear inequality proceeds very much like the solution of a linear equation:

$$2x + 1 > 7$$
$$2x + 1 - \mathbf{1} > 7 - \mathbf{1}$$
$$2x > 6$$
$$x > 3.$$

FIGURE 16

The solution is the interval $(3, +\infty)$. Its graph is shown in Figure 16. The endpoint of the interval $x = 3$ is not a solution of the inequality; to indicate this we mark it with an open circle on the graph.

EXAMPLE 2 Solve $-3 \leq 2x - 5 \leq 11$. Sketch the solution on a coordinate line.

SOLUTION Recall from Section 1.1 that a solution to a compound inequality of the form $a \leq u \leq b$ is a real number that satisfies both $a \leq u$ and $u \leq b$. In this case, we could solve $-3 \leq 2x - 5$ and $2x - 5 \leq 11$. The set of real numbers common to both of these solutions is the solution set of the original inequality. Since the steps are more or less the same, we can solve the inequality in this manner:

$$-3 + 5 \leq 2x - 5 + 5 \leq 11 + 5 \qquad \text{add 5 to each member of the inequality}$$

$$2 \leq 2x \leq 16 \qquad \text{multiply each member of the inequality by } \tfrac{1}{2}$$

$$1 \leq x \leq 8.$$

In interval notation, the solution is $[1, 8]$. The graph is shown in Figure 17. The endpoints of this interval are marked with solid circles because they are part of the set of solutions.

FIGURE 17

EXAMPLE 3 Solve the compound inequality $-\tfrac{3}{2} < 2(5 - x) \leq 12$. Sketch the solution on a coordinate line.

SOLUTION $2\left(-\tfrac{3}{2}\right) < 2[2(5 - x)] \leq 2(12)$

$$-3 < 20 - 4x \leq 24$$

$$-23 < -4x \leq 4$$

$$\tfrac{23}{4} > x \geq -1 \quad \text{or} \quad -1 \leq x < \tfrac{23}{4}.$$

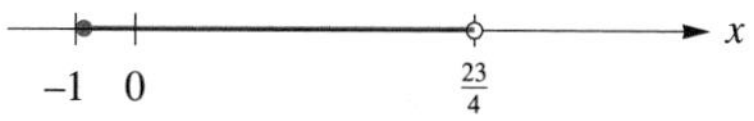

In interval notation, $\left[-1, \tfrac{23}{4}\right)$. See Figure 18.

FIGURE 18

In our discussion of absolute value in Section 1.1, we saw that an inequality such as $|u| < a$ ($a \neq 0$) can be written without absolute value symbols as $-a < u < a$.

Inequalities Involving Absolute Values

> For $a > 0$,
>
> $$|u| < a \quad \Leftrightarrow \quad -a < u < a$$
>
> $$|u| \leq a \quad \Leftrightarrow \quad -a \leq u \leq a$$
>
> $$|u| > a \quad \Leftrightarrow \quad u < -a \text{ or } u > a$$
>
> $$|u| \geq a \quad \Leftrightarrow \quad u \leq -a \text{ or } u \geq a.$$

We will use these facts to reduce an inequality that involves absolute values to one that does not.

FIGURE 19

EXAMPLE 4　Solve the absolute value inequality. Sketch the solution on a coordinate line (Figure 19).

　　a)　$\left|\frac{1}{3}x + 1\right| < 5$ 　　　　　　　　　**b)**　$|2 - 5x| \geq 38$

SOLUTION

a)　This inequality is equivalent to $-5 < \frac{1}{3}x + 1 < 5$. Proceeding as we did in Example 2,

$$-5 < \tfrac{1}{3}x + 1 < 5$$
$$-6 < \tfrac{1}{3}x < 4$$
$$-18 < x < 12.$$

The solution is the interval $(-18, 12)$.

b)　We can express this inequality as

$$2 - 5x \leq -38 \quad \text{or} \quad 2 - 5x \geq 38.$$

Solving,

$$\begin{array}{c|c} -5x \leq -40 & -5x \geq 36 \\ x \geq 8. & x \leq -\tfrac{36}{5}. \end{array}$$

Any real number in the intervals $(-\infty, -\tfrac{36}{5}]$ or $[8, \infty)$ is a solution to the inequality (Figure 20).

FIGURE 20

An inequality can always be expressed in such a way that there is a zero on the right side. If the other side can be factored, then a property of real numbers can be exploited to help solve the inequality.

Zero-Factor Property for Inequalities

> $AB > 0$ if and only if A and B are either both positive or both negative.
>
> $AB < 0$ if and only if A and B are of different signs.

Of course, this depends on whether we can write the left side of an inequality as a product. The next example should shed some light on how this is used.

EXAMPLE 5   Solve the inequality $x^2 < x + 12$. Sketch the solution on a coordinate line.

SOLUTION   First, we add $-x - 12$ to each side to make the right side of the inequality zero:

$$x^2 - x - 12 < 0.$$

Factoring, we get

$$(x + 3)(x - 4) < 0.$$

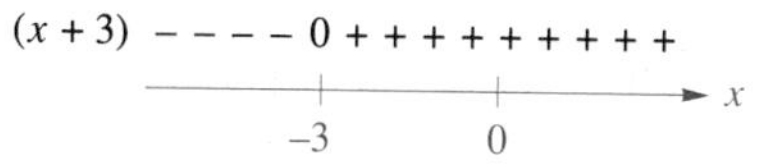

FIGURE 21

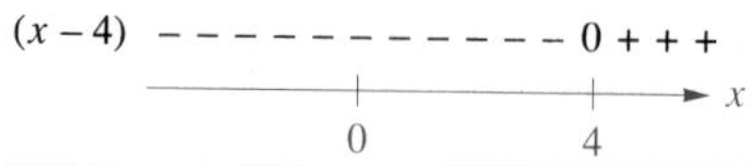

Consider each of the factors separately. The factor $(x + 3)$ is positive for all real numbers x such that $x > -3$ (as Figure 21 indicates). On the other hand, the factor $(x - 4)$ is positive for all real numbers x such that $x > 4$ (Figure 22).

Putting these two together (Figure 23), we can see that the coordinate line is divided into three open intervals: $(-\infty, -3)$, $(-3, 4)$, and $(4, \infty)$.

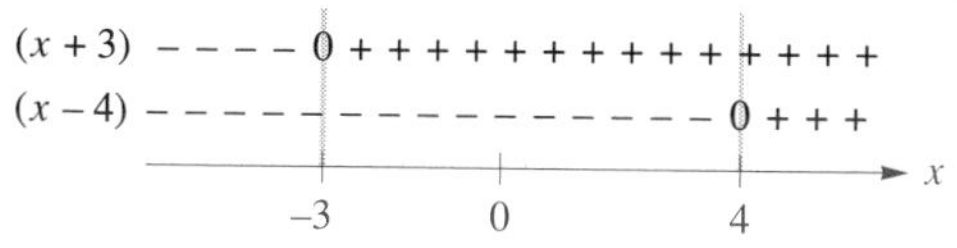

FIGURE 22

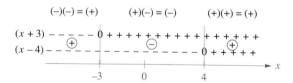

FIGURE 23

Consider the interval $(-\infty, -3)$. For each real number in this interval, both factors $(x + 3)$ and $(x - 4)$ are negative. This implies, of course, that their product $(x + 3)(x - 4)$ is positive. We will indicate all this simply by labeling the interval $(-)(-) = (+)$. The other intervals are appropriately labeled as well (Figure 24).

FIGURE 24

The graph in Figure 24 is called the **sign graph** of this inequality. Since solutions to the inequality are those for which $(x + 3)(x - 4) < 0$, the set of solutions is the interval $(-3, 4)$. See Figure 25.

FIGURE 25

This technique can be generalized to more than two factors, as the next example shows.

EXAMPLE 6   Solve the inequality $x^3 \geq 4x$. Sketch the solution on a coordinate line.

SOLUTION Again, we first express the inequality so that its right side is zero:

$$x^3 - 4x \geq 0.$$

Factoring gives us

$$x(x^2 - 4) \geq 0$$

$$x(x + 2)(x - 2) \geq 0.$$

Constructing the sign graph, we get Figure 26.

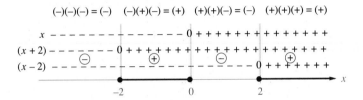

FIGURE 26

The solution is $[-2, 0]$ or $[2, \infty)$.

An inequality involving a rational expression can be solved in much the same manner that we used in the previous example. This is the subject of the next three examples.

EXAMPLE 7   Solve the inequality $\dfrac{x - 5}{x + 2} > 0$. Sketch the solution on a coordinate line.

SOLUTION Using a scheme that is similar to the product in Examples 5 and 6, we construct a sign graph with $(x - 5)$ and $(x + 2)$. From this, we determine the sign of this quotient in each of the intervals shown in Figure 27.

FIGURE 27

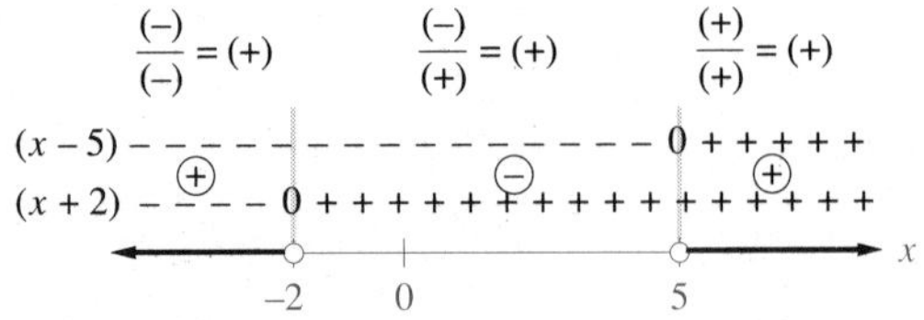

The solution is $(-\infty, -2)$ or $(5, \infty)$.

EXAMPLE 8   Solve the inequality $\dfrac{3x - 1}{2x + 3} \leq 2$. Sketch the solution on a co-ordinate line.

SOLUTION At first glance, you may be tempted to multiply each side of the inequality by $2x + 3$ to clear the fraction. This presents a problem, however. If we multiply each side by a positive value, the sense of the inequality remains the same. If we multiply each side by a negative value, the sense of the inequality changes. We don't know whether $2x + 3$ is positive or negative since it depends on the value of x.

Instead, we start by subtracting 2 from both sides to make the right side zero:

$$\frac{3x - 1}{2x + 3} - 2 \leq 0.$$

Simplifying this expression reduces our problem to one similar to the inequality of Example 7:

$$\frac{(3x - 1) - 2(2x + 3)}{2x + 3} \leq 0$$

$$\frac{-x - 7}{2x + 3} \leq 0.$$

It is to our advantage to multiply each side by -1 (notice that the sense of the inequality changes):

$$\left(\frac{-x - 7}{2x + 3}\right)(-1) \geq 0(-1)$$

$$\frac{x + 7}{2x + 3} \geq 0.$$

From this we construct the sign graph and determine the solution (Figure 28).

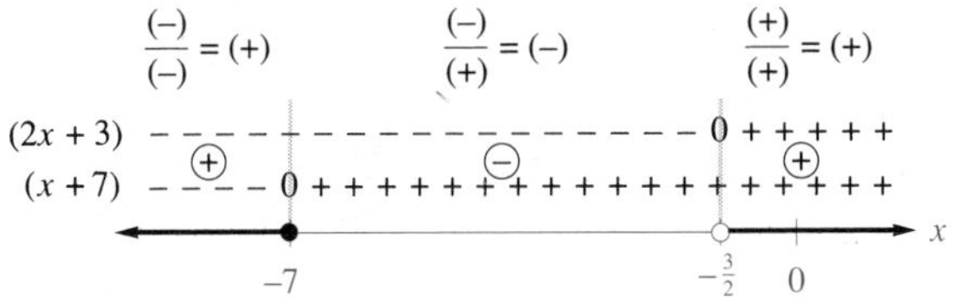

FIGURE 28

Notice that the endpoint $x = -7$ is a solution since it makes the rational expression $\dfrac{x + 7}{2x + 3}$ equal to zero, but the endpoint $x = -\frac{3}{2}$ is not a solution since it makes the denominator zero.

The solution is $(-\infty, -7]$ or $(-\frac{3}{2}, \infty)$.

EXAMPLE 9   Solve the inequality

$$\frac{(x-4)^2(x+7)}{(2x-1)(x+2)} \geq 0.$$

Sketch the solution on a coordinate line.

SOLUTION   We construct a sign graph (Figure 29), accounting for each factor in the numerator and the denominator (look carefully at how the factor $(x-4)^2$ is handled). We leave it to you to verify the sign of the rational expression in each of the intervals.

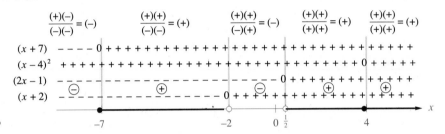

FIGURE 29

Thus the solution is $[-7, -2)$ or $(\frac{1}{2}, \infty)$.

The solution of inequalities will also be discussed in Sections 2.1 and 3.4.

EXERCISE SET 1.5

A

In Problems 1 through 6, solve the inequality. Sketch the solution on a coordinate line.

1. $2(x-3) - 3(x+2) < 5$

2. $x - 4[1 - (2x+1)] \leq 3x - 2$

3. $(x+5)(x-3) \geq x^2 - (2-x)$

4. $4x^2 - 4x + 2 > (2x-2)(2x+3)$

5. $\dfrac{x+2}{3} + \dfrac{x-3}{4} \leq 2$

6. $\dfrac{x-3}{2} - 1 \leq \dfrac{4x+1}{5} + 3$

In Problems 7 through 12, solve the compound inequality. Sketch the solution on a coordinate line.

7. $5 < 3x - 1 < 14$

8. $-3 < 4x + 5 < 13$

9. $4 \leq \dfrac{6+x}{5} \leq 14$

10. $-3 \leq \dfrac{2x+5}{3} \leq 6$

11. $-2 < 8 - 5x < 7$

12. $-5 < 8 - 2(x+3) < 13$

In Problems 13 through 18, solve the quadratic inequality. Sketch the solution on a coordinate line.

13. $(x-2)(x+4) \leq 0$

14. $(x+4)(x-3) > 0$

15. $x^2 + 3x - 10 \leq 0$

16. $x^2 - 11x + 24 > 0$

17. $x^2 - 6 \le x$

18. $5x \ge x^2 + 4$

In Problems 19 through 24, solve the absolute value inequality.
Sketch the solution on a coordinate line.

19. $|x - 1| > 2$

20. $|x + 5| < \frac{1}{2}$

21. $|3x + 2| \ge 3$

22. $|3 - x| < 4$

23. $|4 - 3x| + 4 \le 12$

24. $|x - 8| - 3 \ge 2$

B

In Problems 25 through 48, solve the inequality and sketch
the solution on a coordinate line.

25. $(x - 2)(x + 4)(x + 7) < 0$

26. $(x + 4)(x - 5)x \ge 0$

27. $x^2(x - 16)(x - 3) \le 0$

28. $x(x - 3)^2(x - 7) < 0$

29. $(2 - x)(x - 3)(2x + 5) \le 0$

30. $(5 - x)(2x - 7)(x + 5) \le 0$

31. $\dfrac{x + 5}{x - 4} > 0$

32. $\dfrac{x - 10}{2x + 5} < 0$

33. $\dfrac{x + 7}{x - 5} \le 0$

34. $\dfrac{x - 8}{x + 3} \ge 0$

35. $\dfrac{(x + 6)(x - 2)}{x - 12} < 0$

36. $\dfrac{x}{(x + 5)(x - 1)} \ge 0$

37. $\dfrac{2x}{x^2 - 25} \ge 0$

38. $\dfrac{x + 2}{x^2 - 2x - 48} < 0$

39. $\dfrac{x + 1}{x + 10} \ge 3$

40. $\dfrac{x - 12}{x} < 3$

41. $\dfrac{6}{x + 1} < \dfrac{2}{x - 2}$

42. $\dfrac{10}{3x - 1} < \dfrac{4}{2x - 3}$

43. $\dfrac{2}{x^2 + 10x + 25} \ge \dfrac{3}{x^2 - 25}$

44. $||x| - 2| < 4$

45. $0 < 3 + x < 4x$

46. $0 < 4x < 2x - 1$

47. $0 \le 5 - 2x < x$

48. $0 < -x \le 4x$

49. a) Solve $|2x - 6| < 1.2$. Sketch the solution on a coordinate line.

b) Determine a δ such that $|x - 3| < \delta$ is an equivalent equation to that of part a).

50. a) Solve $|4x - 10| < 0.8$. Sketch the solution on a coordinate line.

b) Determine a δ such that $|x - 2.5| < \delta$ is an equivalent equation to that of part a).

51. a) Solve $|2x - 12| < 0.01$. Sketch the solution on a coordinate line.

b) Determine a δ such that $|x - 6| < \delta$ is an equivalent equation to that of part a).

52. a) Solve $|\frac{1}{2}x - 1| < 1.6$. Sketch the solution on a coordinate line.

b) Determine a δ such that $|x - 2| < \delta$ is an equivalent equation to that of part a).

53. a) Solve $x^3 < 9x$. Sketch the solution on a coordinate line.

b) Solve $x^2 < 9$. Sketch the solution on a coordinate line.

54. a) Solve $2x - 5 < x + 3$. Sketch the solution on a coordinate line.

b) Solve $(2x - 5)^2 < (x + 3)^2$. Sketch the solution on a coordinate line.

55. a) Solve $x^2 - 2x - 8 < 0$. Sketch the solution on a coordinate line.

b) Solve $x - 2 - 8/x < 0$. Sketch the solution on a coordinate line.

56. Prove Property 2 of the Properties of Inequalities listed in this section.

57. Prove Property 3 of the Properties of Inequalities listed in this section.

58. Determine the solution of each of the following by inspection:

a) $\sqrt{x + 2} < 0$

b) $|x^2 - 10| \le -1$

c) $(x - 2)^2 \le 0$

d) $\sqrt{x - 2} > \sqrt{x}$

C

59. Prove that if a and b are nonnegative, then $a > b$ if and only if $a^2 > b^2$. (Hint: Do this by showing that $a - b > 0$ if and only if $a^2 - b^2 > 0$.)

60. Use Problem 59 to solve $|3x - 8| > |2x + 2|$.

61. Use Problem 59 to solve $|x + 2| \le |x^2 - 4|$.

62. Show that $||x| - |y|| \le |x - y|$.

S E C T I O N   1.6   THE APPLICATION OF MATHEMATICS: PROBLEM SOLVING

You have no doubt come to appreciate that to learn or apply mathematical topics you must solve (or at least attempt to solve) problems. In doing so, you not only master the material but also improve your skills as a problem solver. Developing these skills is essential. To this end we will examine various problems and how their solutions evolve.

There is no universal procedure for solving problems. We can, however, offer guidelines to help you improve your problem-solving skills.

Guidelines for Solving Problems

1) You must understand the problem. You may need to read the problem more than once. Draw a figure. Identify the given data and unknowns. Try restating the problem in your own words.

2) Find relations between the given data and the unknown quantities. Introduce variables for the relevant unknown quantities. Label the diagram with these variables. Write any equations relating these quantities. Do you know any formulas or theorems that might be useful?

3) Find an equation involving the unknown quantity that the problem is asking for. Write this equation in terms of one variable; you may be able to use some of the equations you wrote in Step 2. If you have trouble finding this equation, try guessing the solution to the problem. Then scrutinize why your guess does not fulfill the requirements of the problem.

4) Solve the equation.

5) Check your results in the original problem. A solution to the equation may not be a solution to the problem. Interpret your result and state an answer to the original problem. Pay attention to detail such as units of measurement.

EXAMPLE 1   A printed poster is to have 3-inch margins at the top and bottom and 2-inch margins on the sides. The area of the printed portion is 48 square inches, and the area of the poster is 160 square inches. Find the dimensions of the poster.

SOLUTION The first step is to draw a picture (Figure 30). Since the poster has 3-inch margins at the top and bottom, we get

$$\begin{pmatrix} \text{Length} \\ \text{of} \\ \text{poster} \end{pmatrix} = \begin{pmatrix} \text{Length} \\ \text{of printed} \\ \text{portion} \end{pmatrix} + 6.$$

Likewise, because the side margins are 2 inches each,

$$\begin{pmatrix} \text{Width} \\ \text{of} \\ \text{poster} \end{pmatrix} = \begin{pmatrix} \text{Width} \\ \text{of printed} \\ \text{portion} \end{pmatrix} + 4.$$

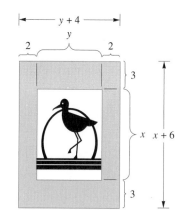

FIGURE 30

If we let x be the length of the printed portion and y be the width of the printed portion, then the length of the poster is $x + 6$, and the width of the poster is $y + 4$. Now,

$$\begin{pmatrix} \text{Area} \\ \text{of} \\ \text{poster} \end{pmatrix} = \begin{pmatrix} \text{Length} \\ \text{of} \\ \text{poster} \end{pmatrix} \begin{pmatrix} \text{Width} \\ \text{of} \\ \text{poster} \end{pmatrix}.$$

We are given that the area of the poster is 160 square inches, so

$$160 = (x + 6)(y + 4).$$

We also know that the area of the printed portion is to be 48 square inches, so

$$48 = xy.$$

If we solve this equation for y, we have

$$y = \frac{48}{x}.$$

Substituting this into the equation for the area of the poster, we obtain

$$160 = (x + 6)(y + 4)$$

$$= (x + 6)\left(\frac{48}{x} + 4\right)$$

$$= 4x + 72 + \frac{288}{x}.$$

Multiply both sides by x, and then solve the resulting quadratic equation:

$$160x = 4x^2 + 72x + 288$$

$$0 = 4x^2 - 88x + 288$$

$$0 = x^2 - 22x + 72$$

$$0 = (x - 18)(x - 4) \Rightarrow x = 18 \text{ or } 4.$$

For each of these values for x, the corresponding values for y are $\frac{48}{18} = \frac{8}{3}$ and $\frac{48}{4} = 12$, respectively. Therefore, there are two pairs of dimensions for the poster. One pair is

$$x + 6 = 18 + 6 = 24, \qquad y + 4 = \tfrac{8}{3} + 4 = 6\tfrac{2}{3},$$

and the other pair is

$$x + 6 = 4 + 6 = 10, \qquad y + 4 = 12 + 4 = 16.$$

The poster can be 24 by $6\frac{2}{3}$ inches, or the poster can be 10 by 16 inches.

EXAMPLE 2   A radiator contains 5 quarts of 30% antifreeze mixture. How much of this mixture must be drained and replaced with pure antifreeze to obtain an 80% antifreeze mixture?

SOLUTION An important general idea for this problem is that

$$\begin{pmatrix} \text{Amount of} \\ \text{antifreeze} \end{pmatrix} = \begin{pmatrix} \text{Percentage of} \\ \text{antifreeze} \end{pmatrix} \begin{pmatrix} \text{Volume of} \\ \text{mixture} \end{pmatrix}$$

We can use this idea for several cases:

$$\begin{pmatrix} \text{Amount of} \\ \text{original antifreeze} \end{pmatrix} = \begin{pmatrix} \text{Percentage of} \\ \text{original antifreeze} \end{pmatrix} \begin{pmatrix} \text{Volume of} \\ \text{original mixture} \end{pmatrix}$$

$$= (0.30)(5)$$

$$= 1.5 \text{ quarts},$$

$$\begin{pmatrix} \text{Amount of} \\ \text{drained antifreeze} \end{pmatrix} = \begin{pmatrix} \text{Percentage of} \\ \text{drained antifreeze} \end{pmatrix} \begin{pmatrix} \text{Volume of} \\ \text{drained mixture} \end{pmatrix}$$

$$= (0.30) \begin{pmatrix} \text{Volume of} \\ \text{drained mixture} \end{pmatrix}$$

$$\begin{pmatrix} \text{Amount of} \\ \text{added antifreeze} \end{pmatrix} = \begin{pmatrix} \text{Percentage of} \\ \text{added antifreeze} \end{pmatrix} \begin{pmatrix} \text{Volume of} \\ \text{added antifreeze} \end{pmatrix}$$

$$= (1.00) \begin{pmatrix} \text{Added volume} \\ \text{of antifreeze} \end{pmatrix}$$

$$\begin{pmatrix} \text{Final amount} \\ \text{of antifreeze} \end{pmatrix} = \begin{pmatrix} \text{Final percentage} \\ \text{of antifreeze} \end{pmatrix} \begin{pmatrix} \text{Final volume} \\ \text{of mixture} \end{pmatrix}$$

$$= (0.80)(5)$$

$$= 4 \text{ quarts}$$

The relationship among these four amounts is:

$$\begin{pmatrix} \text{Amount of} \\ \text{original} \\ \text{antifreeze} \end{pmatrix} - \begin{pmatrix} \text{Amount of} \\ \text{drained} \\ \text{antifreeze} \end{pmatrix} + \begin{pmatrix} \text{Amount of} \\ \text{added} \\ \text{antifreeze} \end{pmatrix} = \begin{pmatrix} \text{Final} \\ \text{amount of} \\ \text{antifreeze} \end{pmatrix}$$

$$1.5 - (0.30)\begin{pmatrix} \text{Volume of} \\ \text{drained mixture} \end{pmatrix} + (1.00)\begin{pmatrix} \text{Volume of} \\ \text{added antifreeze} \end{pmatrix} = 4$$

Since we are replacing the drained mixture with the added mixture, these two mixtures have the same volume. This volume is the only remaining unknown.

Let v be the number of quarts added.

$$1.5 - (0.30)v + (1.00)v = 4$$
$$1.5 + 0.70v = 4$$
$$0.70v = 2.5$$
$$v = \frac{2.5}{0.70} = 3\frac{4}{7}$$

Thus, $3\frac{4}{7}$ quarts must be added.

The next example involves a triangle with a right angle, which is called a **right triangle.** The side of a right triangle that is opposite the right angle is called the **hypotenuse.** The other two sides are called **legs.** No doubt you have seen the next theorem.

A triangle is a right triangle if and only if the sum of the squares of the two shorter sides is equal to the square of the longest side:

$$a^2 + b^2 = c^2$$

Pythagorean Theorem

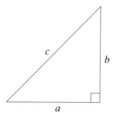

FIGURE 31

EXAMPLE 3   An offshore oil rig is located 5 miles west of a dock on a north-south coast. A pipeline connects the rig to a pumping station on the coast, which is north of the dock. The pipeline then runs north up the coast to a refinery, which is 30 miles north of the dock. The cost of laying the pipeline was $900 per mile under the sea and $200 per mile along the coast, for a total cost of $15,300. Find the distance between the dock and the pumping station.

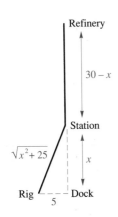

FIGURE 32

SOLUTION   Draw a picture (Figure 32). Let x be the distance between the dock and the station. The rig, dock, and station form a right triangle whose hypotenuse is the segment of pipeline from the rig to the station. By the Pythagorean Theorem,

$$\text{length of pipeline from rig to station} = \sqrt{x^2 + 5^2}$$
$$= \sqrt{x^2 + 25}$$

Also, since the distance from the dock to the refinery is 30 miles,

$$\text{length of pipeline from station to refinery} = 30 - x$$

The cost of the pipeline is

$$\left(\begin{array}{c}\text{Cost of laying}\\\text{pipeline undersea}\end{array}\right) + \left(\begin{array}{c}\text{Cost of laying}\\\text{pipeline on coast}\end{array}\right) = 15{,}300$$

$$\left(\begin{array}{c}\text{Cost per mile}\\\text{for pipeline}\\\text{undersea}\end{array}\right)\left(\begin{array}{c}\text{Number of miles}\\\text{of pipeline}\\\text{undersea}\end{array}\right) + \left(\begin{array}{c}\text{Cost per mile}\\\text{for pipeline}\\\text{on coast}\end{array}\right)\left(\begin{array}{c}\text{Number of miles}\\\text{of pipeline}\\\text{on coast}\end{array}\right) = 15{,}300$$

$$(900)(\sqrt{x^2 + 25}) + (200)(30 - x) = 15{,}300$$

To solve this equation, we simplify and isolate the term with the radical

$$(900)(\sqrt{x^2 + 25}) + 6000 - 200x = 15{,}300$$
$$(900)(\sqrt{x^2 + 25}) = 9300 + 200x$$
$$9(\sqrt{x^2 + 25}) = 93 + 2x$$

Squaring both sides and simplifying leads to

$$81(x^2 + 25) = 4x^2 + 372x + 8649$$
$$77x^2 - 372x - 6624 = 0$$
$$(x - 12)(77x + 552) = 0 \Rightarrow x = 12 \text{ or } -\tfrac{552}{77}.$$

The value $x = -\tfrac{552}{77}$ is not a feasible solution since x represents a distance. The distance from the dock to the station is 12 miles.

Sometimes the solution of a problem involves an inequality.

EXAMPLE 4   A personnel manager finds that it takes new employees an aver-age of $1\frac{1}{3}$ hours to assemble a sewing machine. An experienced employee averages only $\frac{2}{3}$ of an hour to assemble the sewing machine. The union re-quires that the assembly crew must consist of 5 more experienced workers than new workers. How many new workers will be needed on a crew if it must assemble at least 210 machines during an 8-hour shift?

SOLUTION We can relate the time of assembly to the rate of assembly. An experienced worker takes $\frac{2}{3}$ of an hour to assemble a machine. Then the experienced worker's rate is

$$\frac{\text{Number produced}}{\text{Time}} = \frac{1 \text{ machine}}{\frac{2}{3} \text{ hours}} = \frac{3}{2} \text{ machine per hour}$$

Since the crew works for 8 hours, each experienced worker can assemble

$$(\tfrac{3}{2})(8) = 12 \text{ machines per day}$$

Similarly, each new worker's rate is

$$\frac{\text{Number produced}}{\text{Time}} = \frac{1 \text{ machine}}{1\frac{1}{3} \text{ hours}} = \frac{3}{4} \text{ machine per hour}$$

and each new worker can produce

$$(\tfrac{3}{4})(8) = 6 \text{ machines per day}$$

Considering the daily quota, we have the following relationship.

$$\begin{pmatrix} \text{Number of} \\ \text{machines experienced} \\ \text{workers produce} \end{pmatrix} + \begin{pmatrix} \text{Number of} \\ \text{machines new} \\ \text{workers produce} \end{pmatrix} \geq 210$$

$$\begin{pmatrix} \text{Number each} \\ \text{experienced worker} \\ \text{produces per day} \end{pmatrix}\begin{pmatrix} \text{Number of} \\ \text{experienced} \\ \text{workers} \end{pmatrix} + \begin{pmatrix} \text{Number each} \\ \text{new worker} \\ \text{produces per day} \end{pmatrix}\begin{pmatrix} \text{Number} \\ \text{of new} \\ \text{workers} \end{pmatrix} \geq 210$$

Let x be the number of new workers needed. The union rule requires that there must be $x + 5$ experienced workers. Substituting this into the inequality, we have

$$12(x + 5) + 6x \geq 210$$

$$18x + 60 \geq 210$$

$$x \geq 8\tfrac{1}{3}$$

The personnel director needs at least 9 new workers on the crew.

EXERCISE SET 1.6

A

In Problems 1 through 32, write an equation or inequality that models the problem, solve it, and answer the question.

1. A calculator sells for $18.50 less than a textbook. If the cost to buy both is $36, find the cost of each item.

2. A compact disk sells for $3.25 less than twice the cost of a cassette tape. If the cost to buy both is $22.25, find the cost of each item.

3. The length of a rectangle is $4\frac{1}{2}$ inches greater than the width. If the area of the rectangle is 28 square inches, find the dimensions of the rectangle.

4. The area of a triangle is 26 cm². The height is $1\frac{1}{2}$ cm greater than the base. Find the dimensions of the triangle.

5. Mike drives his BMW 135 miles in the same length of time that Sherry takes to drive her Jeep 125 miles. If Mike averages 5 miles per hour faster than Sherry, find their rates.

6. Colleen types 1760 words in the same length of time that it takes Jean to type 1400 words. If Colleen averages 8 words per minute faster than Jean, find their rates.

7. The dimensions of a rectangular lawn are 50 by 80 feet. A border of uniform width is paved around the lawn. As a result, the area of the lawn is reduced to 3619 square feet. Find the width of the border.

8. A strip of uniform width is to be cut from both sides and both ends of a sheet of paper that is 11 by 14 inches. If the remaining area is to be 54 square inches, find the width of the strip.

9. A tank contains 50 gallons of 40% chlorine mixture. How much of this mixture must be drained and replaced with pure chlorine to obtain a 70% chlorine mixture?

10. A bottle of mineral water contains 12 ounces of liquid with a 30% calcium content. How many ounces of this liquid should be poured out and replaced with a 55% calcium solution to obtain a drink that is 40% calcium solution?

11. How many ml of acid must be added to 40 ml of a solution that is 55% acid to obtain a new solution that is 70% acid?

B

12. A candy store sells boxes of candy containing chocolates and caramels. The chocolates cost $0.35 each, and the caramels cost $0.40 each. How many of each should be in the box if it is to contain 20 pieces of candy at a total cost of $7.70?

13. A grocery store owner wants to make a 10-pound mixture of peanuts and cashews to sell for $1.50 per pound. The peanuts sell for $0.90 per pound, and the cashews sell for $1.80 per pound. How many pounds of each should she use?

14. An investor with $25,000 wants to place part of her money in a savings account paying 8% per year and the rest in bonds paying 11% per year. If the total return on her investment is to be $2435, how much should she invest in bonds?

15. A game-show winner split his prize money into two accounts, one paying 9% per year and the other paying 13% per year. The amount invested at 9% is twice as much as the amount invested at 13%. If the combined annual income from these two investments is $1333, how much prize money did he win?

16. A container in the shape of a right circular cylinder is to be constructed. The container is to have an open top and a height of 5 inches. If only 75π square inches of material are available, how long must the radius of the circular base be?

17. A box with an open top is to be constructed so that the base is square and the height is 12 cm. If only 448 cm² of material are available, what are the possible dimensions?

18. A boat is pulled toward a dock by a rope on a pulley (Figure 33). The pulley is 15 feet above the water level. Initially there is 39 feet of rope out. Four seconds later there is 25 feet of rope out. What is the average speed of the boat during this four-second interval?

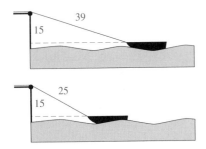

FIGURE 33

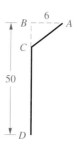

FIGURE 34

19. A Farmall harvester can pick an acre of tomatoes in $2\frac{1}{2}$ hours, and a Blackwelder harvester can pick an acre of tomatoes in 2 hours. A fleet of both types of harvesters is needed to pick 44 acres in 8 hours. If there must be two more Blackwelders than Farmalls in the fleet, how many of each type of harvester are needed?

20. A restaurant manager finds that new employees can make 40 burgers per hour and that an experienced employee can make 55 burgers per hour. Upper management requires that the cooking crew must consist of twice as many experienced workers as new workers. How many new workers will be needed on a cooking crew that can handle the lunchtime demand of at least 770 burgers from noon to 1 P.M.?

21. A pump can fill a pool in $1\frac{1}{2}$ days, and another can fill the same pool in 2 days. How long will it take both pumps to fill the pool?

22. One pipe can fill a tank in 3 hours. The drain for the tank can empty it in 4 hours. How long will it take the pipe to fill the empty tank if the drain was accidentally left open?

23. A buyer for a retail store purchases a truckload of desks for $2760. To furnish new offices in the store, two of these desks are retained. The remainder are sold for $110 above original cost per desk. If the total profit was $640, how many desks were originally purchased?

24. In Figure 34, point A is 6 miles due east of point B. Underground cable is laid from A to C at a cost of $300 per mile, then from C to D at a cost of $450 per mile. The distance from B to D is 50 miles. Find the distance from B to C if the total cost of laying the cable is $21,900.

25. In Figure 35, point A is 7 km due north of point B, which is due west 48 km of point C. Point D is 10 km due north of point C. Starting from A, a ship travels straight to a point E on $\overline{BC}$, and then straight to point D. If the distance travelled from A to E is 1 km less than the distance from E to D, find the distance from B to E.

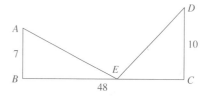

FIGURE 35

26. A plane that can fly 255 miles per hour in still air can travel 150 miles downwind in the same amount of time that it can travel 105 miles upwind. What is the speed of the wind?

27. A boat that can travel 20 miles per hour in still water travels 48 miles upstream and returns downstream in a total of 5 hours. What is the speed of the current of the river?

28. Two planes leave simultaneously from the same airport. One flies due north and the other flies due east. The northbound plane flies 80 miles per hour faster than the eastbound plane, and after 4 hours the planes are 1600 miles apart. Find the ground speed of each plane.

29. An open-top box is to be constructed from a square sheet of cardboard by cutting 3-inch squares from each corner and folding up the sides. If the box is to contain 363 cubic inches, find the size of the original piece of material (Figure 36).

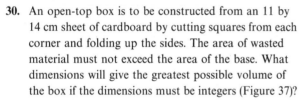

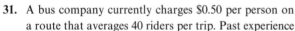

FIGURE 36

30. An open-top box is to be constructed from an 11 by 14 cm sheet of cardboard by cutting squares from each corner and folding up the sides. The area of wasted material must not exceed the area of the base. What dimensions will give the greatest possible volume of the box if the dimensions must be integers (Figure 37)?

31. A bus company currently charges $0.50 per person on a route that averages 40 riders per trip. Past experience

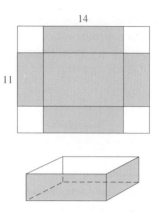

FIGURE 37

indicates that the number of passengers decreases by four people for every $0.10 increase in that rate. If the company needs to receive a revenue of at least $22.40 per run, how much can the company charge?

32. A bridge toll is currently $2.50 per car, and the average number of cars crossing per day is 230. Past experience indicates that the number of cars using the bridge decreases by 10 for every $0.25 increase in that rate. If the bridge needs to receive a revenue of at least $680 per day, how much can the company charge?

SECTION 1.7 **COMPLEX NUMBERS**

The set of integers is sufficient to solve an equation such as $x + 2 = 9$. It is not sufficient, however, when solving equations such as $2x + 9 = 0$; for this we need a larger number system, the set of rational numbers, to accommodate the root $x = -\frac{9}{2}$ of this equation.

Unfortunately, the set of rational numbers is still inadequate when it comes to solving an equation such as $x^2 = 7$. Another number system, the real numbers $\mathbb{R}$, is required to accommodate the roots $x = \pm\sqrt{7}$ of this equation.

Even with the set of real numbers, we have seen equations that still cannot be solved to our satisfaction. A simple equation such as $x^2 = -1$ has no real-number solutions. In Section 1.4, we found that quadratic equations whose discriminants are negative had no real roots.

Apparently, another expansion of the number system is necessary to solve these types of equations satisfactorily. We start by defining a new number that is the square root of -1. Certainly this number is not a real number since the square of any real number is nonnegative. Our new number has no image on the coordinate line. Although the best name for this would be the *imageless unit,* it is conventionally called the **imaginary unit.** It is denoted by the letter i. It follows that $i^2 = -1$.

$$i = \sqrt{-1}.$$

The Imaginary Unit

We can use this nonreal number to construct our extended number system that contains the roots of quadratic equations with negative discriminants. We call this new system the **complex number system.**

A complex number is an expression of the form $a + bi$, where a and b are real numbers, and i is the imaginary unit.

Complex Numbers

A better name for this set of numbers would have been the *complete number system* because this is the last extension that needs to be made to contain the roots of all polynomial equations.

It should seem reasonable that any real number is also a complex number. For example, consider the real number 5. If we express this number as $5 + 0i$, we can see that it qualifies as a complex number.

EXAMPLE 1   Solve $x^2 = -9$ over the set of complex numbers.

SOLUTION   By extracting roots, we get $x = \pm\sqrt{-9}$. The obvious problem is how to interpret $\sqrt{-9}$. It would seem natural that we would want to preserve the laws of radicals from the set of real numbers. So,

$$\sqrt{-9} = \sqrt{9}\sqrt{-1} = 3i.$$

The apparent solutions are $x = -3i$ and $x = 3i$. We can check these directly (notice that we assume that the properties of exponents for real numbers also hold for complex numbers):

$$(-3i)^2 \overset{?}{=} -9 \qquad\qquad (3i)^2 \overset{?}{=} -9$$
$$(-3)^2 i^2 \overset{?}{=} -9 \qquad\qquad (3)^2 i^2 \overset{?}{=} -9$$
$$9(-1) = -9. \qquad\qquad 9(-1) = -9.$$

The check verifies that $x = -3i$ and $x = 3i$ are the roots of this equation.

EXAMPLE 2   Find the complex roots of the quadratic equation

$$(x - 2)^2 = -16.$$

SOLUTION This equation certainly has no real roots since for no real value of x can $(x - 2)^2$ be negative. However, by extracting roots, we get

$$(x - 2)^2 = -16$$
$$x - 2 = \pm\sqrt{-16}$$
$$x - 2 = \pm 4i$$
$$x = 2 \pm 4i.$$

The two complex roots are $x = 2 + 4i$ and $x = 2 - 4i$. We leave the check to you (pause here and do this).

Equality, addition, and multiplication of complex numbers are defined in a natural way. The manipulation of complex numbers is very much like that of polynomials, with one exception. When the quantity i^2 occurs, it can be replaced by -1. Specifically:

$$a + bi = c + di \quad \text{if and only} \quad a = c \text{ and } b = d$$
$$(a + bi) + (c + di) = (a + c) + (b + d)i$$
$$(a + bi)(c + di) = (ac - bd) + (ad + bc)i.$$

EXAMPLE 3   Write each expression in the form $a + bi$.

   a) $(2 + 5i) + (3 - 2i)$ **b)** $(3 + 7i)(1 - 4i)$ **c)** $(2 + i)^2$

SOLUTION Rather than blindly follow the preceding definitions, we will simplify these expressions just as we might simplify a polynomial expression.

   a) $(2 + 5i) + (3 - 2i) = (2 + 3) + (5i - 2i) = 5 + 3i.$

   b) $(3 + 7i)(1 - 4i) = 3 - 5i - 28i^2 = 3 - 5i - 28(-1) = 31 - 5i.$

   c) $(2 + i)^2 = 4 + 4i + i^2 = 4 + 4i + (-1) = 3 + 4i.$

Division of complex numbers is accomplished in a manner that is similar to that of rationalizing the denominator of an algebraic fraction with a radical in the denominator (you may want to pause here and review Example 8 of Section 1.3).

Given a complex number $z = a + bi$, the complex conjugate of z is

$$\bar{z} = a - bi.$$

Furthermore, it follows that

1) $z + \bar{z} = (a + bi) + (a - bi) = 2a,$

2) $z\bar{z} = (a + bi)(a - bi) = a^2 + b^2.$

Complex Conjugate

A bar over any complex expression denotes the complex conjugate of that expression. Also note that the conjugate of a real number is itself.

This second property is the one that we will exploit. The next example demonstrates how the complex conjugate is used to simplify the quotient of two complex numbers.

EXAMPLE 4   Write each expression in the form $a + bi$.

a) $\dfrac{4 - i}{2 + 3i}$

b) $\dfrac{1}{\sqrt{2} - i}$

c) $\dfrac{1 + 3i}{i}$

SOLUTION

a) The plan here is to multiply the numerator and the denominator by the complex conjugate of the denominator. Notice how this operation "realizes" the denominator of the ratio:

$$\frac{4 - i}{2 + 3i} = \frac{4 - i}{2 + 3i} \cdot \frac{2 - 3i}{2 - 3i} = \frac{8 - 14i + 3i^2}{2^2 - (3i)^2} = \frac{8 - 14i - 3}{4 - 9i^2}$$

$$= \frac{5 - 14i}{4 + 9} = \frac{5}{13} - \frac{14}{13}i.$$

b) $\dfrac{1}{\sqrt{2} - i} = \dfrac{1}{\sqrt{2} - i} \cdot \dfrac{\sqrt{2} + i}{\sqrt{2} + i} = \dfrac{\sqrt{2} + i}{\sqrt{2}^2 - i^2} = \dfrac{\sqrt{2} + i}{2 + 1} = \dfrac{\sqrt{2}}{3} + \dfrac{1}{3}i.$

c) The complex conjugate of $i = 0 + i$ is $0 - i = -i$:

$$\frac{1 + 3i}{i} = \frac{1 + 3i}{i} \cdot \frac{-i}{-i} = \frac{-i - 3i^2}{-i^2} = \frac{3 - i}{1} = 3 - i.$$

In certain situations, powers of i need to be simplified. For example, using the fact that $i^2 = -1$, we see that

$$i^3 = (i^2)i = (-1)i = -i \quad \text{and} \quad i^4 = (i^2)(i^2) = (-1)(-1) = 1.$$

EXAMPLE 5   Write each expression in the form $a + bi$.

a) $(1 + 2i)^{-3}$ **b)** i^{-4}

SOLUTION

a) $\dfrac{1}{(1 + 2i)^3} = \dfrac{1}{(1)^3 + 3(1)^2(2i) + 3(1)(2i)^2 + (2i)^3}$

$= \dfrac{1}{1 + 6i + 12i^2 + 8i^3}$

$= \dfrac{1}{1 + 6i - 12 - 8i}$

$= \dfrac{1}{-11 - 2i}$

$= \dfrac{1}{-11 - 2i} \cdot \dfrac{-11 + 2i}{-11 + 2i}$

$= \dfrac{-11 + 2i}{121 + 4} = \dfrac{-11}{125} + \dfrac{2}{125} i.$

b) $i^{-4} = \dfrac{1}{i^4} = \dfrac{1}{1} = 1.$

In our discussion of equations in Section 1.4, we saw that a quadratic equation $ax^2 + bx + c = 0$ has roots of the form

$$x = \frac{-b + \sqrt{b^2 - 4ac}}{2a} \quad \text{and} \quad x = \frac{-b - \sqrt{b^2 - 4ac}}{2a}$$

This implied that a quadratic equation has no real roots, exactly one real root, or two real roots, depending on the sign of the discriminant $b^2 - 4ac$. Specifically, the equation in which the discriminant is negative yielded no real roots. This equation does, however, have complex roots.

EXAMPLE 6   Find the complex roots of the quadratic equation

$$3x^2 + 12x + 14 = 0.$$

(Compare this with part c) of Example 7 in Section 1.4.)

SOLUTION Using the quadratic formula, we get

$$x = \frac{-12 \pm \sqrt{-24}}{6}.$$

Since the discriminant is negative, we know that there are no real roots. However, since

$$\sqrt{-24} = \sqrt{24}\sqrt{-1} = 2\sqrt{6}i,$$

we can proceed as follows:

$$x = \frac{-12 \pm \sqrt{-24}}{6} = \frac{-12 \pm 2\sqrt{6}i}{6} = -2 \pm \frac{\sqrt{6}}{3}i.$$

The complex roots of this equation are

$$-2 + \frac{\sqrt{6}}{3}i \quad \text{and} \quad -2 - \frac{\sqrt{6}}{3}i.$$

Equations of higher degree can be solved in a similar fashion.

EXAMPLE 7   Solve $x^3 - 8 = 0$ over the complex numbers.

SOLUTION We start by factoring:

$$(x - 2)(x^2 + 2x + 4) = 0.$$

So, we obtain

$$x - 2 = 0 \quad \text{or} \quad x^2 + 2x + 4 = 0.$$

The first equation has $x = 2$ as a root. Solving the second equation by quadratic formula yields

$$x = \frac{-2 \pm \sqrt{2^2 - 4(1)(4)}}{2} = \frac{-2 \pm \sqrt{-12}}{2} = \frac{-2 \pm 2\sqrt{3}i}{2} = -1 \pm \sqrt{3}i.$$

The complex roots of the equations are 2, $-1 + \sqrt{3}i$, and $-1 - \sqrt{3}i$.

EXAMPLE 8   Given the equation $x^4 + 13x^2 = 48$:

    a) Solve over the real numbers.

    b) Solve over the complex numbers.

Solution

a) We start by expressing the equation in such a way that the right side is zero:

$$x^4 + 13x^2 - 48 = 0.$$

This factors into

$$(x^2 - 3)(x^2 + 16) = 0.$$

So,

$$x^2 - 3 = 0 \quad \text{or} \quad x^2 + 16 = 0$$

$$x^2 = 3 \qquad\qquad x^2 = -16$$

$$x = \pm\sqrt{3} \quad\bigg|\quad \text{no real roots.}$$

The real roots of the equation are $\sqrt{3}$ and $-\sqrt{3}$.

b) Over the complex numbers, the equation $x^2 + 16 = 0$ does have roots. These are $4i$ and $-4i$. The complex roots of the original equation are $\sqrt{3}$, $-\sqrt{3}$, $4i$, and $-4i$.

We finish this section with a discussion of some of the properties of the complex conjugate.

EXAMPLE 9   Given that $z_1 = a_1 + b_1 i$ and that $z_2 = a_2 + b_2 i$, show that

a) $\bar{z}_1 + \bar{z}_2 = \overline{z_1 + z_2}$ **b)** $(\bar{z}_1)^2 = \overline{(z_1)^2}$.

Solution Our scheme in both part a) and part b) is to express each side of the identity in terms of a and b and to verify that they are equal.

a) Starting with the left side of the identity, we get

$$\bar{z}_1 + \bar{z}_2 = (a_1 - b_1 i) + (a_2 - b_2 i)$$

$$= (a_1 + a_2) - (b_1 + b_2)i.$$

Simplifying the right side gives

$$\overline{z_1 + z_2} = \overline{(a_1 + b_1 i) + (a_2 + b_2 i)}$$

$$= \overline{(a_1 + a_2) + (b_1 + b_2)i}$$

$$= (a_1 + a_2) - (b_1 + b_2)i.$$

This proves the identity to be true.

b) We leave it to you to supply the steps:

$$\bar{z}_1 = a_1 - b_1 i$$

$$(\bar{z}_1)^2 = (a_1 - b_1 i)^2 = a_1{}^2 - 2a_1 b_1 i + (b_1 i)^2 = (a_1{}^2 - b_1{}^2) - 2a_1 b_1 i$$

$$\overline{(z_1)^2} = \overline{(a_1 + b_1 i)^2} = \overline{a_1{}^2 + 2a_1 b_1 i + (b_1 i)^2} = \overline{(a_1{}^2 - b_1{}^2) + 2a_1 b_1 i}$$

$$= (a_1{}^2 - b_1{}^2) - 2a_1 b_1 i.$$

EXERCISE SET 1.7

A

In Problems 1 through 12, solve the equation. Write your answer in the form a + bi.

1. $x^2 = -25$

2. $x^2 = -64$

3. $x^2 = -11$

4. $x^2 = -26$

5. $(x - 3)^2 = -4$

6. $(x + 7)^2 = -16$

7. $(x + 1)^2 = -12$

8. $(x - 5)^2 = -45$

9. $4x^2 = -7$

10. $100x^2 = -25$

11. $9(x - 5)^2 = -16$

12. $4(x - 5)^2 = -49$

In Problems 13 through 24, simplify the complex number expression. Write your answer in the form a + bi.

13. $(2 + i) + (4 - 7i)$

14. $(8 - 2i) + (9 + 5i)$

15. $(4 + 10i) - (12 - 2i)$

16. $(2 + i) - (4 - 7i)$

17. $(2 + 2i)(1 + 3i)$

18. $(3 + i)(4 - 5i)$

19. $i^4 + (4 - 7i)$

20. $(5 + i)^2$

21. $(4 - i)^2$

22. $\dfrac{1 + 2i}{3 - 4i}$

23. $\dfrac{3 + i}{1 - 2i}$

24. $\dfrac{3 + i}{4i}$

B

In Problems 25 through 45, let $z_1 = 2 - i$, $z_2 = 3 + 7i$, and $z_3 = -2i$. Simplify the complex number expression, and write your answer in the form a + bi.

25. $z_1 z_2$

26. $z_3 z_2$

27. $z_1 z_3 + z_2$

28. $z_1 z_3 + z_2 z_3$

29. $z_1 z_3 - z_2 z_3$

30. $(z_1)^3$

31. $(z_2)^3$

32. $(z_1)^{-2}$

33. $(z_3)^{-2}$

34. $\bar{z}_1 z_1$

35. $\bar{z}_2 z_2$

36. $\bar{z}_3 z_3$

37. $\bar{z}_3 z_2$

38. $\bar{z}_1 z_2$

39. $2\bar{z}_1 + 5\bar{z}_2$

40. $4\bar{z}_1 - \bar{z}_2$

41. $(z_1)^{-3}$

42. $(z_3)^{-4}$

43. $\bar{z}_3 \bar{z}_2$

44. $\bar{z}_1 \bar{z}_2$

45. $\bar{z}_1 + z_1$

In Problems 46 through 67, solve equation over the complex numbers. Write nonreal solutions in the form a + bi.

46. $x^2 - 2x + 6 = 0$

47. $x^2 + 4x + 12 = 0$

48. $x^2 + 5x + 7 = 0$

49. $x^2 - 2x + 3 = 0$

50. $2x^2 - 6x + 7 = 0$

51. $3x^2 + x + 1 = 0$

52. $x^2 - 2x - 15 = 0$

53. $x^2 + 4x - 5 = 0$

54. $x^3 - 2x^2 + 6x = 0$

55. $x^3 + 4x^2 + 8x = 0$

56. $x^4 + 13x^2 + 36 = 0$

57. $x^4 + 10x^2 + 9 = 0$

58. $x^4 + 4x^2 + 4 = 0$

59. $x^4 + 6x^2 + 9 = 0$

60. $x^4 - 2x^3 + x - 2 = 0$

61. $x^4 + 5x^3 + x + 5 = 0$

62. $x^3 - 64 = 0$

63. $x^3 - 27 = 0$

64. $8x^3 + 1 = 0$

65. $27x^3 - 64 = 0$

66. $x^6 - 64 = 0$ (Hint: See Problem 94 of Section 1.2.)

67. $x^6 - 1 = 0$ (Hint: See Problem 94 of Section 1.2.)

68. In Example 7, we found the roots of the equation $x^3 - 8 = 0$. Check these solutions.

69. In Example 8, we found the roots of the equation $x^4 + 13x^2 = 48$. Check these solutions.

70. Evaluate:

a) i^{20} **b)** i^{53} **c)** i^{131} **d)** i^{86}

71. Evaluate:

a) i^{-6} **b)** i^{-9} **c)** i^{-19} **d)** i^{-36}

72. Explain why, if a quadratic equation with real coefficients has two nonreal roots, then the roots are a pair of complex conjugates.

73. Use Problem 72 to find a quadratic equation with real coefficients such that $3 - i$ is a root.

74. Use Problem 72 to find a quadratic equation with real coefficients such that $1 + 4i$ is a root.

C

75. Show that $\overline{z_1 + z_2} = \overline{z_1} + \overline{z_2}$ for any complex numbers z_1 and z_2.

76. Show that $a\overline{z_1} + b\overline{z_2} = \overline{az_1 + bz_2}$ for any complex numbers z_1 and z_2.

77. Show that if $\overline{z} = z$, then z is a real number.

78. Show that if $\overline{z} = -z$, then z is of the form bi.

A Note About Proof

One of the most distinguishing characteristics of mathematics is the certainty of its results. When a mathematical statement is proven, it is final and cannot be affected by subsequent empirical discoveries. Unfortunately, there is wide disagreement among mathematicians as to what constitutes a proof. For our purposes, we will define a proof as **an argument or demonstration that convinces the reader.**

When using this text you will not only read proofs, but you will also be asked to write proofs of your own. A fundamental expression in the logic of proofs is the conditional statement, "If . . . then . . .". An example is

$$\text{"if } x < -3 \text{, then } x^2 > 4\text{."}$$

There are other ways to state the same thing, such as

$$\text{"Suppose } x < -3 \text{; this implies } x^2 > 4\text{."}$$
$$\text{"If } x < -3 \text{, it follows that } x^2 > 4\text{."}$$
$$\text{"}x^2 > 4 \text{, if } x < -3\text{."}$$
$$\text{"}x < -3 \Rightarrow x^2 > 4\text{."}$$

Showing that such a conditional statement is true for specific values of x does not constitute a proof of the implication. For example, the following **does not prove** the statement "if $x < -3$, then $x^2 > 4$":

a fallacious proof

$$\text{Let } x = -5 \text{. Then } x^2 = 25 \text{. Since } x < -3$$
$$\text{and } x^2 > 4 \text{, this proves the implication.}$$

To prove this implication requires showing that it is true for all values of x. It is important to understand that the implication

$$\text{"if } x < -3 \text{, then } x^2 > 4\text{"}$$

does not say

$$\text{"if } x^2 > 4 \text{, then } x < -3\text{."}$$

In fact, this last statement is not true (for all values of x). Try the counterexample $x = 5$.

To examine implications in general, let P and Q represent statements.

For the conditional statement,
"if P, then Q,"
the **converse** *is*
"if Q, then P,"
and the **contrapositive** *is*
"if not Q, then not P."

For example, suppose P represents the statement, "it has been raining," and Q represents the statement, "the lawn is wet." The three statements above become

Implication: *"If it has been raining, then the lawn is wet."*
Converse: *"If the lawn is wet, then it has been raining."*
Contrapositive: *"If the lawn is not wet, then it has not been raining."*

Clearly the converse is not necessarily true (there may be dew, or the sprinklers may have been on recently). However, the·contrapositive is true and in fact really says the same thing that the implication says. The following fact can be very useful when reading or doing proofs.

The implication
"if P, then Q"
is equivalent to its contrapositive
"if not Q, then not P."
In other words, proving either of these statements is equivalent to proving the other.

Sometimes a conditional statement is true and its converse is true. For example, the implication
"if $xy = 0$, then $x = 0$ or $y = 0$"
and its converse
"if $x = 0$ or $y = 0$, then $xy = 0$"
are both true. Instead of stating both the implication and its converse, the phrase "if and only if" is used:
"$xy = 0$, if and only if $x = 0$ or $y = 0$"

In general, for statements P and Q,

"P if and only if Q" is actually two statements:

1)   If P then Q, and

2)   If Q then P.

Thus, as a rule, proving "P if and only if Q" requires two proofs.

Archimedes of Syracuse

The Greeks established the existence of irrational numbers in the 5th century B.C., although they did not indicate how irrational numbers fit into the real number system. The greatest mathematician of ancient times was Archimedes of Syracuse (287 to 212 B.C.).

Archimedes is regarded as one of the greatest mechanical geniuses of all time. He founded the science of hydrostatics, and applied the principles of levers to calculate centers of gravity. He is famed for his inventions of war, such as catapults that hurled stones weighing up to a quarter of a ton. Another dramatic account describes a machine that would reach over walls, seize attacking Roman ships, and smash them against the cliffs. He also designed a system of mirrors that focused the rays of the sun on the Roman ships as they approached, causing them to burst into flames. Indeed, his ability to design clever weapons (among other things) became legendary. The historian Plutarch, in his *Lives of the Noble Grecians and Romans,* wrote that the Romans grew so wary that "if they but see a little rope or a piece of wood from the wall, instantly crying out, that there it was again, Archimedes was about to let fly some engine at them, they turned their backs and fled."

However, Archimedes felt that his many achievements in applied mathematics paled in comparison to his contributions to pure mathematics. He computed the value of π to within two decimal places. He discovered general methods for finding planar areas bounded by curves as well as for finding volumes bounded by curved surfaces. The formulas for the area of a circle and for the areas of ellipses are examples. Other examples, of which he admits he was most proud, are the formulas for the volumes of the cone and sphere. As Archimedes explained to his close friend King Hieron II,

. . . mathematics rewards only those who are interested in it not only because of its rewards but also because of itself. Mathematics is like your daughter, Helena, who suspects every time a suitor appears that he is not really in love with her, but is interested in her only because he wants to be the son-in-law of the king. She wants a husband who loves her for her own beauty, her wit and charm, and not for the wealth and power which he can get by marrying her. Similarly, mathematics reveals its secrets only to those who approach it with pure love, for its own beauty. Those who do this are, of course, also rewarded with results of practical importance. But if somebody asks at each step, "What can I profit by this" he will not get far. You remember I told you that the Romans would never be really successful in applying mathematics. Well, now you see why: they are too practical-minded.*

Archimedes was an archetypical mathematician. While doing mathematics, he would become so consumed that he skipped meals and was oblivious to his appearance. One day while Syracuse was under siege by the armies of Marcellus during the second Punic War, the unaware Archimedes was contemplating a geometric problem. When a Roman soldier stepped on his diagrams drawn on the sandy floor, Archimedes exclaimed "Don't disturb my circles!". The angered soldier killed Archimedes.

Archimedes was buried in a simple grave in Sicily, marked only by a stone with the chiseled image of a sphere, cone, and cylinder. The grave was found by Cicero in the first century B.C. but remained undiscovered in modern times until 1965.

* ALFRÉD RÉNYI, "A DIALOGUE ON THE APPLICATION OF MATHEMATICS," FROM *MATHEMATICS PEOPLE, PROBLEMS, RESULTS,* VOL. I, WADSWORTH INTERNATIONAL.

92. a) Solve $|5x - 2| < 0.5$. Sketch the solution on a coordinate line.

b) Determine a δ such that $|x - 0.4| < \delta$ is an equivalent inequality to that of part a).

93. One machine labels 250 packages in ten minutes and a second machine labels the same number of packages in eight minutes. How long will it take the two machines working simultaneously to label 250 packages?

94. How many liters of a 24% alcohol mixture must be added to a 35% alcohol mixture to obtain 15 liters of a 30% alcohol mixture?

95. A traveller goes on a trip averaging 60 miles per hour. She returns along the same route averaging 40 miles per hour. What is her average rate for the entire trip? (It is not 50 miles per hour.)

96. Find all rectangles whose sides are integers and whose area and perimeter are numerically equal.

97. Two boats leave simultaneously from the same port. One heads due north and the other due west. The northbound boat travels 5 miles per hour faster than the westbound vessel, and after 2 hours the boats are 100 miles apart. Find the speed of each boat to the nearest mile per hour.

98. For what values of k are the solutions to $x^2 + 6x + k = 0$ complex, nonreal numbers?

In Problems 99 through 108, perform the indicated operation and write the result in the form $a + bi$.

99. $(6 + 4i) + (-5 + 2i)$ **100.** $(11 + i) - (5 - 7i)$

101. $(3 + 4i)(-2 + 8i)$

102. $\left(\dfrac{\sqrt{6}}{4} - \dfrac{\sqrt{2}}{4}i\right)\left(\dfrac{\sqrt{6}}{4} + \dfrac{\sqrt{2}}{4}i\right)$

103. $(\sqrt{3} + i)^3$ **104.** $\dfrac{4 - i}{i}$

105. $\dfrac{3 - 2i}{4 + 5i}$ **106.** $\dfrac{i}{(1 + 2i)^2}$

107. i^{35} **108.** $(\sqrt{-18} + 2)^2$

In Problems 109 through 112, solve the equation over the complex numbers.

109. $x^2 - 2x + 2 = 0$ **110.** $4x^2 + 2x + 5 = 0$

111. $2x + \dfrac{9}{x} = 5$ **112.** $x^2 = 9(1 + i)^4$

FUNCTIONS AND GRAPHS

COORDINATE GEOMETRY

Our discussion begins with a powerful tool used to represent algebraic concepts in a geometric setting. The idea behind **coordinate geometry** is straightforward, yet it is so powerful that it is no coincidence that the development of calculus and many scientific advances directly followed its widespread introduction.

The basic concepts of coordinate geometry are probably familiar to you. A horizontal axis and a vertical axis (usually called the **x-axis** and the **y-axis**) intersect at right angles at the **origin** and divide the **coordinate plane** into four **quadrants,** as shown in Figure 1. Each of these axes is marked off just as a number line is marked off, with zeros at the origin and the positive directions being to the right for the x-axis and up for the y-axis. Each ordered pair of real numbers (x, y) corresponds to exactly one point on the plane as illustrated in Figure 2. Conversely, each point on the plane corresponds to exactly one ordered pair of real numbers.

The set of all ordered pairs (x, y) of real numbers is sometimes denoted by $\mathbb{R} \times \mathbb{R}$ or $\mathbb{R}^2$. Other names for the coordinate plane include **xy-plane, rectangular plane, Cartesian plane,** and **Euclidean plane.** The ordered pair of numbers corresponding to a point in the coordinate plane are **coordinates** of the point.

The importance of coordinate geometry is that it gives a geometric picture of an algebraic relation. An equation in two unknowns has ordered pairs as solutions; if these points are plotted, the resulting graph gives us a good grasp of the properties of the algebraic equation.

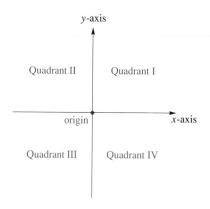

FIGURE 1

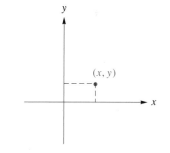

FIGURE 2

Given an equation in two unknowns, x and y, a **solution** of the equation is an ordered pair (x, y) that makes the equation true. The set of points in the coordinate plane that correspond to all such ordered pairs is called the **graph** of the equation.

x	0	1	1	4	4	9	9
y	0	1	-1	2	-2	3	-3

EXAMPLE 1   Sketch the graph of the equation $y^2 - x = 0$.

SOLUTION We can get a good idea of the graph by plotting enough points in the set to get a representative idea of the graph and then sketching a curve through the rest of the points (Figure 3). To determine these points, select a value for either variable and determine the value (or values) of the other variable such that the resulting ordered pair is a solution to the equation.

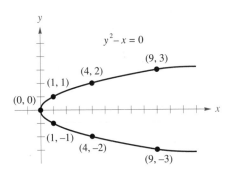

In many circumstances it is important to determine the exact distance between two points whose coordinates are known. The next example looks at a specific case and is followed by a generalization to any two points. The notation $d(P, Q)$ is used to denote the distance from the point P to the point Q.

EXAMPLE 2   Find the distance from point $P(2, 9)$ to point $Q(6, 3)$.

SOLUTION The first step is to draw axes and plot the points (Figure 4). Suppose we plot a point R on the same horizontal line as Q and the same vertical line as P [this point is $(2, 3)$]. Next, we connect R to P and Q by line segments. The triangle formed is a right triangle with a right angle at the point R. The length of the horizontal leg of the triangle is 4, the difference of the x coordinates of the points P and Q (note that this difference is positive since it is a length). Likewise, the length of the vertical leg is 6, the positive difference in the y coordinates. From the Pythagorean theorem, the hypotenuse is

$$d(P, Q) = \sqrt{4^2 + 6^2} = 2\sqrt{13}.$$

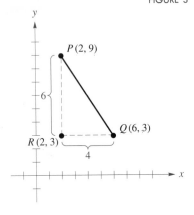

FIGURE 4

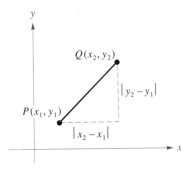

FIGURE 5

We can generalize the procedure for $P(x_1, y_1)$ and $Q(x_2, y_2)$ using Figure 5. The absolute values are used since the length must be nonnegative. So,

$$d(P, Q) = \sqrt{(x_2 - x_1)^2 + (y_2 - y_1)^2}.$$

FIGURE 3

Notice that since the differences are squared, the absolute value symbols are no longer required.

This is a good opportunity to introduce notation that is used in calculus. We define Δx (read as "delta x") as the change in the value of x as we go from one point to another, in this case from point P to point Q. That is,

$$\Delta x = x_2 - x_1.$$

Likewise,

$$\Delta y = y_2 - y_1.$$

Distance Formula

Given the points $P(x_1, y_1)$ and $Q(x_2, y_2)$, the distance from P to Q is given by

$$d(P, Q) = \sqrt{(\Delta x)^2 + (\Delta y)^2}$$

where $\Delta x = x_2 - x_1$ and $\Delta y = y_2 - y_1$.

Even though we now have a formula, the best way to remember this is to think about our picture and apply the Pythagorean theorem.

EXAMPLE 3 Consider the set of all points (x, y) that are equidistant from the given points $A(-3, 1)$ and $B(6, 4)$. For each of these points, determine y as an expression in terms of x.

SOLUTION First, draw the picture (from geometry, you may remember that the points in question are exactly those on the perpendicular bisector of the line segment $\overline{AB}$) (Figure 6). The point $C(x, y)$ represents any point in the set. So,

$$d(A, C) = d(B, C)$$
$$\sqrt{(x + 3)^2 + (y - 1)^2} = \sqrt{(x - 6)^2 + (y - 4)^2}$$
$$x^2 + 6x + 9 + y^2 - 2y + 1 = x^2 - 12x + 36 + y^2 - 8y + 16$$
$$6y = -18x + 42$$
$$y = -3x + 7.$$

From the distance formula, we find that the set in question is the graph of the line $y = -3x + 7$.

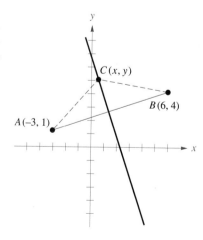

FIGURE 6

EXAMPLE 4   Consider the triangle with vertices

$$A(-3, 1),\ B(3, 4),\ C(-1, -3).$$

a) Determine if this triangle is an isosceles triangle.

b) Determine if this triangle is a right triangle.

SOLUTION

a) The first step is a sketch of the triangle (Figure 7). Use the distance formula to compute the lengths of the sides:

$$d(A, B) = \sqrt{(3 - (-3))^2 + (4 - 1)^2} = \sqrt{45} = 3\sqrt{5}$$
$$d(B, C) = \sqrt{(-1 - 3)^2 + (-3 - 4)^2} = \sqrt{65}$$
$$d(C, A) = \sqrt{(-3 - (-1))^2 + (1 - (-3))^2} = \sqrt{20} = 2\sqrt{5}.$$

Since no two sides of this triangle are the same length, the triangle is not isosceles. You might have guessed this from the picture, but to be completely sure, it is necessary to check it algebraically.

b) We can use the Pythagorean theorem to see whether this is a right triangle. Looking at the distances computed in part a), $\overline{BC}$ must be the hypotenuse (why?). So,

$$[d(A, B)]^2 + [d(C, A)]^2 \overset{?}{=} [d(B, C)]^2$$
$$[2\sqrt{5}]^2 + [3\sqrt{5}]^2 \overset{?}{=} [\sqrt{65}]^2$$
$$20 + 45 = 65.$$

What this means is that this triangle is a right triangle with the right angle at the vertex A. Again, you may have guessed this from the picture, but to be sure that it is a right angle (and not just very close), we must verify it as we just did.

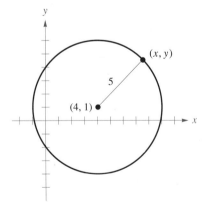

FIGURE 7

One of the direct results of the distance formula is that we can characterize a circle in the coordinate plane with an algebraic equation in two unknowns. For example, consider a circle with center $(4, 1)$ and radius 5 (Figure 8). By the definition of a circle, all points (x, y) are a distance 5 from $(4, 1)$. So, by the distance formula,

$$\sqrt{(x - 4)^2 + (y - 1)^2} = 5.$$

Squaring both sides gives us

$$(x - 4)^2 + (y - 1)^2 = 25.$$

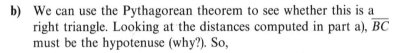

FIGURE 8

This is an equation whose graph is the circle on the coordinate plane with its center at (4, 1) and radius 5. Had we used (h, k) as the center instead of (4, 1) and used r instead of 5 as the radius, we would have arrived at this general result:

An equation whose graph is the circle of radius r and center (h, k) is

$$(x - h)^2 + (y - k)^2 = r^2.$$

Equation of a Circle

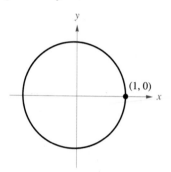

FIGURE 9 *Unit circle*

EXAMPLE 5 Find the equation of the form $(x - h)^2 + (y - k)^2 = r^2$ whose graph is the circle with

a) center at the origin and a radius of 1 unit (Figure 9), and

b) center at the point $(-2, 1)$ and a radius of 6 units (Figure 10).

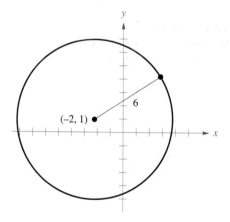

FIGURE 10

SOLUTION

a) Playing the role of (h, k) is the ordered pair (0, 0), and playing the role of r is 1, so

$$(x - 0)^2 + (y - 0)^2 = 1^2$$
$$x^2 + y^2 = 1.$$

This particular circle, the **unit circle,** will be important in our investigation of trigonometry in Chapter 5.

b) In this case (h, k) is $(-2, 1)$ and r is 6, so

$$(x - (-2))^2 + (y - 1)^2 = 6^2$$
$$(x + 2)^2 + (y - 1)^2 = 36.$$

EXAMPLE 6 The graphs of these equations are circles. Sketch them.

a) $x^2 + y^2 + 2x = 15$

b) $4x^2 + 4y^2 + 12x - 24y + 21 = 0$

SOLUTION

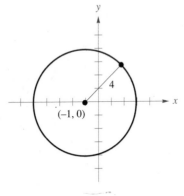

FIGURE 11

a) To get this equation in the form

$$(x - h)^2 + (y - k)^2 = r^2$$

we can complete the square on the x terms in the following manner:

$$(x^2 + 2x + 1) + y^2 = 15 + 1$$
$$(x + 1)^2 + y^2 = 16.$$

Expressing it in the form

$$(x - h)^2 + (x - k)^2 = r^2,$$

we get

$$(x - (-1))^2 + (y - 0)^2 = 4^2.$$

The circle has center at $(-1, 0)$, and its radius is 4 (Figure 11).

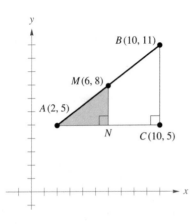

FIGURE 12

b) We proceed by dividing each side of the equation by 4 and then completing the square for both the x terms and the y terms:

$$x^2 + y^2 + 3x - 6y + \tfrac{21}{4} = 0$$
$$(x^2 + 3x + \tfrac{9}{4}) + (y^2 - 6y + 9) = -\tfrac{21}{4} + \tfrac{9}{4} + 9$$
$$(x + \tfrac{3}{2})^2 + (y - 3)^2 = 6.$$

The circle has center $(-\tfrac{3}{2}, 3)$ and radius $\sqrt{6}$ (Figure 12).

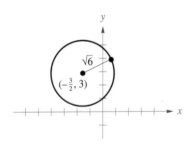

FIGURE 13

Suppose that $A(2, 5)$ and $B(10, 11)$ are the endpoints of a line segment and that M is the midpoint of this line segment. The distance $d(A, B)$ is 10 (you should verify this), so $d(A, M)$ is 5. The shaded triangle AMN and the triangle ABC in Figure 13 are similar (recall that two right triangles are similar if they share a common angle). It follows that the x coordinate of M is the average of the x coordinates of A and B. A similar result follows for the y coordinate of M. This specific case suggests the following formula for determining the coordinates of the midpoint.

The coordinates of the midpoint M of the line segment joining points $P(x_1, y_1)$ and $Q(x_2, y_2)$ are

$$\left(\frac{x_1 + x_2}{2}, \frac{y_1 + y_2}{2} \right).$$

Midpoint Formula

Notice that each of the coordinates of the midpoint is the average of the corresponding coordinates of P and Q. Also notice that while Figure 13 suggests the formula, it does not prove it—that is left for you as an exercise (see Problem 56).

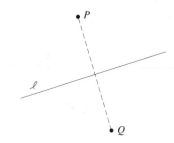

FIGURE 14

EXAMPLE 7   Find the midpoint of the line segment joining $P(2, 5)$ and $Q(-4, 8)$.

SOLUTION Applying the formula gives

$$\left(\frac{x_1 + x_2}{2}, \frac{y_1 + y_2}{2} \right) = \left(\frac{2 + (-4)}{2}, \frac{5 + 8}{2} \right) = \left(-1, \frac{13}{2} \right).$$

We finish this section with a discussion of symmetry. Points P and Q are **symmetric with respect to a line** ℓ if ℓ is the perpendicular bisector of the line segment $\overline{PQ}$ (Figure 14). Points P and Q are **symmetric with respect to a point** M if M is the midpoint of the line segment $\overline{PQ}$ (Figure 15).

In particular, the symmetries of interest for us on the coordinate plane are those with respect to the origin, the x-axis, the y-axis, and the line $y = x$. Here is a summary of these symmetries; also see Figure 16.

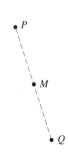

FIGURE 15

FIGURE 16

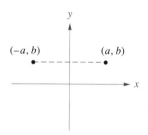

Points are symmetric with respect to the y-axis

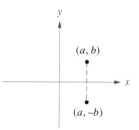

Points are symmetric with respect to the x-axis

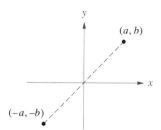

Points are symmetric with respect to the origin

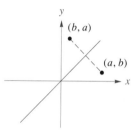

Points are symmetric with respect to the line $y = x$

Symmetry

The points (a, b) and $(-a, b)$ are symmetric with respect to the y-axis.

The points (a, b) and $(a, -b)$ are symmetric with respect to the x-axis.

The points (a, b) and $(-a, -b)$ are symmetric with respect to the origin.

The points (a, b) and (b, a) are symmetric with respect to the line $y = x$.

An important use of symmetry is in graphing equations in two unknowns. Finding that a graph is symmetric to the x-axis, the y-axis, the origin, or the line $y = x$ reduces the amount of work in sketching that graph. What follows are generalizations about these symmetries. With a little practice, you should be able to determine the symmetry of a graph quickly and without difficulty (Figure 17).

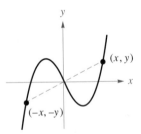

Graph is symmetric with respect to the origin

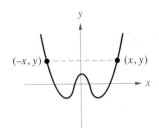

Graph is symmetric with respect to the y-axis

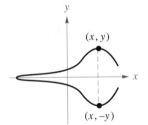

Graph is symmetric with respect to the x-axis

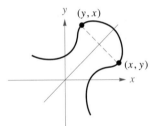

Graph is symmetric with respect to the line $y = x$

FIGURE 17

Symmetry of a Graph

A graph is symmetric with respect to the y-axis if, for any point (x, y) on the graph, the point $(-x, y)$ is also on the graph.

A graph is symmetric with respect to the x-axis if, for any point (x, y) on the graph, the point $(x, -y)$ is also on the graph.

A graph is symmetric with respect to the origin if, for any point (x, y) on the graph, the point $(-x, -y)$ is also on the graph.

A graph is symmetric with respect to the line $y = x$ if, for any point (x, y) on the graph, the point (y, x) is also on the graph.

EXAMPLE 8   Complete the graph in Figure 18 by using symmetry with respect
to the

a) the x-axis,   b) the y-axis,   c) the origin.

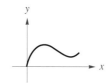

SOLUTION

FIGURE 18

a)   b)

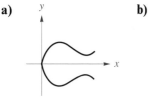

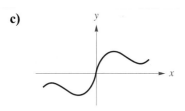

c)

EXAMPLE 9   Discuss the symmetry of the graph of $x + y^2 = 4$ with respect to

a) the x-axis   b) the y-axis   c) the origin

SOLUTION

a) The graph of $x + y^2 = 4$ is symmetric with respect to the x-axis
    if for every point (x, y) on the graph of $x + y^2 = 4$, the point
    $(x, -y)$ is also on the graph. To determine whether $(x, -y)$ is on
    the graph, we can check it in the equation $x + y^2 = 4$. So,

    $$x + (-y)^2 = 4$$

    or

    $$x + y^2 = 4.$$

    Since we ended (after some algebraic simplification) with the
    original equation, $(x, -y)$ is on the graph. The graph is
    symmetric with respect to the x-axis.

b) We can apply a similar procedure to test whether $(-x, y)$ is on
    the graph of the equation $x + y^2 = 4$:

    $$(-x) + y^2 = 4$$

and simplifying,

$$-x + y^2 = 4.$$

Since we do not arrive at an equivalent equation, $(-x, y)$ is not on the graph. The graph is not symmetric with respect to the y-axis.

c) Applying the procedure again to test whether $(-x, -y)$ is on the graph,

$$(-x) + (-y)^2 = 4$$

and simplifying,

$$-x + y^2 = 4.$$

Since we do not arrive at an equivalent equation, $(-x, -y)$ is not on the graph. The graph is not symmetric with respect to the origin.

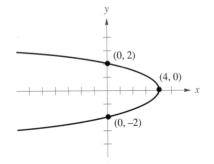

FIGURE 19

The graph of $x + y^2 = 4$ is shown in Figure 19. Notice that it agrees with our findings.

The discussion of symmetries in the last example can be formalized into these tests:

Symmetry Tests

The graph of an equation in x and y is

a) symmetric with respect to the y-axis if and only if an equivalent equation results when (x, y) is replaced by $(-x, y)$ in the equation;

b) symmetric with respect to the x-axis if and only if an equivalent equation is obtained when (x, y) is replaced by $(x, -y)$ in the equation;

c) symmetric with respect to the origin if and only if an equivalent equation is obtained when (x, y) is replaced by $(-x, -y)$ in the equation;

d) symmetric with respect to the line $y = x$ if and only if an equivalent equation is obtained when (x, y) is replaced by (y, x) in the equation.

EXERCISE SET 2.1

A

For Problems 1 through 6, determine the distance between the points and their midpoint.

1. $(2, 4)$ and $(-6, -6)$ **2.** $(0, 0)$ and $(12, -4)$

3. $(3, 8)$ and $(-4, 8)$ **4.** $(2, 6)$ and $(2, -4)$

5. $(\sqrt{3}, \sqrt{2})$ and $\left(\dfrac{2}{\sqrt{3}}, -\sqrt{2}\right)$

6. $(\sqrt{2}, \sqrt{3})$ and $\left(\dfrac{1}{\sqrt{2}}, -\sqrt{3}\right)$

For Problems 7 through 18, sketch the graph of the equation in two unknowns. Note and use any symmetry with respect to the x-axis, y-axis, origin, or the line $y = x$.

7. $y = x + 5$ **8.** $y = -2x$

9. $y = |x + 5|$ **10.** $|y| + x = 3$

11. $y = -2x^2$ **12.** $x = \frac{1}{2}y^2$

13. $xy = 4$ **14.** $xy = -1$

15. $y = \dfrac{1}{x^2}$ **16.** $y = -\dfrac{4}{x^2}$

17. $|x| + |y| = 6$ **18.** $|x| - |y| = 8$

For Problems 19 through 24, the graph of the equation is a circle. Determine the center and the radius.

19. $x^2 + y^2 = 4$ **20.** $x^2 + y^2 = 36$

21. $x^2 + (y + 6)^2 = 36$

22. $(x + 3)^2 + (y + 5)^2 = 30$

23. $(x - 3)^2 + (x + 3)^2 = 25$

24. $(x - 5)^2 + (y - 12)^2 = 169$

For Problems 25 through 30, use the graphs in Figure 20.

25. Complete the graph I using symmetry with respect to the y axis.

26. Complete the graph I using symmetry with respect to the origin.

27. Complete the graph II using symmetry with respect to the x axis.

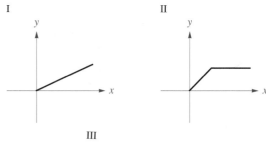

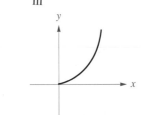

FIGURE 20

28. Complete the graph II using symmetry with respect to the origin.

29. Complete the graph III using symmetry with respect to the y axis.

30. Complete the graph III using symmetry with respect to the origin.

B

For Problems 31 through 36, rewrite each equation (if possible) in the form $(x - h)^2 + (y - k)^2 = r^2$.

31. $x^2 + y^2 - 4x - 6y - 3 = 0$

32. $x^2 + y^2 + 6x - 10y - 2 = 0$

33. $3x^2 + 3y^2 + 6x - 9y + 3 = 0$

34. $2x^2 + 2y^2 - 6x + 4y + 6 = 0$

35. $x^2 + y^2 - 4x + 8y + 20 = 0$

36. $x^2 + y^2 - 2y\sqrt{3} - 2x - 3 = 0$

For Problems 37 through 42, write an equation for the circle described in the standard form $(x - h)^2 + (y - k)^2 = r^2$.

37. The circle in the second quadrant is tangent to the x-axis and to the y-axis. Its radius is 3.

38. The circle has center $(2, -3)$, and it passes through the origin.

39. The circle has its center at the origin and is tangent to the vertical line through $(8, -4)$.

40. The circle has its center at $(5, 2)$ and is tangent to the horizontal line through $(9, -11)$.

41. The circle has its center on the *y*-axis and is tangent to a vertical line at the point $(6, -2)$.

42. A diameter of the circle is the line segment with endpoints $(-2, 2)$ and $(6, 8)$.

For Problems 43 through 55, be sure to draw a picture as part of your solution.

43. Three points A, B, C are collinear (that is, lie on the same line) with B between A and C, if and only if $d(A, B) + d(B, C) = d(A, C)$. Determine whether the points $(3, 5)$, $(6, 9)$, and $(12, 17)$ are collinear.

44. Determine whether the points $P(-4, 12)$, $Q(0, -1)$, and $R(-1, 3)$ are collinear (see Problem 43).

45. Determine whether the triangle with vertices at $M(-1, 2)$, $N(2, 6)$, and $P(4, 2)$ is equilateral, isosceles, or scalene (recall that an equilateral triangle has three sides all the same length, an isosceles triangle has exactly two sides the same length, and a scalene triangle has three sides of different lengths).

46. Determine whether the triangle with vertices at $A(-2, -1)$, $B(\sqrt{3}, 2\sqrt{3})$, and $C(-2, 1)$ is equilateral, isosceles, or scalene (see Problem 45).

47. Determine whether the triangle with vertices at $P(-4, 3)$, $Q(-3, 8)$, and $R(-15, 5)$ is right, acute, or obtuse (for a right triangle, the sum of the squares of the two shortest sides is equal to the square of the longest side; for an acute triangle, the sum of the squares of the two shortest sides is greater than the square of the longest side; and for an obtuse triangle, the sum of the squares of the two shortest sides is less than the square of the longest side).

48. Determine whether the triangle with vertices at $A(-6, 2)$, $B(4, 0)$, and $C(-5, 7)$ is right, acute, or obtuse (see Problem 47).

49. In a right triangle, the midpoint M of the hypotenuse is equidistant from the three vertices (that is, it is the same distance from each vertex). Show that the triangle with vertices at $A(-12, 4)$, $B(-10, 14)$, and $C(8, 0)$ is a right triangle, then show that this statement is true for this triangle.

50. A circle that circumscribes a right triangle (that is, passes through its vertices) has its center at the midpoint M of the right triangle's hypotenuse. Show that the triangle with vertices at $A(8, 7)$, $B(2, 5)$, and $C(6, 9)$, is a right triangle, then find the equation of the circumscribing circle.

51. A quadrilateral is a parallelogram if and only if each pair of opposite sides are of equal length. Use this fact to determine if the quadrilateral with vertices $A(-3, 1)$, $B(3, 3)$, $C(5, 7)$, and $D(-1, 5)$ is a parallelogram.

52. A quadrilateral is a parallelogram if and only if its diagonals bisect each other (that is, the midpoint of one is the midpoint of the other). Use this fact to determine if the quadrilateral with vertices $A(-2, 1)$, $B(7, 1)$, $C(10, 5)$, and $D(1, 5)$ is a parallelogram.

53. A quadrilateral is a rhombus if and only if all sides are of equal length. Use this fact to determine if the quadrilateral with vertices $G(16, 4)$, $H(10, 0)$, $J(3, 4)$, and $K(10, 8)$ is a rhombus.

54. A quadrilateral is a rectangle if and only if its diagonals are of equal length and bisect each other (that is, the midpoint of one is the midpoint of the other). Use this fact to determine if the quadrilateral with vertices $A(0, 5)$, $B(10, 11)$, $C(13, 6)$, and $D(3, 0)$ is a rectangle.

55. A quadrilateral is a square if and only if all four sides are of equal length and the two diagonals are of equal length. Use this fact to determine if the quadrilateral with vertices $P(-5, 2)$, $Q(-2, -2)$, $R(2, 1)$, and $S(-1, 5)$ is a square.

C

56. Prove the midpoint formula, that if M is the midpoint of $P(x_1, y_1)$ and $Q(x_2, y_2)$, then the coordinates of M are

$$\left(\frac{x_1 + x_2}{2}, \frac{y_1 + y_2}{2} \right).$$

To do this, show that P, M, and Q are collinear (see Problem 43) and that $d(P, M) = d(M, Q)$.

57. Prove that the point M on the line segment from $P(x_1, y_1)$ to $Q(x_2, y_2)$ that is one-third of the way from

P to Q has the coordinates

$$\left(\frac{2x_1 + x_2}{3}, \frac{2y_1 + y_2}{3}\right)$$

(see Problem 56).

58. A median of a triangle is the line segment from the midpoint of one of the sides to the vertex opposite the side. Show that the three medians of a triangle with vertices $A(x_1, y_1)$, $B(x_2, y_2)$, and $C(x_3, y_3)$ (Figure 21) are concurrent (that is, they all share a common point) and that the point of concurrency is one-third of the distance from the midpoint to the

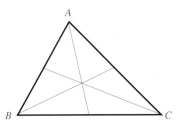

FIGURE 21

vertex on each of the medians (see Problem 57). What are the coordinates of the point?

FUNCTIONS

S E C T I O N   2.2

The concept of a function is of fundamental importance in mathematics and science. There are many instances in which one quantity is dependent on another quantity. For example, the distance a dropped ball falls is dependent on the amount of time that it is allowed to fall. The volume of a cube is dependent on the length of its edge. The pitch of a plucked string is dependent on (among other things) its length. Roughly speaking, a function is a rule that describes how one quantity depends on another.

A **function** is a rule that assigns to each element x of a set A *exactly* one element of a set B.

Function

The important word here is *exactly*. This means that each element in the set A is assigned to one and only one element in the set B.

You can think of a function f as a machine that accepts an input x from the set A and produces an output from the set B (Figure 22). This output of the function f is usually expressed as $f(x)$ (read this as "f of x"); it is also referred to as the **value** of f at x or the **image** of x under f.

The set of all possible inputs to the function f is called the **domain** of f. The set of all possible outputs to the function f is called the **range** of f.

For the present, our concern is with **real functions,** that is, those functions with inputs and outputs that are both real numbers. Unless indicated (or dictated by the situation), the domain of a real function is generally assumed to be the set of those real numbers for which the function is defined. We also say that two functions f and g are **equal** over a given set of real numbers if $f(x) = g(x)$ for all real numbers x in this set.

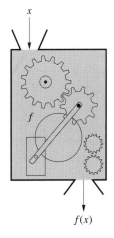

FIGURE 22

Function notation is used to describe the output of a function in terms of a formula. For example, a function g that assigns to each real number its square could be expressed as $g(x) = x^2$. Keep in mind that the function g is a process, that x is the input to the process, and that $g(x)$ is the output from the process when x is input.

We could have just as easily described this squaring function as $f(t) = t^2$, as $f(\Omega) = \Omega^2$, or even as $f(\ ) = (\ )^2$. These all represent the same function; the only difference is the symbol we choose to represent the input.

EXAMPLE 1   Given $f(x) = \dfrac{10x}{x^2 - 4}$.

a) Determine the domain of f. Sketch this set on a coordinate line.

b) Evaluate $f(1)$, $f(3)$, $f(-4)$, $f(-2)$, and $f(0)$.

SOLUTION

a) The domain is the set of all real numbers except for those that make the denominator equal to zero. To find these numbers, we set $x^2 - 4$ equal to zero and solve:

$$x^2 - 4 = 0$$
$$x^2 = 4$$
$$x = \pm 2.$$

FIGURE 23

These values are 2 and -2. The domain is $(-\infty, -2)$ or $(-2, 2)$ or $(2, \infty)$. Its graph is shown in Figure 23.

b) $f(1) = \dfrac{10(1)}{(1)^2 - 4} = -\dfrac{10}{3}$

$f(3) = \dfrac{10(3)}{(3)^2 - 4} = 6$

$f(-4) = \dfrac{10(-4)}{(-4)^2 - 4} = -\dfrac{10}{3}$

$f(-2)$ has no value since -2 is not in the domain of f.

$f(0) = \dfrac{10(0)}{(0)^2 - 4} = 0$

At first glance, it appears that the function in the previous example violates our definition of a function since 1 and -4 are assigned the same value by the function. However, if you read the definition closely (as you should with any definition), you will see that this is perfectly all right. Each of these elements of the domain is assigned exactly one element of the range; they just happen to be the same element.

EXAMPLE 2   Given $h(x) = \sqrt{36 - x^2}$.

   a) Determine the domain.

   b) Find $h(t)$, $h(2x)$, $2h(x)$.

   c) Find $h(t) + 3$, $h(t + 3)$, $h(t) + h(3)$.

SOLUTION

   a) The domain is the set of all real numbers x such that the radicand $36 - x^2$ is nonnegative. So,

$$36 - x^2 \geq 0$$
$$x^2 - 36 \leq 0$$
$$(x - 6)(x + 6) \leq 0.$$

   The sign graph (Figure 24) shows that the domain can be described as the interval $[-6, 6]$.

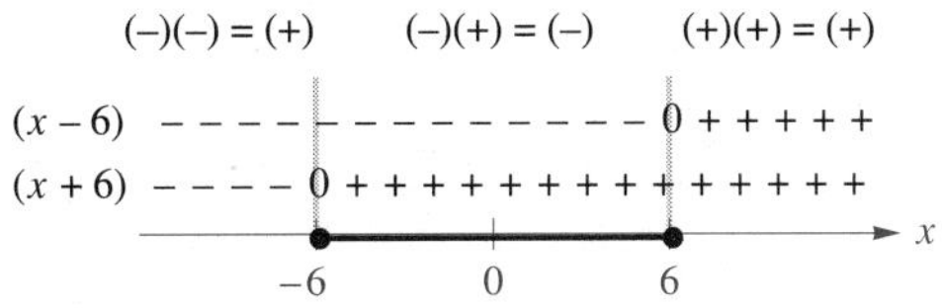

FIGURE 24

   b) $h(t) = \sqrt{36 - t^2}$

$h(2x) = \sqrt{36 - (2x)^2} = 2\sqrt{9 - x^2}$

$2h(x) = 2\sqrt{36 - x^2}$

   c) $h(t) + 3 = \sqrt{36 - t^2} + 3$

$h(t + 3) = \sqrt{36 - (t + 3)^2} = \sqrt{27 - 6t - t^2}$

$h(t) + h(3) = \sqrt{36 - t^2} + \sqrt{36 - (3)^2} = \sqrt{36 - t^2} + 3\sqrt{3}$

EXAMPLE 3   Given $h(x) = \dfrac{x}{1 - x}$.

a) Find $h(1/t)$ and $1/h(t)$.   **b)** Find $[h(t)]^2$ and $h(t^2)$.

SOLUTION

a) $h\left(\dfrac{1}{t}\right) = \dfrac{\dfrac{1}{t}}{1 - \dfrac{1}{t}} = \dfrac{1}{t - 1}$

$\dfrac{1}{h(t)} = \dfrac{1}{\left[\dfrac{t}{1 - t}\right]} = \dfrac{1 - t}{t}$

b) $[h(t)]^2 = \left[\dfrac{t}{1 - t}\right]^2 = \dfrac{t^2}{(1 - t)^2} = \dfrac{t^2}{1 - 2t + t^2}$

$h(t^2) = \dfrac{t^2}{1 - t^2}$

From the two previous examples, you can see it is important to be very careful about the notation: Two expressions that look very similar can have very different results. Note for example that $h(t + 3)$ is not the same as $h(t) + h(3)$ or $h(t) + 3$.

The next example introduces a function that assigns to each real number the greatest integer that is less than or equal to the real number. This function is called the **greatest integer function.** It is denoted by $[\![x]\!]$. For example,

$$[\![2.2]\!] = 2, \qquad [\![4.9]\!] = 4, \qquad [\![-8.1]\!] = -9, \quad \text{and} \quad [\![\pi]\!] = 3.$$

EXAMPLE 4   Determine the domain and range of the following:

a) $f(x) = [\![x]\!]$ **b)** $g(x) = 2[\![x]\!]$

SOLUTION

a) The domain is the set of all real numbers. Since only integers can come out of these functions, the range is the set of integers.

b) The domain is again the set of all real numbers. The range in this case is the set of even integers. This should seem reasonable since any element of the range of the function is twice an integer value.

EXAMPLE 5   Given $f(x) = \begin{cases} 2x & x < 0 \\ x^2 & 0 \le x \le 5. \\ \dfrac{x}{x-3} & x > 5 \end{cases}$

a) Find $f(-5)$ and $f(4)$.   **b)** Find $f(3)$.

SOLUTION First, notice that the function f is defined by different expressions over different intervals in the domain. The expression used to evaluate $f(x)$ depends on which interval of the domain contains x.

a) $f(-5) = 2(-5) = -10$

$f(4) = 4^2 = 16$

b) You might suspect that 3 is not in the domain of $f(x)$ since the expression $\dfrac{x}{x-3}$ is undefined for $x = 3$. But since $0 \le 3 \le 5$,

$$f(3) = 3^2 = 9.$$

The expressions involving functional notation in the next two examples may seem somewhat arbitrary to you. They do, however, play an important role in calculus. It will be worth your while to master the algebraic techniques of simplifying them.

EXAMPLE 6   For the function $g(x) = 2x^2 - 1$, evaluate and simplify

$$\frac{g(x + \Delta x) - g(x)}{\Delta x}.$$

SOLUTION Recall that Δx was defined in Section 2.1; in this case it should be treated as a variable.

$$\frac{g(x + \Delta x) - g(x)}{\Delta x} = \frac{g(x + \Delta x) - g(x)}{\Delta x}$$

$$= \frac{[2(x + \Delta x)^2 - 1] - [2x^2 - 1]}{\Delta x}$$

$$= \frac{[2x^2 + 4x\,\Delta x + 2\,\Delta x^2 - 1] - [2x^2 - 1]}{\Delta x}$$

$$= \frac{4x\,\Delta x + 2\,\Delta x^2}{\Delta x}$$

$$= 4x + 2\,\Delta x.$$

EXAMPLE 7   For the function $f(x) = 1/x$, evaluate and simplify

$$\frac{f(x + h) - f(x)}{h}.$$

SOLUTION

$$\frac{f(x + h) - f(x)}{h} = \frac{\left(\dfrac{1}{x + h}\right) - \left(\dfrac{1}{x}\right)}{h}$$

$$= \frac{\left(\dfrac{1}{x + h}\right) - \left(\dfrac{1}{x}\right)}{h} \cdot \frac{x(x + h)}{x(x + h)}$$

$$= \frac{x - (x + h)}{hx(x + h)}$$

$$= \frac{-h}{hx(x + h)}$$

$$= \frac{-1}{x(x + h)}.$$

Applications

The next three examples look at real situations in which one quantity depends on another and express that relationship as a mathematical function. While there is no specific set of steps for finding these functions, there are some general guidelines that will aid your search.

1) Draw a picture. Label the relevant quantities, known and unknown.

2) Determine which quantity is the input of the function and which quantity is the output of the function. Write down relationships that involve these quantities.

3) Find a relation involving the input and the output of the function.

4) Rewrite this relation as a function.

EXAMPLE 8   Determine the area $A(x)$ of a right isosceles triangle as a function of the length x of its hypotenuse. Use this function to determine the area of a right isosceles triangle with hypotenuse of length 10. What is the domain of the function A?

SOLUTION The first step is to draw a picture of the triangle and label it with the relevant information. The input x of the function A is the length of the hypotenuse; the output is the area $A(x)$ of the triangle.

We need to find a relation involving the area of a triangle and its hypotenuse. The legs of an isosceles triangle are of equal length. Using the Pythagorean Theorem, we obtain

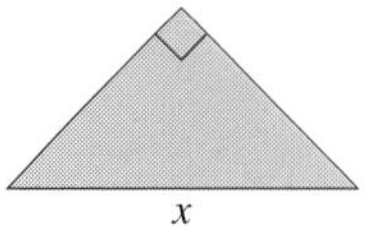

FIGURE 25

$$\begin{pmatrix}\text{length}\\ \text{of leg}\end{pmatrix}^2 + \begin{pmatrix}\text{length}\\ \text{of leg}\end{pmatrix}^2 = \begin{pmatrix}\text{length of}\\ \text{hypotenuse}\end{pmatrix}^2$$

or

$$\begin{pmatrix}\text{length}\\ \text{of leg}\end{pmatrix}^2 = \frac{1}{2}\begin{pmatrix}\text{length of}\\ \text{hypotenuse}\end{pmatrix}^2$$

Now, the area of a right triangle is half the product of its legs, so

$$\begin{aligned}\begin{pmatrix}\text{area of}\\ \text{triangle}\end{pmatrix} &= \frac{1}{2}\begin{pmatrix}\text{length}\\ \text{of leg}\end{pmatrix}^2\\ &= \frac{1}{2}\left[\frac{1}{2}\begin{pmatrix}\text{length of}\\ \text{hypotenuse}\end{pmatrix}^2\right]\end{aligned}$$

so

$$A(x) = \frac{1}{2}\left[\frac{1}{2}x^2\right] = \frac{1}{4}x^2$$

Letting $x = 10$, we get $A(10) = \frac{1}{4}(10)^2 = 25$.

Furthermore, the domain of the function is the set of all positive real numbers. Even though the expression $\frac{1}{4}x^2$ is defined for all real numbers, all values for x must be positive since it is a length.

EXAMPLE 9 A pig farmer wishes to build a rectangular pigpen, using the bank of a river as one of the sides (pigs don't swim well). The farmer has 180 meters of fencing. Find the area of the pen as a function of its width. What is the domain of this function?

SOLUTION Again, draw and label a picture of the situation (Figure 26). Since there is a fixed amount of fencing, it should seem reasonable that once the farmer decides on the width, the length is determined as well. In terms of the rectangle in question,

$$\begin{pmatrix}\text{area of}\\ \text{rectangle}\end{pmatrix} = \begin{pmatrix}\text{width of}\\ \text{rectangle}\end{pmatrix}\begin{pmatrix}\text{length of}\\ \text{rectangle}\end{pmatrix}$$

and

$$\begin{pmatrix}\text{length of}\\ \text{rectangle}\end{pmatrix} = 180 - 2\begin{pmatrix}\text{width of}\\ \text{rectangle}\end{pmatrix}$$

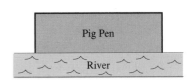

FIGURE 26

So, if we name this function, say, p, and let w represent the width, then

$$p(w) = w(180 - 2w).$$

Similar to the last problem, the domain for our function is determined by the situation. Certainly the width cannot be negative, and it must be no larger than 90 meters (why?). Thus the domain is the interval of $0 < w < 90$. However, if you are willing to consider the ridiculous cases of $w = 0$ and $w = 90$ as pigpens (picture our poor pigs in each of these situations), the domain could be $0 \le w \le 90$.

EXAMPLE 10   A box with a square base is to be constructed from two materials: one for the top and bottom, and one for the sides. The volume of the box is to be 36 ft^3. The cost of the material for the top and bottom is \$3 per ft^2, and the cost of material for the sides is \$5 per ft^2. Determine the cost of the box as a function of its height h.

SOLUTION Of course, the picture is the first step (Figure 27). Next, it should seem reasonable that

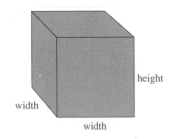

$$\binom{\text{total}}{\text{cost}} = \binom{\text{cost of top}}{\text{and bottom}} + \binom{\text{cost of}}{\text{sides}}$$

$$= 3\binom{\text{area of top}}{\text{and bottom}} + 5\binom{\text{area of}}{\text{sides}}$$

$$= 3\left[2\binom{\text{area of}}{\text{top}}\right] + 5\left[4\binom{\text{area of}}{\text{one side}}\right]$$

$$= 6[(\text{width})^2] + 20[(\text{width})(\text{height})].$$

FIGURE 27

Now, since the base is square and the volume is fixed,

$$(\text{width})^2(\text{height}) = 36.$$

Solving for the width, replacing the height with h, and a little algebra yield

$$(\text{width}) = \frac{6}{\sqrt{h}}.$$

If we name the cost function C, then

$$C(h) = 6\left[\left(\frac{6}{\sqrt{h}}\right)^2\right] + 20\left[\left(\frac{6}{\sqrt{h}}\right)h\right]$$

or

$$C(h) = \frac{216}{h} + 120\sqrt{h}.$$

The domain for C is the set of positive real numbers.

A

For Problems 1 through 6, determine the domain and the range of the function.

1. $f(x) = x^2 - 5$

2. $g(x) = \sqrt{x} + 7$

3. $h(t) = 5 + |t|$

4. $F(x) = x - |x|$

5. $h(x) = x - [\![x]\!]$

6. $G(x) = -3 - |2x|$

For Problems 7 through 12, determine the domain of the function in interval notation. Sketch the domain on a coordinate line.

7. $G(x) = \dfrac{x}{x - 2}$

8. $f(x) = \dfrac{x + 5}{2x + 1}$

9. $G(t) = -\sqrt{5 - 11t}$

10. $h(t) = \sqrt{2t + 1}$

11. $P(w) = \dfrac{5}{\sqrt{w - 3}}$

12. $h(x) = \dfrac{3}{\sqrt{x^2 - 4}}$

For Problems 13 through 18, use

$$f(x) = x - 2|x|, \quad g(x) = x^2 + 3x - 1, \quad and \quad h(x) = \dfrac{1}{x + 6}$$

to evaluate the given expression.

13. a) $f(-2)$
    b) $-f(2)$

14. a) $g(-4)$
    b) $-g(4)$

15. a) $h\left(\dfrac{1}{2}\right)$
    b) $\dfrac{1}{h(2)}$

16. a) $h\left(\dfrac{2}{3}\right)$
    b) $\dfrac{2}{h(3)}$

17. a) $f(1) + 3$
    b) $f(1) + f(3)$

18. a) $g(7) - 2$
    b) $g(7) - g(2)$

For Problems 19 through 24, use

$$f(x) = x - 2|x|, \quad g(x) = x^2 + 3x - 1, \quad and \quad h(x) = \dfrac{1}{x + 6}$$

to simplify the given expression.

19. a) $f(2a)$
    b) $2f(a)$

20. a) $h(x + 2)$
    b) $h(x) + h(2)$

21. a) $g(-a)$
    b) $-g(a)$

22. a) $f(-a^2)$
    b) $-f(a^2)$

23. a) $h(x) + 6$
    b) $h(x + 6)$

24. a) $h(a^2)$
    b) $[h(a)]^2$

B

For Problems 25 through 30, be sure to simplify your answers completely.

25. For $f(x) = 2x^2 - 9$, find $\dfrac{f(3 + h) - f(3)}{h}$.

26. For $g(x) = 3x - x^2$, find $\dfrac{g(2 + h) - g(2)}{h}$.

27. For $f(x) = \dfrac{2}{x}$, find $\dfrac{f(x + h) - f(x)}{h}$.

28. For $g(x) = \dfrac{1}{x + 2}$, find $\dfrac{g(x + h) - g(x)}{h}$.

29. For $f(x) = 2x^2 - 9$, find $\dfrac{f(x) - f(a)}{x - a}$.

30. For $g(x) = 3x - x^2$, find $\dfrac{g(x) - g(a)}{x - a}$.

31. Given

$$h(x) = \begin{cases} -x & x \le 1 \\ \sqrt{x} & 1 < x < 4 \\ \dfrac{x}{2} & x \ge 4 \end{cases}$$

Find $h(-9)$, $h(9)$, and $h(\tfrac{9}{4})$. Determine the domain of h.

32. Given

$$F(x) = \begin{cases} 9 - x^2 & x < -1 \\ \sqrt{x + 1} & x \ge -1 \end{cases}$$

Find $F(-4)$, $F(4)$, and $F(-1)$. Determine the domain of F.

33. Determine the range of the function

$$h(x) = \dfrac{2x}{x - 3}.$$

(Hint: Set the function equal to y; by solving for x, you should be able to determine the range.)

34. Determine the range of the function

$$C(x) = \dfrac{2x}{5 - x}.$$

(Hint: See Problem 33.)

35. Determine the value (or values) of k for the function $f(x) = x^2 - kx + 5$ such that $f(3) = -4$.

36. Determine the value (or values) of k for the function $f(x) = kx^2 - 2kx - 12$ such that $f(6) = 12$.

For Problems 37 through 57, determine the function indicated and its domain. Be sure to draw a picture and label it with information that is important to the problem.

37. Determine the area of a square as a function of its perimeter.

38. Determine the area of a circle as a function of its circumference.

39. A rectangle has a fixed area of 48 m². Find the length as a function of the width. Determine the domain.

40. A rectangle has an area of 54 ft². Express its perimeter as a function of the width.

41. Determine the area of an equilateral triangle as a function of its side.

42. A ladder of length 10 meters leans against a wall (Figure 28). Find the height of the top of the ladder as a function of the distance from the foot of the ladder to the wall.

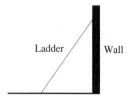

Ladder   Wall

FIGURE 28

43. Sam and Dave leave an intersection at the same time. Sam heads north at a rate of 15 mph while Dave heads east at 20 mph. Find the distance between them as a function of time traveled (in hours).

44. Al walks toward a light post that is 30 feet high. If Al is 6 feet tall, find the length of Al's shadow as a function of the distance that Al is from the lightpost.

45. A 15-m tree and a 10-m tree are 30 m apart (Figure 29). A rope is tied from the top of one tree to a peg in the ground and then to the top of the other tree (assume that the peg and the trees lie in the same plane). Find the length of the entire rope as a function of the distance that the peg is from the base of the taller tree.

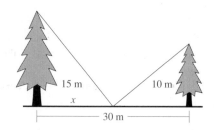

15 m   10 m

x

30 m

FIGURE 29

46. A printed page is to have 400 cm² of printed material with 6-cm margins on top and bottom and 3-cm margins on each side (Figure 30). Find the area of the entire page as a function of the width of the entire page.

FIGURE 30

47. A wire of length 24 cm is cut into two pieces. One piece is bent into a square. The other piece is bent into a circle. Find the total area of the square and the circle as a function of the length used for the square.

48. Equal-sized squares are cut from the corners of a 50-cm × 80-cm sheet of tin. The sides are folded up to make a box with no top (Figure 31). Find the volume

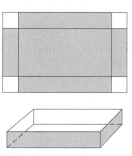

FIGURE 31

of the box as a function of the side of the square cut from each corner.

49. A farmer wishes to construct three rectangular pens as shown in Figure 32, using exactly 80 m of fencing material. Find the area as a function of the lengths of the pens.

FIGURE 32

50. Determine the distance from the point $P(1, 0)$ to a point $Q(x, y)$ on the graph of $y = \sqrt{x}$ as a function of the x-coordinate of the point Q (Figure 33).

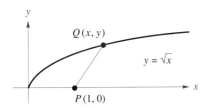

FIGURE 33

51. Determine the distance from the point $A(0, 5)$ to the point $B(x, y)$ on the graph of the line $y = 2x$ as a function of the x-coordinate of the point B (see Problem 50).

52. A wall is to be constructed around a rectangle of 120 m² with a fence running the length as in Figure 34. The fence costs $4 per m to construct and the

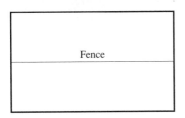

FIGURE 34

wall costs $7 per m to construct. Find the cost of the total project as a function of the length of the fence.

53. A house is on one bank of a river that is 20 m wide. On the other bank, 30 m downstream, is another house. A phone line is to be strung from one house along the bank and then across the river to the other house as in Figure 35. Stringing the line along the bank costs $2 per m; over the water it costs $5 per m. Determine the cost as a function of the distance the wire is strung along the bank.

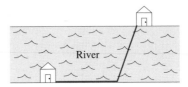

FIGURE 35

54. A water trough 6 ft long is constructed in such a way that the ends are isosceles triangles and the sides are each rectangles that are 6 ft by 3 ft (Figure 36). Determine the volume as a function of the width across the top.

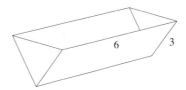

FIGURE 36

C

55. A cylindrical oil can is to hold 58 in³. Express the surface area as a function of the radius of the base.

56. Sand is pouring into a conical pile at a rate of 12 ft³ per minute. The bottom radius of the pile is equal to the height of the pile at all times. Determine the height of the pile as a function of the number of minutes the sand has been pouring into the pile.

57. Water is poured into a conical tank at a rate of 2 m³ per hour. The tank is 6 m high and the top radius is 4 m. Determine the depth of the water as a function of the number of hours that the water is allowed to pour into the tank (Figure 37).

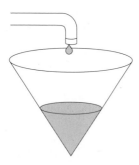

FIGURE 37

58. For $f(x) = \sqrt{x}$, simplify

$$\frac{f(x + \Delta x) - f(x)}{\Delta x}$$

in such a way that there are no radicals in the numerator.

59. For $f(x) = 2/\sqrt{x}$, simplify

$$\frac{f(x + \Delta x) - f(x)}{\Delta x}$$

in such a way that there are no radicals in the numerator.

60. For $g(x) = 1/\sqrt{x + 3}$, simplify

$$\frac{g(x + \Delta x) - g(x)}{\Delta x}$$

in such a way that there are no radicals in the numerator.

61. For $f(x) = \sqrt[3]{x}$, simplify

$$\frac{f(x + \Delta x) - f(x)}{\Delta x}$$

in such a way that there are no radicals in the numerator.

S E C T I O N   2.3   GRAPHS OF FUNCTIONS

Given a function $f(x)$, we can use the graph of the equation $y = f(x)$ to investigate the properties of f. When considering an equation $y = f(x)$, think of x as the input to the function (an element of the domain of f) and y as the output of the function (an element of the range of f). The input variable is called the **independent variable,** and the output variable is called the **dependent variable** (Figure 38).

FIGURE 38

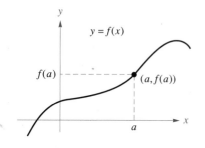

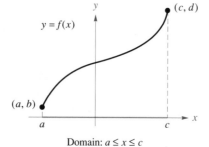

Domain: $a \le x \le c$

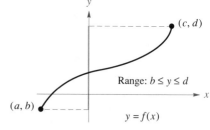

> The **graph of a function** f is the graph of the equation $y = f(x)$. That is, it is the set of points on the coordinate plane with coordinates $(x, f(x))$, where x is in the domain of f.

Graph of a Function

You can distinguish between a graph of a function and a graph that does not represent a function. Recall that for each element of the domain, there is exactly one element of the range that is assigned to it. In the context of the graph of $y = f(x)$, this is equivalent to saying that no two points (x, y) have the same x component yet different y components—that is, no two points on the graph lie on the same vertical line (Figure 39).

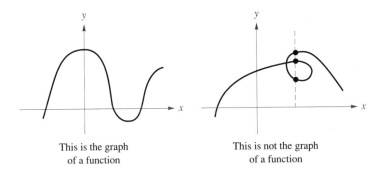

This is the graph   This is not the graph
of a function   of a function   FIGURE 39

> A set of points in a coordinate plane is the graph of a function if and only if no two points lie on the same vertical line. That is, no vertical line intercepts the set at more than one point.

Vertical Line Test

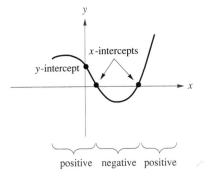

FIGURE. 40

The point at which the graph of a function f crosses the y-axis is the **y-intercept** of f. Likewise, the points at which the graph of f crosses the x-axis are the **x-intercepts** of f. A function f is **positive** at $x = a$ if $f(a) > 0$, and it is **negative** at $x = a$ if $f(a) < 0$. It should be easy to see that if f is positive at $x = a$, then the point $(a, f(a))$ is above the x-axis. Likewise, if f is negative at $x = a$, then the point $(a, f(a))$ is below the x-axis (Figure 40).

Consider an open interval $a < x < b$ in the domain of a function f. The function f is **increasing** on this interval if, for any two real numbers x_1 and x_2 in this interval such that $x_1 < x_2$, $f(x_1) < f(x_2)$. Likewise, the function f is **decreasing** on this interval if, for any two real numbers x_1 and x_2

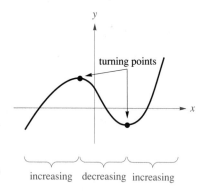

increasing   decreasing   increasing

FIGURE 41

in this interval such that $x_1 < x_2$, $f(x_1) > f(x_2)$ (Figure 41). Imagine a bug walking from left to right on the graph of the function f. If our bug walks over an increasing interval, then it travels uphill. If it walks over a decreasing interval, then it travels downhill.

Now suppose that $x = a$ is in the domain of f. If $x = a$ separates an interval over which a function f is increasing from an interval over which f is decreasing, then the point $(a, f(a))$ is called a **turning point** of f.

EXAMPLE 1 Given the graph of a function f in Figure 42, determine from the graph:

a) the domain and range of f;

b) the intercepts of f;

c) the intervals over which f is positive and the intervals over which f is negative;

d) the intervals over which f is increasing, the intervals over which f is decreasing, and the turning points of f.

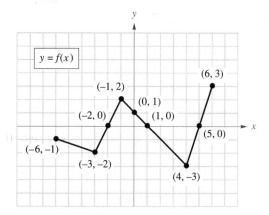

FIGURE 42

SOLUTION

a) The domain of the function is the interval $-6 \le x \le 6$, and the range of the function is $-3 \le y \le 3$.

b) From the graph, the x-intercepts of the graph are $(-2, 0)$, $(1, 0)$, and $(5, 0)$, and the y-intercept of the graph is $(0, 1)$.

c) The function is positive over the interval $-2 < x < 1$ and the interval $5 < x < 6$. It is negative over the interval $-6 < x < -2$ and the interval $1 < x < 5$.

d) The function is increasing over the intervals $-3 < x < -1$ and $4 < x < 6$. It is decreasing over the intervals $-6 < x < -3$ and $-1 < x < 4$.

e) The turning points are $(-3, -2)$, $(-1, 2)$, and $(4, -3)$.

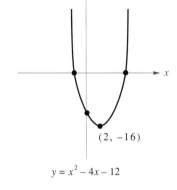

EXAMPLE 2   The graph of $y = x^2 - 4x - 12$ is shown in Figure 43, and the turning point is labeled. Determine:

a) the intercepts of f,

b) the intervals over which f is positive and the intervals over which f is negative,

c) the intervals over which f is increasing and the intervals over which f is decreasing.

$(2, -16)$

$y = x^2 - 4x - 12$

FIGURE 43

SOLUTION

a) The y-intercept is precisely the point on the graph where the x coordinate is zero (Figure 44). Setting $x = 0$,
$y = 0^2 - 4(0) - 12 = -12$. The y-intercept is the point $(0, -12)$. In a similar fashion, the x-intercepts are precisely the points on the graph where the y-coordinate is zero. Setting $y = 0$,

$$0 = x^2 - 4x - 12 \quad \Rightarrow \quad x = -2 \quad \text{or} \quad x = 6.$$

The x-intercepts are $(-2, 0)$ and $(6, 0)$.

b) You can see that the graph is below the x-axis between the x-intercepts, so the function is negative over the interval $-2 < x < 6$. The function is positive over the interval $x < -2$ and over the interval $x > 6$.

c) Again, you can see that the graph is decreasing over the interval $x < 2$ and increasing over the interval $x > 2$.

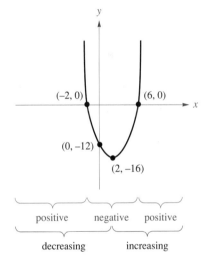

FIGURE 44

In general, finding the exact coordinates of the turning points of a function without using calculus is difficult or even impossible (this gives you something to look forward to). We can, however, make an educated guess once we have a sketch of the graph of the function.

Figure 45 is a catalog of some of the well-known graphs of functions that arise frequently in mathematics courses. You should be able to sketch the graph of each of them quickly and without much difficulty.

Catalog of Graphs

If you are not already completely familiar with the graphs of each of these functions, now is a good time to pause and practice sketching them. You should be able to sketch any of them as quickly as you can write your name.

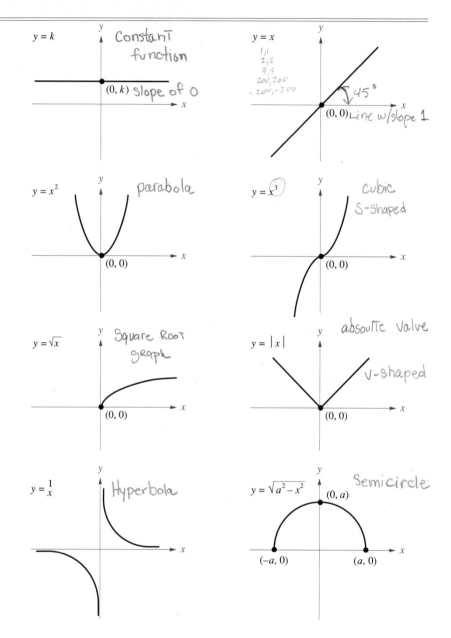

FIGURE 45

Although our catalog contains graphs of only eight functions (so far), these graphs will enable us to determine the graphs that have the same shape and size but are in different positions on the coordinate plane. Such graphs are called **translations.** The idea of translations is demonstrated in the next set of examples.

EXAMPLE 3   Sketch the graph of each of these functions. Label the intercepts and determine the domain.

a) $y = \sqrt{x}$ **b)** $y = \sqrt{x-2}$ **c)** $y = \sqrt{x+4}$

SOLUTION

a) This is in our catalog and can be sketched quickly (Figure 46). A few points on the graph are labeled. The domain is the interval $[0, \infty)$.

b) If we plot a few points and sketch the graph, we get a graph very much like that of $y = \sqrt{x}$. In fact, it is identical except that it is translated 2 units to the right (Figure 47). The domain is the interval $[2, \infty)$.

c) If we plot a few points and sketch the graph, again we get a graph very much like that of $y = \sqrt{x}$. In fact, it too is identical except that it is translated 4 units to the left (Figure 48). The domain is the interval $[-4, \infty)$.

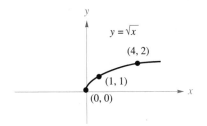

FIGURE 46

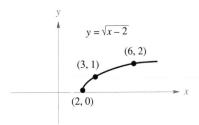

FIGURE 47

In general, the graph of $y = f(x - h)$ is identical to the graph of $y = f(x)$ translated horizontally the **directed** distance of h units. That is, if h is positive, the horizontal translation is to the right, and if h is negative, the horizontal translation is to the le.

In part c) of Example 3, the role of h is played by -4, since

$$\sqrt{x+4} = \sqrt{x-(-4)}.$$

EXAMPLE 4   The graph of a function f is shown in Example 1 (Figure 42). Sketch the graphs of these functions.

a) $y = f(x - 4)$ **b)** $y = f(x + 2)$

SOLUTION

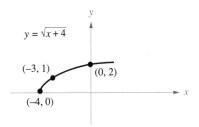

FIGURE 48

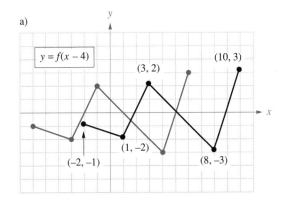

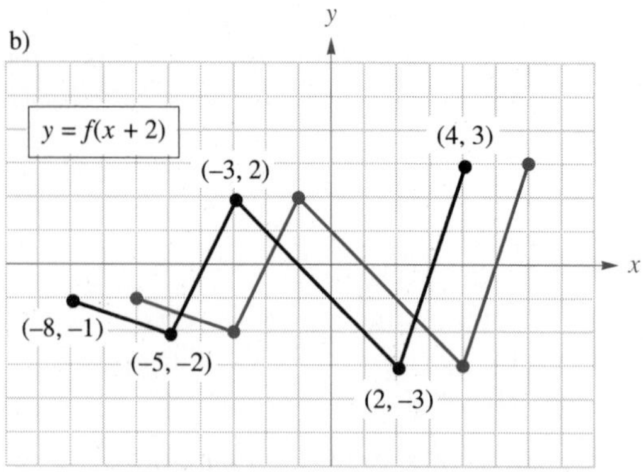
b)
y
x
y = f(x + 2)
(4, 3)
(−3, 2)
(−8, −1)
(−5, −2)
(2, −3)

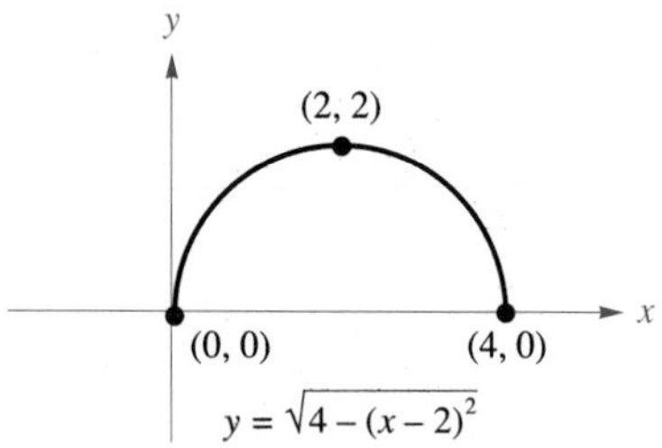
y
x
(2, 2)
(0, 0)
(4, 0)
y = √4 − (x − 2)²

FIGURE 49

EXAMPLE 5 Sketch the graph of $y = \sqrt{4 - (x - 2)^2}$.

SOLUTION If we assign $f(x) = \sqrt{4 - x^2}$, then the graph in question is that of $y = f(x - 2)$. So we take the graph of $f(x)$ (which is in our catalog) and translate it 2 units horizontally in the positive direction (Figure 49).

EXAMPLE 6 Sketch the graphs of these functions. Label the intercepts.

a) $y - 3 = \sqrt{x}$ **b)** $y + 2 = \sqrt{x}$

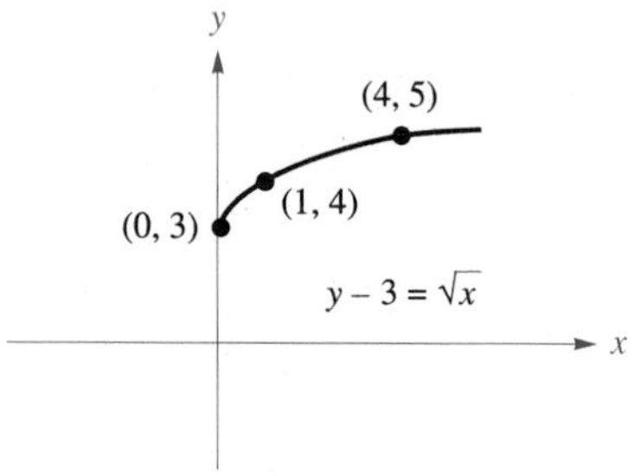
y
x
(4, 5)
(1, 4)
(0, 3)
y − 3 = √x

FIGURE 50

SOLUTION

a) If we plot a few points and sketch the graph, we get a vertical translation of the graph of $y = \sqrt{x}$ (Figure 50). The translation is 3 units up, in the positive direction.

b) If we plot a few points and sketch the graph, we again get a vertical translation of the graph of $y = \sqrt{x}$ (Figure 51). The translation is 2 units down, in the negative direction.

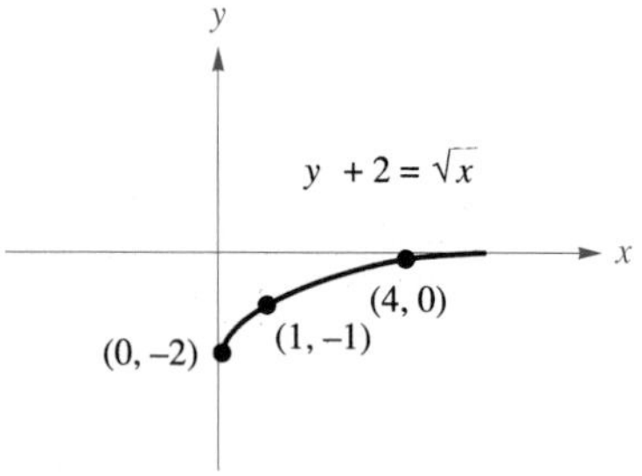
y
x
y + 2 = √x
(4, 0)
(0, −2)
(1, −1)

FIGURE 51

In general, the graph of $y - k = f(x)$ is identical to the graph of $y = f(x)$ translated vertically the **directed** distance of k units.

EXAMPLE 7 The graph of a function f is shown in Example 1 (Figure 42). Sketch the graphs of these functions.

a) $y + 2 = f(x)$ **b)** $y - 4 = f(x)$

SOLUTION

a)

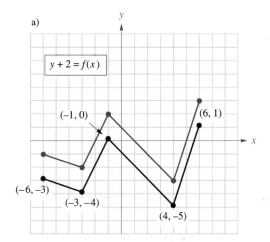

b)

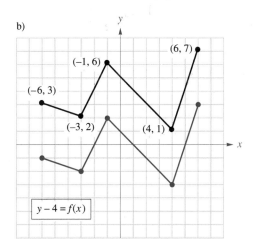

EXAMPLE 8   Sketch the graph of $y = x^3 + 2$. Label the intercepts.

SOLUTION If we rewrite this equation as $y - 2 = x^3$, we can consider this as a translation of $y = x^3$, which is in our catalog. Translating $y = x^3$ up 2 units, we get the graph in Figure 52.

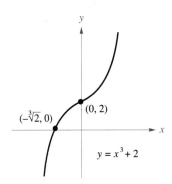

FIGURE 52

If we combine the last two concepts, we can now sketch the graph of $y - k = f(x - h)$ if we know the graph of $y = f(x)$. We just have to translate the graph of $y = f(x)$ horizontally h directed units and vertically k directed units.

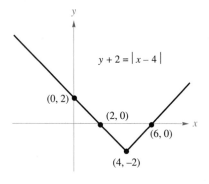

FIGURE 53

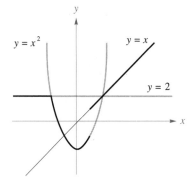

FIGURE 54

Odd and Even Functions

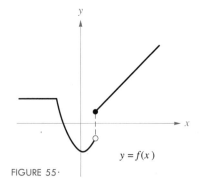

FIGURE 55·

EXAMPLE 9 Sketch the graph of $y + 2 = |x - 4|$. Label the intercepts.

SOLUTION The graph (Figure 53) is a translation of $y = |x|$, with a directed horizontal translation of 4 units and a directed vertical translation of -2 units.

EXAMPLE 10 Sketch the graph of $f(x) = \begin{cases} 2 & x \le -2 \\ x^2 - 2 & -2 \le x < 1 \\ x & x \ge 1 \end{cases}$.

SOLUTION First, we sketch the graphs $y = 2$, $y = x^2 - 2$, and $y = x$ on the same coordinate plane (Figure 54). The graph of $f(x)$ is made up of portions of these graphs, depending on the interval of the domain that includes x (Figure 55). For $x < -2$, the graph of $f(x)$ is the same as the graph of $y = 2$. For $-2 \le x < 1$, the graph of $f(x)$ is the same as the graph of $y = x^2 - 2$. For $x \ge 1$, the graph of $f(x)$ is the same as the graph of $y = x$.

Another aid in sketching the graph of a function is symmetry. A function whose graph is symmetric with respect to the y-axis is called **even**. A function whose graph is symmetric with respect to the origin is called **odd** (completing Problem 58 will give you a hint as to why the names even and odd are appropriate). If a function is determined to be even, or to be odd, the task of sketching its graph is reduced by half since we can exploit the symmetry of the graph. We will use these tests to determine whether a function is even or odd.

> A function $f(x)$ is **even** if for each x in the domain of f, $-x$ is also in the domain of f and $f(-x) = f(x)$. A function $f(x)$ is **odd** if for each x in the domain of f, $-x$ is also in the domain of f and $f(-x) = -f(x)$.

EXAMPLE 11 For each of these functions, determine whether it is even, odd, or neither. Sketch its graph.

    a) $f(x) = x^2 - 4$ **b)** $g(x) = (x - 4)^2$ **c)** $h(x) = x^3$

SOLUTION

    a) Our plan of attack here is to evaluate $f(-x)$, simplify it, and determine if $f(-x)$ is equal to $f(x)$, if $f(-x)$ is equal to $-f(x)$, or neither:

$$f(-x) = (-x)^2 - 4$$
$$= x^2 - 4$$
$$= f(x).$$

This shows that f is an even function. Its graph has symmetry with respect to the y-axis. We can actually sketch the graph easily; it is a vertical translation of the graph of $y = x^2$ (Figure 56).

FIGURE 56

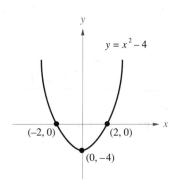

$y = x^2 - 4$

$(-2, 0)$ $(2, 0)$

$(0, -4)$

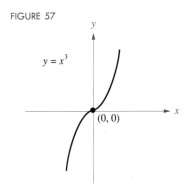

$y = (x - 4)^2$

$(0, 16)$

$(4, 0)$

b) Using our test,

$$g(-x) = (-x - 4)^2$$
$$= (x + 4)^2$$
$$\neq g(x) \quad \text{or} \quad -g(x).$$

This shows that g is neither an even nor an odd function. The sketch of g is a horizontal translation of the graph of $y = x^2$ (Figure 57).

c) The graph of this function is in our catalog of graphs. This graph is apparently symmetric with respect to the origin. Using our test, we obtain

$$h(-x) = (-x)^3$$
$$= -x^3$$
$$= -h(x).$$

This shows that h is an odd function. Its graph has symmetry with respect to the origin (Figure 58).

FIGURE 57

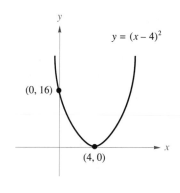

$y = x^3$

$(0, 0)$

FIGURE 58

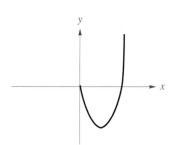

EXAMPLE 12 Figure 59 is a sketch of the graph of $y = x^3 - 4x$ for $x > 0$. Complete the sketch and determine the intercepts.

FIGURE 59

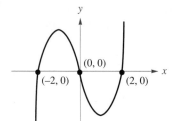

FIGURE 60

SOLUTION First, let $f(x) = x^3 - 4x$. If $f(x)$ is even or odd, then we can use symmetry to complete the graph. So,

$$f(-x) = (-x)^3 - 4(-x)$$
$$= -x^3 + 4x$$
$$= -(x^3 - 4x)$$
$$= -f(x).$$

This means that $f(x)$ is an odd function. We reflect the graph for $x > 0$ through the origin to obtain the complete graph of the function (Figure 60).

EXERCISE SET 2.3

A

For Problems 1 through 6, determine if the graph represents a function.

1.

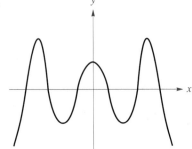

3.

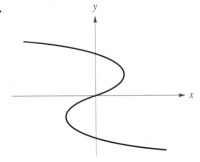

2.

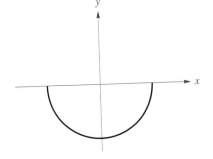

4.

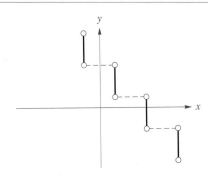

5.

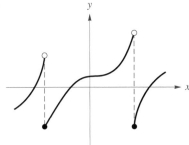

6.

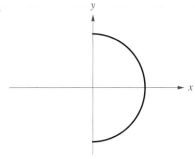

For Problems 7 through 12, from the graph of the function, determine a) the intervals over which the function is positive, the intervals over which the function is negative, and the intercepts; b) the intervals over which the function is increasing, the intervals over which the function is decreasing, and the turning points.

7. $y = x^4 - 2x^2 - 8$

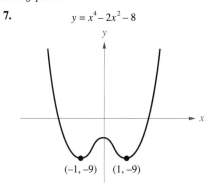

8. $y = x^2 - 2x - 8$

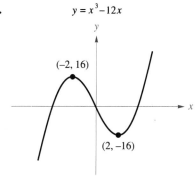

9.
$$y = \begin{cases} -\sqrt{9 - x^2} & x < 0 \\ \sqrt{9 - x^2} & x \geq 0 \end{cases}$$

10. $y = x^3 - 12x$

11.

$$y = \frac{1}{x}$$

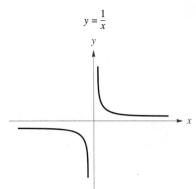

14.

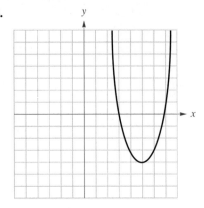

12.

$$y = x^2 - 2|x|$$

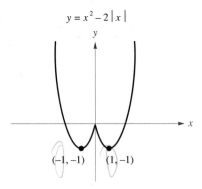

15.

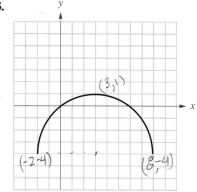

For Problems 13 through 18, the graph shown is a translation of one of the graphs in the catalog. Write an equation for the graph shown.

13.

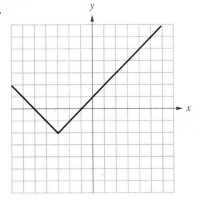

16.

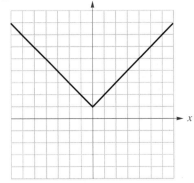

17.

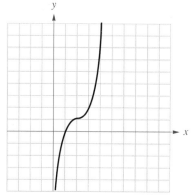

18.

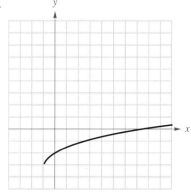

For Problems 19 through 30, sketch the graphs of both equations on the same coordinate plane and label the intercepts.

19. $y = |x|$ and $y = |x - 2|$

20. $y = |x|$ and $y = |x + 3|$

21. $y = \sqrt{x}$ and $y = \sqrt{x - 1}$

22. $y = \sqrt{x}$ and $y = \sqrt{x + 9}$

23. $y = x^2$ and $y = x^2 + 3$

24. $y = x^2$ and $y - 5 = x^2$

25. $y = |x|$ and $y - 2 = |x|$

26. $y = |x|$ and $y = |x| - 4$

27. $y = x^3$ and $y - 5 = (x - 2)^3$

28. $y = x^3$ and $y = (x + 1)^3 - 2$

29. $y = \sqrt{9 - x^2}$ and $y - 1 = \sqrt{9 - (x - 3)^2}$

30. $y = \dfrac{1}{x}$ and $y = \dfrac{1}{(x - 5)} + 2$

B

For Problems 31 through 42, sketch the graph of the given equation by selecting an appropriate graph from the catalog of functions in this section and performing the appropriate translation. Label the intercepts and the turning points. Do not plot points to sketch the graphs.

31. $y = 7 + |x - 5|$ **32.** $y - 3 = |x + 4|$

33. $y = \sqrt{x} - 2$ **34.** $y = \sqrt{x - 3} + 1$

35. $y = 4 + (x - 3)^3$ **36.** $y = (x + 1)^3 + 8$

37. $y = (x + 2)^3 + 7$ **38.** $y - 4 = \sqrt{25 - x^2}$

39. $y = \dfrac{2x - 1}{x - 1}$ (Hint: Use long division.)

40. $y = \dfrac{5 - 2x}{x - 3}$ (Hint: Use long division.)

41. $y = \sqrt{3 - x^2 - 2x}$
(Hint: Complete the square in the radicand.)

42. $y = x^3 + 3x^2 + 3x - 7$
[Hint: $(x + 1)^3 = x^3 + 3x^2 + 3x + 1$.]

For Problems 43 through 48, sketch the graph of the given split-domain function. Label the intercepts. Do not plot points to sketch the graphs.

43. $y = \begin{cases} x^2 & x < 0 \\ x & x \geq 0 \end{cases}$

44. $y = \begin{cases} x & x < 1 \\ \dfrac{1}{x} & x \geq 1 \end{cases}$

45. $y = \begin{cases} \sqrt{25 - x^2} + 2 & x < 3 \\ 6 & x \geq 3 \end{cases}$

46. $y = \begin{cases} x - 3 & x < 0 \\ \dfrac{1}{x + 1} & x \geq 0 \end{cases}$

47. $y = \begin{cases} 3 & x < -3 \\ |x| & -3 \leq x < 1 \\ -x + 2 & x \geq 1 \end{cases}$

48. $y = \begin{cases} 4 - x^2 & x \leq 0 \\ -x + 4 & 0 < x < 4 \\ 2 & x \geq 4 \end{cases}$

For Problems 49 through 54, part of the graph of the function is given. Determine whether the function is even or odd, and complete the graph. Label the intercepts.

49.

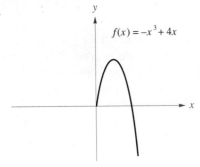

$$f(x) = -x^3 + 4x$$

50.

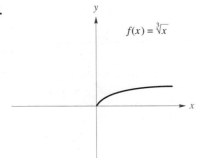

$$f(x) = \sqrt[3]{x}$$

51.

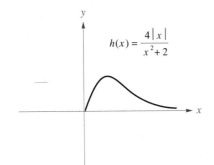

$$h(x) = \frac{4|x|}{x^2 + 2}$$

52.

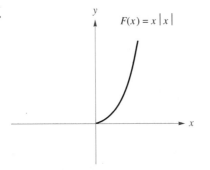

$$F(x) = x|x|$$

53.

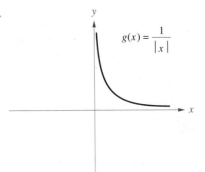

$$g(x) = \frac{1}{|x|}$$

54.

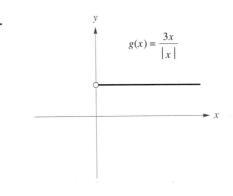

$$g(x) = \frac{3x}{|x|}$$

For Problems 55 through 60, determine whether the function given is even, odd, or neither.

55. $g(x) = \dfrac{2x}{x^2 - 1}$

56. $k(x) = \dfrac{(x + 1)(x - 4)}{x^3} + \dfrac{3}{x^2}$

57. $f(x) = (x^2 - 3x)^2$ **58.** $h(x) = x^5 - 3x^3 + x$

59. $G(x) = |x| - 2x^2$ **60.** $R(x) = |x - 2x^2|$

61. Under what conditions is the function $f(x) = x^n$ an even function? An odd function?

C

62. Define $P(x) = xf(x)$, where $f(x)$ is an even function. Is the function P an even function, an odd function, or neither?

63. Define $Q(x) = f(x) + g(x)$, where $f(x)$ is an even function and $g(x)$ is an odd function. Is the function Q an even function, an odd function, or neither?

64. The graph of $f(x) = x^2 - 2x - 8$ is shown in Exercise 8. Sketch the graphs of $y = |f(x)|$, $y = f(|x|)$, and $y = |f(|x|)|$.

65. The graph of $g(x) = x^2 - 4x - 12$ is shown in Example 2 (Figure 43). Sketch the graphs of $y = -g(x)$, $y = g(-x)$, and $y = -g(-x)$.

LINEAR AND QUADRATIC FUNCTIONS   S E C T I O N   2.4

We turn now from a general discussion of functions and their graphs to two specific, basic functions: linear functions and quadratic functions. We start with linear functions.

> A **linear** function is a function of the form $f(x) = mx + b$ where m and b are constants.

Linear Function

As the name suggests, the graph of a linear function is a straight line (the proof of this is left as an exercise).

The slope of a line plays an important role in calculus and is defined as follows.

> Suppose (x_1, y_1) and (x_2, y_2) are any two distinct points on a line. The slope of a nonvertical line is
> $$\frac{\Delta y}{\Delta x} = \frac{y_2 - y_1}{x_2 - x_1}.$$

Slope of a Line

The vertical directed distance $\Delta y = y_2 - y_1$ is called the **rise**, and the horizontal directed distance $\Delta x = x_2 - x_1$ is called the **run** (Figure 61).

Now consider the linear function $f(x) = mx + b$. Let $(x_1, f(x_1))$ and $(x_2, f(x_2))$ be any two points on the graph. The slope is

$$\frac{y_2 - y_1}{x_2 - x_1} = \frac{f(x_2) - f(x_1)}{x_2 - x_1} = \frac{(mx_2 + b) - (mx_1 + b)}{x_2 - x_1}$$

$$= \frac{m(x_2 - x_1)}{x_2 - x_1}$$

$$= m.$$

Furthermore, the graph crosses the y-axis when $x = 0$, so the y-intercept is $f(0) = m(0) + b = b$.

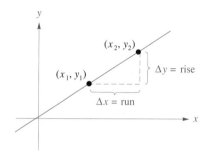

FIGURE 61

The graph of the linear function $f(x) = mx + b$ is a straight line with slope m and y-intercept b.

In applications, the slope of the line through two points on the graph of any function is the average rate of change in y per unit change in x.

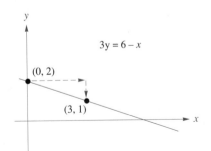

EXAMPLE 1 Sketch the graph of $3y = 6 - x$. Identify the slope and y-intercept.

SOLUTION First, divide both sides by 3 to get the form $y = mx + b$,

$$y = -\tfrac{1}{3}x + 2.$$

So the slope is $-\tfrac{1}{3}$ and the y-intercept is 2. Starting at the point $(0, 2)$, the line must have a run of 3 with a rise of -1.

FIGURE 62

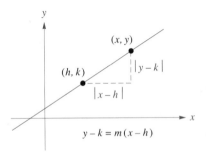

Suppose we know the slope m of a line and a point that it passes through (h, k). Then, if (x, y) is another point on the line (see Figure 63),

$$\frac{y - k}{x - h} = m$$

or, equivalently, $y - k = m(x - h)$. This is the point-slope form of a line.

FIGURE 63

Point-Slope Form

An equation of the line that has slope m that passes through the point (h, k) is

$$y - k = m(x - h).$$

EXAMPLE 2 Find an equation of the line through $(2, -5)$ with a slope of $-\tfrac{4}{3}$.

SOLUTION Substitute $h = 2$, $k = -5$, and $m = -\tfrac{4}{3}$ into the point-slope form:

$$y - k = m(x - h)$$
$$y - (-5) = -\tfrac{4}{3}(x - 2)$$
$$y + 5 = -\tfrac{4}{3}(x - 2).$$

Many functions that arise in the real world are linear. The point-slope form is often used to determine such a function, as the next example illustrates.

EXAMPLE 3   The relationship between Celsius and Fahrenheit is linear. Water freezes at 0° Celsius or 32° Fahrenheit and boils at 100° Celsius or 212° Fahrenheit. Find an equation relating Celsius and Fahrenheit (Figure 64).

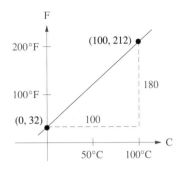

FIGURE 64

SOLUTION   First, the slope of the line is

$$m = \frac{212 - 32}{100 - 0} = \frac{180}{100} = \frac{9}{5}.$$

Now substitute into the point-slope form of a line:

$$F - 32 = \tfrac{9}{5}(C - 0)$$

or, solving for F,

$$F = \tfrac{9}{5}C + 32.$$

Parallel and Perpendicular Lines

Two nonvertical lines are parallel if and only if they have equal slopes. The proof of this is left to you as an exercise (see Problem 57).

The relation between the slopes of two perpendicular lines can be seen in Figure 65. Suppose that two perpendicular lines, one with slope m_1 and one with slope m_2, intersect at point Q. Locate another point P on the first line, as shown in the figure. From this, we can find the slope to be

$$m_1 = \frac{a}{b}.$$

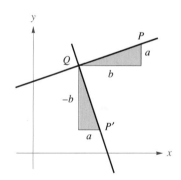

FIGURE 65

Now go back to point Q and locate a point P' on the other line so that the $d(Q, P') = d(Q, P)$. Since the two shaded triangles are congruent (why?), the slope of the line through Q and P' is

$$m_2 = \frac{-b}{a} = -\frac{1}{m_1}$$

or

$$m_1 m_2 = -1.$$

In a fashion similar to this, it can also be shown that if the product of the slopes of two lines is -1, then the lines must be perpendicular.

Two lines with slope m_1 and m_2 are parallel if and only if $m_1 = m_2$.

Two lines with slope m_1 and m_2 are perpendicular if and only if $m_1 m_2 = -1$.

EXAMPLE 4   Find an equation of the perpendicular bisector of the line segment with endpoints $(-3, 1)$ and $(6, 4)$.

SOLUTION The given line segment has a slope

$$m = \frac{4 - 1}{6 - (-3)} = \frac{1}{3}$$

so the perpendicular bisector has a slope of

$$\frac{-1}{m} = -3.$$

Compare with Example 3 in Section 2.1

The perpendicular bisector also passes through the midpoint of the segment

$$\left(\frac{-3 + 6}{2}, \frac{1 + 4}{2} \right) = \left(\frac{3}{2}, \frac{5}{2} \right).$$

Using the point-slope form of a linear equation gives

$$y - \tfrac{5}{2} = -3(x - \tfrac{3}{2})$$

or, simplifying,

$$y = -3x + 7.$$

Vertical and Horizontal Lines

We conclude this discussion of lines with two special cases. An equation of the form $x = h$ (h constant) is a vertical line. This is *not* a linear function and does not have a slope. An equation of the form $y = k$ is a horizontal line. This is a linear function (with $m = 0$ and $b = k$) (Figure 66).

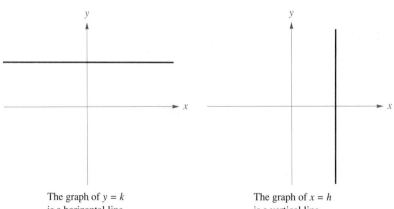

FIGURE 66

The graph of $y = k$ is a horizontal line

The graph of $x = h$ is a vertical line

Quadratic functions play an important role in the application of mathematics. They can be used, for example, in discussing the cost of manufacturing an item as well as the path of a freefalling object. We will see that the graphs of these functions have a characteristic shape.

Quadratic Functions

A **quadratic function** is a function of the form

$$f(x) = ax^2 + bx + c$$

where a, b, and c are constants, and $a \neq 0$.

We start with a few special cases of the form $y = ax^2$ (Figure 67). Note that these are all **even** functions, so we need only use nonnegative values of x to compute ordered pairs (you might need to reflect on this).

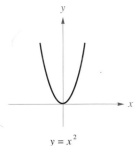

$y = x^2$

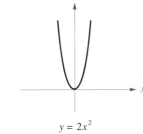

$y = 2x^2$

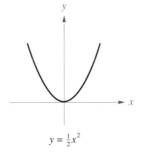

$y = \frac{1}{2}x^2$

FIGURE 67

x	0	1	2	3	4
y	0	1	4	9	16

x	0	1	2	3	4
y	0	2	8	18	32

x	0	1	2	3	4
y	0	$\frac{1}{2}$	2	$\frac{9}{2}$	8

Similarly, graphs for $a < 0$ look like Figure 68.

In general, the graph of $y = ax^2$ is a **parabola** (parabolas are also discussed in Chapter 8). The parabola opens upward if $a > 0$, and it opens

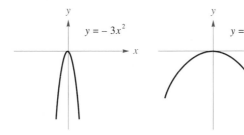

$y = -3x^2$

$y = -\frac{1}{5}x^2$

FIGURE 68

down if $a < 0$. Furthermore, the larger the value of $|a|$, the taller (or thinner) the parabola; likewise, small values of $|a|$ result in a shorter (or wider) parabola.

The turning point of the graph of $y = ax^2$ is called the **vertex.** From the figures, it should be clear that the vertex of $y = ax^2$ is at the origin. If $a > 0$, the vertex is the point with the minimum value of y for the graph. If, on the other hand, $a < 0$, the vertex is the point with the maximum value of y.

It is important for you to be able to graph $y = ax^2$ quickly with a minimal amount of point plotting. Now we are ready to graph the general quadratic function.

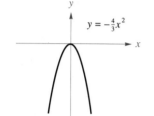

$y = -\frac{4}{3}x^2$

FIGURE 69

EXAMPLE 5   Sketch the graph of $y - 3 = -\frac{4}{3}(x + 1)^2$. Label the vertex and the intercepts.

SOLUTION We start with the graph of $y = -\frac{4}{3}x^2$ (Figure 69). Recall from Section 2.3 that $y - k = f(x - h)$ has the same graph as $y = f(x)$ but is translated h units horizontally and k units vertically. In this case, the basic function is $f(x) = -\frac{4}{3}x^2$. The translation is -1 unit horizontally and 3 units vertically. The vertex is at $(-1, 3)$.

The y-intercept is found by letting $x = 0$ and solving the resulting equation for y:

$$y - 3 = -\tfrac{4}{3}(0 + 1)^2$$
$$y - 3 = -\tfrac{4}{3}$$
$$y = \tfrac{5}{3}.$$

The y-intercept is $(0, \frac{5}{3})$.

The x-intercepts are found by letting $y = 0$ and solving for x:

$$0 - 3 = -\tfrac{4}{3}(x + 1)^2$$
$$-3 = -\tfrac{4}{3}(x + 1)^2$$
$$\tfrac{9}{4} = (x + 1)^2$$
$$\pm\tfrac{3}{2} = x + 1$$
$$x = -1 \pm \tfrac{3}{2}.$$

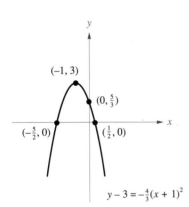

$(-1, 3)$

$(0, \frac{5}{3})$

$(-\frac{5}{2}, 0)$ $(\frac{1}{2}, 0)$

$y - 3 = -\frac{4}{3}(x + 1)^2$

FIGURE 70

Thus the x-intercepts are $(-\frac{5}{2}, 0)$ and $(\frac{1}{2}, 0)$. The graph is shown in Figure 70.

Finding the vertex of a quadratic equation in the form $y - k = a(x - h)^2$ is relatively simple; it is merely the point with coordinates (h, k). The vertex of a quadratic function in the form $y = ax^2 + bx + c$ is found by writing this function in the form $y - k = a(x - h)^2$. This can be accomplished by completing the square, as the next example shows.

EXAMPLE 6   Sketch the graph of $y = 2x^2 - 16x + 22$. Label the vertex and the intercepts.

SOLUTION   First, complete the square to get the form $y - k = a(x - h)^2$:

$$y - 22 = 2x^2 - 16x$$

$$y - 22 = 2(x^2 - 8x)$$

$$y - 22 + \mathbf{32} = 2(x^2 - 8x + \mathbf{16})$$

$$y + 10 = 2(x - 4)^2.$$

The basic function is $f(x) = 2x^2$, a relatively tall parabola that opens upward. Translate it 4 units horizontally and -10 units vertically. The vertex is at $(4, -10)$. To find the y-intercept, let $x = 0$ and determine the corresponding value of y:

$$y = 2(0)^2 - 16(0) + 22 = 22.$$

To find the x-intercepts, set $y = 0$ and solve for x:

$$0 = 2x^2 - 16x + 22.$$

We leave it to you to verify that the roots of this equation are $(4 - \sqrt{5}, 0)$ and $(4 + \sqrt{5}, 0)$. The graph is shown in Figure 71.

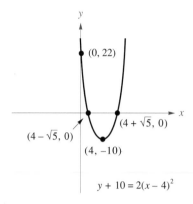

FIGURE 71

We can use the graph of a function to solve an inequality. To do this, we take advantage of the fact that if the point $(a, f(a))$ is above the x-axis, then $f(a) > 0$. Likewise, if the point $(a, f(a))$ is below the x-axis, then $f(a) < 0$. Of course, if $(a, f(a))$ is on the x-axis, then $f(a) = 0$.

For example, consider the graph in Figure 72. The values of x for which the graph of $y = f(x)$ is below the x-axis are (in interval notation)

$$(a, b) \quad \text{or} \quad (c, d).$$

This is exactly the solution to the inequality $f(x) < 0$.

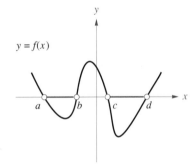

FIGURE 72

EXAMPLE 7   Solve $2x^2 \le 9x + 5$.

SOLUTION   First, rewrite the inequality as

$$2x^2 - 9x - 5 \le 0.$$

Consider the graph of $f(x) = 2x^2 - 9x - 5$. The graph of f is shown in Figure 73. Its x-intercepts are $(-\frac{1}{2}, 0)$ and $(5, 0)$.

The values of x over which the graph of $f(x)$ is below or on the x-axis is the solution to the inequality. This is the interval $[-\frac{1}{2}, 5]$.

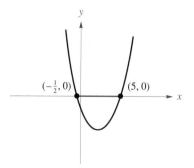

FIGURE 73

A useful formula for the vertex can be obtained if we complete the square for the general quadratic function:

$$y = ax^2 + bx + c$$

$$= a\left(x^2 + \frac{b}{a}x\right) + c$$

$$= a\left(x^2 + \frac{b}{a}x + \frac{b^2}{4a^2}\right) + c - \frac{b^2}{4a}$$

$$= a\left(x + \frac{b}{2a}\right)^2 + \left(c - \frac{b^2}{4a}\right).$$

Or

$$y - \left(c - \frac{b^2}{4a}\right) = a\left(x + \frac{b}{2a}\right)^2.$$

Hence, we have the following:

The Vertex of the Graph of
$f(x) = ax^2 + bx + c$

> The x-coordinate of the vertex of the graph of $f(x) = ax^2 + bx + c$ is $x = -\dfrac{b}{2a}$.

In practice, finding the y-coordinate is best accomplished by evaluating the function at $x = -b/2a$.

Being able to find the vertex has one very important application. It enables us to find the extreme value of a quadratic function. This may be a minimum value (if the parabola opens up) or a maximum value (if the parabola opens down).

EXAMPLE 8   Given $f(x) = \frac{1}{2}(9 - 2x + x^2)$, determine if f has either a maximum value or a minimum value. Find this value.

SOLUTION Rewriting in the form $f(x) = ax^2 + bx + c$, we have

$$f(x) = \tfrac{1}{2}x^2 - x + \tfrac{9}{2}.$$

This opens upward since $\frac{1}{2} > 0$. Computing the x-coordinate of the vertex, we get

$$x = -\frac{b}{2a} = -\frac{(-1)}{2(\frac{1}{2})} = 1.$$

To find the y value, substitute $x = 1$ into the function:

$$f(1) = \tfrac{1}{2}(9 - 2(1) + (1)^2) = 4.$$

The vertex is at the point $(1, 4)$; hence the minimum value is 4, occurring at $x = 1$.

An important application of mathematics is finding optimum values of varying quantities. For example, what is the area of the largest rectangle with a perimeter of 60 cm? What is the smallest surface area of all cylindrical cans that have a volume of 100 in.^{3}? These types of problems are called **optimization problems.** Although the in-depth study of optimization problems is a fundamental part of calculus, we can solve a number of these that can be modeled by quadratic functions. The next example is an illustration.

EXAMPLE 9   A gutter is to be formed from a rectangular piece of tin 8 in. wide by folding up equal rectangles on each side to be at right angles with the base (see Figure 74). Find the dimensions of the gutter that will maximize the flow capacity of the gutter.

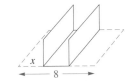

FIGURE 74

SOLUTION The problem is equivalent to finding the largest cross section possible for the gutter. Let x be the length of each segment to be turned up. Then this area is $A = (\text{width})(\text{height})$:

$$A(x) = (8 - 2x)(x)$$
$$= 8x - 2x^2.$$

This is a quadratic function. The graph opens down with a maximum value at $x = -b/2a = -8/2(-2) = 2$. The gutter should be constructed by folding up 2-in. segments for the height. The resulting base is 4 in.

The maximum (or minimum) values of some functions that are not quadratic functions can nonetheless be determined by using the techniques of the last two examples. These types of functions are called **quadratic in form.**

Functions That Are Quadratic in Form

EXAMPLE 10   Find the maximum value or minimum value of

$$f(x) = -x^4 + 36x^2.$$

SOLUTION This is not quadratic in x. However, if we rewrite the equation as

$$f(x) = -(x^2)^2 + 36(x^2)$$

then $f(x)$ is quadratic in x^2. Let $u = x^2$. Then

$$f(x) = -(u)^2 + 36(u).$$

This has a maximum at $u = -b/2a = 18$. Substituting x^2 back for u gives

$$x^2 = 18 \Rightarrow x = \pm\sqrt{18} = \pm 3\sqrt{2}.$$

So the maximum value is $f(\pm 3\sqrt{2}) = 324$.

EXAMPLE 11   Of all the rectangles inscribed in a circle with a diameter of 6, which has the greatest area?

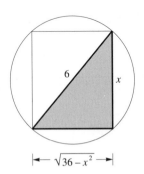

FIGURE 75

SOLUTION The first step of course is to draw a picture (see Figure 75). Draw a diagonal of the rectangle, forming a right triangle. Since the diagonal of the rectangle is also a diameter of the circle, its length is 6. Label one of the sides of the rectangle x. By the Pythagorean theorem, the other dimension of the rectangle is $\sqrt{36 - x^2}$. Now, we want to maximize the area of the rectangle

$$A(x) = x\sqrt{36 - x^2}.$$

At this point we call on a useful fact: The maximum or minimum of a nonnegative function occurs at the same place as the square of the function. Thus, an equivalent problem is to find where the maximum of

$$[A(x)]^2 = x^2(36 - x^2) = -x^4 + 36x^2$$

occurs. This was done in Example 10, except that, since x is a distance, we have a maximum at $3\sqrt{2}$. Hence the rectangle with the maximum area has dimensions $3\sqrt{2}$ by $3\sqrt{2}$ (a square!).

EXERCISE SET 2.4

A

In Problems 1 through 6, sketch the graphs of (a) and (b) on the coordinate plane.

1. a) $y = \frac{3}{4}x$
   b) $y = \frac{3}{4}x - 5$

2. a) $y = -\frac{2}{5}x$
   b) $y = -\frac{2}{5}x + 1$

3. a) $y = 3x$
   b) $y + 2 = 3(x - 4)$

4. a) $y = -2x$
   b) $y - 3 = -2(x + 4)$

5. a) $y = \frac{5}{4}x + 3$
   b) $y = -\frac{4}{5}x - 1$

6. a) $x = 3$
   b) $y + 5 = 0$

In Problems 7 through 14, determine the equation of the line described. Express your answer in the form $y = mx + b$.

7. The line passes through $(2, 1)$ and $(-1, -8)$.

8. The line passes through $(-7, 1)$ and $(8, -2)$.

9. The line passes through $(2, 1)$ and is parallel to the line $3x - 7y = 5$.

10. The line passes through $(6, 2)$ and is parallel to the line $6x - 5y = 0$.

11. The line passes through $(12, -8)$ and is perpendicular to the line $12x + 7y + 10 = 0$.

12. The line passes through $(-2, -2)$ and is perpendicular to the line $2x - y - 4 = 0$.

13. The line is the perpendicular bisector of the line segment with endpoints $(5, 10)$ and $(1, -2)$.

14. The line is the perpendicular bisector of the line segment with endpoints $(2, -2)$ and $(-14, 6)$.

15. Find a linear function f such that $f(-7) = -10$ and $f(3) = 5$.

16. Find a linear function g such that $g(4) = 6$ and $g(12) = 10$.

17. Find a linear function h such that $h(2) = -3$ and $h(-9) = -3$.

18. Find a linear function Q such that $Q(0) = 4$ and $Q(2) = 8$.

In Problems 19 through 24, sketch the graphs of (a) and (b) on the coordinate plane. Label the intercepts and the vertex.

19. a) $y = \frac{2}{3}x^2$ **b)** $y = \frac{1}{9}x^2$

20. a) $y = \frac{1}{5}x^2$ **b)** $y = -\frac{1}{5}x^2$

21. a) $y = -3x^2$ **b)** $y = -\frac{1}{3}x^2$

22. a) $y - 1 = (x + 2)^2$ **b)** $y - 1 = \frac{1}{2}(x + 2)^2$

23. a) $y + 4 = (x + 2)^2$ **b)** $y + 4 = -(x + 2)^2$

24. a) $y - 2 = -\frac{1}{2}(x - 1)^2$ **b)** $y - 2 = -2(x - 1)^2$

B

In Problems 25 through 36, express the function f in the form $f(x) = a(x - h)^2 + k$. Sketch the graph of the function and label the vertex.

25. $f(x) = x^2 - 2x + 1$ **26.** $f(x) = x^2 - 10x + 25$

27. $f(x) = x^2 + 2x + 3$ **28.** $f(x) = x^2 - 4x - 4$

29. $f(x) = 4x - x^2$ **30.** $f(x) = 5x + x^2$

31. $f(x) = -\frac{1}{2}x^2 - 2x$ **32.** $f(x) = 2x^2 + 12x + 9$

33. $f(x) = -\frac{1}{4}x^2 + x + 3$

34. $f(x) = -\frac{2}{3}x^2 + 8x - 19$

35. $f(x) = 2x^2 + 4\sqrt{3}x + 3$

36. $f(x) = \frac{1}{2}x^2 - \sqrt{2}x$

In Problems 37 through 42, determine whether the function has either a maximum value or a minimum value and find that value.

37. $f(x) = 2 - 4x - x^2$ **38.** $f(x) = 8x - 2 + x^2$

39. $f(x) = 4x - x^2$ **40.** $f(x) = x^2 - 8x$

41. $f(x) = 5x - 5 - 2x^2$ **42.** $f(x) = 10 + 3x + \frac{1}{2}x^2$

43. Find two numbers whose sum is 12 and whose product is maximum. (Hint: Let the numbers be x and $12 - x$.)

44. Find two numbers whose difference is 4 and whose product is minimum. (Hint: Let the numbers be $x + 4$ and x.)

45. A projectile is fired upward at time $t = 0$ from ground level. Neglecting air resistance, the height as a function of time (in seconds) is $f(t) = v_0 t - 16t^2$ feet, where v_0 is the initial muzzle velocity. If $v_0 = 1200$ ft/sec, what is the greatest height the projectile reaches?

46. A projectile is fired at time $t = 0$ from ground level at a 30° angle with the ground. Neglecting air resistance, the height as a function of time (in seconds) is $g(t) = \frac{1}{2}v_0 t - 16t^2$, where v_0 is the initial muzzle velocity. If $v_0 = 1200$ ft/sec, what is the greatest height the projectile reaches?

47. A manufacturer of compact disc players has found that the cost in dollars of manufacturing x units per month is given by the function

$$C(x) = 0.1x^2 - 126x + 3457.$$

How many units should be manufactured per month to minimize the cost?

48. A manufacturer of computer printers determines that the profit (or loss) in dollars generated by manufacturing x units per week is given by the function $P(x) = -0.2x^2 + 49.6x - 935.4$.

a) What is the minimum number of units per week that make a profit? What is the maximum?

b) What is the maximum profit?

49. What is the greatest rectangular area that can be enclosed with 300 meters of fencing?

50. A pen is to be constructed by erecting a wall in the shape of a rectangle. Its interior is divided by a chain fence (see Figure 76). Given that the wall costs $10/ft and the fence costs $5/ft, what is the largest pen that can be constructed for a total of $800?

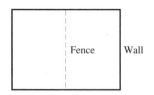

FIGURE 76

51. Three sides of a rectangular pigpen are to be constructed with 600 feet of fencing. (One side is a river and pigs can't swim.) What should the dimensions be if the area is to be a maximum (Figure 77)?

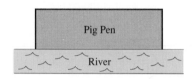

FIGURE 77

52. What is the maximum possible area of a rectangle with one side on the positive x-axis, one side on the positive y-axis, one corner on the line $2x + 3y = 12$, and the opposite corner at the origin (Figure 78)?

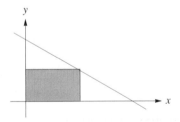

FIGURE 78

53. A rectangle is inscribed in a right isosceles triangle, as shown in Figure 79. The triangle has a hypotenuse of length 12.

a) Show that the area of the rectangle can be expressed as the function

$$A(h) = 12h - 2h^2$$

where h is the height of the rectangle. What is its domain?

b) What is the maximum possible area of the rectangle?

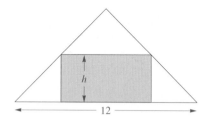

FIGURE 79

54. The monthly rent of a storage shed is a linear function of the square footage. In particular, a 3000-ft^2 shed rents for \$490, and a 1500-ft^2 shed rents for \$320. Find this linear function and use it to determine the monthly rent of a 42,000-ft^2 shed.

55. The income tax levied by a certain state on incomes between \$30,000 and \$60,000 is a linear function of the income. In particular, the tax on \$36,000 is \$790, and the tax on \$42,000 is \$920. Find this linear function and use it to determine the tax on an income of \$53,000.

In Problems 56 through 61, solve the inequality by sketching the graph of a related quadratic function.

56. $x^2 + 2x - 3 < 0$

57. $2x^2 + 9x + 4 > 0$

58. $2x^2 + 5x \geq 12$

59. $11x + 3 \geq 4x^2$

60. $x^2 + 3 \geq x$

61. $2x^2 + 2x + 1 < 0$

C

In Problems 62 through 67, determine whether the function f has either a maximum value or a minimum value and find that value.

62. $f(x) = x^4 - 6x^2 + 1$

63. $f(x) = 4 + 10x^3 - x^6$

64. $f(x) = x - 16\sqrt{x} + 80$

65. $f(x) = \sqrt{2(x^2 - 4x + 10)}$

66. $f(x) = \sqrt[3]{3x(10 - x)}$

67. $f(x) = \dfrac{1}{x^2 + 2x + 3}$

68. An eastbound train, which starts out 25 miles due north of a northbound car, is traveling at 40 mph. If the car is traveling at 50 mph, find the time when the two are closest together (Figure 80).

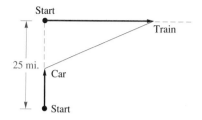

FIGURE 80

69. Prove that the graph of $y = mx + b$ is a straight line. (Hint: Prove that any three points on this line are collinear using the distance formula.)

70. Prove that two nonvertical lines are parallel if and only if they have equal slopes.

71. Consider a line segment with endpoints (x_1, y_1) and

(x_2, y_2) on the graph of the parabola $y = x^2$ (Figure 81).

a) Show that the y-intercept for the line segment is $\sqrt{y_1 y_2}$.

b) Determine the larger of $\sqrt{y_1 y_2}$ and $\frac{1}{2}(y_1 + y_2)$.

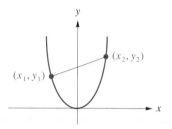

FIGURE 81

MORE ON GRAPHING S E C T I O N   2.5

In the last section we started with the graph of $y = x^2$ and "stretched" and translated it. In this section, we take a closer look at stretching graphs in general.

For example, consider the graph of a function $y = f(x)$ in Figure 82. Suppose that we wish to determine the graphs of $y = 2f(x)$ and $y = \frac{1}{2}f(x)$ (Figure 83). For each value of x, the y-coordinate for the graph of $y = 2f(x)$ is twice the y-coordinate for the graph of the function $y = f(x)$. Similarly, for each x, the y-coordinate for the graph of $y = \frac{1}{2}f(x)$ is half the y-coordinate for the graph of the function $y = f(x)$. The only points that remain fixed are the x-intercepts (why?).

In general, if $c > 1$, the graph of $y = cf(x)$ looks like the graph of $y = f(x)$, expanded away from the x-axis; if $0 < c < 1$, then the graph of $y = cf(x)$

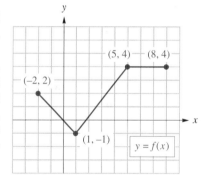

FIGURE 82

FIGURE 83

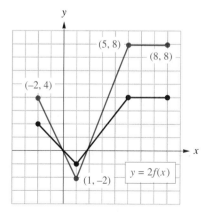

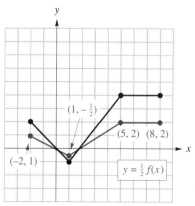

looks like the graph of $y = f(x)$ compressed toward the x-axis. In either case, to graph $y = cf(x)$, replace each point (x, y) on the graph of $y = f(x)$ with (x, cy).

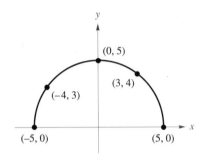

FIGURE 84

EXAMPLE 1   The graph of

$$f(x) = \sqrt{25 - x^2}$$

is shown in Figure 84 (recall that this is a member of our catalog in Section 2.3). Use this to sketch the graphs of

a) $y = \tfrac{3}{5}\sqrt{25 - x^2}$

b) $y = 3\sqrt{25 - x^2}.$

SOLUTION

a)   This is $y = \tfrac{3}{5}f(x)$. Its graph (Figure 85) is a compression of the graph of $y = f(x)$ toward the x-axis. To graph $y = \tfrac{3}{5}f(x)$, we replace each point (x, y) on the graph of $y = f(x)$ with $(x, \tfrac{3}{5}y)$. For example, replace $(0, 5)$ with $(0, 3)$, replace $(3, 4)$ with $(3, \tfrac{12}{5})$, replace $(-4, 3)$ with $(-4, \tfrac{9}{5})$, and so on.

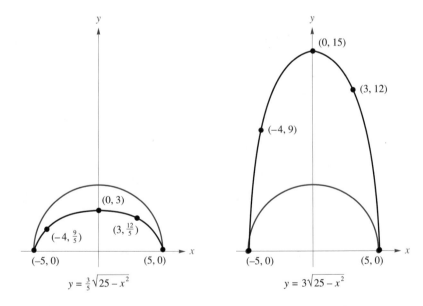

FIGURE 85
$$y = \tfrac{3}{5}\sqrt{25 - x^2} \qquad\qquad y = 3\sqrt{25 - x^2}$$

b)   This is $y = 3f(x)$. Its graph (Figure 85) is an expansion of the graph of $y = f(x)$ away from the x-axis. To graph $y = 3f(x)$, we will replace each point (x, y) on the graph of $y = f(x)$ with $(x, 3y)$. For example, replace $(0, 5)$ with $(0, 15)$, replace $(-4, 3)$ with $(-4, 9)$, and so on.

For the graph of $y = f(cx)$, you may be able to guess that stretching will occur toward or away from the y-axis. Returning to the graph of $y = f(x)$, suppose that we are interested in sketching the graph of $y = f(2x)$ or the graph of $y = f(\frac{2}{3}x)$. Notice from the graph of $y = f(x)$ that $f(-2) = 2$. The point with coordinates $(-1, 2)$ is on the graph of $y = f(2x)$ since $y = f(2(-1)) = f(-2) = 2$. In fact, if the point $(a, f(a))$ is on the graph of $y = f(x)$, then the point with coordinates $(a/2, f(a))$ is on the graph of $y = f(2x)$ since $y = f(2(a/2)) = f(a)$. What this implies is that the graph of $y = f(2x)$ is the graph of $y = f(x)$ compressed toward the y-axis.

In a similar fashion, the graph of $y = f(\frac{2}{3}x)$ is the graph of $y = f(x)$ expanded away from the y-axis by a factor since if the point $(a, f(a))$ is on the graph of $y = f(x)$, then the point with coordinates $(\frac{3}{2}a, f(a))$ is on the graph of $y = f(\frac{2}{3}x)$ since $y = f(\frac{2}{3}(\frac{3}{2}a)) = f(a)$. The graphs of $y = f(2x)$ and $y = f(\frac{2}{3}x)$ are shown in Figure 87.

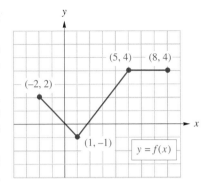

FIGURE 86

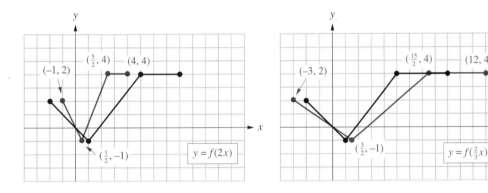

FIGURE 87

From these examples, we can generalize: If $c > 1$, then $y = f(cx)$ will look like $y = f(x)$ compressed toward the y-axis. If $0 < c < 1$, then $y = f(cx)$ will look like $y = f(x)$ expanded away from the y-axis. In either case, to graph $y = f(cx)$, replace each point (x, y) on the graph of $y = f(x)$ with $((1/c)x, y)$.

EXAMPLE 2   Sketch the graph of the following:

a) $y = \sqrt{25 - (\frac{1}{2}x)^2}$ **b)** $y = 3\sqrt{25 - 4x^2}$.

SOLUTION

a) We will use $f(x) = \sqrt{25 - x^2}$ to determine this graph. Since this equation is $y = f(\frac{1}{2}x)$, its graph is an expansion of the graph of $y = f(x)$ away from the y-axis. To graph $y = f(\frac{1}{2}x)$, replace each

FIGURE 88

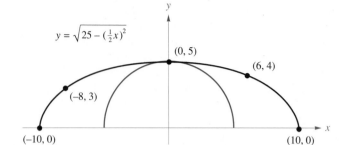

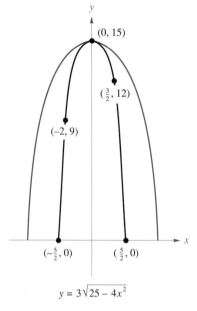

$y = 3\sqrt{25 - 4x^2}$

FIGURE 89

each point (x, y) on the graph of $y = f(x)$ with $(2x, y)$ (Figure 88). For example replace $(5, 0)$ with $(10, 0)$, replace $(3, 4)$ with $(6, 4)$, replace $(-4, 3)$ with $(-8, 3)$, and so on.

b) In this case we use $f(x) = 3\sqrt{25 - x^2}$. Its graph was discussed in part b) of Example 1. Since the equation that we want to graph is $y = f(2x)$, its graph is a compression of the graph of $y = f(x)$ toward the y-axis. Replacing each point (x, y) on the graph of $y = f(x)$ with $(\frac{1}{2}x, y)$ gives the graph in Figure 89.

In Section 2.4, we saw that the graph of $y = -x^2$ is the same as $y = x^2$ reflected through the x-axis. This is a specific case of a general rule. The graph of $y = -f(x)$ is the same as $y = f(x)$ reflected through the x-axis.

On the other hand, $-x$ causes the same functional value for $f(-x)$ as x causes for $f(x)$. This means that the graph of $y = f(-x)$ is the same as the graph of $y = f(x)$ reflected through the y-axis (Figure 90).

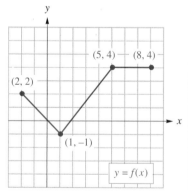
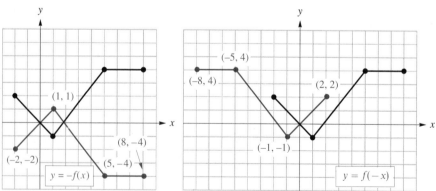

FIGURE 90

EXAMPLE 3 Sketch the graph of $y = \sqrt{x}$. Reflect it about the x-axis, then expand it away from the y-axis by a factor of 3. Give the equation for the final graph.

SOLUTION

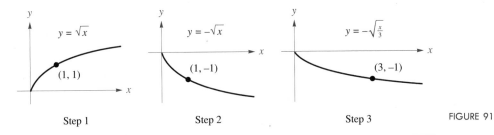

Step 1   Step 2   Step 3   FIGURE 91

EXAMPLE 4 Sketch the graph of

$$f(x) = \begin{cases} x + 1 & x \le 0 \\ \dfrac{1}{x} & x > 0 \end{cases}$$

Reflect through the y-axis, then translate 2 units to the right.

SOLUTION The graph of $y = f(x)$ is shown in Figure 92:

FIGURE 92

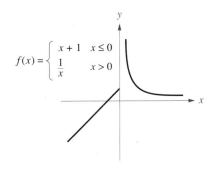

$$f(x) = \begin{cases} x + 1 & x \le 0 \\ \dfrac{1}{x} & x > 0 \end{cases}$$

Step 1 (Reflect)   Step 2 (Translate)

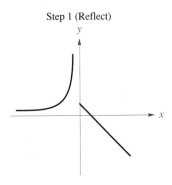
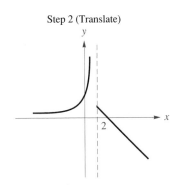

$$y = f(-x) = \begin{cases} -x + 1 & -x \le 0 \\ \dfrac{1}{-x} & -x > 0 \end{cases}$$

$$y = f(-(x - 2)) = \begin{cases} -(x - 2) + 1 & -(x - 2) \le 0 \\ \dfrac{1}{-(x - 2)} & -(x - 2) > 0 \end{cases}$$

The simplified form for the last step is

$$y = \begin{cases} -x + 3 & x \geq 2 \\ \dfrac{1}{-x + 2} & x < 2 \end{cases}.$$

Graphs and Absolute Value

The next example demonstrates how to graph $y = |f(x)|$ or $y = f(|x|)$, given the graph of $y = f(x)$.

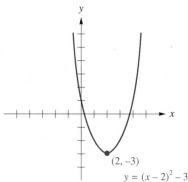

$y = (x - 2)^2 - 3$

FIGURE 93

EXAMPLE 5   Sketch the graph of the following:

a) $y = |(x - 2)^2 - 3|$ **b)** $y = (|x| - 2)^2 - 3$

SOLUTION First, if $f(x) = (x - 2)^2 - 3$, notice that part a) is $y = |f(x)|$ and part b) is $y = f(|x|)$. In each case, start with the graph of $y = f(x)$. This is a translation of the graph of $y = x^2$, shown in Figure 93. Its vertex has the coordinates $(2, -3)$.

a)   For nonnegative values of $f(x)$, the graph of $y = |f(x)|$ has the same y-coordinates for each x as $y = f(x)$. For negative values of $f(x)$, the graph of $y = |f(x)|$ has the opposite (positive) y-coordinate for each x as $y = f(x)$. The result is that all parts of the graph that are below the x-axis are reflected through the x-axis. The remainder of the graph does not change (Figure 94).

FIGURE 94

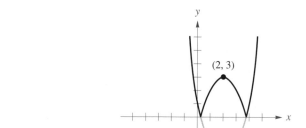

$y = |(x - 2)^2 - 3|$

b)   To graph $y = f(|x|)$, first notice that $f(|x|)$ is an even function. For $x \geq 0$, $|x| = x$. This means that for $x \geq 0$, the graph of $y = f(|x|)$ is identical to the graph of $y = f(x)$. Since $y = f(|x|)$ is even, we complete our graph by using the fact that it must be symmetric with respect to the y-axis (Figure 95).

FIGURE 95

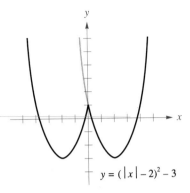

$y = (|x| - 2)^2 - 3$

Summary

Suppose the graph of $y = f(x)$ is known.

To sketch the graph of $y = cf(x)$:
Replace each point (x, y) on the graph of $y = f(x)$ with (x, cy).
If $c > 1$, the resulting graph is an expansion away from the x-axis.
If $0 < c < 1$, the resulting graph is a compression toward the x-axis.

To sketch the graph of $y = f(cx)$:
Replace each point (x, y) on the graph of $y = f(x)$ with $(x/c, y)$.
If $0 < c < 1$, the resulting graph is an expansion away from the y-axis.
If $c > 1$, the resulting graph is a compression toward the y-axis.

To sketch the graph of $y = -f(x)$:
Replace each point (x, y) on the graph of $y = f(x)$ with $(x, -y)$.
The resulting graph is a reflection through the x-axis.

To sketch the graph of $y = f(-x)$:
Replace each point (x, y) on the graph of $y = f(x)$ with $(-x, y)$.
The resulting graph is a reflection through the y-axis.

To sketch the graph of $y = |f(x)|$:
Keep all points on the graph of $y = f(x)$ that are above the x-axis. Reflect all points on the graph of $y = f(x)$ that are below the x-axis through the x-axis.

To sketch the graph of $y = f(|x|)$:
Keep all points on the graph of $y = f(x)$ that are to the right of the y-axis. The graph of $y = f(|x|)$ is always even; complete the graph by noting this symmetry with respect to the y-axis.

EXERCISE SET 2.5

A

In Problems 1 through 9, refer to the graphs of $y = f(x)$ and $y = g(x)$ (Figure 96) to graph the equations.

FIGURE 96

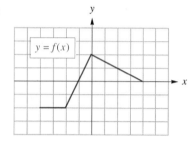

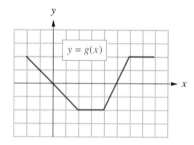

1. a) $y = 3f(x)$ **b)** $y = f(3x)$

2. a) $y = 2g(x)$ **b)** $y = g(2x)$

3. a) $y = f(\frac{1}{2}x)$ **b)** $y = \frac{1}{2}f(x)$

4. a) $y = -f(x)$ **b)** $y = f(-x)$

5. a) $y = -g(x)$ **b)** $y = g(-x)$

6. a) $y = -f(x)$ **b)** $y - 1 = -f(x - 2)$

7. a) $y = 2g(x)$ **b)** $y + 2 = 2g(x + 3)$

8. a) $y = g(-x)$ **b)** $y + 1 = g(-x + 3)$

9. a) $y = f(-x)$ **b)** $y - 3 = f(-x - 1)$

In Problems 10 through 15, use the graph of $y = \sqrt{x}$ to graph the given functions.

10. $y = -2\sqrt{x}$ **11.** $y = -\sqrt{\frac{3}{2}x}$

12. $y = -\sqrt{-x}$ **13.** $y = 2\sqrt{-x}$

14. $y - 2 = \sqrt{-(x + 1)}$ **15.** $y - 3 = \sqrt{-(x - 4)}$

In Problems 16 through 21, use the graph of $y = |x|$ to graph the given functions.

16. $y = -|x|$ **17.** $y = |-x|$

18. $y = |2x| - 4$ **19.** $y = 2|x - 4|$

20. $y - 2 = -\frac{1}{2}|x - 4|$ **21.** $y - 2 = -|-(x - 4)|$

In Problems 22 through 27, use the graph of $y = \sqrt{36 - x^2}$ to graph the given functions.

22. $y = \frac{1}{2}\sqrt{36 - x^2}$ **23.** $y = \frac{2}{3}\sqrt{36 - x^2}$

24. $y = \sqrt{36 - 9x^2}$ **25.** $y = \sqrt{36 - \frac{1}{4}x^2}$

26. $y = \sqrt{36 - (x + 1)^2}$ **27.** $y = \sqrt{36 - (x - 2)^2}$

B

In Problems 28 through 36, sketch the graph of the given functions.

28. a) $y = x$ **b)** $y - 4 = \frac{1}{3}(x + 1)$

29. a) $y = x$ **b)** $y - 2 = (x + 3)$

30. a) $y = x^3$ **b)** $y - 2 = \frac{1}{4}(x + 3)^3$

31. a) $y = x^3$ **b)** $y + 1 = -(x - 4)^3$

32. a) $y = [\![x]\!]$ **b)** $y = -[\![2x]\!]$

33. a) $y = [\![x]\!]$ **b)** $y = [\![\frac{1}{3}x]\!] + 2$

34. a) $y = \dfrac{1}{x}$ **b)** $y = \dfrac{3}{x}$

35. a) $y = \dfrac{1}{x}$ **b)** $y = -\dfrac{1}{2x}$

36. a) $y = \dfrac{1}{x}$ **b)** $y = -\dfrac{1}{x - 3}$

37. Given that

$$f(x) = \begin{cases} x + 2 & x < 0 \\ -x^2 & x \geq 0 \end{cases},$$

sketch:

a) $y = f(x)$ **b)** $y = f(2x)$.

38. Given that

$$g(x) = \begin{cases} x & x < 1 \\ x^2 - 4 & x \geq 1 \end{cases},$$

sketch:

a) $y = g(x)$ **b)** $y - 2 = g(\frac{1}{2}x)$.

39. Given the graph of $y = \sqrt[3]{x}$ (Figure 97), sketch the graph of $y = -\sqrt[3]{x} - 4$. Label the intercepts.

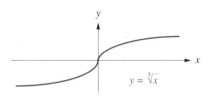

$y = \sqrt[3]{x}$

FIGURE 97

40. Given the graph of $y = \sqrt[3]{x}$ (Figure 97), sketch the graph of $y = -2\sqrt[3]{x} + 4$. Label the intercepts.

41. Sketch the graph of $y = \sqrt{32 - 4x - x^2}$. (Hint: Complete the square.)

42. Sketch the graph of $y = \sqrt{20 + 8x - x^2}$. (Hint: Complete the square.)

In Problems 43 through 48, refer to the graphs of $y = f(x)$ and $y = g(x)$ used in Problems 1–9 (Figure 96) to graph the given equations.

43. **a)** $y = |f(x)|$ **b)** $y = |f(x)| - 3$

44. **a)** $y = g(|x|)$ **b)** $y = g(|x - 2|)$

45. **a)** $y = f(|x|)$ **b)** $y = -2f(|x|)$

46. **a)** $y = |g(x)|$ **b)** $y = \frac{1}{2}|g(x)|$

47. **a)** $y = f(|x|)$ **b)** $y = f(-|x|)$

48. **a)** $y = |g(x)|$ **b)** $y = |g(-x)|$

49. Sketch the graph of **a)** $y = \sqrt{|x|}$ and **b)** $y = \sqrt{|x| - 2}$.

50. Sketch the graph of **a)** $y = |\sqrt{x} - 2|$ and **b)** $y = |\sqrt{x} - 2| - 2$.

C

51. Sketch the graphs of $y = [\![x]\!]$ and $y = [\![\frac{3}{2}x]\!]$ on the same coordinate plane. Use these graphs to solve the equation $[\![x]\!] = [\![\frac{3}{2}x]\!]$ by determining the values of x for which the graphs coincide.

52. Solve the equation $[\![\frac{1}{2}x]\!] = \frac{1}{4}[\![x]\!]$ by using the method described in Problem 51.

53. Sketch the graphs of $y = -\sqrt{2x}$ and $y = |x - 6| - 6$ on the same coordinate plane. Use these graphs to solve the inequality $-\sqrt{2x} > |x - 6| - 6$ by determining the values of x for which the graph of $y = -\sqrt{2x}$ is strictly above the graph of $y = |x - 6| - 6$.

54. Solve the inequality $\sqrt{20 - x^2} \geq |x - 2|$ by using the method described in Problem 53.

55. Sketch the graph of $y = ||x| - 1|$.

56. Sketch the graph of $y = |||x| - 1| - 1|$.

57. Sketch the graph of $y = \sqrt{9 - (|x| - 4)^2}$.

OPERATIONS WITH FUNCTIONS   S E C T I O N   2.6

Two real numbers can be combined by using addition, subtraction, multiplication, or division to yield another real number. In a similar fashion, functions can be combined by using these same four basic operations.

For example, suppose that $f(x) = \sqrt{x}$ and $g(x) = 1/x$. The sum of these two functions f and g is another function $f + g$ defined as

$$(f + g)(x) = f(x) + g(x)$$

$$= \sqrt{x} + \frac{1}{x}.$$

What about the domain of this new function? The domain of f is the set of nonnegative real numbers, and the domain of g is the set of nonzero real numbers. The domain of their sum, therefore, is the *intersection* of these two sets since for a particular value of x, the sum of two functions is defined only if each of these two functions is defined.

The other basic operations on functions can be defined in a like manner.

Operations on Functions

> Given functions f and g, the sum, difference, product, and quotient are given by
>
> $$(f + g)(x) = f(x) + g(x)$$
> $$(f - g)(x) = f(x) - g(x)$$
> $$(fg)(x) = f(x)g(x)$$
> $$\left(\frac{f}{g}\right)(x) = \frac{f(x)}{g(x)}, \qquad g(x) \neq 0.$$
>
> Furthermore, the domain of the sum, difference, and product of f and g is the intersection of the domains of f and the domain of g. The domain of the quotient is also the intersection of the domains of f and the domain of g, with the additional restriction that $g(x) \neq 0$.

EXAMPLE 1   Let $f(x) = \dfrac{3}{x - 5}$ and $g(x) = \dfrac{x + 4}{x}$. Find the following:

a) $f + g$ **b)** fg **c)** The domain for $f + g$ *and* fg

SOLUTION

a) $(f + g)(x) = f(x) + g(x)$

$$= \frac{3}{x - 5} + \frac{x + 4}{x}$$

$$= \frac{3}{x - 5} \cdot \frac{x}{x} + \frac{x + 4}{x} \cdot \frac{x - 5}{x - 5}$$

$$= \frac{3x}{x(x - 5)} + \frac{x^2 - x - 20}{x(x - 5)}$$

$$= \frac{x^2 + 2x - 20}{x(x - 5)}.$$

b) $(fg)(x) = f(x) \cdot g(x)$

$$= \frac{3}{x-5} \cdot \frac{x+4}{x}$$

$$= \frac{3(x+4)}{x(x-5)}.$$

c) The domain for both $f + g$ and fg is the intersection of the domains for f and g. This intersection is the set of all real numbers such that $x \neq 0$ or $x \neq 5$.

To succeed in calculus, it will be crucial that you can reduce complicated expressions to simpler functions. Indeed, an important goal of this section is to develop more sophisticated ways to view complicated functions.

One of the best ways to describe a sum of two functions is geometrically. Consider the function $h(x) = \frac{1}{2}|x| + |x + 3|$. This function h can be described as the sum $f + g$, where $f(x) = \frac{1}{2}|x|$ and $g(x) = |x + 3|$.

Sketch the graph of $y = f(x)$ and $y = g(x)$ on the same axis. Since $h(x) = f(x) + g(x)$, we can add the y values at each value of x to get the graph of $h(x)$.

For example, for $x = 4$, $f(4) = \frac{1}{2}|4| = 2$, and $g(4) = |4 + 3| = 7$. These values are represented in Figure 98 with line segments.

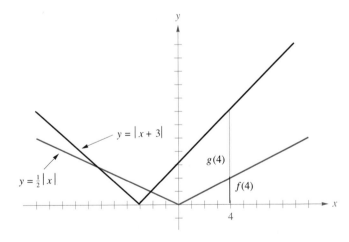

FIGURE 98

To determine the point on the graph of $y = \frac{1}{2}|x| + |x + 3|$ for $x = 4$, we can add these two line segments by placing them end to end. Do this for a few other values of x to complete the graph (Figure 99).

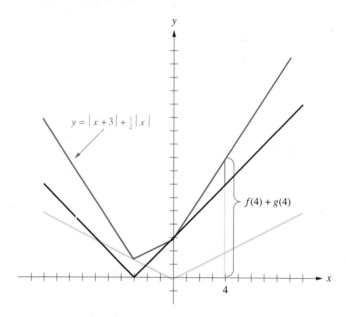

FIGURE 99

EXAMPLE 2   Sketch the graph of $h(x) = \sqrt{x} + 1/x$.

SOLUTION Graph $f(x) = \sqrt{x}$ and $g(x) = 1/x$ on the same axes. Notice that since $f(x) = \sqrt{x}$ is defined only for $x \geq 0$, we only graph $g(x) = 1/x$ for $x > 0$.

By adding the y-coordinates, we can get a good idea of the graph of $h(x)$ (Figure 100). Notice that for larger values of x, the graph of $h(x)$ approaches that of $f(x) = \sqrt{x}$. Similarly, as x approaches 0, the graph of $h(x)$ is very close to that of $g(x) = 1/x$.

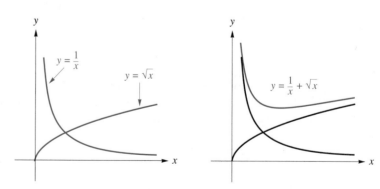

FIGURE 100

A function $h(x)$ is called a **linear combination** of $f(x)$ and $g(x)$ if it can be written as $h(x) = c_1 f(x) + c_2 g(x)$, where c_1 and c_2 are constants.

EXAMPLE 3 Given that $h(x) = \frac{1}{2}(x + 4\sqrt{x - 1}) - x$:

a) Express $h(x)$ as a linear combination of $f(x) = \sqrt{x - 1}$ and $g(x) = x$.

b) Graph $y = h(x)$.

SOLUTION

a) Simplify

$$h(x) = \tfrac{1}{2}(x + 4\sqrt{x - 1}) - x = \tfrac{1}{2}x + 2\sqrt{x - 1} - x$$
$$= 2\sqrt{x - 1} + (-\tfrac{1}{2})x.$$

This has the form $c_1 f(x) + c_2 g(x)$, where $c_1 = 2$ and $c_2 = -\frac{1}{2}$.

b) Graph $y = 2\sqrt{x - 1}$ and $y = -\frac{1}{2}x$ on the same coordinate plane. Then add the y-coordinates where the domains overlap (Figure 101).

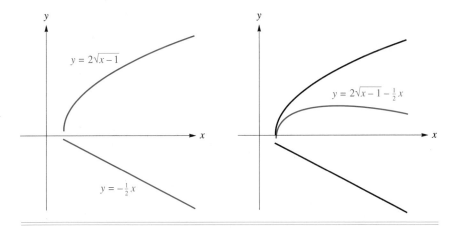

FIGURE 101

An important function in engineering mathematics is $u(x)$, the **unit step function** (Figure 102). This function can be thought of as an on-off switch. Linear combinations of these types of functions are used to describe complicated functions that arise in engineering applications.

$$u(x) = \begin{cases} 0 & x < 0 \\ 1 & x \geq 0 \end{cases}$$

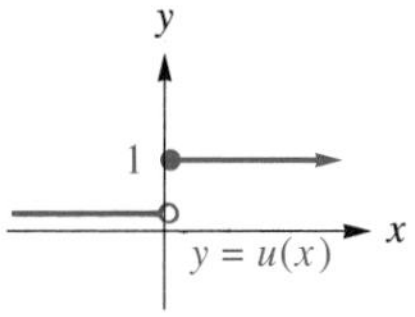

$y = u(x)$

FIGURE 102

EXAMPLE 4 Sketch the graph of the following equations, where $u(x)$ is the unit step function:

a) $y = 2u(x - 1), \; x \geq 0$

b) $y = 2u(x - 1) - 3u(x - 2), \; x \geq 0$

SOLUTION

a) The graph is the same as $y = u(x)$, only translated 1 unit to the right and then expanded away from the x-axis by a factor of 2.

b) First, graph $y = 2u(x - 1)$ and $y = -3u(x - 2)$ on the same axes. Then add y-coordinates (Figure 103).

FIGURE 103

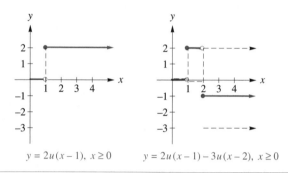

$$y = 2u(x - 1), \ x \geq 0 \qquad y = 2u(x - 1) - 3u(x - 2), \ x \geq 0$$

There is a fifth operation on functions that plays a significant role in calculus. It involves a function that operates successively on the output of another function.

Composition of Functions

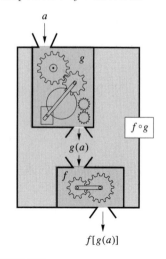

FIGURE 104

Suppose that f and g are functions. The **composition of f with g** is

$$(f \circ g)(x) = f[g(x)].$$

Likewise, the **composition of g with f** is

$$(g \circ f)(x) = g[f(x)].$$

The composition of f with g can be thought of as a black box whose inside components are the functions f and g (Figure 104). If a is the input to the black box it is processed first by g. The output of g, $g(a)$, becomes the input of the function f. The final output is $f[g(a)]$.

The composition $f \circ g$ is defined only if $g(x)$ is in the domain of f. The domain of $(f \circ g)(x)$ is the set of all numbers x such that x is in the domain of g and $g(x)$ is in the domain of f. Figure 105 shows how this works:

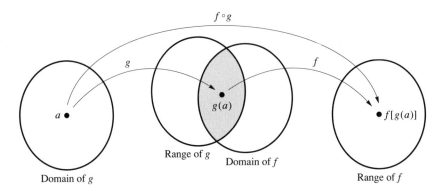

$f \circ g$

g

f

$g(a)$

$f[g(a)]$

a

Range of g

Domain of f

Domain of g

Range of f FIGURE 105

EXAMPLE 5   Let $f(x) = \dfrac{x}{x^2 + 1}$ and $g(x) = 2x^3$. Find

a) $f[g(x)]$ and **b)** $g[f(x)]$.

SOLUTION

a) $f[g(x)] = f(2x^3) = \dfrac{(2x^3)}{(2x^3)^2 + 1} = \dfrac{2x^3}{4x^6 + 1}$.

b) $g[f(x)] = g\left(\dfrac{x}{x^2 + 1}\right) = 2\left(\dfrac{x}{x^2 + 1}\right)^3 = \dfrac{2x^3}{(x^2 + 1)^3}$.

EXAMPLE 6   If $f(x) = \sqrt{x}$ and $g(x) = 4 - x^2$, find

a) $(f \circ g)(x)$,   **b)** $(g \circ f)(x)$, and **c)** the domain of each.

SOLUTION

a) $(f \circ g)(x) = f[g(x)] = f(4 - x^2) = \sqrt{4 - x^2}$.

b) $(g \circ f)(x) = g[f(x)] = g(\sqrt{x}) = 4 - (\sqrt{x})^2 = 4 - x$.

c) Since the domain of f is $x \geq 0$, the domain for $f[g(x)]$ must be all x such that $g(x) \geq 0$: that is, $4 - x^2 \geq 0$, which is equivalent to $-2 \leq x \leq 2$. Therefore, the domain of $(f \circ g)(x)$ is $-2 \leq x \leq 2$. The domain of the function $g \circ f$ is $x \geq 0$ since f, the first process, cannot evaluate negative values. Since the domain for g is any real number, this is the only restriction on their composition.

Notice that, in general, the function $f \circ g$ is not the same as the function $g \circ f$. You should be able to graph both a) and b) quickly. (One is a half circle; the other is a half line.)

The **decomposition** of functions is as important as the composition of functions. The next example illustrates decomposing a function into two simpler functions.

EXAMPLE 7   Write $h(x) = \sqrt{x^2 + 2}$ as the composition of two functions $f(g(x))$. Do not use $g(x) = x$ or $f(x) = x$.

SOLUTION   One way is to let $g(x) = x^2 + 2$ and $f(x) = \sqrt{x}$; then

$$f(g(x)) = f(x^2 + 2) = \sqrt{x^2 + 2}.$$

Another possibility is $f(x) = \sqrt{x + 2}$ and $g(x) = x^2$; then

$$f(g(x)) = f(x^2) = \sqrt{x^2 + 2}.$$

EXAMPLE 8   A watermelon dropped from the top of a 90-foot building has a height of $90 - 16t^2$ feet, t seconds after it is released. A cameraman is crouched (on the sidewalk) 30 feet from the point of impact. Express the distance s between the cameraman and the watermelon as a function of t.

SOLUTION   Let d represent the height of the watermelon (see Figure 106). By the Pythagorean theorem,

$$s^2 = 30^2 + d^2$$

or

$$s = f(d) = \sqrt{900 + d^2}.$$

Since $d = g(t) = 90 - 16t^2$,

$$s = f(d) = f[g(t)] = f(90 - 16t^2) = \sqrt{900 + (90 - 16t^2)^2}.$$

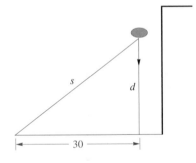

FIGURE 106

EXERCISE SET 2.6

A

In Problems 1 through 6, determine the functions $f + g$, $f - g$, fg, and f/g and state their domains.

1. $f(x) = 2x^4 - x^2$
  $g(x) = (x^2 + 3)^2$

2. $f(x) = 3\sqrt{x} + 2x$
  $g(x) = \sqrt{x}$

3. $f(x) = \dfrac{3}{x}$
  $g(x) = \dfrac{x + 2}{x^2}$

4. $f(x) = \dfrac{x + 2}{x + 1}$
  $g(x) = \dfrac{5}{x}$

5. $f(x) = \dfrac{x}{x + 4}$
  $g(x) = \dfrac{x + 2}{x}$

6. $f(x) = \dfrac{3}{x - 2}$
  $g(x) = \dfrac{2x + 1}{x + 2}$

In Problems 7 through 18, find $f \circ g$ and $g \circ f$.

7. $f(x) = 5x$
  $g(x) = \dfrac{1}{x}$

8. $f(x) = 2x^3 + x$
  $g(x) = \dfrac{1}{3x}$

9. $f(x) = x^2 - \dfrac{2}{x}$

$g(x) = \sqrt{x - 1}$

10. $f(x) = \sqrt{x - 1}$

$g(x) = x^2 + 3$

11. $f(x) = x^2 - 2$

$g(x) = x + \dfrac{1}{x}$

12. $f(x) = 2x^2 + 6$

$g(x) = x - \dfrac{1}{x}$

13. $f(x) = \dfrac{1}{x^3 + 5}$

$g(x) = \sqrt[3]{\dfrac{1 - 5x}{x}}$

14. $f(x) = \sqrt{\dfrac{x + 1}{2}}$

$g(x) = 2x^2 - 1$

15. $f(x) = \frac{1}{2}x - 3$

$g(x) = 2x + 3$

16. $f(x) = \dfrac{1}{x + 1}$

$g(x) = \dfrac{1}{x - 1}$

17. $f(x) = 4$

$g(x) = \dfrac{1}{x^2}$

18. $f(x) = x^2 - 8$

$g(x) = 5$

In Problems 19 through 24, find $f \circ f$.

19. $f(x) = 2x - 1$

20. $f(x) = \frac{2}{3}x - 1$

21. $f(x) = -\sqrt{9 - x^2}$

22. $f(x) = -\sqrt[3]{x^3 + 2}$

23. $f(x) = \dfrac{x - 1}{x + 1}$

24. $f(x) = x + \dfrac{1}{x}$

B

In Problems 25 through 30, use the graph of f, g, and h in Figure 107 to sketch the graph of the given equations.

25. $y = f(x) + h(x)$

26. $y = f(x) + g(x)$

27. $y = f(x) - h(x)$

28. $y = h(x) - g(x)$

29. $y = h(x) + 2f(x)$

30. $y = h(x) + \frac{1}{2}g(x)$

In Problems 31 through 36, sketch the graph of $y = f(x)$ and $y = g(x)$ on the same coordinate plane and then sketch the graph of $y = f(x) + g(x)$ by adding y-coordinates over the intersection of the domains of $f(x)$ and $g(x)$.

31. $f(x) = |x|$

$g(x) = -x$

32. $f(x) = |x - 3|$

$g(x) = x - 1$

33. $f(x) = \frac{1}{2}x^3$

$g(x) = -2x$

34. $f(x) = x^2 - 6$

$g(x) = x$

35. $f(x) = x + 4$

$g(x) = \sqrt{9 - x^2}$

36. $f(x) = -\frac{1}{2}x$

$g(x) = \sqrt{16 - x^2}$

In Problems 37 through 42, sketch the graph of the functions for $x \geq 0$ [$u(x)$ is the unit step function].

37. $y = -2u(x)$

38. $y = 3u(x - 4)$

39. $y = u(x) - 2u(x - 1)$

40. $y = -u(x) + 3u(x - 4)$

41. $y = 2u(x) - u(x - 3) + 2u(x - 5)$

42. $y = u(x) - 4u(x - 2) + 3u(x - 5)$

In Problems 43 through 54, write $h(x)$ as the composition of two functions, $f[g(x)]$. Do not use $g(x) = x$ or $f(x) = x$. (Answers are not unique.)

43. $h(x) = (3x - 1)^2 + (3x - 1) + 6$

44. $h(x) = |1 - 2x|$

45. $h(x) = \dfrac{1}{(x - 1)^2}$

FIGURE 107

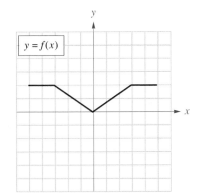
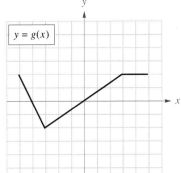
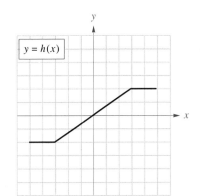

46. $h(x) = \left(\dfrac{2}{x-5}\right)^2 + 3\left(\dfrac{2}{x-5}\right) + 1$

47. $h(x) = \sqrt{8 - x^3}$

48. $h(x) = \dfrac{|x| - 1}{|x|}$

49. $h(x) = \sqrt{25 - (2x + 1)^2}$

50. $h(x) = \dfrac{1}{1 + (x - 4)^2}$

51. $h(x) = \dfrac{2}{5 - (1 - x)^3}$

52. $h(x) = \sqrt{5 - x} + \dfrac{1}{x - 5}$

53. $h(x) = |x - 2| + \dfrac{3}{(2 - x)^3}$

54. $h(x) = \dfrac{7x^2 + 3}{7x^2 + 1}$

55. The radius of a balloon is initially 6 in. and grows at a rate of 3 in./sec.

    a) Determine a function $r(t)$ that gives the radius at t seconds.

    b) Use the $r(t)$ from part a) to find a function $V(t)$ that gives the volume at t seconds.

56. A car travels north from an intersection at a rate of 50 mph. One hundred miles on a road due east from the intersection, there is a town (Figure 108).

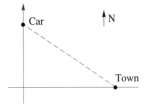

FIGURE 108

    a) Determine a function $f(t)$ that gives the distance from the car to the intersection t hours after the car passes the intersection.

    b) Determine a function $D(t)$ that gives the distance from the car to the town t hours after the car passes the intersection.

C

In Problems 57 through 59, write an equation for the graph given as a linear combination of unit step functions of the form $u(x - a)$.

57.

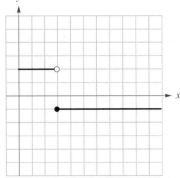

58.

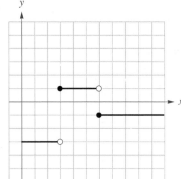

59.

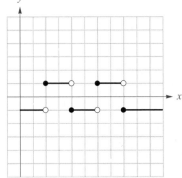

60. Use the graphs of the functions f and g used for Problems 25–30 (Figure 107) to sketch the graphs of $y = f[g(x)]$ and $y = g[f(x)]$.

INVERSE FUNCTIONS

S E C T I O N   2.7

A true mathematician cannot discuss what a process does without giving some thought to what undoes it. For example, subtraction undoes addition; the process of squaring a natural number is undone by taking a square root. In this section, we will discuss undoing as it applies to functions.

Consider two functions f and g such that a is a number in the domain of f and b is a number in the range of f with $f(a) = b$. Now suppose that b is also in the domain of g and that $g(b) = a$. What this actually says is that the function g undoes what the function f does. We call the function g the **inverse** of the function f. Figure 109 shows this relationship.

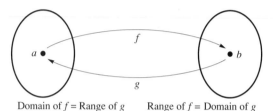

FIGURE 109

Domain of f = Range of g Range of f = Domain of g

You can see from the picture that if the function g is the inverse of the function f, then f is the inverse of g. Furthermore, the domain of f is the range of g, and the range of f is the domain of g.

Two functions f and g are **inverses** of each other if

$$g[f(a)] = a \qquad \text{for all } a \text{ in the domain of } f, \text{ and}$$

$$f[g(b)] = b \qquad \text{for all } b \text{ in the domain of } g.$$

Inverse Functions

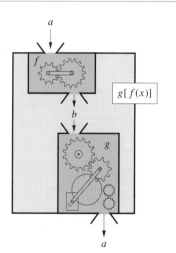

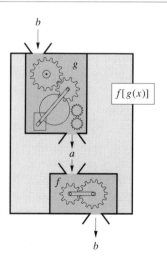

FIGURE 110

In other words, the composition of the function and its inverse is a function whose output is the same as the input (this function is called the **identity function**). See Figure 110.

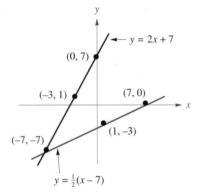

FIGURE 111

EXAMPLE 1   Show that the functions

$$f(x) = 2x + 7 \quad \text{and} \quad g(x) = \tfrac{1}{2}(x - 7)$$

are inverse functions.

SOLUTION To show that these two functions are inverses of each other, we need to verify the two conditions in the preceding definition. So,

$$g[f(a)] = g[2a + 7] = \tfrac{1}{2}((2a + 7) - 7) = a$$

and

$$f[g(b)] = f[\tfrac{1}{2}(b - 7)] = 2(\tfrac{1}{2}(b - 7)) + 7 = b.$$

This shows that f and g are inverse functions. The graphs of $f(x) = 2x + 7$ and $g(x) = \tfrac{1}{2}(x - 7)$ shown together in Figure 111. A few selected points are plotted.

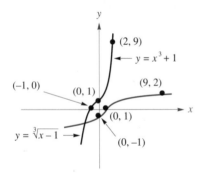

FIGURE 112

EXAMPLE 2   Show that the functions $F(x) = x^3 + 1$ and $H(x) = \sqrt[3]{x - 1}$ are inverse functions.

SOLUTION

$$H[F(a)] = H[a^3 + 1] = \sqrt[3]{(a^3 + 1) - 1} = \sqrt[3]{a^3} = a$$
$$F[H(b)] = H[\sqrt[3]{b - 1}] = (\sqrt[3]{b - 1})^3 + 1 = b - 1 + 1 = b.$$

This shows that H and F are inverse functions. The graphs $y = x^3 + 1$ and $y = \sqrt[3]{x - 1}$ are shown together in Figure 112.

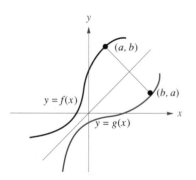

FIGURE 113

The two examples point out a connection between the graph of a function and the graph of its inverse. Suppose that the function f and the function g are inverses of each other; that is, $f(a) = b$ if and only if $g(b) = a$. Now consider the graph of $y = f(x)$ and the graph of $y = g(x)$ (Figure 113). Since $f(a) = b$, the point (a, b) is on the graph of $y = f(x)$. Since $g(b) = a$, the point (b, a) is on the graph of $y = g(x)$. These two points are **symmetric** with respect to the line of $y = x$ (recall our discussion of symmetry in Section 2.1).

Suppose that the function f and the function g are inverses of each other. Then the graph of $y = g(x)$ is the reflection of the graph of $y = f(x)$ through the line $y = x$.

The Graph of an Inverse Function

EXAMPLE 3   Determine whether the functions $f(x) = x^2$ and $g(x) = \sqrt{x}$ are inverses of each other.

SOLUTION At first glance, you might think that these functions are inverses of each other. However, consider this counterexample. If $g(x)$ is to be the inverse function of $f(x)$, then $g[f(a)] = a$ for all a in the domain of f. Letting $a = -2$ for example,

$$g[f(-2)] = g(4) = \sqrt{4} = 2.$$

But, by our definition of an inverse function, $g[f(-2)]$ should be -2, not 2. This shows that these two functions are not inverses of each other.

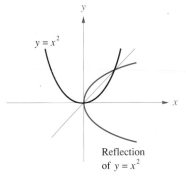

Reflection of $y = x^2$

FIGURE 114

As a matter of fact, the function $f(x) = x^2$ cannot have an inverse function. To see this, suppose that there is such a function g. The graph of the inverse function g must be the reflection of the function f, as shown in Figure 119. However, the reflection does not represent a function at all because it fails the vertical line test. Thus, no such inverse function g exists.

By now, two fundamental questions have probably occurred to you. First, for a given function, does an inverse function exist? And second, if an inverse function does exist, how do we find it?

To answer the first question, we need to define a special property of certain functions.

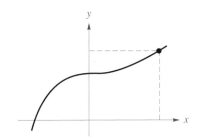

FIGURE 115

A function f is **one-to-one** if, for any a and b in the domain of f such that $a \neq b$, then $f(a) \neq f(b)$. That is, no two elements in the domain of f are assigned to the same element of the range.

One-to-One Functions

In other words, if a function f is one-to-one, then there are no two distinct numbers a and b such that $f(a) = f(b)$.

Consider the graph of a function f (Figure 115). To say the function f is one-to-one is to say that for each value of y in the range of f, there is only one point on its graph with that y-coordinate. What this means is that no two points on the graph of f lie on the same horizontal line (Figure 116).

A function is one-to-one if each value of y has only one value of x assigned to it.

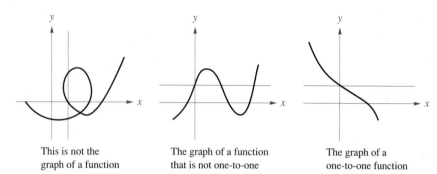

FIGURE 116

This is not the
graph of a function

The graph of a function
that is not one-to-one

The graph of a
one-to-one function

Horizontal Line Test

> The graph of a function is the graph of a one-to-one function if
> and only if no two points lie on the same horizontal line. That is,
> no horizontal line intercepts the graph at more than one point.

EXAMPLE 4 Show that $f(x) = -x^3 + 3$ is one-to-one.

SOLUTION Suppose that there is a pair of numbers in the domain of f, say a
and b, such that $f(a) = f(b)$. If we can show that $a = b$, then we have shown
that the function is one-to-one. So,

$$f(a) = f(b)$$
$$-a^3 + 3 = -b^3 + 3$$
$$a^3 = b^3$$
$$a = b.$$

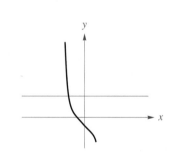

FIGURE 117

What this shows (in an indirect way) is that if $a \neq b$, then $f(a) \neq f(b)$. Therefore, f is one-to-one.

   If we apply the horizontal line test to the graph of $y = -x^3 + 3$, we
see (Figure 117) that the function f is one-to-one.

EXAMPLE 5 Show that $h(x) = (x + 3)^2$ is not one-to-one.

SOLUTION Our plan of attack here is to find two distinct numbers a and b
in the domain of h such that $h(a) = h(b)$. So,

$$(a + 3)^2 = (b + 3)^2$$
$$(a + 3) = \pm(b + 3)$$
$$a + 3 = b + 3 \quad \text{or} \quad a + 3 = -b - 3$$
$$a = b \quad \text{or} \quad a = -b - 6.$$

The second result $a = -b - 6$ implies that, for example, $b = 1$ and $a = -7$ are both assigned to the same value by h, namely 16. This shows that the function cannot be one-to-one.

If we apply the horizontal line test to the graph of $y = (x + 3)^2$, we see that the function h is not one-to-one (Figure 118).

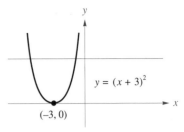

FIGURE 118

We can now answer the question of whether a given function has an inverse function. If the graph of a function passes the horizontal line test, then the reflection through the line $y = x$ of this graph passes the vertical line test. This justifies the following statement.

| A function f has an inverse function if and only if f is one-to-one. |

Which Functions Have Inverse Functions?

Now that we know whether a given function has an inverse, we need to find the inverse if it does exist. Here we introduce the conventional notation f^{-1} for the inverse function of the function f. This should not be confused with the *reciprocal* of the function f. This is one instance where the conventional notation is not as clear as it could be; be sure to exercise caution when using it.

EXAMPLE 6   Find the inverse function of $f(x) = \frac{1}{2}x + 6$.

SOLUTION By our definition, $f[f^{-1}(x)] = x$ for any x in the domain of f^{-1}. For simplicity, we let $y = f^{-1}(x)$. So,

$$f(y) = x$$

or

$$\tfrac{1}{2}y + 6 = x.$$

If we solve this for y, we find $f^{-1}(x)$:

$$\tfrac{1}{2}y = x - 6$$
$$y = 2(x - 6).$$

Thus, $f^{-1}(x) = 2(x - 6)$. You should verify that $f^{-1}[f(a)] = a$ and $f[f^{-1}(b)] = b$.

We can generalize the procedure in Example 6 into the following steps.

Finding the Inverse of a Function

Suppose that f is a function with inverse function f^{-1}. To find f^{-1}:

1) Set $f(y) = x$.

2) Solve the resulting equation for y.

3) Replace y with $f^{-1}(x)$. This is the inverse function of f.

4) Verify that $f^{-1}[f(a)] = a$ and $f[f^{-1}(b)] = b$.

EXAMPLE 7   Find the inverse function of $G(x) = \dfrac{2x}{x+3}$.

SOLUTION We start by setting $G(y) = x$ and solving the resulting equation for y:

$$\frac{2y}{y+3} = x$$

$$2y = x(y+3)$$

$$2y = xy + 3x$$

$$2y - xy = 3x$$

$$(2-x)y = 3x$$

$$y = \frac{3x}{2-x}.$$

Next, we set $G^{-1}(x) = \dfrac{3x}{2-x}$ and verify that it is the inverse of $G(x)$:

$$G^{-1}[G(a)] = \frac{3\left[\dfrac{2a}{a+3}\right]}{2 - \left[\dfrac{2a}{a+3}\right]} = \frac{6a}{2(a+3) - 2a} = a$$

and

$$G[G^{-1}(b)] = \frac{2\left[\dfrac{3b}{2-b}\right]}{\left[\dfrac{3b}{2-b}\right] + 3} = \frac{6b}{3b + 3(2-b)} = b.$$

Thus, $G^{-1}(x) = \dfrac{3x}{2-x}$.

Unfortunately, not all the functions for which we would like to find inverse functions are one-to-one. One way to allow an inverse function is to **restrict the domain** of the function so that it is one-to-one. For example, the squaring function, $f(x) = x^2$, is not one-to-one, but it is important to have an inverse function for this special function. If we restrict the domain to $x \geq 0$, the function is one-to-one and does have an inverse, namely $f^{-1}(x) = \sqrt{x}$ (Figure 119).

Restricting the domain so that $x \leq 0$ also would result in a function that is one-to-one. The inverse in this case would be $f^{-1}(x) = -\sqrt{x}$.

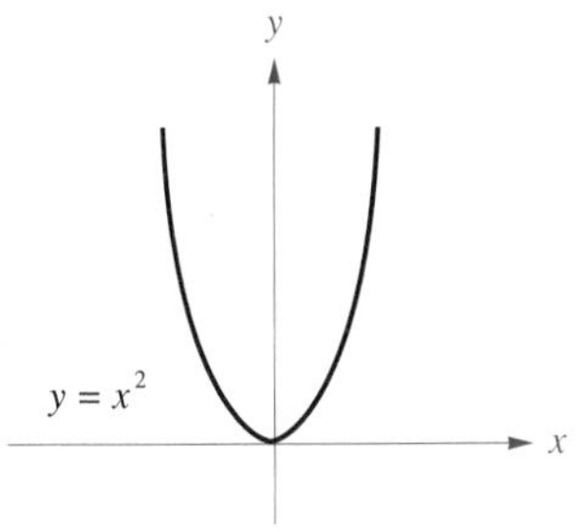

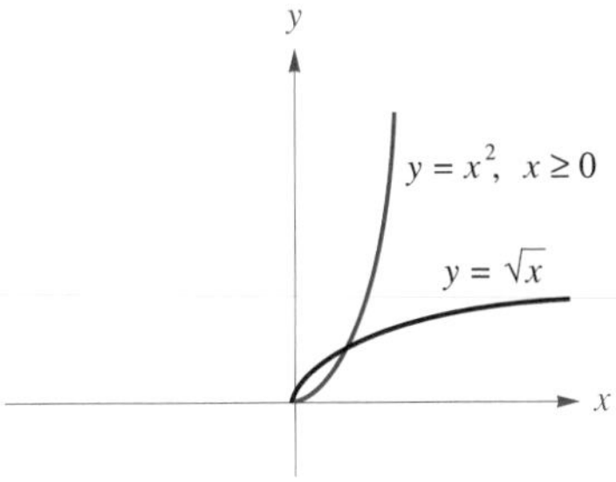

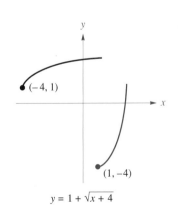

FIGURE 119

EXAMPLE 8   Given $h(x) = x^2 - 2x - 3$, $x \geq 1$. Show that h is one-to-one by means of its graph and find h^{-1}.

SOLUTION If we rewrite the function h as $h(x) = (x - 1)^2 - 4$, we can see that the graph is a parabola opening upward with a vertex at $(1, -4)$. If we consider only $x \geq 1$, this graph passes the horizontal-line test (Figure 120). This means that h is one-to-one and h has an inverse. Now, to find h^{-1}:

$$(y - 1)^2 - 4 = x \qquad \text{(set } h(y) = x\text{)}$$

$$(y - 1)^2 = x + 4 \qquad \text{(now solve for } y\text{)}$$

$$y - 1 = \pm\sqrt{x + 4}$$

$$y = 1 \pm \sqrt{x + 4}.$$

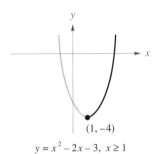

FIGURE 120

$$y = x^2 - 2x - 3, \ x \geq 1$$

Since our restriction on the domain of h required that $x \geq 1$, the corresponding restriction on the range of h^{-1} is such that $h^{-1}(x) \geq 1$. It follows that $h^{-1}(x) = 1 + \sqrt{x + 4}$ (Figure 121). We leave it to you to verify that $h^{-1}[h(a)] = a$ and that $h[h^{-1}(b)] = b$.

$$y = 1 + \sqrt{x + 4}$$

FIGURE 121

EXERCISE SET 2.7

A

In Problems 1 through 6, determine whether $f(x)$ and $g(x)$ are inverse functions by checking if $g[f(a)] = a$ and $f[g(b)] = b$. Be sure to consider all a in the domain of f and all b in the domain of g.

1. $f(x) = 4x + 2$, $g(x) = \frac{1}{4}(x - 2)$

2. $f(x) = \frac{1}{2}x - 7$, $g(x) = 2x + 7$

3. $f(x) = \frac{4}{5}x + 2$, $g(x) = \frac{5}{4}x - \frac{1}{2}$

4. $f(x) = \frac{4}{5}x + 2$, $g(x) = \frac{5}{4}x - \frac{5}{2}$

5. $f(x) = \frac{1}{2}\sqrt{25 - 4x^2}$, $g(x) = \frac{1}{2}\sqrt{25 - 4x^2}$

6. $f(x) = \dfrac{x}{x + 3}$, $g(x) = \dfrac{x - 3}{x}$

In Problems 7 through 12, sketch the graph of the function given and determine if it is one-to-one.

7. $f(x) = x^2 - 4x$

8. $f(x) = x^3 + 3$

9. $h(x) = \dfrac{-2}{x - 2}$

10. $f(x) = \dfrac{1}{x^2}$

11. $f(x) = 4$

12. $f(x) = [\![x]\!]$

In Problems 13 through 18, copy the graph and use the horizontal line test to determine if it is the graph of a one-to-one function. If the function is one-to-one, sketch the graph of the inverse function. If the function is not one-to-one, show where it fails the horizontal line test.

13.

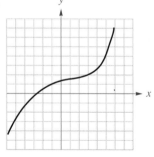

14.

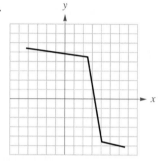

15.

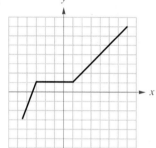

16.

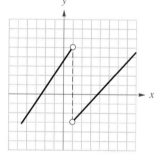

17.

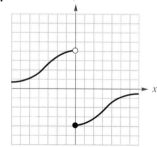

18.

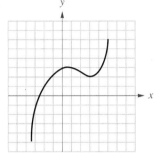

35. $H(x) = x^2 - 2x - 8, \; x \geq 1$

36. $G(x) = x^2 - 2x - 8, \; x \leq 1$

In Problems 37 through 42, a function is given along with its graph. Determine the inverse of the function, and sketch the graph of the inverse function.

37.

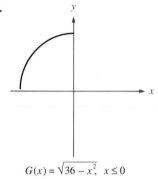

$$G(x) = \sqrt{36 - x^2}, \; x \leq 0$$

In Problems 19 through 30, find the inverse of the given functions. You may assume that the given functions are one-to-one.

19. $h(x) = 5(x - 3)$

20. $F(x) = \frac{2}{3}x + 7$

21. $G(x) = -2(x - 3)$

22. $N(x) = \frac{2}{3}(x + 5)$

23. $d(x) = x^3 + 4$

24. $H(x) = x^5 - 7$

25. $p(x) = \sqrt[3]{x + 4}$

26. $q(x) = \sqrt[3]{x} - 6$

27. $R(x) = \dfrac{2}{x - 3}$

28. $S(x) = \dfrac{4}{2 - x}$

29. $S(x) = \dfrac{x}{x + 4}$

30. $f(x) = \dfrac{2x}{x - 6}$

38.

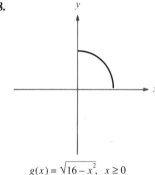

$$g(x) = \sqrt{16 - x^2}, \; x \geq 0$$

B

In Problems 31 through 36, a function is given with a restricted domain. Sketch the graph of this function, determine the inverse of the function, and sketch the graph of the inverse function.

31. $f(x) = x^2 + 4, \; x \geq 0$

32. $g(x) = x^2 + 4, \; x \leq 0$

33. $g(x) = |x - 1|, \; x \leq 1$

34. $H(x) = |x + 3|, \; x \leq -3$

39.

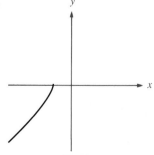

$$D(x) = -\sqrt{x^2 - 9}, \; x \leq -3$$

40.

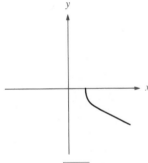

$f(x) = -\frac{1}{2}\sqrt{x^2 - 4}, \quad x \geq 2$

41.

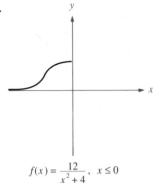

$f(x) = \dfrac{12}{x^2 + 4}, \quad x \leq 0$

42.

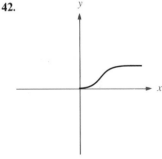

$F(x) = 4 - \dfrac{4}{x^2 + 1}, \quad x \geq 0$

43. Verify that the functions f and g are inverse functions:

$$f(x) = \sqrt[3]{(x-1)(x^2 + x + 1)}$$
$$g(x) = \sqrt[3]{(x+1)(x^2 - x + 1)}$$

44. Verify that the functions $f(x) = x^5 + 3x^{10/3} + 3x^{5/3} + 1$ and $g(x) = (x^{1/3} - 1)^{3/5}$ are inverse functions.

45. The function $F(C) = \frac{9}{5}C + 32$ is used to determine the temperature in degrees Fahrenheit $F(C)$ given the temperature in degrees Celcius C. For example, if the temperature in degrees Celcius is 25°C, then the temperature in degrees Fahrenheit is 77°F since $F(25) = 77$. Find the inverse of the function F and describe what it can be used for.

46. The function $N(f) = 4f - 160$ can be used to determine the number of times that a cricket will chirp per minute when the temperature is f degrees Fahrenheit. For example, if the temperature is 70°F, then $N(70) = 120$. Find the inverse of the function N and describe what is can be used for.

47. Sketch the graph of

$$f(x) = \begin{cases} -x^2 & x < 0 \\ x^2 & x \geq 0 \end{cases}$$

and sketch the graph of $y = f^{-1}(x)$. From the graph, determine $f^{-1}(x)$.

48. Sketch the graph of

$$f(x) = \begin{cases} -\dfrac{1}{2}x + 6 & x \leq 0 \\ -2x + 3 & 0 < x < 3 \\ -x & x \geq 3 \end{cases}$$

and sketch the graph of $y = f^{-1}(x)$. From the graph, determine $f^{-1}(x)$.

C

49. Suppose that $f_1(x)$ and $f_2(x)$ are one-to-one functions and define $F(x) = f_1(x) + f_2(x)$. Is $F(x)$ a one-to-one function? Support your answer.

50. Suppose that $f_1(x)$ and $f_2(x)$ are one-to-one functions and define $F(x) = f_1[f_2(x)]$. Show that $F(x)$ is a one-to-one function and find $F^{-1}(x)$.

51. Given

$$f(x) = \frac{x+2}{x-5},$$

determine $f^{-1}(x)$, $f(x^{-1})$, and $[f(x)]^{-1}$.

52. Show that

$$f(x) = \frac{x}{x-1}$$

is its own inverse. Sketch the graph of f. What can you generalize about the graph of a function that is its own inverse?

*C*ogito, ergo sum ("I think, therefore I am"). This famous statement belongs to René Descartes (1596–1650) in his *Discourse on the Method of Reasoning Well and Seeking Truth in the Sciences.* In the first appendix to that publication, Descartes made the simple and profound connection between an algebraic equation and a geometric curve in a coordinate plane. (The name of this plane, the *Cartesian* plane, is derived from *Cartesius,* the latinized form of *Descartes.*) This was the birth of analytic geometry, a unification of algebra and geometry that exploited the strengths of both disciplines to overcome their separate weaknesses. It was also at this time that Descartes introduced the word *function,* although his idea of a function was quite rudimentary compared with the modern concept. Pierre de Fermat, also a French mathematician, considered a union of algebra and geometry, but he did not publish his ideas until 1679. Even though both men independently developed the tools of coordinate geometry to solve important mathematical problems, Descartes is given the primary credit for its discovery.

Descartes also made significant contributions in optics, astronomy, and biology, but he is best known as the first great modern philosopher. His was an exciting period of history, the advent of the Age of Reason, during the time of Galileo, Kepler, Newton, Shakespeare, and Spinoza. In his *Discourse,* he developed the principles of scientific knowledge and reason that are used today.

As a youngster, René was often ill. He was allowed to spend his mornings in bed. This became his custom in later life, and it was during these long hours in bed that he did his most important thinking. The son of a French lawyer, Descartes himself obtained a law degree at the age of 20. He served as a mercenary in several armies for the next nine years. After this period, Descartes was a wanderer, seeking solitude to think, traveling often between the scientific activity of France and the intellectual freedom of Holland. At the age of 54, he accepted an appointment to the royal court of Queen Christine of Sweden. Unfortunately, the climate proved too harsh for his frail health; he died shortly after his arrival.

Descartes was a rather eccentric individual. He was often seen walking the streets of Paris dressed in foppish attire, sampling its pleasures and comforts. There is one account of his encounter with a drunkard in a sword fight for the honor of a lady of the evening. Descartes was the superior swordsman and made short work of his adversary. He spared the drunkard's life, however, since "a beautiful lady should not witness the slaying of one so vile."

The importance of Descartes' contributions is without measure. Without the creation of analytic geometry, calculus would not have developed to its full extent. His work in science and philosophy were instrumental in the development of Western civilization. Most importantly, though, René Descartes showed others how to think and solve problems.

There are some men who are counted great because they represent the actuality of their own age, and mirror it as it is. . . . But there are other men who attain greatness because they embody the potentiality of their day and magically reflect the future. They express the thoughts which will be everybody's two or three centuries after them. Such a one was Descartes.

—Thomas Huxley

René Descartes

MISCELLANEOUS EXERCISES

In Problems 1 through 3, determine the distance between the given points and their midpoint.

1. $(6, -12)$ $(-1, 12)$ **2.** $(2, \frac{1}{6})$ $(0, \frac{5}{3})$

3. $\left(\sqrt{2}, \frac{1}{\sqrt{2}}\right)$ $\left(\sqrt{8}, \frac{3}{\sqrt{2}}\right)$

In Problems 4 through 9, express the equation of the circle described in the form $(x - h)^2 + (y - k)^2 = r^2$

4. The circle with radius 5 and center at the origin.

5. The circle with radius 2 and center at $(2, 6)$.

6. The circle with equation $x^2 + y^2 + 4x - 2y - 31 = 0$.

7. The circle with equation $x^2 + y^2 = 8x - 15y$.

8. The circle with radius 4 and tangent to both the negative x-axis and the negative y-axis.

9. The circle with center on the graph of $y = \frac{1}{2}x + 2$, radius 5, and tangent to the positive x-axis.

In Problems 10 through 15, express the equation of the line described in the form $y = mx + b$.

10. The line passing through $(22, 4)$ and $(-1, 16)$.

11. The line passing through $(4, 7)$ and with x-intercept of 8.

12. The line parallel to the graph of $3x + 4y = 16$ and passing through $(0, -2)$.

13. The line perpendicular to the graph of $3x + 4y = 16$ and passing through $(0, -2)$.

14. The line tangent to the circle with the equation $x^2 + y^2 = 52$ at the point $(-4, 6)$. (Hint: A line tangent to a circle is perpendicular to a radius of the circle at the point of tangency.)

15. The line tangent the circle with the equation $(x - 4)^2 + y^2 = 25$ at the point $(7, 4)$. (See Problem 14.)

In Problems 16 through 21, use the symmetry tests to tell whether the graph of the equation is symmetric with respect to the x-axis, y-axis, the line $y = x$, or the origin.

16. $2x^2 + 5y^2 = 12$ **17.** $x^2 + xy + 2y^2 = 7$

18. $x^2 y^2 = 16$ **19.** $y = 4x$

20. $y = 18 - x$ **21.** $y^2 = x^3 - \dfrac{1}{x}$

In Problems 22 through 27, determine the domain of the function.

22. $f(x) = 2x^4 - 14$ **23.** $g(x) = 12 - x^5$

24. $h(x) = 3 + \sqrt{x - 3}$ **25.** $f(x) = \dfrac{1}{2x - 6}$

26. $B(x) = \dfrac{3x}{\sqrt{x^2 - 9}}$ **27.** $r(x) = \sqrt{x - \dfrac{1}{x}}$

28. Determine the range of the function given in Problem 22.

29. Determine the range of the function given in Problem 24.

30. Determine the range of the function given in Problem 25.

In Problems 31 through 40, determine $\dfrac{g(x + h) - g(x)}{h}$.

31. $g(x) = 12x - 5$ **32.** $g(x) = 4 - 3x$

33. $g(x) = 12 - x^2$ **34.** $g(x) = x^2 + 3$

35. $g(x) = x^3$ **36.** $g(x) = 1 + 3x - 4x^2$

37. $g(x) = -\dfrac{1}{x}$ **38.** $g(x) = \dfrac{1}{x - 1}$

39. $g(x) = 4 + \dfrac{1}{x}$

In Problems 40 through 42, determine $\dfrac{f(x + h) - f(x)}{h}$ such that there are no radicals in the numerator.

40. $f(x) = \sqrt{x + 2}$ **41.** $f(x) = \sqrt{2x}$

42. $f(x) = 1 + \sqrt{x}$

In Problems 43 through 66, sketch the graph of the equation. Label the intercepts.

43. $y = 2x + 1$ **44.** $y = -\frac{2}{3}x + 6$

45. $y = x^2 - 16$ **46.** $y = 4 - x^2$

47. $y = 8 - x^3$ **48.** $y = |x - 6|$

49. $y = |x| - 9$ **50.** $y - 5 = (x - 2)^2$

51. $y + 2 = (x + 1)^2$

52. $y = \frac{1}{4}x^2$

53. $y = -\frac{2}{5}x^2$

54. $y = x^2 - 6x$

55. $y = x^2 + 4x$

56. $y + 5 = |x - 2|$

57. $y - 3 = |x + 4| - 5$

58. $y = \frac{2}{3}|x|$

59. $y = -\frac{5}{2}|x|$

60. $y + 2 = \sqrt{x - 4}$

61. $y - 3 = \sqrt{x + 5}$

62. $y - 8 = -\sqrt{x}$

63. $y + 3 = \frac{3}{2}\sqrt{x}$

64. $y = 2 + \sqrt{16 - x^2}$

65. $y + 3 = \sqrt{9 - (x - 3)^2}$

66. $y = \frac{1}{2}\sqrt{36 - 9x^2}$

In Problems 67 through 69, sketch the graph of the equation.

67. $y = \begin{cases} 4 - x, & x < 0 \\ \frac{1}{2}x, & x \geq 0 \end{cases}$

68. $y = \begin{cases} 3, & x \leq 0 \\ |x|, & 0 < x < 3 \\ |x - 6|, & x \geq 3 \end{cases}$

69. $y = \begin{cases} 4 - x^2, & x \leq 0 \\ 4, & 0 < x < 5 \\ x - 5, & x \geq 5 \end{cases}$

In Problems 70 through 75, sketch the graph of the equation by adding y-coordinates.

70. $y = x^2 + \dfrac{1}{x}$

71. $y = |x| + x$

72. $y = |2x| - |x|$

73. $y = |x - 6| - |x + 2|$

74. $y = \sqrt{x + 2} - \sqrt{2 - x}$

75. $y = \sqrt{4 - x^2} + \sqrt{4 - (x - 2)^2}$

In Problems 76 through 81, determine whether the function is an even function, an odd function, or neither.

76. $f(x) = x^4 - 8$

77. $f(x) = (x - 8)^5$

78. $f(x) = 5|x| - 3x^2$

79. $f(x) = \dfrac{4x}{x^2 - 2}$

80. $f(x) = x^{2/5} - 4x^{1/3}$

81. $f(x) = \dfrac{x^{3/5}}{(x^{4/3} + 3)(x^2 - 1)}$

In Problems 82 through 93, express the given function $h(x)$ as a composition of these functions:

$$p(x) = x^2 \qquad q(x) = \frac{4}{x} \qquad r(x) = \sqrt{x} \qquad u(x) = x + 2.$$

82. $h(x) = (x + 2)^2$

83. $h(x) = x^2 + 2$

84. $h(x) = \sqrt{x + 2}$

85. $h(x) = \sqrt{x} + 2$

86. $h(x) = \dfrac{4}{x^2}$

87. $h(x) = \dfrac{2}{\sqrt{x}}$

88. $h(x) = \dfrac{16}{x^2}$

89. $h(x) = \dfrac{4}{\sqrt{x}}$

90. $h(x) = x^4$

91. $h(x) = x$

92. $h(x) = \sqrt[4]{x}$

93. $h(x) = x + 4$

In Problems 94 through 96, express the functions $F(x)$ and $G(x)$ as the composition of three of the functions given for Problems 73–84.

94. a) $F(x) = \dfrac{4}{\sqrt{x + 2}}$

b) $G(x) = \dfrac{4}{\sqrt{x} + 2}$

95. a) $F(x) = \dfrac{4}{|x|}$

b) $G(x) = \dfrac{2}{|x|}$

96. $F(x) = \sqrt{x^2 + 2}$

b) $G(x) = |x| + 2$

Problems 97 through 111 refer to the graphs shown in Figure 122 on p. 156.

97. Determine the domain and range of the function f from its graph.

98. Determine the domain and range of the function g from its graph.

99. Are either of these functions odd? even?

100. Determine the value of

a) $f(-2)$ **b)** $f(0)$ **c)** $f(2)$

101. Determine the value of

a) $g(-2)$ **b)** $g(2)$ **c)** $g(4)$

102. Sketch the graph of

a) $y = f(x - 1)$ **b)** $y = f(x) + 3$

c) $y = f(x - 1) + 3$

103. Sketch the graph of

a) $y = g(x + 1)$ **b)** $y = g(x) - 4$

c) $y = g(x + 1) - 4$

104. Sketch the graph of

a) $y = 2f(x)$ **b)** $y = f(2x)$

c) $y = \frac{1}{2}f(x)$

105. Sketch the graph of

a) $y = -g(x)$ **b)** $y = g(-x)$

c) $y = -g(-x)$

FIGURE 122

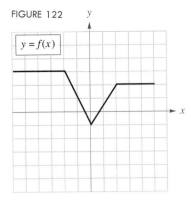

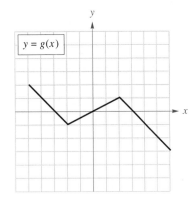

106. Sketch the graph of

    a) $y = g(|x|)$ **b)** $y = |g(x)|$

    c) $y = -|g(x)|$

107. Sketch the graph of

    a) $y = f(|x|)$ **b)** $y = |f(x)|$

    c) $y = -|f(x)|$

108. Sketch the graph of

    a) $y = f(x) + g(x)$ **b)** $y = f(x) - g(x)$

    c) $y = g(x) - f(x)$

109. Sketch the graph of

    a) $y = \frac{1}{2}(f(x) + |f(x)|)$ **b)** $y = \frac{1}{2}(f(x) - |f(x)|)$

110. Sketch the graph of

    a) $y = \frac{1}{2}(g(x) + |g(x)|)$ **b)** $y = \frac{1}{2}(g(x) - |g(x)|)$

111. Consider any function F. Based on your answers to Problems 109 and 110, make a conjecture about the graph of $y = \frac{1}{2}(F(x) + |F(x)|)$ and the graph of $y = \frac{1}{2}(F(x) - |F(x)|)$.

In Problems 112 through 117, determine whether the function f is one-to-one. If it is, find its inverse. If it is not, find two distinct values x_1 and x_2 such that $f(x_1) = f(x_2)$.

112. $f(x) = 3x^2 - 2$ **113.** $f(x) = 7(x - 4)$

114. $f(x) = 5 + x^3$ **115.** $f(x) = \dfrac{4}{x - 1}$

116. $f(x) = \sqrt[3]{x - 8}$ **117.** $f(x) = |5 - x|$

In Problems 118 through 120, a function f is given with a restricted domain such that the function is one-to-one. Find f^{-1}, the inverse of the function, and sketch both $y = f(x)$ and $y = f^{-1}(x)$ on the same coordinate plane.

118. a) $f(x) = 4 - x^2, x \geq 0$

    b) $f(x) = 4 - x^2, x \leq 0$

119. a) $f(x) = \frac{1}{2}x^2 - 4x + 8, x \geq 4$

    b) $f(x) = \frac{1}{2}x^2 - 4x + 8, x \leq 4$

120. a) $f(x) = -\sqrt{16 - x^2}, 0 \leq x \leq 4$

    b) $f(x) = -\sqrt{16 - x^2}, -4 \leq x \leq 0$

In Problems 121 through 126, determine whether the function has either a maximum or minimum value and find that value.

121. $f(x) = x^2 - 2$ **122.** $g(x) = x^2 + 8x$

123. $P(x) = 2x^2 - 8x + 1$ **124.** $Q(x) = \frac{1}{2}x^2 + x - 3$

125. $F(x) = \dfrac{4}{x^2 + 6}$

126. $F(x) = \sqrt{x^2 - 4x + 13}$

127. The Kumquat Computer Company finds that the revenue (in dollars) generated by selling x computer printers in a certain market is given by $R(x) = 480x - 0.4x^2$. The company's cost of producing each printer is 320 dollars.

    a) Determine the number of units to be sold to maximize revenue.

    b) Determine the profit (in dollars) generated by selling x units. (Hint: profit is the difference between total revenue and total cost.)

    c) Determine the number of units to be sold to maximize profit.

128. A farmer wants to construct a rectangular pen using 220 feet of fencing and two sides of her barn (Figure 123).

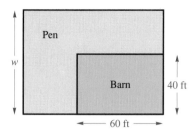

FIGURE 123

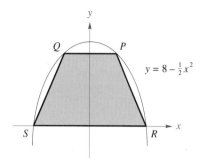

FIGURE 124

a) Determine the area of the pen as a function of its width w.

b) What is the maximum area possible for the pen?

129. Show that the midpoints of the sides of a quadrilateral are the vertices of a parallelogram. (Hint: Consider the quadrilateral with vertices $A(x_1, y_1)$, $B(x_2, y_2)$, $C(x_3, y_3)$, and $D(x_4, y_4)$; find the midpoints of the sides and use the test for a parallelogram given in Problem 52 in Section 2.1.)

130. A right isosceles triangle has a hypotenuse with endpoints $(2, 1)$ and $(9, 2)$. Find the third vertex of the triangle. (Hint: there are two possibilities.)

131. Three vertices of a parallelogram are $(0, 0)$, $(a, 0)$, and $(0, b)$, where a and b are positive real numbers. Find the fourth vertex. (Hint: there are three possibilities; one is a square.)

132. John is speeding directly north across the Nevada desert in his sports car at 150 ft/sec. Tillie, flying her helicopter at an altitude of 210 ft directly east at 120 ft/sec, passes directly over John. Find the distance $d(t)$ in feet between John and Tillie as a function of t, the number of seconds after Tillie passes over John. Find the domain and range of this function.

133. A trapezoid is inscribed in the graph of $y = 8 - \frac{1}{2}x^2$ as shown (Figure 124). Find the area of the trapezoid as a function of the x-coordinate of the point P. What is the domain of this function?

134. A cone has a height of 5 cm and a base with diameter 10 cm. A cylinder is placed in the cone as shown (Figure 125). Find the volume of the cylinder as a function of the radius of the cylinder. What is the domain of this function?

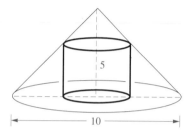

FIGURE 125

135. A 16-ft ladder leans against a wall so that the bottom of the ladder is 4 ft from the wall (Figure 126). Scott climbs a ladder at a rate of 1 ft/sec. When Scott gets 6 ft up the ladder, Geoffrey begins to pull the bottom of the ladder from the wall at a rate of 1.5 ft/sec. Find Scott's height above the floor as a function of t, the number of seconds after Geoffrey begins to pull the ladder.

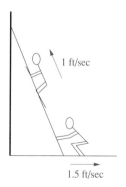

FIGURE 126

CHAPTER 3

ALGEBRAIC FUNCTIONS

POLYNOMIAL FUNCTIONS

The linear functions and the quadratic functions that were the subject of our investigations in Section 2.4 are specific examples of the more general set of **polynomial functions.** In this section, we will examine these functions and their graphs.

> A **polynomial function** is a function that can be expressed in the form
>
> $$P(x) = a_n x^n + a_{n-1} x^{n-1} + a_{n-2} x^{n-2} + \cdots + a_1 x + a_0$$
>
> where the **degree** n is a nonnegative integer, the **coefficients** a_n, $a_{n-1}, a_{n-2}, \ldots, a_1, a_0$ are real numbers, and $a_n \neq 0$.

Polynomial Functions

A function of the form $f(x) = k$ (k is a nonzero constant) is the trivial case of a polynomial function; its degree is zero. Since any linear function can be written $f(x) = mx + b$, it is a polynomial function of degree one. Likewise, any quadratic function $f(x) = ax^2 + bx + c$ is a polynomial function of degree two.

To investigate properties of the graphs of polynomial functions, we will start by looking at **monomial functions**—polynomial functions with only one term. The next three examples involve this type of polynomial function.

EXAMPLE 1   Sketch the graphs of $P_1(x) = x$, $P_3(x) = x^3$, and $P_5(x) = x^5$.

SOLUTION We can construct a table to determine points on each of the graphs.

x	-10	-2	-1	0	$\frac{1}{2}$	1	2	10
$P_1(x)$	-10	-2	-1	0	$\frac{1}{2}$	1	2	10
$P_3(x)$	$-1{,}000$	-8	-1	0	$\frac{1}{8}$	1	8	$1{,}000$
$P_5(x)$	$-100{,}000$	-32	-1	0	$\frac{1}{32}$	1	32	$100{,}000$

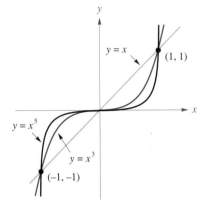

FIGURE 1

The graphs are shown in Figure 1. Notice that all three pass through the points $(-1, -1)$, $(0, 0)$, and $(1, 1)$. Each of these graphs is symmetric with respect to the origin.

EXAMPLE 2   Sketch the graphs of $P_2(x) = x^2$, $P_4(x) = x^4$, and $P_6(x) = x^6$.

SOLUTION   We proceed as in Example 1.

x	-10	-2	-1	0	$\frac{1}{2}$	1	2	10
$P_2(x)$	100	4	1	0	$\frac{1}{4}$	1	4	100
$P_4(x)$	$10{,}000$	16	1	0	$\frac{1}{16}$	1	16	$10{,}000$
$P_6(x)$	$1{,}000{,}000$	64	1	0	$\frac{1}{64}$	1	64	$1{,}000{,}000$

The graphs are shown in Figure 2. Notice that all three pass through the points $(-1, 1)$, $(0, 0)$, and $(1, 1)$. Each of these graphs is symmetric with respect to the y-axis.

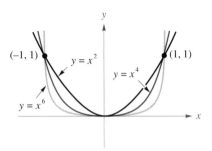

FIGURE 2

From these two examples, we can generalize about the shapes of graphs of monomial functions. First, though, we introduce some new notation. The symbols $x \rightarrow +\infty$ are used to indicate that we are allowing x to "approach positive infinity"; that is, we are allowing x to grow without bound in the positive direction. Likewise, the symbols $x \to -\infty$ indicate that x is allowed to grow without bound in the negative direction.

Consider the graph of $y = x^n$ for $x \geq 0$ (Figure 3). First, we can generalize that $y \to +\infty$ as $x \to +\infty$, and that for larger values of n, y increases more rapidly. Also note that for any n, the graph of $y = x^n$ passes through the points $(0, 0)$ and $(1, 1)$.

To complete the graph of $f(x) = x^n$, consider two cases. First, if n is odd (as in Example 1), then

$$f(-x) = (-x)^n = -x^n = -f(x).$$

This implies that f is an odd function (this may suggest why odd functions are called *odd*) and that the graph is symmetric with respect to the origin. In a similar fashion, if n is even (as in Example 2), then the function is even and its graph is symmetric to the y-axis (Figure 4).

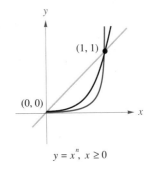

$y = x^n$, $x \geq 0$

FIGURE 3

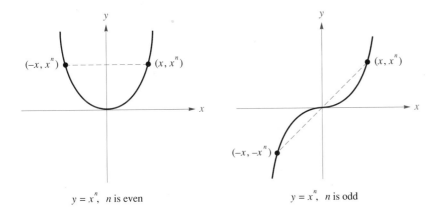

$y = x^n$, n is even

$y = x^n$, n is odd

FIGURE 4

EXAMPLE 3   Sketch the graphs of $P(x) = x^3$, $Q(x) = \frac{1}{2}x^3$, and $R(x) = -2x^3$.

SOLUTION The graph of $P(x)$ is a familiar one. Since $Q(x) = \frac{1}{2}P(x)$, the graph of $Q(x)$ is a vertical compression of the graph of $P(x)$. Likewise, the graph of $R(x)$ is a reflection and a vertical expansion of $P(x)$ since $R(x) = -2P(x)$.

From Example 3, we can generalize about the graph of the monomial function $y = ax^n$ (a is a constant). Suppose that a is positive. Then $y = ax^n$ is a vertical expansion or compression of $y = x^n$ (depending on whether $a > 1$ or $a < 1$); $y = ax^n$ behaves in a similar fashion to $y = x^n$ at the extreme values of x. If, on the other hand, a is negative, then $y = ax^n$ is a reflection about the x-axis as well as a vertical expansion or compression of $y = x^n$. The graph of $y = ax^n$ behaves in an opposite fashion to $y = x^n$ at the extreme values of x. All the possibilities are shown in the following figure.

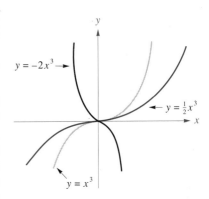

$y = -2x^3$

$y = \frac{1}{2}x^3$

$y = x^3$

FIGURE 5

Behavior of the Graph of
$y = ax^n$ *at the Extremes*

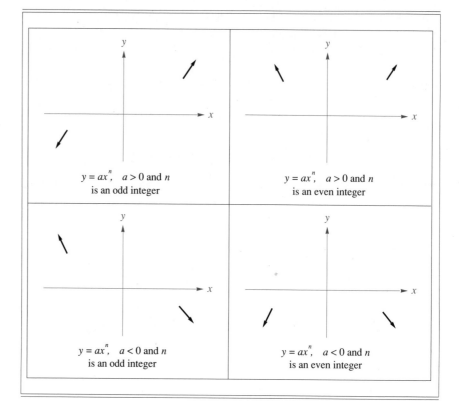

$y = ax^n$, $a > 0$ and n is an odd integer

$y = ax^n$, $a > 0$ and n is an even integer

$y = ax^n$, $a < 0$ and n is an odd integer

$y = ax^n$, $a < 0$ and n is an even integer

$y = x^5$

FIGURE 6

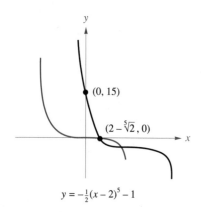

$(0, 15)$

$(2 - \sqrt[5]{2}, 0)$

$y = -\frac{1}{2}(x - 2)^5 - 1$

FIGURE 7

EXAMPLE 4   Sketch the graphs of these functions.

a) $f(x) = -\frac{1}{2}x^5$ **b)** $g(x) = -\frac{1}{2}(x - 2)^5 - 1$

SOLUTION

a) The graph of $f(x)$ is a vertical compression and reflection about the x-axis of the graph of $y = x^5$ (Figure 6). (The graph of $y = x^5$ is also shown for reference.)

b) The graph of $g(x)$ is a translation of the graph of $f(x)$ from part a) (Figure 7). (The graph of $f(x)$ is also shown for reference.) To find the x-intercept, we set $y = 0$ and solve for x:

$$0 = -\frac{1}{2}(x - 2)^5 - 1$$
$$\frac{1}{2}(x - 2)^5 = -1$$
$$(x - 2)^5 = -2$$
$$x - 2 = -\sqrt[5]{2} \Rightarrow x = 2 - \sqrt[5]{2}.$$

To find the y-intercept, we set $x = 0$ and solve for y:

$$y = -\tfrac{1}{2}(0 - 2)^5 - 1 = 15.$$

Sketching the graph of a constant function, linear function, quadratic function, or monomial function is relatively straightforward. Unfortunately, this is not true for polynomial functions in general. The best we can do in most cases is to get a good approximation of the graph.

There are, however, some things that we can generalize about the graph of a polynomial. These ideas will be justified and discussed in greater detail in calculus, but for now we assume them to be true.

1) The graph of a polynomial function is **continuous;** there are no breaks or jumps in the graph. In essence, this means that the graph of a polynomial can be sketched over any part of its domain without lifting the pencil (Figure 8).

2) The graph of a polynomial function of degree n can have *at most* $n - 1$ *turning points.* We have already seen that a linear function (degree one) has no turning points and that a quadratic function (degree two) has one turning point. A direct result of this is that a polynomial of degree n can have at most n intercepts.

3) The graph of a polynomial function is **smooth.** It has no corners or abrupt changes of direction (Figure 9).

4) The graph of the polynomial function

$$P(x) = a_n x^n + a_{n-1} x^{n-1} + a_{n-2} x^{n-2} + \cdots + a_1 x + a_0$$

behaves as its leading term $a_n x^n$ for the extreme values of x. What this means is that as $x \to \infty$ or $x \to -\infty$, the function $P(x)$ must do one of two things: Either $P(x) \to +\infty$ or $P(x) \to -\infty$, depending on the behavior of the leading term $a_n x^n$ at the extremes (Figure 10).

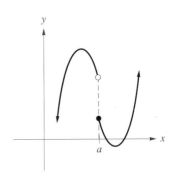

FIGURE 8 *There is a break in the graph at $x = a$. This is not the graph of a polynomial function.*

Note that the graph of a polynomial function need not have $n - 1$ turning points. For example, the graph of $P(x) = x^3$ has no turning points.

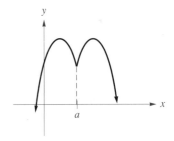

FIGURE 9 *There is a "corner" in the graph at $x = a$. This is not the graph of a polynomial function.*

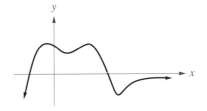

FIGURE 10 *This is not the graph of a polynomial function since it does not grow without bound at the extreme values of x.*

EXAMPLE 5 Sketch the graph of $f(x) = x^3 - 5x^2 - 4x + 20$ and label the intercepts.

SOLUTION Before we start plotting points, note that our graph can have at most two turning points since $f(x)$ is of degree three. Also, the leading term is x^3, so $f(x)$ behaves like $y = x^3$ at the extreme values of x:

$$y \to +\infty \quad \text{as} \quad x \to +\infty$$

and

$$y \to -\infty \quad \text{as} \quad x \to -\infty.$$

To determine the y-intercept, we evaluate $f(0)$:

$$f(0) = (0)^3 - 5(0)^2 - 4(0) + 20 = 20.$$

The y-intercept of $f(x)$ is $(0, 20)$.

In general, the x-intercepts are not so easy to find. In later sections, we will encounter other techniques to assist in finding the x-intercepts.

To determine the x-intercepts, we set $f(x) = 0$ and solve this equation for x. Notice the factoring by grouping:

$$x^3 - 5x^2 - 4x + 20 = 0$$
$$x^2(x - 5) - 4(x - 5) = 0$$
$$(x^2 - 4)(x - 5) = 0$$
$$(x - 2)(x + 2)(x - 5) = 0$$
$$x - 2 = 0 \quad \text{or} \quad x + 2 = 0 \quad \text{or} \quad x - 5 = 0$$
$$x = 2 \quad \text{or} \quad x = -2 \quad \text{or} \quad x = 5.$$

The x-intercepts of the graph of $f(x)$ are $(2, 0)$, $(-2, 0)$, and $(5, 0)$.

x	-3	-1	3	6
$f(x)$	-40	18	-10	32

Next, we compute additional points on the graph by evaluating $f(x)$ at selected values of x. Choosing integers simplifies the computation (however, it is important to select at least one value for x between each x-intercept). We also use arrows to indicate the behavior of the graph at the extreme values of x (Figure 11).

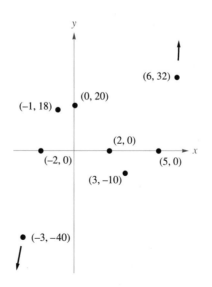

FIGURE 11

Finally, we sketch a smooth, continuous curve through the plotted points and subject to all of the conditions of our analysis (Figure 12).

Notice that the *y*-axis is scaled to allow large values of *y*. This is often necessary to get the "big picture" when graphing polynomials.

It may bother you that we didn't take great pains to make our graph extremely accurate in the last example. Remember that our goal is to get an overall picture of the graph to better understand the function. All we really need is a rough sketch that shows us the overall picture of the function. Also, notice that we had to scale the *y*-axis so that we could show the important parts of the graph. This is common for the graphs of polynomial functions.

EXAMPLE 6   Sketch the graph $Q(x) = (2x^2 - 2)(x^2 - x - 12)$ and label the intercepts.

SOLUTION If we were to simplify by multiplying, the leading term would be $(2x^2)(x^2) = 2x^4$. This means that Q is a polynomial function of degree 4 and that $y \to +\infty$ as $x \to +\infty$ and $y \to +\infty$ as $x \to -\infty$. It also implies that there are no more than three turning points.

To determine the *y*-intercept, evaluate $Q(0)$:

$$Q(0) = (2(0)^2 - 2)((0)^2 - (0) - 12).$$

The *y*-intercept of $f(x)$ is $(0, 24)$.

To determine the *x*-intercepts, we set $Q(x) = 0$ and solve for *x*:

$$(2x^2 - 2)(x^2 - x - 12) = 0$$

$$2(x - 1)(x + 1)(x - 4)(x + 3) = 0$$

$$x = 1, -1, 4, -3.$$

The *x*-intercepts of the graph of $f(x)$ are $(1, 0)$, $(-1, 0)$, $(4, 0)$, and $(-3, 0)$. These are plotted along with some additional points (computed by evaluating $f(x)$ at selected values of *x*) on the graph in Figure 13.

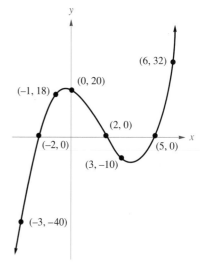

FIGURE 12

x	-2	2	3
$Q(x)$	-36	-60	-96

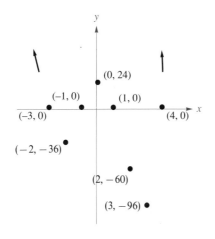

FIGURE 13

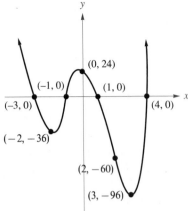

FIGURE 14

Finally, we sketch a smooth, continuous curve through the plotted points (Figure 14).

EXAMPLE 7   Show that $H(x) = -x^4 + 3x^2 + 4$ is an even function. Sketch its graph and label the intercepts.

SOLUTION To show that $H(x)$ is an even function, we need to verify that $H(-x) = H(x)$:

$$H(-x) = -(-x)^4 + 3(-x)^2 + 4 = -x^4 + 3x^2 + 4 = H(x).$$

Since H is an even function, we only need to sketch its graph for $x \geq 0$ and reflect this portion across the y-axis to complete the graph.

The leading term of $H(x)$ is $-x^4$. The graph of H has at most three turning points, and

$$y \to -\infty \quad \text{as} \quad x \to +\infty, \quad \text{and} \quad y \to -\infty \quad \text{as} \quad x \to -\infty.$$

To determine the y-intercept, we evaluate $H(0)$:

$$H(0) = -(0)^4 + 3(0)^2 + 4 = 4.$$

The y-intercept is $(0, 4)$.

To find the x-intercepts, we set $H(x) = 0$ and solve for x:

$$-x^4 + 3x^2 + 4 = 0$$
$$x^4 - 3x^2 - 4 = 0$$
$$(x^2 + 1)(x^2 - 4) = 0$$
$$x^2 + 1 = 0 \quad \text{or} \quad x^2 - 4 = 0$$
$$x^2 = -1 \qquad\qquad x^2 = 4$$
$$\text{no real solutions} \qquad x = \pm 2.$$

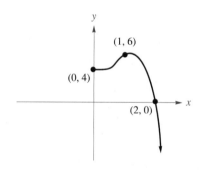

FIGURE 15

The x-intercepts are $(2, 0)$ and $(-2, 0)$.

Next, we plot a few additional points for $x \geq 0$ and sketch a smooth graph through them (Figure 15). Reflecting this across the y-axis gives us the completed graph (Figure 16).

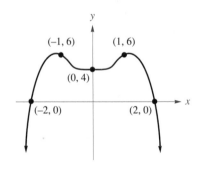

FIGURE 16

Just as we used the graph of a quadratic function to solve a quadratic inequality, we can use the graph of a polynomial function to solve a polynomial inequality (you may wish to pause here and review Section 2.4).

EXAMPLE 8   Solve $(2x^2 - 2)(x^2 - x - 12) < 0$.

SOLUTION   In Example 6, we sketched the graph of $f(x) = (2x^2 - 2)(x^2 - x - 12)$. We can use this sketch now to determine the intervals over which the graph of $f(x)$ is below the x-axis.

   Plotting a few more points and sketching gives us the graph in Figure 17. The values of x over which the graph of $f(x)$ is below the x-axis are in the set $(-3, -1)$ or $(1, 4)$. This is the solution to our inequality.

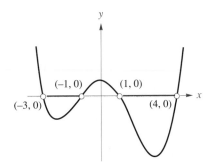

FIGURE 17

EXERCISE SET 3.1

A

In Problems 1 through 6, determine whether the graph given could be the graph of a polynomial function. If it is not a polynomial function, state why not. If it could be a polynomial function, state whether the leading coefficient of the polynomial function would be positive or negative, and whether the degree would be even or odd.

1.

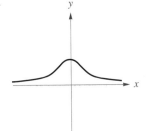

2.

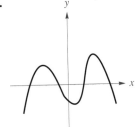

3.

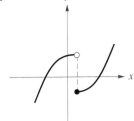

4.

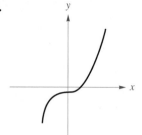

5.

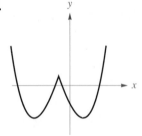

6.

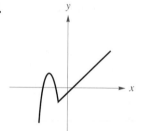

In Problems 7 through 12, the graph of one of the six polynomial functions I through VI is given:

   I. $P_1(x) = 5x^4 - 10x^3 - 40x^2 + 9x + 45$
  II. $P_2(x) = -x^6 + 14x^4 - 49x^2 + 36$
 III. $P_3(x) = -x^5 + 7x$
 IV. $P_4(x) = x^4 - 16x^2 + 2$
  V. $P_5(x) = x^3 - 9x$
 VI. $P_6(x) = x^4 + 4x^3 - 4x^2 - 16x$

7.

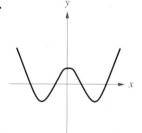

8.

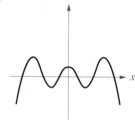

9.

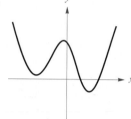

10.

11.

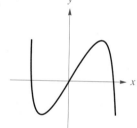

12.

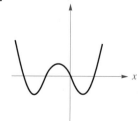

In Problems 13 through 21, sketch the graph of the given monomial function in a). On the same axes, sketch the graph and label the intercepts of the function given in b).

13. **a)** $y = x^3$ **b)** $y = x^3 - 8$

14. **a)** $y = x^4$ **b)** $y = x^4 - 1$

15. **a)** $y = x^4$ **b)** $y = (x - 2)^4$

16. **a)** $y = x^3$ **b)** $y = (x + 3)^3$

17. **a)** $y = \frac{1}{2}x^5$ **b)** $y = \frac{1}{2}x^5 + 16$

18. **a)** $y = \frac{1}{3}x^4$ **b)** $y = \frac{1}{3}x^4 - 27$

19. **a)** $y = \frac{1}{2}x^4$ **b)** $y = \frac{1}{2}(x - 2)^4 - 8$

20. **a)** $y = \frac{1}{6}x^5$ **b)** $y = \frac{1}{6}(x + 2)^5 - 16$

21. **a)** $y = -x^6$ **b)** $y = -(x - 1)^6 - 5$

In Problems 22 through 33, roughly sketch the graph of the given polynomial functions. Label the intercepts.

22. $P(x) = x(x + 3)(x - 5)$

23. $Q(x) = x(x + 2)(x - 8)$

24. $f(x) = (x - 5)(x + 2)(x - 8)$

25. $f(x) = (x - 1)(x + 2)(x - 2)$

26. $p(x) = -(x + 1)(x + 3)(x - 4)$

27. $q(x) = -(x + 4)(x - 5)(x - 1)$

28. $R(x) = (x^2 + 8x + 12)(x - 2)(x + 5)$

29. $w(x) = (x^2 + 5x + 4)(x^2 - 10x)$

30. $f(x) = -2(3x - 1)(x + 4)(x - 7)$

31. $h(x) = -3(2x - 1)(x + 1)(x + 5)$

32. $h(x) = \frac{2}{3}(x^2 - 4)(x - 7)$

33. $D(x) = -\frac{1}{2}(2x - 15)(x^2 - 9)$

B

In Problems 34 through 45, roughly sketch the graph of the given polynomial function. Label the intercepts.

34. $p(x) = x^3 - 4x^2 - 8x$

35. $g(x) = x^3 - 5x^2 + 4x$

36. $P(x) = x^4 - 2x^3 - 8x^2$

37. $Q(x) = -x^4 + 4x^3 + 21x^2$

38. $R(x) = x^3 - 5x^2$

39. $n(x) = 4x^2 - \frac{1}{2}x^3$

40. $H(x) = -2x^3 + 54x$

41. $y(x) = 8x^2 - 0.5x^4$

42. $f(x) = x^3 + 6x^2 - 2x - 12$

43. $g(x) = x^3 - 4x^2 - x + 4$

44. $f(x) = 2x^3 - 10x^2 - 6x + 30$

45. $g(x) = -x^3 - 3x^2 + 7x + 21$

In Problems 46 through 51, the polynomial function is either an even function or an odd function. Sketch the graph of the function for which $x \geq 0$. If the function is odd, complete the graph by reflecting this portion through the origin. If the function is even, complete the graph by reflecting this portion through the y-axis.

46. $p(x) = x^3 - 9x$

47. $g(x) = x^3 - 16x$

48. $P(x) = x^4 - 8x^2$

49. $Q(x) = -\frac{1}{2}x^4 + 4x^2$

50. $R(x) = x^4 - 22x^2 + 40$

51. $n(x) = x^4 - 9x^2 + 20$

In Problems 52 through 57, use the graph of a polynomial function to solve the inequality.

52. $(x - 1)(x + 2)(x - 2) \leq 0$
    (Hint: Refer to Problem 25.)

53. $-3(2x - 1)(x + 1)(x + 5) \geq 0$
    (Hint: Refer to Problem 31.)

54. $4x^3 + 21x^2 > x^4$ *(Hint: Refer to Problem 37.)*

55. $8x^2 > 0.5x^4$ *(Hint: Refer to Problem 41.)*

56. $x^3 \leq 16x$ *(Hint: Refer to Problem 47.)*

57. $x^4 \geq 9x^2 - 20$ *(Hint: Refer to Problem 51.)*

In Problems 58 through 60, determine whether the polynomial function is even or odd and sketch its graph over the given interval. Each of these functions has applications in advanced physics and mathematics.

58. Third-degree **Legendre** polynomial:

$$p_3(x) = \frac{1}{2}(5x^3 - 3x) \quad \text{interval: } [-1, 1].$$

59. Third-degree **Tchebyshev** polynomial:

$$T_3(x) = 4x^3 - 3x \quad \text{interval: } [-1, 1].$$

60. Fourth-degree **Hermite** polynomial:

$$H_3(x) = x^4 - 6x^2 + 3 \quad \text{interval: } (-\infty, +\infty).$$

61. Sketch the graphs $f(x) = \frac{1}{6}x^3$ and $g(x) = -2x + 1$ on the same axes. Use the method of adding y-coordinates

(Section 2.6) to sketch the graph of
$h(x) = \frac{1}{6}x^3 - 2x + 1$.

62. Sketch the graphs of $F(x) = \frac{1}{6}x^4$ and $G(x) = -3x^2 + 3$ on the same axes. Use the method of adding y-coordinates (Section 2.6) to sketch the graph of $H(x) = \frac{1}{6}x^4 - 3x^2 + 3$.

63. Sketch the graph of $P(x) = x^3 - 3x^2$. Use the fact that $Q(x) = x^3 - 3x^2 - 2$ is a translation of $P(x)$ to sketch the graph of $Q(x)$.

64. Sketch the graph of $p(x) = x^4 - 9x^2$. Use the fact that $q(x) = x^4 - 9x^2 + 4$ is a translation of $p(x)$ to sketch the graph of $q(x)$.

65. Show that the monomial function $y = ax^n$ is an even function if n is an even positive integer and that it is an odd function if n is an odd positive integer.

C

66. In Example 6 of this section, the graph of $Q(x) = (2x^2 - 2)(x^2 - x - 12)$ is shown. Use this picture to sketch the graphs of:

   a) $y = |Q(x)|$

   b) $y = Q(|x|)$

   c) $y = |Q(|x|)|$

67. In Example 7, we discussed the graph of $H(x) = -x^4 + 3x^2 + 4$. Explain why we can be sure that there is a turning point at $(0, 4)$. Generalize to all even polynomial functions.

68. Given the third-degree polynomial function $P(x) = ax^3 + bx^2 + cx + d$, it can be shown (using calculus) that the x-coordinates of the turning points of its graph are the solutions of the quadratic equation $3ax^2 + 2bx + c = 0$. Show that the x-coordinates of the turning points are

$$x = \frac{-b + \sqrt{b^2 - 3ac}}{3a} \quad \text{and} \quad x = \frac{-b - \sqrt{b^2 - 3ac}}{3a}.$$

In Problems 69 through 72, sketch the graph of the third-degree polynomial function. Use Problem 68 to find its turning points.

69. $f(x) = x^3 - 12x$

70. $g(x) = 4x^3 - 27x$

71. $f(x) = 2x^3 - 9x^2 - 24x + 108$

72. $g(x) = (x^2 - 6x + 16)(x + 2)$

73. Use Problem 68 to find the maximum volume of the box described in Problem 48 of Section 2.2.

S E C T I O N 3.2 VALUES OF POLYNOMIAL FUNCTIONS

One of the problems with sketching the graph of an nth-degree polynomial function $P(x)$ is that many points must be plotted to do an adequate job. Another problem with sketching the graph of $P(x)$ is that of finding its x-intercepts.

We have seen that $(c, 0)$ is on the graph of a function f if and only if $f(c) = 0$. These values c are called the **zeros** of the function f and are found by setting $f(x) = 0$ and finding the real solutions of this equation. Finding the zeros of nth-degree polynomial function $P(x)$ means finding the real solutions of $P(x) = 0$, a polynomial equation of degree n. For $n > 2$, this solution is usually a difficult or impossible task.

In this section, we will look at a way to evaluate a polynomial function at a value of x. We will also investigate methods of determining the x-intercepts of its graph.

We begin by reviewing polynomial division in Example 1. (Recall this was discussed in Section 1.3; you may wish to pause here and refer to it.)

EXAMPLE 1   Use polynomial division to find the quotient and the remainder of

$$\frac{2x^4 - x^3 - 5x^2 + 6x - 11}{x^2 - 4}.$$

SOLUTION The division tableau follows:

$$
\begin{array}{r}
2x^2 - x + 3 \\
x^2 - 4 \overline{\smash{\big)}\ 2x^4 - x^3 - 5x^2 + 6x - 11} \\
\underline{2x^4 \qquad\quad -8x^2} \\
-x^3 + 3x^2 + 6x \\
\underline{-x^3 \qquad\quad +4x} \\
3x^2 + 2x - 11 \\
\underline{3x^2 \qquad -12} \\
2x + 1
\end{array}
$$

The quotient is $2x^2 - x + 3$, and the remainder is $2x + 1$.

To check our answer, multiply the quotient by the divisor and add the remainder. We should end up with the dividend:

$$(x^2 - 4)(2x^2 - x + 3) + (2x + 1) \overset{?}{=} 2x^4 - x^3 - 5x^2 + 6x - 11$$

$$(2x^4 - x^3 - 5x^2 + 4x - 12) + (2x + 1) \overset{?}{=} 2x^4 - x^3 - 5x^2 + 6x - 11$$

$$2x^4 - x^3 - 5x^2 + 6x - 11 = 2x^4 - x^3 - 5x^2 + 6x - 11.$$

This verifies that our answer is correct.

The check we performed illustrates this property:

Suppose that $P(x)$ and $D(x)$ are nonconstant polynomial functions such that the degree of $P(x)$ is greater than or equal to the degree of $D(x)$. There exist two unique polynomials $Q(x)$ and $R(x)$ such that the degree of $R(x)$ is less than the degree of $D(x)$ and

Division Algorithm

$$\frac{P(x)}{D(x)} = Q(x) + \frac{R(x)}{D(x)}$$

or, equivalently,

$$P(x) = D(x)Q(x) + R(x).$$

$D(x)$ is the **divisor,** $Q(x)$ is the **quotient,** and $R(x)$ is the **remainder.**

EXAMPLE 2　Let $P(x) = 3x^3 - 8x^2 - x + 19$. Find the quotient $Q(x)$ and the remainder $R(x)$ such that

$$P(x) = (x - 2)Q(x) + R(x).$$

SOLUTION

$$
\begin{array}{r}
3x^2 - 2x - 5 \\
x - 2 \overline{\smash{\big)}\ 3x^3 - 8x^2 - x + 19} \\
\underline{3x^3 - 6x^2} \\
-2x^2 - x \\
\underline{-2x^2 + 4x} \\
-5x + 19 \\
\underline{-5x + 10} \\
9
\end{array}
$$

The quotient is $Q(x) = 3x^2 - 2x - 5$, and the remainder is $R(x) = 9$. So,

$$3x^3 - 8x^2 - x + 19 = (x - 2)(3x^2 - 2x - 5) + 9.$$

Notice that the remainder is a constant value. This should seem perfectly reasonable since the divisor is of degree one, and the remainder must be of degree less than the divisor.

Consider again the polynomial function P in Example 2. Suppose that we wish to evaluate $P(2)$. We could use the representation

$$P(x) = 3x^3 - 8x^2 - x + 19,$$

so that

$$P(2) = 3(2)^3 - 8(2)^2 - (2) + 19 = 9.$$

On the other hand, we could use the representation we determined in Example 2, namely,

$$P(x) = (x - 2)(3x^2 - 2x - 5) + 9$$
$$P(2) = (2 - 2)(3(2)^2 - 2(2) - 5) + 9 = (0)(3) + 9 = 9.$$

We get $P(2) = 9$ using either representation, but notice something special about the second representation. Since the first term has a factor of zero, the value of $P(2)$ is precisely 9, the remainder in our division. As you may suspect, a generalization follows.

If a polynomial $P(x)$ is divided by $(x - c)$, then the remainder is $P(c)$. That is, if

$$P(x) = (x - c)Q(x) + R$$

then

$$P(c) = R.$$

Remainder Theorem

This is not difficult to show to be true, for if

$$P(x) = (x - c)Q(x) + R$$

then

$$P(c) = (c - c)Q(c) + R = (0)Q(c) + R = R.$$

EXAMPLE 3   Let $f(x) = 2x^3 + 6x - 7$. Use the remainder theorem to evaluate $f(-4)$.

SOLUTION First, we write $f(x)$ in the form

$$f(x) = (x - (-4))Q(x) + R = (x + 4)Q(x) + R$$

for some $Q(x)$ and R. Finding R entails polynomial division:

$$
\begin{array}{r}
2x^2 - 8x + 38 \\
x + 4 \overline{\smash{)}\ 2x^3 + 0x^2 + \ 6x - 7} \\
\underline{2x^3 + 8x^2} \\
-8x^2 + \ 6x \\
\underline{-8x^2 - 32x} \\
38x - \quad 7 \\
\underline{38x + 152} \\
-159
\end{array}
$$

Using the remainder theorem, $f(-4) = -159$.

You may argue at this point that using the remainder theorem to evaluate a polynomial function at a value of its domain is not much easier than evaluating it directly by inserting that value into the function. Since it involves tedious polynomial division, it actually seems to be more work. Our discussion now turns to a streamlined division algorithm—**synthetic division**—that makes the remainder theorem the more efficient method of evaluating polynomials.

Look back at the long-division tableau in Example 3. Notice that there is a great deal of unnecessary writing. If we eliminate the variables and preserve the coefficients in place, we still have the essence of the algorithm.

$$\begin{array}{r} 2x^2 - 8x + 38 \\ x + 4\overline{\smash{\big)}\ 2x^3 + 0x^2 + 6x - 7} \\ \underline{2x^3 + 8x^2} \\ -8x^2 + 6x \\ \underline{-8x^2 - 32x} \\ 38x - 7 \\ \underline{38x + 152} \\ -159 \end{array}$$

Now examine the coefficients and constants; many are repetitions (they are boxed here). These can also be eliminated without loss to the algorithm. We compress the tableau vertically to save space.

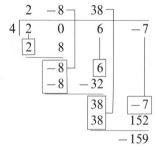

Finally, we eliminate the source of the most common mistake in polynomial long division: subtraction of polynomial expressions. By changing the signs on the divisor and the second line of the tableau, we change the subtraction operations to addition. The final format of our algorithm, synthetic division, is shown.

$$\begin{array}{r} 2 \quad -8 \quad\quad 38 \\ 4\,|\ 2 \quad\ 0 \quad\ 6 \quad\ -7 \\ \underline{\quad\ 8\quad\quad\quad} \\ \quad\quad\ -32 \\ \underline{\quad\quad\quad\quad\quad} \\ \quad\quad\quad\quad\ -152 \\ \quad\quad\quad\quad\ -159 \end{array}$$

$$\boxed{\begin{array}{r} -4\,|\ 2 \quad\ 0 \quad\ 6 \quad\ -7 \\ \underline{\quad\ -8\quad 32\ -152} \\ 2\ -8\quad 38\,|\ -159 \end{array}}$$

If you feel a bit unsure about this development of the synthetic division algorithm, don't despair. It is more important that you be able to perform synthetic division quickly and accurately. The following are the steps of the algorithm:

Synthetic Division Algorithm

To divide $P(x) = 5x^4 - 13x^3 - 10x - 1$ by $x - 3$ using synthetic division:

1) Enter the coefficients of $P(x)$ into the tableau (be sure to account for all coefficients, including the zero coefficients). Bring the leading coefficient 5 down below the line. Enter 3 in the upper left corner.

$$\underline{3}\,|\ 5 \quad -13 \quad 0 \quad -10 \quad -1$$
$$\overline{\hphantom{xx}5}$$

2) Multiply 5 by 3 and enter the product 15 under the coefficient -13. Add the coefficient -13 and the product 15 and write the sum 2 under the line as shown.

$$\underline{3}\,|\ 5 \quad -13 \quad 0 \quad -10 \quad -1$$
$$\boxed{15}$$
$$5 \quad\ 2$$

3) Multiply 2 by 3 and enter the product 6 under the coefficient 0. Add again and enter the sum 6 under the line. Continue this process for all the coefficients.

$$\underline{3}\,|\ 5 \quad -13 \quad 0 \quad -10 \quad -1$$
$$15 \quad 6 \quad\ \ 18 \quad\ \ 24$$
$$\overline{5 \quad\ \ 2 \quad 6 \quad\ \ 8\ |\ 23}$$

4) The first three values along the bottom row are the coefficients of the quotient $Q(x) = 5x^3 + 2x^2 + 6x + 8$; note that the quotient is of degree one less than $P(x)$. The value 23 in the bottom right corner is the remainder R.
Keep in mind that the direct result of the remainder theorem is that $P(3) = 23$.

The next example shows the value of the remainder theorem and synthetic division. We now have a way to quickly evaluate a polynomial function at a given value of its domain.

EXAMPLE 4   Sketch the graph of $f(x) = x^5 - 5x^3 + 3x + 2$. Use synthetic division to plot selected points.

Using a calculator speeds up this process considerably. With practice, you may be able to determine ordered pairs without writing down the synthetic division tableau. A programmable calculator or a personal computer is even more of an asset.

SOLUTION We start by noting that $y \to -\infty$ as $x \to -\infty$ and $y \to +\infty$ as $x \to +\infty$. Using synthetic division, we obtain

$$\begin{array}{r|rrrrrr} -3 & 1 & 0 & -5 & 0 & 3 & 2 \\ & & -3 & 9 & -12 & 36 & -117 \\ \hline & 1 & -3 & 4 & -12 & 39 & -115 \end{array} \Rightarrow (-3, -115)$$

$$\begin{array}{r|rrrrrr} -2 & 1 & 0 & -5 & 0 & 3 & 2 \\ & & -2 & 4 & 2 & -4 & 2 \\ \hline & 1 & -2 & -1 & 2 & -1 & 4 \end{array} \Rightarrow (-2, 4)$$

In a similar manner we also get $(-1, 3)$, $(0, 2)$, $(1, 1)$, $(2, 0)$, and $(3, 119)$ (pause here and verify these ordered pairs). Figure 18 shows our results so far: It is still difficult to get a sense of the graph of the function, so we continue to determine and plot ordered pairs:

$$\begin{array}{r|rrrrrr} -1.5 & 1 & 0 & -5 & 0 & 3 & 2 \\ & & -1.5 & 2.25 & 4.125 & -6.1875 & 4.78125 \\ \hline & 1 & -1.5 & -2.75 & 4.125 & -3.1875 & 6.78125 \end{array}$$

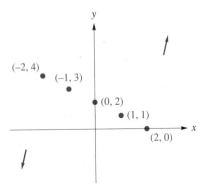

FIGURE 18

The y-coordinates of these ordered pairs have been rounded to the nearest 0.1.

This gives us the ordered pair $(-1.5, 6.8)$. Similar computations yield the ordered pairs $(-0.5, 1.1)$, $(0.5, 2.9)$, and $(1.5, -2.8)$. Adding these points to our picture gives us a better sense of the graph (Figure 19).

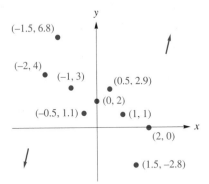

FIGURE 19

The graph is smooth and continuous. It has no more than four turning points and five x-intercepts (Figure 20).

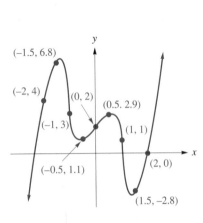

FIGURE 20

In Example 4, we found $f(2) = 0$. This enabled us to plot an x-intercept $(2, 0)$. By the synthetic division in Example 4, we can express $f(x)$ as

$$f(x) = (x - 2)(x^4 + 2x^3 - x^2 - 2x - 1) + 0$$

or just

$$f(x) = (x - 2)(x^4 + 2x^3 - x^2 - 2x - 1).$$

In other words, $(x - 2)$ is a factor of $f(x)$.

In general, suppose that we examine a polynomial function

$$P(x) = (x - c)Q(x) + R \quad \text{where } R = P(c).$$

If $P(c) = 0$, then $P(x) = (x - c)Q(x)$, implying that $(x - c)$ is a factor of $P(x)$. On the other hand, if $(x - c)$ is a factor of $P(x)$, then $P(x) = (x - c)Q(x)$, and the remainder theorem implies that $P(c) = 0$. We have just proved the **factor theorem.**

Given a polynomial function P, a real number c is a zero of P if and only if $(x - c)$ is a factor of $P(x)$.

Factor Theorem

EXAMPLE 5 Show that $x + \frac{4}{3}$ is a factor of $3x^3 - 11x^2 + x + 28$ and rewrite this polynomial as

$$3x^3 - 11x^2 + x + 28 = (x + \tfrac{4}{3})Q(x) \quad \text{for some } Q(x).$$

SOLUTION We use synthetic division to find the quotient $Q(x)$:

$$
\begin{array}{r|rrrr}
-\frac{4}{3} & 3 & -11 & 1 & 28 \\
             &   & -4 & 20 & -28 \\
\hline
             & 3 & -15 & 21 & 0 \\
\end{array}
$$

The fact that the remainder is zero is sufficient to show that $x + \frac{4}{3}$ is a factor of $3x^3 - 11x^2 + x + 28$. Recall that the coefficients of $Q(x)$ are shown in the bottom row of the tableau. So,

$$3x^3 - 11x^2 + x + 28 = (x + \tfrac{4}{3})(3x^2 - 15x + 21).$$

A direct result of the factor theorem is that $(c, 0)$ is an x-intercept of the graph of a polynomial function $P(x)$ if and only if $(x - c)$ is a factor of $P(x)$. This is used in the next example.

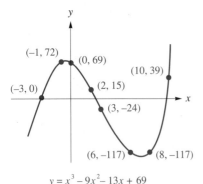

$y = x^3 - 9x^2 - 13x + 69$

FIGURE 21

EXAMPLE 6   Show that $x = -3$ is an x-intercept of the graph of $f(x) = x^3 - 9x^2 - 13x + 69$. Sketch the graph of $f(x)$.

SOLUTION To verify that $(-3, 0)$ is an x-intercept, we use synthetic division.

$$
\begin{array}{r|rrrr}
-3 & 1 & -9 & -13 & 69 \\
   &   & -3 & 36 & -69 \\
\hline
   & 1 & -12 & 23 & 0
\end{array}
$$

Since the remainder is 0, $(-3, 0)$ is an x-intercept of the graph of $f(x)$. We plot a few selected points and sketch the graph (Figure 21).

EXERCISE SET 3.2

A

In Problems 1 through 6, a polynomial function P and a divisor D(x) are given. Find the quotient Q(x) and remainder R(x). Rewrite P as P(x) = D(x)Q(x) + R(x).

1. $P(x) = 2x^3 + x^2 - 10x + 2$
   $D(x) = 2x + 1$

2. $P(x) = 3x^3 - 2x^2 - 12x - 4$
   $D(x) = 3x - 2$

3. $P(x) = x^4 - x^3 - 8x^2 + 6x - 4$
   $D(x) = x^2 + 2x - 1$

4. $P(x) = 2x^4 + x^3 - 5x^2 + 14x + 7$
   $D(x) = 2x^2 + x - 5$

5. $P(x) = x^5 - x^3 - x^2 + 8x + 4$
   $D(x) = x^2 - 3$

6. $P(x) = 6x^4 - 2x^3 + 29x^2 - 7x + 28$
   $D(x) = 2x^2 + 7$

In Problems 7 through 15, use synthetic division to evaluate P(c) for the polynomial P and value of c given. Rewrite P as P(x) = (x - c)Q(x) + P(c).

7. $P(x) = 3x^3 - 2x^2 + x - 7$; $c = 2$

8. $P(x) = x^3 + 8x^2 - 9x + 2$; $c = -1$

9. $P(x) = x^4 - 4x^3 - 17x^2 + 2$; $c = 6$

10. $P(x) = 6x^4 - x^3 - 3x^2 + 5x + 7$; $c = \frac{1}{2}$

11. $P(x) = 12x^4 - 2x^3 - 7x^2 - x + 5$; $c = \frac{2}{3}$

12. $P(x) = x^4 - 4x^3 - 20x^2 + 9$; $c = 8$

13. $P(x) = \frac{1}{2}x^5 + 7x^4 - 2x^3 + x^2 - 5x + 9$; $c = 4$

14. $P(x) = \frac{2}{3}x^5 - 2x^4 + 4x^3 + x + 5$; $c = 6$

15. $P(x) = 6x^8 - 6x^3 + x - 9$; $c = 2$

In Problems 16 through 24, use synthetic division to show that $(x - c)$ is a factor of P(c). Rewrite P as P(x) = (x - c)Q(x).

16. $P(x) = x^3 - 6x^2 + 15x - 14$; $c = 2$

17. $P(x) = 2x^3 - 11x^2 + 17x - 6$; $c = 3$

18. $P(x) = 5x^4 + 35x^3 - 3x^2 - 17x + 28$; $c = -7$

19. $P(x) = 3x^4 + 7x^3 + 2x^2 - 6x - 12$; $c = -2$

20. $P(x) = 2x^4 - 5x^3 - 10x^2 + 4x + 1$; $c = \frac{1}{2}$

21. $P(x) = 6x^4 + 4x^3 - 18x^2 + 9x + 14$; $c = -\frac{2}{3}$

22. $P(x) = 8x^6 + 38x^5 + 23x^4 - 82x^3 - 190x^2 - 115x + 210$; $c = 2$

23. $P(x) = 2x^3 + 7x^2 - 10x - 35$; $c = -\sqrt{5}$

24. $P(x) = x^3 + 3x^2 - 3x - 9$; $c = \sqrt{3}$

B

In Problems 25 through 30, use synthetic division to determine which of the given values for x are roots of the equation.

25. $x^3 - 7x + 6 = 0$; $x = -2, 1, 2, 3$

26. $2x^3 - 10x^2 - 28x + 60 = 0$; $x = 0, -3, -1, 7$

27. $(x^3 + 4x^2 - x - 4)(x^3 + 3) = 0$; $x = -7, -4, -1, 5$

28. $(x^2 + x - 4)(x^3 - 2x + 1) = 0$; $x = -1, 1, 3$

29. $2x^4 - 7x^3 = 18x^2 - 49x - 28$; $x = -\frac{1}{2}, \sqrt{7}, 1, 4$

30. $-3x^4 = 11x^3 - 29x^2 - 33x + 60$; $x = -\sqrt{3}, 3, \sqrt{2}, \frac{4}{3}$

In Problems 31 through 36, sketch the graph of the polynomial function. Use synthetic division to determine points of the graph. To give you a head start, one or more of the x-intercepts is given.

31. $f(x) = x^3 - 7x + 6$; $(1, 0)$

32. $f(x) = -2x^3 - 5x^2 + 28x + 15$; $(3, 0)$

33. $f(x) = x^3 + 12x^2 + 12x - 80$; $(2, 0)$

34. $f(x) = x^3 + 5x^2 - 9x - 45$; $(3, 0)$

35. $f(x) = x^4 - 4x^3 - 5x^2 + 12x + 6$;
$(\sqrt{3}, 0)$ and $(-\sqrt{3}, 0)$

36. $f(x) = -x^4 + 8x^3 + 22x^2 - 160x - 40$;
$(2\sqrt{5}, 0)$ and $(-2\sqrt{5}, 0)$

In Problems 37 through 42, use the graph of a polynomial function to solve the inequality.

37. $x^3 - 7x + 6 < 0$ *(Hint: See Problem 31.)*

38. $-2x^3 - 5x^2 + 28x + 15 > 0$ *(Hint: See Problem 32.)*

39. $x^3 + 12x^2 + 12x - 80 \geq 0$ *(Hint: See Problem 33.)*

40. $x^3 + 5x^2 - 9x - 45 \leq 0$ *(Hint: See Problem 34.)*

41. $x^4 < 4x^3 + 5x^2 - 12x - 6$ *(Hint: See Problem 35.)*

42. $x^4 - 8x^3 \leq 22x^2 - 160x - 40$ *(Hint: See Problem 36.)*

43. Show that $x - 1$ is a factor of $x^{76} - 4x^{54} + x^{14} + 2$.

44. Show that $x + 1$ is a factor of $x^{94} + 6x^{23} - 23x^{12} + x^{14} + 27$.

45. Suppose that n is a positive integer. Use the factor theorem to explain why $x - a$ is a factor of $x^n - a^n$.

46. Suppose that n is an even positive integer. Use the factor theorem to explain why $x + a$ is a factor of $x^n - a^n$.

47. Suppose that f is a continuous function on an interval containing a and b, and that $f(a)$ and $f(b)$ are of different signs (one is positive and one is negative). Explain why f must have a zero between a and b. (This result is called the *location theorem*.)

In Problems 48 through 51, use the location theorem of Problem 47 to show that the function f has a zero in the given interval.

48. $f(x) = -x^3 - 3x^2 + 2x + 2$; $[-1, 0]$

49. $f(x) = -x^3 - 3x^2 + 2x + 2$; $[-4, -3]$

50. $f(x) = x^4 - 2x^3 - x^2 - 6x - 12$; $[3, 4]$

51. $f(x) = x^4 - 2x^3 - x^2 - 6x - 12$; $[-2, -1]$

52. Use the location theorem of Problem 47 to approximate the $\sqrt[3]{5}$ to the nearest 0.01. Do this by approximating the zero of the function $f(x) = x^3 - 5$.

53. Use the location theorem of Problem 47 to approximate the $\sqrt[4]{11}$ to the nearest 0.01. Do this by approximating a zero of the function $f(x) = x^4 - 11$.

C

*One method of approximating zeros of a continuous function is **false position**. In Figure 22, the graph of a function f is shown. Since f(a) is negative and f(b) is positive, the location theorem of Problem 47 guarantees a zero in the interval $[a, b]$. Now, suppose that we do this:*

1. Draw a line from a to b.

2. Find the x-intercept (r, 0) of the line.

3. Compute f(r). If it is within our given tolerance, then we are done; otherwise we replace either a or b with r and perform another iteration.

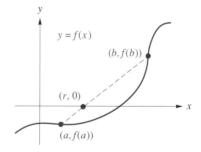

FIGURE 22

54. Show that in Figure 22

$$r = \frac{af(b) - bf(a)}{f(b) - f(a)}.$$

(Hint: Find the equation of the line using the point-slope form. Find the x-intercept.)

55. Use the method of false position to approximate the zero of $f(x) = -x^3 - 3x^2 + 2x + 2$ in the interval $[-1, 0]$. Use two iterations.

56. Use the method of false position to approximate the zero of $f(x) = -x^3 - 3x^2 + 2x + 2$ in the interval $[-4, -3]$. Use two iterations.

57. Show that $x = 1$ is a zero of $f(x) = -x^3 - 3x^2 + 2x + 2$ and use this fact to find the other two zeros exactly. Compare your answers with the approximations found in Problems 55 and 56.

S E C T I O N   3.3   **ZEROS OF POLYNOMIAL FUNCTIONS**

One of our concerns in Sections 3.1 and 3.2 was to find the zeros of a polynomial function. We saw that this is equivalent to solving a polynomial equation and that this solution was not always an easy task.

In this section, we will again be concerned with the zeros of polynomial functions. While we still won't be able to find all the zeros of just any given polynomial function, we will be able to determine specific information about these zeros and to make some educated guesses about them. It is important to keep in mind that any information we discover about the zeros of the polynomial function P also tells us something about the roots of the polynomial equation $P(x) = 0$ and something about the x-intercepts of the graph of P.

We need to make distinctions about types of zeros: rational zeros, real zeros, and complex zeros. (Recall that these sets of numbers were discussed in Chapter 1; you may wish to pause here and refer to it.) Example 1 illustrates these distinctions of the zeros of a polynomial function.

EXAMPLE 1   Given the polynomial function

$$P(x) = (x^2 + x + 1)(2x - 1)(x - \sqrt{2})$$

find the rational zeros, the real zeros, and the complex zeros.

SOLUTION Finding the zeros of P is equivalent to solving the equation $P(x) = 0$, so

$$(x^2 + x + 1)(2x - 1)(x - \sqrt{2}) = 0$$

$$x^2 + x + 1 = 0 \quad \text{or} \quad 2x - 1 = 0 \quad \text{or} \quad x - \sqrt{2} = 0$$

$$x = -\frac{1}{2} \pm \frac{\sqrt{3}}{2}i \quad \bigg| \quad x = \frac{1}{2} \quad \bigg| \quad x = \sqrt{2}.$$

Thus, the rational zero is $\frac{1}{2}$; the real zeros are $\sqrt{2}$ and $\frac{1}{2}$; and the complex zeros are

$$\sqrt{2}, \frac{1}{2}, -\frac{1}{2} + \frac{\sqrt{3}}{2}i, \text{ and } -\frac{1}{2} - \frac{\sqrt{3}}{2}i.$$

Before we start searching for the zeros of a given polynomial function, it will be to our advantage to know if indeed they do exist. If they do, it will also be to our advantage to determine how many and of what type (rational, real, or complex). The theorems of this section address these questions.

Our first theorem, the fundamental theorem of algebra, guarantees the existence of at least one complex zero for any polynomial function (keep in mind that integers, rational numbers, and real numbers are also considered complex numbers). Unfortunately, it doesn't tell us how to find this zero; it only tells us that the zero exists. In mathematics, this type of a theorem is called an existence theorem.

The fundamental theorem of algebra was suspected to be true for many years, but it was not actually proven until Karl Friedrich Gauss (1777–1855) accomplished the task at the age of 20 in 1797. As you might guess from its name, this is an important theorem in mathematics. The proof entails calculus and the advanced theory of complex numbers, so we will be content to state the theorem without proof.

| If P is a polynomial function of degree $n > 0$, then P has at least one complex zero. | *Fundamental Theorem of Algebra* |

Of course, a direct result of this is that the polynomial equation $P(x) = 0$ has at least one complex root.

Now, suppose that we have a polynomial function

$$P(x) = a_n x^n + a_{n-1} x^{n-1} + a_{n-2} x^{n-2} + \cdots + a_1 x + a_0$$

and that c_1 is a zero of P. By the factor theorem of Section 3.2, this implies that $(x - c_1)$ is a factor of P and that we can write P as

$$P(x) = (x - c_1)Q_{n-1}(x)$$

where Q_{n-1} is a polynomial function of degree $n - 1$. Now, by the fundamental theorem of algebra, Q_{n-1} in turn has a zero c_2, and Q_{n-1} can be written as

$$Q_{n-1}(x) = (x - c_2)Q_{n-2}(x)$$

where Q_{n-2} is a polynomial function of degree $n - 2$. This implies that

$$P(x) = (x - c_1)(x - c_2)Q_{n-2}(x).$$

If we continue this process, we eventually arrive at

$$P(x) = (x - c_1)(x - c_2) \cdots (x - c_n)Q_0(x)$$

where Q_0 is a polynomial function of degree 0, namely, a constant function. We leave it to you as an exercise to show that $Q_0(x) = a_n$, the leading coefficient of P. So,

$$P(x) = a_n(x - c_1)(x - c_2) \cdots (x - c_n).$$

It is important that you realize that, by the factor theorem, each of these $c_i (i = 1, \ldots, n)$ is a zero of P. Furthermore, these are the only zeros of P. To see this, suppose that $x = k$ is a zero of P, distinct from the zeros $c_i (i = 1, \ldots, n)$. Then,

$$P(k) = a_n(k - c_1)(k - c_2) \ldots (k - c_n).$$

But none of these factors can be zero (why?), so $P(k) \neq 0$. We are inescapably forced to the conclusion that no such k exists. Our results are summarized in this theorem.

Linear Factors of a Polynomial Function

If P is a polynomial function of degree $n > 0$ and leading coefficient a_n, then

$$P(x) = a_n(x - c_1)(x - c_2) \cdots (x - c_n)$$

where $c_1, c_2, c_3, \ldots, c_n$ are the complex zeros of P.

Nothing in the discussion of this theorem guaranteed that these complex zeros $c_i (i = 1, \ldots, n)$ are necessarily distinct. Example 2 is a case in point.

EXAMPLE 2   Let $f(x) = 2x^3 - 5x^2 - 4x + 12$. Find the zeros of f and write f in its linear factored form, given that $x = 2$ is a zero of f.

SOLUTION Since 2 is a zero of f, then f can be written as

$$f(x) = (x - 2)Q(x)$$

for some polynomial function $Q(x)$ of degree two. Recall that we can find $Q(x)$ by using synthetic division:

$$\begin{array}{r|rrrr} 2 & 2 & -5 & -4 & 12 \\ & & 4 & -2 & -12 \\ \hline & 2 & -1 & -6 & 0 \end{array}$$

So $f(x) = (x - 2)(2x^2 - x - 6)$. If we factor $(2x^2 - x - 6)$, we get $(x - 2)(2x + 3)$. Our conclusion is that the zeros of f are 2, 2, and $-\frac{3}{2}$, and the linear factored form of f is

$$f(x) = 2(x - 2)(x - 2)(x - (-\tfrac{3}{2}))$$
$$= 2(x - 2)^2(x + \tfrac{3}{2}).$$

In Example 2, the linear factored form of f has two factors $(x - 2)$. In this case we say that 2 is a zero of **multiplicity two** of P. In general, if the linear factored form of a polynomial function P has n factors $(x - c)$, then we say that c is a zero of **multiplicity n** of P.

EXAMPLE 3 Determine polynomial function Q of degree 4 and leading coefficient 1 with these zeros: 0, -2, 4 (multiplicity two). Sketch the graph of Q.

SOLUTION We can write

$$Q(x) = (x - 0)(x - (-2))(x - 4)(x - 4)$$
$$= x(x + 2)(x - 4)^2.$$

Simplifying yields

$$Q(x) = x^4 - 6x^3 + 32x.$$

If we plot the x-intercepts (recall that each zero of the function gives us an intercept) and a few more selected points, we can sketch the graph (Figure 23). Notice the behavior of the graph at the x-intercept $(4, 0)$.

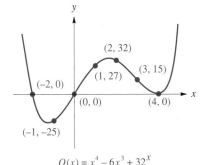

$Q(x) = x^4 - 6x^3 + 32^x$

FIGURE 23

EXAMPLE 4 Given the polynomial equation $x^4 - 18 = 7x^2$, solve the following:

a) The polynomial equation over the real numbers

b) The polynomial equation over the complex numbers

SOLUTION

a) First we rewrite the equation as $x^4 - 7x^2 - 18 = 0$. Since this is in essence a quadratic equation in x^2, we can factor and solve:

$$(x^2)^2 - 7(x^2) - 18 = 0$$
$$(x^2 - 9)(x^2 + 2) = 0$$
$$x^2 - 9 = 0 \quad \text{or} \quad x^2 + 2 = 0$$
$$x^2 = 9 \qquad\qquad x^2 = -2$$
$$x = \pm 3 \qquad \text{no real roots.}$$

The real roots of the equation are -3 and 3.

b) If we allow nonreal, complex roots, then

$$x^2 + 2 = 0$$
$$x^2 = -2$$
$$x = \pm i\sqrt{2}$$

The complex roots of the equation are -3, 3, $i\sqrt{2}$, and $-i\sqrt{2}$.

The fact that we found four roots should not be surprising; after all, it is a fourth-degree polynomial equation.

Example 4 points out that while an nth-degree polynomial function has n linear factors, it does not necessarily have n real zeros. The zeros may not be distinct, and some may be nonreal complex numbers. However, we can say something about the maximum number of real zeros of a polynomial function (and therefore the number of x-intercepts of its graph).

Number of Real Zeros of a Polynomial Function

> If P is a polynomial function of degree $n > 0$, then P has at most n real zeros. The graph of P has at most n x-intercepts.

Next, we turn our attention to the existence of rational zeros of polynomial functions (recall that a rational number is a number that can be expressed as the ratio p/q of two integers p and q). Suppose that we wish to examine the possible rational zeros of

$$f(x) = 2x^4 + 3x^3 - 12x^2 - 15x + 10.$$

Assume that p/q is a zero of f and that p/q is reduced to lowest terms (that is, p and q have no common factors). Then, by definition of a zero, $f(p/q) = 0$, or

$$2(p/q)^4 + 3(p/q)^3 - 12(p/q)^2 - 15(p/q) + 10 = 0.$$

Simplifying and multiplying through by q^4 gives

$$2p^4 + 3p^3q - 12p^2q^2 - 15pq^3 + 10q^4 = 0$$

or, subtracting $10q^4$ from both sides,

$$2p^4 + 3p^3q - 12p^2q^2 - 15pq^3 = -10q^4.$$

Now, if we factor p from the left side, then

$$p(2p^3 + 3p^2q - 12pq^2 - 15q^3) = -10q^4.$$

Think about what this implies. Both sides are the products of integers, and p is a factor of the left side. Since the right side is equal to the left side, p is also a factor of the $-10q^4$. However, remember that p and q have no common factors, so p must be a factor of -10 (the constant term of f). Possible candidates for p are the divisors of -10; they are $-1, 1, -2, 2, -5, 5, -10,$ and 10.

In a similar fashion, it can be shown that q must be a factor of 2 (the leading coefficient of f). Possible candidates for q are $-1, 1, -2,$ or 2. The next theorem is a generalization of this discussion.

If $P(x) = a_n x^n + a_{n-1} x^{n-1} + a_{n-2} x^{n-2} + \ldots + a_1 x + a_0$ is a polynomial function of degree $n > 0$ with integer coefficients and if p/q is a rational zero (reduced to lowest terms), then p is a factor of a_0 and q is a factor of a_n.

Rational Root Theorem

EXAMPLE 5   Find the rational zeros p/q of the polynomial function

$$f(x) = 2x^4 + 3x^3 - 12x^2 - 15x + 10.$$

(This is the function in our previous discussion.)

SOLUTION We have already determined the candidates for both p and q. From these, we can determine potential rational roots of f by considering all possible combinations of p and q:

$$\frac{p}{q} = \frac{\pm 1, \pm 2, \pm 5, \pm 10}{\pm 1, \pm 2}$$

$$= -1, 1, -2, 2, -5, 5, -10, 10, -\frac{1}{2}, \frac{1}{2}, -\frac{5}{2}, \frac{5}{2}.$$

There are 12 potential rational zeros, not all of which can actually be zeros. At best, only four of these can be zeros, since f is of fourth degree.

Next, we use synthetic division to determine which (if any) are really zeros of f:

$$
\begin{array}{r|rrrrr}
1 & 2 & 3 & -12 & -15 & 10 \\
  &   & 2 & 5 & -7 & -22 \\
\hline
  & 2 & 5 & -7 & -22 & \boxed{-12} \Rightarrow 1 \text{ is not a zero of } f
\end{array}
$$

$$
\begin{array}{r|rrrrr}
-1 & 2 & 3 & -12 & -15 & 10 \\
   &   & -2 & -1 & 13 & 2 \\
\hline
   & 2 & 1 & -13 & -2 & \boxed{12} \Rightarrow -1 \text{ is not a zero of } f
\end{array}
$$

Similarly, 2, 5, and -5 also fail to be zeros of f. However, the potential zero -2 does prove to be a zero:

$$
\begin{array}{r|rrrrr}
-2 & 2 & 3 & -12 & -15 & 10 \\
   &   & -4 & 2 & 20 & -10 \\
\hline
   & 2 & -1 & -10 & 5 & \boxed{0} \Rightarrow -2 \text{ is a zero of } f.
\end{array}
$$

It follows that

$$f(x) = (x + 2)(2x^3 - x^2 - 10x + 5).$$

Any of the remaining zeros of f must also be zeros of the **reduced function**

$$2x^3 - x^2 - 10x + 5$$

so we perform the same procedure with the reduced function. The potential rational zeros of the reduced function are

$$\frac{p}{q} = \frac{\pm 1, \pm 2, \pm 5}{\pm 1, \pm 2}.$$

We need not check the potential zeros that previously failed. This leaves only -2, $\pm\frac{1}{2}$, and $\pm\frac{5}{2}$ to try. We find that $\frac{1}{2}$ is a zero:

$$
\begin{array}{r|rrrr}
\frac{1}{2} & 2 & -1 & -10 & 5 \\
            &   & 1 & 0 & -5 \\
\hline
            & 2 & 0 & -10 & \boxed{0} \Rightarrow \tfrac{1}{2} \text{ is a zero of the reduced function}
\end{array}
$$

and $f(x) = (x + 2)(x - \frac{1}{2})(2x^2 - 10)$. The zeros of our new reduced function can be found by solving the quadratic equation

$$2x^2 - 10 = 0$$
$$x^2 = 5$$
$$x = \sqrt{5} \quad \text{or} \quad x = -\sqrt{5}.$$

The zeros $\sqrt{5}$ and $-\sqrt{5}$ were not in our list of potential zeros since they are not rational numbers.

We can conclude that the rational zeros of f are -2 and $\frac{1}{2}$. We have discovered all four zeros; the linear factored form of f is

$$f(x) = 2(x + 2)(x - \tfrac{1}{2})(x - \sqrt{5})(x + \sqrt{5}).$$

EXAMPLE 6   Solve $x^3 - \frac{1}{3}x^2 - \frac{9}{2}x + \frac{3}{2} = 0$.

SOLUTION Since the rational root theorem applies only to equations and functions with integer coefficients, we will have to make an adjustment. Multiplying each side of the equation by 6 to clear the fractions, we get

$$6x^3 - 2x^2 - 27x + 9 = 0.$$

The potential rational roots are

$$\frac{p}{q} = \pm \frac{1, 3, 9}{1, 2, 3, 6}$$

$$= \pm 1, \pm 3, \pm 9, \pm\frac{1}{2}, \pm\frac{3}{2}, \pm\frac{9}{2}, \pm\frac{1}{3}, \pm\frac{1}{6}.$$

After checking a few of these potential rational roots using synthetic division, we finally discover that $\frac{1}{3}$ is a root and that our equation becomes

$$(x - \tfrac{1}{3})(6x^2 - 27) = 0.$$

Setting $6x^2 - 27 = 0$, we get the irrational roots $-3\sqrt{2}/2$ and $3\sqrt{2}/2$.
    The roots of the equation are $\frac{1}{3}$, $3\sqrt{2}/2$, and $-3\sqrt{2}/2$.

Our next task is to examine the complex zeros of a polynomial function with real coefficients. First, recall from our discussion in Section 1.7 that the **complex conjugate** of a complex number $z = a + bi$ is the complex number $\bar{z} = a - bi$ (recall that a line over a complex variable or an complex expression indicates the conjugate of that particular quantity). In particular, a real number is its own conjugate. Now suppose that the complex number $z = a + bi$ is a zero of a polynomial function P with real coefficients. Then,

$$P(x) = a_n x^n + a_{n-1} x^{n-1} + a_{n-2} x^{n-2} + \cdots + a_1 x + a_0.$$

Now, $P(z) = 0$, so $\overline{P(z)} = 0$ or

$$\overline{a_n z^n + a_{n-1} z^{n-1} + a_{n-2} z^{n-2} + \cdots + a_1 z + a_0} = 0$$

$$\overline{a_n z^n} + \overline{a_{n-1} z^{n-1}} + \overline{a_{n-2} z^{n-2}} + \cdots + \overline{a_1 z} + \overline{a_0} = 0$$

$$a_n (\bar{z})^n + a_{n-1}(\bar{z})^{n-1} + a_{n-2}(\bar{z})^{n-2} + \cdots + a_1(\bar{z}) + a_0 = 0$$

or

$$P(\bar{z}) = 0.$$

We have proven the following theorem for polynomial functions with real coefficients.

Conjugate Root Theorem

> If $P(x) = a_n x^n + a_{n-1} x^{n-1} + a_{n-2} x^{n-2} + \cdots + a_1 x + a_0$ is a polynomial function of degree $n > 0$ with real coefficients, and if $a + bi$ is a zero of P, then $a - bi$ is also a zero of P.

EXAMPLE 7   Find the zeros of

$$g(x) = 2x^4 - 8x^3 + 11x^2 - 6x - 35$$

given that $1 - 2i$ is a zero of g.

SOLUTION We need to find four zeros, not necessarily distinct. From the conjugate root theorem, the fact that $1 - 2i$ is a zero tells us that its conjugate $1 + 2i$ is a zero of g as well. Once we pare off these two zeros, the reduced function is of second degree, and it yields to the quadratic formula. So,

$$
\begin{array}{r|rrrrl}
1 - 2i & 2 & -8 & 11 & -6 & -35 \\
       &   & 2 - 4i & -14 + 8i & 13 + 14i & 35 \\
\hline
1 + 2i & 2 & -6 - 4i & -3 + 8i & 7 + 14i & \boxed{0} \Rightarrow 1 - 2i \text{ is a zero of } g \\
       &   & 2 + 4i & -4 + 8i & -7 - 14i & \\
\hline
       & 2 & -4 & -7 & & \boxed{0} \Rightarrow 1 + 2i \text{ is a zero of } g
\end{array}
$$

It follows that

$$g(x) = (x - (1 - 2i))(x - (1 + 2i))(2x^2 - 4x - 7).$$

The zeros of the reduced function are $1 + \frac{3}{2}\sqrt{2}$ and $1 - \frac{3}{2}\sqrt{2}$. The zeros of g are $1 - 2i$, $1 + 2i$, $1 + \frac{3}{2}\sqrt{2}$, and $1 - \frac{3}{2}\sqrt{2}$. The linear factored form of g is

$$g(x) = (x - (1 - 2i))(x - (1 + 2i))(x - (1 + \tfrac{3}{2}\sqrt{2}))(x - (1 - \tfrac{3}{2}\sqrt{2})).$$

EXAMPLE 8   Find g, a polynomial function (of minimal degree) with real coefficients, leading coefficient of 2, and with the following zeros: $-2 + i$ and 1 (multiplicity two).

SOLUTION Since $-2 + i$ is zero, then $-2 - i$ must also be a zero by the conjugate root theorem. The polynomial function that we are seeking is of degree four. So,

$$g(x) = 2(x - 1)^2(x - (-2 - i))(x - (-2 + i))$$

$$g(x) = 2(x^2 - 2x + 1)(x^2 + 4x + 5)$$

$$g(x) = 2x^4 + 4x^3 - 4x^2 - 12x + 10.$$

EXERCISE SET 3.3

A

In Problems 1 through 6, find the rational zeros, the real zeros, and the complex zeros of the given function. Write the function in linear factored form.

1. $P(x) = (x^2 - 7x + 12)(x^2 - 4x - 8)$

2. $Q(x) = (3x^2 + 8x + 1)(2x^2 - 11x + 5)$

3. $f(x) = (x^3 - 8)(x^2 + 4x + 4)$

4. $g(x) = (x^3 + 64)(x^2 - 8x + 16)$

5. $H(x) = (x^2 - 4)(x^4 - 7x^2 + 12)$

6. $G(x) = (x^2 - 36)(2x^4 + x^2 - 1)$

In Problems 7 through 12, determine the real polynomial function P that is described. Write your answer in the form

$$P(x) = a_n x^n + a_{n-1} x^{n-1} + a_{n-2} x^{n-2} + \cdots + a_1 x + a_0.$$

7. The leading coefficient of P is -2; degree is 3; zeros are 1, -3 (multiplicity 2).

8. The leading coefficient of P is 5; degree is 4; zeros are $\frac{1}{5}$, 2, -7 (multiplicity 2).

9. The leading coefficient of P is 1; the coefficients of P are real numbers; degree is 4; zeros are -4 (multiplicity 2), $1 + i\sqrt{3}$.

10. The leading coefficient of P is 2; the coefficients of P are real numbers; degree is 5; zeros are -1 (multiplicity 3), $2i$.

11. The zeros of P are 1, -3 (multiplicity 2); degree is 3; and $P(-4) = -5$.

12. The zeros of P are 3, 1 (multiplicity 3); degree is 4; and $P(2) = 4$.

In Problems 13 through 18, solve the given equation over a) the rational numbers, b) the real numbers, and c) the complex numbers.

13. $(2x^2 - 3)(x - 4)(3x - 8) = 0$

14. $(x^2 - 8)(7x - 4)(x + 7) = 0$

15. $(2x^2 - 4x + 5)(x^2 + 32) = 0$

16. $(x^2 + 12)(x^2 - 4x + 8) = 0$

17. $(x^2 + 12x + 20)(x^2 + 2x - 6)(x + 8) = 0$

18. $(x^3 + 64)(x^2 - 4x - 10)(2x - 5) = 0$

In Problems 19 through 24, list all the potential zeros of the function according to the rational root theorem and determine which of these are actually zeros of the function.

19. $P(x) = 2x^3 - 7x^2 + 13x - 5$

20. $P(x) = 2x^3 - 13x^2 + 17x - 3$

21. $P(x) = 2x^3 - 5x^2 + 5x - 6$

22. $P(x) = 3x^3 - 4x^2 - 6x + 8$

23. $P(x) = 2x^4 - 7x^2 + 3$

24. $P(x) = x^4 - x^3 - 7x^2 + 5x + 10$

B

In Problems 25 through 30, find all the complex zeros of the given function. You will need to use the rational root theorem. Rewrite the function in linear factored form.

25. $P(x) = x^3 + 6x^2 - x - 30$

26. $Q(x) = x^3 + 2x^2 - 41x - 42$

27. $g(x) = x^4 - 2x^3 - 38x^2 + 6x + 105$

28. $f(x) = 2x^4 - x^3 - 13x^2 + 5x + 15$

29. $R(x) = 2x^5 - 7x^4 - 44x^3 + 97x^2 + 204x - 396$

30. $q(x) = 2x^5 + 5x^4 - 70x^3 - 360x^2 - 540x - 189$

In Problems 31 through 36, sketch the graph of the polynomial function and label its intercepts.

31. $P(x) = x^3 + 6x^2 - x - 30$ *(Hint: See Problem 25.)*

32. $Q(x) = x^3 + 2x^2 - 41x - 42$ *(Hint: See Problem 26.)*

33. $g(x) = x^4 - 2x^3 - 38x^2 + 6x + 105$ *(Hint: See Problem 27.)*

34. $f(x) = 2x^4 - x^3 - 13x^2 + 5x + 15$ *(Hint: See Problem 28.)*

35. $R(x) = 2x^5 - 7x^4 - 44x^3 + 97x^2 + 204x - 396$
(Hint: See Problem 29.)

36. $q(x) = 2x^5 + 5x^4 - 70x^3 - 360x^2 - 540x - 189$
(Hint: See Problem 30.)

In Problems 37 through 42, find all the zeros of the polynomial function, given one or more zeros.

37. $P(x) = x^4 - 16x^2 - 64x - 64$; a zero is $-2 - 2i$.

38. $Q(x) = x^4 - 4x^3 + 4x^2 - 36$; a zero is $1 - i\sqrt{5}$.

39. $f(x) = x^5 - 2x^4 - 5x^3 + 26x^2 - 80$; a zero is $2 - 2i$.

40. $g(x) = x^5 - x^4 + 6x^2 - 4x - 8$; a zero is $1 + i\sqrt{3}$.

41. $H(x) = x^6 - 3x^5 - 4x^4 - x^2 + 3x + 4$; zeros are $-i$ and 1.

42. $F(x) = x^6 - 5x^5 - 8x^4 - 16x^2 + 80x + 128$; zeros are $2i$ and -2.

43. Find the complex cube roots of 8. *(Hint: These are the three zeros of the function $f(x) = x^3 - 8$; two of them are nonreal complex numbers.)*

44. Find the complex cube roots of 64. *(Hint: See Problem 43.)*

45. Find the complex cube roots of -27. *(Hint: See Problem 43.)*

46. Find the complex cube roots of 1. *(Hint: See Problem 43.)*

C

47. Suppose that c_1 and c_2 are the zeros of the quadratic function $f(x) = x^2 + ax + b$. Show that $c_1 + c_2 = -a$ and $c_1 c_2 = b$.

48. Suppose that c_1, c_2, and c_3 are the zeros of the cubic function $f(x) = x^3 + px^2 + qx + r$. Show that $c_1 + c_2 + c_3 = -p$, $c_1 c_2 + c_1 c_3 + c_2 c_3 = q$, and $c_1 c_2 c_3 = -r$.

49. Solve the equation $x^3 + 8x^2 - 3x - 24 = 0$ by using the fact that two of the roots r_1 and r_2 are such that $r_1 + r_2 = 0$. (Hint: Use the results of Problem 48 to find the third root.)

50. Solve the equation $x^3 - 6x^2 + 13x - 12 = 0$ by using the fact that two of the roots r_1 and r_2 are such that $r_1 r_2 = 4$. (Hint: Use the results of Problem 48 to find the third root.)

51. Given a polynomial function

$$P(x) = a_n x_n{}^n + a_{n-1} x^{n-1} + a_{n-2} x^{n-2} + \cdots + a_1 x + a_0$$

with linear factored form

$$P(x) = a(x - c_1)(x - c_2) \cdots (x - c_n)$$

show that $a_n = a$.

SECTION 3.4 RATIONAL FUNCTIONS

In our discussion in Section 2.6, we saw that the quotient of two functions f and g determined another function f/g. If the two functions are polynomial functions, then the function formed by their quotient is called a **rational function.** The relation between rational functions and polynomial functions is very much like the relation between rational numbers and integers; just as a rational number is a ratio of two integers, a rational function is a ratio of two polynomial functions.

> A **rational function** is a function that can be expressed as the ratio of two polynomial functions. That is, $Q(x)$ is a rational function if there exist polynomial functions $N(x)$ and $D(x)$ such that
>
> $$Q(x) = \frac{N(x)}{D(x)}.$$

Rational Function

The domain of $Q(x)$ is the set of all x such that the denominator $D(x)$ is not zero (otherwise the ratio would be undefined). The zeros of $Q(x)$ are those zeros of the numerator $N(x)$ that are in the domain of $Q(x)$ since a ratio is zero if and only if its numerator is zero.

If the degree of the numerator $N(x)$ is less than the degree of the denominator $D(x)$, then $Q(x)$ is a **proper** rational function. If, on the other hand, the degree of the numerator $N(x)$ is larger than or equal to the degree of denominator $D(x)$, then $Q(x)$ is an **improper** rational function.

Just as an improper fraction can be rewritten as the sum of an integer and a proper fraction (for example, $\frac{17}{5} = 3 + \frac{2}{5}$), we can rewrite an improper rational function as the sum of a polynomial function and a proper rational function.

EXAMPLE 1   Let $f(x) = \dfrac{2x^2 + 3x + 1}{x - 4}$.

a) Determine the domain of f.

b) Determine the zeros of f.

c) Rewrite f as the sum of a polynomial function and a proper rational function.

SOLUTION

a) To determine the domain of f, we need to find the zeros of the denominator $x - 4$. Since 4 is the zero of the denominator, the domain of f is the set of all real numbers except 4.

b) To determine the zeros of f, we need to find the zeros of the numerator $2x^2 + 3x + 1$. So,

$$2x^2 + 3x + 1 = 0$$

$$(2x + 1)(x + 1) = 0$$

$$2x + 1 = 0 \quad \text{or} \quad x + 1 = 0$$

$$x = -\tfrac{1}{2} \qquad x = -1.$$

The zeros of f are $-\tfrac{1}{2}$ and -1.

c) We start by performing polynomial division. Since the divisor is of the form $x - c$, we can use synthetic division:

$$
\begin{array}{r|rrr}
4 & 2 & 3 & 1 \\
  &   & 8 & 44 \\
\hline
  & 2 & 11 & 45
\end{array}
$$

Interpreting this, we get

$$f(x) = (2x + 11) + \frac{45}{x - 4}.$$

Just as a rational number can be written with no common factors in the numerator and denominator, a rational function can also be **reduced to lowest terms.**

EXAMPLE 2   Let

$$g(x) = \frac{2x^2 - 10x + 12}{x - 3}.$$

Determine the domain of g and the zeros of g. Sketch the graph of g.

SOLUTION The domain of g is the set of all real numbers except 3. Factoring the numerator of g, we get

$$g(x) = \frac{2(x - 3)(x - 2)}{x - 3}.$$

Since there is a factor $(x - 3)$ in both the numerator and the denominator, we can reduce this fraction:

$$g(x) = \frac{2(x - 3)(x - 2)}{x - 3} = 2(x - 2) = 2x - 4.$$

For all values of x except for 3, $g(x) = 2x - 4$. At $x = 3$, the function g is undefined. The graph of g is the same as the graph of $y = 2x - 4$ except that there is a break in the graph at $x = 3$. The open circle on the graph of g (Figure 24) shows this break in the graph. The zero is $x = 2$.

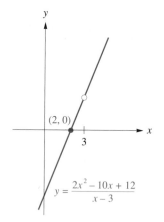

FIGURE 24

In the next two examples, we examine rational functions of the form $f(x) = 1/x^n$. This will give us a foothold on the more formidable task of sketching the graph of a rational function in general.

EXAMPLE 3   Sketch graphs of $f_1(x) = 1/x$ and $f_3(x) = 1/x^3$.

SOLUTION The domain of each of these functions is the set of nonzero real numbers. Each of these is an odd function:

$$f_1(-x) = \frac{1}{(-x)} = -\frac{1}{x} = -f_1(x)$$

$$f_3(-x) = \frac{1}{(-x)^3} = -\frac{1}{x^3} = -f_3(x).$$

We need only plot points for $x > 0$, sketch the graph for $x > 0$, and then reflect that graph through the origin to obtain the complete graph. We start with a table of values.

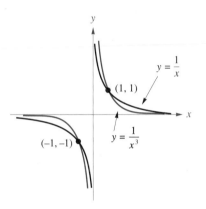

x	0.01	0.1	0.5	1	2	5	10	100	1000
$f_1(x)$	100	10	2	1	$\frac{1}{2}$	$\frac{1}{5}$	$\frac{1}{10}$	$\frac{1}{100}$	$\frac{1}{1000}$
$f_3(x)$	10^6	1000	8	1	$\frac{1}{8}$	$\frac{1}{125}$	$\frac{1}{1000}$	10^{-6}	10^{-9}

The graphs of these rational functions are shown in Figure 25.

FIGURE 25

EXAMPLE 4   Sketch the graphs of $f_2(x) = 1/x^2$ and $f_4(x) = 1/x^4$.

SOLUTION Again, we generate a table of values. Each of these functions is even, so for each we only need to sketch the graph for $x > 0$ and reflect it through the y-axis to obtain the complete graph.

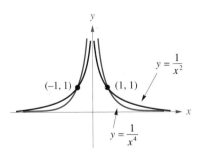

x	0.01	0.1	0.5	1	2	5	10	100	1000
$f_2(x)$	10,000	100	4	1	$\frac{1}{4}$	$\frac{1}{25}$	$\frac{1}{100}$	$\frac{1}{10,000}$	10^{-6}
$f_4(x)$	10^8	10,000	16	1	$\frac{1}{16}$	$\frac{1}{625}$	$\frac{1}{10,000}$	10^{-8}	10^{-12}

The graphs of these rational functions are shown in Figure 26.

FIGURE 26

We can generalize about the behavior of the graphs of $f(x) = 1/x^n$. For example, as $x \to +\infty$, $y \to 0^+$. (We have introduced some new notation here; we write $t \to a^+$ to indicate that t approaches a from the positive direction; if t were to approach a from the negative side, we would write $t \to a^-$.) Also, larger values for n cause the graph to decrease more rapidly. Furthermore, as $x \to 0^+$, $y \to +\infty$. The graph includes the point $(1, 1)$. We can complete the graphs by reflection as we did in each of the examples.

The graphs of many rational functions can be sketched by translating, reflecting, compressing, or expanding the graph of $f(x) = 1/x^n$ for some n.

EXAMPLE 5   Sketch the graph of

$$f(x) = \frac{3x + 11}{x + 4}.$$

SOLUTION Since f is an improper rational function (the degree of the numerator and the degree of the denominator are both one), we can perform the division and rewrite the function as

$$f(x) = 3 - \frac{1}{x + 4}.$$

This tells us that the graph of f is a reflection and a translation of the graph of $y = 1/x$. The line $x = -4$ plays the same role for the graph of f as the y-axis plays for the graph of $y = 1/x$. Likewise, the line $y = 3$ plays the same role for the graph of f as the x-axis plays for the graph of $y = 1/x$.

The zero of the numerator is $-\frac{11}{3}$, so the x-intercept is $(-\frac{11}{3}, 0)$. The graphs of both f and $y = 1/x$ are shown (Figure 27). The lines $x = -4$ and $y = 3$ are shown as dotted lines.

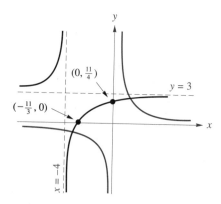

FIGURE 27

In general, sketching the graph of a rational function requires more thought and investigation than sketching the graph of a polynomial function. Rational functions, like polynomial functions, are smooth, without corners and abrupt changes of direction. Unlike polynomial functions, however, they may have breaks (called **discontinuities**), which occur at the zeros of the denominator. A zero of the denominator causes a hole in the graph at the zero if it is also a zero of the numerator (as in Example 2).

A zero of the denominator can also cause the graph to grow without bound and approach a vertical line (such as the line $x = -4$ in Example 5). To see this, suppose that x approaches a zero of the denominator, say $x = a$. As $x \to a$ from either side, the denominator tends toward a small number. This causes the ratio to grow without bound. The vertical line $x = a$ is called a **vertical asymptote.**

The graphs of rational functions also differ from the graphs of polynomial functions at the extreme values of x. Instead of $y \to \pm\infty$ as $x \to \pm\infty$, the value of y may approach a finite value b. This causes the graph of the function to approach a horizontal line (such as the line $y = 3$ in Example 5). The line $y = b$ is called a **horizontal asymptote.**

Suppose that $f(x) = N(x)/D(x)$ is a rational function:

a) If f is a proper rational function (that is, the degree of N is less than the degree of D), then $y = 0$ is the horizontal asymptote of f.

b) If the degree of N is equal to the degree of D, then

$$f(x) = q + \frac{R(x)}{D(x)}$$

where q is a real number, and R is a polynomial function with degree less than the degree of D (this is a result of the discussion in Section 3.2). Then $y = q$ is a horizontal asymptote. It can be shown that q is the ratio of the leading coefficient of N and the leading coefficient of D.

c) If the degree of N is greater than the degree of D, then

$$f(x) = Q(x) + \frac{R(x)}{D(x)}$$

where $Q(x)$ is a nonconstant polynomial, and R is a polynomial function with degree less than the degree of D. Then f has no horizontal asymptote. The behavior of f at the extreme values of x is the same as the behavior of $Q(x)$ at the extreme values of x.

Finding the Horizontal Asymptotes of a Rational Function

Trying to sketch the graph of a rational function by plotting points is tedious at best. Even with a large number of points, it is difficult to determine the graph, and this is especially true near the horizontal and vertical asymptotes. We offer these guidelines for sketching the graph of a rational function.

Sketching the Graph of a
Rational Function

To sketch the graph of $f(x) = N(x)/D(x)$:

1) Factor the $N(x)$ and $D(x)$ into real polynomial factors, if possible. Only those values of x for which $D(x) \neq 0$ are in the domain of $f(x)$. Reduce $f(x)$ to lowest terms.

2) Use the remaining zeros of the denominator to determine the vertical asymptotes of the graph. Use the remaining zeros of the numerator to determine the x-intercepts of the graph.

3) Determine the horizontal asymptotes as discussed before.

4) Determine the intervals over which the function is positive and the intervals over which the function is negative. Use the zeros of the denominator and the numerator as break points.

5) Take advantage of symmetry if $f(x)$ is an odd or even function.

EXAMPLE 6   Sketch the graph of $H(x) = \dfrac{x}{x^2 - 16}$.

SOLUTION We rewrite

$$H(x) = \frac{x}{(x - 4)(x + 4)}.$$

The domain of H includes all real numbers except for $x = \pm 4$. The function H is already in lowest terms. The vertical asymptotes are therefore $x = 4$ and $x = -4$. Since H is a proper rational function, the horizontal asymptote is $y = 0$. If we examine the individual factors of both the numerator and the denominator, we can construct a sign graph for the function (such as we did to solve rational inequalities in Section 1.5). Once we know, for example, that $H(x)$ must be positive on the interval $(-4, 0)$, we know that $y \to \infty$ as $x \to -4^+$ (why?). In a similar manner,

$$y \to -\infty \quad \text{as} \quad x \to -4^-$$

$$y \to +\infty \quad \text{as} \quad x \to 4^+$$

$$y \to -\infty \quad \text{as} \quad x \to 4^-.$$

Also, considering the horizontal asymptote $y = 0$:

$$y \to 0^+ \quad \text{as} \quad x \to +\infty$$

$$y \to 0^- \quad \text{as} \quad x \to -\infty.$$

All of this information is compiled in Figure 28. As a bonus, we make note that H is an odd function and that the graph of H is symmetric with respect to the origin.

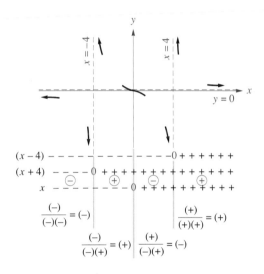

FIGURE 28

We sketch three smooth, continuous curves, separated by the vertical asymptotes, according to our investigation (Figure 29).

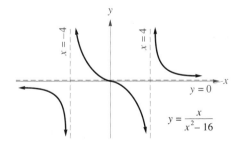

FIGURE 29

EXAMPLE 7 Sketch the graph of

$$f(x) = \frac{(x^2 - 36)(x^2 - 3)}{(x + 5)(x - 3)^2(x - 6)}.$$

SOLUTION First we factor over the real numbers and reduce f to lowest terms:

$$f(x) = \frac{(x - 6)(x + 6)(x - \sqrt{3})(x + \sqrt{3})}{(x + 5)(x - 3)^2(x - 6)}$$

$$= \frac{(x + 6)(x - \sqrt{3})(x + \sqrt{3})}{(x + 5)(x - 3)^2}.$$

The graph of f has a hole at $x = 6$ and vertical asymptotes at $x = -5$ and $x = 3$. The x-intercepts are $(-6, 0)$, $(-\sqrt{3}, 0)$ and $(\sqrt{3}, 0)$.

Doing the long division to determine the horizontal asymptote looks like quite a task. However, inspection tells us that the degree of the numera-

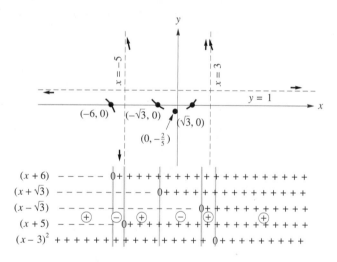

FIGURE 30

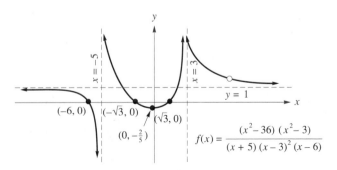

FIGURE 31

tor and the degree of the denominator are both four and that the leading coefficients are both one. From this, we get a horizontal asymptote at $y = 1$.

Next we construct a sign graph for the function and examine the behavior of f about the asymptotes and the intercepts. The results of our research are shown in Figure 30.

Now we sketch four smooth, continuous curves, separated by the vertical asymptotes and the discontinuity at $x = 6$. The hole occurs at $x = 6$ since it was not in our original domain (Figure 31).

EXAMPLE 8 Sketch the graph of $G(x) = \dfrac{x^3}{2x^2 + 4}$.

SOLUTION The numerator of G has a zero (of multiplicity three) at $x = 0$. The denominator has no real zeros, so the graph of G has no vertical asymptotes; its domain is the set of all real numbers. The function is an odd function [you should verify that $G(-x) = -G(x)$]. Since the denominator is always positive, it should seem reasonable that $G(x) > 0$ for $x > 0$ and that $G(x) < 0$ for $x < 0$. Because G is an improper rational function, we perform the polynomial division and arrive at

$$G(x) = \frac{1}{2}x - \frac{x}{x^2 + 2}.$$

There is no horizontal asymptote. As $x \to \infty$, the proper fraction

$$\frac{x}{x^2 + 2}$$

tends to zero. This means that as $x \to \infty$, G behaves just as the linear function $Q(x) = \frac{1}{2}x$ behaves. In fact, for $x > 0$, $G(x) < \frac{1}{2}x$ and, for $x < 0$, $G(x) > \frac{1}{2}x$ (why?). The graph of $y = \frac{1}{2}x$ is shown as a dotted line in Figure 32.

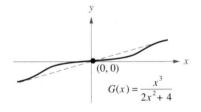

$$G(x) = \frac{x^3}{2x^2 + 4}$$

FIGURE 32

The line $y = \frac{1}{2}x$ acts very similar to a horizontal asymptote except that, of course, it is not horizontal. This type of situation occurs whenever the quotient of the polynomial division is a linear function. We call such a line an **oblique asymptote**. The function in the next example also has such an asymptote.

EXAMPLE 9 Sketch the graph of $f(x) = \dfrac{x^2 - 9}{x - 2}$.

SOLUTION The denominator has one zero at $x = 2$, so the graph of f has a vertical asymptote at $x = 2$. The numerator of f has zeros $x = \pm 3$, so

FIGURE 33

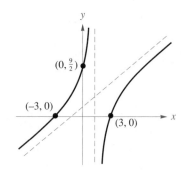

the graph of f has x-intercepts at $x = 3$ and $x = -3$. Performing the long division

$$f(x) = (x + 2) - \frac{5}{x - 2}$$

we get the oblique asymptote $y = x + 2$. You should finish the steps to complete the graph (Figure 33).

EXERCISE SET 3.4

A

In Problems 1 through 6, determine the domain and the real zeros of the rational function.

1. $P(x) = \dfrac{(x - 8)(x + 3)}{(x - 3)(x + 6)(x + 1)}$

2. $Q(x) = \dfrac{(x - 5)(x + 7)}{(x - 1)(x + 2)(x + 4)}$

3. $H(x) = \dfrac{x^2 + 3x - 40}{x^2 - 9x + 20}$

4. $R(x) = \dfrac{x^2 - 4x - 12}{x^2 - 5x - 14}$

5. $f(x) = \dfrac{x^3 - 1}{x^2 - 2x - 8}$

6. $g(x) = \dfrac{x^3 - 8}{x^2 + 4x - 12}$

In Problems 7 through 12, reduce the rational function and sketch its graph.

7. $r(x) = \dfrac{(x - 3)(x + 5)}{(x - 3)}$

8. $g(x) = \dfrac{(x + 1)(x - 2)}{(x - 2)}$

9. $f(x) = \dfrac{x^2 - 25}{x + 5}$

10. $f(x) = \dfrac{x^2 - 9}{x - 3}$

11. $P(x) = \dfrac{(x + 4)^2(x - 3)}{x^2 + x - 12}$

12. $Q(x) = \dfrac{(x + 2)(x - 1)^2}{x^2 + x - 2}$

In Problems 13 through 18, sketch the graphs of the two equations on the same coordinate plane. Notice that the first is of the form $y = 1/x^n$ and the second is a translation, reflection, expansion, or compression of the first. Label the asymptotes and intercepts of the second.

13. a) $y = \dfrac{1}{x^2}$ **b)** $y = -\dfrac{2}{x^2}$

14. a) $y = \dfrac{1}{x^3}$ **b)** $y = -\dfrac{4}{x^3}$

15. a) $y = \dfrac{1}{x^2}$ **b)** $y = 2 - \dfrac{1}{(x - 3)^2}$

16. a) $y = \dfrac{1}{x^3}$ **b)** $y = 2 + \dfrac{1}{(x + 1)^3}$

17. a) $y = \dfrac{1}{x}$ **b)** $y = \dfrac{2x - 3}{x - 1}$

18. a) $y = \dfrac{1}{x}$ **b)** $y = \dfrac{3x - 2}{x - 1}$

In Problems 19 through 24, select from functions I through VI the one that correctly describes the given graph.

I $y = \dfrac{x}{2 - x}$ **II** $y = \dfrac{x}{x + 2}$ **III** $y = \dfrac{x}{x^2 - 4}$

IV $y = \dfrac{2}{x(x - 2)}$ **V** $y = \dfrac{x^2}{x^2 - 4}$ **VI** $y = \dfrac{4}{x - 2}$

19.

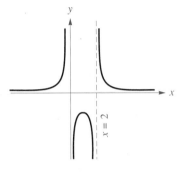

20.

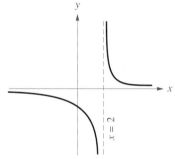

21.

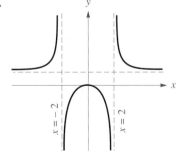

22.

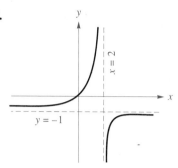

23.

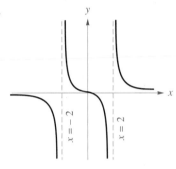

24.

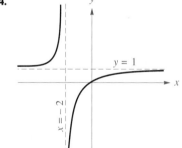

In Problems 25 through 36, sketch the graphs of the rational function. Label intercepts and asymptotes.

25. $g(x) = \dfrac{x}{x^2 - 16}$ **26.** $f(x) = \dfrac{x - 2}{x^2 - 25}$

27. $H(x) = \dfrac{x^2 + 3x - 40}{x^2 - x - 20}$ **28.** $R(x) = \dfrac{x^2 + 4x - 12}{x^2 - 5x - 14}$

29. $P(x) = \dfrac{2x^2}{x^2 - 9}$ **30.** $G(x) = \dfrac{2x^2 - 8}{x^2 + 5x}$

31. $f(x) = \dfrac{x - 7}{x^2 - 16}$ **32.** $H(x) = \dfrac{x + 8}{x^2 - 4}$

33. $Q(x) = \dfrac{x(x-4)}{(x+3)(x-7)}$

34. $P(x) = \dfrac{2x(x-2)}{(x+1)(x-8)}$

35. $f(x) = \dfrac{x+3}{x^2 - 5x + 4}$

36. $g(x) = \dfrac{x+3}{x^2 - 3x - 10}$

B

In Problems 37 through 48, sketch the graphs of the rational function. Label the intercepts and the asymptotes.

37. $f(x) = \dfrac{(x-4)^2(x+1)}{3(x-2)^2(x+3)}$

38. $f(x) = \dfrac{2x(x+1)^2}{(x+5)^2(x-1)}$

39. $G(x) = \dfrac{x^2 - 16}{x^2 + 4}$

40. $p(x) = \dfrac{2x^2 - 18}{x^2 + 25}$

41. $F(x) = \dfrac{x^2 - 9}{x - 6}$

42. $G(x) = \dfrac{x^2 - 4}{x + 8}$

43. $v(x) = \dfrac{x^3 - 4x}{x^2 + 1}$

44. $H(x) = \dfrac{x^3 - 9x}{2x^2 + 8}$

45. $W(x) = \dfrac{x^3 - 9x}{x^2 - 16}$

46. $D(x) = \dfrac{x^3 - 25x}{2x^2 - 18}$

47. $P(x) = \dfrac{x^3 - 1}{x^2 - 9}$

48. $Q(x) = \dfrac{x^3 + 1}{x^2 - 16}$

49. Use the graph of Problem 25 to solve the inequality

$$\frac{x}{x^2 - 16} < 0.$$

50. Use the graph of Problem 29 to solve the inequality

$$\frac{2x^2}{x^2 - 9} > 0.$$

51. Use the graph of Problem 35 to solve the inequality

$$\frac{x+3}{x^2 - 5x + 4} > 0.$$

52. Use the graph of Problem 39 to solve the inequality

$$\frac{x^2 - 16}{x^2 + 4} \leq 0.$$

53. Use the graph of Problem 41 to solve the inequality

$$\frac{x^2 - 9}{x - 6} \geq 0.$$

54. Use the graph of Problem 45 to solve the inequality

$$\frac{x^3 - 9x}{x^2 - 16} \leq 0.$$

55. The function in Example 8,

$$G(x) = \frac{x^3}{2x^2 + 4},$$

was shown to be the same as

$$G(x) = \frac{1}{2}x - \frac{x}{x^2 + 2}$$

by polynomial division. Sketch

$$y = \tfrac{1}{2}x \quad \text{and} \quad y = \frac{x}{x^2 + 2}$$

on the same coordinate plane. Use the technique of adding y-coordinates to arrive at the graph of $G(x)$.

56. The function in Example 9,

$$f(x) = \frac{x^2 - 9}{x - 2},$$

was shown to be the same as

$$f(x) = x + 2 - \frac{5}{x - 2}$$

by polynomial division. Sketch

$$y = x + 2 \quad \text{and} \quad y = \frac{-5}{x - 2}$$

on the same coordinate plane. Use the technique of adding y-coordinates to arrive at the graph of $G(x)$.

57. Rewrite the rational function

$$f(x) = \frac{x^4 + 1}{2x^2}$$

as the sum of a polynomial function P and a proper rational function Q. Sketch the graph of $y = f(x)$, $y = P(x)$, and $y = Q(x)$ on the same coordinate plane. Describe the behavior of the graph of f at the extremes and near its vertical asymptote relative to the other graphs.

58. Rewrite the rational function

$$f(x) = \frac{x^3 + 3}{4x}$$

as the sum of a polynomial function and a proper rational function. Sketch the graph of f, the polynomial function, and the proper rational function on the same coordinate plane. Describe the behavior of the graph of f at the extremes and near its vertical asymptote, relative to the other graphs.

C

59. Sketch the graph of the rational function

$$Q(x) = \frac{2(x - 1)(x - 2)}{x^2}.$$

Notice that the graph crosses the line $y = 2$, its horizontal asymptote. The point at which it crosses the asymptote can be determined by setting $Q(x) = 2$ and solving this equation for x. Do this to find the point of intersection.

60. In general it is difficult to find a turning point of a rational function without calculus. However, finding a turning point is relatively easy in certain instances. Sketch the graph of the rational function

$$Q(x) = \frac{2(x - 1)(x - 2)}{x^2}.$$

Notice that the graph has a turning point between the x-intercepts, $x = 1$ and $x = 2$. Suppose that we also add the horizontal line $y = k$ ($k \le 0$) to the sketch. If this line is above the turning point, it intersects the graph at two points. If this line is below the turning point, it does not intersect the graph. If this line passes through the turning point, it intersects the graph at one point. This means that if we set $Q(x) = k$ and determine the value of k for which this equation has exactly one root, we can discover the turning point. Do this to find the turning point.

GRAPHING ALGEBRAIC FUNCTIONS S E C T I O N 3.5

Many of the applications and examples encountered in calculus involve a set of functions called algebraic functions. Loosely speaking, an **algebraic function** is a function that involves only the operations of addition, subtraction, multiplication, division, and taking roots. In fact, almost all the functions we have discussed up to this point have been algebraic.

Polynomial functions and rational functions certainly are algebraic functions. Also, a function such as

$$f(x) = x - 2\sqrt{x}$$

qualifies as an algebraic function, even though it is neither a polynomial function nor a rational function.

EXAMPLE 1 Determine the domain of the function:

a) $f(x) = \dfrac{\sqrt{x + 2}}{x - 4}$ **b)** $g(x) = -3(x - 2)^{3/4}$

SOLUTION

a) The domain of the function f is such that the radicand $x + 2$ is nonnegative and the denominator $x - 4$ is nonzero. Using interval notation, the intersection of these two sets is

$$[-2, 4) \quad \text{or} \quad (4, +\infty).$$

b) Think of the function as $g(x) = -3\sqrt[4]{(x-2)^3}$. Since this is a radical with an even index, the domain is such that the radicand is nonnegative. The domain is $[2, +\infty)$.

Many of the sketching techniques, skills, and tricks that we acquired in our previous investigations of polynomial functions and rational functions will apply to our task at hand. We will also acquire some new methods of discovering these graphs. Keep in mind as you examine these sketches that we are only concerned with obtaining the big picture of the function.

What we will not do is plot a set of points and play "connect-the-dots." This is an inefficient and inelegant method of sketching, which we hope that you now use only as a last resort. We may, however, plot a few selected points to finish off a sketch.

We start with an examination of graphs of functions of the type $f(x) = x^{p/q}$, where p and q are positive integers. We already have investigated some functions of this type; for example, $f(x) = x$, $f(x) = x^2$, and $f(x) = \sqrt{x}$ are all members of this set, and we know their graphs (Figure 34) from our previous investigations.

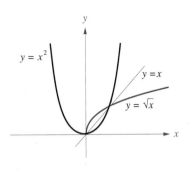

FIGURE 34

EXAMPLE 2 Sketch the graph of the following:

a) $y = x^{2/3}$
b) $y = x^{3/2}$

SOLUTION

a) The domain of the function $f(x) = x^{2/3}$ is the set of all real numbers. Since it is an even function, we need only sketch the graph for $x > 0$ and reflect it through the y-axis. The graph can be sketched easily if we use the graphs of $y = \sqrt{x}$ and $y = x$ as guides:

$$0 < x < 1 \Rightarrow x < x^{2/3} < \sqrt{x},$$
$$x = 1 \Rightarrow x = x^{2/3} = \sqrt{x}$$
$$x > 1 \Rightarrow \sqrt{x} < x^{2/3} < x.$$

Our results are shown (Figure 35).

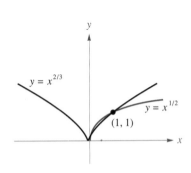

FIGURE 35

b) The domain of the function $f(x) = x^{3/2}$ is the set of nonnegative real numbers, so we need only sketch the graph for $x \geq 0$. The graph can be sketched easily if we use the graphs of $y = x$ and $y = x^2$ as guides:

$$0 < x < 1 \Rightarrow x^2 < x^{3/2} < x$$

$$x = 1 \Rightarrow x = x^{3/2} = x^2$$

$$x > 1 \Rightarrow x < x^{3/2} < x^2$$

Our results are shown in Figure 36.

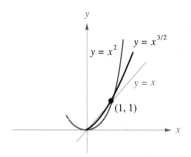

FIGURE 36

From Example 2, we can generalize about the overall shapes of the graphs of $f(x) = x^{p/q}$ for $x > 0$. If $p/q < 1$, then the graph curves down (as $y = \sqrt{x}$ does); such a graph is called **concave down** (Figure 37). If, on the other hand, $p/q > 1$, then the graph curves up (as $y = x^2$ does); this curve is **concave up.** Although you must wait until calculus to discuss the formal definitions of these terms, we can use them in an informal sense to describe the graphs of functions.

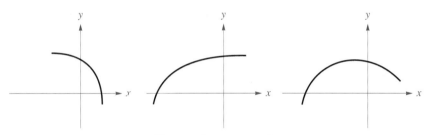

These graphs are concave down

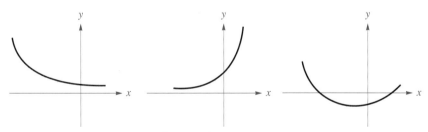

These graphs are concave up

FIGURE 37

We can now make some observations about the graphs of $f(x) = x^{p/q}$. Each of these functions is an increasing function for $x > 0$, and each passes through the point $(0, 0)$ and $(1, 1)$. The following box summarizes the various possibilities for the shape of the graph, depending on the values of p and q.

Sketching the Graph of $y = x^{p/q}$

To sketch the graph of $f(x) = x^{p/q}$ (assume that p and q have no common factors):

1)   Sketch $y = f(x)$ for $x \geq 0$.

If $p > q$, then $y = f(x)$ is concave up   If $p < q$, then $y = f(x)$ is concave down

2)   If q is even, then the domain of f is $x \geq 0$. The graph is complete, as in part b) of Example 2. If q is odd, then the domain of f is the set of all real numbers. To complete the graph in this case, we can exploit the symmetry of the graph. If p is an even integer, then f is an even function; we complete the graph by reflecting about the y-axis, as we did in part a) of Example 2. On the other hand, if p is an odd integer, then f is an odd function; we complete the graph by reflecting about the origin.

For $p > q$:

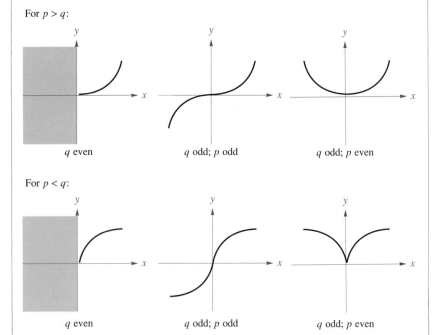

EXAMPLE 3   Sketch the graphs of the following:

a) $g(x) = x^{3/5}$ **b)** $h(x) = -2(x + 3)^{3/5} + 2$

Determine the intercepts.

SOLUTION

a) Since $\frac{3}{5} < 1$, the graph of g for $x > 0$ is concave down. Also, since the denominator 5 is odd, the domain is the set of all real numbers. Observing that the numerator 3 is also odd, we complete the graph by reflecting through the origin (Figure 38).

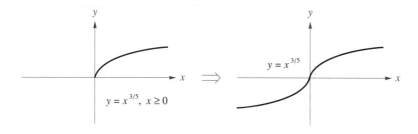

FIGURE 38

b) The graph of $h(x)$ is a reflection, expansion, and translation of the graph of $y = x^{3/5}$. To arrive at the graph of h, we expand the graph of $y = x^{3/5}$ by a factor of 2 from the x-axis, reflect it through the x-axis, and translate -3 units horizontally and 2 units vertically.

To find the y-intercept (Figure 39):

$$h(0) = -2(0 + 3)^{3/5} + 2 = -2(3)^{3/5} + 2.$$

To the nearest 0.0001, this is $y = -1.8664$.
To find the x-intercept:

$$0 = -2(x + 3)^{3/5} + 2 \Rightarrow x = -2.$$

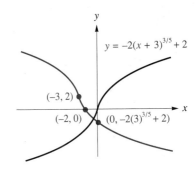

FIGURE 39

EXAMPLE 4   Sketch the graph of $h(x) = \frac{1}{2}x + x^{2/3}$.

SOLUTION Since we know the graph of the linear function $y = \frac{1}{2}x$ and the graph of $y = x^{2/3}$ (this is the function discussed in Example 2 of this section), we can sketch the graph of h by adding y-coordinates (Figure 40).

We leave it to you to verify that the x-intercepts are $(-8, 0)$ and $(0, 0)$.

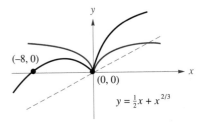

FIGURE 40

Quite often, the algebraic function of interest is the square root of a polynomial or rational function. The next two examples deal with this sort of function.

EXAMPLE 5   Sketch the graph of $g(x) = \sqrt{x^2 - 6x - 7}$.

SOLUTION The function is the square root of a quadratic function. The graph of $y = x^2 - 6x - 7$ is shown in Figure 41. Since the radicand must be non-negative, the domain of g is precisely the set of values of x for which the graph of $y = x^2 - 6x - 7$ is above the x-axis (this excluded region is shaded). The x-intercepts of g are the same as the x-intercepts of $y = x^2 - 6x - 7$.

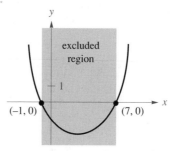

FIGURE 41

On the intervals over which the value of $x^2 - 6x - 7$ is between 0 and 1, the graph of g is above the graph of $y = x^2 - 6x - 7$ (why is this true?). Likewise, on the intervals over which the value of $x^2 - 6x - 7$ is greater than 1, the graph of g is below the graph of $y = x^2 - 6x - 7$. Figure 42 shows these results.

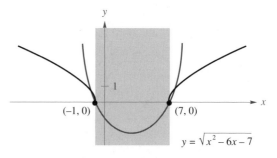

$$y = \sqrt{x^2 - 6x - 7}$$

FIGURE 42

In general, if the graph of $y = f(x)$ is one we know, we can use it to discover the graph of $y = \sqrt{f(x)}$. What follows is a description of how this can be accomplished.

To sketch the graph of $y = \sqrt{f(x)}$:

1) Sketch the graph of $y = f(x)$. Label the x-intercepts. These x-intercepts are also the x-intercepts of the graph of $y = \sqrt{f(x)}$.

2) Determine the domain of $y = \sqrt{f(x)}$ by examining the graph of $y = f(x)$. Include only those values of x for which the graph of $y = f(x)$ is above or on the x-axis.

3) Sketch the graph of $y = \sqrt{f(x)}$, using $y = f(x)$ as a guide:

$$0 < f(x) < 1 \Rightarrow \sqrt{f(x)} > f(x)$$

$\Rightarrow$ the graph of $y = \sqrt{f(x)}$ lies *above* the graph of $y = f(x)$

$$f(x) = 1 \Rightarrow \sqrt{f(x)} = f(x)$$

$\Rightarrow$ the graph of $y = \sqrt{f(x)}$ *intersects* the graph of $y = f(x)$

$$f(x) > 1 \Rightarrow \sqrt{f(x)} < f(x)$$

$\Rightarrow$ the graph of $y = \sqrt{f(x)}$ lies *below* the graph of $y = f(x)$.

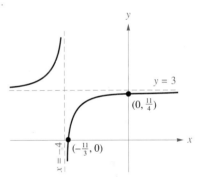

FIGURE 43

EXAMPLE 6   Sketch the graph of

$$f(x) = \sqrt{\frac{3x + 11}{x + 4}}.$$

SOLUTION   The function is the square root of a function we sketched in Example 5 in Section 3.4 (pause here and refer to this example). The graph of

$$y = \frac{3x + 11}{x + 4}$$

is shown in Figure 43. We apply the preceding steps to arrive at the graph of f. Notice that the graph of f has a horizontal asymptote at $y = \sqrt{3}$. To see this, rewrite f as

$$f(x) = \sqrt{3 - \frac{1}{x + 4}}.$$

As $x \to \infty$, the value of $1/(x + 4)$ approaches 0. The radicand approaches 3, and so the radical approaches $\sqrt{3}$ (Figure 44).

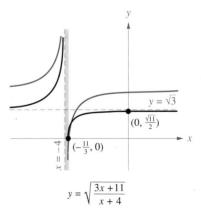

$$y = \sqrt{\frac{3x + 11}{x + 4}}$$

FIGURE 44

Evariste Galois

The problem of finding solutions to polynomial equations occupied the great mathematical minds of the 18th and 19th centuries. While many specific types of these equations were conquered by some, the basic question of which equations can be solved remained to be answered conclusively by Evariste Galois during this short, tragic life.

Born in 1811 in a village near Paris, Evariste was educated at home by his mother in his early years. It is obvious now that he was gifted; but when he entered school at the age of twelve, he neglected his studies of rhetoric, history, and Latin, which he found boring, and devoted his time to his passion, mathematics. Because of this, his teachers thought Evariste to be at best an average student.

Not discouraged, Galois applied to the École Polytechnique, the great school of French mathematics, science, and engineering. Unfortunately, he was rejected after his oral examination; he knew the answers, but his examiners could not understand his explanations. He reapplied later but was refused again after striking an examiner with a chalk eraser.

At the age of 17, Galois submitted a paper of his original work to the French Academy of Science. The paper so impressed Augustin-Louis Cauchy, the premiere French mathematician of the age, that he agreed to present it to a meeting of the society. However, Cauchy either forgot or he lost the paper. Galois submitted another paper two years later, but the Academy's secretary, Joseph Fourier, died shortly after he accepted it from Galois. The paper was never found again. In 1831, Galois submitted yet another paper, but this one was not given serious consideration and was returned as "incomprehensible." Had any of these three papers been given its proper due, Galois would have been properly recognized as a mathematical genius.

Rebuked by the academic community, Evariste turned to politics. He spent the next year and a half in jail due to his outspoken republican views. Later, Galois fell in love with the mistress of an aristocrat who challenged him to a pistol duel. His adversary was a trained marksman, and Evariste knew that he was outmatched. Galois spent the entire night before the duel feverishly scribbling down his mathematical ideas. Galois was mortally wounded at dawn and died two days later. He was only 21.

It is mostly from these scribblings of Galois' last night that we now know the magnitude of our loss at his early death. His notes were deciphered by the mathematician Liouville and published 15 years later. Galois set the foundations for much of modern day mathematics.

In France, about 1830, a new star of unimaginable brightness appeared in the heavens of pure mathematics . . . Evariste Galois.

Felix Klein

In general, the discovery of the graph of an algebraic function is not easy. However, many of the tools that we have acquired in our discussions of polynomial functions and rational functions apply to these graphs.

While there is no one specific set of instructions for discovering the graph of a general algebraic function, there are some general guidelines that simplify this task. We offer a few now.

1) Determine the domain. Mark off the excluded regions of the plane by shading.

2) Check for symmetry. Recall that an even function is symmetric with respect to the *y*-axis and that an odd function is symmetric with respect to the origin.

3) Check the behavior of the function at extremes and near any discontinuities.

4) If the graph is the sum of two or more simpler functions, use the graphs of these simpler functions and the method of adding *y*-coordinates (recall that this method was discussed back in Section 2.6).

5) Take advantage of any other information your investigation may uncover. Be clever.

EXAMPLE 7   Sketch the graph of $f(x) = x/\sqrt{4 - x^2}$.

SOLUTION We write the radicand in factored form:

$$f(x) = \frac{x}{\sqrt{(2 - x)(2 + x)}}.$$

This tells us two things. First, the domain is the interval $(-2, 2)$ since the radicand must be positive. Second, there are asymptotes at $x = 2$ and $x = -2$ since those are zeros of the denominator. As a matter of fact,

$$y \to +\infty \quad \text{as} \quad x \to 2^-$$

and

$$y \to -\infty \quad \text{as} \quad x \to -2^+.$$

This is an odd function, so we can also use its symmetry in sketching the graph. Notice that the graph passes through the origin. This is sufficient information to draw a sketch of the graph (Figure 45).

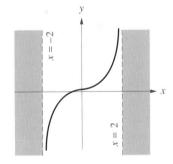

FIGURE 45

EXAMPLE 8   Sketch the graph of

$$F(x) = \frac{x - 3}{\sqrt{x^2 - 4}}.$$

SOLUTION The domain is such that $x^2 - 4 > 0$. This implies that $x < -2$ or $x > 2$. Also, as $x \to 2^+$, $y \to -\infty$; and as $x \to 2^-$, $y \to -\infty$ (you should verify this). This gives us vertical asymptotes at $x = -2$ and $x = 2$. There is an x-intercept at $(3, 0)$.

Now consider the behavior of F at the extreme values of x. As $x \to \pm\infty$, the numerator $x - 3$ behaves as x does, and the denominator $\sqrt{x^2 - 4}$ behaves as $\sqrt{x^2}$ does. Thus, since $\sqrt{x^2} = |x|$, we get

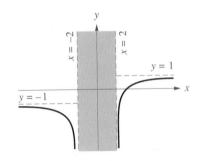

$$x \to +\infty \Rightarrow \frac{x - 3}{\sqrt{x^2 - 4}} \to \frac{x}{|x|} = 1$$

and

$$x \to -\infty \Rightarrow \frac{x - 3}{\sqrt{x^2 - 4}} \to \frac{x}{|x|} = -1.$$

FIGURE 46

Figure 46 shows these results.

EXERCISE SET 3.5

A

In Problems 1 through 6, find the domain and the zeros of the algebraic function. Sketch the graph of the domain on a coordinate line.

1. $f(x) = \dfrac{1}{\sqrt{x^2 - 1}}$

2. $g(x) = \dfrac{6}{\sqrt{25 - x^2}}$

3. $R(x) = \sqrt{x - 2} - \sqrt{2 + x}$

4. $Q(x) = \dfrac{\sqrt{x - 2}}{\sqrt{5 - x}}$

5. $H(x) = \dfrac{2x - 5}{x\sqrt{8 - x^2}}$

6. $S(x) = \dfrac{x^2 - 6}{(x - 2)\sqrt{x}}$

In Problems 7 through 18, sketch the graphs of the two equations given in parts a) and b) on the same coordinate plane. Label the intercepts of the equation given in part b).

7. **a)** $y = x^{2/3}$ **b)** $y = x^{2/3} - 4$

8. **a)** $y = x^{3/5}$ **b)** $y = x^{3/5} - 8$

9. **a)** $y = x^{3/4}$ **b)** $y = (x - 1)^{3/4}$

10. **a)** $y = x^{2/5}$ **b)** $y = (x + 1)^{2/5}$

11. **a)** $y = -x^{4/3}$ **b)** $y = -(x - 2)^{4/3}$

12. **a)** $y = -\frac{2}{3}x^{3/5}$ **b)** $y = -\frac{2}{3}x^{3/5} + 18$

13. **a)** $y = -\frac{1}{2}x^{1/4}$ **b)** $y = -\frac{1}{2}(x - 1)^{1/4}$

14. **a)** $y = \frac{1}{2}x^{3/7}$ **b)** $y = \frac{1}{2}(x - 1)^{3/7}$

15. **a)** $y = -x^{7/4}$ **b)** $y = -x^{7/4} + 1$

16. **a)** $y = -\frac{1}{2}x^{4/3}$ **b)** $y = -\frac{1}{2}x^{4/3} + 8$

17. **a)** $y = \frac{1}{2}x^{2/3}$ **b)** $y = \frac{1}{2}(x - 4)^{2/3}$

18. **a)** $y = x^{5/3}$ **b)** $y = \frac{1}{2}(x - 8)^{5/3} - \frac{1}{2}$

B

In Problems 19 through 30, the algebraic function is the square root of a polynomial function or a rational function. First sketch the radicand, and then sketch the algebraic function on the same coordinate plane. Determine the asymptotes and the intercepts.

19. $f(x) = \sqrt{x^2 - 2x - 8}$

20. $g(x) = \sqrt{3x^2 - 4x - 15}$

21. $H(x) = \sqrt{x^3 - 16x}$

22. $f(x) = \sqrt{8x - x^3}$

23. $U(x) = \sqrt{-x^4 + 5x^2 - 4}$

24. $N(x) = \sqrt{9x^2 - x^4}$

25. $f(x) = \sqrt{\dfrac{1}{x - 4}}$

26. $f(x) = \sqrt{\dfrac{4}{x + 2}}$

27. $Q(x) = \sqrt{\dfrac{x + 4}{x + 3}}$

28. $F(x) = \sqrt{\dfrac{5 - x}{x - 4}}$

29. $G(x) = \sqrt{\dfrac{4x}{x^2 + 2}}$

30. $M(x) = \sqrt{\dfrac{4 - x^2}{x^2 + 1}}$

In Problems 31 through 36, sketch the algebraic function by adding y-coordinates. Label any asymptotes and intercepts.

31. $f(x) = x^{3/5} + \dfrac{1}{x}$

32. $g(x) = \frac{1}{2}x - x^{3/5}$

33. $T(x) = \frac{1}{2}x + x^{2/3}$

34. $R(x) = x^{3/5} - \frac{1}{3}x^2$

35. $f(x) = x^{2/3} - \dfrac{1}{x}$

36. $Q(x) = x^{3/5} - \dfrac{1}{x}$

In Problems 37 through 48, sketch the algebraic function. Label any asymptotes and intercepts.

37. $f(x) = \dfrac{2}{\sqrt{x-1}}$

38. $R(x) = \dfrac{8}{\sqrt{4-x}}$

39. $g(x) = \dfrac{x}{\sqrt{x+4}}$

40. $T(x) = \dfrac{x}{\sqrt{9-x}}$

41. $P(x) = \dfrac{3}{\sqrt{9-x^2}}$

42. $V(x) = \dfrac{16}{\sqrt{4-x^2}}$

43. $g(x) = \dfrac{2x-10}{\sqrt{x^2-9}}$

44. $f(x) = -\dfrac{x+4}{\sqrt{x^2-1}}$

45. $H(x) = \dfrac{x-1}{\sqrt{4-x^2}}$

46. $F(x) = \dfrac{2x+6}{\sqrt{16-x^2}}$

47. $H(x) = \dfrac{\sqrt{4-x^2}}{x+1}$

48. $W(x) = \dfrac{\sqrt{9+x}}{x}$

49. Use the graph of Problem 35 to solve the inequality

$$x^{2/3} - \frac{1}{x} > 0.$$

50. Use the graph of Problem 41 to solve the inequality

$$\frac{3}{\sqrt{9-x^2}} \geq 0.$$

C

51. Are the functions

$$f_1(x) = \frac{\sqrt{x-2}}{\sqrt{x+3}} \quad \text{and} \quad f_2(x) = \sqrt{\frac{x-2}{x+3}}$$

the same function? Determine the domain of real numbers of each.

52. Suppose that we want to sketch the graph of the equation

$$y^2 = \frac{x}{x-4}$$

(since this is not a function, we cannot directly use the techniques of this chapter to find the graph).

a) Solve the equation for y in terms of x to get

$$y = \pm\sqrt{\frac{x}{x-4}}.$$

b) Sketch each of the equations in part a) on the same coordinate plane using the techniques of this section. The union of these two graphs is the graph of the equation.

53. Sketch the graph of $y^2(4-x) = x^2$, using the direction of Problem 52.

54. Sketch the graph of $5x^2 - 4xy + 4y^2 = 36$, using the direction of Problem 52. (Hint: Use the quadratic formula to solve for y.)

MISCELLANEOUS EXERCISES

In Problems 1 through 18, sketch the graph of the function.

1. $y = x^4$

2. $y = x^5$

3. $y = x^{-3}$

4. $y = x^{-4}$

5. $y = x^{2/3}$

6. $y = x^{3/7}$

7. $y = x^4 - 16$

8. $y = x^5 - \frac{243}{32}$

9. $y = (x+8)^{2/3}$

10. $y = (x-4)^{5/2}$

11. $y = (x+2)^{-4}$

12. $y = (x+3)^{-3}$

13. $y = (x-2)^4 - 1$

14. $y = (x-8)^{4/3} - 4$

15. $y = 4 - (x+1)^{-4}$

16. $y = 18 - \frac{1}{2}(x-4)^{5/2}$

17. $y = -2(x+1)^{5/4} + 2$

18. $y = -3(x-1)^{3/7}$

In Problems 19 through 30, sketch the graph of the polynomial function and label its intercepts.

19. $y = (x-2)(x+4)(x-7)$

20. $y = x(x-5)(x+2)$

21. $y = x^2(x-2)(x+7)$

22. $y = x(4x-3)(x+5)^2$

23. $y = \frac{1}{4}(x-2)^2(x^2-36)$

24. $y = 2(x^2 - 5x - 4)(x - 10)$

25. $y = x^3 - 36x$

26. $y = x^4 - 16x^2$

27. $y = 20x^2 - x^4$

28. $y = 18x - 2x^3$

29. $y = x(x - 2)^2(x + 4)^2$

30. $y = x^4 - 6x^2 + 5$

In Problems 31 through 48, sketch the graph of the rational function and label its intercepts and asymptotes.

31. $P(x) = \dfrac{3x + 5}{2x - 10}$

32. $P(x) = \dfrac{2x - 4}{x}$

33. $P(x) = \dfrac{-5(x^2 - 9)}{x^2 + 3x - 10}$

34. $P(x) = \dfrac{16 - x^2}{3x^2 - 9x}$

35. $P(x) = \dfrac{x(x - 5)(x + 2)}{(x - 3)(x + 4)(x - 7)}$

36. $P(x) = \dfrac{(x^2 - 6x + 8)(x + 7)}{x(x^2 + x - 12)}$

37. $P(x) = \dfrac{x(x - 6)}{(x^2 + 1)}$

38. $P(x) = \dfrac{(x + 2)(x - 6)}{(x^2 + 4)}$

39. $P(x) = \dfrac{x^2(x - 3)}{(x^2 - 36)}$

40. $P(x) = \dfrac{x(x - 6)^2}{(x^2 - 4)}$

41. $P(x) = \dfrac{x^4 - 17x^2 + 16}{2x^3 - 8x}$

42. $P(x) = \dfrac{x^4 - 29x^2 + 100}{x^3 - 9x}$

43. $P(x) = \dfrac{x^2 - 3x}{x^3 - 9x}$

44. $P(x) = \dfrac{x^2 - 4x + 4}{x^3 - 6x^2 + 8x}$

45. $P(x) = \left[\dfrac{x + 1}{x^2 - x - 12}\right]^2$

46. $P(x) = \left[\dfrac{x^2 - 16}{x^2 - x - 2}\right]^2$

47. $P(x) = \left[\dfrac{x + 2}{x^2 - 4}\right]^2$

48. $P(x) = \left[\dfrac{x^2 - 3x + 2}{x - 2}\right]^2$

In Problems 49 through 54, sketch the graph of the algebraic function and label its intercepts and asymptotes.

49. $f(x) = \sqrt{16 - x^4}$

50. $f(x) = \sqrt{27 - x^3}$

51. $f(x) = \dfrac{\sqrt{9 - x^2}}{x}$

52. $f(x) = x\sqrt{x - 4}$

53. $f(x) = \dfrac{\sqrt{x^2 - 1}}{x - 4}$

54. $f(x) = x\sqrt{x + 2}$

In Problems 55 through 60, sketch the graph of the equation by adding y-coordinates. Label its intercepts.

55. $y = \dfrac{1}{x} - \dfrac{1}{4}x^3$

56. $y = \frac{1}{2}x - x^{1/3}$

57. $y = x^{2/3} - \frac{1}{2}x$

58. $y = \frac{1}{4}x^2 - x^{2/3}$

59. $y = \frac{1}{8}x^3 - x^{3/5}$

60. $y = 2x^{2/5} - x^{3/5}$

In Problems 61 through 66, use synthetic division to evaluate $P(c)$, for the polynomial P and value of c given. Rewrite $P(x)$ as $P(x) = (x - c)Q(x) + P(c)$.

61. $P(x) = 2x^3 - 3x^2 + x - 4$; $c = 4$

62. $P(x) = x^3 + 6x^2 - 3x + 2$; $c = -5$

63. $P(x) = x^4 - 2x^3 - 12x^2 + 2$; $c = 6$

64. $P(x) = 6x^5 - 23x^4 + 21x^3 + 6x^2 - 63x + 107$; $c = \frac{3}{2}$

65. $P(x) = 12x^4 - 2x^3 - 7x^2 - x + 2$; $c = \frac{2}{3}$

66. $P(x) = x^4 - 10x^2 + 16$; $c = \sqrt{2}$

In Problems 67 through 72, use the rational root theorem to solve these equations over the real numbers.

67. $x^3 - 5x + 2 = 0$

68. $3x^3 + 8x^2 - 5 = 0$

69. $2x^4 + 46x = 15x^2 - 5x^3 + 20$

70. $3x^4 + 17x^2 = 11x^3 + 13x - 4$

71. $6x^4 + 27x + 945 = x^3 + 197x^2$

72. $-13x^3 - 9x^2 + 133x - 110 = x^4$

In Problems 73 through 84, first sketch the graph of the equation in a) then use this graph to solve the inequality given in part b).

73. a) $y = x^3 - 4x$ **b)** $x^3 - 4x < 0$

74. a) $y = x^3 - 36x$ **b)** $x^3 - 36x \geq 0$

75. a) $y = x^4 - 12x^2 - 13$ **b)** $x^4 > 12x^2 + 13$

76. a) $y = x^4 - 4x^2 - 12$ **b)** $x^4 - 4x^2 \leq 12$

77. a) $y = x^4 - 20x^2$ **b)** $x^4 - 20x^2 \geq 0$

78. a) $y = -x^3 + 5x^2$ **b)** $x^3 \geq 5x^2$

79. a) $y = \dfrac{2x}{x - 3}$ **b)** $\dfrac{2x}{x - 3} > 0$

80. a) $y = \dfrac{x - 3}{x + 6}$ **b)** $\dfrac{x - 3}{x + 6} < 0$

81. a) $y = \dfrac{x - 2}{x^2 - 5x}$ **b)** $\dfrac{x - 2}{x^2 - 5x} \geq 0$

82. a) $y = \dfrac{x^2 - 2}{2x - 5}$ **b)** $\dfrac{x^2 - 2}{2x - 5} \le 0$

83. a) $y = \dfrac{x}{\sqrt{4 - x^2}}$ **b)** $\dfrac{x}{\sqrt{4 - x^2}} > 0$

84. a) $y = \dfrac{x^2}{\sqrt{25 - x^2}}$ **b)** $\dfrac{x^2}{\sqrt{25 - x^2}} \le 0$

85. Sketch the graph of

a) $y = \dfrac{|x|}{x - 5}$,   **b)** $y = \left|\dfrac{x}{x - 5}\right|$.

86. Sketch the graph of

a) $y = \dfrac{|x|}{x^2 - 16}$,   **b)** $y = \left|\dfrac{x}{x^2 - 16}\right|$.

87. Determine the real polynomial function P of the form
$$P(x) = a_n x^n + a_{n-1} x^{n-1} + a_{n-2} x^{n-2} + \cdots + a_1 x + a_0$$
such that the leading coefficient of P is -4, its degree is 3, and its zeros are 7 and -1 (multiplicity 2).

88. Determine the real polynomial function P of the form
$$P(x) = a_n x^n + a_{n-1} x^{n-1} + a_{n-2} x^{n-2} + \cdots + a_1 x + a_0$$
such that the leading coefficient of P is -4, its degree is 4, and its zeros are $1 + i$ and 0 (multiplicity 2).

89. Determine the real polynomial function P of the form
$$P(x) = a_n x^n + a_{n-1} x^{n-1} + a_{n-2} x^{n-2} + \cdots + a_1 x + a_0$$
such that its degree is 4, $P(\sqrt{5}) = 12$, and its zeros are i, -2, and 2.

CHAPTER 4

EXPONENTIAL AND LOGARITHMIC FUNCTIONS

EXPONENTIAL FUNCTIONS

The functions in Chapter 3 belong to a category of functions known as algebraic functions. A function that is not algebraic is called a **transcendental** function. One important example is the **exponential** function.

Before we can define the exponential function, we need to give meaning to expressions with irrational exponents. As an example, we will discuss how the value of $5^{\sqrt{2}}$ is defined. We start with a succession of rational numbers that approach the irrational number $\sqrt{2}$:

$$1, 1.4, 1.41, 1.414, 1.4142, 1.41421, \ldots \rightarrow \sqrt{2}.$$

Now consider the values

$$5^1, 5^{1.4}, 5^{1.41}, 5^{1.414}, 5^{1.4142}, 5^{1.41421}, \ldots.$$

It can be shown that this succession approaches a real number. This is precisely the value that we will assign to $5^{\sqrt{2}}$:

$$5^1, 5^{1.4}, 5^{1.41}, 5^{1.414}, 5^{1.4142}, 5^{1.41421}, \ldots \rightarrow 5^{\sqrt{2}}.$$

In the same way, for $b > 0$, we will assume b^r can be obtained for any real number r. The restriction $b > 0$ is necessary to avoid expressions like $(-9)^{1/2}$. Moreover, we will assume that the laws of exponents hold for all real-valued exponents.

Before the availability of the hand-held calculator, finding a good approximation for expressions like $5^{\sqrt{2}}$ was a tedious task. The next example illustrates how to perform such calculations.

EXAMPLE 1   Using a calculator with a y^x key, approximate:

a) $5^{\sqrt{2}}$ **b)** $2^{-\pi^2}$

SOLUTION

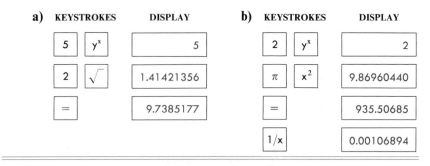

a) KEYSTROKES	DISPLAY
5 yˣ	5
2 √	1.41421356
=	9.7385177

b) KEYSTROKES	DISPLAY
2 yˣ	2
π x²	9.86960440
=	935.50685
1/x	0.00106894

If $b > 0$, then for every real value of x, the expression b^x designates a unique real number. This observation leads to the following definition.

Exponential Function

> An **exponential** function is a function of the form
> $$f(x) = b^x$$
> where $b > 0$ and x is any real number.

Notice that the base b is constant, and the exponent x varies. For example,

$$f(x) = 2^x \quad \text{and} \quad g(x) = x^2$$

are different kinds of functions. The function $f(x)$ is an exponential function. The function $g(x)$ is not an exponential function; it is a quadratic function.

One of the best ways to get acquainted with a new function is to sketch its graph.

EXAMPLE 2   Sketch the graph of $f(x) = 2^x$.

SOLUTION The following table shows some of the ordered pairs:

x	3	2	1	$\frac{1}{2}$	0	$-\frac{1}{2}$	-1
2^x	8	4	2	$\sqrt{2} \approx 1.4$	1	$1/\sqrt{2} \approx 0.7$	$\frac{1}{2}$

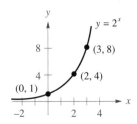

FIGURE 1

This is enough to get an idea of the graph (Figure 1).

By connecting the ordered pairs with a continuous curve, we are using not only rational values but also irrational values for x. It is important to realize that ordered pairs like $(\sqrt{3}, 2^{\sqrt{3}})$ and $(\pi, 2^{\pi})$ are included on the graph.

To get a feeling of how typical our previous graph is, we will look at $f(x) = b^x$ for various values of $b > 0$.

EXAMPLE 3   Sketch the graphs $f(x) = b^x$ for $b = 10, 3, \frac{3}{2},$ and 1.

SOLUTION The table shows several ordered pairs for each value of b:

x	-2	-1	0	1	2	3
10^x	$\frac{1}{100}$	$\frac{1}{10}$	1	10	100	1000
3^x	$\frac{1}{9}$	$\frac{1}{3}$	1	3	9	27
$(\frac{3}{2})^x$	$\frac{4}{9}$	$\frac{2}{3}$	1	$\frac{3}{2}$	$\frac{9}{4}$	$\frac{27}{8}$
1^x	1	1	1	1	1	1

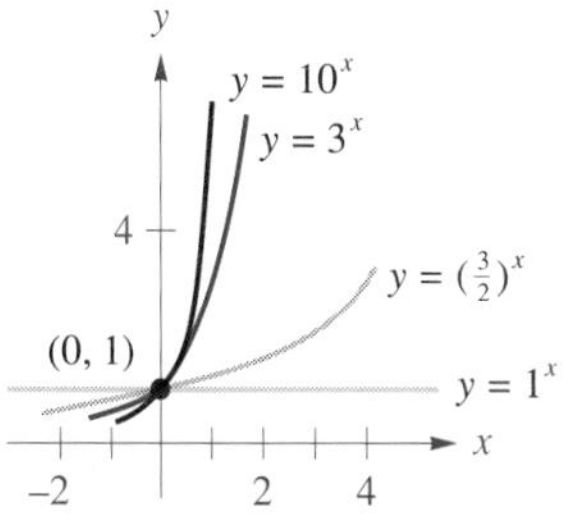

The respective graphs are shown in Figure 2. The value $b = 1$ appears to be critical, and it is natural to wonder about values $0 < b < 1$. Looking at the previous progression, you might anticipate the answer.

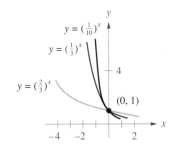

FIGURE 2

EXAMPLE 4   Graph $f(x) = b^x$ for $b = \frac{2}{3}, \frac{1}{3},$ and $\frac{1}{10}$.

SOLUTION The following table shows several ordered pairs for each value. The respective graphs are shown in Figure 3.

x	-3	-2	-1	0	1	2
$(\frac{2}{3})^x$	$\frac{27}{8}$	$\frac{9}{4}$	$\frac{3}{2}$	1	$\frac{2}{3}$	$\frac{4}{9}$
$(\frac{1}{3})^x$	27	9	3	1	$\frac{1}{3}$	$\frac{1}{9}$
$(\frac{1}{10})^x$	1000	100	10	1	$\frac{1}{10}$	$\frac{1}{100}$

FIGURE 3

In Section 2.5, we saw that the relationship between the graphs of $y = f(x)$ and $y = f(-x)$ is that one is the reflection of the other through the

y-axis. The graphs of $y = b^x$ and $y = (1/b)^x$ display this reflective relationship. This suggests taking advantage of the fact that

$$b^{-x} = \left(\frac{1}{b}\right)^x.$$

For example, to graph $f(x) = (\frac{1}{3})^x$, we could first graph $g(x) = 3^x$. Since

$$g(-x) = 3^{-x} = (\frac{1}{3})^x$$

the graph of $f(x)$ is the reflection of the graph of $g(x)$ through the y-axis.

The graphs of $f(x) = b^x$ for the two cases $b > 1$ and $0 < b < 1$ are used frequently in mathematics. You should learn their general shape without relying on plotting points.

Graph of $f(x) = b^x$

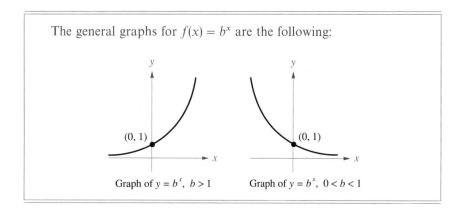

The general graphs for $f(x) = b^x$ are the following:

Graph of $y = b^x$, $b > 1$ Graph of $y = b^x$, $0 < b < 1$

You should commit these graphs to memory because they illustrate some important facts about the exponential function:

The graph is smooth, continuous, and concave up over its domain.

The domain is $\mathbb{R}$; that is, b^x is defined for all real x.

The range is all positive real numbers; that is, $b^x > 0$.

A horizontal line crosses the graph no more than once; that is, $f(x) = b^x$ is a **one-to-one** function. This means that if $b^r = b^s$, then $r = s$.

If $b > 1$, then as $x \to \infty$, $y \to \infty$; as $x \to -\infty$, $y \to 0+$. If $b < 1$, then as $x \to \infty$, $y \to 0+$; as $x \to -\infty$, $y \to \infty$.

Thus, the x-axis is a horizontal asymptote. However, the graph approaches the x-axis only on one side of the y-axis (depending on b). On the other side of the y-axis the graph increases very rapidly.

These properties can be useful in a number of situations. They should be kept in mind when solving equations that have exponential functions.

EXAMPLE 4   Solve $5^t = \sqrt{5}$.

SOLUTION The right side of the equation can be expressed with a rational exponent:

$$5^t = 5^{1/2}.$$

Since $f(x) = 5^x$ is one-to-one, this implies that

$$t = \tfrac{1}{2}.$$

The previous example used the fact that the exponential function is one-to-one. In the next example we use the fact that an exponential function is always positive.

EXAMPLE 5   Solve for x: $7^x(x - 3) = 0$.

SOLUTION The product can be zero if and only if one of the factors is zero:

$$7^x = 0 \quad \text{or} \quad x - 3 = 0.$$

But since $7^x > 0$, the only solution is $x = 3$.

We will discuss equations with exponential functions in detail in Section 4.4.

Knowing the general shape of the graph of $y = b^x$, we can use the techniques discussed in Chapter 2 to graph more complicated functions that involve exponential functions.

EXAMPLE 6   Graph $y - 2 = -3^x$.

SOLUTION First, the graph of $y = -3^x$ is the same as $y = 3^x$ reflected through the x-axis. (See Section 2.5 to review how the graphs of $y = f(x)$ and $y = -f(x)$ are related.) To graph $y - 2 = -3^x$, translate the graph of $y = -3^x$ two units up (Figure 4).

The y-intercept $(0, 1)$ is obtained by setting $x = 0$ in the equation $y - 2 = -3^x$ and solving for y.

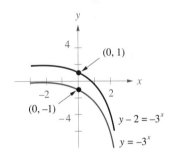

FIGURE 4

EXAMPLE 7   Sketch the graph of $y = 2^{x/3}$.

SOLUTION We could write $2^{x/3} = (2^{1/3})^x = (\sqrt[3]{2})^x$ and construct a table of values. Instead, we will quickly graph $f(x) = 2^x$ and recall from Section 2.5

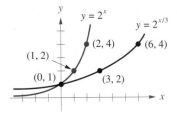

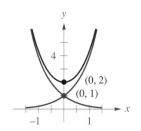

FIGURE 5

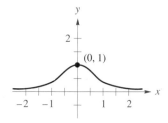

FIGURE 6

that the graph of $f(\frac{1}{3}x) = 2^{x/3}$ is an expansion away from the y-axis by a factor of 3 (Figure 5).

EXAMPLE 8 Sketch the graph of $f(x) = 4^x + 4^{-x}$.

SOLUTION One way to start is to graph $y = 4^x$ and $y = 4^{-x}$ on the same axes and add y-coordinates. This gives the graph of $f(x)$ (Figure 6). The function is even, since

$$f(-x) = 4^{(-x)} + 4^{-(-x)} = 4^{-x} + 4^x = f(x).$$

This graph is an example of a type of curve called a **catenary.** (Catenaries have an important application in real life: A flexible cable suspended from two fixed points corresponds to a catenary.)

EXAMPLE 9 Sketch the graph of $f(x) = 2.5^{-x^2}$.

SOLUTION This function is even since $f(x) = f(-x)$; so we form a table of values for $x \geq 0$. Using the symmetry with respect to the y-axis, we obtain the graph in Figure 7.

FIGURE 7

x	y
0	$2.5^0 = 1$
$\frac{1}{2}$	$2.5^{-1/4} \approx 0.8$
1	$2.5^{-1} = 0.4$
$\frac{3}{2}$	$2.5^{-9/4} \approx 0.1$
2	$2.5^{-4} \approx 0.03$

This type of curve is encountered in the study of probability and statistics. It is sometimes called a bell-shaped curve.

Applications

Exponential functions arise in a number of real-world situations. We will focus on two situations, one involving growth and the other involving decay.

Suppose $5000 is invested at a 10% annual interest rate compounded monthly. This means that the rate per month is $(\frac{1}{12})(10\%)$ or 0.10/12. After the first month the interest earned will be

$$(\text{principal})(\text{rate}) = 5000\left(\frac{0.10}{12}\right).$$

If the original $5000 remains invested, the total amount accrued is

$$(\text{principal})(\text{interest rate}) + \text{principal} = 5000\left(\frac{0.10}{12}\right) + 5000$$

$$= 5000\left(1 + \frac{0.10}{12}\right).$$

This becomes the new principal for the next month. After the second month the amount accumulated will be

$$(\text{principal})(\text{interest rate}) + \text{principal}$$

$$= \left[5000\left(1 + \frac{0.10}{12}\right)\right]\left(\frac{0.10}{12}\right) + 5000\left(1 + \frac{0.10}{12}\right).$$

If we factor $5000(1 + 0.10/12)$ from both terms and simplify, this can be written as

$$5000\left(1 + \frac{0.10}{12}\right)\left[\left(\frac{0.10}{12}\right) + 1\right] = 5000\left(1 + \frac{0.10}{12}\right)^2.$$

After three months the total amount will be

$$5000\left(1 + \frac{0.10}{12}\right)^3.$$

Generally, after x months the accumulated amount invested A will be

$$A = 5000\left(1 + \frac{0.10}{12}\right)^x.$$

After t years, the total amount will be

$$A = 5000\left(1 + \frac{0.10}{12}\right)^{12t}.$$

The ideas for this problem can be generalized.

Suppose P dollars are invested at an interest rate r compounded n times per year. The accumulated amount A, after t years will be

$$A = P\left(1 + \frac{r}{n}\right)^{nt}.$$

Compound Interest

EXAMPLE 10   If $8500 is invested at an interest rate of 8% compounded quarterly, what is the accumulated amount after 1 year? 5 years?

SOLUTION Use the compound interest formula with $P = 8500$, $r = 0.08$, and $n = 4$:

$$A = 8500\left(1 + \frac{0.08}{4}\right)^{4t} = 8500(1.02)^{4t}.$$

After 1 year, there will be $8500(1.02)^4 = \$9200.67$. After 5 years, there will be $8500(1.02)^{20} = \$12630.55$.

Compound interest is an example of **exponential growth.** Another application of an exponential function involves **exponential decay.** Certain substances, such as carbon-14 and radioactive isotopes of other elements, decay at a rate that can be described with an exponential function. The concept of half-life is important in many problems of decay. The half-life of a substance is the time it takes for one-half of an initial amount to disintegrate.

Half-Life

Suppose there is an initial amount Q_0 of a substance with a half-life of h. The amount of remaining Q after a time period of t will be

$$Q = Q_0 \left(\frac{1}{2}\right)^{t/h}.$$

The units of measure for t and h must be the same.

EXAMPLE 11   The radioactive isotope of lead, Pb-209, has a half-life of approximately 3.3 hours. If there are 5 grams of lead initially, how many grams will remain after 6.6 hours? After 1 day?

SOLUTION In this case $Q_0 = 5$ (grams) and $h = 3.3$ (hours), giving

$$Q = 5\left(\frac{1}{2}\right)^{t/3.3}.$$

After 6.6 hours

$$Q = 5\left(\frac{1}{2}\right)^{6.6/3.3} = 5\left(\frac{1}{2}\right)^{2} = 1.25$$

grams will remain. After 1 day,

$$Q = 5\left(\frac{1}{2}\right)^{24/3.3} \approx 0.032$$

grams will remain.

EXERCISE SET 4.1

A

In Problems 1 through 6, use a calculator to approximate the expression to the nearest 0.0001.

1. 3^π
2. $\pi^{\sqrt{2}}$
3. 2.718^π
4. $4^{2+\sqrt{5}}$
5. $3^{1-\sqrt{6}}$
6. $5^{-\sqrt{3}}$

In Problems 7 through 14, solve each equation.

7. a) $3^x = 3^4$ b) $2^x = \sqrt{2}$
8. a) $5^x = 5^8$ b) $7^x = \sqrt{7}$
9. a) $5^x = \frac{1}{5}$ b) $3^x = 1$
10. a) $4^x = \frac{1}{16}$ b) $6^x = 1$
11. a) $3^x = 0$ b) $4^x = -12$
12. a) $5^x = 0$ b) $5^x = -25$
13. $2^x(4x + 1) = 0$
14. $3^x(2x - 5) = 0$

In Problems 15 through 26, sketch the graphs of a and b on the same axes. Identify asymptotes.

15. a) $y = 4^x$ b) $y = (\frac{1}{4})^x$
16. a) $y = (\frac{5}{2})^x$ b) $y = (\frac{2}{5})^x$
17. a) $y = 4^x$ b) $y = -4^{-x}$
18. a) $y = 3^x$ b) $y = -3^{-x}$
19. a) $y = (\frac{1}{5})^x$ b) $y = (\frac{1}{5})^{x/2}$
20. a) $y = 10^x$ b) $y = 10^{x/4}$
21. a) $y = 2^x$ b) $y - 3 = 2^x$
22. a) $y = 5^x$ b) $y + 1 = 5^x$
23. a) $y = 6^x$ b) $y = 6^{x-2}$
24. a) $y = 9^x$ b) $y = 9^{x+3}$
25. a) $y = (\frac{1}{4})^x$ b) $y - 1 = (\frac{1}{4})^{x-3}$
26. a) $y = (\frac{1}{3})^x$ b) $y + 4 = (\frac{1}{3})^{x-2}$

B

In Problems 27 through 32, write h(x) as the composition of two functions, f[g(x)]. Do not use f(x) = x or g(x) = x.

27. $h(x) = 10^{3x-1}$
28. $h(x) = 3^{x+5}$
29. $h(x) = (\frac{1}{2})^{|x|}$
30. $h(x) = 5^{(x-1)/2}$
31. $h(x) = \sqrt{2^x}$
32. $h(x) = \sqrt[3]{6^x}$

Graph the functions given in Problems 33 through 44. Identify asymptotes.

33. a) $y = 7^{-x}$ b) $y = 7^{2-x}$
34. a) $y = (\frac{1}{3})^{-x}$ b) $y = (\frac{1}{3})^{3-x}$
35. a) $y = 8^{(x-2)/3}$ b) $y = 8^{2(x-2)}$
36. a) $y = 5^{(x+1)/2}$ b) $y = 5^{3(x+1)}$
37. a) $y = 2^{-x^2}$ b) $y = 2^{-(x-2)^2}$
38. a) $y = 3^{-x^2}$ b) $y = 3^{-(x-4)^2}$
39. a) $y = \frac{1}{2}(2^x + 2^{-x})$ b) $y = \frac{1}{2}(2^x - 2^{-x})$
40. a) $y = \frac{1}{2}(3^x + 3^{-x})$ b) $y = \frac{1}{2}(3^x - 3^{-x})$
41. a) $y = 3^{|x|}$ b) $y = (\frac{1}{3})^{|x|}$
42. a) $y = 2^{|x|}$ b) $y = (\frac{1}{2})^{|x|}$
43. a) $x = 10^y$ b) $x = 10^{-y}$
44. a) $x = 2^y$ b) $x = 2^{-y}$

45. If $7000 is invested at an interest rate of 6% compounded twice a year, what is the accumulated amount at the end of 3 years? $4\frac{1}{2}$ years?

46. Repeat Problem 45 with interest compounded monthly.

47. A savings account pays an interest rate of 9% compounded quarterly. How much invested now will amount to $8000 in 2 years?

48. Two parents want to invest enough money in a savings account to pay for their daughter's college education. If the account pays 10% interest compounded monthly, how much should they deposit to accumulate $12,000 in 4 years?

49. Is it better to invest your money at 7% compounded weekly or at 7.50% compounded annually?

50. All living wood contains a fixed concentration of carbon-14, which has a half-life of about 5700 years in dead wood. If the present amount of carbon-14 is 140 grams, how many grams will remain after 100 years?

51. Gold-196 has a half-life of about 6 days. If we initially have 2 pounds, how many pounds will we have in 2 days? in 20 days?

52. Radium has a half-life of about 1600 years. Suppose we presently have 10 grams of radium. How many grams were present 50 years ago?

53. A certain one-celled animal divides in two every half-hour. If 100 are present now, how many will be present in 3 hours? How long will it take until there are 25,600?

54. Find a formula describing the growth of the one-celled animals in Problem 53.

55. Suppose $1 is invested at a 100% interest rate for 1 year. Find the accumulated amount if compounded **a)** monthly **b)** weekly **c)** daily **d)** hourly.

56. If the annual rate of inflation is 7% per year, then the equation $P = P_0(1.07)^t$ yields the expected price P in t years of an item that presently costs P_0. Find the expected price of each of the following items for the indicated number of years from now:

   a) a $1.74 loaf of bread in 4 years.

   b) a $36.00 book in 3 years.

   c) a $12,000 car in 5 years.

   d) a $112,000 house in 6 years.

57. Some drugs are eliminated from the body only through the kidneys. For many of these cases, an exponential function is used to model the amount of the drug remaining in the body. If a person initially consumes Q_0 milligrams of caffeine, then the amount $Q(t)$ in the body t hours later is approximated by $Q(t) = Q_0(0.9)^t$. Suppose a student drinks some coffee containing 40 milligrams of caffeine.

a) How many milligrams are remaining in the student's body after 3 hours?

b) To the nearest hour, how long does it take the body to eliminate half the caffeine?

58. The amount present y after the principal p is invested for x years at an interest rate of r compounded annually is $y = p(1 + r)^x$. Graph this function for $50 deposited at a rate of 8% compounded annually with $0 \le x \le 15$.

59. If P dollars are borrowed at an interest rate r for t years, the monthly payment for a fully amortized loan is

$$m = \frac{P\left(\dfrac{r}{12}\right)}{1 - \left(1 + \dfrac{r}{12}\right)^{-12t}}.$$

What is the monthly payment for a 4-year car loan of $8,000 at a 14% interest rate?

60. What is the monthly payment for a 30-year home loan of $110,000 at 11% interest? Refer to the formula given in Problem 59.

61. For $f(x) = x^2$ and $g(x) = 2^x$, answer true or false:

   a) If $f(a) = f(b)$, then $a = b$.

   b) If $g(a) = g(b)$, then $a = b$.

62. For $f(x) = |x|$ and $g(x) = 3^x$, answer true or false:

   a) If $f(a) = f(b)$, then $a = b$.

   b) If $g(a) = g(b)$, then $a = b$.

63. Graph $f(x) = x^2$ and $g(x) = 2^x$ on the same set of axes for $-2 \le x \le 5$. How many solutions does $x^2 = 2^x$ have?

64. Graph $f(x) = x^3$ and $g(x) = 3^x$ on the same set of axes for $x > 0$. As $x \to \infty$, both functions grow without bound. Which function grows more rapidly?

65. Sketch the graph of $f(x) = \frac{1}{2}x^2$ and $g(x) = (\frac{1}{2})^{x^2}$ on the same set of axes. For what values of x is $\frac{1}{2}x^2 \le (\frac{1}{2})^{x^2}$?

C

66. Let $c(x) = \frac{1}{2}(2^x + 2^{-x})$ and $s(x) = \frac{1}{2}(2^x - 2^{-x})$. Show:

   a) $c^2(x) - s^2(x) = 1$ **b)** $s(2x) = 2s(x)c(x)$.

67. **a)** Find a function $f(x)$ such that $f(a + b) = f(a)f(b)$.

   b) Find a function $f(x)$ such that $3f(x) = f(x + 1)$.

68. Sketch the graph of $y = (1 + x)^{1/x}$ for $x > 0$.

THE NATURAL EXPONENTIAL FUNCTION   S E C T I O N   4.2

Suppose a biology experiment requires that we place 100 rodents in a supportive environment (one with sufficient food and appropriate climate), and record the population growth. Let $f(t)$ represent the number of rodents after t days. The following table and graph (Figure 8) are typical of the data we might obtain.

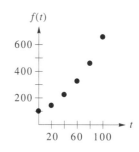

Days	t	0	20	40	60	80	100
Rodents	$f(t)$	100	146	214	314	460	673

FIGURE 8

The graph indicates that the larger the population, the more rapid the population growth. We can formulate the average rate of population growth in terms of $f(t)$. Since $f(t)$ represents the number of rodents after t days, then h days later, the number of rodents will be $f(t + h)$. The (average) growth rate will then be

$$\text{rate of growth} = \frac{\text{change in number of rodents}}{\text{time interval}}$$

$$= \frac{f(t + h) - f(t)}{h} \quad \text{(rodents per day)}$$

To get a good approximation of this growth rate at (or near) time t, h must be small (near zero).

As we observed earlier, the population grows more rapidly as the population increases. More precisely, the rate of growth is proportional to the number present. To analyze this relationship and mathematically model similar situations, mathematicians have asked the question, "Is there a function such that the rate of growth is equal to the amount (number) present at any time t?" In other words, find $f(t)$ such that

$$\frac{f(t + h) - f(t)}{h} \approx f(t).$$

Suppose $f(t)$ is an exponential function, that is, $f(t) = b^t$ (this supposition is inspired by our graph of the rodent population). Then the preceding equation would take the form

$$\frac{b^{t+h} - b^t}{h} \approx b^t \quad \text{or} \quad \frac{b^t(b^h - 1)}{h} \approx b^t.$$

Multiplying both sides by h/b^t and solving for b, we have

$$b^h - 1 \approx h \Rightarrow b^h \approx 1 + h \Rightarrow b \approx (1 + h)^{1/h}.$$

Remember that we want h to be small; in fact, the closer h is to zero, the better the approximation. The table illustrates $b = (1 + h)^{1/h}$ for values of h near zero:

h	$b = (1 + h)^{1/h}$
1	$2^1 = 2$
0.1	$(1.1)^{10} \approx 2.59374$
0.01	$(1.01)^{100} \approx 2.70481$
0.001	$(1.001)^{1000} \approx 2.71692$
0.0001	$(1.0001)^{10000} \approx 2.71815$

It can be shown that as $h \to 0$,

$$(1 + h)^{1/h} \to 2.71828182845904523 \ldots .$$

This is an irrational number that is so important in mathematics, science, and engineering that it is given its own symbol, e.

Definition of the Number e

> As $h \to 0$, $(1 + h)^{1/h} \to e = 2.71828182845904523 \ldots$. Equivalently, as $k \to \infty$, $(1 + 1/k)^k \to e$.

Futhermore, the function we were originally seeking is $f(t) = e^t$. This exponential function is referred to as the **natural exponential function.** The constant e is referred to as the **natural base.**

> $f(x) = e^x$ is the natural exponential function.

Many calculators have an e^x key. If you have trouble locating it, consult the owner's manual.

EXAMPLE 1   Using a calculator with an e^x key, approximate:

a) $e^{(0.023)(11)}$ **b)** $16e^{-5/8}$

SOLUTION

a)

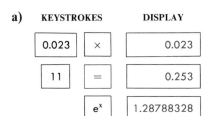

KEYSTROKES		DISPLAY
0.023	×	0.023
11	=	0.253
	e^x	1.28788328

b)

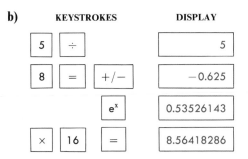

KEYSTROKES			DISPLAY
5	÷		5
8	=	+/−	−0.625
		e^x	0.53526143
×	16	=	8.56418286

EXAMPLE 2 Write $h(x) = e^{3/(2x)}$ as the composition of two functions $f[g(x)]$. Do not use $f(x) = x$ or $g(x) = x$.

SOLUTION One way is to let

$$g(x) = \frac{3}{2x} \quad \text{and} \quad f(x) = e^x.$$

Then

$$f[g(x)] = f\left[\frac{3}{2x}\right] = e^{3/(2x)}.$$

EXAMPLE 3 Sketch the graph of $f(x) = e^x$.

SOLUTION Since $2 < e < 3$, the graph of $f(x) = e^x$ is between $y = 2^x$ and $y = 3^x$ (Figure 9).

EXAMPLE 4 Graph $y = 2e^{-x}$.

SOLUTION We start with the graph of $y = e^x$ (see Example 3). If we reflect this graph through the y-axis, the resulting curve is the graph of $y = e^{-x}$ (Figure 10). To graph $y = 2e^{-x}$, we expand the graph of $y = e^{-x}$ from the x-axis by a factor of 2 (Figure 11).

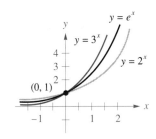

FIGURE 9

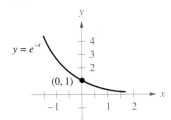

FIGURE 10

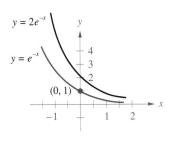

FIGURE 11

Keep in mind that the natural exponential function e^x is a special case of the general exponential function b^x. All of the properties of exponential functions that we discussed in Section 4.1 apply to the natural exponential function.

When a certain quantity is tending toward a value related to e, there are techniques for analyzing that exact value the quantity is approaching.

EXAMPLE 5   As $h \to 0$, what does $(1 + h)^{4/h}$ approach?

SOLUTION First, to get an estimate, let h take on values close to zero. Using a calculator with a y^x key:

h	$(1 + h)^{4/h}$
0.1	$(1.1)^{40} \approx 45.26$
0.01	$(1.01)^{400} \approx 53.52$
0.001	$(1.001)^{4000} \approx 54.49$
-0.01	$(0.99)^{-400} \approx 55.71$
-0.001	$(0.999)^{-4000} \approx 54.71$

To get the exact value, first note that

$$(1 + h)^{4/h} = [(1 + h)^{1/h}]^4.$$

Then, as $h \to 0$,

$$[(1 + h)^{1/h}]^4 \to [e]^4.$$

So the exact value is e^4. Using a calculator with an e^x key gives

$$e^4 \approx 54.598.$$

EXAMPLE 6   As $n \to \infty$, what does $(1 + 2/n)^{3n}$ approach?

SOLUTION To get an estimate, let n take on large values.

n	$\left(1 + \dfrac{2}{n}\right)^{3n}$
1000	$(1.002)^{3000} \approx 401$
50000	$(1.00004)^{150,000} \approx 403$
National deficit	403

To find the exact value, use the fact that as $k \to \infty$

$$\left(1 + \frac{1}{k}\right)^k \to e.$$

Comparing this with the base of our problem

$$\left(1 + \frac{2}{n}\right)$$

suggests letting $2/n = 1/k$. This means that $k = n/2$ and that $n = 2k$. So,

$$\left(1 + \frac{2}{n}\right)^{3n} = \left(1 + \frac{2}{(2k)}\right)^{3(2k)} = \left(1 + \frac{1}{k}\right)^{6k} = \left[\left(1 + \frac{1}{k}\right)^k\right]^6.$$

Because $k = n/2$, $n \to \infty$ if and only if $k \to \infty$. Therefore,

$$\left(1 + \frac{2}{n}\right)^{3n} = \left[\left(1 + \frac{1}{k}\right)^k\right]^6 \to [e]^6.$$

The exact value is e^6. A calculator approximation, using the e^x key, is $e^6 \approx$ 403.4288.

In Section 4.1 we introduced the compound interest formula

$$A = P\left(1 + \frac{r}{n}\right)^{nt}$$

for P dollars invested at an interest rate r for t years compounded n times per year. Suppose P, r, and t remain fixed, but n increases.

For example, suppose $1000 is invested at 6% for 7 years. The table shows several situations.

Compounded	Amount
Quarterly ($n = 4$)	$A = 1000\left(1 + \dfrac{0.06}{4}\right)^{(4)(7)} = \$1517.22$
Monthly ($n = 12$)	$A = 1000\left(1 + \dfrac{0.06}{12}\right)^{(12)(7)} = \$1520.37$
Daily ($n = 365$)	$A = 1000\left(1 + \dfrac{0.06}{365}\right)^{(365)(7)} = \$1521.91$
Every second ($n = 525,600$)	$A = 1000\left(1 + \dfrac{0.06}{525600}\right)^{(525600)(7)} = \$1521.96$

We could continue to increase n, but the accumulated total would not exceed $1,521.96 (rounded to the nearest penny).

Some saving institutions advertise interest **compounded continuously,** meaning that n increases without bound. For our example, $1000 invested at 6% interest for 7 years compounded continuously would amount to $1,521.96. That is, as $n \to \infty$,

$$A = 1000 \left(1 + \frac{0.06}{n} \right)^{7n} \to 1521.96.$$

To get a more general understanding of continuously compounded interest, we need to look at the behavior of $P(1 + r/n)^{nt}$ as $n \to \infty$.

We want to use the fact that as $k \to \infty$, $(1 + 1/k)^k \to e$. This suggests letting $1/k = r/n$. This is equivalent to $n = kr$. We have

$$A = P \left(1 + \frac{r}{n} \right)^{nt} = P \left(1 + \frac{r}{(kr)} \right)^{(kr)t} = P \left(1 + \frac{1}{k} \right)^{krt} = P \left[\left(1 + \frac{1}{k} \right)^{k} \right]^{rt}.$$

Now, $n \to \infty$ if and only if $k \to \infty$. And, as $n \to \infty$,

$$P \left(1 + \frac{r}{n} \right)^{nt} = P \left[\left(1 + \frac{1}{k} \right)^{k} \right]^{rt} \to Pe^{rt}.$$

The result is summarized in the following statement.

Continuously Compounded Interest

> If P dollars is invested for t years at an interest rate r compounded continuously, the amount accrued is
>
> $$A = Pe^{rt}.$$

EXAMPLE 7   Suppose $6000 is deposited into an account that pays 8% interest. Find the amount in the account after 3 years if the interest is compounded (a) semiannually and (b) continuously.

SOLUTION

a)   Use the formula $A = P(1 + r/n)^{nt}$ with $P = 6000$, $r = 0.08$, $n = 2$, and $t = 3$.

$$A = 6000 \left(1 + \frac{0.08}{2} \right)^{(2)(3)} = 7591.91.$$

The amount in the account after three years with semiannual compounding is \$7,591.91.

b) Use the formula $A = Pe^{rt}$ with $P = 6000$, $r = 0.08$, and $t = 3$.

$$A = 6000e^{(0.08)(3)} = 6000e^{0.24} = 7627.49.$$

The amount in the account after three years with continuous compounding is \$7,627.49.

At the beginning of this section we considered the population growth of rodents. It appeared that the rate of growth of the population is proportional to the size of the population. This idea can be applied to many other situations. It should seem reasonable that, under favorable conditions, the rate of growth of a population is proportional to the size of the population. The following mathematical model is based on this assumption.

Population Growth

> A formula used to model population growth states that the population $P(t)$ at time t is
>
> $$P(t) = P_0 e^{kt}$$
>
> where P_0 is the initial population and k is a constant.

EXAMPLE 8   Suppose that for a certain culture of bacteria the function

$$Q(t) = 25000e^{0.02t}$$

gives the number of bacteria present after t hours. Find **a)** the initial number of bacteria, and **b)** the number of bacteria after 2 days.

SOLUTION

a) The initial number of bacteria corresponds to $Q(t)$ when $t = 0$:

$$Q(0) = 25000e^{(0.02)(0)} = 25,000.$$

b) The number of bacteria after 2 days corresponds to $Q(t)$ when $t = 48$ hours:

$$Q(48) = 25000e^{(0.02)(48)} \approx 65,292.$$

EXAMPLE 9   In 1930 the population of the western United States (the states in Pacific and Mountain time zones) was 12,320,000. In 1950 the population was 20,190,000. According to the model for population growth, **a)** predict the population of the west in the year 2000, and **b)** find and compare the population predicted by this model in 1980 to the actual population of 42,410,000.

SOLUTION   We start with the population growth model

$$P(t) = P_0 e^{kt}.$$

Our plan is to solve for e^k. Then we will be able to determine the required values of $P(t)$. Our initial record of the population is in 1930. This will correspond to $t = 0$. The initial population is

$$P_0 = 12{,}320{,}000.$$

Substituting this into the growth model gives us

$$P(t) = (12320000)e^{kt}.$$

We are also given that the population 20 years later (in 1950) is 20,190,000. In terms of our model, this means that

$$P(20) = 20190000.$$

If we substitute this into the growth model, we have

$$20190000 = P(20) = (12320000)e^{k(20)}.$$

We divide both sides of this equation by 12320000:

$$\frac{20190000}{12320000} = e^{k(20)}$$

$$\frac{2019}{1232} = (e^k)^{20}.$$

Now to solve for e^k we raise both sides of the last equation to the power $\frac{1}{20}$.

$$\left(\frac{2019}{1232}\right)^{1/20} = e^k.$$

Therefore our growth model is

$$P(t) = 12320000[e^k]^t$$

$$= 12320000\left[\left(\frac{2019}{1232}\right)^{1/20}\right]^t.$$

a) The year 2000 is 70 years after 1930. The population then should be

$$P(70) = 12320000\left[\left(\frac{2019}{1232}\right)^{1/20}\right]^{70}$$

$$= 12320000\left(\frac{2019}{1232}\right)^{7/2}$$

Using a calculator, we estimate that

$$12320000\left(\frac{2019}{1232}\right)^{7/2} \approx 69{,}410{,}000.$$

b) The year 1980 is 50 years after 1930. According to the model, the population should have been

$$P(50) = 12320000\left[\left(\frac{2019}{1232}\right)^{1/20}\right]^{50}$$

$$= 12320000\left(\frac{2019}{1232}\right)^{5/2}.$$

With a calculator, we get

$$12320000\left(\frac{2019}{1232}\right)^{5/2} \approx 42{,}360{,}000.$$

Comparing this with the actual population, our error is

$$\frac{\text{actual population} - \text{predicted population}}{\text{actual population}}$$

$$= \frac{42{,}410{,}000 - 42{,}360{,}000}{42{,}410{,}000} = 0.00117\ldots.$$

This is an error of less than 0.12%.

Consider the rate at which an object cools. It seems reasonable that the warmer the object is (compared with the surrounding temperature), the faster it will cool. More precisely, suppose an object is at temperature T_0. At time $t = 0$ it is placed into a surrounding temperature A. Let $T(t)$ denote the temperature of the object at time t. **Newton's law of cooling** says the difference $T(t) - A$ decreases at a rate proportional to itself. This leads to the following model.

Newton's Law of Cooling

An object at temperature T_0, surrounded by air temperature A, cools to the temperature $T(t)$, t minutes later, according to the formula

$$T(t) = A + (T_0 - A)e^{kt}.$$

The constant k depends on the particular object.

EXAMPLE 10   A soft drink at a temperature of $80°$ is placed in a refrigerator where the temperature is $36°$. Ten minutes later, the temperature of the soft drink is $68°$. Use Newton's law of cooling to predict the temperature 25 minutes after it was placed in the refrigerator.

SOLUTION Our strategy is similar to the solution for Example 9. We begin with the model

$$T(t) = A + (T_0 - A)e^{kt}.$$

The surrounding temperature is $A = 36°$, and the initial temperature is $T_0 = 80°$. Thus we have

$$T(t) = 36 + (80 - 36)e^{kt}$$
$$= 36 + 44e^{kt}.$$

Since we are given that the temperature ten minutes later is $68°$, we have

$$68 = 36 + 44e^{k(10)}.$$

We now solve for e^k. First subtract 36 from both sides of this equation and then divide both sides by 44:

$$32 = 44e^{k(10)}$$

$$\frac{32}{44} = (e^k)^{10}$$

$$\frac{8}{11} = (e^k)^{10}.$$

Raising both sides of the last equation to the power $\frac{1}{10}$, we get

$$\left(\frac{8}{11}\right)^{1/10} = e^k.$$

−56.9264

Substituting this into our model for $T(t)$, we have

$$T(t) = 36 + 44\left[\left(\frac{8}{11}\right)^{1/10}\right]^t.$$

The temperature of the soft drink after being in the refrigerator for 25 minutes is

$$T(25) = 36 + 44\left[\left(\frac{8}{11}\right)^{1/10}\right]^{25}.$$

Using a calculator, this can be approximated as

$$36 + 44\left(\frac{8}{11}\right)^{2.5} \approx 56°.$$

EXERCISE SET 4.2

A

In Problems 1 through 3, use a calculator to approximate $(1 + h)^{1/h}$ to the nearest 0.0001, for the indicated values of h.

1. $h = 0.05$ **2.** $h = 0.002$

3. $h = -0.0004$

In Problems 4 through 9, use a calculator to approximate each expression to the nearest 0.0001.

4. $e^{3.2}$ **5.** $e^{1.55}$ **6.** $18e^{-(1.8)(4)}$

7. $-6e^{(0.75)(3)}$ **8.** $e^{3-\sqrt{2}}$ **9.** $e^{2+\sqrt{5}}$

In Problems 10 through 15, sketch the graph of the equation. Identify asymptotes.

10. $y = (\frac{1}{2})e^x$ **11.** $y = 3e^x$

12. $y + 1 = e^{x-3}$ **13.** $y - 2 = e^{x-1}$

14. $y = e^{-x} + 2$ **15.** $y = e^{-x} - 4$

In Problems 16 through 21, write h(x) as the composition of two functions, $f[g(x)]$. Do not use $f(x) = x$ or $g(x) = x$.

16. $h(x) = e^{2x+5}$ **17.** $h(x) = e^{3-x}$

18. $h(x) = e^{\sqrt{x}}$ **19.** $h(x) = e^{|x|}$

20. $h(x) = e^{-x^2}$ **21.** $h(x) = e^{(x-5)/4}$

In Problems 22 through 24, solve for x.

22. $5xe^x + 9e^x = 0$ **23.** $2e^x - 3xe^x = 0$

24. $x^2e^x - 4xe^x = 0$

B

In Problems 25 through 30, find the value that $f(x)$ approaches.

25. $f(x) = (1 + x)^{3/x}$ as $x \to 0$

26. $f(x) = (1 + x)^{1/(2x)}$ as $x \to 0$

27. $f(x) = \left(1 + \frac{2}{x}\right)^{5x}$ as $x \to \infty$

28. $f(x) = \left(1 + \frac{3}{2x}\right)^{4x}$ as $x \to \infty$

29. $f(x) = x^{1/(x-1)}$ as $x \to 1$

30. $f(x) = (x - 4)^{2/(x-5)}$ as $x \to 5$

In Problems 31 through 38, sketch the graph of the equations. Identify asymptotes.

31. $y = -e^{2x}$ **32.** $y = -2e^x$ **33.** $y = e^{|x|}$

34. $y = e^{-x^2}$ **35.** $y = e^{3-x}$ **36.** $y = e^{2-x}$

37. $y = e^x + e^{-x}$ **38.** $y = e^x - e^{-x}$

39. Suppose $15,000 is deposited at a 12% rate compounded continuously. This will grow to what amount after 4 years?

40. Suppose $9,000 is deposited at a 10.5% rate compounded continuously. This will grow to what amount after 5 years?

41. Is it better to invest at 8% compounded quarterly or 7.75% compounded continuously?

42. Suppose a two-year certificate of deposit pays interest at a 10% rate compounded continuously. If the certificate is worth $9,500 at maturity, what was it originally purchased for two years ago?

43. Suppose that for a certain culture of bacteria, the function $Q(t) = 25,000e^{0.2t}$ gives the number of bacteria present after t hours. Find **a)** the initial number of bacteria and **b)** the number of bacteria after 2 days.

44. Suppose that for a certain city, the function $Q(t) = 45,000e^{0.06t}$ yields the population t years after 1988. Find **a)** population in 1988 and **b)** the population in 1995 (7 years after 1988).

45. In 1940 the population of the southern United States was 41,666,000. In 1960 the population was 54,973,000. According to the model for population growth, **a)** predict the population of the south in the year 2000 and **b)** find and compare the population predicted by this model in 1980 to the actual population of 75,370,000.

46. In 1930 the population of the northeastern United States was 34,427,000. In 1950 the population was 39,478,000. According to the model for population growth, **a)** predict the population of the northeast in the year 2000 and **b)** find and compare the population predicted by this model in 1980 to the actual population of 49,135,000.

47. A culture of bacteria is observed to have a population of about 10,000 at 3:00 P.M. Two hours later the number of bacteria has grown to 25,000. Find the population $P(t)$, where t is the number of hours after 3:00 P.M. Assume the population growth model.

48. Based on the function $P(t)$ in Problem 47, predict the number of bacteria at 8:00 P.M. (5 hours after the initial observation).

49. A cake at a temperature of 350° is removed from an oven and placed in a room where the temperature is a constant 75°. Twenty minutes later the cake has cooled to 210°. Using Newton's law of cooling, **a)** find the cooling function $T(t)$ and **b)** predict the temperature of the cake 30 minutes after the cake is removed from the oven.

50. A cup of coffee initially at 140° cools to 125° after ten minutes in a room where the temperature is a constant 72°. Using Newton's law of cooling, **a)** find the cooling function $T(t)$ and **b)** predict the temperature of the coffee after 20 minutes of cooling.

51. Deforestation is a major problem facing sub-Saharan Africa. The fuel-wood consumption per year (in million cubic meters) in Sudan is approximated by the function $C(t) = 76e^{0.3t}$, where $t = 0$ corresponds to the year 1980. Predict the consumption in the year 1994.

52. In an isolated population, a virus is introduced by one infected person. A formula used to model the spread of the virus is

$$n(t) = \frac{a}{1 + (a - 1)e^{akt}}$$

where $n(t)$ is the number of infected people, a is the population, and t is the number of days after the virus is introduced. Suppose an isolated farming community with a population of 1600 is infected by a person with the flu. Suppose it is further observed that after 5 days, $n(5) = 20$. How many people will have the flu after 7 days?

53. A mathematical model known as the **learning curve,** which is used to model a learning process, is $N(t) = c - ce^{-kt}$, where $N(t)$ is the number of tasks mastered after t units of time, and k and c are positive constants. Suppose a new employee must learn the phone numbers of her 30 colleagues. She learned 8 of them in two days. Use the given mathematical model $N(t) = 30 - 30e^{-kt}$, where t is in days, to predict how many phone numbers she will have learned in 5 days.

54. Experiments have shown that sales of a new product, under relatively stable conditions, with no promotional activities, declines according to the function $S(t) = S_0e^{-kt}$, where S_0 is the initial number of units sold. The value of k varies from product to product but is a constant throughout the years for a particular product. Suppose that a new product has been on the market for 30 days. Suppose further that on the 30th day the number of units sold was only one-third of the number sold on the first day. Predict the sales on the 45th day.

C

55. In calculus, you will see that e^x can be approximated by the polynomial function

$$P(x) = 1 + x + \tfrac{1}{2}x^2 + \tfrac{1}{6}x^3.$$

Use $P(x)$ to approximate the value of $e^{0.75}$. Compare your result by using your calculator to approximate $e^{0.75}$ directly.

56. Graph $y = x^{1/x}$ for $0 < x \le 5$. Make a conjecture about where the maximum occurs.

57. Simplify $\dfrac{e^x(2e^{-2x} + 1) - (e^{2x} + 1)e^{-x}}{e^{4x}}$.

58. What does the following continued fraction seem to approach?

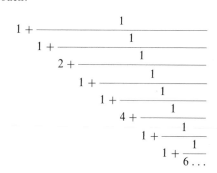

LOGARITHMIC FUNCTIONS

SECTION 4.3

Consider the equation

$$x^3 = 5.$$

The notation for the solution is

$$x = \sqrt[3]{5}.$$

The most common name for this number is "the cube root of 5". Perhaps a better name would have been "the number whose cube is 5".

Consider the equation

$$3^x = 5.$$

For this equation, we introduce the notation for the solution as

$$x = \log_3 5.$$

The most common way to say this is "the **logarithm** base 3 of 5." Perhaps a better version would be "the **exponent** to which 3 must be raised to get 5." Try not to worry about the value of this symbol—we will get to that soon enough. For the time being, concentrate on its meaning.

The Logarithm

> For $a > 0$, $b > 0$, and $b \neq 1$, the solution to the equation $b^t = a$ is denoted as
>
> $$t = \log_b a.$$
>
> This solution is called the logarithm base b of a.

Always keep in mind that a logarithm is an exponent.

EXAMPLE 1   Compute the following:

    a) $\log_4 16$ **b)** $\log_5(\frac{1}{5})$ **c)** $\log_{1/2} 8$

SOLUTION

    a) $\log_4 16$ represents the solution of $4^t = 16$. Knowing that $16 = 4^2$, the answer is $\log_4 16 = 2$.

    b) $\log_5(\frac{1}{5})$ represents the solution of $5^t = \frac{1}{5}$. Knowing that $5^{-1} = \frac{1}{5}$, the answer is $\log_5(\frac{1}{5}) = -1$.

    c) $\log_{1/2} 8$ represents the solution of $(\frac{1}{2})^t = 8$. Recognizing that $(\frac{1}{2})^{-3} = 8$, the answer is $\log_{1/2} 8 = -3$.

In many cases, finding the value of $\log_b a$ may not be easy. However, we should be able to estimate a given logarithm between two consecutive integers.

EXAMPLE 2   For each of the following logarithms, find two consecutive integers such that the logarithm is between them.

    a) $\log_3 2$ **b)** $\log_{10} 0.004$ **c)** $\log_5 425$

SOLUTION

    a) Since $3^0 < 2 < 3^1$, $0 < \log_3 2 < 1$.

    b) Since $10^{-3} < 0.004 < 10^{-2}$, $-3 < \log_{10} 0.004 < -2$.

    c) Since $5^3 < 425 < 5^4$, $3 < \log_5 425 < 4$.

Consider the statement

$$3^4 = 81.$$

This is equivalent to saying that

$$\log_3 81 = 4.$$

The first equation is called the exponential form of the statement, and the second equation is the logarithmic form. This equivalence can be generalized.

For $a > 0$, $b > 0$, and $b \neq 1$,

$$b^r = a \text{ is equivalent to } \log_b a = r.$$

The equation

$$b^r = a$$

is the **exponential form.** The equation

$$\log_b a = r$$

is the **logarithmic form.**

EXAMPLE 3

    a) Write $4^{3/2} = 8$ in logarithmic form.

    b) Write $\log_{25} 5 = \frac{1}{2}$ in exponential form.

SOLUTION

    a) The equation $4^{3/2} = 8$ is equivalent to $\log_4 8 = \frac{3}{2}$.

    b) The equation $\log_{25} 5 = \frac{1}{2}$ is equivalent to $25^{1/2} = 5$.

The rule $b^r b^s = b^{r+s}$ can be stated as: "The resulting exponent (logarithm) of a product is the sum of the individual exponents (logarithms)." More precisely, suppose P and Q are positive real numbers. If we write

$$P = b^r \quad \text{and} \quad Q = b^s$$

then the logarithmic forms of these equations are

$$r = \log_b P \quad \text{and} \quad s = \log_b Q.$$

If we multiply P and Q, we have

$$PQ = b^r b^s = b^{r+s}.$$

Putting this in logarithmic form gives

$$r + s = \log_b(PQ).$$

Since $r = \log_b P$ and $s = \log_b Q$, we also have

$$r + s = \log_b P + \log_b Q.$$

Therefore,

$$\log_b(PQ) = \log_b P + \log_b Q.$$

The rules $b^r/b^s = b^{r-s}$ and $(b^r)^t = b^{rt}$ can also be stated using logarithms.

Properties of Logarithms

For all positive real numbers P, Q, and b with $b \neq 1$,

1) $\log_b(PQ) = \log_b P + \log_b Q$ because $b^r b^s = b^{r+s}$.

2) $\log_b\left(\dfrac{P}{Q}\right) = \log_b P - \log_b Q$ because $\dfrac{b^r}{b^s} = b^{r-s}$.

3) $\log_b(P^t) = t \log_b P$ because $(b^r)^t = b^{rt}$.

The proofs of properties 2 and 3 are left as exercises. These three important properties are used extensively. They should be committed to memory.

EXAMPLE 4   Express the following in terms of one logarithm.

a) $\log_b x + 2 \log_b y$ **b)** $\log_b(u + 2) - \log_b(u - 4)$

c) $\log_b(x^4) - \frac{1}{2} \log_b(9x^6) + \log_b(xy)$

SOLUTION

a) $\log_b x + 2 \log_b y = \log_b x + \log_b(y^2)$ Using Property 3

$ = \log_b(xy^2)$ Using Property 1

b) $\log_b(u + 2) - \log_b(u - 4) = \log_b\left(\dfrac{u + 2}{u - 4}\right)$ Using Property 2

c) Using Property 3, the second term simplifies as

$$\tfrac{1}{2} \log_b(9x^6) = \log_b(9x^6)^{1/2} = \log_b(3x^3).$$

So the expression that we want to simplify is

$$\log_b(x^4) - \tfrac{1}{2}\log_b(9x^6) + \log_b(xy) = \log_b(x^4) - \log_b(3x^3) + \log_b(xy)$$

$$= \log_b\left(\frac{x^4}{3x^3} \cdot xy\right)$$

$$= \log_b\left(\frac{x^2 y}{3}\right).$$

EXAMPLE 5   Use the properties of logarithms to express each of the following as a sum or difference having no exponents or radicals.

a) $\log_b\left(\dfrac{t}{p^3}\right)$ b) $\log_b b\sqrt{x}$ c) $\log_b(5b^t)$

SOLUTION

a) $\log_b\left(\dfrac{t}{p^3}\right) = \log_b t - \log_b(p^3) = \log_b t - 3\log_b p$

b) $\log_b b\sqrt{x} = \log_b b + \log_b \sqrt{x} = 1 + \log_b(x^{1/2}) = 1 + \tfrac{1}{2}\log_b x$

c) $\log_b(5b^t) = \log_b 5 + \log_b(b^t) = \log_b 5 + t$

EXAMPLE 6   Assume that $5^{0.43} \approx 2$ and $5^{0.68} \approx 3$. Approximate the following:

a) $\log_5 2$ b) $\log_5 3$ c) $\log_5(\tfrac{3}{2})$

d) $\log_5 \sqrt{3}$ e) $\log_5 10$

SOLUTION

a) $5^{0.43} \approx 2$ is equivalent to $\log_5 2 \approx 0.43$.

b) $5^{0.68} \approx 3$ is equivalent to $\log_5 3 \approx 0.68$.

c) Using the results of a) and b), we have

$$\log_5(\tfrac{3}{2}) = \log_5 3 - \log_5 2$$

$$\approx 0.68 - 0.43$$

$$= 0.25.$$

d) Using the results of a), we have

$$\log_5 \sqrt{3} = \log_5(3^{1/2})$$

$$= \tfrac{1}{2}\log_5 3$$

$$\approx \tfrac{1}{2}(0.68)$$

$$= 0.34.$$

e)
$$\log_5 10 = \log_5 (2)(5)$$
$$= \log_5 (2) + \log_5 (5)$$
$$\approx 0.43 + 1$$
$$= 1.43.$$

The three properties of logarithms are not the only relationships we will need. What follows are some additional facts about logarithms that we will occasionally use.

Useful facts

For all real numbers a, b, and r, with $a > 0$, $b > 0$, and $b \neq 1$,

$$b^{\log_b a} = a \qquad\qquad \log_b b^r = r$$

$$\log_b \frac{1}{a} = -\log_b a \qquad \log_b 1 = 0$$

We will prove three of these. The fourth is left as an exercise.

To prove $b^{\log_b a} = a$, recall that

$$\log_b a = r \text{ is equivalent to } b^r = a.$$

Substituting $\log_b a$ for r in this last statement, we have

$$\log_b a = \log_b a \text{ is equivalent to } b^{\log_b a} = a.$$

Since the first equation in this statement is true, the second equation must be true also.

To prove $\log_b b^r = r$, we again exploit the fact that

$$\log_b a = r \text{ is equivalent to } b^r = a.$$

Substituting b^r for a, the statement becomes

$$\log_b b^r = r \text{ is equivalent to } b^r = b^r.$$

Since the second equation in this statement is true, the first equation must be true also.

To prove $\log_b 1 = 0$, we once again appeal to the fact that

$$\log_b a = r \text{ is equivalent to } b^r = a.$$

Letting $a = 1$ and $r = 0$ in this statement establishes that $\log_b 1 = 0$.

Recall that the exponential function $f(x) = b^x$ is a one-to-one function. Therefore we know it has an inverse function, $g(x)$. We start with

$$y = g(x).$$

Since f and g are inverses, we must have

$$f(y) = f[g(x)] = x.$$

The function f is the exponential function, so this last equation is

$$b^y = x.$$

The logarithmic form of this equation is

$$y = \log_b x.$$

Since we originally designated $y = g(x)$, we have

$$g(x) = \log_b x.$$

Logarithmic Function

A **logarithmic function** is a function of the form

$$g(x) = \log_b x$$

where $b > 0$, and $b \neq 1$. It is the inverse of the function $f(x) = b^x$.

Recall from Section 2.7 how the graph of a function $y = f(x)$ is related to the graph of its inverse $y = g(x)$. The graph of $y = g(x)$ is the reflection of the graph of $y = f(x)$ through the line $y = x$.

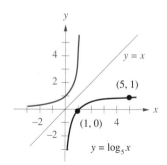

FIGURE 12

EXAMPLE 7   Graph $y = 5^x$. Use the fact that $f(x) = 5^x$ and $g(x) = \log_5 x$ are inverses to sketch the graph of $y = \log_5 x$.

SOLUTION We start with the graph of $y = 5^x$. Then we reflect the graph through the line $y = x$ (Figure 12).

EXAMPLE 8   Sketch the graph of

    a) $y = \log_{(1/5)} x$ **b)** $y = -\log_5 x$.

SOLUTION

    a) $y = \log_{(1/5)} x$ is the inverse function of $y = (\frac{1}{5})^x$.
        First we sketch the graph of $y = (\frac{1}{5})^x$. Then we reflect the graph through the line $y = x$ (Figure 13).

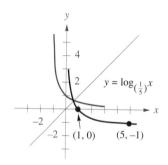

FIGURE 13

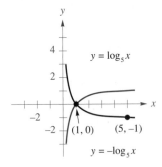

FIGURE 14

The Graph of $y = \log_b x$

b) First, sketch the graph of $y = \log_5 x$. The graph of $y = -\log_5 x$ is a reflection through the x-axis (Figure 14).

Notice from the graphs of Figures 13 and 14 that $\log_{(1/5)} x = -\log_5 x$.

The previous examples are typical of the general graphs of logarithmic functions. The graphs of $f(x) = \log_b x$ for the two cases $b > 1$ and $0 < b < 1$ follow.

The general graphs for $f(x) = \log_b x$ are as follows:

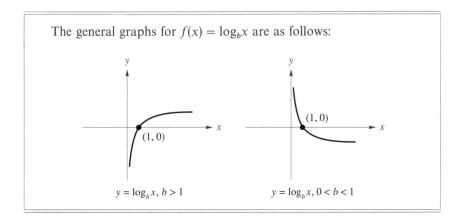

You should commit these graphs to memory because they illustrate some useful facts about logarithms:

The domain is all positive real numbers. This means that if $x \leq 0$ then $\log_b x$ does not exist.

As $x \to 0^+$, $\log_b x \to \pm\infty$. The y-axis is a vertical asymptote.

$f(x) = \log_b x$ is a one-to-one function. This means that if $\log_b p = \log_b q$, then $p = q$.

EXAMPLE 9   Find the domain of each function:

a) $f(x) = \log_4(3 - x)$ **b)** $g(x) = \log_{10}(x^2 - 1)$

SOLUTION

a) $\log_4(3 - x)$ is defined if and only if $3 - x > 0$. Solving for x, the domain is $x < 3$.

b) $\log_{10}(x^2 - 1)$ is defined if and only if $x^2 - 1 > 0$. Solving for x, the domain is $x < -1$ or $x > 1$.

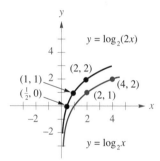

FIGURE 15

Knowing the general shape of the graph of $y = \log_b x$, we can use the techniques discussed in Chapter 2 to graph more complicated functions that involve logarithmic functions.

EXAMPLE 10 Sketch the graph of $y = \log_2(2x)$.

SOLUTION First, sketch the graph of $y = \log_2 x$; then recall from Section 2.5 that the graph of $f(2x)$ is a compression of $f(x)$ toward the y-axis (Figure 15).

ALTERNATE SOULTION Using a property of logarithms gives

$$\log_2(2x) = \log_2 2 + \log_2 x = 1 + \log_2 x.$$

So $y = \log_2(2x)$ can be written as $y = 1 + \log_2 x$ or $y - 1 = \log_2 x$. Now, sketch $y = \log_2 x$ and then translate the curve up 1 unit, (Figure 16).

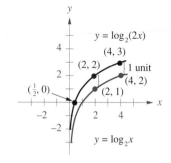

FIGURE 16

EXAMPLE 11 Sketch $y = \log_3\sqrt{x}$.

SOLUTION Using a property of logarithms gives

$$\log_3\sqrt{x} = \log_3(x^{1/2}) = \tfrac{1}{2}\log_3 x.$$

So $y = \log_3\sqrt{x}$ can be written as $y = \tfrac{1}{2}\log_3 x$. The graph of $y = \tfrac{1}{2}\log_3 x$ is a compression of $y = \log_3 x$ toward the x-axis (Figure 17).

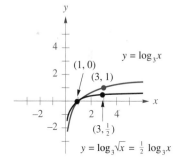

FIGURE 17

We know that the inverse of the exponential function b^x is the logarithmic function $\log_b x$. For the case in which the base is e, the logarithm has a special notation.

Natural Logarithm

The inverse of e^x is $\log_e x$. The function

$$f(x) = \log_e x$$

is called the **natural logarithm** and is usually written as $\ln x$:

$$f(x) = \log_e x = \ln x.$$

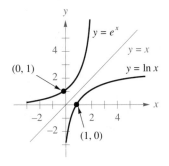

FIGURE 18

The graph of $y = \ln x$ is obtained by reflecting the graph of $y = e^x$ through the line $y = x$ (Figure 18). Furthermore, the graph of $y = \ln x$ lies between the graphs of $y = \log_2 x$ and $y = \log_3 x$ because $2 < e < 3$ (Figure 19).

The next three examples are graphs of functions that are typically used in calculus.

EXAMPLE 12 Sketch the graph of $y = \ln(x + 4)$.

SOLUTION First, sketch the graph of $y = \ln x$. To graph $y = \ln(x + 4)$, translate the graph of $y = \ln x$ four units to the left (Figure 20). The line $x = -4$ is a vertical asymptote.

EXAMPLE 13 Sketch the graph of $y = \ln(3 - x)$.

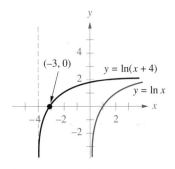

FIGURE 19

SOLUTION First, we sketch the graph of $y = \ln(-x)$, which is a reflection about the y-axis of the graph of $y = \ln x$ (Figure 21).

Note that

$$\ln(3 - x) = \ln[-(x - 3)].$$

Therefore, to graph $y = \ln(3 - x)$, we translate the graph of $y = \ln(-x)$ three units to the right. The vertical line $x = 3$ is an asymptote. The domain of the function is $x < 3$ (Figure 22).

EXAMPLE 14 Sketch the graph of $y = \ln|x|$.

SOLUTION We start with the graph of $y = \ln x$ and use the graphing technique for $y = f(|x|)$ discussed in Section 2.5. There we saw that the graph of $y = f(|x|)$ is the same graph as $y = f(x)$ to the right of the y-axis. To the left of the y-axis, the graph of $y = f(|x|)$ is the reflection of the right "half" of $y = f(x)$ (Figure 23).

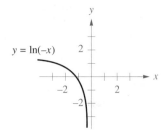

FIGURE 20

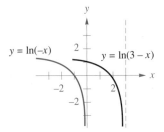

FIGURE 21

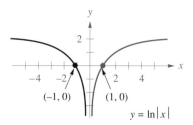

FIGURE 22   FIGURE 23

The natural logarithm is used so often in applications that most scientific calculators have an ln key.

EXAMPLE 15   Find a numerical approximation for $\ln(\frac{5}{3})$.

SOLUTION   The keystrokes are shown below:

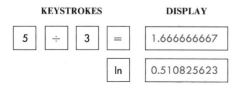

KEYSTROKES	DISPLAY
5 ÷ 3 =	1.666666667
ln	0.510825623

An alternate way uses a property of logarithms:

$$\ln(\tfrac{5}{3}) = \ln 5 - \ln 3.$$

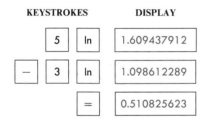

KEYSTROKES	DISPLAY
5 ln	1.609437912
− 3 ln	1.098612289
=	0.510825623

EXERCISE SET 4.3

A

Express each equation in Problems 1 through 9 in logarithmic form.

1. $5^3 = 125$ **2.** $2^6 = 64$

3. $36^{1/2} = 6$ **4.** $10^{-1} = 0.1$

5. $4^{-2} = 0.0625$ **6.** $2.25 = 1.5^2$

7. $5^x = 12$ **8.** $3^t = \frac{1}{2}$

9. $5 = (2.7)^k$

Express each equation in Problems 10 through 18 in exponential form.

10. $\log_{10}1 = 0$ **11.** $\log_9 3 = \frac{1}{2}$

12. $\log_9(\frac{1}{3}) = -\frac{1}{2}$ **13.** $\log_{10}1000 = 3$

14. $\log_8\sqrt{512} = 1.5$ **15.** $\log_2 1.4 = 5t$

16. $\log_7 30 = 3x$ **17.** $\log_{2.7}2 = 3t^2$

18. $\log_{10}1.3 = c^5$

Without using a calculator, find two consecutive integers that each logarithm in Problems 19 through 24 is between.

19. $\log_5 80$ **20.** $\log_3 50$ **21.** $\log_{10}3250$

22. $\log_{10}0.03$ **23.** $\log_3(\frac{1}{2})$ **24.** $\log_2(\frac{1}{5})$

In Problems 25 through 30, find the inverse function.

25. $f(x) = 10^x$ **26.** $f(x) = 9^x$

27. $f(x) = \log_4 x$ **28.** $f(x) = \log_6 x$

29. $f(x) = \log_3(x + 5)$ **30.** $f(x) = 5^{x-3}$

In Problems 31 through 36 find the domain of each function.

31. $f(x) = \log_3(2x + 1)$ **32.** $f(x) = \log_4(5 - 2x)$

33. $g(x) = \log_{2.5}(x^2)$ **34.** $g(x) = \log_2(x^2 - 4)$

35. $h(x) = \log_5(x^3 - 2x^2)$

36. $h(x) = \log_5(3x^3 - 4x^2)$

In Problems 37 through 55, sketch the graph of $y = f(x)$. Identify any asymptotes.

37. $f(x) = \ln(2x)$ **38.** $f(x) = \ln(\frac{3}{2}x)$

39. $f(x) = \log_2(x + 3)$ **40.** $f(x) = \log_5(x - 2)$

41. $f(x) = \log_4(-x)$ **42.** $f(x) = \log_5(4 - x)$

43. $f(x) = \ln(2 - x)$ **44.** $f(x) = \ln(x^3)$

45. $f(x) = \ln(x^5)$ **46.** $f(x) = \log_6(2x - 4)$

47. $f(x) = \log_2(\frac{1}{3}x + 1)$ **48.** $f(x) = \ln(\sqrt[3]{x})$

49. $f(x) = \ln(\sqrt[4]{x})$ **50.** $f(x) = -2 \log_{10}x$

51. $f(x) = -3 \log_{10}x$

52. $f(x) = \log_9(x^2)$ (*Hint: $f(x)$ is even.*)

53. $f(x) = \log_8(x^4)$ (*Hint: $f(x)$ is even.*)

54. $f(x) = 3 - \log_2(x - 1)$

55. $f(x) = 4 - \log_5(x + 2)$

B

Express Problems 56 through 64 as a single logarithm.

56. $\log_b a + \log_b 2a$

57. $\log_b \dfrac{a^2}{x} + \log_b 5a$

58. $\log_b(2x + 3) - \log_b(x - 1)$

59. $\log_b(u + 2) - 2 \log_b(u - 3)$

60. $\log_b x + \log_b 4 - 3 \log_b 3x$

61. $5 \log_b(2x) - \log_b x - \log_b(4x^2)$

62. $2 \log_b x + \log_b(\sqrt[3]{x})$

63. $\log_b \cdot \sqrt{3x} - 4 \log_b x$

64. $\log_b(y^2 - 4) - 3 \log_b(y + 2) + \log_b 1$

Use the properties of logarithms to expand Problems 65 through 70.

65. $\log_b \left(\dfrac{x^2}{y}\right)$ **66.** $\log_b \left(\dfrac{3}{m^4}\right)$

67. $\log_b A \sqrt[3]{B}$ **68.** $\log_b K \sqrt{D}$

69. $\log_b \left(\dfrac{5x^7 \sqrt[3]{y}}{\sqrt{z}}\right)$ **70.** $\log_b \left(\dfrac{2x^3(\sqrt{y})^5}{\sqrt[3]{z}}\right)$

In Problems 71 through 79, assume that $10^{0.301} \approx 2$ and $10^{0.477} \approx 3$. Use these to approximate the logarithm.

71. $\log_{10}2$ **72.** $\log_{10}3$ **73.** $\log_{10}(\frac{2}{3})$

74. $\log_{10}9$ **75.** $\log_{10}6$ **76.** $\log_{10}12$

77. $\log_{10}30$ **78.** $\log_{10}0.2$ **79.** $\log_{10}5$

80. Graph $y = \ln|x - 4|$.

81. Graph $y = \ln|x + 1|$.

82. Montgomery Ward determined the following model to approximate sales in terms of the allocation of catalog advertising.

$$\ln S = \ln k + 0.56 \ln x_1 + 0.56 \ln x_2$$
$$+ 0.23 \ln x_3 + 0.46 \ln x_4$$

where S is the total sales of a line of merchandise (in thousands of dollars), x_1 is the number of catalogs distributed (in thousands), x_2 is the number of pages devoted to the line, x_3 is the number of items in the line, and x_4 is a color factor. Express the right side of the equation as a single logarithm.

83. Allometry is the study of the relative growth of a part in relation to an entire organism. If A and B are the sizes (in cubic units) of two organs of a particular animal, then A and B are related by the allometric equation $\ln B - k \ln A = \ln C$, where k and C depend only on the organs that are measured and are constant among animals belonging to the same species. Express the left side of the equation as a single logarithm.

84. It was found in a study that the average walking speed s (in feet per second) of people that live in a city of population p (in thousands) is given by

$$s(p) = 0.37 \ln p + 0.05.$$

a) The population of Santa Rosa, California, is 100,000. What is the average walking speed of a Santa Rosan?

b) The population of Chicago, Illinois, is 3,100,000. What is the average walking speed of a Chicagoan?

85. The number of years $N(r)$ since two independently evolving languages split off from a common ancestral language is approximated by $N(r) = -5000 \ln r$. The value of r is the proportion of the words from the ancestral language that are common to both languages now. It has been approximately 1300 years since the north and south French dialects split. Use this to estimate the proportion of words from the ancestral language that are common to both dialects now.

86. The population p of blue whales in the southern hemisphere t years after 1988 is related by the equation

$$t = 21 \ln \frac{p}{8000}.$$

To the nearest year and month, when will the population reach 12,000?

C

87. Evaluate $\log_{10}(\frac{1}{2}) + \log_{10}(\frac{2}{3}) + \log_{10}(\frac{3}{4}) + \cdots + \log_{10}(\frac{98}{99}) + \log_{10}(\frac{99}{100})$.

88. Prove that $\log_b(P/Q) = \log_b P - \log_b Q$.

89. Prove that $\log_b P^t = t \log_b P$.

90. Prove or disprove that $\log_b(P + Q) = \log_b P + \log_b Q$.

91. Prove or disprove that $(\log_b P)(\log_b Q) = \log_b PQ$.

92. If p and q are the roots of $2x^2 - mx + 1 = 0$, what is the value of $\log_2 p + \log_2 q$?

93. Simplify $3^{\log_9 x}$.

EXPONENTIAL AND LOGARITHMIC EQUATIONS

S E C T I O N 4.4

Any positive number except 1 can be used as the base for a logarithm. One of the most practical bases, from a computational point of view, is base 10. This is because our decimal system uses powers of 10.

> The logarithm base 10 is called the **common logarithm** and is written $\log x$. In other words, the base for $\log x$ is understood to be 10.

Common Logarithm

For 300 years, computations with logarithms were done using tables of common "logs." Now, hand-held calculators and computers have made computations with tables unnecessary.

For example, suppose we want to approximate $\log 732.26$. Since $10^2 < 732.26 < 10^3$, it follows that $2 < \log 732.26 < 3$. We can obtain a much better approximation using a calculator:

KEYSTROKES	DISPLAY
732.26 | log |	2.864665311

We emphasize that this is an approximation. Calculators do not yield exact values in general. For example, many calculators approximate $\log(10^{0.025})$ as

$$\log(10^{0.025}) \approx 0.024999999.$$

Of course, we know that the exact answer is

$$\log(10^{0.025}) = 0.025.$$

The properties of logarithms and a calculator are useful tools in solving equations.

EXAMPLE 1   Solve $\log x = 3.54$ for x. Also give an approximation to the nearest 0.0001.

SOLUTION First we write the equation in exponential form:

$$x = 10^{3.54}.$$

This is the exact solution. With a calculator, we get the approximation $x \approx 3467.368505$. To the nearest .0001 this is

$$x = 3467.3685.$$

EXAMPLE 2   Solve $\log_7(a - 2) = \log_7 a - \log_7 8$ for a.

SOLUTION Using the property $\log_b P - \log_b Q = \log_b(P/Q)$, we have

$$\log_7(a - 2) = \log_7\left(\frac{a}{8}\right).$$

Because $\log_7 x$ is a one-to-one function,

$$a - 2 = \frac{a}{8}$$

$$8a - 16 = a$$

$$7a = 16$$

$$a = \frac{16}{7}.$$

EXAMPLE 3   Solve $\log_{16} x + \log_{16}(10x + 3) = \frac{1}{2}$ for x.

SOLUTION Using the property $\log_b P + \log_b Q = \log_b(PQ)$, we have

$$\log_{16}[x(10x + 3)] = \frac{1}{2}.$$

This can be written in exponential form as

$$x(10x + 3) = 16^{1/2}.$$

This is a quadratic equation:

$$10x^2 + 3x - 4 = 0$$

$$(5x + 4)(2x - 1) = 0 \Rightarrow x = -\tfrac{4}{5} \quad \text{or} \quad \tfrac{1}{2}.$$

We must reject the value $x = -\tfrac{4}{5}$, however, because this calls for the logarithm of a negative number when it is substituted into the original equation. The only solution is $x = \tfrac{1}{2}$.

EXAMPLE 5   Solve $2 \log^2 x - \log(x^3) = 2$ for x. Also, give approximations to the nearest 0.0001.

Solution Using a property of logarithms, we have

$$2 \log^2 x - 3 \log x = 2.$$

This is a quadratic equation in $\log x$:

$$2 \log^2 x - 3 \log x - 2 = 0$$

$$(2 \log x + 1)(\log x - 2) = 0$$

$$\log x = -\tfrac{1}{2} \quad \text{or} \quad \log x = 2$$

$$x = 10^{-1/2} \quad \text{or} \quad x = 10^2$$

$$x \approx 0.3162 \quad \text{or} \quad x = 100.$$

The notation $\log^2 x$ means $(\log x)^2$

We often need to solve an equation in which the unknown is in the exponent. This type of equation is called an **exponential equation.** In most cases, we must use logarithms to solve to exponential equations.

For example, suppose we want to solve

$$3^x = 2.$$

We know the answer is between 0 and 1 because

$$3^0 < 2 < 3^1.$$

There are two ways to approach the solution.

Theoretical Approach We could write $3^x = 2$ in the equivalent logarithmic form:

$$x = \log_3 2.$$

We have roughly estimated this to be between 0 and 1. Unfortunately, calculators don't have a $\log_3$ key. We would need to resort to trial-and-error, substituting values into the original equation until we finally find one that approximates the solution to the precision we desire.

Practical Approach Apply the common log to both sides:

$$\log 3^x = \log 2.$$

Using a property of logarithms we have

$$x \log 3 = \log 2.$$

Solving for x yields

$$x = \frac{\log 2}{\log 3}.$$

Using the log key on a calculator gives a numerical approximation, $x \approx 0.63093$. Comparing the theoretical answer with the practical answer, we must have

$$\log_3 2 = \frac{\log 2}{\log 3}$$

since the original equation $3^x = 2$ can have only one solution. If we follow the same line of reasoning to solve the general exponential equation $b^x = c$, we obtain the following useful formula.

Change of Base Formula

> If a and b are real numbers such that $a > 0$, $b > 0$, $a \neq 1$, $b \neq 1$, then
>
> $$\log_b c = \frac{\log_a c}{\log_a b}.$$

In practice, the value of a in this formula is often 10 or e:

$$\log_b c = \frac{\log c}{\log b} \quad \text{or} \quad \log_b c = \frac{\ln c}{\ln b}.$$

This is because the logarithm keys on a calculator are log (base 10) and ln (base e).

EXAMPLE 6   Solve $150 = 5(2^x)$.

SOLUTION First divide both sides of the equation by 5, then write the exponential equation as a logarithmic equation:

$$30 = 2^x \Rightarrow x = \log_2 30.$$

Using the change of base formula (with $a = 10$),

$$x = \log_2 30 = \frac{\log 30}{\log 2} \approx 4.9068906.$$

ALTERNATE SOLUTION Apply log to both sides and use the properties of logarithms:

$$\log 150 = \log[5(2)^x]$$

$$\log 150 = \log 5 + \log 2^x$$

$$\log 150 = \log 5 + x \log 2$$

$$\frac{\log 150 - \log 5}{\log 2} = x$$

$$x \approx 4.9068906.$$

Before using the calculator, we could have simplified the expression

$$\frac{\log 150 - \log 5}{\log 2}$$

using properties of logarithms as follows:

$$\frac{\log 150 - \log 5}{\log 2} = \frac{\log[\frac{150}{5}]}{\log 2} = \frac{\log 30}{\log 2}.$$

Notice that this agrees with the answer in the first solution.

EXAMPLE 7 Solve $6^{3x} = 11(7^{x+2})$.

SOLUTION Apply log to both sides and use the properties of logarithms:

$$\log 6^{3x} = \log 11(7)^{x+2}$$

$$3x \log 6 = \log 11 + \log 7^{x+2}$$

$$3x \log 6 = \log 11 + (x+2)\log 7$$

$$3x \log 6 = \log 11 + x \log 7 + 2 \log 7$$

$$3x \log 6 - x \log 7 = \log 11 + 2 \log 7$$

$$x(3 \log 6 - \log 7) = \log 11 + 2 \log 7$$

$$x = \frac{\log 11 + 2 \log 7}{3 \log 6 - \log 7} \approx \frac{2.7315888}{1.4893557} \approx 1.8340741.$$

EXAMPLE 8 Solve $\frac{1}{2}(3^x + 3^{-x}) = 2$.

SOLUTION Multiplying both sides by $2(3^x)$, we get

$$3^{2x} + 1 = 4(3^x).$$

Writing 3^{2x} as $(3^x)^2$, we see the equation is a quadratic equation in 3^x:

$$(3^x)^2 - 4(3^x) + 1 = 0.$$

Since this doesn't factor, we apply the quadratic formula:

$$3^x = \frac{-(-4) \pm \sqrt{(-4)^2 - 4(1)(1)}}{2(1)} = \frac{4 \pm \sqrt{12}}{2} = 2 \pm \sqrt{3}.$$

The real roots are $x = \log_3(2 + \sqrt{3})$ and $x = \log_3(2 - \sqrt{3})$. These roots can be approximated using the change of base formula and a calculator:

$$x = \frac{\log(2 + \sqrt{3})}{\log 3} \approx 1.1987467$$

or

$$x = \frac{\log(2 - \sqrt{3})}{\log 3} \approx -1.1987467.$$

EXAMPLE 9   The half-life of plutonium is 50 years. How long will it take a certain amount of plutonium to decay to 30% of its original amount? Use the formula

$$Q = Q_0(\tfrac{1}{2})^{t/h}$$

where Q_0 is the original amount, h is the half-life, t is time, and Q is the present amount. (See Section 4.1.)

SOLUTION   Applying the formula, we have

$$0.30Q_0 = Q_0(\tfrac{1}{2})^{t/50}.$$

Dividing both sides by Q_0 and then using logarithmic notation, we have

$$0.30 = (\tfrac{1}{2})^{t/50}$$

$$\frac{t}{50} = \log_{1/2}(0.30)$$

$$t = 50 \log_{1/2}(0.30)$$

$$t = 50 \frac{\log(0.30)}{\log(\tfrac{1}{2})} \approx 87 \text{ years}$$

EXAMPLE 10   How long will it take a principal to double if it is invested at 9% interest and compounded monthly? Use the formula

$$A = P\left(1 + \frac{r}{n}\right)^{nt}$$

where A is the accrued amount after t years of a principal P invested at a rate r compounded n times per year. (See Section 4.1.)

SOLUTION Applying the formula, we get

$$2P = P\left(1 + \frac{0.09}{12}\right)^{12t}.$$

Dividing both sides by P gives

$$2 = \left(1 + \frac{0.09}{12}\right)^{12t}$$

$$2 = (1.0075)^{12t}$$

$$12t = \log_{1.0075}2$$

$$t = \frac{\log 2}{12 \log 1.0075} \approx 7.73 \text{ years.}$$

The principal will double in approximately 7 years, 9 months.

EXAMPLE 11   The Richter scale for measuring the magnitude M of earthquakes was adopted in 1935. If we are given the amplitude A of the seismic wave, the magnitude is given by

$$M = \log \frac{A}{A_0}$$

where A_0 is the amplitude of the smallest recorded wave at that time. The amount of energy released E (measured in ergs) is approximated by

$$\log E = 11.4 + 1.5M.$$

a) How much energy was released by the 1964 Alaskan earthquake, which was 8.4 on the Richter scale?

b) Suppose two earthquakes have magnitudes that differ by 3. How many times greater is the amplitude of the larger earthquake than that of the smaller?

SOLUTION

a) Substituting 8.4 for M in the appropriate equation, we get

$$\log E = 11.4 + 1.5M \Rightarrow \log E = 11.4 + 1.5(8.4).$$

If we simplify and solve for E, we get

$$\log E = 24$$

$$E = 10^{24} \text{ ergs.}$$

b) Let a be the amplitude for the larger earthquake and let b represent the amplitude for the smaller earthquake. The difference

Leonhard Euler

The most productive mathematician of all time was probably Leonhard Euler (pronounced "oiler"). It has been estimated that he averaged 800 pages of mathematical works per year during his life. Even more impressive than the quantity of his writings is the quality and diversity. Euler was truly a universal mathematician; his research ranged from arithmetic and elementary algebra to the most advanced fields of mathematics. He also wrote books and articles on astronomy, hydraulics, life expectancy, annuities, and the mathematics of lotteries. He even presented a theory of music, of which P.H. von Fuss said was "too mathematical for musicians and too musical for mathematicians."

Euler was born in Basel, Switzerland, in 1707. His father was a minister and mathematician who studied under the great mathematician Jakob Bernoulli. Due to his father's influence, Leonhard developed a deep religious conviction and studied theology at the University of Basel for the purpose of becoming a pastor. But he also continued to study mathematics, and his father, recognizing his son's genius, was eventually persuaded to let him leave theology and concentrate on mathematics. In 1733, Euler was appointed to the Academy at St. Petersburg, one of the centers of mathematical and scientific research of the time. It was said that he worked with such zeal and devotion that he developed a fever from overwork that resulted in the loss of sight in his right eye. Undaunted, Euler continued his prolific research. He earned an excellent reputation throughout Europe and was persuaded to leave St. Petersburg to join the Berlin Academy of Sciences in 1741.

An interesting event illustrates the high regard that Euler enjoyed. During his tenure at the Berlin Academy, the Russians, who were at war with the Prussians, devastated a farm near Berlin. When they learned that it belonged to Euler, he was promptly overcompensated for his loss. Probably due to this kind of royal treatment by the Russians and the personal differences that Euler had with certain members and the leadership of the Academy at Berlin, Catherine the Great induced him to return to St. Petersburg in 1766. A few years later, Euler lost the sight in his left eye. Total blindness did not slow his machinelike pace one bit. He continued to dictate mathematical papers and books to secretaries, students, and his sons. He remained in St. Petersburg until his death at age 77.

His remarkable composure after losing his sight is indicative of his faith and lifelong religious convictions. His ability to continue mathematical research in total darkness is testimony to his legendary mental capacity. He could recite from memory all of Virgil's *Aenid,* which he read as a boy. He could even identify the first and last lines on any page of the book. He was known to do mental calculations involving as many as 50 decimal places and once recited the first six powers of all the integers from 1 to 20. His powers of concentration were also impressive. Euler loved children and often did mathematics while his children played around him. Francois Arago, a French academician said Euler could calculate effortlessly, "just as men breathe, as eagles sustain themselves in the air."

Among Euler's most significant achievements is the order that he brought to mathematics and its notation. As one example, he masterfully took a mass of scattered trigonometric formulas and codified them into a systematic development of trigonometry that has since needed little improvement. He similarly summarized calculus and analytic geometry into presentations that remain the models of present-day college mathematics. As a result, Euler's writings were very popular, and the notation he used has lasted. He was the first to use the notation $f(x)$ for a function of x. He adopted the use of π for the ratio of circumference to diameter of a circle, i for $\sqrt{-1}$, and e for the base of the natural exponential function. Although John Napier is credited as the first to use the idea of logarithms, Euler was the first to introduce the modern notion of the logarithm as a root of an exponential equation. One of the great mathematicians of all time, Karl F. Gauss, said

The study of Euler's works will remain the best school for the different fields of mathematics and nothing else can replace it.

The great mathematician P.S. Laplace told younger mathematicians:

Read Euler, read Euler, he is the master of us all!

in magnitudes is

$$3 = \log \frac{a}{A_0} - \log \frac{b}{A_0}.$$

Utilizing the property $\log P - \log Q = \log(P/Q)$, we have

$$3 = \log \frac{\dfrac{a}{A_0}}{\dfrac{b}{A_0}} = \log \frac{a}{b}.$$

Thus $a/b = 10^3$ or $a = 1000b$. The larger earthquake had an amplitude that is 1000 times greater than the smaller.

EXERCISE SET 4.4

A

Using a calculator and the properties of logarithms, if necessary, approximate Problems 1 through 9. Approximate all answers to the nearest 0.0001.

1. $\log 43.15$

2. $\log 0.00312$

3. $\log_9 0.00312$

4. $\log_8 52.389$

5. $\log(2.34 \times 10^{45})$

6. $\log(1.06 \times 10^{-12})$

7. $\log(4.255 \times 10^{-236})$

8. $\log(3.081 \times 10^{-407})$

9. $\log(3.2 \times 10^{455})$

Solve the equations in Problems 10 through 18. In addition, approximate all answers to the nearest 0.0001.

10. $\log 3x = 2$

11. $\log 4x = 3$

12. $\log x^3 = -1$

13. $\ln 6 + \ln 2x = 3$

14. $\log 3x - \log 4 = 2$

15. $\log x + \log(x + 3) = 1$

16. $\log x + \log(x - 15) = 2$

17. $3 \log^2 x - \log x^2 = 1$

18. $\ln^2 x - \ln x^5 + 4 = 0$

B

Solve the equations in Problems 19 through 39. In addition, approximate each solution to the nearest 0.0001.

19. $10^x = 15$

20. $10^x = 1850$

21. $e^x = 19$

22. $e^x = 871$

23. $112 = 9^t$

24. $74 = 5^t$

25. $2.5^{3t} = 24$

26. $1.8^{4t} = 15$

27. $90 = 3(12^t)$

28. $45 = 9(8^t)$

29. $6^x = 8(5^x)$

30. $11^x = 6(2.5^x)$

31. $3^x = 5^{x-2}$

32. $2^{x+3} = 9^x$

33. $\frac{1}{2}(7^x + 7^{-x}) = 2$

34. $\dfrac{10^x - 10^{-x}}{2} = 4$

35. $\frac{1}{2}(3^x + 3^{-x}) = 4$

36. $3^{2x} - 7(3^x) = 8$

37. $2^{2x} - 5(2^x) = 6$

38. $3xe^x + 5e^x = 0$

39. $x^2 e^x - 9e^x = 0$

Solve the equations in Problems 40 through 42. In addition, approximate each solution to the nearest 0.0001.

40. $\log_x 5 = 4 \log_x 8 + 1$

41. $2 \log_6 x = 4 \log_3 x - 1$

42. $\log_7 x = \log_2 x - 3$

43. How many years will it take \$18,000 invested at 10% compounded quarterly to grow to \$25,000?

44. How many years will it take for the investment of \$18,000 in Problem 43 to triple?

45. The half-life for krypton is about 11 years. How long will it take 25 grams of krypton to decay to only 10 grams?

46. The half-life for a certain substance is 3 days. How long will it take the substance to decay to only 10% of the original amount?

47. Initially there were 80 milligrams of a radioactive substance present. After 5 hours the mass decreased to 73 milligrams. What is the half-life of this substance?

48. How many times greater is the amplitude of an earthquake measuring 6.0 on the Richter scale than one measuring 4.5?

49. Find the magnitude of an earthquake whose amplitude is $1500A_0$. Round your answer to the nearest 0.1.

50. Find the energy released (in ergs) for the two earthquakes in Problem 48.

51. A bacteriocide is introduced into a bacteria culture. The number of bacteria remaining $D(t)$ after t minutes is given by $D(t) = 80,000e^{-0.01t}$. After how many minutes will there be only 5000 bacteria remaining?

52. If P dollars are borrowed at an interest rate r for t years, the monthly payment for a fully amortized loan is

$$m = \frac{P\left(\dfrac{r}{12}\right)}{1 - \left(1 + \dfrac{r}{12}\right)^{-12t}}.$$

Suppose we want to obtain a $20,000 loan at a 10% interest rate. If we can afford monthly payments of $250, what should be the length of time for paying off the loan (to the nearest month)?

53. The pH scale is a logarithmic scale used to determine the acidity of a chemical. The hydrogen ion concentration (in moles per liter) is denoted $[H^+]$. The pH is defined as $pH = -\log[H^+]$.

a) If the pH of orange juice is 5.8, what is the hydrogen ion concentration?

b) If the hydrogen ion concentration of wine is 4.3×10^{-4}, what is the pH?

c) Does a higher pH mean a higher concentration of hydrogen ions or a lower concentration of hydrogen ions?

54. The voltage gain G (in dB) of an amplifier is computed as

$$G = 10 \log\left(\frac{P}{P_0}\right)$$

where P_0 is the input signal (in watts) and P is the output power (in watts) of the amplifier. If an amplifier produces 60 watts when driven by an input signal of 0.15 watt, what is the voltage gain?

55. Refer to the formula in Problem 54. If you are to design an amplifier that has a voltage gain of 28 dB with an input signal of 0.10 watt, what must the output power be?

56. The magnitude m of a star is the measure of its brightness B given by

$$m = -2.5 \log\left(\frac{B}{B_0}\right)$$

where B_0 is the intensity of an "average star." Find the magnitude of a star that **a)** has one-half the brightness of an average star and **b)** has one-tenth the brightness of an average star.

57. The decibel scale is used for measuring sound intensity. The number of decibels (dB) is computed from the sound intensity I:

$$dB = 10 \log\left(\frac{I}{I_0}\right)$$

where I_0 is the intensity of the least audible sound. A sustained intensity level of over 120 dB is considered potentially damaging to the ear. Approximately how many times louder than I_0 must I be to attain 120 dB?

58. A measure of egg quality is H, the Haugh unit:

$$H = 100 \log(h - 1.70m^{0.37} + 7.57),$$

where m is the mass of the egg (in grams) and h is the height of the albumen (in mm) when the egg is broken on a flat surface.

a) Calculate H for an egg whose mass is 22 grams and whose albumen height is 14 mm

b) What must h be for an egg that is 30 grams to obtain a Haugh unit of 106?

59. A company has determined that the number of units sold, N, is related to the dollars spent in advertising, a, by the following formula: $N = 2000 + 400 \ln a$. If the company wants to sell 7000 units, how much money will it have to spend in advertising?

60. The **barometric pressure,** measured in centimeters of mercury, is given for any altitude of h kilometers (above sea level) by the formula

$$P = 76e^{-h/(30T + 8)}$$

where T is the air temperature in degrees Celsius. Suppose that on a certain day the barometric pressure measures 40 centimeters of mercury. Find the height when the air temperature is $10°$ Celsius.

61. The electric current I in amperes flowing in a series circuit with an inductance L henrys, a resistance R ohms, and an electromotive force E volts, is given by

$$I = \frac{E}{R}(1 - e^{-Rt/L})$$

where t is the time in seconds after the current begins to flow. If $E = 6$ volts, $L = 0.02$ henry, and $R = 4$ ohms, find the amount of time it will take for the current to reach 1.1 amperes.

62. A rocket of structural mass m_1 contains fuel of initial mass m_2. It travels straight up from the surface of the earth by burning fuel at a constant rate a and expelling the exhaust at a constant velocity b. Neglecting all external forces except gravity, the burnout velocity of the rocket is given by

$$V = b \log\left(1 + \frac{m_2}{m_1}\right) - \frac{gm_2}{a}$$

where g is a constant (the acceleration due to gravity). Solve this equation for m_1.

63. In 1920 the population of the north central United States was 34,020,000. In 1950 the population was 44,461,000. According to the model for population growth in Section 4.2, **a)** predict the population in the year 2000, and **b)** predict the year that the population should reach 85,000,000?

64. Some populations are constrained by environmental factors. A formula for inhibited growth that assumes constraints have a greater effect as time passes is

$$P = \frac{C}{1 + C_0 e^{-kt}}$$

where C and C_0 are constants. Solve this equation for t.

65. In determining the pollution of a factory's effluent, an important measurement is the amount of oxidizable organic matter M, given by $M = M_0 10^{-kt}$. If a lake initially has $M = 160$ milligrams per liter, and 6 days later $M = 110$ milligrams per liter, when will the water be at a drinkable level of $M = 0.4$ milligrams per liter?

C

66. Without using a calculator, evaluate

$$\frac{1}{\log_2 36} + \frac{1}{\log_3 36}.$$

67. Find all values satisfying $x^{\log x} = x^3/100$.

MISCELLANEOUS EXERCISES

In Problems 1 through 6, use a calculator to approximate each expression to the nearest 0.0001.

1. $5^{\sqrt{3}/2}$

2. $56e^{(0.021)(14)}$

3. $\frac{1}{2}(e^{3/2} + e^{-3/2})$

4. $\ln 76.4$

5. $\log \sqrt{87}$

6. $\log_6 165$

In Problems 7 through 18, sketch the graph of $y = f(x)$. Identify asymptotes.

7. $f(x) = 7^x$

8. $f(x) = (\frac{3}{5})^x$

9. $f(x) = 2^{x+2}$

10. $f(x) = e^x + 1$

11. $f(x) = e^{2x}$

12. $f(x) = 3^{x/2}$

13. $f(x) = 2^{-x^2}$

14. $f(x) = e^{-\sqrt{x}}$

15. $f(x) = 3e^{x-5}$

16. $f(x) = \frac{1}{2}e^{-x}$

17. $f(x) = -e^{x-3}$

18. $f(x) = e^{2-x}$

In Problems 19 through 30, sketch the graph of $y = f(x)$. *Identify asymptotes.*

19. $f(x) = \log_5(x)$

20. $f(x) = 4\log_3(x)$

21. $f(x) = \log_4(x + 5)$

22. $f(x) = \log_5(x) + 2$

23. $f(x) = -\ln(x)$

24. $f(x) = \ln(-x)$

25. $f(x) = \ln(x^2)$

26. $f(x) = \ln|x|$

27. $f(x) = \ln|3 + x|$

28. $f(x) = 8^{\log_8 x}$

29. $f(x) = \ln(e^x)$

30. $f(x) = \ln\left(\dfrac{x}{e}\right)$

In Problems 31 through 33, write h(x) as the composition of two functions, $f[g(x)]$. *Do not use* $f(x) = x$ *or* $g(x) = x$.

31. $h(x) = e^{\sqrt{x+3}}$

32. $h(x) = \ln\left(2 + \dfrac{1}{x}\right)$

33. $h(x) = \sqrt[4]{\log x}$

Without using a calculator, find two consecutive integers that each logarithm in Problems 34 through 36 is between.

34. $\log_5 12$

35. $\log 2323$

36. $\log_6(\frac{1}{10})$

37. What value does $f(x) = (1 + 3/x)^x$ approach as $x \to \infty$?

In Problems 38 through 40, find the inverse function.

38. $f(x) = e^{x+4}$

39. $f(x) = \log_7(2x - 3)$

40. $f(x) = \ln\sqrt{x}$

Express Problems 41 through 44 as a single logarithm.

41. $\log_b(2t) + \log_b(t^2)$

42. $\log_b(6y^5) - 3\log_b(y)$

43. $\ln\left(\dfrac{\sqrt{x}}{y^4}\right) - \ln(1) + \dfrac{1}{2}\ln(9xy^2)$

44. $\frac{1}{5}\log(a^3) - \log(a^{8/5}b) + 2$ (Hint: $2 = \log 100$.)

45. Express

$$\ln\left(\dfrac{3z^4\sqrt{x}}{\sqrt{y^3}}\right)$$

as terms involving $\ln x$, $\ln y$, and $\ln z$.

Solve the equations in Problems 46 through 65. In addition, approximate all answers to the nearest 0.0001.

46. $5\log x = 3$

47. $\log_6 2x = -1$

48. $\frac{1}{2} = \log_x 5$

49. $\log_5(x + 7) + \log_5(13 - 2x) = 2$

50. $\log_3(x^2 + 2) = 1 + \log_3 x$

51. $\ln(x - 1) = 2 + \ln x$

52. $\ln^2 x = 3 + \ln x^2$

53. $\ln|x - 5| = 2$

54. $2\ln 5x = \ln(x + 2)$

55. $\ln x = 0.7506$

56. $e^x = 44.302$

57. $4xe^x - 3e^x = 0$

58. $\dfrac{2xe^x - x^2e^x}{e^{2x}} = 0$

59. $46.6 = 7^{2x}$

60. $184 = 12e^{3x}$

61. $120 = 72 + 210e^{-0.2x}$

62. $3550 = 1200\left(1 + \dfrac{0.06}{12}\right)^{12x}$

63. $3^x = 12(8^x)$

64. $2^x = 5^{x-4}$

65. $e^x - e^{-x} = 3$

66. Solve for y: $-\frac{1}{2}\ln(y + 3) + \frac{1}{2}\ln(y - 3) = x + 1$

67. Suppose that $f(x) = e^x + ce^x$, where c is a constant to be determined. Given that $f(0) = 5$, determine c and $f(x)$.

68. The position of the top of a certain shock absorber at time t is given by $h(t) = -4e^{-t} + 12e^{-2t}$. Graph this function for $t \geq 0$.

69. A commonly used function to approximate the amount of a drug in the body is $C(t) = C_0e^{-kt}$, where C_0 is the initial amount introduced into the system, and k is a constant that depends on the specific drug. If 15 mg of the drug codeine is introduced into the body, 7.5 mg will remain after 2.4 hours. Predict the amount of codeine remaining after 4 hours.

70. Refer to the function in Problem 69. Solatol is a drug used to treat hypertension or angina. If 240 mg of solatol is taken, 120 mg will remain in the body after 13 hours. Predict the amount remaining after 18 hours.

71. The age t (in years) of young sperm whales can be related to the body length L (in feet) by the equation $L = 13 + 6.5\ln(1 + t)$. **a)** How long should a 4-year-old whale be? **b)** Solve the equation for t. Use this to determine the age of a 28-foot sperm whale.

72. Suppose $6,000$ is invested at a rate of 10% compounded six times per year. This will grow to what amount after 3 years?

73. Suppose \$8,500 is deposited at a rate of 9% compounded continuously. What will the principal be after 7 years?

74. How long will it take an investment to double at 12% compounded continuously?

75. For a biology experiment, 1000 fruit flies are put in a supportive environment. Twenty days later, there are 1200 of them. Find a function of the form $n(t) = n_0 b^t$, where $n(t)$ is the number of flies after t days, that is consistent with the given information. Use the resulting function to predict the number of fruit flies after sixty days.

76. A lightbulb at a temperature of 300° is switched off and begins cooling in a room where the temperature is a constant 68°. Ten minutes later the bulb has cooled to 220°. Based on Newton's law of cooling, **a)** determine the cooling function $T(t)$, and **b)** predict the temperature of the lightbulb 40 minutes after it is switched off.

77. Initially, there were 62 mg of a radioactive substance. After 15 minutes, there were 54 mg left. What is the half-life of this substance? Use the formula

$$Q = Q_0 (\tfrac{1}{2})^{t/h}$$

given in Section 4.1, where t and h are in minutes.

78. A fungicide is applied to a fungus culture, and after t minutes the number of fungi remaining is given by $N(t) = 100,000 e^{-0.06t}$. **a)** How many fungi will remain after 20 minutes? **b)** After how many minutes will there be only 15,000 fungi remaining?

79. An electric current of I amperes flowing in a series circuit with an inductance of L henrys, a resistance of R ohms, and an electromotive force of E volts is given by

$$I = \frac{E}{R}(1 - e^{-Rt/L})$$

where t is the time in seconds after the current begins to flow. If $E = 12$ volts, $L = 0.1$ henry, and $R = 2$ ohms, find the amount of time it takes for the current to reach 5 amperes.

80. Suppose that a saturated solution of a certain substance is obtained by dissolving 60 grams of the substance in 100 grams of water. Suppose that 20 grams of the substance is put into 100 grams of water. The amount (in grams) that is undissolved, x, after t hours is given by

$$\ln\left[\frac{3x}{x + 40}\right] = kt$$

where k is a constant. If 10 grams of the substance have dissolved after 3 hours, what amount remains undissolved after 2 additional hours?

TRIGONOMETRIC FUNCTIONS

TRIGONOMETRY OF THE RIGHT TRIANGLE

The development of *trigonometry,* as its name suggests, arose historically from the study of triangles. Specifically, it is the study of a set of functions that relate the angles of a right triangle and its sides.

The present importance of these trigonometric functions, however, lies far beyond the study of triangles. As functions on the real numbers, they are used to describe periodic phenomena such as light and sound waves, vibrating strings, and the orbits of planets and electrons. You will find that these real functions play an important role in solving certain fundamental problems that you will encounter in calculus.

We start our investigation in this section by considering two similar right triangles (Figure 1). Suppose that we examine the ratio of the length of the side opposite the angle θ and the length of the hypotenuse in each of these triangles. For the smaller right triangle,

You should recall that the corresponding angles of two similar triangles are equal and that their corresponding sides are in constant proportion. The verification that these two triangles are right triangles and that they are similar is left to you as an exercise.

$$\frac{\text{length of opposite side}}{\text{length of hypotenuse}} = \frac{10}{26} = \frac{5}{13}.$$

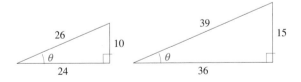

FIGURE 1

For the larger right triangle,

$$\frac{\text{length of opposite side}}{\text{length of hypotenuse}} = \frac{15}{39} = \frac{5}{13}.$$

It should be no surprise that the ratios are the same; after all, they are the ratios of corresponding parts of similar right triangles. As a matter of fact,

265

any other right triangle that is similar to these two will have the same acute angle θ and would yield a like result.

The ratio we computed here is not dependent on the particular right triangle but rather on the angle θ. In other words, this ratio is strictly a function of the measure of the angle θ (pause here and mull this over until you truly appreciate this point). This ratio is called the **sine** of the angle θ, usually written sin θ. It is one of six **trigonometric functions,** which are defined as follows.

Trigonometric Functions of Acute Angles

These definitions should be committed to memory.

Consider a right triangle with an acute angle θ and with sides labeled *adj, opp,* and *hyp* (for adjacent, opposite, and hypotenuse, respectively). Then the six trigonometric functions **sine, cosine, tangent, cotangent, secant,** and **cosecant** of θ (abbreviated **sin θ, cos θ, tan θ, cot θ, sec θ,** and **csc θ,** respectively) are the ratios

$$\sin \theta = \frac{\text{opp}}{\text{hyp}} \qquad \cos \theta = \frac{\text{adj}}{\text{hyp}} \qquad \tan \theta = \frac{\text{opp}}{\text{adj}}$$

$$\csc \theta = \frac{\text{hyp}}{\text{opp}} \qquad \sec \theta = \frac{\text{hyp}}{\text{adj}} \qquad \cot \theta = \frac{\text{adj}}{\text{opp}}$$

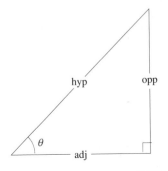

EXAMPLE 1 Consider the triangle in Figure 2. In the preceding discussion, we determined that

$$\sin \theta = \tfrac{5}{13}.$$

Find the values of the five other trigonometric functions of the angle θ.

SOLUTION In this triangle, the hypotenuse is 39. The length of the side opposite the angle θ is 15, and the length of the side adjacent to the angle θ is

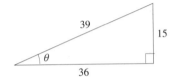

FIGURE 2

36. So,

$$\cos \theta = \frac{\text{adj}}{\text{hyp}} = \frac{36}{39} = \frac{12}{13} \Rightarrow \text{the cosine of } \theta \text{ is } \frac{12}{13}$$

$$\tan \theta = \frac{\text{opp}}{\text{adj}} = \frac{15}{36} = \frac{5}{12} \Rightarrow \text{the tangent of } \theta \text{ is } \frac{5}{12}$$

$$\cot \theta = \frac{\text{adj}}{\text{opp}} = \frac{36}{15} = \frac{12}{5} \Rightarrow \text{the cotangent of } \theta \text{ is } \frac{12}{5}$$

$$\sec \theta = \frac{\text{hyp}}{\text{adj}} = \frac{39}{36} = \frac{13}{12} \Rightarrow \text{the secant of } \theta \text{ is } \frac{13}{12}$$

$$\csc \theta = \frac{\text{hyp}}{\text{opp}} = \frac{39}{15} = \frac{13}{5} \Rightarrow \text{the cosecant of } \theta \text{ is } \frac{13}{5}.$$

Look at the definitions of the trigonometric functions and the results of Example 1. Notice that the functions cosecant, secant, and cotangent are reciprocals of the functions sine, cosine, and tangent, respectively. These observations can be summarized in the form of identities.

$$\csc \theta = \frac{1}{\sin \theta} \qquad \sec \theta = \frac{1}{\cos \theta} \qquad \cot \theta = \frac{1}{\tan \theta}$$

Reciprocal Identities for Acute Angles

We will prove the first of these identities; the rest are left to you as an exercise. By the definition of the cosecant function,

$$\csc \theta = \frac{\text{hyp}}{\text{opp}} = \frac{1}{\text{opp/hyp}} = \frac{1}{\sin \theta}.$$

Also, it is left to you as an exercise to derive the following two identities from the definitions of the trigonometric functions:

$$\tan \theta = \frac{\sin \theta}{\cos \theta} \qquad \cot \theta = \frac{\cos \theta}{\sin \theta}$$

Ratio Identities for Acute Angles

The reciprocal and ratio identities should be committed to memory.

EXAMPLE 2 Given that θ is an acute angle such that $\sin \theta = \frac{3}{8}$, determine the values of the other five trigonometric functions of θ. Approximate these values to the nearest 0.0001.

SOLUTION The angle θ is an acute interior angle of any right triangle such that the ratio of the side opposite θ and the hypotenuse is $\frac{3}{8}$. There are an infinite number of possibilities here, but the most obvious choice is the right triangle with side of length 3 and hypotenuse of length 8. The remaining leg, the side adjacent to the angle θ, can be computed using the Pythagorean theorem (Figure 3):

$$\text{adj} = \sqrt{8^2 - 3^2} = \sqrt{64 - 9} = \sqrt{55}.$$

So,

$$\cos \theta = \frac{\text{adj}}{\text{hyp}} = \frac{\sqrt{55}}{8}.$$

Instead of computing the other three functions directly from the definitions, as we did in Example 1, we can exploit the reciprocal and ratio identities to determine their values:

$$\csc \theta = \frac{1}{\sin \theta} = \frac{1}{\left(\dfrac{3}{8}\right)} = \frac{8}{3} \qquad \sec \theta = \frac{1}{\cos \theta} = \frac{1}{\left(\dfrac{\sqrt{55}}{8}\right)} = \frac{8}{\sqrt{55}}$$

$$\tan \theta = \frac{\sin \theta}{\cos \theta} = \frac{\left(\dfrac{3}{8}\right)}{\left(\dfrac{\sqrt{55}}{8}\right)} = \frac{3}{\sqrt{55}} \qquad \cot \theta = \frac{\cos \theta}{\sin \theta} = \frac{\left(\dfrac{\sqrt{55}}{8}\right)}{\left(\dfrac{3}{8}\right)} = \frac{\sqrt{55}}{3}.$$

Approximating to the nearest 0.0001, we obtain

$$\sin \theta = 0.3750 \qquad \cos \theta = 0.9270 \qquad \tan \theta = 0.4045$$

$$\csc \theta = 2.6667 \qquad \sec \theta = 1.0787 \qquad \cot \theta = 2.4721.$$

EXAMPLE 3 Consider the right triangle with legs of length 4, and 6 and acute angles α and β (Figure 4).

a) Compute the length of the hypotenuse of the triangle.

b) Determine the values of the trigonometric functions of α.

c) Determine the values of the trigonometric functions of β.

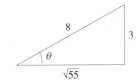

FIGURE 3

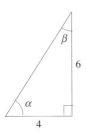

FIGURE 4

SOLUTION

a) Using the Pythagorean theorem, the hypotenuse (Figure 5) is

$$\sqrt{4^2 + 6^2} = \sqrt{16 + 36}$$
$$= \sqrt{52}$$
$$= 2\sqrt{13}.$$

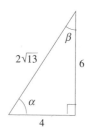

FIGURE 5

b) Relative to the angle α, the opposite side is of length 6 and the adjacent side is 4, so

$$\sin \alpha = \frac{6}{2\sqrt{13}} = \frac{3}{\sqrt{13}} \qquad \csc \alpha = \frac{2\sqrt{13}}{6} = \frac{\sqrt{13}}{3}$$

$$\cos \alpha = \frac{4}{2\sqrt{13}} = \frac{2}{\sqrt{13}} \qquad \sec \alpha = \frac{2\sqrt{13}}{4} = \frac{\sqrt{13}}{2}$$

$$\tan \alpha = \frac{6}{4} = \frac{3}{2} \qquad \cot \alpha = \frac{4}{6} = \frac{2}{3}.$$

c) Relative to the angle β, the opposite side is of length 4 and the adjacent side is 6, so

$$\sin \beta = \frac{4}{2\sqrt{13}} = \frac{2}{\sqrt{13}} \qquad \csc \beta = \frac{2\sqrt{13}}{4} = \frac{\sqrt{13}}{2}$$

$$\cos \beta = \frac{6}{2\sqrt{13}} = \frac{3}{\sqrt{13}} \qquad \sec \beta = \frac{2\sqrt{13}}{6} = \frac{\sqrt{13}}{3}$$

$$\tan \beta = \frac{4}{6} = \frac{2}{3} \qquad \cot \beta = \frac{6}{4} = \frac{3}{2}.$$

Compare the results of parts a) and b) of Example 3. Notice in this particular case that

$$\cos \beta = \sin \alpha \qquad \sin \beta = \cos \alpha$$
$$\cot \beta = \tan \alpha \qquad \tan \beta = \cot \alpha$$
$$\sec \beta = \csc \alpha \qquad \csc \beta = \sec \alpha.$$

In general, if two angles α and β are the acute interior angles of a right triangle with sides a, b, and c, then $\beta = 90° - \alpha$. The angles α and β are complementary (recall that two angles are complementary if their sum is a right

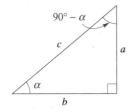

FIGURE 6

angle). Then (Figure 6),

$$\cos(90° - \alpha) = \cos \beta = \frac{\text{side adjacent to } \beta}{\text{hypotenuse}}$$

$$= \frac{a}{c} = \frac{\text{side opposite } \alpha}{\text{hypotenuse}} = \sin \alpha.$$

This is the proof of the first of six **cofunction identities** (the others are left as exercises):

Cofunction Identities for Acute Angles

$\cos(90° - \alpha) = \sin \alpha$	$\sin(90° - \alpha) = \cos \alpha$
$\cot(90° - \alpha) = \tan \alpha$	$\tan(90° - \alpha) = \cot \alpha$
$\sec(90° - \alpha) = \csc \alpha$	$\csc(90° - \alpha) = \sec \alpha$

The six trigonometric functions can be thought of as three pairs of **cofunc**tions: sine and **co**sine, tangent and **cot**angent, secant and **co**secant. The function of an angle α is equal to the **co**function of its **co**mplement $90° - \alpha$.

Special Triangles

There are two special right triangles which play an important role in our discussion of trigonometry. They are the **45°-45° right triangle** and the **30°-60° right triangle.**

Consider an isosceles triangle with sides of length a, hypotenuse of length c, and acute angles of 45° (Figure 7). By the Pythagorean theorem, we get

$$c = \sqrt{a^2 + a^2} = \sqrt{2a^2} = \sqrt{2}a.$$

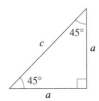

FIGURE 7

The hypotenuse of a 45°-45° right triangle is $\sqrt{2}$ times the length of each leg.

The triangle formed by placing two 30°-60° right triangles as shown in Figure 8 is an equilateral triangle since each of its interior angles is 60°. From this we find that

$$c = 2a.$$

Using the Pythagorean theorem to find b, we obtain

$$b = \sqrt{c^2 - a^2} = \sqrt{(2a)^2 - a^2} = \sqrt{3a^2} = \sqrt{3}a.$$

The length of the hypotenuse of a 30°-60° right triangle is twice the length of the shorter leg, and the longer leg is $\sqrt{3}$ times the shorter leg.

For easy reference, the following box summarizes our discussion of these two types of triangles. The important idea is the ratios of the sides of the triangles; however, we illustrate with frequently used specific cases.

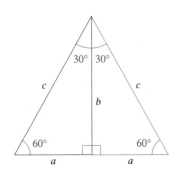

FIGURE 8

45°-45° right triangles:

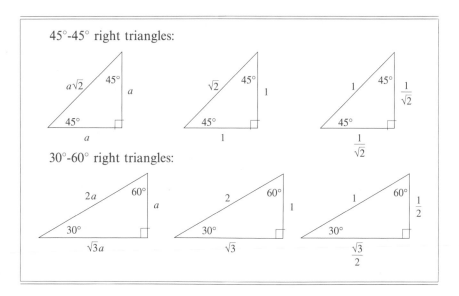

30°-60° right triangles:

EXAMPLE 4 Determine the exact value of $\sin \theta$, $\cos \theta$, and $\tan \theta$ for

a) $\theta = 45°$ **b)** $\theta = 30°$ **c)** $\theta = 60°$

SOLUTION

a) $\sin 45° = \dfrac{a}{a\sqrt{2}} = \dfrac{1}{\sqrt{2}}$ $\cos 45° = \dfrac{a}{a\sqrt{2}} = \dfrac{1}{\sqrt{2}}$

$\tan 45° = \dfrac{a}{a} = 1$

b) $\sin 60° = \dfrac{a\sqrt{3}}{2a} = \dfrac{\sqrt{3}}{2}$ $\cos 60° = \dfrac{a}{2a} = \dfrac{1}{2}$

$\tan 60° = \dfrac{a\sqrt{3}}{a} = \sqrt{3}$

c) $\sin 30° = \dfrac{a}{2a} = \dfrac{1}{2}$ $\cos 30° = \dfrac{a\sqrt{3}}{2a} = \dfrac{\sqrt{3}}{2}$

$\tan 30° = \dfrac{a}{a\sqrt{3}} = \dfrac{1}{\sqrt{3}}$

The exact values of the trigonometric functions arise frequently enough in mathematics to warrant your memorization of the accompanying table.

The other functions of these values can then be determined quickly by using the reciprocal identities. For example, to recall csc 45° quickly:

$$\csc 45° = \frac{1}{\sin 45°} = \frac{1}{\left[\dfrac{1}{\sqrt{2}}\right]} = \sqrt{2}.$$

θ	$\sin \theta$	$\cos \theta$	$\tan \theta$
30°	$\dfrac{1}{2}$	$\dfrac{\sqrt{3}}{2}$	$\dfrac{1}{\sqrt{3}}$
45°	$\dfrac{1}{\sqrt{2}}$	$\dfrac{1}{\sqrt{2}}$	1
60°	$\dfrac{\sqrt{3}}{2}$	$\dfrac{1}{2}$	$\sqrt{3}$

Using Your Calculator

The exact evaluation of the trigonometric functions is difficult and usually impossible for most angles θ. Even the approximation of these values within a specified tolerance involves the use of advanced mathematics (you may see some of these in calculus). These approximations have been collected in tables, and, until recently, these tables were the primary tool used to determine these values. With the advent of the inexpensive, hand-held calculator, the approximation of these trigonometric values has become much easier, faster, and more precise.

Using the calculator to approximate trigonometric functions is very easy. You need only enter the measure of the angle and press the appropriate function button. For example, suppose that you wish to evaluate sin 36°. First, your calculator must be set in *degree mode* (consult your owner's manual to see how this is accomplished). The keystrokes and output are:

KEYSTROKES   **DISPLAY**

3   6   sin   0.587785252

Rounded to the nearest 0.0001, sin 36° = 0.5878.

Now suppose that you wish to approximate the value of csc 36°. More than likely, you do not have a csc button on your calculator. You can, however, use the sin button and the reciprocal identity $\csc \theta = 1/\sin \theta$:

KEYSTROKES   **DISPLAY**

$\boxed{3}$ $\boxed{6}$ $\boxed{\sin}$ $\boxed{1/x}$ $\boxed{1.701301617}$

EXAMPLE 5   Given the right triangle in Figure 9, determine the lengths of the sides a and b. Approximate these to the nearest 0.1 using your calculator.

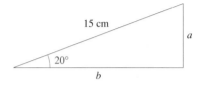

SOLUTION Given the length of the hypotenuse and the measure of an acute interior angle, we need to determine the length of the side opposite the given angle and the length of the side adjacent to the given angle:

$$\sin 20° = \frac{a}{15} \Rightarrow a = 15 \sin 20° = 5.1$$

$$\cos 20° = \frac{b}{15} \Rightarrow b = 15 \cos 20° = 14.1.$$

FIGURE 9

You can also use your calculator to approximate the measure of an angle whose sine (or cosine or tangent) is known. For example, suppose that we wish to approximate the measure of the angle θ in Example 1. Precisely,

$$\theta = \text{angle whose sine is } \tfrac{5}{13}.$$

This is abbreviated as

$$\theta = \arcsin(\tfrac{5}{13}) \qquad \text{(read as "arcsine of } \tfrac{5}{13}\text{")}$$

or

$$\theta = \sin^{-1}(\tfrac{5}{13}) \qquad \text{(read as "inverse sine of } \tfrac{5}{13}\text{").}$$

Use your calculator to approximate θ:

KEYSTROKES   **DISPLAY**

To the nearest degree, the angle θ is $23°$.

We could have arrived at the same approximation by considering this angle as $\cos^{-1}(\tfrac{12}{13})$, or $\arccos(\tfrac{12}{13})$, or by considering this angle as $\tan^{-1}(\tfrac{5}{12})$, or $\arctan(\tfrac{5}{12})$. We leave it to you to verify both of these using your calculator.

The keystrokes involved may be different depending on the calculator, but most calculators are similar. You may have a button marked *2nd* or *arc* instead of *inv*.

A word of caution is in order. Don't confuse the notation $\sin^{-1}x$, which denotes the inverse sine function, and $1/\sin x$, which denotes the reciprocal of the sine function (namely, $\csc x$).

Applications of Right Triangle Trigonometry

Many practical problems encountered in science and mathematics can be reduced to the solution of a right triangle. It is definitely to your advantage to begin the solution of these types of applications with a sketch of the situation.

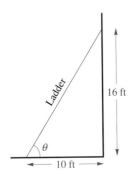

FIGURE 10

EXAMPLE 6   A ladder leans against a wall so that the top of the ladder is 16 ft from the floor and the bottom of the ladder is 10 ft from the wall (Figure 10). Find the angle (to the nearest degree) of the angle ϕ formed by the ladder and the floor.

SOLUTION This problem reduces to finding an acute interior angle of a right triangle, given the lengths of the sides opposite and the side adjacent to this angle. So,

$$\tan \phi = \tfrac{16}{10} \Rightarrow \phi = \tan^{-1}(\tfrac{16}{10}) = 58°.$$

Many practical applications involve the measure of the angle formed by a direct line-of-sight to an object and a horizontal line. If the object in question is above the horizontal line, this angle is called an **angle of elevation.** If, on the other hand, the object is below the horizontal line, the angle is called an **angle of depression** (Figure 11).

FIGURE 11

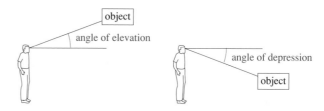

EXAMPLE 7   From a point A on the ground, an observer notes that the angle of elevation to the top of the Sears Tower in Chicago is 64.6° (Figure 12). After moving to a point B, 100 m closer to the building, the observer notes the angle of elevation to be 76.0°. Find the height of the Sears Tower (to the nearest meter).

SOLUTION The height of the building h is the side opposite the angle DAC, so

$$\tan 64.6° = \frac{h}{100 + x} \Rightarrow h = (100 + x)\tan 64.6°.$$

The obvious problem here is that we do not know the distance x from point B to point C. But,

$$\tan 76.0° = \frac{h}{x} \Rightarrow x = \frac{h}{\tan 76.0°}$$

so

$$h = \left(100 + \frac{h}{\tan 76.0°}\right)\tan 64.6°$$

or

$$h = 100\tan 64.6° + \frac{h}{\tan 76.0°}\tan 64.6°.$$

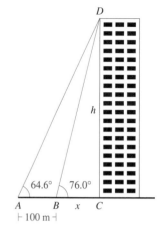

FIGURE 12

Solving for h gives

$$h\tan 76.0° = 100\tan 64.6°\tan 76.0° + h\tan 64.6°$$

$$h\tan 76.0° - h\tan 64.6° = 100\tan 64.6°\tan 76.0°$$

$$h(\tan 76.0° - \tan 64.6°) = 100\tan 64.6°\tan 76.0°$$

$$h = \frac{100\tan 64.6°\tan 76.0°}{\tan 76.0° - \tan 64.6°}$$

$$= 443 \text{ (to the nearest meter).}$$

The height of the Sears Tower is 443 m.

The keystrokes for this computation are as shown (notice the use of the parentheses):

KEYSTROKES

| 100 | × | 64.6 | tan | × | 76.0 | tan | ÷ |

| (| 76.0 | tan | − | 64.6 | tan |) | = |

The next example involves the concept of a **compass bearing.** The bearing from point A to point B is the acute angle formed by the line AB and the north-south line. The angle is also described as being measured either east or west of either north or south. For example in Figure 13, the bearing of point B from point A is N18°E, and the bearing of point C from point A is S71°W.

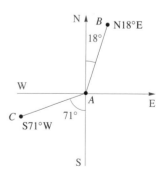

FIGURE 13

EXAMPLE 8 An observer in a lookout tower detects a fire directly south of the tower. An observer in an airplane 13.0 miles directly east of the tower sees the fire to be at a bearing of S34.2°W. To the nearest 0.1 mile, how far from the tower is the fire?

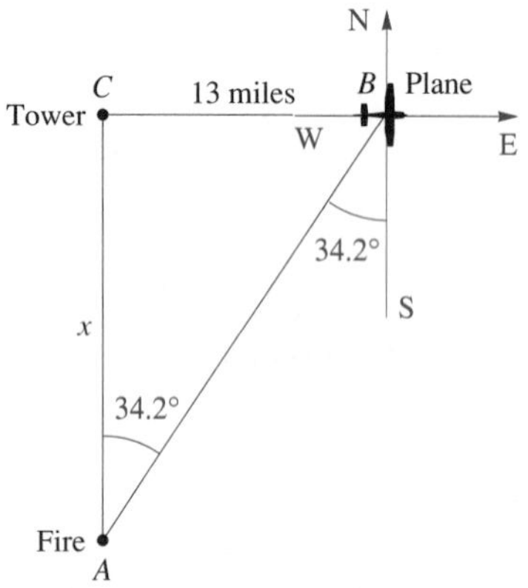

SOLUTION In Figure 14, the angle CAB is also 34.2° (why?). So,

$$\tan 34.2° = \frac{13}{x}$$

or

$$x = \frac{13}{\tan 34.2°}$$

$$= 19.1.$$

The fire is 19.1 miles from the lookout tower.

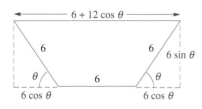

FIGURE 14

Many of the applications of right triangle trigonometry in calculus involve determining a mathematical function as a model for a particular situation (these types of models were discussed in Section 2.2).

EXAMPLE 9 A rain gutter is to be constructed in such a way that its cross section is a trapezoid (Figure 15). Determine the area of the cross section as a function of the angle θ.

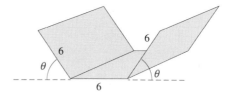

FIGURE 15

SOLUTION The area A of a trapezoid is

$$A = \tfrac{1}{2}[\text{Height}][\text{Sum of the bases}].$$

From Figure 16, we see that the height of the trapezoid is $6 \sin \theta$, the upper base is $6 + 12 \cos \theta$, and the lower base is 6. So,

$$A(\theta) = \tfrac{1}{2}[6 \sin \theta][(6 + 12 \cos \theta) + (6)]$$

or

$$A(\theta) = 36 \sin \theta + 36 \sin \theta \cos \theta.$$

FIGURE 16

EXERCISE SET 5.1

A

*In Problems 1 through 6, determine the values of the trigono-
metric functions of θ.*

1.

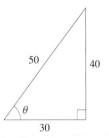

2.

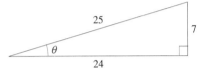

3.

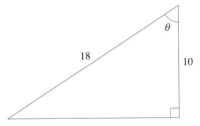

4.

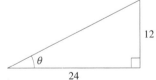

5.   **6.**

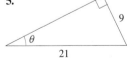   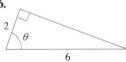

*In Problems 7 through 9, use your calculator to determine the
value of the function (to the nearest 0.0001).*

7. a) sin 29°   **b)** cos 29°   **c)** sin 61°   **d)** cos 61°

8. a) sin 21°   **b)** cos 21°   **c)** tan 21°   **d)** cot 69°

9. a) csc 32°   **b)** sec 58°   **c)** cot 49°   **d)** tan 41°

*In Problems 10 through 12, use your calculator to determine
the value of the function (to the nearest 0.01°).*

10. a) $\sin^{-1}(\frac{1}{3})$ **b)** $\cos^{-1}(\frac{1}{3})$

c) $\tan^{-1}(\frac{1}{2})$ **d)** $\tan^{-1}(2)$

11. a) $\sin^{-1}(0.12)$ **b)** $\cos^{-1}(0.12)$

c) $\sin^{-1}(0.87)$ **d)** $\tan^{-1}(1.27)$

12. a) $\arcsin(\frac{1}{5})$ **b)** $\arccos(\frac{1}{5})$

c) $\arccos(\frac{2}{3})$ **d)** $\arctan(\frac{5}{3})$

*In Problems 13 through 18, use your calculator to determine
the value of x in the figure (to the nearest 0.1).*

13.

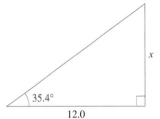

14.

15.

16.

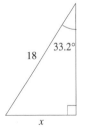

17.

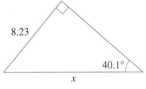

18.

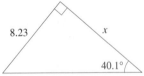

In Problems 19 through 24, given the value of one of the trigo-nometric functions of θ, determine the values of the remaining functions of θ.

19. $\sin \theta = \frac{2}{3}$ **20.** $\cos \theta = \frac{7}{9}$

21. $\tan \theta = \frac{7}{2}$ **22.** $\cot \theta = \frac{4}{9}$

23. $\sec \theta = \dfrac{\sqrt{5}}{2}$ **24.** $\csc \theta = \dfrac{4}{\sqrt{11}}$

B

25. Prove the two remaining reciprocal identities of this section using the definitions of the trigonometric functions.

26. Prove the two ratio identities of this section using the definitions of the trigonometric functions.

27. Prove the five remaining cofunction identities of this section using the definitions of the trigonometric functions.

28. At the beginning of this section, the two triangles in Figure 17 were discussed. **a)** Use the Pythagorean theorem to prove that each of these triangles is a right triangle. **b)** Prove that the two triangles are similar by showing that corresponding sides are in constant proportion.

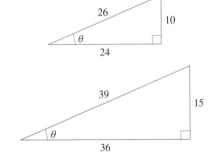

FIGURE 17

In Problems 29 through 34, determine the value of the six trig-onometric functions for θ in terms of x from the figure. (Hint: First determine the remaining side in terms of x using the Pythagorean theorem.)

29.

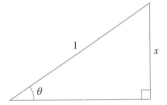

30.

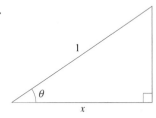

31.

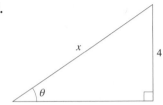

32.

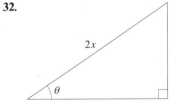

33.

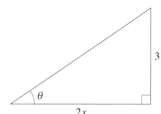

34.

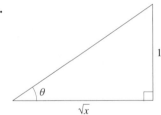

35. From the trailhead, Nancy walks due south for 3 miles and then walks N37°W until she is directly west of the trailhead. How far is she from the trailhead (nearest 0.1 mile)?

36. Eric hikes cross country 6 miles directly north, then 3 miles directly west, and finally 4 miles at a heading of S12°E. How far is he from the trailhead (nearest 0.1 mile)?

37. From an intersection, John walks 100 m directly north. From the same intersection, Marleen walks 70 m directly east. What is the compass bearing from John to Marleen (nearest degree)?

38. From an intersection, Dave runs 2 km at a heading of S42°W. From the same intersection, Inés runs 7 km at a heading of N48°W. What is the compass bearing from Dave to Inés (nearest degree)?

39. The angle of depression from a 100-ft lighthouse to a rowboat on the sea is 40.2° (see Figure 18). How far is the boat from the base of the light (nearest ft)?

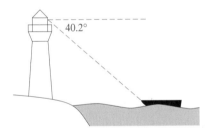

FIGURE 18

40. Find the angle of elevation of the sun (to the nearest degree) if a person 6.0-ft tall casts a shadow that is 3.8-ft long.

41. A helicopter is hovering 320 m from the Eiffel Tower in Paris. The angle of depression from the helicopter to the base of the tower is 20.0°. The angle of elevation from the helicopter to the top of the tower is 30.7°. Find the height of the tower (to the nearest meter).

42. While standing in his backyard, Craig observes a plane directly overhead at an elevation of 200 m. Two seconds later, he notes the angle of elevation to the plane to be 32.6°. Planes passing over this area are not to exceed 150 m/sec. Should Craig call the Airport Complaint Center to report a speeding plane?

43. While standing in her backyard, Liz notices a plane flying at a speed of 120 m/sec at an angle of elevation

of 28.2°. Three seconds later, the plane is directly overhead. Planes passing over this area are not to fly lower than 160 m. Should Liz call the Airport Complaint Center to report a low-flying plane?

44. A balloon for a sales promotion is tethered by two ropes 100 ft apart (see Figure 19). The angle that the ropes make with the horizontal are 23.4° and 51.2°. Find the height of the balloon above the ground (to the nearest foot).

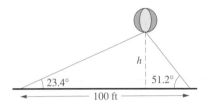

FIGURE 19

45. A pyramid is composed of a square base (10 ft on a side) and four equilateral triangles (see Figure 20). Find the height of the pyramid.

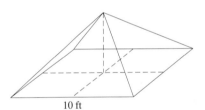

FIGURE 20

46. A pentagon with sides of equal length is inscribed in a circle with radius 12 (see Figure 21). Determine the area of the pentagon.

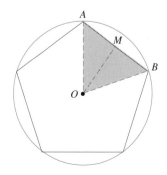

FIGURE 21

47. Determine the area of the right triangle in Figure 22 as a function of the measure of the angle θ.

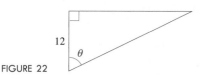

12

θ

FIGURE 22

48. A parallelogram with sides of length 10 in. and 14 in. has an acute interior angle of measure θ (see Figure 23). Determine the area of the parallelogram as a function of the measure of the angle θ.

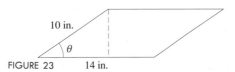

10 in.

θ

FIGURE 23   14 in.

49. A rocket climbs vertically after launch so that at t seconds after launch it is at an altitude of $38t^2$ ft (see Figure 24). A camera at ground level tracks the rocket. Find the angle of elevation as a function of t, the number of seconds after launch.

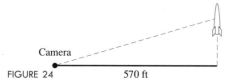

Camera

FIGURE 24   570 ft

50. An oversized gorilla (with a bad attitude) starts to climb the Empire State Building at 10 ft/sec (see Figure 25). Meanwhile, a terrorized bystander stands 200 ft away observing the gorilla. Find the angle of elevation from the bystander to the gorilla as a function of t, the number of seconds after he starts to climb.

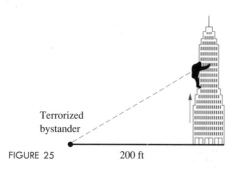

Terrorized
bystander

FIGURE 25   200 ft

C

51. A painting 3-ft high is hung on a wall so that the base of the painting is 8 ft above the floor (see Figure 26). An observer (5-ft tall) stands x ft from the wall to view the painting. Determine the angle θ as a function of x.

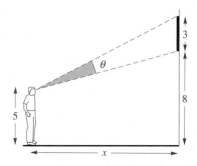

3

θ

8

5

x

FIGURE 26

52. A right triangle with an interior angle of 50° and hypotenuse of 9 in. is rotated about the leg opposite this angle to form a cone (see Figure 27). Find the volume and lateral surface area of the cone.

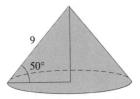

9

50°

FIGURE 27

53. In Figure 28, determine the measure of $\angle EGB$, $\angle DGB$, and $\angle HGB$.

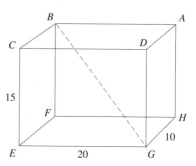

B   A

C   D

15

F

H

10

E   20   G

FIGURE 28

54. One method that surveyors use to determine the height h of an inaccessible object is illustrated in Figure 29. From a point A, measure the angle of elevation to the top of the object. Next, travel a specified distance m to a point B along a heading that is at a right angle to the heading from A to the base of the object. From point B, measure the angle β. From these measurements, the height h can be determined. Derive an expression for h in terms of α, β, and m.

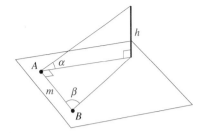

FIGURE 29

ANGLES AND THE UNIT CIRCLE

S E C T I O N   5.2

It is important to note that the discussion of the trigonometric functions of the previous section applies only to acute angles, that is, those angles θ such that $0 < \theta < 90°$. To expand our discussion of these functions, we need to define angles in a more general manner.

Angle and its Measure

An **angle** is formed by rotating a ray about its vertex. The position of the ray before the rotation is the **initial side** of the angle, and the position of the ray after the rotation is the **terminal side** of the angle (see figure). The amount of this rotation is the **measure** of the angle. If the rotation is counterclockwise, then the measure is **positive**. If the rotation is clockwise, then the measure is **negative**.

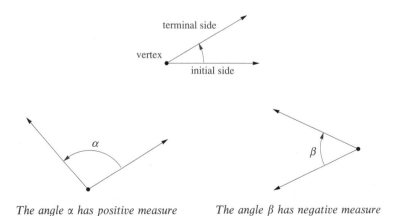

The angle α has positive measure   *The angle β has negative measure*

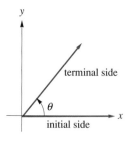

FIGURE 30 *The angle θ is in standard position.*

If we impose a coordinate plane on an angle so that the vertex is on the origin and the initial side falls along the positive *x*-axis, then the angle is in **standard position** (Figure 30).

EXAMPLE 1   Sketch each angle in standard position:

a) $-150°$ **b)** $630°$ **c)** $60°$ **d)** $-450°$

SOLUTION

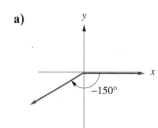

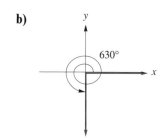

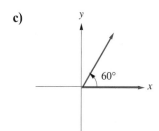

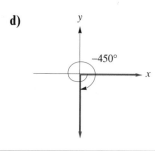

FIGURE 31 *The angles ψ and λ are coterminal.*

If two angles in standard position have terminal sides that coincide, then these angles are said to be **coterminal** (Figure 31). In Example 1, you can see that $630°$ and $-450°$ are coterminal.

Coterminal Angles

> Given an angle with measure $θ$, any angle with measure $θ \pm 360n°$ ($n = 1, 2, 3, \ldots$) is coterminal with the given angle.

It is important that you realize that *coterminal* does not mean *equal*. Two angles are equal if and only if their measures are the same.

EXAMPLE 2 Determine the smallest positive angle that is coterminal with each of the angles given in Example 1.

SOLUTION

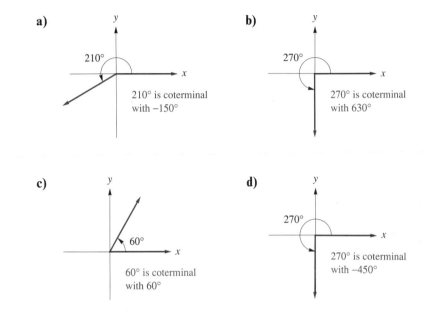

a) 210° is coterminal with −150°

b) 270° is coterminal with 630°

c) 60° is coterminal with 60°

d) 270° is coterminal with −450°

Since degree measurement is so widely accepted, you might think that there is some natural reason why, for example, a right angle has a measure of 90°. On the contrary, the reason we use this system of assigning measure to an angle is mainly historical. Another system of measuring an angle is used in some European countries. In that system, a right angle is divided into 100 *new degrees* or *grades*. A grade is divided into 100 *new minutes*. Each new minute, in turn, is divided into 100 *new seconds*. Your calculator may even support this mode of angle measurement.

Both of these methods of measurement are basically arbitrary. For calculus, we need a different way to measure angles, one that will allow us to consider the trigonometric functions as real functions.

Consider the graph of the equation $x^2 + y^2 = 1$. From Section 2.1, we know that this is a circle of radius 1 with center at $(0, 0)$. We call this the **unit circle.**

Suppose that a bug starts at the point $(1, 0)$ on the unit circle and walks along this circle a distance of one unit (the length of the radius) (see Figure 32). The central angle that intercepts this path of arc is the angle about which our natural method of measure will be determined.

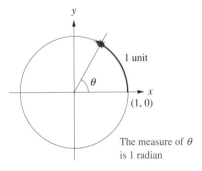

FIGURE 32 *The measure of θ is 1 radian.*

Radian Measure

An angle with measure **one radian** is the central angle of a unit circle that intercepts an arc of circle of length one (the length of the radius). An angle with measure *t* **radians** is the central angle of the unit circle that intercepts an arc of *t* units (see figure).

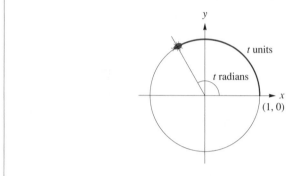

The circumference of the unit circle is 2π units. So, to get completely around the unit circle, our bug must walk 2π units. From this it follows that one complete revolution has a measure of 2π radians, and that one half of a revolution has a measure of π radians. Thus,

$$180° = \pi \text{ radians.}$$

From this equation, it follows that 1 radian is $180/\pi$ degrees, or approximately $57.2958°$.

On occasion, it will be necessary to change from one mode of angle measurement to another.

EXAMPLE 3

  a) Find the radian measure of the angle α if the degree measure of α is $70°$.

  b) Find the degree measure of the angle β if the radian measure of β is $7\pi/8$.

SOLUTION

  a) Since $180° = \pi$ radians, we can multiply $70°$ by π radians/180 degrees (this is in essence the same as multiplying by 1) and determine the corresponding radian measure (notice how the dimensions cancel):

$$70 \text{ degrees} = 70 \, \cancel{\text{degrees}} \cdot \frac{\pi \text{ radians}}{180 \, \cancel{\text{degrees}}} = \frac{7\pi}{18} \text{ radians.}$$

The radian measure of α is $7\pi/18$ radians (approximately 1.2217 radians).

b) In a similar fashion,

$$\frac{7\pi}{8} \text{ radians} \cdot \frac{180 \text{ degrees}}{\pi \text{ radians}} = 157.5 \text{ degrees}.$$

The degree measure of β is $157.5°$.

Since radian measure is such a natural form of angle measurement, it is conventional to write radian measure without units such as degrees. For example, $\theta = 2$ implies that θ is two radians (not two degrees).

Many angles occur frequently enough in calculus to merit special consideration, especially those that are multiples of $30°$ ($\pi/6$ radians) or $45°$ ($\pi/4$ radians). It is definitely to your advantage to be able to rapidly change modes for such angles. The table below shows the corresponding measures for some of these angles.

0°	30°	45°	60°	90°	120°	135°	150°	180°	210°	225°	240°	270°	300°	315°	330°	360°
0	$\frac{\pi}{6}$	$\frac{\pi}{4}$	$\frac{\pi}{3}$	$\frac{\pi}{2}$	$\frac{2\pi}{3}$	$\frac{3\pi}{4}$	$\frac{5\pi}{6}$	π	$\frac{7\pi}{6}$	$\frac{5\pi}{4}$	$\frac{4\pi}{3}$	$\frac{3\pi}{2}$	$\frac{5\pi}{3}$	$\frac{7\pi}{4}$	$\frac{11\pi}{6}$	2π

These should be committed to memory.

From now on, an angle measure is assumed to be a radian measure unless the degree symbol is used.

EXAMPLE 4   For each angle, sketch in standard position:

    a) $-\pi/2$ **b)**   4   **c)** $7\pi/8$ **d)** $-13\pi/6$

SOLUTION

a)

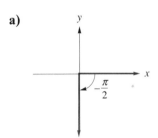

b)

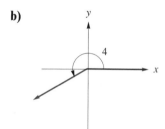

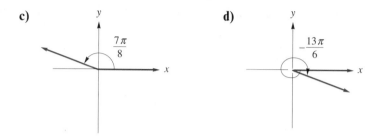

c)   d)

EXAMPLE 5   Determine the smallest positive angle θ that is coterminal with an angle of radian measure -2.

SOLUTION   The first step is to sketch the given angle in standard position, as well as the angle in question. From Figure 33 it should be clear that the angle θ is $2\pi - 2$. This is approximately 4.283 radians.

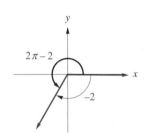

FIGURE 33

The importance of radian measure is this: The measure of arc length that our bug travels along the unit circle from the point $(1, 0)$ is exactly the same as the radian measure of the central angle that subtends the arc. Thus we can talk about the measure of the arc and the radian measure of the angle interchangeably (stop here and ponder this statement until you truly appreciate what it says).

We can use radian measure to determine arc length on any circle of radius r. It should seem reasonable that the ratio of the arc length to the circumference of the circle is equal to the ratio of the measure of the central angle to the measure of a complete revolution:

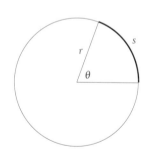

FIGURE 34

$$\frac{\left(\begin{array}{c}\text{length of} \\ \text{the arc}\end{array}\right)}{\left(\begin{array}{c}\text{circumference} \\ \text{of the circle}\end{array}\right)} = \frac{\left(\begin{array}{c}\text{measure of the} \\ \text{central angle}\end{array}\right)}{\left(\begin{array}{c}\text{measure of} \\ \text{one revolution}\end{array}\right)}.$$

Writing this proportion, using radian measure (see Figure 34), we get

$$\frac{s}{2\pi r} = \frac{\theta}{2\pi}.$$

Solving for the arc length s, we get the following result:

Arc Length

> In a circle of radius r, the length s of an arc that is subtended by a central angle θ (in radian measure) is given by
>
> $$s = r\theta.$$

It is important to note that this formula holds only when the central angle θ is measured in radians.

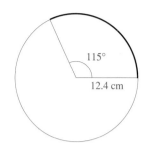

FIGURE 35

EXAMPLE 6   Find the arc length of a circle with radius 12.4 cm that is inter-cepted by an angle of 115°. Express the answer to the nearest 0.1 cm (see Figure 35).

SOLUTION To use the arc length formula, the measure of the central angle must be in radians:

$$115° = 115 \cancel{\text{degrees}} \left(\frac{\pi \text{ radians}}{180 \cancel{\text{degrees}}} \right) = \frac{23\pi}{36} \text{ radians.}$$

So,

$$s = 12.4 \left(\frac{23\pi}{36} \right) = 24.9 \text{ cm.}$$

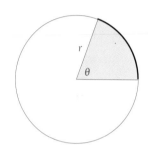

FIGURE 36

A similar result can be derived for the area of a **sector** intercepted by a central angle of a circle with radius r (this is the shaded region in Figure 36). Since it is very much like that for arc length, we will state the result and leave its derivation as an exercise.

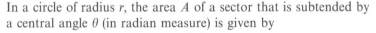

> In a circle of radius r, the area A of a sector that is subtended by a central angle θ (in radian measure) is given by
>
> $$A = \tfrac{1}{2}r^2\theta.$$

Area of a Sector

EXAMPLE 7   Find the radius of a circle, given that a sector of the circle with central angle $\frac{3}{2}$ radians has an area of 24 in.².

SOLUTION In the context of our formula for the area of a sector, we are given two values (the central angle and the area), and we are asked to find the third (radius of the circle) (Figure 37). By substituting the values for the central angle and the area in the formula, and solving for r, the radius:

$$24 = \tfrac{1}{2}r^2(\tfrac{3}{2})$$
$$24 = \tfrac{3}{4}r^2$$

or

$$r = 4\sqrt{2} \text{ in.}$$

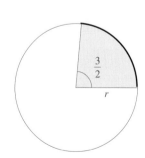

FIGURE 37

E X E R C I S E   S E T   5.2

A

In Problems 1 through 6, determine (in degree measure) the smallest positive angle that is coterminal with the given angle.

1. 500°   **2.** 647°   **3.** −240°

4. −110°   **5.** 1020°   **6.** −824°

In Problems 7 through 12, determine (in radian measure) the smallest positive angle that is coterminal with the given angle. Express this exactly and also approximate it to the nearest 0.1 radian.

7. $\dfrac{7\pi}{2}$ **8.** $\dfrac{10\pi}{3}$ **9.** 9.8

10. −5.7   **11.** π^2 **12.** $\dfrac{1}{\pi}$

In Problems 13 through 18, express the angle in radian measure exactly and also approximately to the nearest 0.01.

13. 120°   **14.** 13.02°   **15.** −200°

16. −1000°   **17.** 100π°   **18.** $(\pi^e)°$

In Problems 19 through 24, express the angle in degree measure. If necessary, round to the nearest degree.

19. $\dfrac{7\pi}{2}$ **20.** $-\dfrac{9\pi}{8}$ **21.** 3

22. −0.5   **23.** $\sqrt{2}$ **24.** π^e

B

25. Find the arc length of a circle with radius 3.7 cm that is intercepted by a central angle of 45°. Express the answer to the nearest 0.1 cm.

26. Find the arc length of a circle with radius 8.2 in. that is intercepted by an angle of 120°. Express the answer to the nearest 0.1 in.

27. Find the radius of a circle such that a central angle of 110° intercepts an arc of length 13.7 in. Express the answer to the nearest 0.1 in.

28. Find the radius of a circle such that a central angle of 220° intercepts an arc of length 5.9 cm. Express the answer to the nearest 0.1 cm.

29. Find the measure of a central angle that intercepts an arc of length 4.1 m in a circle with radius 2.4 m. Express the answer to the nearest degree.

30. Find the measure of a central angle that intercepts an arc of length 12 ft in a circle with diameter 4.3 ft. Express the answer to the nearest degree.

In Problems 31 through 36, suppose that the angle θ is such that $0 < \theta < \pi/2$ and the point (x, y) is on the terminal side of θ (see Figure 38). Determine (in terms of θ) the smallest positive angle with terminal side that passes through the given point.

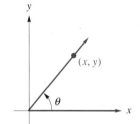

FIGURE 38

31. $(-x, y)$ **32.** $(x, -y)$ **33.** $(-x, -y)$

34. (y, x) **35.** $(-y, x)$ **36.** $(y, -x)$

37. The earth is approximately 93 million miles from the sun. Assuming that the earth travels in a circle at a constant speed and that it completes one revolution in $365\frac{1}{4}$ days, determine the angle θ swept out by the radial line from the sun to the earth over the course of 10 days (to the nearest degree), and determine the distance that the earth travels relative to the sun over this period (to the nearest 0.1 million miles) (Figure 39).

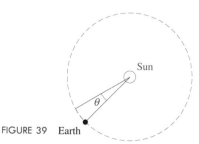

FIGURE 39   Earth

38. The moon is approximately 240 thousand miles from the earth. Assume that the moon travels in a circle at a constant speed and that it completes one revolution in 29 days. Determine the angle θ swept out by the radial line from the earth to the moon over the course of 20 days (to the nearest degree), and determine the distance that the moon travels relative to the earth over this period (to the nearest 0.1 thousand miles).

39. Eratosthenes, the librarian of the great library at Alexandria, Egypt observed that exactly at noon (on the same day) a vertical stick cast a shadow of 7.2° in the library courtyard whereas at Syene (site of present-day Aswan) a vertical stick cast no shadow at all (see Figure 40). He also knew that the distance from Alexandria to Syene is 5000 stadia (1 stadion is about 0.11 miles). From this information, he was able to approximate the radius and the circumference of the earth. Show how he might have done this and determine these distances.

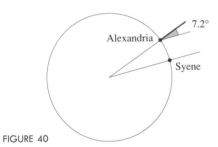

FIGURE 40

40. Determine the area of the segment of a circle of radius 1.2 in. between the arc $\overset{\frown}{AB}$ intercepted by a central angle of 60° and the chord $\overline{AB}$ (this segment is the area that is shaded in Figure 41).

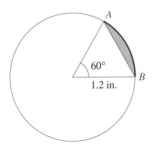

FIGURE 41

41. A nautical mile is the distance between two points on the equator whose longitudes differ by 1 minute (recall that 1 minute = 1/60 degree). Given that the radius of the earth at the equator is approximately 3960 miles, determine the length of one nautical mile in terms of feet (see Figure 42). (Actually, there are several definitions of a nautical mile, so your answer may not agree with other sources.)

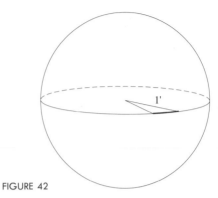

FIGURE 42

42. Find the angle (to the nearest degree) between the hour hand and the minute hand of a clock at 10:30 P.M.

43. Find the angle (to the nearest degree) between the hour hand and the minute hand of a clock at 4:45 A.M.

44. Derive the formula for the area of a sector given in this section.

45. Determine a formula to find the area of a sector of a circle in terms of its radius r and its arc s.

46. A bug starts at the point $(1, 0)$ on the unit circle and walks along the arc of the circle in a counterclockwise direction for 4 units. Determine the quadrant in which the bug ends its trip.

47. A bug starts at the point $(1, 0)$ on the unit circle and walks along the arc of the circle in a clockwise direction for 5 units. Determine the quadrant in which the bug ends its trip.

48. A bug starts at the point $(0, 1)$ on the unit circle and walks along the arc of the circle in a counterclockwise direction for 16 units. Determine the quadrant in which the bug ends its trip.

49. A bug starts at the point $(0, -1)$ on the unit circle and walks along the arc of the circle in a counterclockwise direction for 9 units. Determine the quadrant in which the bug ends its trip.

C

50. A sector with central angle of 45° is cut from a circular piece of paper with radius 24 in., and the edges are taped together to form a cone (see Figure 43). Determine the volume of the cone formed.

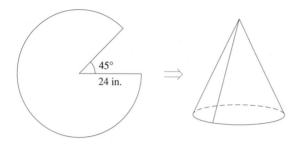

FIGURE 43

51. A region inside of a circle is bounded by the diameter $\overline{AB}$, the chord $\overline{AC}$, and the arc $\overset{\frown}{CB}$. The angle at A is 30°, and the length of $\overline{AB}$ is 12 m (the region is shaded in Figure 44). Find the area of the region.

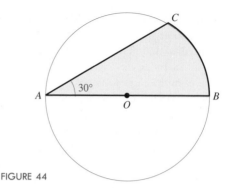

FIGURE 44

52. Find the maximum area of a sector with a perimeter of 8 cm (Figure 45).

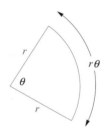

FIGURE 45

S E C T I O N   5.3   **THE TRIGONOMETRIC FUNCTIONS**

The definitions of the six trigonometric functions in Section 5.1 applied to acute angles only since they were based on right triangles. In Section 5.2 the concept of angle was extended to include general angles of any size, positive or negative. The stage is now set to extend the definitions of the six trigonometric functions for this general angle.

Consider any angle θ in standard position. For the sake of our discussion, we will consider an angle with its terminal side in the third quadrant (Figure 46). Suppose that we select two points on the terminal side of θ as shown. The point $P_1(x_1, y_1)$ is r_1 units from the origin, and the point $P_2(x_2, y_2)$ is r_2 units from the origin. Since the right triangles OP_1Q_1 and OP_2Q_2 share the interior angle at O, they are similar. It follows that

$$\frac{y_1}{r_1} = \frac{y_2}{r_2}.$$

For any point on the terminal side of θ, this ratio y/r is not dependent on the particular point that we select. Stated more precisely, this ratio is a func-

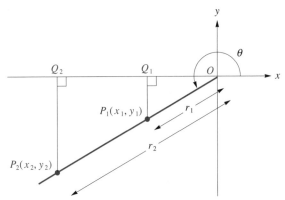

FIGURE 46

tion of the angle θ. The same can be said for the ratios x/r, y/x, x/y, r/y, and r/x.

Let θ be an angle in standard position, let P be a point on the terminal side of θ with coordinates (x, y), and let $r = \sqrt{x^2 + y^2}$ be the distance from P to the origin. Then the **trigonometric functions** of θ are

$$\sin \theta = \frac{y}{r} \qquad \cos \theta = \frac{x}{r} \qquad \tan \theta = \frac{y}{x}$$

$$\csc \theta = \frac{r}{y} \qquad \sec \theta = \frac{r}{x} \qquad \cot \theta = \frac{x}{y}$$

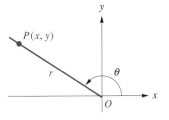

Trigonometric Functions

A little thought should convince you that these six ratios are the only ratios of x, y, and r.

EXAMPLE 1 Suppose that the terminal side of an angle θ passes through the point $P(-8, 15)$ and $0 < \theta < \pi$. Sketch θ and determine the values of the six trigonometric functions of θ.

SOLUTION The sketch of the angle θ is in Figure 47. We find that the distance from the origin O to the point P is

$$r = \sqrt{(-8)^2 + 15^2} = \sqrt{289} = 17.$$

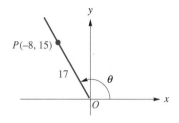

$P(-8, 15)$

17

θ

O

FIGURE 47

So,

$$\sin \theta = \frac{y}{r} = \frac{15}{17} \qquad\qquad \csc \theta = \frac{r}{y} = \frac{17}{15}$$

$$\cos \theta = \frac{x}{r} = \frac{-8}{17} = -\frac{8}{17} \qquad \sec \theta = \frac{r}{x} = \frac{17}{-8} = -\frac{17}{8}$$

$$\tan \theta = \frac{y}{x} = \frac{15}{-8} = -\frac{15}{8} \qquad \cot \theta = \frac{x}{y} = \frac{-8}{15} = -\frac{8}{15}.$$

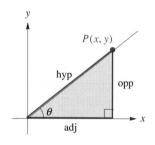

$P(x, y)$

hyp

opp

θ

adj

FIGURE 48

It is important that you appreciate that these definitions are merely an extension of the definitions of the trigonometric functions for acute angles, discussed earlier. To see this, consider Figure 48. A coordinate system has been imposed on the right triangle in such a way that the acute angle θ is in standard position. Notice that the hypotenuse coincides with the terminal side of θ and that $x = $ adj, $y = $ opp, and $r = $ hyp.

You should also notice from these definitions of the trigonometric functions that the reciprocal identities and the ratio identities of acute angles are also true for angles in general. For example,

$$\sin \theta = \frac{y}{r} = \frac{1}{\left[\dfrac{r}{y}\right]} = \frac{1}{\csc \theta}.$$

The proofs of the rest are left to you as exercises. We state these identities again for easy reference.

Reciprocal Identities
for Angles

$$\csc \theta = \frac{1}{\sin \theta} \qquad \sec \theta = \frac{1}{\cos \theta} \qquad \cot \theta = \frac{1}{\tan \theta}$$

Ratio Identities for Angles

$$\tan \theta = \frac{\sin \theta}{\cos \theta} \qquad \cot \theta = \frac{\cos \theta}{\sin \theta}$$

The reciprocal identities enable us to focus our investigation of the trigonometric functions on the sine, cosine, and tangent functions. The other three functions can be determined directly from these.

EXAMPLE 2   Suppose that the terminal side of an angle θ passes through the point $P(2, -8)$.

a) Sketch θ and determine the values of sin θ, cos θ, and tan θ from the sketch.

b) Use part (a) and the reciprocal identities to determine the values of csc θ, sec θ, and cot θ.

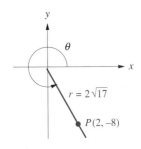

FIGURE 49

SOLUTION

a) Sketch θ (see Figure 49) and determine the values of sin θ, cos θ, and tan θ. We need to determine the distance from the origin to P:

$$r = \sqrt{2^2 + (-8)^2} = 2\sqrt{17}.$$

So,

$$\sin \theta = \frac{y}{r} = \frac{-8}{2\sqrt{17}} = -\frac{4}{\sqrt{17}}$$

$$\cos \theta = \frac{x}{r} = \frac{2}{2\sqrt{17}} = \frac{1}{\sqrt{17}}$$

$$\tan \theta = \frac{y}{x} = \frac{-8}{2} = -4.$$

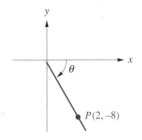

b) By the reciprocal identities,

$$\csc \theta = \frac{1}{\sin \theta} = \frac{1}{\left[-\dfrac{4}{\sqrt{17}}\right]} = -\frac{\sqrt{17}}{4}$$

$$\sec \theta = \frac{1}{\cos \theta} = \frac{1}{\left[\dfrac{1}{\sqrt{17}}\right]} = \sqrt{17}$$

$$\cot \theta = \frac{1}{\tan \theta} = \frac{1}{-4} = -\frac{1}{4}.$$

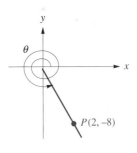

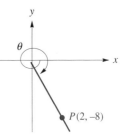

You may object to the way the angle was sketched in Example 2. After all, there are an infinite number of angles whose terminal sides pass through the point given. Three other possibilities are shown in Figure 50. However, since the values of the six trigonometric values of an angle are actually

FIGURE 50

dependent on the terminal side of the angle, our particular choice of angle makes no difference. The following statement should seem reasonable (recall that two angles are coterminal if they share a common terminal side).

Trigonometric Functions of Coterminal Angles

If angles α and β are coterminal, then

$$\sin \alpha = \sin \beta \qquad \cos \alpha = \cos \beta \qquad \tan \alpha = \tan \beta$$

$$\csc \alpha = \csc \beta \qquad \sec \alpha = \sec \beta \qquad \cot \alpha = \cot \beta.$$

The next four examples deal with actually determining the values of the sine, cosine, and tangent functions for specific angles.

EXAMPLE 3   Determine $\sin 135°$, $\cos 135°$, and $\tan 135°$.

SOLUTION First, we sketch the angle in standard position. Consider an isosceles right triangle with both legs of length 1 unit and hypotenuse of $\sqrt{2}$ units, as shown in Figure 51. (This triangle was discussed in Section 5.1.) If we were to superimpose this triangle on our sketch of the angle, we can use it to locate a point on the terminal side, namely, $(-1, 1)$. We can also determine the distance $r = \sqrt{2}$ from our triangle. Thus,

$$\sin 135° = \frac{1}{\sqrt{2}} \qquad \cos 135° = \frac{-1}{\sqrt{2}} = -\frac{1}{\sqrt{2}} \qquad \tan 135° = \frac{-1}{1} = -1.$$

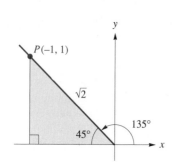

FIGURE 51

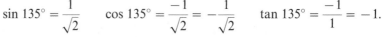

EXAMPLE 4   Determine $\sin 240°$, $\cos 240°$, and $\tan 240°$.

SOLUTION First, we sketch the angle in standard position. In this case we consider a 30°-60° right triangle with legs of lengths 1 and $\sqrt{3}$ and hypotenuse of length 2. Superimposing this triangle on our sketch of the angle (see Figure 52), we can determine a point $(-1, -\sqrt{3})$ on the terminal side and find the distance $r = 2$. So,

$$\sin 240° = \frac{-\sqrt{3}}{2} = -\frac{\sqrt{3}}{2}$$

$$\cos 240° = \frac{-1}{2} = -\frac{1}{2}$$

$$\tan 240° = \frac{-\sqrt{3}}{-1} = \sqrt{3}.$$

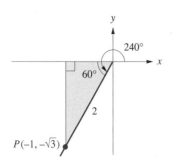

FIGURE 52

EXAMPLE 5 Determine $\sin 11\pi/6$, $\cos 11\pi/6$, and $\tan 11\pi/6$.

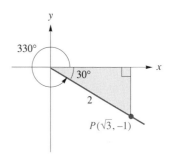

FIGURE 53

SOLUTION The equivalent degree measure for $11\pi/6$ radians is $330°$. Sketching the angle in standard position (see Figure 53), and using the same $30°$-$60°$ right triangle as in Example 4, we can find a point on the terminal side of the angle, $(\sqrt{3}, -1)$. The value of r is 2. So,

$$\sin\frac{11\pi}{6} = \frac{-1}{2} = -\frac{1}{2} \qquad \cos\frac{11\pi}{6} = \frac{\sqrt{3}}{2} \qquad \tan\frac{11\pi}{6} = \frac{-1}{\sqrt{3}} = -\frac{1}{\sqrt{3}}.$$

EXAMPLE 6 Determine the values of the six trigonometric functions of $3\pi/2$.

SOLUTION First, we sketch the angle in standard position (Figure 54). Suppose that we choose the point $(0, -3)$ on the terminal side of the angle. Then $r = 3$ (the distance r is always positive), and we can find the values of the functions:

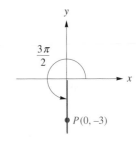

FIGURE 54

$$\sin\frac{3\pi}{2} = \frac{-3}{3} = -1 \qquad\qquad \csc\frac{3\pi}{2} = \frac{3}{-3} = -1$$

$$\cos\frac{3\pi}{2} = \frac{0}{3} = 0 \qquad\qquad \sec\frac{3\pi}{2} = \frac{3}{0} \quad \text{(undefined)}$$

$$\tan\frac{3\pi}{2} = \frac{-3}{0} \quad \text{(undefined)} \qquad \cot\frac{3\pi}{2} = \frac{0}{-3} = 0.$$

Whether the value of a trigonometric function of θ is positive or negative depends on the quadrant in which the terminal side of θ lies. For example, suppose that θ is an angle whose terminal side lies in the second quadrant and the point $P(x, y)$ is a point on its terminal side. Then, since P is in the second quadrant, $x < 0$ and $y > 0$ (recall that r, a distance, is always positive). This implies that $\sin \theta > 0$, $\cos \theta < 0$, and $\tan \theta < 0$.

A similar analysis can be made for each of the three remaining quadrants. Figure 55 summarizes the results in all quadrants.

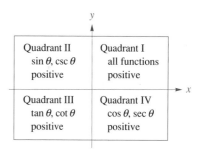

FIGURE 55

Notice that in Examples 3, 4, and 5, a right triangle played a role in the evaluation of the six trigonometric functions of an angle θ. This triangle, the **reference triangle** of θ, is formed by the x-axis, the terminal side of the angle, and a vertical line passing through the selected point P on the terminal side. More importantly, you may have noticed that the acute interior angle of the reference triangle at the origin, the **reference angle** of θ, is related to the values of the trigonometric functions of a general angle.

Reference Angle

For an angle θ in standard position, the **reference angle** θ' for θ is the acute angle that the terminal side of θ makes with the x-axis. Let $f(\theta)$ represent any of the six trigonometric functions. If θ' is the reference angle for the angle θ, then either

$$f(\theta) = f(\theta') \quad \text{or} \quad f(\theta) = -f(\theta')$$

depending on the function and the quadrant in which the terminal side of θ falls.

EXAMPLE 7   Determine the reference angle θ' for the given angle θ.

a) $300°$ **b)** $-510°$ **c)** $9\pi/4$ **d)** 4

SOLUTION

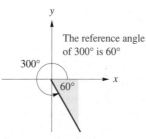

The reference angle of $300°$ is $60°$

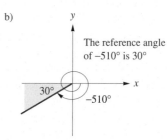

The reference angle of $-510°$ is $30°$

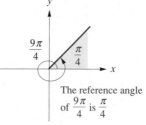

The reference angle of $9\pi/4$ is $\pi/4$.

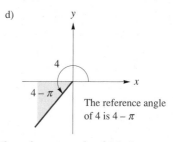

The reference angle of 4 is $4 - \pi$.

EXAMPLE 8   Determine:

a) $\sin 300°$ **b)** $\tan(-510°)$ **c)** $\cos 9\pi/4$.

SOLUTION

a) The terminal side of 300° is in the fourth quadrant. The sine
function is negative in this quadrant. Using the reference angle
found in Example 7, we obtain

$$\sin 300° = -\sin 60° = -\frac{\sqrt{3}}{2}.$$

b) Again, from Example 7, $-510°$ is in the third quadrant. Since
the tangent function is positive in the third quadrant,

$$\tan(-510°) = +\tan 30° = \frac{1}{\sqrt{3}}.$$

c) Because $9\pi/4$ is in the first quadrant,

$$\cos\frac{9\pi}{4} = +\cos\frac{\pi}{4} = \frac{1}{\sqrt{2}}.$$

EXAMPLE 9 Suppose that for some angle θ, $\cos\theta = -\frac{5}{13}$ and $\sin\theta < 0$.
Sketch the terminal side of θ and determine the values of the other five
trigonometric functions of θ.

SOLUTION Because both $\cos\theta$ and $\sin\theta$ are negative, the terminal side of the
angle θ must fall in the third quadrant (Figure 56). Since $\cos\theta = -\frac{5}{13}$, there
exists a point P on the terminal side of θ whose x-coordinate is -5 and distance from the origin is 13. We still need to find the y-coordinate of this
point P. Now, recall that

$$x^2 + y^2 = r^2.$$

So,

$$(-5)^2 + y^2 = (13)^2$$

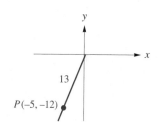

FIGURE 56

or

$$y = \pm\sqrt{(13)^2 - (-5)^2} = \pm 12.$$

Since the angle is in the third quadrant, we choose $y = -12$. The values of the trigonometric functions follow:

$$\sin\theta = -\frac{12}{13} \qquad \tan\theta = \frac{12}{5} \qquad \cot\theta = \frac{5}{12}$$

$$\csc\theta = -\frac{13}{12} \qquad \sec\theta = -\frac{13}{5}$$

Although all the values of the table don't necessarily need to be memorized, you should be able to determine any of these values quickly and without much difficulty. Committing those in the shaded box to memory will be to your advantage; they are often used as reference angles.

The accompanying table shows many of the common trigonometric values that occur frequently in application. Pause here for at least five minutes and examine the table for the patterns caused by the reciprocal and ratio identities, as well as the concept of reference angle. Eventually, you should be able to reproduce this table without much effort.

deg	rad	$\sin \theta$	$\cos \theta$	$\tan \theta$	$\cot \theta$	$\sec \theta$	$\csc \theta$
0°	0	0	1	0	undef	1	undef
30°	$\pi/6$	1/2	$\sqrt{3}/2$	$1/\sqrt{3}$	$\sqrt{3}$	$2/\sqrt{3}$	2
45°	$\pi/4$	$1/\sqrt{2}$	$1/\sqrt{2}$	1	1	$\sqrt{2}$	$\sqrt{2}$
60°	$\pi/3$	$\sqrt{3}/2$	1/2	$\sqrt{3}$	$1/\sqrt{3}$	2	$2/\sqrt{3}$
90°	$\pi/2$	1	0	undef	0	undef	1
120°	$2\pi/3$	$\sqrt{3}/2$	$-1/2$	$-\sqrt{3}$	$-1/\sqrt{3}$	-2	$2/\sqrt{3}$
135°	$3\pi/4$	$1/\sqrt{2}$	$-1/\sqrt{2}$	-1	-1	$-\sqrt{2}$	$\sqrt{2}$
150°	$5\pi/6$	1/2	$-\sqrt{3}/2$	$-1/\sqrt{3}$	$-\sqrt{3}$	$-2/\sqrt{3}$	2
180°	π	0	-1	0	undef	-1	undef
210°	$7\pi/6$	$-1/2$	$-\sqrt{3}/2$	$1/\sqrt{3}$	$\sqrt{3}$	$-2/\sqrt{3}$	-2
225°	$5\pi/4$	$-1/\sqrt{2}$	$-1/\sqrt{2}$	1	1	$-\sqrt{2}$	$-\sqrt{2}$
240°	$4\pi/3$	$-\sqrt{3}/2$	$-1/2$	$\sqrt{3}$	$1/\sqrt{3}$	-2	$-2/\sqrt{3}$
270°	$3\pi/2$	-1	0	undef	0	undef	-1
300°	$5\pi/3$	$-\sqrt{3}/2$	1/2	$-\sqrt{3}$	$-1/\sqrt{3}$	2	$-2/\sqrt{3}$
315°	$7\pi/4$	$-1/\sqrt{2}$	$1/\sqrt{2}$	-1	-1	$\sqrt{2}$	$-\sqrt{2}$
330°	$11\pi/6$	$-1/2$	$\sqrt{3}/2$	$-1/\sqrt{3}$	$-\sqrt{3}$	$2/\sqrt{3}$	-2
360°	2π	0	1	0	undef	1	undef

Just as with trigonometric functions of acute angles, the exact evaluation of trigonometric functions of general angles is difficult if not impossible for most values of θ. And, just as with acute angles, one can obtain a good approximation of these functions by using a hand-held calculator. Care must exercised, however, to insure that your calculator knows that you mean *degrees* when you want *degrees* and that you mean *radians* when you want *radians*.

For example(suppose that we wish to approximate tan 18°; the keystrokes are

KEYSTROKES		DISPLAY
18	tan	0.324919696

The calculator must be in the degree mode to get a proper approximation.

On the other hand, suppose that we wish to approximate tan 18, the tangent of 18 radians. The calculator must be set in **radian mode.** The keystrokes are similar to those for tan 18°, but the output is quite different:

KEYSTROKES		DISPLAY
18	tan	−1.137313722

Approximating cot θ by calculator is a bit more difficult since most calculators don't have a cot button. To evaluate, say, cot 2 (recall that this implies 2 radians), we first evaluate tan 2 and then its reciprocal:

KEYSTROKES			DISPLAY
2	tan	1/x	−0.457657553

Approximating sec θ or csc θ involves similar procedures, using the cos or sin buttons, respectively. We leave these for you to develop.

Using Your Calculator

EXAMPLE 10   Using a calculator, determine the value of sin $3\pi/8$ to the nearest 0.0001.

SOLUTION The keystrokes and the output are shown:

KEYSTROKES						DISPLAY
3	×	π	÷	8	=	sin	0.923879532

To the nearest 0.0001, sin $3\pi/8 = 0.9239$.

One final note on calculators: Even though they are faster and more precise than other means of approximating trigonometric functions, they do not yield *exact* values in general. For example, being able to approximate $\sin(-240°)$ as 0.866025403 is no substitute in many contexts for knowing the exact value of $\sin(-240°)$, namely, $\sqrt{3}/2$.

EXERCISE SET 5.3

A

In Problems 1 through 12, the terminal side of the angle θ passes through the given point. Determine the values of the sin θ, cos θ and tan θ.

1. $(3, 4)$ **2.** $(5, 13)$

3. $(-3, 4)$ **4.** $(5, -13)$

5. $(-3, -4)$ **6.** $(-5, -13)$

7. $(-12, 8)$ **8.** $(-3, -9)$

9. $(\frac{1}{2}, -\frac{3}{4})$ **10.** $(-\frac{1}{4}, -\frac{3}{2})$

11. $(\sqrt{3}, -1)$ **12.** $(-\sqrt{5}, \sqrt{5})$

In Problems 13 through 18, an angle θ is given. Find the reference angle of θ in degrees, and determine sin θ, cos θ, and tan θ without using a calculator.

13. $-300°$ **14.** $-120°$ **15.** $570°$

16. $-150°$ **17.** $720°$ **18.** $540°$

In Problems 19 through 24, an angle θ is given. Find the reference angle of θ in radians, and determine sin θ, cos θ, and tan θ without using a calculator.

19. $-\dfrac{2\pi}{3}$ **20.** $\dfrac{15\pi}{4}$ **21.** $\dfrac{11\pi}{4}$

22. $-\dfrac{13\pi}{6}$ **23.** 13π **24.** -12π

In Problems 25 through 36, use a calculator to evaluate the function (to the nearest 0.0001).

25. $\sin 110°$ **26.** $\cos(-130°)$ **27.** $\cos 92.145°$

28. $\sin 21.145°$ **29.** $\sin\left(-\dfrac{2\pi}{5}\right)$ **30.** $\cos\left(-\dfrac{3\pi}{7}\right)$

31. $\cos 13$ **32.** $\sin 6$ **33.** $\cos 13°$

34. $\sin 6°$ **35.** $\cos 10\pi°$ **36.** $\sin(e^4)$

B

37. Given that $\cos\theta = \frac{3}{5}$ and $3\pi/2 < \theta < 2\pi$, determine the other five trigonometric functions.

38. Given that $\sin\theta = -\frac{12}{13}$ and $\pi/2 < \theta < 3\pi/2$, determine the other five trigonometric functions.

39. Given that $\tan\theta = -\frac{2}{7}$ and $0 < \theta < \pi$, determine the other five trigonometric functions.

40. Given that $\cot\theta = \sqrt{7}/5$ and $0 < \theta < \pi$, determine the other five trigonometric functions.

41. Given that $\sec\theta = 7/\sqrt{17}$ and $\pi < \theta < 2\pi$, determine the other five trigonometric functions.

42. Given that $\cos\theta = \frac{2}{5}$ and $0 < \theta < \pi$, determine the other five trigonometric functions.

43. Prove the reciprocal identities by using the definitions of this section.

44. Prove the ratio identities by using the definitions of this section.

45. Suppose that α is an acute angle and that $\sin\alpha = k$. Determine (in terms of k):

    a) $\sin(\pi - \alpha)$ **b)** $\sin(\alpha + \pi)$

    c) $\sin(-\alpha)$ **d)** $\sin(-\pi - \alpha)$

46. Suppose that α is an acute angle and that $\cos\alpha = r$. Determine (in terms of r):

    a) $\cos(\alpha - \pi)$ **b)** $\cos(\pi + \alpha)$

    c) $\cos(2\pi - \alpha)$ **d)** $\cos(2\pi + \alpha)$

C

47. In calculus, you will see that $\sin t$ (t in radians) can be approximated by the polynomial function

$$s(t) = t - \tfrac{1}{6}t^3 + \tfrac{1}{120}t^5$$

with a maximum error of $\frac{1}{5040}t^7$. Use $s(t)$ and your calculator to approximate the value of sin 0.2 and to determine the maximum error in this approximation.

48. In calculus, you will see that cos t (t in radians) can be approximated by the polynomial function

$$c(t) = 1 - \tfrac{1}{2}t^2 + \tfrac{1}{24}t^4$$

with a maximum error of $\frac{1}{720}t^6$. Use $c(t)$ and your calculator to approximate the value of $\cos(-\tfrac{3}{4})$ and to determine the maximum error in this approximation.

49. In calculus, you will see that tan t (t in radians) can be approximated by the polynomial function

$$T(t) = t + \tfrac{1}{3}t^3 + \tfrac{2}{15}t^5$$

for $-\pi/2 \le t \le \pi/2$. Use $T(t)$ to approximate the value of tan $\tfrac{1}{2}$. Check your result by using your calculator to approximate tan $\tfrac{1}{2}$ directly.

50. In Figure 57 of a unit circle, each of the line segments listed below corresponds to one of the six trigonometric functions of θ. Determine which one matches each line segment: $\overline{PN}$, $\overline{PM}$, $\overline{AR}$, $\overline{BQ}$, $\overline{OR}$, and $\overline{OQ}$.

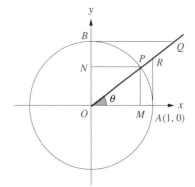

FIGURE 57 *Unit circle*

TRIGONOMETRIC FUNCTIONS OF REAL NUMBERS

SECTION 5.4

This is a good point in our development of the trigonometric functions to step back and take a look at the big picture. In Section 5.1, we defined the six trigonometric functions for acute angles in terms of right triangles. After introducing the concept of the general angle in Section 5.2, we expanded these functions in Section 5.3 to include all angles, positive and negative.

Our task at hand is to extend these functions so that they are not merely functions of angles but functions of real numbers. It is in this context that these functions are most important in mathematics and science.

Recall from Section 5.2 that if a bug travels t units along an arc of the unit circle from the point (1, 0), the central angle that intercepts this arc has a measure of t radians. Consider the point at which the bug ends up after traveling t units. For the sake of this discussion, we label this point $P(t)$ (Figure 58). Since this point $P(t)$ is on the terminal side of this angle with measure t radians, we can use it to determine sin t and cos t. The corresponding distance r is 1. So, if $P(t)$ has coordinates (x, y), then

$$\cos t = \frac{x}{1} = x \quad \text{and} \quad \sin t = \frac{y}{1} = y.$$

Simply stated, the x-coordinate of $P(t)$ is cos t, and the y-coordinate is sin t. There is an important point to be made here; the sine function and the

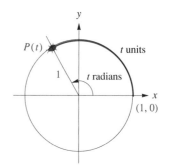

FIGURE 58

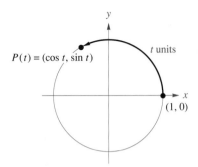

FIGURE 59

cosine function can be considered as functions not only of angles but also of real numbers (Figure 59). This real number is the distance traveled by the bug along the unit circle (ponder this point until you truly appreciate it).

Because of this alternate way to think of these trigonometric functions (using the unit circle), the sine and cosine functions are sometimes called **circular functions.**

Sines and Cosines of Real Numbers: Circular Functions

Consider a real number t and its corresponding point $P(t)$ on the unit circle. The **sin t** is the y-coordinate of the point $P(t)$. The **cos t** is the x-coordinate of $P(t)$.

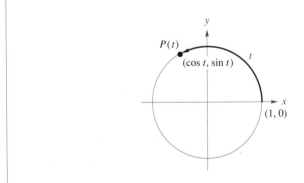

Once these two functions have been determined, the tangent, cotangent, secant, and cosecant functions can be derived by using the reciprocal and ratio identities.

EXAMPLE 1   Plot on the unit circle and find the coordinates of the following:

     a) $P(\pi/2)$ **b)** $P(2\pi/3)$ **c)** $P(-2\pi/3)$.

SOLUTION

a)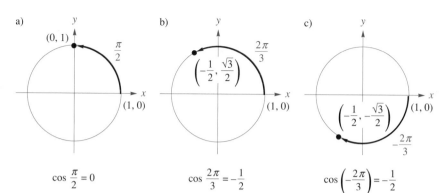

$$\cos \frac{\pi}{2} = 0$$

$$\sin \frac{\pi}{2} = 1$$

b)

$$\cos \frac{2\pi}{3} = -\frac{1}{2}$$

$$\sin \frac{2\pi}{3} = \frac{\sqrt{3}}{2}$$

c)

$$\cos \left(-\frac{2\pi}{3}\right) = -\frac{1}{2}$$

$$\sin \left(-\frac{2\pi}{3}\right) = -\frac{\sqrt{3}}{2}$$

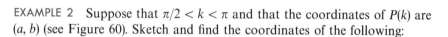

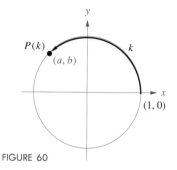

FIGURE 60

EXAMPLE 2   Suppose that $\pi/2 < k < \pi$ and that the coordinates of $P(k)$ are (a, b) (see Figure 60). Sketch and find the coordinates of the following:

a) $P(-k)$ b) $P(k + 2\pi)$ c) $P(k + \pi)$ d) $P(\pi - k)$

SOLUTION

a) From Figure 61, it should be apparent that $P(k)$ and $P(-k)$ are symmetric with respect to the x-axis. As you recall from our discussion of symmetry in Section 2.1, the coordinates of $P(-k)$ are $(a, -b)$. Thus,

$$\sin(-k) = -\sin k \quad \text{and} \quad \cos(-k) = \cos k.$$

b) Since the circumference of the unit circle is 2π units, the points $P(k)$ and $P(k + 2\pi)$ coincide; the coordinates of $P(k + 2\pi)$ are also (a, b) (Figure 62). Thus,

$$\sin(k + 2\pi) = \sin k \quad \text{and} \quad \cos(k + 2\pi) = \cos k.$$

c) It should be apparent that $P(k)$ and $P(k + \pi)$ are symmetric with respect to the origin (Figure 63). This implies that the coordinates of $P(k + \pi)$ are $(-a, -b)$. Thus,

$$\sin(k + \pi) = -\sin k \quad \text{and} \quad \cos(k + \pi) = -\cos k.$$

d) Think of this as instructing our bug to travel along the unit circle π units, then backing up k units. From the figure, we can see that $P(k)$ and $P(\pi - k)$ are symmetric with respect to the

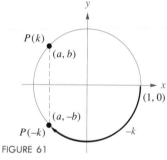

FIGURE 61

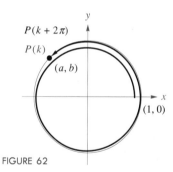

FIGURE 62

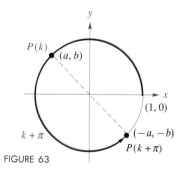

FIGURE 63

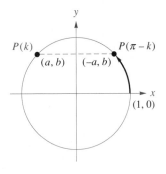

FIGURE 64

x-axis (Figure 64). The coordinates of $P(\pi - k)$ are $(-a, b)$. Thus,

$$\sin(\pi - k) = \sin k \quad \text{and} \quad \cos(\pi - k) = -\cos k.$$

We can derive a few facts about the sine and cosine functions. For example, since the point $P(t)$ is on the unit circle, the *x*-coordinate and the *y*-coordinate of $P(t)$ must be in the interval $[-1, 1]$. Stated more precisely:

For any real number t,
$$-1 \leq \cos t \leq 1 \quad \text{and} \quad -1 \leq \sin t \leq 1.$$

The notation $\sin^2 t$ is used to denote the square of the sine of t. That is, $\sin^2 t = (\sin t)^2$.

Also, since the point $P(t)$ is on the unit circle $x^2 + y^2 = 1$, the coordinates $(\cos t, \sin t)$ should make this equation true. So,

$$(\cos t)^2 + (\sin t)^2 = 1$$

or

$$\cos^2 t + \sin^2 t = 1.$$

This identity is a **Pythagorean identity.** There are two other Pythagorean identities; the proof of the other two are left as exercises.

Pythagorean Identities

$$\cos^2 t + \sin^2 t = 1 \qquad \tan^2 t + 1 = \sec^2 t \qquad 1 + \cot^2 t = \csc^2 t$$

The results of parts b) and c) of Example 1 suggest another pair of properties of the sine and cosine functions. For a real number t, consider the points $P(t)$ with the coordinates $(\cos t, \sin t)$ and $P(-t)$ with the coordinates $(\cos(-t), \sin(-t))$. These two points are symmetric with respect to the *x*-axis (see Figure 65), and because of this symmetry, $\cos(-t) = \cos t$ and $\sin(-t) = -\sin t$. In other words, the cosine function is an even function, and the sine function is an odd function.

What about the tangent function? Is this odd or even? Recall that by a ratio identity the tangent function is the quotient of the sine and cosine functions, so

$$\tan(-t) = \frac{\sin(-t)}{\cos(-t)} = \frac{-\sin t}{\cos t} = -\tan t.$$

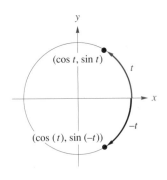

FIGURE 65

Thus, the tangent function is an odd function.

These and other results are summarized as identities:

Odd-Even Identities

The cosine function is an even function. The sine function is an odd function. That is,

$$\sin(-t) = -\sin t \quad \text{and} \quad \cos(-t) = \cos t$$

for any real number t. Furthermore,

$$\tan(-t) = -\tan t \qquad \cot(-t) = -\cot t$$

$$\sec(-t) = \sec t \qquad \csc(-t) = -\csc t.$$

Again, the proof of the rest of these identities are left as exercises.

We can also make a generalization based on our result in part b) of Example 2. It should seem reasonable that for any real number t, the points $P(t)$ and $P(t + 2\pi)$ coincide. Thus, for any real number t,

$$\sin(t + 2\pi) = \sin t \quad \text{and} \quad \cos(t + 2\pi) = \cos t.$$

This implies that the values of these functions repeat after any interval of 2π units. We say that the sine and cosine functions are **periodic** functions, and that each has **period** of 2π.

From part c) of Example 2, we can generalize that

$$\sin(t + \pi) = -\sin t \quad \text{and} \quad \cos(t + \pi) = -\cos t,$$

so

$$\tan(t + \pi) = \frac{\sin(t + \pi)}{\cos(t + \pi)} = \frac{-\sin t}{-\cos t} = \tan t.$$

This implies that the period of the tangent function is π. The periods of the other functions are left as exercises.

Graphs of the Trigonometric Functions

Any discussion of a function would be incomplete without investigating its graph. We start by sketching one period of the graph over the interval $[0, 2\pi]$. In Figure 66a, points $P(t)$ for selected values of t are plotted on the unit circle. For each, a vertical line segment representing the value of the sine for that value of t is shown. Using these line segments, we can determine the graph of $y = \sin t$ (Figure 66b). This is a periodic function.

Once we determine the portion of the graph over the interval $[0, 2\pi]$, we can copy this to the intervals $[2\pi, 4\pi]$, $[4\pi, 6\pi]$, and so on. We complete the entire graph by exploiting the symmetry of this odd function:

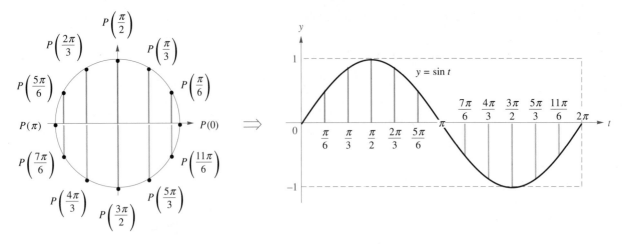

FIGURE 66

The Graph of $y = \sin t$

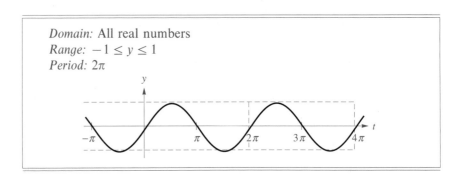

Domain: All real numbers
Range: $-1 \leq y \leq 1$
Period: 2π

It is very important that you can sketch one period of this graph quickly and use this period to complete the graph.

The cosine function can be sketched in the same manner using the unit circle. In Figure 67a, horizontal line segments are used to represent values of the cosine function for selected values of t. These line segments are used to construct one period of the cosine curve (Figure 67b).

Using this one period over the interval $[0, 2\pi]$ and the symmetry of this even function, we can complete the graph of the cosine function.

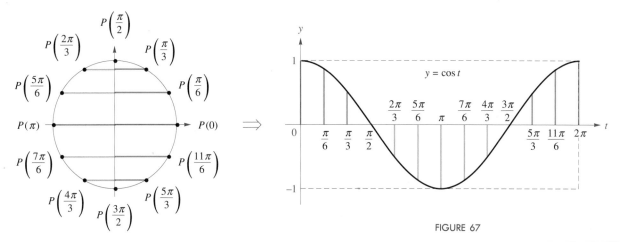

FIGURE 67

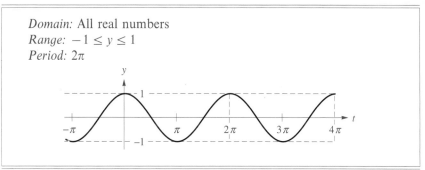

Domain: All real numbers
Range: $-1 \le y \le 1$
Period: 2π

The Graph of $y = \cos t$

The graphing techniques of Chapter 2—translations, reflections, expansions, and compressions—allow us to use the graphs of $y = \sin t$ and $y = \cos t$ to sketch the graphs of more general sine and cosine graphs.

EXAMPLE 3   Sketch the graph of $y = 2 + \sin(t - \pi/2)$ for $-2\pi \le t \le 4\pi$.

SOLUTION This is a translation of the graph of $y = \sin t$. We shift that graph up 2 units and to the right $\pi/2$ units (Figure 68).

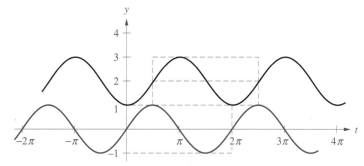

FIGURE 68

EXAMPLE 4   Sketch the graph of $y = -3 - \cos t$ for $-2\pi \le t \le 4\pi$.

SOLUTION This is a reflection through the t-axis and vertical translation of three units down of the graph of $y = \cos t$ (Figure 69).

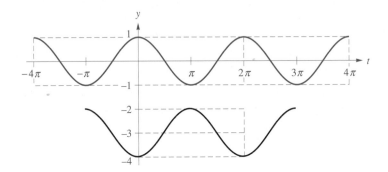

FIGURE 69

Examine the graphs of $y = \sin t$ and $y = \cos t$; they appear to be more or less the same graph. Looking at the unit circle, you can see that the x-coordinate and the y-coordinate of $P(t)$ vary in the same manner except for their initial values and direction (increasing or decreasing). It turns out that if we reflect the graph of $y = \sin t$ through the y-axis and then translate $\pi/2$ units to the right, we end up with the graph of $y = \cos t$. That is,

$$\sin\left(\frac{\pi}{2} - t\right) = \cos t \quad \text{for all } t.$$

Similarly,

$$\cos\left(\frac{\pi}{2} - t\right) = \sin t \quad \text{for all } t.$$

Astute readers will remember these as the cofunction identities for acute angles of Section 5.1. These identities are also true for angles in general, but their proofs in this context will be deferred until Chapter 6.

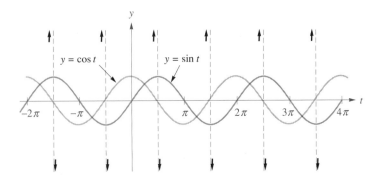

FIGURE 70

We can use the ratio identities and the techniques developed in Chapter 3 for sketching the graphs of rational functions to investigate the graphs of $y = \tan t$ and $y = \cot t$. Since $\tan t = \sin t/\cos t$, the graph of $y = \tan t$ has the same t-intercepts as the graph of $y = \sin t$ (Figure 70). They are 0, $\pm\pi$, $\pm 2\pi$, $\pm 3\pi, \ldots, \pm n\pi, \ldots$ (n is an integer). It also has vertical asymptotes for every value of t such that $\cos t = 0$, namely, $t = n\pi/2$, for odd integers n. The graph of $y = \tan t$ is positive over all intervals in which $\sin t$ and $\cos t$ have the same sign, and the graph is negative over the intervals in which these two functions have different signs.

We now complete the sketch of $y = \tan t$:

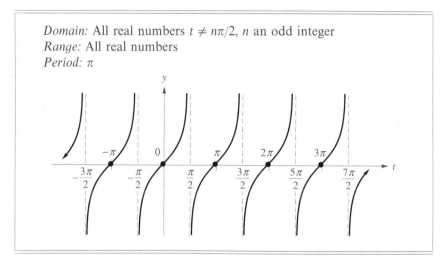

Domain: All real numbers $t \neq n\pi/2$, n an odd integer
Range: All real numbers
Period: π

The Graph of $y = \tan t$

In a similar fashion, we can sketch the graph of $y = \cot t$ using $\cot t = \cos t/\sin t$.

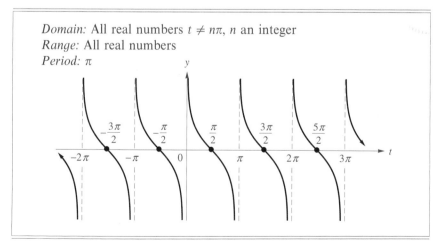

Domain: All real numbers $t \neq n\pi$, n an integer
Range: All real numbers
Period: π

The Graph of $y = \cot t$

EXAMPLE 5   Sketch the graph of $f(t) = \tan 2t$.

SOLUTION   Starting with the graph of $y = \tan t$, we can arrive at the graph of $f(t)$ by shrinking it horizontally by a factor of two. The vertical asymptotes are $t = n\pi/4$ (n is an odd integer) (Figure 71).

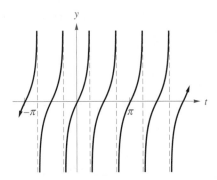

FIGURE 71

To investigate the graphs of $y = \sec t$ and $y = \csc t$, we can use the reciprocal identities. Since $\sec t = 1/\cos t$, each t-intercept of the graph of $y = \cos t$ corresponds to a vertical asymptote of the graph of $y = \sec t$. Likewise, the graph of $y = \sec t$ is positive in any interval over which the graph of $y = \cos t$ is positive, and it is negative in any interval over which the graph of $y = \cos t$ is negative.

The Graph of $y = \sec t$

Domain: All real numbers $t \neq n\pi/2$, n an odd integer
Range: $y \leq -1$ or $y \geq 1$
Period: 2π

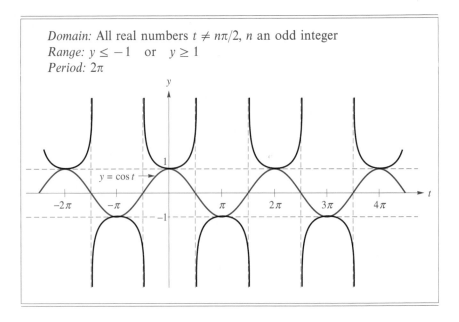

The sketch of the graph $y = \csc t$ can be found in a similar manner using the reciprocal identity $\csc t = 1/\sin t$.

The Graph of $y = \csc t$

Domain: All real numbers $t \neq n\pi$, n an integer
Range: $y \leq -1$ or $y \geq 1$
Period: 2π

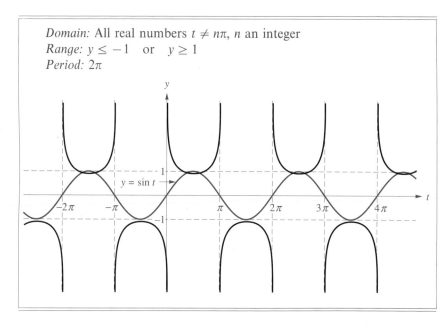

EXAMPLE 6   Sketch the graphs of $y = \cos(t + \pi/3)$ and $y = \sec(t + \pi/3)$ on the same coordinate plane.

SOLUTION These functions are horizontal translations of $y = \cos t$ and $y = \sec t$. The asymptotes of the graph of $y = \sec(t + \pi/3)$ are $t = n\pi/2 - \pi/3$ (n is an odd integer) (Figure 72).

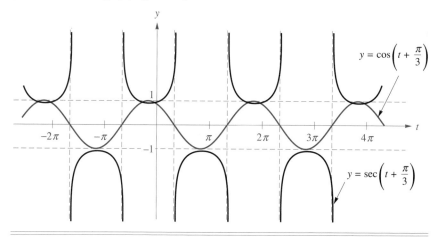

FIGURE 72

EXERCISE SET 5.4

A

In Problems 1 through 12, for each point P(t), plot the point on a unit circle and determine its coordinates exactly.

1. **a)** $P(\pi)$ **b)** $P(-\pi)$
2. **a)** $P(3\pi/2)$ **b)** $P(-\pi/2)$
3. **a)** $P(-3\pi/4)$ **b)** $P(3\pi/4)$
4. **a)** $P(2\pi)$ **b)** $P(-2\pi)$
5. **a)** $P(3\pi)$ **b)** $P(-3\pi)$
6. **a)** $P(\pi/6)$ **b)** $P(5\pi/6)$
7. **a)** $P(-4\pi/3)$ **b)** $P(-2\pi/3)$
8. **a)** $P(\pi/3)$ **b)** $P(2\pi/3)$
9. **a)** $P(10\pi)$ **b)** $P(11\pi)$
10. **a)** $P(24\pi)$ **b)** $P(13\pi)$
11. **a)** $P(-11\pi/2)$ **b)** $P(13\pi/2)$
12. **a)** $P(7\pi/6)$ **b)** $P(-\pi/6)$

In Problems 13 through 18, use the graph of the point P(k) and its coordinates (a, b) to plot the point given and find its coordinates, using a) $P(k - \pi)$, b) $P(k + 2\pi)$, c) $P(k + \pi)$, and d) $P(\pi - k)$.

13.

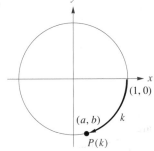

14.

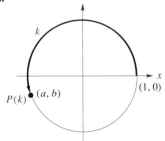

15.

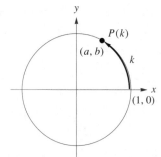

16.

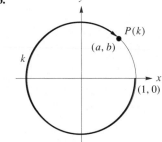

17.

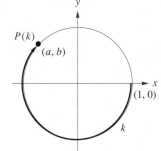

18.

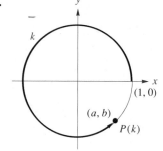

B

In Problems 19 through 30, use the appropriate techniques of translation, scaling, and reflection on the graphs of $y = \sin t$ and $y = \cos t$ to sketch the graph of the equation over the given interval.

19. $y = \sin(t - \pi)$ $-2\pi \le t \le 2\pi$

20. $y = \cos\left(t + \dfrac{\pi}{2}\right)$ $-\pi \le t \le \pi$

21. $y = 3 + \cos\left(t - \dfrac{2\pi}{3}\right)$ $-\pi \le t \le \pi$

22. $y = 4 + \sin(t - \pi)$ $0 \le t \le 4\pi$

23. $y = 5 - \sin\left(t - \dfrac{\pi}{3}\right)$ $-2\pi \le t \le 2\pi$

24. $y = -2 - \cos\left(t + \dfrac{\pi}{6}\right)$ $-2\pi \le t \le 2\pi$

25. $y = 2\sin t$ $-2\pi \le t \le 4\pi$

26. $y = 3\cos t$ $-2\pi \le t \le 4\pi$

27. $y = \sin 2t$ $0 \le t \le 4\pi$

28. $y = \cos 3t$ $0 \le t \le 4\pi$

29. $y = -\sin(t - \pi)$ $0 \le t \le 4\pi$

30. $y = -\cos\frac{1}{2}t$ $0 \le t \le 4\pi$

In Problems 31 through 36, sketch the graph of the equations all on the same coordinate plane over the given interval.

31. $y = \sin 2t,\; y = \cos 2t,\; y = \tan 2t;$ $-\pi \le t \le \pi$

32. $y = \sin 2t,\; y = \cos 2t,\; y = \cot 2t;$ $-\pi \le t \le \pi$

33. $y = \sin\frac{1}{2}t,\; y = \csc\frac{1}{2}t;$ $-2\pi \le t \le 2\pi$

34. $y = \sin\frac{2}{3}t,\; y = \csc\frac{2}{3}t;$ $-3\pi \le t \le 3\pi$

35. $y = \sin\left(t - \dfrac{\pi}{4}\right),\; y = \csc\left(t - \dfrac{\pi}{4}\right);$ $-2\pi \le t \le 2\pi$

36. $y = \cos\left(t + \dfrac{\pi}{3}\right),\; y = \sec\left(t + \dfrac{\pi}{3}\right);$ $-2\pi \le t \le 4\pi$

In Problems 37 through 42, sketch the graph of the equation over the interval $-2\pi \le t \le 2\pi$.

37. $y = |\cos t|$ **38.** $y = \sin|t|$ **39.** $y = |\tan t|$

40. $y = \tan|t|$ **41.** $y = \cot|t|$ **42.** $y = |\sec t|$

In Problems 43 through 51, the graph of the given function is the same as the graph of one of the following: $y = \sin t$,
$y = \cos t$, $y = -\sin t$, $y = -\cos t$. *Use the techniques of translation and reflection to determine which one.*

43. $y = \sin(t + \pi)$ **44.** $y = \cos(t - 2\pi)$

45. $y = \sin\left(-t - \dfrac{\pi}{2}\right)$ **46.** $y = \cos\left(-t - \dfrac{\pi}{2}\right)$

47. $y = \cos\left(t + \dfrac{3\pi}{2}\right)$ **48.** $y = \sin(3\pi - t)$

49. $y = \cos(4\pi + t)$ **50.** $y = -\cos(-t)$

51. $y = -\sin(t + \pi)$

In Problems 52 through 57, the graph of the given function is the same as the graph of one of the following: $y = \tan t$, $y = -\tan t$, $y = \cot t$, $y = -\cot t$. Use the techniques of translation and reflection to determine which one.

52. $y = \tan(t + \pi)$ **53.** $y = \cot(t - 2\pi)$

54. $y = \cot\left(-t - \dfrac{\pi}{2}\right)$ **55.** $y = \tan\left(-t - \dfrac{\pi}{2}\right)$

56. $y = \tan\left(t + \dfrac{3\pi}{2}\right)$ **57.** $y = \cot(3\pi - t)$

58. A bug starts at the point $(1, 0)$ on the unit circle and walks along the arc of the circle in a counterclockwise direction for 4 units. Estimate (to the nearest 0.0001) the coordinates (x, y) of the point at which the bug ends its trip.

59. A bug starts at the point $(1, 0)$ on the unit circle and walks along the arc of the circle in a clockwise direction for 5 units. Estimate (to the nearest 0.0001) the coordinates (x, y) of the point at which the bug ends its trip.

60. A bug starts at the point $(0, 1)$ on the unit circle and walks along the arc of the circle in a counterclockwise direction for 16 units. Estimate (to the nearest 0.0001) the coordinates (x, y) of the point at which the bug ends its trip.

61. A bug starts at the point $(0, -1)$ on the unit circle and walks along the arc of the circle in a counterclockwise direction for 9 units. Estimate (to the nearest 0.0001) the coordinates (x, y) of the point at which the bug ends its trip.

62. Use a ratio identity, $\sin(-t) = -\sin t$, and $\cos(-t) = \cos t$ to show that the cotangent function is an odd function.

63. Use a reciprocal identity, and $\cos(-t) = \cos t$ to show that the secant function is an even function.

64. Use a reciprocal identity, and $\sin(-t) = -\sin t$ to show that the cosecant function is an odd function.

65. Use the identity, $\sin^2 t + \cos^2 t = 1$ to prove the identity $\tan^2 t + 1 = \sec^2 t$. (Hint: Multiply each side of $\sin^2 t + \cos^2 t = 1$ by $1/\cos^2 t$.)

66. Use the identity, $\sin^2 t + \cos^2 t = 1$ to prove the identity $1 + \cot^2 t = \csc^2 t$. (Hint: Multiply each side of $\sin^2 t + \cos^2 t = 1$ by $1/\sin^2 t$.)

C

67. A bug starts at the point $(0, 2)$ on the graph of the circle $x^2 + y^2 = 4$ and walks along the arc of the circle in a counterclockwise direction for $11\pi/2$ units. Determine the coordinates (x, y) of the point at which the bug ends its trip.

68. A bug starts at the point $(-3, 0)$ on the graph of the circle $x^2 + y^2 = 9$ and walks along the arc of the circle in a clockwise direction for 5π units. Determine the coordinates (x, y) of the point at which the bug ends its trip.

S E C T I O N   5.5   **GENERAL TRIGONOMETRIC GRAPHS**

Many of the natural phenomena that arise in the study of science are periodic. They can be modeled by functions that repeat their values over a specified interval (this interval is called the period of the function). For example, sound and electromagnetic waves, the movement of a pendulum of a grandfather clock, and the phases of the moon are all periodic phenomena.

Models of these periodic situations involve functions whose graphs are translations, scalings, and reflections of the graph of the prototype sine function $y = \sin t$. The graphs of these functions are called **sinusoids.** Of course, the cosine function $y = \cos t$ is also a sinusoid since $\cos t = \sin(\pi/2 - t)$.

EXAMPLE 1   Sketch these graphs on the same axes over the interval $-2\pi \leq t \leq 2\pi$.

a) $y = \sin t$ and $y = 2 \sin t$ **b)** $y = \sin t$ and $y = -\frac{1}{2} \sin t$

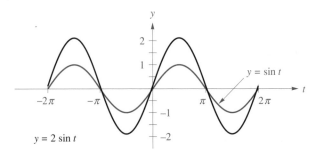

FIGURE 73

SOLUTION

a) The graph of $y = 2 \sin t$ is a vertical expansion of the graph of $y = \sin t$. Since the range of $y = \sin t$ is $[-1, 1]$, the range of $y = 2 \sin t$ is $[-2, 2]$ (Figure 73).

b) The graph of $y = -\frac{1}{2} \sin t$ is a vertical compression and reflection through the t-axis of the graph of $y = \sin t$. Since the range of $y = \sin t$ is $[-1, 1]$, the range of $y = -\frac{1}{2} \sin t$ is $[-\frac{1}{2}, \frac{1}{2}]$ (Figure 74).

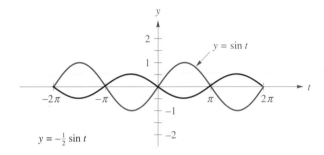

FIGURE 74

The **amplitude** of a sinusoid $y = A \sin t$ or $y = A \cos t$ is the maximum vertical displacement from the t-axis.

Amplitude

It should be fairly obvious that the amplitude of the graph of $y = \sin t$ and the graph of $y = \cos t$ is 1.

Now, consider the relation between the graph of $y = f(t)$ and $y = Af(t)$. (This was discussed in Section 2.5; you may wish to pause here and review it.) If A is positive, then the graph of $y = Af(t)$ is the same as the graph of $y = f(t)$ scaled vertically by a factor of A. If A is negative, then the graph of $y = Af(t)$ is the same as the graph of $y = f(t)$ reflected through the t-axis and scaled vertically by the positive factor of $-A$.

If we apply this idea to the graphs of $y = \sin t$ and $y = A \sin t$, it becomes reasonable (as we see in Example 1) that the amplitude of the graph is $|A|$ (the absolute value is needed since the amplitude is a distance; the value of A may be either positive or negative).

The amplitude of $y = A \sin t$ or $y = A \cos t$ is $|A|$.

Determining the Amplitude of a Sinusoid

EXAMPLE 2   Sketch these graphs on the same axes over the interval $-2\pi \le t \le 2\pi$.

a) $y = \sin t$ and $y = \sin 2t$ **b)** $y = \sin t$ and $y = \sin(-\tfrac{1}{2}t)$

SOLUTION

a) The graph of $y = \sin 2t$ is a horizontal compression of the graph of $y = \sin t$. Since the period of $y = \sin t$ is 2π, the period of $y = \sin 2t$ is π (Figure 75).

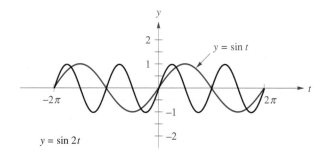

FIGURE 75 $y = \sin 2t$

b) Since the sine function is an odd function,

$$\sin(-\tfrac{1}{2}t) = -\sin(\tfrac{1}{2}t).$$

This means that the graph of $y = \sin(-\tfrac{1}{2}t)$ is the same as the graph of $y = -\sin(\tfrac{1}{2}t)$. The graph of $y = -\sin(\tfrac{1}{2}t)$ is a horizontal expansion and reflection through the t-axis of the graph of $y = \sin t$ (Figure 76). Notice that there is one complete period over the interval $[-2\pi, 2\pi]$.

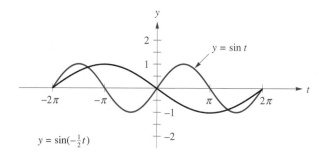

FIGURE 76 $y = \sin(-\tfrac{1}{2}t)$

The connection between the graphs of $y = f(t)$ and $y = f(Bt)$ was also discussed in Section 2.5. Recall that the graph of $y = f(Bt)$ is a horizontal scaling of the graph of $y = f(t)$. In part a) of Example 2, the role of B is played by 2, and the graph of $y = \sin 2t$ is scaled so that the period is π, half the period 2π of the graph of $y = \sin t$. Likewise, in part b) the period of $y = \sin(-\frac{1}{2}t)$ is 4π, twice the period of $y = \sin t$. This suggests a general process for finding the period of a sinusoid.

For $B > 0$, the **period** of the graph of $y = A \sin Bt$ or $y = A \cos Bt$ is $2\pi/B$.

Determining the Period of a Sinusoid

There are four types of curves for the graphs of $y = A \sin Bt$ or $y = A \cos Bt$ with $B > 0$. One period of each is sketched in Figure 77.

These concepts of amplitude and period of a sinusoid enable us to sketch the graph of a sinusoid quickly and without much difficulty. These steps are outlined below.

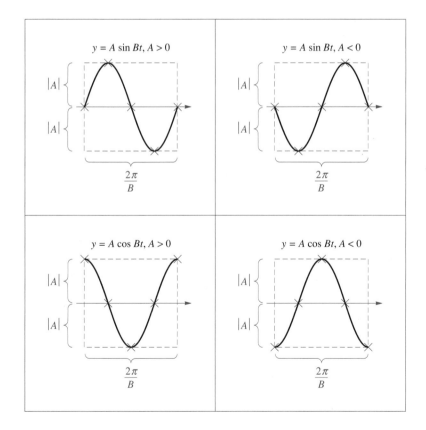

Practice each of these graphs until you can sketch them without difficulty. Notice that the endpoints, the halfpoints, and the quarterpoints are marked with an ×. In each case, locate these five points and sketch the smooth curve through them.

FIGURE 77

Sketching the Graph of y =
A sin Bt or y = A cos Bt

To sketch the graph of $y = A \sin Bt$ or $y = A \cos Bt$ $(B > 0)$:

1) Determine the amplitude $|A|$. Sketch horizontal lines above and below the t-axis to mark the maximum vertical displacement of the graph:

2) Determine the period $2\pi/B$. Sketch vertical lines to mark each increment of the period over the interval of the domain of the sketch:

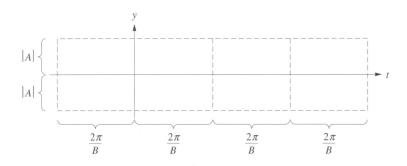

3) Sketch one period of the appropriate curve, cosine or sine, in each box formed:

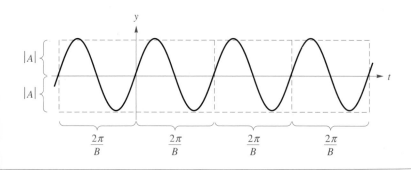

EXAMPLE 3   Sketch the graph of $y = \frac{3}{2} \cos \pi t$ over the interval $-4 \le t \le 6$.

SOLUTION   The amplitude of the graph is $\frac{3}{2}$. The period of the graph is

$$\frac{2\pi}{B} = \frac{2\pi}{\pi} = 2.$$

In Figure 78, the horizontal lines indicate the maximum displacement of the graph, and the vertical lines indicate each period. In each box formed by these lines, we sketch a cosine curve.

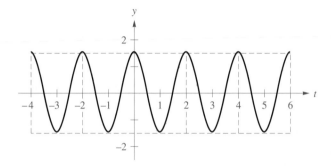

FIGURE 78

EXAMPLE 4   Sketch the graph of $y - 1 = 2 \sin 3(x - \pi/3)$, $-2\pi \le x \le 2\pi$.

SOLUTION   The graph in question is a translation of the graph of $y = 2 \sin 3x$. This graph has amplitude 2 and period $2\pi/3$. The sketch of this function is shown in Figure 79.

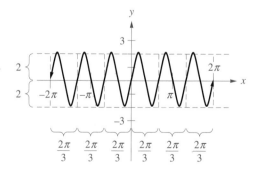

FIGURE 79

The graph of $y - 1 = 2 \sin 3(x - \pi/3)$ can be obtained from the previous graph by translating it $\pi/3$ units to the right and 1 unit up (Figure 80).

FIGURE 80

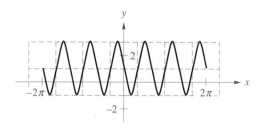

EXAMPLE 5   Sketch the graph of $y = 1 - 2\cos(-x + \frac{1}{2}\pi)$, $-2\pi \le x \le 3\pi$.

SOLUTION Since the cosine is an even function,

$$\cos(-x + \tfrac{1}{2}\pi) = \cos(-(x - \tfrac{1}{2}\pi)) = \cos(x - \tfrac{1}{2}\pi).$$

Thus our graph is the same as the graph of $y = 1 - 2\cos(x - \frac{1}{2}\pi)$, which in turn is a reflection and translation (through the x-axis) of the graph of $y = 2\cos x$. Playing the role of (h, k) is the point $(\frac{1}{2}\pi, 1)$. The amplitude is 2 and the period is 2π. We start at the point $(\frac{1}{2}\pi, 1)$ instead of the origin (Figure 81).

FIGURE 81

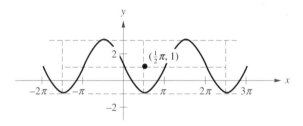

EXAMPLE 6   Assuming stable climatic conditions, the temperature variation at a particular place over the course of one day can be modeled approximately by a sinusoid of the form

$$T(t) = A\cos B(t - h) + k \qquad A > 0 \text{ and } B > 0.$$

One winter day in Santa Rosa, California, the low temperature was 20°. The high temperature of 48° occurred at 4 P.M. Find a function $T(t)$ that describes the temperature at time t hours after midnight and sketch its graph for $0 \le t \le 24$.

SOLUTION The period of the function is 24 (why?). Since the high point of the graph $y = A\cos Bx$ is at the beginning of its period, the horizontal translation h is 16 (4 P.M. is 16 hours after midnight). The vertical translation k is 34°, the average of the high and the low. The amplitude, is follows, is 14.

Using all this information, we write the function as

$$T(t) = 14 \cos\left[\frac{\pi}{12}(t - 16)\right] + 34.$$

To sketch the graph, we consider the period of the graph directly before and directly after $t = 4$ (Figure 82).

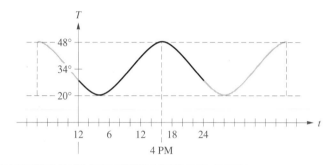

FIGURE 82

One of the important graphing techniques of Chapter 2 is that of graphing by adding y-coordinates. Many of the periodic functions encountered in nature can be described by the sum of sinusoids. In fact, virtually all periodic functions can be approximated (to any required degree of precision) by a sum of sinusoids.

Sum of Sinusoids

EXAMPLE 7 Sketch the graph of $y = \sin 2t + \cos t$, $0 \le t \le 2\pi$.

SOLUTION As a first step, we sketch the graphs of $y = \sin 2t$ and $y = \cos t$ over $[0, 2\pi]$. Next, we plot a few points by adding y-coordinates. Finally, we sketch a smooth curve through these points (Figure 83).

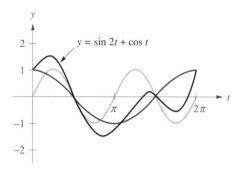

FIGURE 83

EXAMPLE 8   Sketch the graph of $y = t + \sin t$, $-2\pi \leq t \leq 2\pi$.

SOLUTION

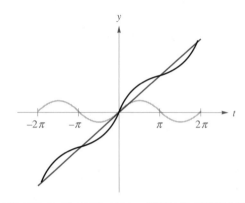

The Graphs of $y = A(x) \sin Bx$ and $y = A(x) \cos Bx$

Many applications of calculus and differential equations involve the functions of the form $y = A(x) \sin Bx$ or $y = A(x) \cos Bx$, where $A(x)$ is an algebraic function such as $x^2 + 1$ or a transcendental function such as e^{2x}.

Sketching the graph of $y = A(x) \sin Bx$ is similar to sketching the graphs of $y = A \sin Bx$. Since $|\sin Bx| \leq 1$, it follows that

$$|A(x) \sin Bx| \leq |A(x)|$$

or

$$-|A(x)| \leq A(x) \sin Bx \leq |A(x)|.$$

This inequality suggests that the graph of $y = A(x) \sin Bx$ lies between the graph of $y = -|A(x)|$ and the graph of $y = |A(x)|$ (there is a similar result for the graph of $y = A(x) \cos Bx$). The next two examples demonstrate how this inequality assists in sketching the graphs of these functions.

These functions are generally not periodic, but they do still have a oscillatory nature. We shall call an interval of one complete oscillation a **quasiperiod.**

EXAMPLE 9   Sketch the graph of $y = (\frac{1}{8}x^2 + 1) \sin x$, $-3\pi \leq x \leq 3\pi$.

SOLUTION   First, we sketch $y = |\frac{1}{8}x^2 + 1|$ and $y = -|\frac{1}{8}x^2 + 1|$ (actually, these are the same as $y = \frac{1}{8}x^2 + 1$ and $y = -\frac{1}{8}x^2 - 1$). The quasiperiod of the function is 2π. The function is an odd function, so the graph is symmetric with respect to the origin. These results are shown in Figure 84.

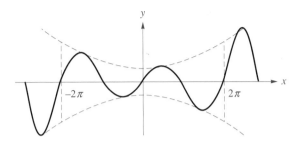

FIGURE 84

EXAMPLE 10   Sketch the graph of $y = x \sin \pi x$, $-3 \le x \le 3$.

SOLUTION First, we sketch the graphs of $y = |x|$ and $y = -|x|$ (Figure 85). The function is even, so we can exploit its symmetry to simplify the task. The quasiperiod is 2.

FIGURE 85

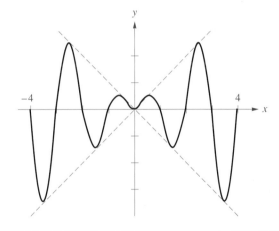

EXAMPLE 11   A weight is attached to a spring hanging from the ceiling of a room. The weight is displaced 1 ft below its position of rest and allowed to oscillate up and down (in this application consider down as the positive direction, as shown in Figure 86). A function that gives the vertical position at time x is

$$H(x) = (1.5)^{-x} \cos(2.3x)$$

Sketch the graph of H and describe its behavior as $x \to \infty$.

SOLUTION First we sketch $y = (1.5)^{-x}$ and $y = -(1.5)^{-x}$. The quasiperiod of H is $2\pi/2.3 = 2.7$ (to the nearest 0.1). From this, we sketch the graph in Figure 87.

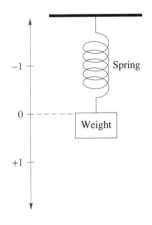

FIGURE 86

FIGURE 87

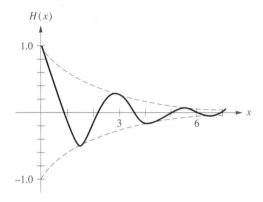

In our discussion of exponential functions in Chapter 4, we saw that functions such as $A(x) = (1.5)^{-x}$ tend to zero as $x \to \infty$. Since $0 \le |H(x)| \le (1.5)^{-x}$, it follows that $H(x) \to 0$ as both the sketch in Figure 87 and the situation seem to suggest.

EXERCISE SET 5.5

A

In Problems 1 through 12, sketch the graph of the equation over one period. Identify the amplitude, the period, and the translation (if any).

1. $y = 2 \sin 3t$

2. $y = \frac{2}{3} \cos 2t$

3. $y = \frac{1}{2} \cos 2t$

4. $y = \frac{2}{3} \sin(\frac{1}{2}t)$

5. $y = 2 \sin 3\left(t - \frac{\pi}{3}\right)$

6. $y = 3 \cos\left(t - \frac{\pi}{6}\right)$

7. $y = \frac{5}{2} \cos \pi t$

8. $y = 6 \sin \pi t$

9. $y = \sin\left(2t - \frac{\pi}{2}\right)$

10. $y = \cos(3t - \pi)$

11. $y = -3 \cos\left(t - \frac{3\pi}{2}\right) + 3$

12. $y = 4 - \sin\left(t - \frac{\pi}{2}\right)$

In Problems 13 through 24, sketch the graph of the equation over the given interval. Identify the amplitude, the period, and the translation.

13. $y = \frac{3}{2} \sin 2t$, $-2\pi \le t \le 2\pi$

14. $y = \frac{2}{3} \sin 4t$, $-\pi \le t \le \pi$

15. $y = \sin 2t + 1$, $0 \le t \le 4\pi$

16. $y = \frac{2}{3} \sin(\frac{1}{2}t) - 2$, $0 \le t \le 2\pi$

17. $y = 2 \sin 3\left(t - \frac{\pi}{3}\right)$, $0 \le t \le 4\pi$

18. $y = 3 \cos 2\left(t - \frac{\pi}{2}\right)$, $-2\pi \le t \le 2\pi$

19. $y = \frac{5}{2} \cos \pi t$, $-4 \le t \le 5$

20. $y = 4 \sin \pi t$, $2 \le t \le 6$

21. $y = 2 \sin\left(2t - \frac{\pi}{2}\right)$, $-\pi \le t \le 2\pi$

22. $y = 2 \cos(3t - \pi)$, $-\pi \le t \le 2\pi$

23. $y = -3 \sin\left(t - \dfrac{\pi}{2}\right) + 3,$ $-\pi \le t \le 3\pi$

24. $y = 2 - 2 \sin\left(t - \dfrac{\pi}{2}\right),$ $-2\pi \le t \le 2\pi$

In Problems 25 through 30, select two functions from I through XII that correctly describe the graph shown.

I $y = -3 \sin \pi x$

II $y = 2 \cos\left[\frac{1}{4}\pi(x + 2)\right]$

III $y = -2 \sin(\frac{1}{4}\pi x)$

IV $y = 3 \cos \pi x$

V $y = 1 - \cos x$

VI $y = 2 \sin 2x$

VII $y = -3 \cos \pi(x + 3)$

VIII $y = 3 \sin \pi(x - 1)$

IX $y = 1 + \cos(x + \pi)$

X $y = -2 \sin\left(x - \dfrac{\pi}{2}\right) - 2$

XI $y = 2 \cos(2x - \frac{1}{2}\pi)$

XII $y = -2 + 2 \cos x$

25.

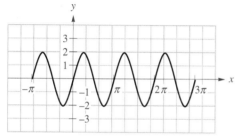

26.

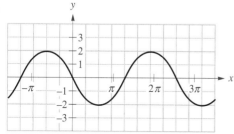

27.

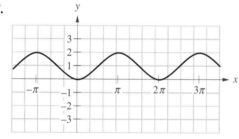

28.

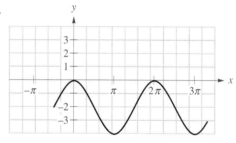

29.

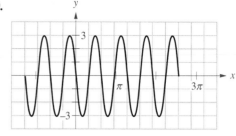

30.

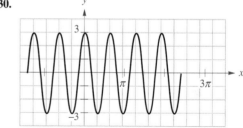

In Problems 31 through 42, use the technique of adding y-coordinates to sketch the graph of the given function.

31. $y = \cos t + 3 \sin t$

32. $y = 2 \cos t + \sin t$

33. $y = t - \sin t$

34. $y = \cos t + t$

35. $y = 2 \cos t + \sin 2t$

36. $y = \cos t - \sin 2t$

37. $y = \cos \frac{1}{2}t + \sin t$

38. $y = \sin \frac{1}{2}t - \sin t$

39. $y = \ln t - \sin \pi t$

40. $y = \log_2 t + \cos\left(\dfrac{\pi}{2}t\right)$

41. $y = |t| + \cos t - 1$

42. $y = \dfrac{1}{t} + \sin t$

B

In Problems 43 through 54, sketch the graph of the given function. Identify the function as either an odd function, an even function, or neither.

43. $y = x \cos x$

44. $y = |x| \sin x$

45. $y = \frac{1}{2}x \sin x$

46. $y = -\frac{2}{3}x \cos x$

47. $y = e^{-x} \cos 2x$

48. $y = \frac{1}{2}e^{-x} \sin x$

49. $y = \sqrt{x} \sin x$

50. $y = \sqrt[3]{x} \cos x$

51. $y = \dfrac{4}{x^2 + 1} \sin x$

52. $y = \dfrac{6}{x^2 + 2} \cos x$

53. $y = (2^{-x} + 1) \cos 2x$

54. $y = (2^{-x} + 1) \sin 2x$

55. Determine the domain of the $f(x) = \tan x \cot x$. Sketch the graph of this function.

56. Determine the domain of the $f(x) = \sec x \cos x$. Sketch the graph of this function.

57. Determine a value for B such that $y = 3 \sin Bx$ has a period of 4 units.

58. Determine a value for B such that $y = -6 \cos Bx$ has a period of $\frac{1}{2}$ unit.

59. A function $f(x) = A \sin B(x - h) + k$ has a maximum value of 20 at $x = 7$ and a minimum value of 4 at $x = 15$ (there are no other extreme points between 7 and 15). Draw a rough sketch of f and determine the function such that $4 \leq h \leq 18$ and $A > 0$.

60. A function $g(x) = A \cos B(x - h) + k$ has a maximum value of -2 at $x = -2$ and a minimum value of -3 at $x = 10$ (there are no other extreme points between -2 and 10). Draw a rough sketch of f and determine the function such that $20 \leq h \leq 30$.

61. The average low temperature (°F) on day of the year t ($t = 0$ corresponds to January 1) can be approximated by the function

$$F(t) = 36 \sin \frac{2\pi}{365} (t - 101) + 14.$$

Sketch the graph of F. Determine the warmest day and the coldest day of the year based on this function.

62. The number of hours of daylight on a given day t of a non-leap year at latitude θ in the northern hemisphere can be approximated (to the nearest 0.2

hour) by the function

$$L(t) = A \sin \frac{2\pi}{365} (t - 79) + 12$$

where $A = (24/\pi)\sin^{-1}(\tan 23.45° \tan \theta)$. Use this to approximate the number of hours of daylight in Belmont, California (latitude 37.5°) on February 25, 1990.

63. Approximate the number of hours of daylight in London, England (latitude 51.5°) on February 25, 1990 (see Problem 62).

C

64. Sketch the graph of

$$f(t) = \frac{\sin t}{t}.$$

Note that there is a discontinuity at $t = 0$. Use your calculator to evaluate the function for a few values of t in the interval $[0, 0.5]$. Deduce the limiting value of

$$\frac{\sin t}{t} \qquad \text{as } t \to 0^+.$$

To complete the sketch, you can use the fact that this is an even function. (Be sure that your calculator is in radian mode.)

65. Sketch the graph of $y = \sin(1/t)$. Describe what happens to the graph as $t \to \pm 0$.

66. **a)** Sketch the graph of

$$h(x) = \begin{cases} -\pi & -1 \leq x \leq 0 \\ \pi & 0 \leq x \leq 1 \end{cases}$$

on a full-sized sheet of paper. Allow for values of y between -4 to 4.

b) Sketch $y = 4 \sin \pi x$, $-1 \leq x \leq 1$, on the same coordinate plane.

c) Sketch $y = \frac{4}{3} \sin 3\pi x$, $-1 \leq x \leq 1$, on the same coordinate plane and use this and a) to sketch the graph of $y = 4 \sin \pi x + \frac{4}{3} \sin 3\pi x$, $-1 \leq x \leq 1$.

d) Sketch $y = 4 \sin \pi x + \frac{4}{3} \sin 3\pi x + \frac{4}{5} \sin 5\pi x$, $-1 \leq x \leq 1$, on the same coordinate plane.

e) Make a conjecture about the general shape of the graph of $y = 4 \sin \pi x + \frac{4}{3} \sin 3\pi x + \frac{4}{5} \sin 5\pi x + \frac{4}{7} \sin 7\pi x + \cdots + (4/99)\sin 99\pi x$, $-1 \leq x \leq 1$.

I n the third century B.C., the seat of all western knowledge was the city of Alexandria in Egypt. Built by Alexander the Great to be the capital of his empire, it was the cosmopolitan center of the civilization and scholarship. The great library here housed works of mathematics, science, politics, and the arts. It was during this period that the library was under the direction of Eratosthenes of Cyrene (c. 276–194 B.C.).

Eratosthenes was an accomplished mathematician, historian, astronomer, philosopher, geographer, poet, and athlete. He earned his nickname, Beta (the second letter of the Greek alphabet), because his range and depth of abilities were that of a second Plato. His other nickname, Pentathlus, was due to his unprecedented five championships in the city athletic games. He is thought by many to be the originator of academic research.

Eratosthenes' genius was in applying the knowledge of antiquity to the problems of his time. He developed the modern calendar of 365 days with an extra day every fourth year. He constructed the first maps of the entire known world that could be used for accurate navigation. Through research and ingenuity (see Problem 39 of Section 5.2),

he was able to determine that the earth is round and compute its circumference to within four percent. (By comparison, the best estimates during the time of Columbus were 17,000 miles; had he known what Eratosthenes had known over 1500 years earlier, Columbus probably would have thought the distance to China far too long to attempt.)

During Eratosthenes' tenure as director, the library amassed over half a million volumes. The great minds of the western world flocked to Alexandria. The municipal resources were devoted not to war and conquest, but to the acquisition of knowledge and pursuit of scholarship. Eratosthenes remained at his post for over 40 years. When blindness prevented him from pursuing his research, he killed himself by voluntary starvation.

His library flourished for another 600 years. Here the lore of trigonometry, geometry, and astronomy grew to become academic disciplines. It was burned to the ground in the first century A.D. Its destruction was a harbinger of the Dark Ages.

Eratosthenes was a scientist, and his musings on the commonplaces changed the world; in a way, they made the world.
CARL SAGAN

Eratosthenes

MISCELLANEOUS EXERCISES

In Problems 1 through 6, determine the value of the trigonometric functions of θ.

1.

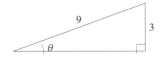

2.

3.

4.

c) $\sec\left(-\dfrac{11\pi}{6}\right)$ **d)** $\sec\dfrac{8\pi}{3}$

5.

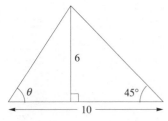

6.

7. Use your calculator to determine the value of θ to the nearest degree in Problem 1.

8. Use your calculator to determine the value of θ to the nearest degree in Problem 2.

9. Use your calculator to determine the value of θ to the nearest degree in Problem 3.

In Problems 10 through 15, determine the exact value of the trigonometric function without the aid of a calculator.

10. a) $\sin 150°$ **b)** $\cos(-30°)$

    c) $\sin(-135°)$ **d)** $\cos 330°$

11. a) $\tan(-60°)$ **b)** $\tan 240°$

    c) $\cot(-315°)$ **d)** $\cot 540°$

12. a) $\sec(-45°)$ **b)** $\csc 135°$

    c) $\csc 150°$ **d)** $\sec(-30°)$

13. a) $\sin\dfrac{7\pi}{4}$ **b)** $\cos\left(-\dfrac{5\pi}{3}\right)$

    c) $\sin\left(-\dfrac{5\pi}{6}\right)$ **d)** $\cos\dfrac{13\pi}{2}$

14. a) $\tan\dfrac{11\pi}{3}$ **b)** $\cot\left(-\dfrac{2\pi}{3}\right)$

    c) $\tan\left(-\dfrac{5\pi}{3}\right)$ **d)** $\cot\dfrac{10\pi}{3}$

15. a) $\csc\dfrac{9\pi}{2}$ **b)** $\csc\left(-\dfrac{7\pi}{3}\right)$

In Problems 16 through 18, determine the coordinates of the point P(t) coordinates exactly.

16. a) $P\left(\dfrac{\pi}{4}\right)$ **b)** $P(3\pi)$

    c) $P\left(-\dfrac{3\pi}{2}\right)$ **d)** $P\left(\dfrac{3\pi}{2}\right)$

17. a) $P\left(-\dfrac{\pi}{4}\right)$ **b)** $P\left(-\dfrac{13\pi}{3}\right)$

    c) $P\left(\dfrac{11\pi}{6}\right)$ **d)** $P\left(-\dfrac{17\pi}{3}\right)$

18. a) $P\left(\dfrac{125\pi}{2}\right)$ **b)** $P(88\pi)$

    c) $P\left(-\dfrac{155\pi}{6}\right)$ **d)** $P(-21\pi)$

In Problems 19 through 42, sketch the graph of the equation over the given interval. Identify the amplitude, period, and translation (if any).

19. $y = \sin 3t$ $-2\pi \le t \le 2\pi$

20. $y = \cos 2t$ $-2\pi \le t \le 2\pi$

21. $y = 2\cos 2t$ $0 \le t \le 4\pi$

22. $y = \frac{5}{2}\sin\frac{1}{2}t$ $-2\pi \le t \le 2\pi$

23. $y = 2\sin t - 2$ $-2\pi \le t \le 4\pi$

24. $y = 3 - 2\cos t$ $-2\pi \le t \le 4\pi$

25. $y = \cos\frac{1}{2}t - 4$ $-2\pi \le t \le 4\pi$

26. $y = 3 + \sin\frac{2}{3}t$ $-3\pi \le t \le 3\pi$

27. $y = 2\sin\left(t - \dfrac{\pi}{2}\right)$ $-2\pi \le t \le 2\pi$

28. $y = 3\cos\left(t + \dfrac{\pi}{2}\right)$ $-2\pi \le t \le 2\pi$

29. $y = 3 + \cos\left(t - \dfrac{\pi}{3}\right)$ $-2\pi \le t \le 2\pi$

30. $y = 2 + \sin\left(t + \dfrac{\pi}{4}\right)$ $-2\pi \le t \le 2\pi$

31. $y = 4 - 4\sin(t + \pi)$ $0 \le t \le 4\pi$

32. $y = 2 + 2\cos(t - \pi)$ $0 \le t \le 4\pi$

33. $y = \cos(3t + \pi)$ $-2\pi \le t \le 2\pi$

34. $y = \sin(2t - \pi)$ $-2\pi \le t \le 2\pi$

35. $y = \sin(3t + 3\pi)$ $-2\pi \le t \le 2\pi$

36. $y = \cos(4t - 2\pi)$ $-2\pi \le t \le 2\pi$

37. $y = 2\sin(\pi t + 3\pi)$ $-2 \le t \le 4$

38. $y = -\cos(\pi t - \pi)$ $-2 \le t \le 4$

39. $y = \tan \frac{2}{3}t$ $-3\pi \le t \le 3\pi$

40. $y = \cot \frac{1}{3}t$ $-3\pi \le t \le 3\pi$

41. $y = \tan\left(\dfrac{\pi t}{2}\right)$ $-2 \le t \le 2$

42. $y = \cot\left(\dfrac{2\pi t}{3}\right)$ $-2 \le t \le 2$

In Problems 43 through 48, the equation is of the form $y = f(x)\sin x$. Sketch the graph of the equation.

43. $y = \frac{1}{3}x \sin x$

44. $y = |x| \cos \pi x$

45. $y = \ln(x + 1)\cos x, \quad x \ge 0$

46. $y = \sqrt{x} \sin \pi x$

47. $y = \frac{1}{2}(2^x - 2^{-x})\sin x$

48. $y = -\frac{1}{2}(2^x + 2^{-x})\cos x$

In Problems 49 through 54, sketch the graph of the function over the given interval using the technique of adding y-coordinates.

49. $y = \cos t - 2\sin t$ $-2\pi \le t \le 2\pi$

50. $y = 3\sin 2t + \cos t$ $-2\pi \le t \le 2\pi$

51. $t = \cos \frac{1}{2}\pi t - \sin \pi t$ $0 \le t \le 4$

52. $y = \sin \frac{1}{2}\pi t + \sin \pi t$ $-2 \le t \le 2$

53. $y = \log_2 t - \sin \pi t$ $0 \le t \le 8$

54. $y = t^{-1} - \cos t$ $0 \le t \le 4\pi$

In Problems 55 through 60, sketch the graph of the function (recall that $[\![u]\!]$ is used to indicate the greatest integer less than or equal to the real number u).

55. $y = \frac{1}{2}(\sin x + |\sin x|)$ $-2\pi \le x \le 2\pi$

56. $y = \frac{1}{2}(\tan x + |\tan x|)$ $-2\pi \le x \le 2\pi$

57. $y = \frac{1}{2}(\cos 2x - |\cos 2x|)$ $-2\pi \le x \le 2\pi$

58. $y = \frac{1}{2}(\cot \frac{1}{2}\pi x - |\cot \frac{1}{2}\pi x|)$ $-2 \le x \le 2$

59. $y = [\![2\sin x]\!]$ $-2\pi \le x \le 2\pi$

60. $y = \sin\left(\dfrac{\pi}{2}[\![x]\!]\right)$ $-4 \le x \le 4$

61. Find the arc length of a circle with radius 16.2 in. that is intercepted by an angle of 80°. Express the answer to the nearest 0.1 in.

62. Find the radius of a circle such that a central angle of 240° intercepts an arc of length 26.2 cm. Express the answer to the nearest 0.1 cm.

63. Find the measure of a central angle that intercepts an arc of length 74 ft in a circle with radius 42 ft. Express the answer to the nearest degree.

64. From atop a smaller building, the angle of elevation to the top of a larger building is 20.0°, and the angle of depression to its base is 46.9°. If the distance between the buildings is 107 m, find the heights of the buildings. (See Figure 88.)

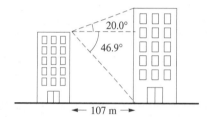

FIGURE 88

65. From an observation point A, the angle of elevation to a point 15.0 ft up a vertical flagpole is 20.0°. The angle of elevation from A to the top of the pole is 46.9°. Determine the height of the flagpole. (See Figure 89.)

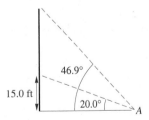

FIGURE 89

66. From the trailhead, Pham walks due east for 5 km, then walks N37°W until he is directly north of the trailhead. How far is he from the trailhead (nearest 0.1 km)?

67. From her cabin, Juanita hikes cross country first 2 miles directly west, then 3 miles directly south, and finally at a heading of S12°E until she is directly south of the trailhead. How far is she from her cabin (to the nearest 0.1 mile)?

68. Given that $\sin \theta = \frac{21}{29}$ and $3\pi/2 < \theta < 5\pi/2$, sketch the angle and determine the other five trigonometric functions.

69. Given that $\cos \theta = 3/\sqrt{10}$ and $\pi < \theta < 2\pi$, sketch the angle and determine the other five trigonometric functions.

70. Given that $\cot \theta = \sqrt{2}/11$ and $0 < \theta < \pi$, sketch the angle and determine the other five trigonometric functions.

71. A bug starts at the point $(1, 0)$ on the unit circle and walks along the arc of the circle in a counterclockwise direction for 18 units. Determine the quadrant in which the bug ends his trip.

72. A bug starts at the point $(0, 1)$ on the unit circle and walks along the arc of the circle in a clockwise direction for 7 units. Determine the quadrant in which the bug ends his trip.

73. In Figure 90, the center of the smaller circle is point C and the center of the larger circle is point A. Given that $\angle BCD = 90°$, $\angle BAD = 60°$, and that the radius of the smaller circle is 12, find a) the radius of the larger circle, and b) the area in common to both circles (this area is shaded in Figure 90).

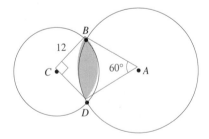

FIGURE 90

74. A right trapezoid has bases of lengths 8 cm and 6 cm (Figure 91). Find the area of the trapezoid as a function of its acute interior angle θ.

FIGURE 91

75. For the trapezoid in Problem 74, find the perimeter as a function of its acute interior angle θ.

76. A right triangle ABC with legs of length 3 and 4 is revolved about a line passing through the point B and perpendicular to the line segment AB. Find the volume of the solid generated (Figure 92).

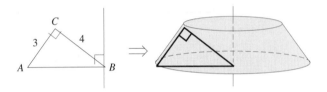

FIGURE 92

77. A right triangle ABC with legs of length 3 in. and 4 in. is revolved about a line passing through the point D that is on the line AB and perpendicular to the line AB. The distance from point A to point D is 9 in. Find the volume of the solid generated (Figure 93).

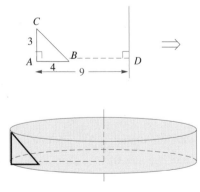

FIGURE 93

78. A regular polygon is a polygon such that all sides are of equal length and all interior angles are of equal measure. Consider a circle of diameter 1 with a regular n-sided polygon inscribed in it and a regular n-sided polygon circumscribed about it (Figure 94 shows the case for $n = 5$).

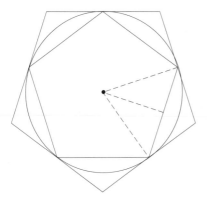

FIGURE 94

a) Show that the perimeter of the inscribed polygon is $2n \sin(\pi/n)$.

b) Show that the perimeter of the circumscribed polygon is $2n \tan(\pi/n)$.

c) Use your calculator and parts a) and b) to complete this table.

Number of sides	Perimeter of inscribed polygon	Perimeter of circumscribed polygon
3		
5		
10		
50		
100		
1000		
10000		

Explain why these values for $n = 10000$ are not unexpected.

CHAPTER 6

TRIGONOMETRIC IDENTITIES

PROVING AND USING TRIGONOMETRIC IDENTITIES

Our investigations of trigonometry in Chapter 5 yielded five sets of fundamental trigonometric identities: reciprocal identities, quotient identities, Pythagorean identities, odd-even identities, and cofunction identities. These identities are used frequently in calculus in the evaluation and simplification of trigonometric expressions. They are used to such an extent in mathematics that you should be as familiar with them as you are with, say, basic multiplication facts.

These fundamental identities are listed below and on the back inside cover of the book.

Reciprocal Identities:

$$\sec \theta = \frac{1}{\cos \theta} \qquad \text{(R1)}$$

$$\csc \theta = \frac{1}{\sin \theta} \qquad \text{(R2)}$$

$$\cot \theta = \frac{1}{\tan \theta} \qquad \text{(R3)}$$

Quotient Identities:

$$\tan \theta = \frac{\sin \theta}{\cos \theta} \qquad \text{(Q1)}$$

$$\cot \theta = \frac{\cos \theta}{\sin \theta} \qquad \text{(Q2)}$$

Pythagorean Identities:

$$\sin^2\theta + \cos^2\theta = 1 \qquad \text{(P1)}$$

$$\tan^2\theta + 1 = \sec^2\theta \qquad \text{(P2)}$$

$$1 + \cot^2\theta = \csc^2\theta \qquad \text{(P3)}$$

Fundamental Trigonometric Identities

333

Odd-Even Identities:

$$\sin(-\theta) = -\sin\theta \qquad \text{(OE1)}$$

$$\cos(-\theta) = \cos\theta \qquad \text{(OE2)}$$

$$\tan(-\theta) = -\tan\theta \qquad \text{(OE3)}$$

Cofunction Identities:

$$\sin\left(\frac{\pi}{2} - \theta\right) = \cos\theta \qquad \text{(C1)}$$

$$\cos\left(\frac{\pi}{2} - \theta\right) = \sin\theta \qquad \text{(C2)}$$

$$\tan\left(\frac{\pi}{2} - \theta\right) = \cot\theta \qquad \text{(C3)}$$

For the sake of this discussion, we have stated the functions in terms of the variable θ. However, these identities are true for acute angles, for general angles, and for real numbers. Also note that next to each of these identities there is an abbreviation which we will use for quick reference in this section.

These identities arise in various forms. For example, the Pythagorean identity $\sin^2\theta + \cos^2\theta = 1$ can also take the form of $1 - \sin^2\theta = \cos^2\theta$, $1 - \cos^2\theta = \sin^2\theta$, or even $\cos\theta = \pm\sqrt{1 - \sin^2\theta}$ (the sign in the last form of the identity depends on the quadrant in which the terminal side of θ falls). Familiarity with the fundamental identities will help you recognize them in their alternate forms.

EXAMPLE 1   Simplify the given expressions.

a) $\sin^2\theta \csc\theta$

b) $\sin^2 2\alpha + \cos^2 2\alpha$

c) $\sec x - \sin x \tan x$

d) $4\cos^2 A(1 + \tan^2 A)$

SOLUTION

a) Since $\csc\theta$ is the reciprocal of $\sin\theta$,

$$\sin^2\theta \csc\theta = \sin^2\theta \frac{1}{\sin\theta} \qquad \text{(using R2)}$$

$$= \sin\theta.$$

b) Simplifying this expression is a direct result of one of the Pythagorean identities (with the role of θ being played by 2α):

$$\sin^2 2\alpha + \cos^2 2\alpha = 1. \qquad \text{(using P1)}$$

c) Again, we rewrite the expression in terms of the sine and cosine function. Notice the alternate form of the identity P1.

$$\sec x - \sin x \tan x = \frac{1}{\cos x} - \sin x \cdot \frac{\sin x}{\cos x} \qquad \text{(using R1 and Q1)}$$

$$= \frac{1 - \sin^2 x}{\cos x}$$

$$= \frac{\cos^2 x}{\cos x} \qquad \text{(using P1)}$$

$$= \cos x.$$

d) $$4 \cos^2 A(1 + \tan^2 A) = 4 \cos^2 A(\sec^2 A) \qquad \text{(using P2)}$$

$$= 4 \cos^2 A \frac{1}{\cos^2 A} \qquad \text{(using R1)}$$

$$= 4.$$

Simplifying trigonometric expressions often requires algebraic manipulations such as factoring, finding common denominators, or reducing fractions. The next two examples involve such techniques.

EXAMPLE 2   Simplify $\dfrac{\sec^4 \theta - \tan^4 \theta}{1 + \sin^2 \theta}$.

SOLUTION Recall that the expression $a^4 - b^4$ can be partially factored as $(a^2 + b^2)(a^2 - b^2)$. We use this factoring form to get started. We leave it to you to fill in the identities used.

$$\frac{\sec^4 \theta - \tan^4 \theta}{1 + \sin^2 \theta} = \frac{(\sec^2 \theta + \tan^2 \theta)(\sec^2 \theta - \tan^2 \theta)}{1 + \sin^2 \theta}$$

$$= \frac{(\sec^2 \theta + \tan^2 \theta) \cdot 1}{1 + \sin^2 \theta}$$

$$= \frac{\left[\dfrac{1}{\cos^2 \theta} + \dfrac{\sin^2 \theta}{\cos^2 \theta} \right]}{1 + \sin^2 \theta}$$

$$= \frac{1 + \sin^2 \theta}{\cos^2 \theta(1 + \sin^2 \theta)}$$

$$= \frac{1}{\cos^2 \theta}.$$

EXAMPLE 3　Simplify

$$\frac{1}{\sec t - 1} + \frac{1}{\sec t + 1}.$$

State the answer in terms of sines and cosines.

SOLUTION　Again, we leave it to you to supply the steps.

$$\frac{1}{\sec t - 1} + \frac{1}{\sec t + 1} = \frac{1}{\sec t - 1} \cdot \frac{\sec t + 1}{\sec t + 1} + \frac{1}{\sec t + 1} \cdot \frac{\sec t - 1}{\sec t - 1}$$

$$= \frac{\sec t + 1}{\sec^2 t - 1} + \frac{\sec t - 1}{\sec^2 t - 1}$$

$$= \frac{2 \sec t}{\sec^2 t - 1}$$

$$= \frac{2 \sec t}{\tan^2 t}$$

$$= \frac{2 \dfrac{1}{\cos t}}{\left[\dfrac{\sin^2 t}{\cos^2 t}\right]}$$

$$= \frac{2}{\cos t}\left[\frac{\cos^2 t}{\sin^2 t}\right]$$

$$= \frac{2 \cos t}{\sin^2 t}.$$

Proving Trigonometric Identities

One of the applications of the fundamental identities is to provide a foundation for proving other, more specific identities. The practical experience obtained in verifying these identities helps to build the skills used to manipulate trigonometric expressions encountered in calculus. Unlike the fundamental identities, the specific identities of the next three examples and the exercise set of this section are not worth the effort of memorizing.

Identities are expressed as two expressions with an equal sign between them. To "prove" or "verify" a trigonometric identity in essence means to use the fundamental identities and algebraic manipulations to show that the two expressions are the same. This is done in three ways:

1)　Show that the left side is the same as the right side.

2)　Show that the right side is the same as the left side.

3)   Show independently that the left side and the right side are both equal to the same expression (this is usually a last resort after the first two methods fail).

EXAMPLE 4   Prove the identity

$$2 \csc^2\phi - 1 \overset{?}{=} 2 \cot^2\phi + 1.$$

SOLUTION   We start with the expression on the left side and show that it is equal to the right side:

$$2 \csc^2\phi - 1 = 2(\cot^2\phi + 1) - 1 \qquad \text{(using P3)}$$

$$= 2 \cot^2\phi + 2 - 1$$

$$= 2 \cot^2\phi + 1.$$

Since we end up with the right side of the given identity, we have verified the identity.

There is no algorithm to verify a trigonometric identity. There are, however, some general guidelines and tricks-of-the-trade that can be of some help.

Guidelines for Proving Identities

1)   If one side of the identity appears more complicated than the other, start with the more complicated side and attempt to show that it is equivalent to the simpler side.

2)   If one side is in terms of one function, try to rewrite the other side in terms of that function.

3)   Consider using an appropriate Pythagorean identity to replace a term that is the square of a trigonometric function.

4)   Multiplying both numerator and denominator by the conjugate of either the numerator or denominator may help to simplify an expression.

5)   Expanding products or factoring common factors from terms may help.

6)   If nothing else seems to work, try rewriting all functions in terms of sines and cosines and simplifying.

Above all, the most important guideline is to do *something*. Don't be afraid to make a mistake. You may have to try four or five approaches before you finally reach your goal.

EXAMPLE 5   Prove the identity

$$\csc \beta \sec \beta \overset{?}{=} \tan \beta + \cot \beta.$$

SOLUTION In this case, we start with the right side and show that it is equivalent to the left side.

$$\tan \beta + \cot \beta = \frac{\sin \beta}{\cos \beta} + \frac{\cos \beta}{\sin \beta} \qquad \text{(using Q1 and Q2)}$$

$$= \frac{\sin^2 \beta}{\sin \beta \cos \beta} + \frac{\cos^2 \beta}{\sin \beta \cos \beta}$$

$$= \frac{\sin^2 \beta + \cos^2 \beta}{\sin \beta \cos \beta}$$

$$= \frac{1}{\sin \beta \cos \beta} \qquad \text{(using P1)}$$

$$= \frac{1}{\sin \beta} \cdot \frac{1}{\cos \beta}$$

$$= \csc \beta \sec \beta. \qquad \text{(using R1 and R2)}$$

EXAMPLE 6   Prove the identity

$$\frac{\tan A}{\csc A} \overset{?}{=} \sec A - \cos A.$$

SOLUTION Examine the solution that follows. Starting with the left side, we arrive at an equivalent expression $\sin^2 A / \cos A$. Since this seems to be a dead end, we go back to the right side and eventually find that it is also equivalent to $\sin^2 A / \cos A$. This verifies the identity.

$$\frac{\tan A}{\csc A} = \frac{\left[\dfrac{\sin A}{\cos A} \right]}{\left[\dfrac{1}{\sin A} \right]} \qquad\qquad \sec A - \cos A = \frac{1}{\cos A} - \cos A$$

$$= \frac{\sin A}{\cos A} \sin A \qquad\qquad\qquad\qquad = \frac{1}{\cos A} - \frac{\cos^2 A}{\cos A}$$

$$= \frac{\sin^2 A}{\cos A} \qquad\qquad\qquad\qquad\qquad = \frac{1 - \cos^2 A}{\cos A}$$

$$\qquad\qquad\qquad\qquad\qquad\qquad\qquad = \frac{\sin^2 A}{\cos A}$$

These are the same

The Pythagorean identities play an important role in the types of manipulations that arise in calculus. It is with these identities that great care must be taken in the algebraic steps.

EXAMPLE 7   Given that $\pi/2 < t < 3\pi/2$ and $\sin t = x$, use the fundamental identities to determine the other trigonometric functions in terms of x.

SOLUTION We start with finding $\cos t$ in terms of x:

$$\sin^2 t + \cos^2 t = 1$$
$$x^2 + \cos^2 t = 1$$
$$\cos^2 t = 1 - x^2$$
$$\cos t = \pm\sqrt{1 - x^2}.$$

Now, consider t as a general angle in standard position. Since the terminal side of t lies in the second or third quadrant, it follows that $\cos t < 0$. So,

$$\cos t = -\sqrt{1 - x^2}.$$

Using the reciprocal and quotient identities, the other functions can be written in terms of $\sin t$ and $\cos t$:

$$\tan t = \frac{\sin t}{\cos t} = \frac{x}{-\sqrt{1 - x^2}} = -\frac{x}{\sqrt{1 - x^2}}$$

$$\cot t = \frac{\cos t}{\sin t} = \frac{-\sqrt{1 - x^2}}{x} = -\frac{\sqrt{1 - x^2}}{x}$$

$$\sec t = \frac{1}{\cos t} = \frac{1}{-\sqrt{1 - x^2}} = -\frac{1}{\sqrt{1 - x^2}}$$

$$\csc t = \frac{1}{\sin t} = \frac{1}{x}.$$

In calculus, certain problems can be solved by changing the form of an algebraic expression to an expression involving trigonometric functions. This process is called **trigonometric substitution.**

EXAMPLE 8   Use the trigonometric substitution

$$u = a \tan\theta, \qquad -\frac{\pi}{2} < \theta < \frac{\pi}{2}, \quad \text{and} \quad a > 0$$

to rewrite (and simplify) the algebraic expression $\sqrt{a^2 + u^2}$ in terms of θ.

SOLUTION Substituting $u = a \tan \theta$ in the expression gives

$$\sqrt{a^2 + u^2} = \sqrt{a^2 + (a \tan \theta)^2}$$
$$= \sqrt{a^2 + a^2\tan^2\theta}$$
$$= \sqrt{a^2(1 + \tan^2\theta)}$$
$$= \sqrt{a^2\sec^2\theta}$$
$$= a \sec \theta.$$

One might object to the last step since, in general $\sqrt{x^2} = |x|$, not x. This step is justified, however, since $a > 0$ and $\sec \theta > 0$ for $-\pi/2 < \theta < \pi/2$ (recall that the secant function is positive for an angle θ whose terminal side falls in the first or fourth quadrant).

EXERCISE SET 6.1

A

In Problems 1 through 18, simplify the given expressions.

1. $\dfrac{\tan \theta}{\sec \theta}$

2. $\dfrac{\csc \alpha}{\cot \alpha}$

3. $\tan^2 3\beta - \sec^2 3\beta$

4. $\csc^2\left(\dfrac{\gamma}{3}\right) - \cot^2\left(\dfrac{\gamma}{3}\right)$

5. $(\sec^2 t - \tan^2 t)^3$

6. $\tan^5 A \cot^3 A$

7. $\dfrac{1 + \cot^2\theta}{\cot^2\theta}$

8. $\dfrac{1 + \tan^2\theta}{\sec^2\theta}$

9. $\dfrac{\cos^4 x - \cos^2 x}{\sin x}$

10. $\dfrac{\sin \theta - \sec \theta}{\tan \theta}$

11. $\dfrac{\sec \alpha + 1}{\csc \alpha}$

12. $\dfrac{\cos^3\alpha - \sin^3\alpha}{\cos \alpha - \sin \alpha}$

13. $\dfrac{\tan^3\alpha + 1}{\tan \alpha + 1}$

14. $\dfrac{\cos t}{1 - \sin t} - \dfrac{\cos t}{1 + \sin t}$

15. $\dfrac{1}{1 - \cos x} - \dfrac{1}{1 + \cos x}$

16. $\left(\dfrac{1 - \tan \alpha}{1 + \tan \alpha}\right)\left(\dfrac{1 + \cot \alpha}{1 - \cot \alpha}\right)$

17. $\sqrt{16 \tan^2 x + 16}$

18. $\sqrt{4 \csc^2 x - 4}$

In Problems 19 through 30, prove the identity.

19. $\csc \alpha \cos \alpha \overset{?}{=} \cot \alpha$

20. $\dfrac{\sec A}{\csc A} \overset{?}{=} \tan A$

21. $\dfrac{\tan^2\gamma}{\sec \gamma - 1} \overset{?}{=} \sec \gamma + 1$

22. $(1 + \sin \theta)(1 - \sin \theta) \overset{?}{=} \cos^2\theta$

23. $\dfrac{\cos t}{\sec t} + \dfrac{\sin t}{\csc t} \overset{?}{=} 1$

24. $\dfrac{\sin w}{\sec w} + \dfrac{\cos w}{\csc w} \overset{?}{=} 2 \sin w \cos w$

25. $\cos \theta(\sec \theta - \cos \theta) \overset{?}{=} \sin^2\theta$

26. $\sin \theta(\cot \theta + \tan \theta) \overset{?}{=} \sec \theta$

27. $\sin^2\alpha - \sec^2\alpha + \cos^2\alpha \overset{?}{=} -\tan^2\alpha$

28. $\csc^2\alpha - \sec^2\alpha \overset{?}{=} \cot^2\alpha - \tan^2\alpha$

29. $\cos^2\beta - \sin^2\beta \overset{?}{=} 2 \cos^2\beta - 1$

30. $\cos^2\beta - \sin^2\beta \overset{?}{=} 1 - 2 \sin^2\beta$

B

In Problems 31 through 48, prove the identity.

31. $(\csc \beta - \cot \beta)(\sec \beta + 1) \overset{?}{=} \tan \beta$

32. $(\sec \alpha - \tan \alpha)(\sin \alpha + 1) \overset{?}{=} \cos \alpha$

33. $\dfrac{\cos \psi - \sin \psi}{\cos \psi + \sin \psi} \overset{?}{=} \dfrac{\csc \psi - \sec \psi}{\csc \psi + \sec \psi}$

34. $\dfrac{1 + \cos x}{1 - \cos x} - \dfrac{1 - \cos x}{1 + \cos x} \overset{?}{=} 4 \csc x \cot x$

35. $\dfrac{\sin x \tan x}{\tan x - \sin x} \overset{?}{=} \dfrac{\tan x + \sin x}{\sin x \tan x}$

36. $1 - \cos^2 y \overset{?}{=} \sin^2 y \cos^2 y + \sin^4 y$

37. $\tan^4 \lambda - 2 \sec^2 \lambda \tan^2 \lambda + \sec^4 \lambda \overset{?}{=} 1$

38. $\csc^4 \alpha - \cot^4 \alpha \overset{?}{=} 2 \cot^2 \alpha + 1$

39. $\dfrac{\csc x + 1}{\csc x - 1} \overset{?}{=} (\tan x + \sec x)^2$

40. $\dfrac{\cot A + 1}{\cot A - 1} \overset{?}{=} \dfrac{\csc^2 A + 2 \cot A}{\csc^2 A - 2}$

41. $\dfrac{\tan^3 \theta - \cot^3 \theta}{\tan \theta - \cot \theta} \overset{?}{=} \sec^2 \theta + \cot^2 \theta$

42. $\dfrac{\sin^3 \theta - \cos^3 \theta}{\sin \theta - \cos \theta} \overset{?}{=} \sin \theta \cos \theta + 1$

43. $\left(\dfrac{\sin \theta - \sin \theta \cos^2 \theta}{\tan^3 \theta} \right)^2 \left(\dfrac{\csc \theta}{\cot \theta} \right)^6 \overset{?}{=} 1$

44. $\left(\dfrac{\sin^3 \theta}{\tan^3 \theta} \right)^4 \overset{?}{=} \left(\dfrac{\cos \theta \tan^2 \theta + \cos \theta}{\sec^5 \theta} \right)^3$

45. $\dfrac{\sin \theta + \cos \theta}{\cos \theta - \sin \theta} - \dfrac{\tan \theta - 1}{\tan \theta + 1} \overset{?}{=} \dfrac{2}{\cos^2 \theta - \sin^2 \theta}$

46. $\dfrac{\sec \theta + 1}{\tan^2 \theta + \sec \theta + 1} \overset{?}{=} \dfrac{\cos^2 \theta + \sin \theta \cos \theta}{\sin \theta + \cos \theta}$

47. $1 + \sin \theta + \cos \theta \overset{?}{=} \dfrac{2 \cos \theta \sin \theta}{\cos \theta + \sin \theta - 1}$

48. $\dfrac{2 \cot^2 \theta + 2 \cot \theta \csc \theta}{\cot \theta + \csc \theta - 1} \overset{?}{=} \cot \theta + \csc \theta + 1$

In Problems 49 through 60, show by counterexample that the equation is not an identity (do this by finding a value of θ for which the equation is false).

49. $\sin(\theta + \pi/2) \overset{?}{=} \sin \theta + \sin \pi/2$

50. $\cos(\theta - \pi) \overset{?}{=} \cos \theta - \cos \pi$

51. $\tan(\pi/4 + \theta) \overset{?}{=} \pi/4 + \tan \theta$

52. $\sec(\pi - \theta) \overset{?}{=} \pi - \sec \theta$

53. $1 + \tan \theta \overset{?}{=} \sec \theta$

54. $1 - \sin \theta \overset{?}{=} \cos \theta$

55. $2 \sin \theta \overset{?}{=} \sin 2\theta$

56. $\frac{1}{2} \cos \theta \overset{?}{=} \cos \frac{1}{2}\theta$

57. $\dfrac{\sin \theta}{\cos \theta + \sin \theta} \overset{?}{=} \tan \theta + 1$

58. $\dfrac{1}{\csc \theta - \sec \theta} \overset{?}{=} \sin \theta - \cos \theta$

59. $\sqrt{\sin^2 \theta} \overset{?}{=} \sin \theta$

60. $\sqrt{1 + \tan^2 \theta} \overset{?}{=} -\sec \theta$

In Problems 61 through 66, prove the identity (you may wish to review the properties of logarithms and exponents first).

61. $\ln(1 - \cos \theta) + \ln(1 + \cos \theta) \overset{?}{=} \ln(\sin^2 \theta)$

62. $\ln(\sec \theta - \tan \theta) \overset{?}{=} -\ln(\sec \theta + \tan \theta)$

63. $(3^{\sin^2 t})(3^{\cos^2 t}) \overset{?}{=} 3$

64. $2^{\sec^2 t} \cdot \left(\frac{1}{2}\right)^{\tan^2 t} \overset{?}{=} 2$

65. $\cos t \cdot \log(100^{\sec t}) \overset{?}{=} 2$

66. $e^{\ln(\cos t) - \ln(1 - \sin t)} \overset{?}{=} \sec t + \tan t$

In Problems 67 through 72, prove the identities (note that there are trigonometric functions of both α and β).

67. $\dfrac{\cos \alpha \sin \beta + \sin \alpha \cos \beta}{\cos \alpha \cos \beta - \sin \alpha \sin \beta} \overset{?}{=} \dfrac{\tan \alpha + \tan \beta}{1 - \tan \alpha \tan \beta}$

68. $\dfrac{\cos \alpha \cos \beta + \sin \alpha \sin \beta}{\cos \alpha \sin \beta - \sin \alpha \cos \beta} \overset{?}{=} \dfrac{\cot \alpha \cot \beta + 1}{\cot \alpha - \cot \beta}$

69. $(\cos \alpha - \cos \beta)^2 + (\sin \alpha - \sin \beta)^2 \overset{?}{=}$
$\qquad\qquad 2 - 2(\cos \alpha \cos \beta + \sin \alpha \sin \beta)$

70. $(\tan \alpha - 1)^2 - (\sec \alpha - 1)^2 \overset{?}{=} 2 \sec \alpha - 2 \tan \alpha - 1$

71. $(\sin \alpha \cot \alpha \sin \beta + \cos \alpha \tan \beta \cos \beta) \csc \alpha \csc \beta \overset{?}{=}$
$\qquad 2 \cot \alpha$

72. $(\sin \alpha \cos \alpha \sin^2 \beta + \cos^2 \alpha \tan \alpha \cos^2 \beta) \sec \alpha \csc \alpha \overset{?}{=}$
$\qquad (\sec^3 \alpha \cos \alpha \sec^2 \beta - \tan^2 \alpha \sec^2 \beta) \cos^2 \beta$

In Problems 73 through 78, a trigonometric function of θ is given in terms of x. Determine the other trigonometric functions in terms of x over the given interval of θ.

73. $\cos \theta = x, \qquad 0 < \theta < \pi$

74. $\tan \theta = x, \qquad -\pi/2 < \theta < \pi/2$

75. $\cot \theta = x, \qquad 0 < \theta < \pi$

76. $\sec \theta = x, \qquad 0 < \theta < \pi/2$

77. $\cos \theta = 1/x, \qquad -\pi/2 < \theta < 0$

78. $\sin \theta = -x, \qquad -\pi/2 < \theta < \pi/2$

C

In Problems 79 through 84, use the trigonometric substitution

$$u = a \sin \theta, \qquad -\frac{\pi}{2} < \theta < \frac{\pi}{2}, \qquad a > 0$$

to rewrite the given algebraic expression in terms of θ, and simplify.

79. $\dfrac{1}{\sqrt{a^2 - u^2}}$

80. $\dfrac{1}{u}\sqrt{a^2 - u^2}$

81. $\dfrac{(a^2 - u^2)^2}{u^4}$

82. $\dfrac{u}{\sqrt{a^2 - u^2}}$

83. $(a^2 - u^2)^{-3/2}$

84. $u^{-4}(a^2 - u^2)^4$

In Problems 85 through 90, use the trigonometric substitution

$$u = a \tan \theta, \qquad -\frac{\pi}{2} < \theta < \frac{\pi}{2}, \qquad a > 0$$

to rewrite the given algebraic expression in terms of θ, and simplify.

85. $\sqrt{u^2 + a^2}$

86. $\dfrac{1}{\sqrt{u^2 + a^2}}$

87. $\dfrac{u^2}{u^2 + a^2}$

88. $\dfrac{1}{u}\sqrt{u^2 + a^2}$

89. $(a^2 + u^2)^{5/2}$

90. $u^3(a^2 + u^2)^{-3/2}$

In Problems 91 through 96, use the trigonometric substitution

$$u = a \sec \theta, \qquad 0 < \theta < \frac{\pi}{2}, \qquad a > 0$$

to rewrite the given algebraic expression in terms of θ, and simplify.

91. $\sqrt{u^2 - a^2}$

92. $\dfrac{1}{(u^2 - a^2)^2}$

93. $\dfrac{1}{u\sqrt{u^2 - a^2}}$

94. $\dfrac{1}{u}\sqrt{u^2 - a^2}$

95. $(u^2 - a^2)^{3/2}$

96. $u^5(u^2 - a^2)^{-5/2}$

S E C T I O N 6.2 IDENTITIES FOR CALCULUS, PART I

In our investigations, we have seen many examples that show, in general, $f(x - y) \neq f(x) - f(y)$. In Chapter 1, we saw that $\sqrt{x - y} \neq \sqrt{x} - \sqrt{y}$. In Chapter 4, we saw that $\log_b(x - y) \neq \log_b x - \log_b y$, and that $b^{x-y} \neq b^x - b^y$. As you might suspect, such is also the case for the trigonometric functions. For example, for the cosine function,

$$\cos(\alpha - \beta) \neq \cos \alpha - \cos \beta \qquad \text{for all } \alpha \text{ and } \beta.$$

This assertion is easy to verify. We pick two values for α and β, say $\alpha = \pi/2$ and $\beta = \pi/3$:

$$\cos\left(\frac{\pi}{2} - \frac{\pi}{3}\right) \overset{?}{=} \cos\frac{\pi}{2} - \cos\frac{\pi}{3}$$

$$\cos\left(\frac{\pi}{6}\right) \overset{?}{=} \cos\frac{\pi}{2} - \cos\frac{\pi}{3}$$

$$\frac{\sqrt{3}}{2} \neq 0 - \frac{1}{2}.$$

Since it is now apparent that $\cos(\alpha - \beta) \neq \cos \alpha - \cos \beta$ for all α and β, the obvious question arises: Is there an identity by which $\cos(\alpha - \beta)$ can be expressed strictly in terms of trigonometric functions of α and of β? How about $\cos(\alpha + \beta)$, $\sin(\alpha + \beta)$, and $\sin(\alpha - \beta)$? Indeed, there are such identities.

$$\cos(\alpha - \beta) = \cos \alpha \cos \beta + \sin \alpha \sin \beta$$

$$\cos(\alpha + \beta) = \cos \alpha \cos \beta - \sin \alpha \sin \beta$$

$$\sin(\alpha + \beta) = \sin \alpha \cos \beta + \cos \alpha \sin \beta$$

$$\sin(\alpha - \beta) = \sin \alpha \cos \beta - \cos \alpha \sin \beta$$

$$\tan(\alpha + \beta) = \frac{\tan \alpha + \tan \beta}{1 - \tan \alpha \tan \beta}$$

$$\tan(\alpha - \beta) = \frac{\tan \alpha - \tan \beta}{1 + \tan \alpha \tan \beta}$$

Addition-Subtraction Identities

We start with proving the identity for $\cos(\alpha - \beta)$. Consider two angles α and β in standard position and the graph of $x^2 + y^2 = 1$, the unit circle on a coordinate plane. Now, from our discussion in Chapter 5, the point of intersection A of the terminal side of α and the unit circle in Figure 1 has coordinates $(\cos \alpha, \sin \alpha)$. Likewise, the point of intersection B of the terminal side of β and the unit circle has coordinates $(\cos \beta, \sin \beta)$.

Now, define an angle θ as the difference between α and β, (that is, $\theta = \alpha - \beta$) and place θ in standard position (Figure 2). The point P of intersection of the terminal side of θ and the unit circle has coordinates $(\cos \theta, \sin \theta)$.

Now, convince yourself that $d(A, B)$, the distance from A to B in the first figure, is the same as $d(P, Q)$ the distance from P to Q in the second figure. These distances can be computed from their coordinates using the

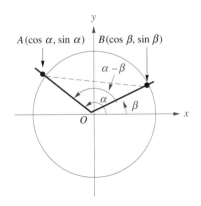

FIGURE 1   FIGURE 2

distance formula:

$$d(A, B) = \sqrt{(\cos\alpha - \cos\beta)^2 + (\sin\alpha - \sin\beta)^2}$$
$$= \sqrt{\cos^2\alpha - 2\cos\alpha\cos\beta + \cos^2\beta + \sin^2\alpha - 2\sin\alpha\sin\beta + \sin^2\beta}$$
$$= \sqrt{(\cos^2\alpha + \sin^2\alpha) + (\cos^2\beta + \sin^2\beta) - 2(\cos\alpha\cos\beta + \sin\alpha\sin\beta)}$$
$$= \sqrt{2 - 2(\cos\alpha\cos\beta + \sin\alpha\sin\beta)}$$

and

$$d(P, Q) = \sqrt{(\cos\theta - 1)^2 + (\sin\theta - 0)^2}$$
$$= \sqrt{\cos^2\theta - 2\cos\theta + 1 + \sin^2\theta}$$
$$= \sqrt{(\cos^2\theta + \sin^2\theta) + 1 - 2\cos\theta}$$
$$= \sqrt{2 - 2\cos\theta}.$$

So, setting these two distances equal and solving for $\cos\theta$, we get

$$d(A, B) = d(P, Q)$$
$$\sqrt{2 - 2\cos\theta} = \sqrt{2 - 2(\cos\alpha\cos\beta + \sin\alpha\sin\beta)}$$
$$2 - 2\cos\theta = 2 - 2(\cos\alpha\cos\beta + \sin\alpha\sin\beta)$$
$$-2\cos\theta = -2(\cos\alpha\cos\beta + \sin\alpha\sin\beta)$$
$$\cos\theta = \cos\alpha\cos\beta + \sin\alpha\sin\beta.$$

Because we defined $\theta = \alpha - \beta$, we arrive at

$$\cos(\alpha - \beta) = \cos\alpha\cos\beta + \sin\alpha\sin\beta.$$

This proves the first addition-subtraction identity.

The addition-subtraction identity for $\cos(\alpha + \beta)$ is a direct result of this identity. Recall that the cosine function is an even function and the sine is an odd function. So,

$$\cos(\alpha + \beta) = \cos[\alpha - (-\beta)]$$
$$= \cos\alpha\cos(-\beta) - \sin\alpha\sin(-\beta)$$
$$\cos(\alpha + \beta) = \cos\alpha\cos\beta - \sin\alpha\sin\beta.$$

Using these identities, we can add to the list of special angles for which we can derive exact values of the trigonometric functions.

EXAMPLE 1   Determine the exact value of

a) $\cos 15°$ b) $\cos 7\pi/12$.

SOLUTION

a) Using the subtraction identity and the fact that $15° = 45° - 30°$, we get

$$\cos 15° = \cos(45° - 30°)$$
$$= \cos 45° \cos 30° + \sin 45° \sin 30°$$
$$= \frac{\sqrt{2}}{2}\frac{\sqrt{3}}{2} + \frac{\sqrt{2}}{2}\frac{1}{2} = \frac{\sqrt{6} + \sqrt{2}}{4}.$$

b) Using the addition identity and the fact that $7\pi/12 = \pi/4 + \pi/3$, we get

$$\cos\left(\frac{7\pi}{12}\right) = \cos\left(\frac{\pi}{4} + \frac{\pi}{3}\right)$$
$$= \cos\frac{\pi}{4}\cos\frac{\pi}{3} - \sin\frac{\pi}{4}\sin\frac{\pi}{3}$$
$$= \frac{\sqrt{2}}{2}\frac{1}{2} - \frac{\sqrt{2}}{2}\frac{\sqrt{3}}{2} = \frac{\sqrt{6} - \sqrt{2}}{4}.$$

One of the most important uses of these addition-subtraction identities is to reduce a trigonometric function of a sum or difference to a linear combination of functions (linear combinations of functions were discussed in Section 2.6).

EXAMPLE 2   Verify these identities.

a) $\cos(\alpha - \pi) \overset{?}{=} -\cos\alpha$

b) $\cos(\pi/3 + \phi) \overset{?}{=} \frac{1}{2}(\cos\phi - \sqrt{3}\sin\phi).$

SOLUTION

a) Starting with the left side,

$$\cos(\alpha - \pi) = \cos\alpha\cos\pi + \sin\alpha\sin\pi$$
$$= \cos\alpha(-1) + \sin\alpha(0)$$
$$= -\cos\alpha.$$

This verifies the identity.

b) Again, starting with the left side,

$$\cos\left(\frac{\pi}{3} + \phi\right) = \cos\frac{\pi}{3}\cos\phi - \sin\frac{\pi}{3}\sin\phi$$

$$= \left(\frac{1}{2}\right)\cos\phi - \left(\frac{\sqrt{3}}{2}\right)\sin\phi$$

$$= \tfrac{1}{2}(\cos\phi - \sqrt{3}\sin\phi).$$

This verifies the identity.

The identity for $\cos(\alpha - \beta)$ can be used to prove the cofunction identities from Chapter 5 (recall that the proofs of these were deferred until this chapter). We start with the proof of $\cos(\pi/2 - \theta) = \sin\theta$:

$$\cos\left(\frac{\pi}{2} - \theta\right) = \cos\frac{\pi}{2}\cos\theta + \sin\frac{\pi}{2}\sin\theta$$

$$= (0)\cos\theta + (1)\sin\theta$$

$$\cos\left(\frac{\pi}{2} - \theta\right) = \sin\theta.$$

The proof of $\sin(\pi/2 - \theta) = \cos\theta$ also follows directly from this identity:

$$\sin\left(\frac{\pi}{2} - \theta\right) = \cos\left[\frac{\pi}{2} - \left(\frac{\pi}{2} - \theta\right)\right]$$

$$= \cos\left[\frac{\pi}{2} - \frac{\pi}{2} + \theta\right] = \cos\theta.$$

Using these two cofunction identities, we can now prove the addition-subtraction identity for $\sin(\alpha + \beta)$:

$$\sin(\alpha + \beta) = \cos\left[\frac{\pi}{2} - (\alpha + \beta)\right]$$

$$= \cos\left[\left(\frac{\pi}{2} - \alpha\right) - \beta\right]$$

$$= \cos\left(\frac{\pi}{2} - \alpha\right)\cos\beta + \sin\left(\frac{\pi}{2} - \alpha\right)\sin\beta$$

$$\sin(\alpha + \beta) = \sin\alpha\cos\beta + \cos\alpha\sin\beta.$$

The proofs of the remaining three addition-subtraction identities are left as exercises.

EXAMPLE 3   Verify these identities.

a) $\sin(3\pi/2 + \theta) \overset{?}{=} -\cos\theta$

b) $\sqrt{2}\sin(\phi - \pi/4) \overset{?}{=} \cos\phi - \sin\phi$

SOLUTION

a) $\sin(3\pi/2 + \theta) = \sin 3\pi/2 \cos\theta + \cos 3\pi/2 \sin\theta$

$$= (-1)\cos\theta + (0)\sin\theta$$

$$= -\cos\theta.$$

b) $\sqrt{2}\sin(\phi - \pi/4) = \sqrt{2}[\sin\phi\cos\pi/4 - \cos\phi\sin\pi/4]$

$$= \sqrt{2}[\sin\phi(1/\sqrt{2}) - \cos\phi(1/\sqrt{2})]$$

$$= \sin\phi - \cos\phi.$$

EXAMPLE 4   Consider two acute angles α and β. Suppose that $\sin\alpha = \frac{12}{13}$ and $\cos\beta = \frac{3}{5}$. Find the exact value of (a) $\sin(\alpha + \beta)$, (b) $\cos(\alpha - \beta)$, and (c) $\tan(\beta + \pi/4)$.

SOLUTION   Each of these computations depends on first finding the values of the sine, cosine, and tangent of both α and β. Since $\sin\alpha = \frac{12}{13}$, we can choose (5, 12) as a point on the terminal side of α (Figure 3). Likewise, a point on the terminal side of β is (3, 4) (Figure 4). From this, we get

$$\cos\alpha = \tfrac{5}{13} \qquad \sin\beta = \tfrac{4}{5} \qquad \tan\beta = \tfrac{4}{3}.$$

a) $\sin(\alpha + \beta) = \sin\alpha\cos\beta + \cos\alpha\sin\beta$

$$= (\tfrac{12}{13})(\tfrac{3}{5}) + (\tfrac{5}{13})(\tfrac{4}{5}) = \tfrac{56}{65}.$$

b) $\cos(\alpha - \beta) = \cos\alpha\cos\beta + \sin\alpha\sin\beta$

$$= (\tfrac{5}{13})(\tfrac{3}{5}) + (\tfrac{12}{13})(\tfrac{4}{5}) = \tfrac{63}{65}.$$

c) $\tan\left(\beta + \dfrac{\pi}{4}\right) = \dfrac{\tan\beta + \tan\dfrac{\pi}{4}}{1 - \tan\beta\tan\dfrac{\pi}{4}} = \dfrac{\dfrac{4}{3} + 1}{1 - \dfrac{4}{3}\cdot 1} = -7.$

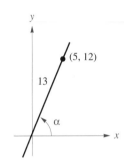

FIGURE 3

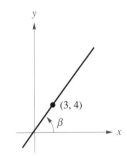

FIGURE 4

If a, b, and B are real constants, then for any real number t,

$$a\cos Bt + b\sin Bt = A\cos(Bt - \phi)$$

where $A = \sqrt{a^2 + b^2}$ and ϕ is such that the terminal side of ϕ passes through the point (a, b).

Linear Combinations of $\cos Bt$ *and* $\sin Bt$

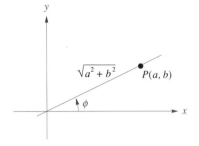

FIGURE 5

To prove this identity, we start (of course) with a sketch of an angle ϕ (Figure 5). There is a point P on the terminal side of ϕ with coordinates (a, b). Starting with the right side,

$$A \cos(Bt - \phi) = \sqrt{a^2 + b^2} \cos(Bt - \phi)$$

$$= \sqrt{a^2 + b^2} \left[\cos Bt \cos \phi + \sin Bt \sin \phi \right]$$

$$= \sqrt{a^2 + b^2} \left[\cos Bt \, \frac{a}{\sqrt{a^2 + b^2}} + \sin Bt \, \frac{b}{\sqrt{a^2 + b^2}} \right]$$

$$= a \cos Bt + b \sin Bt.$$

EXAMPLE 5   Given the function $f(t) = \cos 2t - \sqrt{3} \sin 2t$:

a) Rewrite the function f in the form $A \cos(2t - \phi)$.

b) Sketch the graphs of $y = \cos 2t$, and $y = -\sqrt{3} \sin 2t$ over tne interval $[-\pi, 2\pi]$. Also, sketch the graph of f over the interval $[-\pi, 2\pi]$ using part a).

SOLUTION

a) Since $a = 1$ and $b = -\sqrt{3}$, we get

$$A = \sqrt{(1)^2 + (-\sqrt{3})^2} = 2.$$

The terminal side of the angle ϕ passes through the point $(1, -\sqrt{3})$ (Figure 6). This implies that $\phi = -\pi/3$ (why?). Thus, $f(t) = 2 \cos(2t + \pi/3)$.

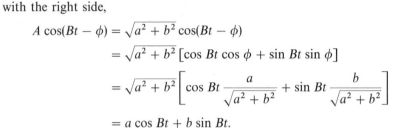

b) The graph of $y = \cos 2t$, has amplitude 1 and period π. The graph of $y = -\sqrt{3} \sin 2t$, has amplitude $\sqrt{3}$ and period π. From part a) the graph of $f(t) = 2 \cos(2t + \pi/3)$ is a cosine curve with amplitude 2, period π, and horizontal translation of $-\pi/6$. The graphs of all three are shown in Figure 7. Notice that the graph of $f(t) = 2 \cos(2t + \pi/3)$ appears to be the same as the result of adding y-coordinates to determine the graph of $y = \cos 2t - \sqrt{3} \sin 2t$.

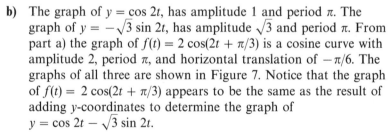

FIGURE 6

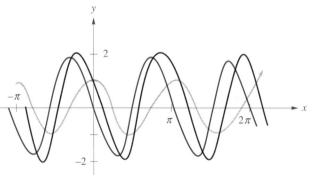

FIGURE 7

In the addition-subtraction identities, special situations arise when α and β are the same angle. The identities derived in these cases are called the double-angle identities. They are listed below.

$$\cos 2\theta = \cos^2\theta - \sin^2\theta$$
$$= 2\cos^2\theta - 1$$
$$= 1 - 2\sin^2\theta$$
$$\sin 2\theta = 2\sin\theta\cos\theta$$
$$\tan 2\theta = \frac{2\tan\theta}{1 - \tan^2\theta}$$

Double-Angle Identities for Sine, Cosine, and Tangent Functions

We prove the first of these identities (the rest are left as exercises). Starting with the left side,

$$\cos 2\theta = \cos(\theta + \theta)$$
$$= \cos\theta\cos\theta - \sin\theta\sin\theta$$
$$= \cos^2\theta - \sin^2\theta.$$

Two alternate forms of this identity exist:

$$\cos 2\theta = 2\cos^2\theta - 1 \quad \text{and} \quad \cos 2\theta = 1 - 2\sin^2\theta.$$

The proofs of these are also left as exercises.

EXAMPLE 6   Given that $\cos\theta = -\frac{5}{13}$ and $\sin\theta < 0$, determine the values of the trigonometric functions of 2θ.

SOLUTION   Back in Example 9 of Section 5.3, we found that $\sin\theta = -\frac{12}{13}$ and $\tan\theta = \frac{12}{5}$. Using the double-angle identities,

$$\cos 2\theta = \cos^2\theta - \sin^2\theta = \left(-\frac{5}{13}\right)^2 - \left(-\frac{12}{13}\right)^2 = -\frac{119}{169}$$

$$\sin 2\theta = 2\sin\theta\cos\theta = 2\left(-\frac{12}{13}\right)\left(-\frac{5}{13}\right) = \frac{120}{169}$$

$$\tan 2\theta = \frac{2\tan\theta}{1 - \tan^2\theta} = \frac{2(\frac{12}{5})}{1 - (\frac{12}{5})^2} = -\frac{120}{119}.$$

Alternatively, $\tan 2\theta$ could have been determined by

$$\tan 2\theta = \frac{\sin 2\theta}{\cos 2\theta} = \frac{(\frac{120}{169})}{(-\frac{119}{169})} = -\frac{120}{119}.$$

The remaining three functions can be found by using the reciprocal identities:

$$\cot 2\theta = -\frac{120}{119} \qquad \sec 2\theta = -\frac{169}{119} \qquad \csc 2\theta = \frac{169}{120}.$$

EXAMPLE 7 Verify the identity $\cos 4\alpha \overset{?}{=} 8\cos^4\alpha - 8\cos^2\alpha + 1$.

SOLUTION We start with the left side. It is helpful here to think of 4α as $2(2\alpha)$. Then,

$$\cos 4\alpha = \cos 2(2\alpha)$$
$$= 2\cos^2(2\alpha) - 1$$
$$= 2(2\cos^2\alpha - 1)^2 - 1$$
$$= 2(4\cos^4\alpha - 4\cos^2\alpha + 1) - 1$$
$$= 8\cos^4\alpha - 8\cos^2\alpha + 1.$$

This verifies the identity.

EXERCISE SET 6.2

A

In Problems 1 through 6, find the exact value of the expression. Do not use a calculator.

1. a) $\sin 15°$
   b) $\csc 15°$

2. a) $\tan 165°$
   b) $\cot 165°$

3. a) $\cos(-75°)$
   b) $\sec(-75°)$

4. a) $\sin 195°$
   b) $\csc 195°$

5. a) $\tan\left(-\frac{5\pi}{12}\right)$
   b) $\cot\left(-\frac{5\pi}{12}\right)$

6. a) $\cos\left(\frac{11\pi}{12}\right)$
   b) $\sec\left(\frac{11\pi}{12}\right)$

In Problems 7 through 12, find the exact value of the expression. Do not use a calculator.

7. a) $\cos\left(\frac{3\pi}{4} - \frac{\pi}{6}\right)$
   b) $\cos\frac{3\pi}{4} - \cos\frac{\pi}{6}$

8. a) $\sin\left(\frac{\pi}{4} + \frac{5\pi}{6}\right)$
   b) $\sin\frac{\pi}{4} + \sin\frac{5\pi}{6}$

9. a) $\cos 17° \cos 43° - \sin 17° \sin 43°$
   b) $\sin 17° \cos 43° + \cos 17° \sin 43°$

10. a) $\sin 75° \cos 75°$
    b) $2\sin(-15°)\cos 15°$

11. a) $\cos^2\left(\frac{\pi}{8}\right) - \sin^2\left(\frac{\pi}{8}\right)$
    b) $\cos^2\left(\frac{\pi}{8}\right) + \sin^2\left(\frac{\pi}{8}\right)$

12. a) $\dfrac{\tan 20° + \tan 25°}{1 - \tan 20° \tan 25°}$
    b) $\dfrac{1 - \tan^2 67.5°}{2\tan 67.5°}$

In Problems 13 through 18, use the addition-subtraction identities to verify the identities.

13. $\cos\left(\theta + \dfrac{\pi}{2}\right) \overset{?}{=} -\sin\theta$

14. $\sin\left(\theta + \dfrac{\pi}{2}\right) \overset{?}{=} \cos\theta$

15. $\cos(\theta - \pi) \overset{?}{=} -\cos\theta$

16. $\tan(\pi - \theta) \overset{?}{=} -\tan\theta$

17. $2\cos\left(\theta + \dfrac{\pi}{4}\right) \stackrel{?}{=} \sqrt{2}(\cos\theta - \sin\theta)$

18. $4\sin\left(\theta + \dfrac{\pi}{3}\right) \stackrel{?}{=} 2\sin\theta + 2\sqrt{3}\cos\theta$

In Problems 19 through 24, use the addition-subtraction identities to write the expression as a linear combination of $\sin\theta$ *and* $\cos\theta$.

19. $2\sin\left(\theta + \dfrac{\pi}{6}\right) - \cos\theta$

20. $2\sin\left(\theta - \dfrac{\pi}{3}\right) - \sin\theta$

21. $\sin\left(\theta + \dfrac{\pi}{2}\right) - \cos\left(\theta + \dfrac{\pi}{3}\right)$

22. $\cos\left(\theta + \dfrac{\pi}{2}\right) + \cos\left(\theta - \dfrac{3\pi}{2}\right)$

23. $2\cos\left(\theta + \dfrac{\pi}{4}\right) + \sin(-\theta + \pi)$

24. $4\sin\left(\theta - \dfrac{\pi}{4}\right) - 3\cos(-\theta)$

B

In Problems 25 through 36, verify the identities.

25. $\tan\left(\theta + \dfrac{\pi}{4}\right) \stackrel{?}{=} \dfrac{1 + \tan\theta}{1 - \tan\theta}$

26. $\dfrac{1 - \tan\alpha\tan\beta}{\tan\alpha - \tan\beta} \stackrel{?}{=} \dfrac{\cos(\alpha + \beta)}{\sin(\alpha - \beta)}$

27. $\tan\lambda + \tan\phi \stackrel{?}{=} \dfrac{\sin(\lambda + \phi)}{\cos\lambda\cos\phi}$

28. $\dfrac{\sin\left(\theta + \dfrac{\pi}{4}\right)}{\cos\left(\theta - \dfrac{\pi}{4}\right)} \stackrel{?}{=} 1$

29. $\cot\left(\theta + \dfrac{\pi}{4}\right) \stackrel{?}{=} \dfrac{\cot\theta - 1}{1 + \cot\theta}$

30. $\cot\left(\theta - \dfrac{\pi}{3}\right) \stackrel{?}{=} \dfrac{\sqrt{3} + \cot\theta}{1 - \sqrt{3}\cot\theta}$

31. $\sin(x + y)\sin(x - y) \stackrel{?}{=} \sin^2 x - \sin^2 y$

32. $\cos(x + y)\cos(x - y) \stackrel{?}{=} \cos^2 x - \sin^2 y$

33. $\sin(x + y) + \sin(x - y) \stackrel{?}{=} 2\sin x\cos y$

34. $\sin(x + y) - \sin(x - y) \stackrel{?}{=} 2\cos x\sin y$

35. $\cos(x + y) + \cos(x - y) \stackrel{?}{=} 2\cos x\cos y$

36. $\cos(x - y) - \cos(x + y) \stackrel{?}{=} 2\sin x\sin y$

37. Consider two angles α and β such that $0 < \alpha < \pi/2$ and $0 < \beta < \pi$. Given that $\sin\alpha = \frac{2}{3}$ and $\cos\beta = \frac{5}{7}$, find the exact value of the expressions.

  a) $\sin(\alpha + \beta)$ **b)** $\sin(\alpha - \beta)$

  c) $\cos(\alpha + \beta)$ **d)** $\cos(\alpha - \beta)$

38. Consider two angles α and β such that $\pi < \alpha < 3\pi/2$ and $-\pi/2 < \beta < 0$. Given that $\sin\alpha = -1/\sqrt{10}$ and $\cos\beta = 2/\sqrt{5}$, find the exact value of the expressions.

  a) $\tan(\alpha + \beta)$ **b)** $\tan(\alpha - \beta)$

  c) $\cos 2\alpha$ **d)** $\sin 2\beta$

39. Consider two angles α and β such that $\pi/2 < \alpha < \pi$ and $0 < \beta < \pi/2$. Given that $\cot\alpha = -\frac{7}{24}$ and $\tan\beta = \frac{3}{4}$, find the exact value of the expressions.

  a) $\sin(\alpha + \beta)$ **b)** $\cos(\alpha - \beta)$

  c) $\tan(\alpha - \beta)$ **d)** $\tan(\alpha + \beta)$

40. Consider two angles α and β such that $-\pi/2 < \alpha < 0$ and $0 < \beta < \pi/2$. Given that $\sin\alpha = -1/\sqrt{26}$ and $\cos\beta = 1/\sqrt{5}$, find the exact value of the expressions.

  a) $\tan(\alpha + \beta)$ **b)** $\tan(\alpha - \beta)$

  c) $\cos 2\alpha$ **d)** $\sin 2\beta$

41. Consider an angle θ such that $0 < \theta < \pi$. Given that $\tan\theta = 2$, find the exact value of the expressions.

  a) $\tan 2\theta$ **b)** $\cos\left(\dfrac{\pi}{3} + \theta\right)$

  c) $\sin\left(\theta + \dfrac{\pi}{4}\right)$ **d)** $\tan\left(\theta - \dfrac{\pi}{4}\right)$

42. Consider an angle θ such that $\pi < \theta < 2\pi$. Given that $\cos\theta = -\frac{2}{5}$, find the exact value of the expressions.

  a) $\cos 2\theta$ **b)** $\csc 2\theta$

  c) $\cos\left(\theta + \dfrac{\pi}{6}\right)$ **d)** $\tan\left(\theta + \dfrac{3\pi}{4}\right)$

43. Use the addition identity for the sine function and the fact that the cosine is an even function and the sine is an odd function to prove the subtraction identity for the sine function.

44. a) Prove the identity

$$\tan(\alpha + \beta) = \frac{\sin\alpha\cos\beta + \cos\alpha\sin\beta}{\cos\alpha\cos\beta - \sin\alpha\sin\beta}.$$

b) Use the result in a) to prove the addition identity for the tangent function. *Hint: Multiply the numerator and the denominator by*

$$\frac{1}{\cos \alpha \cos \beta}.$$

c) Use the result of part b) and the fact that the tangent is an odd function to prove the subtraction identity for the tangent function.

45. a) Prove the double-angle identity
$\cos 2\theta = 2 \cos^2\theta - 1$. *(Hint: Use*
$\cos 2\theta = \cos^2\theta - \sin^2\theta.)$

b) Prove the double-angle identity
$\cos 2\theta = 1 - 2 \sin^2\theta$.

46. a) Prove the double-angle identity
$\sin 2\theta = 2 \sin \theta \cos \theta$ by using the addition identity for $\sin(\alpha + \beta)$.

b) Prove the double-angle identity

$$\tan 2\theta = \frac{2 \tan \theta}{1 - \tan^2\theta},$$

using the addition identity for $\tan(\alpha + \beta)$.

47. Rewrite the function $f(t) = \cos t + \sin t$ in the form of $A \cos(Bt - \phi)$.

48. Rewrite the function $f(t) = 2 \cos t - 2 \sin t$ in the form of $A \cos(Bt - \phi)$.

49. Rewrite the function $f(t) = \sqrt{3} \cos t + \sin t$ in the form of $A \cos(Bt - \phi)$.

50. Rewrite the function $f(t) = 3 \cos 2t - \sqrt{3} \sin 2t$ in the form of $A \cos(Bt - \phi)$.

51. Rewrite the function $f(t) = \sin 5t - \cos 5t$ in the form of $A \cos(Bt - \phi)$.

52. Rewrite the function $f(t) = 2\sqrt{3} \sin 4t - 2 \cos 4t$ in the form of $A \cos(Bt - \phi)$.

53. Prove: If a, b, and B are real constants, then for any real number t,

$$a \cos Bt + b \sin Bt = A \sin(Bt + \lambda),$$

where $A = \sqrt{a^2 + b^2}$ and λ is such that the terminal side of λ passes through the point (b, a). *(Note: this is an alternate approach to that in the section for rewriting linear combinations of* $\sin Bt$ *and* $\cos Bt$.)

C

In Problems 54 through 63, verify the identity.

54. $\sin(\alpha + \beta + \gamma) \overset{?}{=} \sin \alpha \cos \beta \cos \gamma + \cos \alpha \sin \beta \cos \gamma$
$+ \cos \alpha \cos \beta \sin \gamma - \sin \alpha \sin \beta \sin \gamma$

55. $\sin 3\theta \overset{?}{=} 3 \sin \theta \cos^2\theta - \sin^3\theta$ (Hint: Use the result of Problem 54.)

56. $\sin 4x \overset{?}{=} 4 \cos x(\sin x - 2 \sin^3 x)$

57. $\cos 3x \overset{?}{=} 4 \cos^3 x - 3 \cos x$

58. $\cos 5x \overset{?}{=} 16 \cos^5 x - 20 \cos^3 x + 5 \cos x.$

59. $\tan 4x \overset{?}{=} \dfrac{4 \tan x - 4 \tan^3 x}{1 - 6 \tan^2 x + \tan^4 x}$

60. $\dfrac{\sin(x + \Delta x) - \sin x}{\Delta x}$

$$\overset{?}{=} \sin x \left(\frac{\cos \Delta x - 1}{\Delta x} \right) + \cos x \left(\frac{\sin \Delta x}{\Delta x} \right)$$

(Recall that Δx is to be treated as a single variable.)

61. $\dfrac{\cos(x + \Delta x) - \cos x}{\Delta x}$

$$\overset{?}{=} \cos x \left(\frac{\cos \Delta x - 1}{\Delta x} \right) - \sin x \left(\frac{\sin \Delta x}{\Delta x} \right)$$

(Recall that Δx is to be treated as a single variable.)

62. $\sin(n\alpha) \overset{?}{=} 2 \sin[(n - 1)\alpha]\cos \alpha - \sin[(n - 2)\alpha]$

63. $\cos(n\alpha) \overset{?}{=} 2 \cos[(n - 1)\alpha]\cos \alpha - \cos[(n - 2)\alpha]$

SECTION 6.3 IDENTITIES FOR CALCULUS, PART II

From the double-angle identities of the last section, we can derive four sets of identities to use for the manipulation of trigonometric expressions that are encountered in calculus.

If we take the identities

$$\cos 2\theta = 1 - 2\sin^2\theta \quad \text{and} \quad \cos 2\theta = 2\cos^2\theta - 1$$

and solve them for $\sin^2\theta$ and $\cos^2\theta$, respectively, we obtain the following identities, which are useful in simplifying expressions involving powers of functions.

$$\sin^2\theta = \frac{1 - \cos 2\theta}{2} \qquad \cos^2\theta = \frac{1 + \cos 2\theta}{2}$$

Square-of-Function Identities

EXAMPLE 1   Using the square-of-functions identities, find an equivalent expression to $\sin^4\theta$ in terms of first powers of trigonometric functions.

SOLUTION We can rewrite $\sin^4\theta$ as the square of $\sin^2\theta$ and use the square-of-function identity:

$$\sin^4\theta = (\sin^2\theta)^2$$

$$= \left(\frac{1 - \cos 2\theta}{2}\right)^2 \qquad \text{(using the identity for } \sin^2\theta)$$

$$= \tfrac{1}{4} - \tfrac{1}{2}\cos 2\theta + \tfrac{1}{4}\cos^2 2\theta$$

$$= \frac{1}{4} - \frac{1}{2}\cos 2\theta + \frac{1}{4}\left(\frac{1 + \cos 4\theta}{2}\right) \qquad \begin{array}{l}\text{(using the identity} \\ \text{for } \cos^2\theta)\end{array}$$

$$= \tfrac{3}{8} - \tfrac{1}{2}\cos 2\theta + \tfrac{1}{8}\cos 4\theta.$$

In this type of derivation, the possibility of error is high. As a "reality check," we can choose a value for θ, say $\theta = \pi/4$, and substitute this value into both sides of our new-found identity:

$$\sin^4\left(\frac{\pi}{4}\right) \overset{?}{=} \frac{3}{8} - \frac{1}{2}\cos 2\left(\frac{\pi}{4}\right) + \frac{1}{8}\cos 4\left(\frac{\pi}{4}\right)$$

$$\left(\frac{1}{\sqrt{2}}\right)^4 \overset{?}{=} \frac{3}{8} - \frac{1}{2}\cos\left(\frac{\pi}{2}\right) + \frac{1}{8}\cos \pi$$

$$\tfrac{1}{4} \overset{?}{=} \tfrac{3}{8} - \tfrac{1}{2}(0) + \tfrac{1}{8}(-1)$$

$$\tfrac{1}{4} = \tfrac{1}{4}.$$

The fact that the equation seems to be true for the value we chose does not guarantee the correctness of the identity, but it does give us some confidence that the identity is indeed true.

The preceding identity for $\sin^2\theta$ is true for all θ. In particular, it is true for say, $\alpha/2$. So,

$$\sin^2\frac{\alpha}{2} = \frac{1 - \cos 2\left(\dfrac{\alpha}{2}\right)}{2} = \frac{1 - \cos \alpha}{2}.$$

Solving for $\sin \alpha/2$ yields this identity:

$$\sin \frac{\alpha}{2} = \pm\sqrt{\frac{1 - \cos \alpha}{2}}.$$

Similar half-angle identities can be derived for the cosine and tangent functions. Notice that there are two forms for the tangent function. The proof of these are left as exercises.

Half-Angle Identities

$$\sin \frac{\alpha}{2} = \pm\sqrt{\frac{1 - \cos \alpha}{2}}$$

$$\cos \frac{\alpha}{2} = \pm\sqrt{\frac{1 + \cos \alpha}{2}}$$

$$\tan \frac{\alpha}{2} = \frac{1 - \cos \alpha}{\sin \alpha} = \frac{\sin \alpha}{1 + \cos \alpha}$$

The choice of sign of $\sin(\alpha/2)$ and $\cos(\alpha/2)$ depends, of course, on the quadrant in which the terminal side of $\alpha/2$ lies.

EXAMPLE 2   Using an appropriate half-angle identity, determine the exact value of these expressions.

a) $\cos 15°$ **b)** $\sin(-\pi/8)$ **c)** $\tan 5\pi/12$

SOLUTION

a) Since $15°$ is half of $30°$,

The terminal side of $15°$ is in the first quadrant, and the cosine function is positive in that quadrant.

$$\cos 15° = \cos \frac{1}{2}(30°)$$

$$= +\sqrt{\frac{1 + \cos 30°}{2}}$$

$$= \sqrt{\frac{1 + \dfrac{\sqrt{3}}{2}}{2}} = \frac{\sqrt{2 + \sqrt{3}}}{2}.$$

b) Since $-\pi/8$ is half of $-\pi/4$,

$$\sin\left(-\frac{\pi}{8}\right) = \sin\frac{1}{2}\left(-\frac{\pi}{4}\right)$$

$$= -\sqrt{\frac{1 - \cos\left(-\dfrac{\pi}{4}\right)}{2}}$$

$$= -\sqrt{\frac{1 - \left(\dfrac{\sqrt{2}}{2}\right)}{2}} = -\frac{\sqrt{2 - \sqrt{2}}}{2}.$$

The terminal side of $-\pi/8$ is in the fourth quadrant, and the sine function is negative in that quadrant.

c) Since $5\pi/12$ is half of $5\pi/6$,

$$\tan\frac{5\pi}{12} = \tan\frac{1}{2}\left(\frac{5\pi}{6}\right)$$

$$= \frac{1 - \cos\dfrac{5\pi}{6}}{\sin\dfrac{5\pi}{6}}$$

$$= \frac{1 - \left(-\dfrac{\sqrt{3}}{2}\right)}{\left(\dfrac{1}{2}\right)} = 2 + \sqrt{3}.$$

EXAMPLE 3   Given that $\tan\alpha = \frac{7}{24}$ and that $\pi < \alpha < 2\pi$, find the exact values of $\sin\frac{1}{2}\alpha$, $\cos\frac{1}{2}\alpha$, and $\tan\frac{1}{2}\alpha$.

SOLUTION The sketch of α in standard position is shown in Figure 8. Examination of the half-angle identities shows that we need first to find the values of $\sin\alpha$ and $\cos\alpha$. From the figure we see that $\sin\alpha = -\frac{7}{25}$ and that $\cos\alpha = -\frac{24}{25}$. Also, since $\pi < \alpha < 2\pi$, it follows that $\pi/2 < \frac{1}{2}\alpha < \pi$, implying of course that the terminal side of $\frac{1}{2}\alpha$ is in the second quadrant. Hence $\sin\frac{1}{2}\alpha > 0$ and $\cos\frac{1}{2}\alpha < 0$. So,

$$\sin\frac{\alpha}{2} = +\sqrt{\frac{1 - \left(-\frac{24}{25}\right)}{2}} = \frac{7\sqrt{2}}{10}$$

$$\cos\frac{\alpha}{2} = -\sqrt{\frac{1 + \left(-\frac{24}{25}\right)}{2}} = -\frac{\sqrt{2}}{10}$$

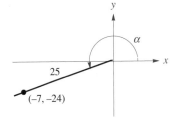

FIGURE 8

and

$$\tan\frac{\alpha}{2} = \frac{1 - \left(-\frac{24}{25}\right)}{\left(-\frac{7}{25}\right)} = -7.$$

Certain applications that arise in the study of mathematics and engineering involve the products of sine and cosine functions. It is to your advantage to write such expressions as sums. The identities that follow can be used to accomplish this task. They are a direct result of the addition identities of Section 6.2.

Product-to-Sum Identities

$$\sin \alpha \cos \beta = \tfrac{1}{2}[\sin(\alpha + \beta) + \sin(\alpha - \beta)]$$

$$\cos \alpha \cos \beta = \tfrac{1}{2}[\cos(\alpha + \beta) + \cos(\alpha - \beta)]$$

$$\sin \alpha \sin \beta = \tfrac{1}{2}[\cos(\alpha - \beta) - \cos(\alpha + \beta)]$$

As usual, we prove the first identity and leave the rest as exercises. Starting with the right side, we can use the addition-subtraction identities for $\sin(\alpha + \beta)$ and $\sin(\alpha - \beta)$:

$$\tfrac{1}{2}[\sin(\alpha + \beta) + \sin(\alpha - \beta)] = \tfrac{1}{2}[(\sin \alpha \cos \beta + \cos \alpha \sin \beta)$$
$$+ (\sin \alpha \cos \beta - \cos \alpha \sin \beta)]$$
$$= \tfrac{1}{2}[2 \sin \alpha \cos \beta]$$
$$= \sin \alpha \cos \beta.$$

This verifies the identity.

EXAMPLE 4   Rewrite the product as a sum:

a) $\sin 4\theta \cos 3\theta$ **b)** $2 \sin(x + \pi/4) \sin(5x - \pi/4)$

SOLUTION

a) $\sin 4\theta \cos 3\theta = \tfrac{1}{2}[\sin(4\theta + 3\theta) + \sin(4\theta - 3\theta)]$
$$= \tfrac{1}{2}[\sin 7\theta + \sin \theta]$$

b) $2 \sin(x + \pi/4) \sin(5x - \pi/4) = 2\{\tfrac{1}{2}[\cos((x + \pi/4) - (5x - \pi/4))$
$$- \cos((x + \pi/4) + (5x - \pi/4))]\}$$
$$= \cos(\pi/2 - 4x) - \cos 6x$$
$$= \sin 4x - \cos 6x.$$

(using the cofunction identity)

EXAMPLE 5   Using the product-to-sums identities, find an equivalent expression to $\cos^3\theta$ in terms of first powers of trigonometric functions.

SOLUTION

$$\cos^3\theta = \cos\theta[\cos^2\theta]$$

$$= \cos\theta[\tfrac{1}{2}(1 + \cos 2\theta)]$$

(using the square-of-functions identity)

$$= \tfrac{1}{2}(\cos\theta + \cos 2\theta \cos\theta)$$

$$= \tfrac{1}{2}(\cos\theta + (\tfrac{1}{2}\cos 3\theta + \tfrac{1}{2}\cos\theta))$$

(using the product-to-sum identity)

$$= \tfrac{3}{4}\cos\theta + \tfrac{1}{4}\cos 3\theta.$$

Conversely, there are instances in which it is to your advantage to rewrite a sum of sines or of cosines as a product. These identities are the "inverses" of the sum-to-product identities.

$$\sin\alpha + \sin\beta = 2\sin\left(\frac{\alpha+\beta}{2}\right)\cos\left(\frac{\alpha-\beta}{2}\right)$$

$$\sin\alpha - \sin\beta = 2\cos\left(\frac{\alpha+\beta}{2}\right)\sin\left(\frac{\alpha-\beta}{2}\right)$$

$$\cos\alpha + \cos\beta = 2\cos\left(\frac{\alpha+\beta}{2}\right)\cos\left(\frac{\alpha-\beta}{2}\right)$$

$$\cos\alpha - \cos\beta = -2\sin\left(\frac{\alpha+\beta}{2}\right)\sin\left(\frac{\alpha-\beta}{2}\right)$$

Sum-to-Product Identities

The proofs of these identities are straightforward. For example, starting with the right side of the first identity,

$$2\sin\left(\frac{\alpha+\beta}{2}\right)\cos\left(\frac{\alpha-\beta}{2}\right) = 2\left\{\frac{1}{2}\left[\sin\left(\frac{\alpha+\beta}{2}+\frac{\alpha-\beta}{2}\right)+\sin\left(\frac{\alpha+\beta}{2}-\frac{\alpha-\beta}{2}\right)\right]\right\}$$

$$= \sin\alpha + \sin\beta.$$

The proofs of the remaining three identities are similar.

EXAMPLE 6   Rewrite as a product:

    a) $\sin 5u - \sin u$ **b)** $\cos \theta - \cos 3\theta$

SOLUTION

    a) $\sin 5u - \sin u = 2 \cos[\frac{1}{2}(5u + u)] \sin[\frac{1}{2}(5u - u)]$

                              $= 2 \cos 3u \sin 2u.$

    b) $\cos \theta - \cos 3\theta = -2 \sin[\frac{1}{2}(\theta + 3\theta)] \sin[\frac{1}{2}(\theta - 3\theta)]$

                           $= -2 \sin 2\theta \sin(-\theta)$

                           $= 2 \sin 2\theta \sin \theta.$

    (Recall that the sine function is an odd function.)

EXAMPLE 7   Verify the identity

$$\frac{\sin \alpha - \sin \beta}{\cos \alpha + \cos \beta} \stackrel{?}{=} \tan \tfrac{1}{2}(\alpha - \beta).$$

SOLUTION We start with the left side and use the sum-to-product identities:

$$\frac{\sin \alpha - \sin \beta}{\cos \alpha + \cos \beta} = \frac{2 \cos[\frac{1}{2}(\alpha + \beta)] \sin[\frac{1}{2}(\alpha - \beta)]}{2 \cos[\frac{1}{2}(\alpha + \beta)] \cos[\frac{1}{2}(\alpha - \beta)]}$$

$$= \frac{\sin \frac{1}{2}(\alpha - \beta)}{\cos \frac{1}{2}(\alpha - \beta)}$$

$$= \tan \tfrac{1}{2}(\alpha - \beta).$$

EXERCISE SET 6.3

A

In Problems 1 through 9, use the half-angle identities to find the exact value of the expression. Do not use a calculator.

1. **a)** $\sin 15°$

    b) $\csc 15°$

3. **a)** $\cos(-157.5°)$

    b) $\sec(-157.5°)$

2. **a)** $\tan 67.5°$

    b) $\cot 67.5°$

4. **a)** $\sin 195°$

    b) $\csc 195°$

5. **a)** $\tan\left(-\dfrac{5\pi}{8}\right)$

    b) $\cot\left(-\dfrac{5\pi}{8}\right)$

7. **a)** $\cos\dfrac{1}{2}\left(\dfrac{3\pi}{4}\right)$

    b) $\dfrac{1}{2}\cos\left(\dfrac{3\pi}{4}\right)$

6. **a)** $\cos\left(\dfrac{11\pi}{8}\right)$

    b) $\sec\left(\dfrac{11\pi}{8}\right)$

8. **a)** $\sin\dfrac{1}{2}\left(\dfrac{5\pi}{6}\right)$

    b) $\dfrac{1}{2}\sin\left(\dfrac{5\pi}{6}\right)$

9. a) $\tan\frac{1}{2}\left(-\frac{3\pi}{4}\right)$

    b) $\frac{1}{2}\tan\left(-\frac{3\pi}{4}\right)$

In Problems 10 through 15, write the product as a sum using the product-to-sum identities. Simplify if possible.

10. $\cos 8\theta \cos 3\theta$ **11.** $\cos 4t \cos 6t$

12. $\sin 3\phi \cos 2\phi$ **13.** $\cos\frac{2}{3}\alpha \sin\frac{4}{3}\alpha$

14. $\cos\left(2x - \frac{\pi}{4}\right)\cos\left(4x + \frac{\pi}{4}\right)$

15. $\cos\left(x + \frac{\pi}{4}\right)\cos\left(3x + \frac{\pi}{4}\right)$

In Problems 16 through 21, write the sum as a product using the sum-to-product identities. Simplify if possible.

16. $\cos 3\theta - \cos \theta$ **17.** $\cos 2x + \cos 3x$

18. $\sin 4y - \sin 3y$ **19.** $\cos 2\theta - \cos \theta$

20. $\sin\left(x - \frac{\pi}{2}\right) - \sin\left(x + \frac{\pi}{2}\right)$

21. $\cos\left(x - \frac{3\pi}{2}\right) + \cos\left(\frac{\pi}{2} - x\right)$

B

In Problems 22 through 33, verify the identities.

22. $\tan\frac{1}{2}A \overset{?}{=} \csc A - \cot A$

23. $\left(\sin\frac{\theta}{2} + \cos\frac{\theta}{2}\right)^2 \overset{?}{=} 1 + \sin \theta$

24. $2\sin^2\frac{x}{2} \overset{?}{=} \sin x \tan\frac{x}{2}$

25. $\sin^2\frac{\alpha}{2} - \cos^2\frac{\alpha}{2} \overset{?}{=} -\cos \alpha$

26. $\frac{1 + \tan\frac{1}{2}\theta}{1 - \tan\frac{1}{2}\theta} \overset{?}{=} \tan \theta + \sec \theta$

27. $\cos t \overset{?}{=} \cos^4\left(\frac{t}{2}\right) - \sin^4\left(\frac{t}{2}\right)$

28. $\tan 5t \overset{?}{=} \frac{\sin 3t + \sin 7t}{\cos 3t + \cos 7t}$

29. $\cot 4\theta \overset{?}{=} \frac{\sin 2\theta - \sin 6\theta}{\cos 6\theta - \cos 2\theta}$

30. $\frac{\cos(r - t)}{\cos(r + t)} \overset{?}{=} \frac{\cot r + \tan t}{\cot r - \tan t}$

31. $\frac{\tan(x + y)}{\tan(x - y)} \overset{?}{=} \frac{\sin 2x + \sin 2y}{\sin 2x - \sin 2y}$

32. $\sin \theta \sin 2\theta \cos 2\theta \overset{?}{=} \frac{1}{4}(\cos 3\theta - \cos 5\theta)$

33. $\frac{\sin \alpha + \sin 3\alpha + \sin 5\alpha}{\cos \alpha + \cos 3\alpha + \cos 5\alpha} \overset{?}{=} \tan 3\alpha$

34. Consider an angle θ such that $0 < \theta < \pi$ and $\cos \theta = \frac{3}{5}$. Find the exact value of the expressions.

    a) $\sin\frac{1}{2}\theta$ **b)** $\cos\frac{1}{2}\theta$ **c)** $\tan\frac{1}{2}\theta$

35. Consider an angle θ such that $\pi/2 < \theta < \pi$ and $\cos \theta = -\frac{7}{25}$. Find the exact value of the expressions.

    a) $\sin\frac{1}{2}\theta$ **b)** $\cos\frac{1}{2}\theta$ **c)** $\tan\frac{1}{2}\theta$

36. Consider an angle α such that $0 < \alpha < \pi/2$ and $\sin \alpha = 3/\sqrt{10}$. Find the exact value of the expressions.

    a) $\sin\frac{1}{2}\alpha$ **b)** $\sin 2\alpha$ **c)** $\csc\frac{1}{2}\alpha$

37. Consider an angle ϕ such that $\pi < \phi < 2\pi$ and $\cos \phi = -2/\sqrt{5}$. Find the exact value of the expressions.

    a) $\cos\frac{1}{2}\phi$ **b)** $\cos 2\phi$ **c)** $\sec\frac{1}{2}\phi$

38. Consider an angle β such that $\pi < \beta < 2\pi$ and $\tan \beta = -2$. Find the exact value of the expressions.

    a) $\tan\frac{1}{2}\beta$ **b)** $\tan 2\beta$ **c)** $\cot\frac{1}{2}\beta$

39. Consider an angle β such that $\pi < \beta < 2\pi$ and $\tan \beta = \frac{24}{7}$. Find the exact value of the expression.

    a) $\sin\frac{1}{2}\beta$ **b)** $\cos\frac{1}{2}\beta$ **c)** $\cot\frac{1}{2}\beta$

In Problems 40 through 45, write the function as a sum or difference and use this to sketch the graph of the function over the interval $[-2\pi, 4\pi]$.

40. $f(t) = \sin 2t \cos t$ **41.** $f(t) = \cos 2t \cos t$

42. $f(t) = \cos\frac{1}{2}t \sin t$ **43.** $f(t) = \sin\frac{1}{2}t \sin t$

44. $f(t) = -\cos 2t \cos t$ **45.** $f(t) = \sin\frac{1}{4}t \cos\frac{3}{4}t$

In Problems 46 through 51, use the square-of-functions identities and the product-to-sum identities to find an equivalent expression in terms of first powers of trigonometric functions.

46. $\cos^4\theta$ **47.** $\sin^6\theta$

48. $\cos^6\theta$ **49.** $\cos^3\theta$

50. $\sin^3\theta$ **51.** $\cos^5\theta$

52. Use the square-of-functions identities to prove the half-angle identities.

53. Prove the product-to-sum identities for $\cos \alpha \cos \beta$ and $\sin \alpha \sin \beta$.

54. Prove the following identity (recall that Δx is to be treated as a single variable).

$$\frac{\sin(x + \Delta x) - \sin x}{\Delta x} \stackrel{?}{=} \left(\frac{\sin(\frac{1}{2}\Delta x)}{\frac{1}{2}\Delta x}\right)\cos(x + \tfrac{1}{2}\Delta x).$$

55. Prove the following identity (recall that Δx is to be treated as a single variable).

$$\frac{\cos(x + \Delta x) - \cos x}{\Delta x} \stackrel{?}{=} -\sin(x + \tfrac{1}{2}\Delta x)\left(\frac{\sin(\frac{1}{2}\Delta x)}{\frac{1}{2}\Delta x}\right).$$

In calculus, a particular method of substitution is used to simplify rational expressions involving $\sin \theta$ and $\cos \theta$. Problems 56 through 62 deal with this substitution.

56. a) Let $u = \tan \frac{1}{2}\theta$. Show that

$$\cos\left(\frac{\theta}{2}\right) = \frac{1}{\sqrt{1 + u^2}} \quad \text{and} \quad \sin\left(\frac{\theta}{2}\right) = \frac{u}{\sqrt{1 + u^2}}.$$

b) Let $u = \tan \frac{1}{2}\theta$. Using a) and the double-angle identities, show that

$$\cos \theta = \frac{1 - u^2}{1 + u^2} \quad \text{and} \quad \sin \theta = \frac{2u}{1 + u^2}.$$

Using the substitution described in part b) of Problem 56, write the expressions in Problems 57 through 62 in terms of u and simplify:

57. $\dfrac{1}{1 - \sin \theta}$

58. $\dfrac{2}{4 \sin \theta - 3 \cos \theta}$

59. $\dfrac{\cos \theta - \sin \theta}{\sin \theta}$

60. $\dfrac{\cot \theta}{1 - \tan \theta}$

61. $\dfrac{\sec \theta}{\tan \theta - \csc \theta}$

62. $\dfrac{\sin 2\theta}{1 - \cos \theta}$

C

63. a) Use the square-of-functions identities to sketch the graph of $f(t) = \sin^2 t$ and $g(t) = \cos^2 t$ over the interval $0 < t < 2\pi$.

b) By adding y-coordinates, sketch the graph of $y = f(t) + g(t)$.

SECTION 6.4 INVERSE TRIGONOMETRIC FUNCTIONS

In Section 5.1, the inverse sine of a real number, $\sin^{-1}x$, was defined as an acute angle θ such that $\sin \theta = x$. Since then, we have expanded the idea of the trigonometric ratios of acute angles to trigonometric functions of general angles and of real numbers. Our task at hand is to investigate the inverses of these functions in these expanded concepts. The **inverse trigonometric functions** are solutions to many fundamental problems that you will encounter in calculus.

Section 2.7 addressed the ideas of the inverse function, which functions have inverses, and how inverse functions are determined. One of the problems encountered in discussing the inverse of a function is that the original function may not be one-to-one. Our way of sidestepping this problem, as you recall, is to restrict the domain of the original function to a convenient interval so that the new, restricted function is one-to-one.

EXAMPLE 1   Determine all values of θ such that $\sin \theta = -\frac{1}{2}$.

SOLUTION Since $\sin \theta < 0$, the terminal side of θ must fall in either the third or fourth quadrant. The reference angle θ' of θ is $\pi/6$ (recall that $\sin \pi/6 = \frac{1}{2}$). This implies that in the third quadrant, θ is $7\pi/6$ or coterminal with $7\pi/6$. Likewise, in the fourth quadrant θ is $11\pi/6$ or coterminal with $11\pi/6$ (Figure 9). In summary,

$$\theta = 7\pi/6 \pm 2\pi n \quad \text{or} \quad \theta = 11\pi/6 \pm 2\pi n \qquad (n = 0, 1, 2, 3, \ldots).$$

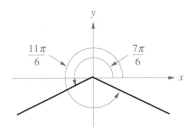

FIGURE 9

Example 1 points out that the sine function is certainly not one-to-one. As a matter of fact, things couldn't be worse: For any particular element in its range, there are an infinite number of elements in its domain which correspond to that element! To solve this problem, we restrict the domain to a convenient interval so that $y = \sin \theta$ is one-to-one over that interval (Figure 10). Our convenient interval is $[-\pi/2, \pi/2]$.

The domain and range of the inverse function are the same as the range and domain, respectively, of the restricted function. The graph of the inverse is the reflection through the line $y = x$ of the restricted function. These results are summarized below.

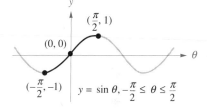

FIGURE 10

The **inverse sine function,** denoted **$\sin^{-1}$**, is defined by the following relation:

$$y = \sin^{-1}x \quad \Leftrightarrow \quad x = \sin y \text{ and } -\frac{\pi}{2} \leq y \leq \frac{\pi}{2}.$$

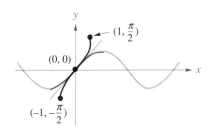

Domain: $[-1, 1]$ Range: $[-\pi/2, \pi/2]$

Inverse Sine Function

The inverse sine function is also known in some circles as the **arcsine** function. The notation arcsin x is used interchangeably with $\sin^{-1}x$.

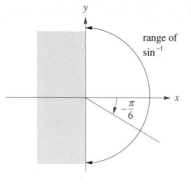

range of sin^{-1}

FIGURE 11

EXAMPLE 2   Determine $\sin^{-1}(-\frac{1}{2})$.

SOLUTION The problem here is to find the value of θ such that

$$\sin \theta = -\tfrac{1}{2} \quad \text{and} \quad -\frac{\pi}{2} \le \theta \le \frac{\pi}{2}.$$

Notice that this limits the value of the θ in question to the first and fourth quadrants. In Example 1, we found all the values of θ such that $\sin \theta = -\frac{1}{2}$. The one in particular that is between $-\pi/2$ and $\pi/2$ is $-\pi/6$ (Figure 11). So,

$$\sin^{-1}\left(-\frac{1}{2}\right) = -\frac{\pi}{6}.$$

Someone who is not paying close attention might incorrectly suspect that the answer in Example 2 could also be $11\pi/6$ since $\sin(11\pi/6) = -\frac{1}{2}$. The problem with $11\pi/6$ is that it is not in the interval $[-\pi/2, \pi/2]$. It bears repeating that the only value for $\sin^{-1}(-\frac{1}{2})$ is $-\pi/6$. If we were to allow any other value, we would violate the definition of a function, namely, that only one element of the range of $\sin^{-1}$ is assigned to each element of the domain of $\sin^{-1}$.

EXAMPLE 3   Determine the following.

    a) $\sin^{-1}(-1)$ **b)** $\arcsin(1/\sqrt{2})$

SOLUTION

    a) By the definition, $\theta = \sin^{-1}(-1)$ if and only if $\sin y = -1$ and $-\pi/2 \le \theta \le \pi/2$. Using what we know about the sine function and its values, we deduce that θ must be $-\pi/2$. That is,

$$\sin^{-1}(-1) = -\frac{\pi}{2}.$$

    b) Again, $\theta = \arcsin(1/\sqrt{2})$ if and only if $\sin \theta = 1/\sqrt{2}$ and $-\pi/2 \le \theta \le \pi/2$. Of course, $\sin \pi/4 = 1/\sqrt{2}$. Thus,

$$\arcsin\left(\frac{1}{\sqrt{2}}\right) = \frac{\pi}{4}.$$

Now consider an inverse of the cosine function. Just as with the sine function, the cosine function in not one-to-one. The domain is conventionally restricted to the interval $[0, \pi]$ to allow an inverse function (Figure 12).

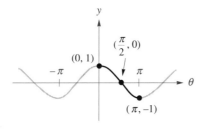

$y = \cos \theta, 0 \le \theta \le \pi$

FIGURE 12

The **inverse cosine function,** denoted **cos^{-1}**, is defined by the following relation:

$$y = \cos^{-1}x \iff x = \cos y \text{ and } 0 \le y \le \pi.$$

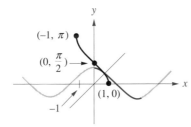

Domain: $[-1, 1]$ Range: $[0, \pi]$

Inverse Cosine Function

Just as you might suspect, the inverse cosine function is also known as the **arccosine** function, and the notations $\cos^{-1}x$ and arccos x are used interchangeably.

EXAMPLE 4 Determine:

    a) $\cos^{-1}(-\frac{1}{2})$ **b)** arccos(1)   **c)** $\cos^{-1}(13/6)$

SOLUTION

    a) By the definition, $\theta = \cos^{-1}(-\frac{1}{2})$ if and only if $\cos \theta = -\frac{1}{2}$ and $0 \le \theta \le \pi$. Using what we know about the cosine function and its values, we deduce that θ must be $2\pi/3$. That is,

$$\cos^{-1}\left(-\frac{1}{2}\right) = \frac{2\pi}{3}.$$

b) We start by letting $\theta = \arccos 1$. Thus, $\cos \theta = 1$ and $0 \leq \theta \leq \pi$. Of course, $\cos 0 = 1$. So,

$$\arccos 1 = 0.$$

c) The value $\frac{13}{6}$ is not in the domain of the inverse cosine function. That is, there is no real number that has a cosine of $\frac{13}{6}$. The expression $\cos^{-1}(\frac{13}{6})$ should be considered as meaningless as, say, $\frac{7}{0}$ or $\tan \pi/2$.

Care must be taken with the range of each of these two inverse functions. This is the moral of the next example.

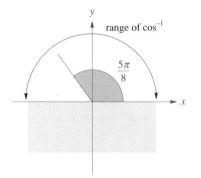

FIGURE 13

EXAMPLE 5   Determine:

a) $\cos^{-1}(\cos 5\pi/8)$ b) $\sin^{-1}(\sin 5\pi/8)$.

SOLUTION

a) The essence of the problem is this: Find a number in the interval $[0, \pi]$ whose cosine is the same as $\cos 5\pi/8$ (repeat this last sentence to yourself until you understand it completely). The obvious answer is $5\pi/8$ (Figure 13). So,

$$\cos^{-1}\left(\cos \frac{5\pi}{8}\right) = \frac{5\pi}{8}.$$

b) This expression appears to be similar to the expression of a). We need to find a number in the interval $[-\pi/2, \pi/2]$ whose sine is the same as $\sin 5\pi/8$. The difference, however, is that $5\pi/8$ is not in the interval $[-\pi/2, \pi/2]$. The value in the interval $[-\pi/2, \pi/2]$ that does satisfy the requirements is $3\pi/8$ (Figure 14). This angle has the same reference angle and therefore the same sine function value as $5\pi/8$ (recall that the sine function has positive values for both the first and second quadrants). Thus,

$$\sin^{-1}\left(\sin \frac{5\pi}{8}\right) = \frac{3\pi}{8}.$$

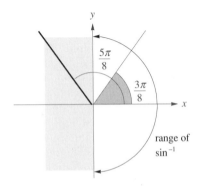

FIGURE 14

The inverse tangent function is defined by restricting the tangent function to the interval $(-\pi/2, \pi/2)$ and then considering the inverse of this restricted function (Figure 15).

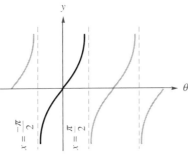

$$y = \tan \theta, \frac{-\pi}{2} < \theta < \frac{\pi}{2}$$

FIGURE 15

Inverse Tangent Function

The **inverse tangent function,** denoted **tan⁻¹,** is defined by the following relation:

$$y = \tan^{-1}x \quad \Leftrightarrow \quad x = \tan y \text{ and } -\frac{\pi}{2} < y < \frac{\pi}{2}.$$

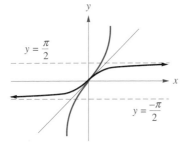

Domain: $(-\infty, \infty)$ Range: $\left(-\frac{\pi}{2}, \frac{\pi}{2}\right)$

EXAMPLE 6   Determine the exact value:

a) $\tan^{-1}(-\sqrt{3})$

b) $\arctan(1)$.

SOLUTION

a) By the definition, $\theta = \tan^{-1}(-\sqrt{3})$ if and only if $\tan\theta = -\sqrt{3}$ and $-\pi/2 < \theta < \pi/2$. From our experience with the tangent function, we deduce that $\tan^{-1}(-\sqrt{3}) = -\pi/3$.

b) Again, $\theta = \arctan(1)$ if and only if $\tan \theta = 1$ and $-\pi/2 < \theta < \pi/2$. Thus, $\arctan(1) = \pi/4$.

Expressions such as $\tan(\sin^{-1}(\frac{4}{7}))$ or $\sin^{-1}(\cos(3\pi/5))$ often occur in calculus. These generally can be evaluated exactly without a calculator.

EXAMPLE 7   Determine the exact value:

a) $\tan(\sin^{-1}(\frac{4}{7}))$ **b)** $\sin^{-1}(\cos(3\pi/5))$

SOLUTION

a) First, let $\theta = \sin^{-1}(\frac{4}{7})$, the angle whose sine is $\frac{4}{7}$. This reduces the problem to finding $\tan \theta$, where $\sin \theta = \frac{4}{7}$ and $-\pi/2 \le \theta \le \pi/2$. This implies that the terminal side of θ is in the first quadrant, and a point P is on the terminal side such that the y-coordinate of P is 4 (Figure 16). The distance from the origin to P is 7 (recall that $\sin \theta = y/r$). We can compute the x-coordinate of P:

$$x = \sqrt{7^2 - 4^2} = \sqrt{33}.$$

From the figure, we can determine that the $\tan \theta$ is $4/\sqrt{33}$. So,

$$\tan\left(\sin^{-1}\left(\frac{4}{7}\right)\right) = \frac{4}{\sqrt{33}}.$$

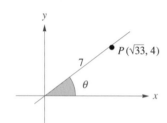

FIGURE 16

b) Recall that $\cos \theta = \sin(\pi/2 - \theta)$. We can use this to rewrite $\cos(3\pi/5)$ as $\sin(\pi/2 - 3\pi/5)$ or $\sin(-\pi/10)$. So,

$$\sin^{-1}\left(\cos\left(\frac{3\pi}{5}\right)\right) = \sin^{-1}\left(\sin\left(-\frac{\pi}{10}\right)\right) = -\frac{\pi}{10}.$$

The identities of Section 6.2 and Section 6.3 are often useful for simplifying expressions.

EXAMPLE 8   Determine the exact value:

a) $\tan[2 \tan^{-1}(-\frac{1}{3})]$ **b)** $\cos[\pi/6 + \sin^{-1}(5/13)]$

SOLUTION

a) First, let $\theta = \tan^{-1}(-\frac{1}{3})$, the angle whose tangent is $-\frac{1}{3}$. So, using the double-angle identity for the tangent function, we get

$$\tan[2\tan^{-1}(-\tfrac{1}{3})] = \tan 2\theta$$

$$= \frac{2\tan\theta}{1 - \tan^2\theta}$$

$$= \frac{2(-\frac{1}{3})}{1 - (-\frac{1}{3})^2} = -\frac{3}{4}.$$

Thus, $\tan[2\tan^{-1}(-\frac{1}{3})] = -\frac{3}{4}$.

b) First, let $\theta = \sin^{-1}(\frac{5}{13})$, the angle whose sine is $\frac{5}{13}$. Since the sine of θ is positive, it should seem reasonable that θ is in the interval $(0, \pi/2)$. This reduces the expression in question to $\cos(\pi/6 + \theta)$. Using the addition identity for the cosine function, we obtain

$$\cos\left[\frac{\pi}{6} + \sin^{-1}\left(\frac{5}{13}\right)\right] = \cos\left(\frac{\pi}{6} + \theta\right)$$

$$= \cos\frac{\pi}{6}\cos\theta - \sin\frac{\pi}{6}\sin\theta$$

$$= \left(\frac{\sqrt{3}}{2}\right)\cos\theta - \left(\frac{1}{2}\right)\sin\theta.$$

From the sketch of θ in Figure 17, we find that $\cos\theta = \frac{12}{13}$. So,

$$\cos\left[\frac{\pi}{6} + \sin^{-1}\left(\frac{5}{13}\right)\right] = \left(\frac{\sqrt{3}}{2}\right)\left(\frac{12}{13}\right) - \left(\frac{1}{2}\right)\left(\frac{5}{13}\right)$$

$$= \frac{12\sqrt{3} - 5}{26}.$$

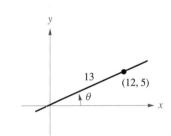

FIGURE 17

Thus,

$$\cos\left[\frac{\pi}{6} + \sin^{-1}\left(\frac{5}{13}\right)\right] = \frac{12\sqrt{3} - 5}{26}.$$

On certain occasions, the composition or the sum of trigonometric and inverse trigonometric functions can be simplified to a simpler algebraic function. The next two examples demonstrate how this can be done.

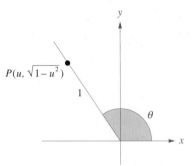

$P(u, \sqrt{1-u^2})$

1

θ

FIGURE 18

EXAMPLE 9   Simplify the function $f(u) = \sin(\cos^{-1}u)$.

SOLUTION   Let $\theta = \cos^{-1}u$. Then $\cos\theta = u$ with $0 \le \theta \le \pi$, and $f(u) = \sin\theta$. Now, think of θ as an angle with its terminal side in either the first or second quadrant (Figure 18 shows the terminal side in the second quadrant). The terminal side passes through the point P with x-coordinate u and a distance 1 from the origin. Its y-coordinate therefore is $\sqrt{1-u^2}$; the y-coordinate is positive since the point is in the first or second quadrant. From the figure we see that $\sin\theta = \sqrt{1-u^2}$, so $f(u) = \sqrt{1-u^2}$.

EXAMPLE 10   Simplify the expression

$$\sin^{-1}x + \cos^{-1}x$$

(this expression arises frequently in calculus).

SOLUTION   Let $\theta = \sin^{-1}x$, which in turn, implies that

$$\sin\theta = x \quad \text{and} \quad -\frac{\pi}{2} \le \theta \le \frac{\pi}{2}.$$

By the cofunction identity,

$$x = \sin\theta = \cos\left(\frac{\pi}{2} - \theta\right)$$

and it follows from $-\pi/2 \le \theta \le \pi/2$ that $0 \le (\pi/2 - \theta) \le \pi$, so

$$\cos^{-1}x = \frac{\pi}{2} - \theta.$$

Thus,

$$\sin^{-1}x + \cos^{-1}x = \theta + \left(\frac{\pi}{2} - \theta\right)$$

or

$$\sin^{-1}x + \cos^{-1}x = \frac{\pi}{2}.$$

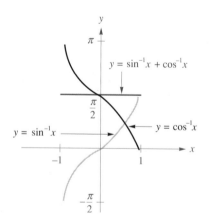

$y = \sin^{-1}x + \cos^{-1}x$

$y = \sin^{-1}x$

$y = \cos^{-1}x$

FIGURE 19

Sketching the graph (Figure 19) of $y = \sin^{-1}x + \cos^{-1}x$ suggests this identity; however it does not prove it. Our discussion above is the actual verification for this identity.

Just as the functions cot θ, sec θ, and csc θ play a lesser role in the discussion of trigonometric functions, their inverses $\cot^{-1}x$, $\sec^{-1}x$, and $\csc^{-1}x$ do as well. They are summarized here.

Inverse Functions of Cotangent, Secant, and Cosecant

Inverse Cotangent Function

$$y = \cot^{-1}x \quad \Leftrightarrow \quad x = \cot y \text{ and } 0 < y < \pi.$$

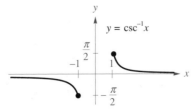

$y = \pi$

$y = \cot^{-1}x$ $(0, \frac{\pi}{2})$

Domain: $(-\infty, \infty)$ Range: $(0, \pi)$

Inverse Secant Function

$$y = \sec^{-1}x \quad \Leftrightarrow \quad x = \sec y \text{ and either } 0 \le y < \frac{\pi}{2} \text{ or } \frac{\pi}{2} < y \le \pi.$$

$(-1, \pi)$ $y = \sec^{-1}x$

$y = \frac{\pi}{2}$

-1 $(1, 0)$

Domain: $(-\infty, -1]$ or $[1, \infty)$ Range: $[0, \pi/2)$ or $(\pi/2, \pi]$

Inverse Cosecant Function

$$y = \csc^{-1}x \Leftrightarrow x = \csc y \text{ and either } -\frac{\pi}{2} \le y < 0 \text{ or } 0 < y \le \frac{\pi}{2}.$$

$y = \csc^{-1}x$

$\frac{\pi}{2}$

-1 1

$-\frac{\pi}{2}$

Domain: $(-\infty, -1]$ or $[1, \infty)$ Range: $[-\pi/2, 0)$ or $(0, \pi/2]$

Using Your Calculator to Approximate the Values of Inverse Trigonometric Functions

Just as with trigonometric functions, the exact evaluation of the inverse trigonometric functions is impossible for almost all values in their domains. However, in most practical applications an approximation to some degree of precision is all that it is required. This can be done easily with a hand-held calculator.

Suppose, for example, we wish to approximate the value of $\tan^{-1}(1.3)$ in radians. Your calculator should be set in radian mode. The keystrokes and output are

KEYSTROKES   **DISPLAY**

| 1.3 | inv | tan | | 0.9151007 |

The keystrokes involved may vary depending on the calculator, but most calculators are similar. You may have a button marked *2nd* or *arc* instead of *inv*.

On the other hand, suppose that your calculator is set in degree mode. The same keystrokes yield an output of 52.43140797. This is actually the same quantity found before: 0.9151007 radians is equivalent to 52.43140797 degrees.

Using your calculator to approximate the values of $\cot^{-1}$, $\sec^{-1}$, or $\csc^{-1}$ is a bit more involved. Suppose that we wish to determine the value of $\sec^{-1}(2.6)$ in radians. Since there typically is no secant button on a calculator, we need to exploit the relationship between the secant function and the cosine function. Recall that the secant of a number is the same as the cosine of its reciprocal. In essence, we seek the number in the interval $[0, \pi]$ whose secant is 2.6. Equivalently, its cosine is the reciprocal of 2.6, namely, 1/2.6. This line of reasoning suggests these keystrokes:

KEYSTROKES   **DISPLAY**

| 2.6 | 1/x | inv | cos | | 1.176005207 |

EXERCISE SET 6.4

A

1. a) Find all values of θ such that $\sin \theta = \frac{1}{2}$.

  b) Evaluate $\sin^{-1}(\frac{1}{2})$.

2. a) Find all values of θ such that $\cos \theta = \frac{1}{2}$.

  b) Evaluate $\cos^{-1}(\frac{1}{2})$.

3. a) Find all values of θ such that $\tan \theta = \dfrac{1}{\sqrt{3}}$.

  b) Evaluate $\tan^{-1}\left(\dfrac{1}{\sqrt{3}}\right)$.

4. a) Find all values of θ such that $\sin \theta = 1$.

  b) Evaluate $\sin^{-1}(1)$.

5. a) Find all values of θ such that $\cos \theta = -\dfrac{1}{\sqrt{2}}$.

  b) Evaluate $\cos^{-1}\left(-\dfrac{1}{\sqrt{2}}\right)$.

6. a) Find all values of θ such that $\tan \theta = 0$.

  b) Evaluate $\tan^{-1}(0)$.

In Problems 7 through 18, evaluate the given expression without using a calculator. If the expression is meaningless, explain why.

7. $\sin^{-1}(0)$

8. $\cos^{-1}(1)$

9. $\arccos\left(-\dfrac{\sqrt{3}}{2}\right)$

10. $\arcsin\left(-\dfrac{\sqrt{3}}{2}\right)$

11. $\tan^{-1}(\sqrt{3})$

12. $\tan^{-1}\left(-\dfrac{1}{\sqrt{3}}\right)$

13. $\operatorname{arccot} 1$

14. $\arctan(-1)$

15. $\cos^{-1}(\sqrt{3})$

16. $\sin^{-1}(-\sqrt{2})$

17. $\csc^{-1}(-1)$

18. $\operatorname{arcsec}\left(-\dfrac{2\sqrt{3}}{3}\right)$

In Problems 19 through 30, approximate (to the nearest 0.01 radian) the given expression using a calculator.

19. $\sin^{-1}(0.24)$

20. $\cos^{-1}(0.57)$

21. $\arccos(-0.56)$

22. $\arcsin(-0.91)$

23. $\arctan(2.1)$

24. $\tan^{-1}(-0.32)$

25. $\tan^{-1}(500)$

26. $\tan^{-1}(-480)$

27. $\sin^{-1}(3.4)$

28. $\cos^{-1}(4.1)$

29. $\csc^{-1}(1.3)$

30. $\operatorname{arcsec}(-1.3)$

B

In Problems 31 through 39, evaluate the given expression without using a calculator. It will be helpful to sketch an angle in standard position.

31. a) $\sin^{-1}\left(\sin\dfrac{\pi}{5}\right)$

32. a) $\cos^{-1}\left(\cos\dfrac{2\pi}{7}\right)$

b) $\sin^{-1}\left(\sin\dfrac{4\pi}{5}\right)$

b) $\cos^{-1}\left(\cos\left(-\dfrac{\pi}{7}\right)\right)$

33. a) $\tan^{-1}\left(\tan\dfrac{2\pi}{9}\right)$

34. a) $\cot^{-1}\left(\cot\dfrac{\pi}{5}\right)$

b) $\cot^{-1}[\cot(-2)]$

b) $\tan^{-1}(\tan 4)$

35. a) $\sin[\cos^{-1}(\frac{3}{5})]$

36. a) $\sec[\cos^{-1}(\frac{6}{11})]$

b) $\cos[\sin^{-1}(\frac{5}{13})]$

b) $\csc[\sin^{-1}(\frac{12}{17})]$

37. a) $\cot[\tan^{-1}(-\frac{2}{7})]$

38. a) $\csc[\cos^{-1}(\frac{12}{13})]$

b) $\tan[\cot^{-1}(\frac{9}{14})]$

b) $\sec[\sin^{-1}(\frac{8}{17})]$

39. a) $\tan[\cos^{-1}(\frac{5}{13})]$

b) $\tan[\sin^{-1}(\frac{15}{17})]$

In Problems 40 through 45, evaluate the given expression using the identities of Sections 6.2 and 6.3.

40. $\cos\left[\dfrac{\pi}{3} - \sin^{-1}\left(\dfrac{1}{3}\right)\right]$

41. $\sin\left[\dfrac{\pi}{4} + \cos^{-1}\left(\dfrac{2}{7}\right)\right]$

42. $\cos[2\cos^{-1}(-\frac{3}{5})]$

43. $\sin[2\sin^{-1}(\frac{7}{25})]$

44. $\tan[\frac{1}{2}\sin^{-1}(\frac{3}{5})]$

45. $\sin[\frac{1}{2}\cos^{-1}(\frac{2}{7})]$

In Problems 46 through 57, sketch the graph of the function. Use the techniques of translation, reflection, expansion, and compression.

46. $y = 2\cos^{-1}x$

47. $y = 3\sin^{-1}x$

48. $y = \sin^{-1}2x$

49. $y = \cos^{-1}\left(\dfrac{x}{2}\right)$

50. $y = \dfrac{\pi}{2} - \cos^{-1}x$

51. $y = \dfrac{\pi}{2} - \sin^{-1}x$

52. $y = \sin^{-1}(x - 1)$

53. $y = \cos^{-1}(x + 1)$

54. $y = 2\arctan x$

55. $y = 2\operatorname{arccot} x$

56. $y = \dfrac{\pi}{2} - \cot^{-1}x$

57. $y = \dfrac{\pi}{2} - \tan^{-1}x$

In Problems 58 through 63, write an algebraic expression for the given trigonometric expression.

58. $\sin(\cos^{-1}2x)$

59. $\cos(\sin^{-1}\sqrt{u})$

60. $\sin(\csc^{-1}2n)$

61. $\cos(\sec^{-1}\frac{2}{3}n)$

62. $\tan(\sin^{-1}5t)$

63. $\cot(\cos^{-1}8t)$

64. Sketch $y = \tan^{-1}x + \cot^{-1}x$. Determine an identity suggested by the graph and prove it. *(Hint: See Example 10.)*

65. Sketch $y = \sec^{-1}x + \csc^{-1}x$. Determine an identity suggested by the graph and prove it. *(Hint: See Example 10.)*

66. Use your calculator to approximate the value of $\cot^{-1}(-3.1)$ to the nearest 0.0001 radian. *(Hint: Be sure that your answer is in the range for the inverse cotangent function.)*

67. a) Explain why $0 < \tan^{-1}\left(\dfrac{1}{3}\right) + \tan^{-1}\left(\dfrac{1}{2}\right) < \dfrac{\pi}{2}$.

b) Show that $\tan[\tan^{-1}(\frac{1}{3}) + \tan^{-1}(\frac{1}{2})] = 1$.

c) Explain why parts a) and b) imply that $\tan^{-1}(\frac{1}{3}) + \tan^{-1}(\frac{1}{2}) = \pi/4$.

68. a) Explain why $0 < \tan^{-1}(\frac{1}{5}) + \tan^{-1}(\frac{2}{3}) < \pi/2$.

   b) Show that $\tan[\tan^{-1}(\frac{1}{5}) + \tan^{-1}(\frac{2}{3})] = 1$.

   c) Explain why parts a) and b) imply that
$\tan^{-1}(\frac{1}{5}) + \tan^{-1}(\frac{2}{3}) = \pi/4$.

C

69. a) Evaluate $\tan^{-1}(x) + \tan^{-1}\left(\dfrac{1}{x}\right)$ for all $x > 0$.

   b) Evaluate $\tan^{-1}(x) + \tan^{-1}\left(\dfrac{1}{x}\right)$ for all $x < 0$.

70. Sketch the graphs of $y = \sin(\sin^{-1}x)$ and $y = \sin^{-1}(\sin x)$. (Hint: Be sure to consider the domain and range of each.)

71. Sketch the graphs of $y = \cos(\cos^{-1}x)$ and $y = \cos^{-1}(\cos x)$. (Hint: Be sure to consider the domain and range of each.)

72. a) Show that $\tan 4\theta = \dfrac{4 \tan \theta - 4 \tan^3\theta}{1 - 6 \tan^2\theta + \tan^4\theta}$.

   b) Evaluate $\tan[4 \tan^{-1}(\frac{1}{5})]$.

73. Show that $\pi/4 = 4 \tan^{-1}(\frac{1}{5}) - \tan^{-1}(\frac{1}{239})$. *(Hint: Use Problem 72.)*

SECTION 6.5   TRIGONOMETRIC EQUATIONS

Not all equations involving trigonometric functions are identities. Those equations that are true for only selected values of the variable, as you recall from Chapter 1, are called conditional equations. In this section, we will investigate methods of solving conditional trigonometric equations.

    One of the differences between the solutions of algebraic equations and trigonometric equations is that trigonometric equations generally have an infinite number of real-number solutions. Quite often, however the context of the problem requires only solutions from a specific domain, such as the interval $[0, 2\pi)$ or $[-\pi, \pi]$.

EXAMPLE 1   Solve the equation $2 \sin\left(x + \dfrac{\pi}{6}\right) = -1$

   a)   over the set of all real numbers.

   b)   over the interval $[0, 2\pi)$

   c)   over the interval $[-\pi, \pi]$.

SOLUTION

   a)   The plan here is first to find the value of $\sin(x + \pi/6)$, to use this value to determine all possible values of $(x + \pi/6)$, and finally to use this to solve for all possible values of x.

         Dividing both sides of the equation by 2 yields

$$\sin\left(x + \frac{\pi}{6}\right) = -\frac{1}{2}.$$

From Example 1 of Section 6.4, we saw that the solutions for $\sin \theta = -\frac{1}{2}$ are

$$\theta = \frac{7\pi}{6} \pm 2\pi n \quad \text{or} \quad \theta = \frac{11\pi}{6} \pm 2\pi n \qquad (n = 0, 1, 2, 3, \ldots).$$

So,

$$x + \frac{\pi}{6} = \frac{7\pi}{6} \pm 2\pi n \quad \text{or} \quad x + \frac{\pi}{6} = \frac{11\pi}{6} \pm 2\pi n \qquad (n = 0, 1, 2, 3, \ldots).$$

Subtracting $\pi/6$ from both sides of each equation, we get

$$x = \pi \pm 2\pi n \quad \text{or} \quad x = \frac{5\pi}{3} \pm 2\pi n \qquad (n = 0, 1, 2, 3, \ldots)$$

or

$$\{\ldots, -3\pi, -\pi, \pi, 3\pi, 5\pi, \ldots\} \text{ or } \left\{\ldots, -\frac{7\pi}{3}, -\frac{\pi}{3}, \frac{5\pi}{3}, \frac{11\pi}{3}, \frac{17\pi}{3}, \ldots\right\}.$$

b) From the results of part a), solutions to the equation in the interval $[0, 2\pi)$ are

$$x = \pi, \frac{5\pi}{3}.$$

c) Again, from the results of part a), solutions to the equation in the interval $[-\pi, \pi]$ are

$$x = -\pi, -\frac{\pi}{3}, \pi.$$

EXAMPLE 2   Solve the equation $3 \tan^2 x = 1$ over the interval $[0, 2\pi)$.

SOLUTION   Solving this equation for $\tan x$:

$$3 \tan^2 x = 1$$
$$\tan^2 x = \tfrac{1}{3}$$
$$\tan x = \pm \frac{1}{\sqrt{3}}$$
$$\tan x = \frac{1}{\sqrt{3}} \quad \text{or} \quad \tan x = -\frac{1}{\sqrt{3}}.$$

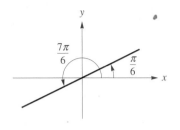

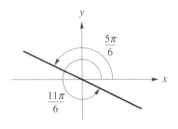

FIGURE 20 *Solutions to*
$\tan x = 1/\sqrt{3}$ *over* $[0, 2\pi)$
Solutions to
$\tan x = -1/\sqrt{3}$ *over* $[0, 2\pi)$

The reference angle for each of the solutions of these equations is $\pi/6$. From Figure 20, we get

$$\tan x = \frac{1}{\sqrt{3}} \Rightarrow x = \frac{\pi}{6}, \frac{7\pi}{6}$$

$$\tan x = -\frac{1}{\sqrt{3}} \Rightarrow x = \frac{5\pi}{6}, \frac{11\pi}{6}.$$

The solutions of $3 \tan^2 x = 1$ over the interval $[0, 2\pi)$ are $\pi/6$, $5\pi/6$, $7\pi/6$, and $11\pi/6$.

EXAMPLE 3   Solve the equation $\sqrt{2} \cos 3\theta - 1 = 0$ over the interval $[0, 2\pi)$.

SOLUTION Solving for $\cos 3\theta$ we get $\cos 3\theta = 1/\sqrt{2}$. Now, since we are seeking all the solutions θ such that $0 \leq \theta < 2\pi$, this means that the possible range for 3θ is such that $0 \leq 3\theta < 6\pi$. The reference angle for 3θ is $\pi/4$, so

$$3\theta = \underbrace{\frac{\pi}{4}, \frac{7\pi}{4}}_{0 \leq 3\theta \leq 2\pi}, \quad \underbrace{\frac{9\pi}{4}, \frac{15\pi}{4}}_{2\pi \leq 3\theta \leq 4\pi}, \quad \underbrace{\frac{17\pi}{4}, \frac{23\pi}{4}}_{4\pi \leq 3\theta \leq 6\pi}.$$

Dividing each of these by 3, we obtain these solutions to the original equation:

$$\theta = \frac{\pi}{12}, \frac{7\pi}{12}, \frac{3\pi}{4}, \frac{5\pi}{4}, \frac{17\pi}{12}, \frac{23\pi}{12}.$$

Many of the skills and techniques that are used to solve algebraic equations are also used to solve trigonometric equations.

EXAMPLE 4   Solve the equation $2 \cos^2 x - 3 \cos x = 2$ over the interval $[0, 2\pi)$.

SOLUTION The important intermediate step here is to find the value of $\cos x$. The equation is a quadratic equation in $\cos x$. Solving this equation for $\cos x$ is equivalent to solving the equation $2u^2 - 3u = 2$ for u. This equation can be solved by factoring:

$$2 \cos^2 x - 3 \cos x = 2$$

$$2 \cos^2 x - 3 \cos x - 2 = 0$$

$$(\cos x - 2)(2 \cos x + 1) = 0$$

$$\cos x - 2 = 0 \quad \text{or} \quad 2 \cos x + 1 = 0$$

$$\cos x = 2 \qquad\qquad \cos x = -\tfrac{1}{2}.$$

Compare the solution with the solution of this equivalent equation:

$$2u^2 - 3u = 2$$

$$2u^2 - 3u - 2 = 0$$

$$(u - 2)(2u + 1) = 0$$

$$u - 2 = 0 \quad \text{or} \quad 2u + 1 = 0$$

$$u = 2 \qquad\qquad u = \tfrac{1}{2}.$$

Consider these cases one at a time, starting with $\cos x = 2$. Since the range of the cosine function is $[-1, 1]$, the equation $\cos x = 2$ has no real solutions.

On the other hand, the equation $\cos x = -\frac{1}{2}$ has a solution in the second quadrant and one in the third quadrant. Each of these solutions has a reference angle of $\pi/3$ (Figure 21).

Thus, $x = 2\pi/3$ and $4\pi/3$ are the only solutions to the equation in the interval $[0, 2\pi)$. Below is the check for the first solution (we leave the second for you):

$$2 \cos^2 \left(\frac{2\pi}{3}\right) - 3 \cos\left(\frac{2\pi}{3}\right) \stackrel{?}{=} 2$$

$$2\left(-\tfrac{1}{2}\right)^2 - 3\left(-\tfrac{1}{2}\right) \stackrel{?}{=} 2$$

$$\tfrac{1}{2} - \left(-\tfrac{3}{2}\right) = 2.$$

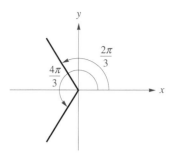

FIGURE 21

The solutions of many trigonometric equations rely on the trigonometric identities of this chapter. Example 5 shows how these identities can be used to solve these types of equations.

EXAMPLE 5   Solve the equation $-2 \sin t = \tan t$ over the interval $[-\pi, \pi]$.

SOLUTION Our solution depends on using a quotient identity to rewrite $\tan t$ in terms of $\sin t$ and $\cos t$:

$$-2 \sin t = \frac{\sin t}{\cos t}$$

or

$$-2 \sin t \cos t = \sin t.$$

Next, we subtract $\sin t$ from both sides and factor:

$$-2 \sin t \cos t - \sin t = 0$$

$$-\sin t(2 \cos t + 1) = 0$$

$$\sin t = 0 \qquad \text{or} \quad 2 \cos t - 1 = 0$$

$$t = -\pi, 0, \pi \qquad\qquad \cos t = \tfrac{1}{2}$$

$$t = -\frac{\pi}{3}, \frac{\pi}{3}.$$

The solutions are $t = -\pi, -\pi/3, 0, \pi/3, \pi$.

Quite often, trigonometric equations involve linear combinations of sine and cosine functions. The next example shows how we can exploit the linear combination identity of Section 6.2 to solve these types of equations.

EXAMPLE 6   Solve the equation $\cos 2t - \sqrt{3} \sin 2t = 1$ over the set of real numbers.

SOLUTION From Example 5 of Section 6.2, we know that the linear combination $\cos 2t - \sqrt{3} \sin 2t$ can be written as $2 \cos(2t + \pi/3)$. So, the equation becomes

$$2 \cos\left(2t + \frac{\pi}{3} \right) = 1$$

$$\cos\left(2t + \frac{\pi}{3} \right) = \frac{1}{2}.$$

Thus,

$$2t + \frac{\pi}{3} = \frac{\pi}{3} \pm 2\pi n \quad \text{or} \quad 2t + \frac{\pi}{3} = \frac{5\pi}{3} \pm 2\pi n$$

$$2t = \pm 2\pi n \quad \text{or} \quad 2t = \frac{4\pi}{3} \pm 2\pi n$$

$$t = \pm \pi n \quad \text{or} \quad t = \frac{2\pi}{3} \pm \pi n$$

or

$$\{\ldots, -\pi, 0, \pi, 2\pi, 3\pi, \ldots\} \quad \text{or} \quad \left\{\ldots, -\frac{\pi}{3}, \frac{2\pi}{3}, \frac{5\pi}{3}, \frac{8\pi}{3}, \frac{11\pi}{3}, \ldots\right\}.$$

The sum-to-product identities can also be used to solve equations involving certain linear combinations of sine and cosine functions

EXAMPLE 7   Solve the equation $\cos \theta - \cos 3\theta = 0$ over the interval $[0, 2\pi)$.

SOLUTION In part b) of Example 6 in Section 6.3, we showed that $\cos \theta - \cos 3\theta = 2 \sin 2\theta \sin \theta$, so the equation can be written

$$2 \sin 2\theta \sin \theta = 0$$

implying that either

$$\sin 2\theta = 0 \qquad \text{or} \quad \sin \theta = 0$$

$$2\theta = 0, \pi, 2\pi, 3\pi \qquad \theta = 0, \pi.$$

$$\theta = 0, \frac{\pi}{2}, \pi, \frac{3\pi}{2}$$

The solutions are $\theta = 0, \pi/2, \pi, 3\pi/2$.

The solutions of some trigonometric equations cannot be found exactly. In these cases, a hand-held calculator can be employed to approximate the solutions.

EXAMPLE 8 Solve the equation $\sec^2 t - 6 = 4 \tan t$ over the interval $[0°, 360°)$.

SOLUTION Replacing $\sec^2 t$ by $1 + \tan^2 t$ gives a quadratic equation in $\tan t$:

$$\sec^2 t - 6 = 4 \tan t$$

$$(1 + \tan^2 t) - 6 = 4 \tan t$$

$$\tan^2 t - 5 = 4 \tan t$$

$$\tan^2 t - 4 \tan t - 5 = 0$$

$$(\tan t + 1)(\tan t - 5) = 0$$

$$\tan t + 1 = 0 \qquad \text{or} \quad \tan t - 5 = 0$$

$$\tan t = -1 \quad \text{or} \qquad \tan t = 5.$$

The solutions of $\tan t = -1$ in the interval $[0, 360°)$ are $135°$ and $315°$. The best we can do with $\tan t = 5$, on the other hand, is state the solutions in terms of $\tan^{-1} 5$, namely, $\tan^{-1} 5$ and $180° + \tan^{-1} 5$ (Figure 22). To the nearest $1°$, these are $79°$ and $259°$. Thus, the solutions are $135°$, $79°$, $315°$, and $259°$.

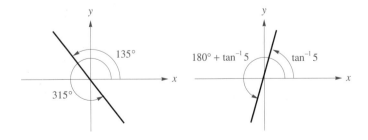

FIGURE 22 *Solutions of* $\tan t = -1$
Solutions of $\tan t = 5$

EXAMPLE 9   Solve the equation $\tan^2 x - 2 \tan x - 2 = 0$ over the interval $[-\pi/2, \pi/2]$. Approximate the solutions to the nearest 0.1 radians.

SOLUTION Solving this equation for $\tan x$ is equivalent to finding the solution of the equation $u^2 - 2u - 2 = 0$. This equation does not yield to solution by factoring (go ahead and try it), so we use the quadratic formula to find the values for $\tan x$:

$$\tan x = \frac{-(-2) \pm \sqrt{(-2)^2 - 4(1)(-2)}}{2(1)} = 1 \pm \sqrt{3}.$$

So,

$$\tan x = 1 + \sqrt{3} \Rightarrow x = \tan^{-1}(1 + \sqrt{3})$$
$$\tan x = 1 - \sqrt{3} \Rightarrow x = \tan^{-1}(1 - \sqrt{3}).$$

Using a calculator to approximate these to the nearest 0.1 radians, we get 1.2 and -0.6.

We finish the section with two applications, each requiring the solution of a trigonometric equation.

EXAMPLE 10   Find the x-intercepts of the graph of $y = \sin 2t + \cos t$ over the interval $[0, 2\pi]$ (the graph of this equation was discussed in Example 7 in Section 5.5).

SOLUTION To find the x-intercepts of a graph, we set the $y = 0$ and solve the resulting equation for x. Doing this, we obtain

$$\sin 2t + \cos t = 0.$$

We must be careful; notice that there is a trigonometric function of t and of $2t$. However, using the double-angle identity $\sin 2t = 2 \sin t \cos t$ reduces this equation to

$$2 \sin t \cos t + \cos t = 0.$$

Solving,

$$\cos t(2 \sin t + 1) = 0$$

$$\cos t = 0 \qquad \text{or} \quad 2 \sin t + 1 = 0$$

$$t = \frac{\pi}{2}, \frac{3\pi}{2} \qquad\qquad\qquad t = \frac{7\pi}{6}, \frac{11\pi}{6}.$$

The t-intercepts of the graph (Figure 23) are $(\pi/2, 0)$, $(7\pi/6, 0)$, and $(11\pi/6, 0)$.

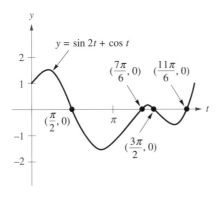

FIGURE 23

EXAMPLE 11   The temperatures on a particular day in Santa Rosa, California, are given by the function

$$T(t) = 14 \cos\left[\frac{\pi}{12}(t - 16)\right] + 34.$$

(See Example 6 of Section 5.5.) During what time interval is the temperature at least 40°?

SOLUTION The graph of $y = T(t)$ is in Figure 24. The problem reduces to finding for what values of t is the graph above or on the line $y = 40°$. The points of intersection can be found by solving the equation over the interval $[0, 24]$:

$$14 \cos\left[\frac{\pi}{12}(t - 16)\right] + 34 = 40.$$

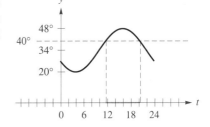

FIGURE 24

Solving for the cosine function gives

$$\cos\left[\frac{\pi}{12}(t - 16)\right] = \frac{3}{7}.$$

From Figure 25, it should seem reasonable that there are two solutions to the equation $\cos\theta = \frac{3}{7}$, namely, $\cos^{-1}(\frac{3}{7})$ and $-\cos^{-1}(\frac{3}{7})$:

$$\frac{\pi}{12}(t - 16) = \cos^{-1}\left(\frac{3}{7}\right)$$

or

$$\frac{\pi}{12}(t - 16) = -\cos^{-1}\left(\frac{3}{7}\right).$$

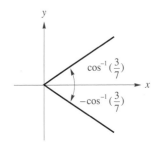

FIGURE 25 *Solutions of* $\cos\theta = \frac{3}{7}$

Solving the first for t, we get

$$\frac{\pi}{12}(t - 16) = \cos^{-1}\left(\frac{3}{7}\right)$$

$$t - 16 = \frac{12}{\pi}\left[\cos^{-1}\left(\frac{3}{7}\right)\right]$$

$$t = \frac{12}{\pi}\left[\cos^{-1}\left(\frac{3}{7}\right)\right] + 16 \approx 20.3.$$

Solving the second equation, we obtain

$$\frac{\pi}{12}(t - 16) = -\cos^{-1}\left(\frac{3}{7}\right)$$

$$t - 16 = \frac{12}{\pi}\left[-\cos^{-1}\left(\frac{3}{7}\right)\right]$$

$$t = -\frac{12}{\pi}\left[\cos^{-1}\left(\frac{3}{7}\right)\right] + 16 \approx 11.7.$$

The temperature is at least $40°$ from approximately 11:42 A.M. to approximately 8:18 P.M.

EXERCISE SET 6.5

A

In Problems 1 through 6, solve the equation exactly over the given interval.

1. **a)** $2 \sin x = \sqrt{3}$, $(-\infty, +\infty)$
   b) $2 \sin x = \sqrt{3}$, $[0, 2\pi)$

2. **a)** $\tan x = 1$, $(-\infty, +\infty)$
   b) $\tan x = 1$, $[-\pi, \pi]$

3. **a)** $2 \cos x = 1$, $[0, 2\pi)$
   b) $2 \cos x = -1$, $[0, 2\pi)$

4. **a)** $\tan x = 1$, $[-\pi, \pi]$
   b) $\tan x = -1$, $[-\pi, \pi]$

5. **a)** $\sqrt{2} \sec x = -2$, $[0, 2\pi)$
   b) $\sqrt{2} \sec x = 2$, $[0, 2\pi)$

6. **a)** $\cot x = \sqrt{3}$, $[-\pi, \pi]$
   b) $\cot x = -\sqrt{3}$, $[-\pi, \pi]$

In Problems 7 through 18, solve the equation exactly over the interval $[0, 2\pi)$.

7. $6 \sec x - 4\sqrt{3} = 0$ 8. $4 \csc x + 8 = 0$

9. $2 \sin^2 x - 1 = 0$

10. $4 \cos^2 x - 1 = 0$

11. $(\tan x - 1) \sin x = 0$

12. $(\cos x - 1)(2 \sin x - 1) = 0$

13. $2 \sin^2 x - \sin x = 0$ 14. $\cos^2 x + 2 \cos x = 0$

15. $6 - 2 \tan^2 x = 0$ 16. $5 - 15 \cot^2 x = 0$

17. $3 \sin x - 5 = 1$ 18. $4 \csc x + 2 = 0$

B

In Problems 19 through 30, solve the equation exactly over the given interval.

19. $2 \sin\left(x + \frac{\pi}{6}\right) = 1$, $[0, 2\pi)$

20. $\cos\left(x - \frac{\pi}{4}\right) = -\frac{1}{\sqrt{2}}$, $[-\pi, \pi]$

21. $\cos\left(x + \frac{\pi}{3}\right) = -1$, $[-\pi, \pi]$

22. $\sin\left(x - \frac{2\pi}{3}\right) = 0$, $[0, 2\pi)$

23. $\tan\left(x - \frac{\pi}{3}\right) = -1$, $[-\pi, \pi]$

24. $\sqrt{3}\cot\left(x + \frac{\pi}{6}\right) = -1$, $[0, 2\pi)$

25. $2\sin 2x = 1$, $[0, 2\pi)$

26. $4\cos 4x + 2 = 0$, $[0, \pi]$

27. $\tan 3x = -\sqrt{3}$, $[0, \pi]$

28. $\cot 3x = -1$, $[\pi, 2\pi)$

29. $\sqrt{2}\sec 2x = 2$, $[-\pi, \pi]$

30. $4\csc 3x - 8 = 0$, $[0, \pi]$

In Problems 31 through 48, solve the equation over the given interval. You may wish first to review the identities of this chapter. Give the solution as an exact value when possible (otherwise, to the nearest 0.1 radian).

31. $2\cos^2 x = 3\sin x$, $[0, 2\pi)$

32. $2\tan^2 x - \cos^2 x = 1 + \sin^2 x$, $[-\pi, \pi]$

33. $\tan^2 x = \frac{\sin x}{1 - \sin x}$, $[0, 2\pi)$

34. $\tan x - \frac{3}{\tan x} = 0$, $[-\pi, \pi]$

35. $2\cot^2 x - 2\cot^2 x \cos^2 x = 1$, $[0, 2\pi)$

36. $2\csc^2 x - 2\cot^2 x \cos^2 x = 1$, $[0, 2\pi)$

37. $\cos 2t = 2\cos t$, $[-\pi, \pi]$

38. $6\cos^2\phi - 5\cos\phi + 1 = 0$, $[0, 2\pi)$

39. $12\cos y + 5 = 2\sec y$, $[-\pi, \pi]$

40. $2\tan x \sin x - \sec x = 2\cos x$ $[-\pi, \pi]$

41. $4\cos t \sin t = 1$, $[0, \pi]$

42. $6\cos^2 u - \sin^2 u = 1$, $[0, 2\pi)$

43. $4\tan^3 x - 4\tan^2 x + \tan x = 0$, $\left[-\frac{\pi}{2}, \frac{\pi}{2}\right]$

44. $\tan^2 x = \frac{\sin x}{1 + \sin x}$, $[0, 2\pi)$

45. $\sin x - \sqrt{3}\cos x = 1$, $[0, 2\pi)$

46. $\tan\theta + \sec\theta = 0$, $[-\pi, \pi]$

47. $\cos 3x + \cos 5x = 0$, $\left[-\frac{\pi}{2}, \frac{\pi}{2}\right]$

48. $\sin 3x + \sin 5x = 0$, $[0, \pi]$

49. Solve the equation $\ln(\sin t) - \ln(\cos t) = 0$ over the interval $[-\pi/2, \pi/2]$. (Be sure to check your prospective solutions.)

50. Solve the equation $\log_2(\tan^2 t) + \log_2(\cos^2 t) = -2$ over the interval $[0, 2\pi)$. (Be sure to check your prospective solutions.)

51. In Example 4 of Section 5.5, the graph of $y - 1 = 2\sin 3(x - \pi/3)$ over the interval $-2\pi \leq x \leq 2\pi$ was discussed. Determine the coordinates of the x-intercepts by solving an appropriate equation.

52. In Example 5 of Section 5.5, the graph of $y = 1 - 2\cos(-x + \frac{1}{2}\pi)$ over the interval $-2\pi \leq x \leq 3\pi$ was discussed. Determine the x-intercepts by solving an appropriate equation. Copy the graph and label the x-intercepts.

C

53. The t-coordinates of the turning points of the graph of $y = \sin 2t + \cos t$ are the solutions of the equation $2\cos 2t - \sin t = 0$. Solve this equation over the interval $[0, 2\pi)$ to approximate the coordinates of the turning points of the graph to the nearest 0.001. Copy the graph and label the turning points (see Example 9 of this section and Example 7 of Section 5.5).

54. The x-coordinates of the turning points of the graph of $y = e^{-x}\sin 2x$ are the solutions of the equation $e^{-x}(2\cos 2x - \sin 2x) = 0$. Solve this equation over the interval $[0, 2\pi)$ to approximate the coordinates of the turning points of the graph to the nearest 0.001. Sketch the graph of $y = e^{-x}\sin 2x$ and label the turning points (you may wish to review Section 5.5).

55. The x-coordinates of the turning points of the graph of $y = e^{-x}\cos 2x$ are the solutions of the equation $-e^{-x}(\cos 2x + 2\sin 2x) = 0$. Solve this equation over the interval $[0, 2\pi)$ to approximate the coordinates of the turning points of the graph to the nearest 0.001. Sketch the graph of $y = e^{-x}\cos 2x$ and label the turning points (you may wish to review Section 5.5).

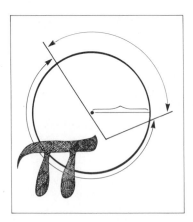

The Story of π

The story of π is truly an enigma wrapped in mystery. By definition, the value of this real-number constant is the ratio of the circumference of a circle to its diameter. The use of the symbol π was popularized by the great mathematician Euler in the 18th century, but the idea behind the symbol goes back to antiquity. In spite of its long history, there have always been unanswered questions about this number.

Almost every society has invested effort in answering these questions. The Rhind Papyrus, written in Egypt in the 17th century B.C., gives $(\frac{16}{9})^2 \approx 3.1605$ as the value of π. The Bible refers to a 10th century B.C. approximation of π as three:

And he made a molten sea, ten cubits from one brim to another, it was round all about, . . . and a line of thirty cubits did compass it round about.

I KINGS 7:23

In the 5th century B.C., the Chinese mathematician Tsu Ch'ung-chih applied π using $\frac{355}{113} \approx 3.141592$. About the same time, Hindu mathematicians approximated π as $\frac{62832}{20000} \approx 3.1416$. Archimedes was able to determine that π is in the interval $(3\frac{10}{71}, 3\frac{1}{7})$ by measuring the perimeters of polygons; he may have been the first to suspect that this elusive value is irrational.

In the 17th century, the development of calculus yielded a handful of formulas that made possible the approximation of π to any degree of accuracy. For example, John Wallis, an English mathematician, found this approximation:

$$\pi \approx 2\left(\frac{2 \cdot 2 \cdot 4 \cdot 4 \cdot 6 \cdot 6 \cdot 8 \cdot 8}{1 \cdot 3 \cdot 3 \cdot 5 \cdot 5 \cdot 7 \cdot 7 \cdot 9}\right).$$

This approximation improves if the pattern suggested by this ratio is extended (try this on your calculator). Later, John Bernoulli was able to show that

$$\pi^2 \approx 6\left(\frac{1}{1^2} + \frac{1}{2^2} + \frac{1}{3^2} + \frac{1}{4^2} + \frac{1}{5^2}\right).$$

Again, continuing this pattern yields more accurate approximations.

In 1706, John Machin computed the value of π to 100 decimal places using this formula (see Problems 72 and 73 of Section 6.4):

$$\pi = 4[4 \tan^{-1}(\tfrac{1}{5}) - \tan^{-1}(\tfrac{1}{239})].$$

William Shanks used Machin's formula to compute over 700 places in 1829. This was a herculean task; Shanks did not have the advantage of modern computing machinery. Unfortunately, there was an error in the 528th decimal place of his computation, which invalidated all of the subsequent digits. Much of his labor was in vain. Presently, high-speed computers have been used to determine millions of digits of π

Throughout history, there have been well-intentioned attempts to simplify this constant. In 1892, the New York *Tribune* published an article that declared the true value of π to be exactly 3.2. In 1897, the General Assembly of a midwestern state passed a bill to redefine the value of π. The bill incorrectly assumed that if the perimeter of a square is equal to the circumference of a circle, then the square and the circle have the same area. This implied that the value of π is four. Later, in the 1920s, another state legislature seriously discussed a bill to restore π to its biblical value of three.

What fascinates mathematicians about π is its frequent appearance in unexpected situations. For example, consider this experiment. Suppose that you drop a number of needles (or toothpicks) onto a lined surface such as a planked floor. The distance between the lines should be greater than the length of the needles. Next, count the number of needles that happen to land across one of the lines. Finally, compute this expression:

$$\frac{2\left(\begin{array}{c}\text{length}\\\text{of pin}\end{array}\right)\left(\begin{array}{c}\text{number of}\\\text{pins dropped}\end{array}\right)}{\left(\begin{array}{c}\text{distance}\\\text{between lines}\end{array}\right)\left(\begin{array}{c}\text{number of pins}\\\text{crossing line}\end{array}\right)}.$$

For a large number of needles, this solution is remarkably close to π!

MISCELLANEOUS EXERCISES

In Problems 1 through 12, simplify the expression to one of the following: $\sin\theta$, $\cos\theta$, $\tan\theta$, $\cot\theta$, $\sec\theta$, *or* $\csc\theta$.

1. $\sin\theta\tan\theta + \cos\theta$

2. $\tan^3\theta(\csc^2\theta - 1)$

3. $\dfrac{\sec\theta}{\sin\theta} - \dfrac{\sin\theta}{\cos\theta}$

4. $\sin\left(\dfrac{\pi}{2} - \theta\right)\csc\theta$

5. $\cos\left(\dfrac{\pi}{2} - \theta\right)\cot\theta$

6. $\cos\theta\csc\theta\left(\dfrac{\sec^2\theta}{\sin\theta} - \csc\theta\right)$

7. $\dfrac{\sin 2\theta}{2\cos\theta}$

8. $\dfrac{\cot\theta\cos\theta - \sin\theta}{\cos 2\theta}$

9. $\dfrac{\csc\theta(1+\cos\theta)(1-\cos\theta)}{\cos\theta}$

10. $\dfrac{\sin\theta\cot\theta + \cos\theta}{2\cot\theta}$

11. $\sin 2\theta\csc\theta - \cos\theta$

12. $\sin\theta + \cos\theta\cot\theta$

In Problems 13 through 42, prove the identity.

13. $\cos\theta(\sec\theta - \cos\theta) \stackrel{?}{=} \sin^2\theta$

14. $\tan^2\theta\csc^2\theta - 1 \stackrel{?}{=} \sin^2\theta\sec^2\theta$

15. $\csc^2\theta + \cot^2\theta + 1 \stackrel{?}{=} \dfrac{2}{\sin^2\theta}$

16. $\cos\theta\cot\theta + \sin\theta \stackrel{?}{=} \csc\theta$

17. $\dfrac{\sin\theta}{\sin\theta + \cos\theta} \stackrel{?}{=} \dfrac{\tan\theta}{\tan\theta + 1}$

18. $(1 + \cos^2\theta)\csc^2\theta \stackrel{?}{=} \csc^4\theta - \cot^4\theta$

19. $\dfrac{\sin\theta + \cos\theta}{\sin\theta - \cos\theta} \stackrel{?}{=} \dfrac{\tan\theta + 1}{\tan\theta - 1}$

20. $\dfrac{\sin^2\theta + 2}{\sin^2\theta - 2} \stackrel{?}{=} \dfrac{1 - 3\sec^2\theta}{1 + \sec^2\theta}$

21. $\cos^4\theta - \sin^4\theta \stackrel{?}{=} \cos 2\theta$

22. $(\sin\theta + \cos\theta)^2 + (\cot^2\theta - \csc^2\theta) \stackrel{?}{=} \sin 2\theta$

23. $\csc\theta + \cot\theta = \dfrac{\sin\theta}{1 - \cos\theta}$

24. $\tan\theta + \cot\theta \stackrel{?}{=} \sec\theta\csc\theta$

25. $\dfrac{\sin^3\theta + \cos^3\theta}{1 - \sin\theta\cos\theta} \stackrel{?}{=} \sin\theta + \cos\theta$

26. $\dfrac{\cot^4\theta - 1}{\cot\theta - 1} \stackrel{?}{=} (\cot\theta + 1)\csc^2\theta$

27. $\cos\left(\theta + \dfrac{3\pi}{2}\right) \stackrel{?}{=} \sin\theta$

28. $\tan(\theta - 5\pi) \stackrel{?}{=} \cot\left(\theta + \dfrac{3\pi}{2}\right)$

29. $\dfrac{\cos\theta + \sin\theta}{\cos\theta - \sin\theta} \stackrel{?}{=} \dfrac{\cos 2\theta}{1 - \sin 2\theta}$

30. $\csc\theta\cot\theta \stackrel{?}{=} (1 + \cot^2\theta)\cos\theta$

31. $\dfrac{3\cos\theta + \cos 3\theta}{3\sin\theta - \sin 3\theta} \stackrel{?}{=} \cot^3\theta$

32. $\cos^6\theta - \sin^6\theta \stackrel{?}{=} (1 - \sin^2\theta\cos^2\theta)\cos 2\theta$

33. $\dfrac{\sin(2\alpha - \beta) + \sin\beta}{\cos(2\alpha - \beta) + \cos\beta} \stackrel{?}{=} \tan\alpha$

34. $\dfrac{\cos\alpha - \cos\beta}{\cos\alpha + \cos\beta} \stackrel{?}{=} -\tan\left(\dfrac{\alpha - \beta}{2}\right)\tan\left(\dfrac{\alpha + \beta}{2}\right)$

35. $\tan\left(\dfrac{\pi}{4} - \theta\right) \stackrel{?}{=} \dfrac{\cot\theta - 1}{\cot\theta + 1}$

36. $\dfrac{\sin\left(\theta - \dfrac{\pi}{3}\right)}{\cos\left(\theta + \dfrac{\pi}{6}\right)} \stackrel{?}{=} -1$

37. $\dfrac{\cos\frac{1}{2}\theta + \sin\frac{1}{2}\theta}{\cos\frac{1}{2}\theta - \sin\frac{1}{2}\theta} \stackrel{?}{=} \dfrac{1 + \sin\theta}{\cos\theta}$

38. $\tan 2\theta - \tan\theta \stackrel{?}{=} \tan\theta\sec 2\theta$

39. $\sin^2 2\theta + 4\cos^4\theta \stackrel{?}{=} 4\cos^2\theta$

40. $\sin 4\theta \stackrel{?}{=} 4\cos^3\theta\sin\theta - 2\sin 2\theta\sin^2\theta$

41. $\cot\dfrac{\theta}{2} - \tan\dfrac{\theta}{2} \stackrel{?}{=} 2\cot\theta$

42. $\cot\dfrac{\theta}{2} + \tan\dfrac{\theta}{2} \stackrel{?}{=} 2\csc\theta$

In Problems 43 through 48, prove the identity given that α, β, *and* γ *are the interior angles of a triangle.* *(Hint:* $\alpha + \beta + \gamma = \pi$.)

43. $\tan\gamma \stackrel{?}{=} -\tan(\alpha + \beta)$

44. $\cos\gamma \stackrel{?}{=} -\cos(\alpha + \beta)$

45. $\sin\beta \stackrel{?}{=} -\cos(\frac{1}{2}\alpha + \frac{3}{2}\beta + \frac{1}{2}\gamma)$

46. $\sin\alpha \stackrel{?}{=} \cos(\frac{1}{2}\alpha - \frac{1}{2}\beta - \frac{1}{2}\gamma)$

47. $\tan \frac{1}{2}\beta \overset{?}{=} \cot \frac{1}{2}(\alpha + \gamma)$

48. $\sin \frac{1}{2}(\alpha + \beta) \overset{?}{=} \cos \frac{1}{2}\gamma$

49. Consider two angles α and β such that $0 < \alpha < \pi$ and $0 < \beta < \pi$. Given that $\cos \alpha = \frac{2}{5}$ and $\cos \beta = -\frac{5}{8}$, find the exact value of the expression:

   a) $\sin(\alpha + \beta)$ b) $\sin(\alpha - \beta)$

   c) $\cos(\alpha + \beta)$ d) $\cos(\alpha - \beta)$

50. Consider two angles α and β such that $0 < \alpha < \pi$ and $0 < \beta < \pi$. Given that $\cos \alpha = -3/\sqrt{10}$ and $\cos \beta = 1/\sqrt{5}$, find the exact value of the expression:

   a) $\tan(\alpha + \beta)$ b) $\tan(\alpha - \beta)$

   c) $\cos 2\alpha$ d) $\sin 2\beta$

51. Consider two angles α and β such that $-\pi < \alpha < 0$ and $\pi < \beta < 2\pi$. Given that $\cos \alpha = \frac{2}{5}$ and $\cos \beta = \frac{2}{5}$, find the exact value of the expression:

   a) $\cos \frac{1}{2}\alpha$ b) $\cos \frac{1}{2}\beta$ c) $\cos 2\alpha$ d) $\cos 2\beta$

52. Given that θ is an acute angle such that $\sin \theta = x$, determine (in terms of x):

   a) $\csc \theta$ b) $\cos \theta$ c) $\tan \theta$ d) $\sec \theta$

53. Given that θ is an acute angle such that $\cos \theta = x$, determine (in terms of x):

   a) $\sin \theta$ b) $\cos\left(\dfrac{\pi}{2} - \theta\right)$

   c) $\cot \theta$ d) $\cos(\pi - \theta)$

54. Rewrite the function $g(t) = 2 \cos t + \sin t$ in the form of $A \cos(Bt - \phi)$.

55. Rewrite the function $f(t) = 3 \cos t - 4 \sin t$ in the form of $A \cos(Bt - \phi)$.

56. Consider an angle θ such that $0 < \theta < \pi$ and $\cos \theta = \frac{7}{13}$. Find the exact value of the expression:

   a) $\sin \frac{1}{2}\theta$ b) $\cos \frac{1}{2}\theta$

   c) $\tan \frac{1}{2}\theta$ d) $\sqrt{2} \cos\left(\dfrac{1}{2}\theta - \dfrac{\pi}{4}\right)$

57. Consider an angle θ such that $\pi/2 < \theta < \pi$ and $\cos \theta = -\frac{7}{25}$. Find the exact value of the expression:

   a) $\sin \frac{1}{2}\theta$ b) $\cos \frac{1}{2}\theta$

   c) $2 \cos \frac{1}{2}\left(\theta - \dfrac{\pi}{3}\right)$ d) $2 \sin \frac{1}{2}\left(\theta - \dfrac{\pi}{2}\right)$

In Problems 58 through 60, evaluate the given expression without using a calculator. If the expression is meaningless, explain why.

58. a) $\sin^{-1}(0)$ b) $\cos^{-1}(0)$

    c) $\cos^{-1}(\sqrt{2})$ d) $\tan^{-1}(\sqrt{3})$

59. a) $\sin^{-1}(-1)$ b) $\cos^{-1}(-1)$

    c) $\tan^{-1}(1)$ d) $\tan^{-1}(-1)$

60. a) $\arccos(2)$ b) $\arcsin\left(-\frac{1}{2}\right)$

    c) $\arcsin\left(-\dfrac{\sqrt{3}}{2}\right)$ d) $\arctan\left(-\dfrac{1}{\sqrt{3}}\right)$

In Problems 61 through 66, evaluate the given expression using the identities of Sections 6.2 and 6.3

61. $\cos\left[\dfrac{\pi}{2} - \sin^{-1}\left(\dfrac{3}{5}\right)\right]$ 62. $\sin\left[\dfrac{\pi}{3} + \cos^{-1}\left(\dfrac{4}{5}\right)\right]$

63. $\sin[2 \sin^{-1}(-\frac{1}{2})]$ 64. $\cos[2 \cos^{-1}(\frac{15}{17})]$

65. $\sin[\frac{1}{2} \cos^{-1}(\frac{3}{5})]$ 66. $\cos[\frac{1}{2} \cos^{-1}(\frac{2}{7})]$

In Problems 67 through 84, solve the equation for θ over the interval. Determine an exact value for θ when possible (otherwise, approximate θ to the nearest 0.1 radian).

67. $2 \cos \theta - 1 = 0$, $[0, \pi]$

68. $\sqrt{3} \cot \theta - 1 = 0$, $[\pi, 2\pi]$

69. $4 \sin^2\theta - 3 = 0$, $[0, \pi]$

70. $3 \tan^2 \theta + 5 = 6$, $[-\pi, 0]$

71. $2 \cos \theta \tan \theta - \tan \theta = 0$, $[0, \pi]$

72. $\dfrac{1 - \sin \theta}{\sin \theta} = \cos \theta \tan \theta$, $[0, \pi]$

73. $\cos^2\theta - \sin^2\theta = -1$, $[-\pi, \pi]$

74. $2 \tan^2\theta - \sec^2\theta = 0$, $\left[-\dfrac{\pi}{2}, \dfrac{\pi}{2}\right]$

75. $6 \cos^2\theta = \cos \theta + 1$, $[-\pi, \pi]$

76. $\tan^2\theta - 2 \tan \theta = 4$, $\left[-\dfrac{\pi}{2}, \dfrac{\pi}{2}\right]$

77. $\sin \theta - \sin(\frac{1}{2}\theta) = 0$, $[0, \pi]$

78. $\sin 2\theta \cos \theta + \cos 2\theta \sin \theta = 1$, $[0, 2\pi]$

79. $\tan 2\theta = \cot \theta$, $[-\pi, \pi]$

80. $\cos 2\theta = 2 \sin \theta \cos \theta$, $[0, 2\pi]$

81. $3 \cos \theta - 4 \sin \theta = 5$, $[0, 2\pi]$

82. $\cos 2\theta - \cos 3\theta = 0$, $[0, \pi]$

83. $\sin 2\theta + \sin \theta = 1 + 2 \cos \theta$, $\qquad [-\pi, \pi]$

84. $\tan(\frac{1}{2}\theta) + \sin 2\theta = \csc \theta$, $\qquad [-\pi, \pi]$

Problems 85 through 87 involve the following situation. A soccer ball is kicked from rest on the ground with an initial velocity of v_0 ft/sec. The ball leaves the ground at an angle of θ with the horizontal (Figure 26). Its path is approximated by the graph of

$$y = (\tan \theta)x - \left(\frac{16}{v_0^2 \cos^2\theta}\right)x^2.$$

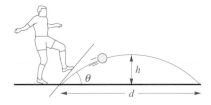

FIGURE 26

85. Show that the distance d (in feet) in Figure 26 is given by

$$d = \frac{v_0^2 \sin 2\theta}{32}.$$

86. Explain why the distance d is maximized when $\theta = 45°$.

87. Show that the maximum height h is given by

$$h = \frac{v_0^2 \sin^2\theta}{64}.$$

(Hint: This is the vertex of the parabola.)

88. In Figure 27, triangles ABC and CDA are isosceles triangles, $\overline{AB} = r$, $\overline{BC} = r$, and $\angle CDA = 90°$. Show

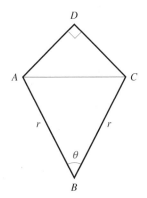

FIGURE 27

that the area of the quadrilateral $ABCD$ is given by $K(\theta) = \frac{1}{2}r^2[\sin \theta - \cos \theta + 1]$, where θ is the measure of $\angle ABC$.

89. Given the function in Problem 88, find the value of θ if $r = 12$ cm and the area of quadrilateral $ABCD$ is 72 in^2.

90. Show that $\tan(2 \tan^{-1}x) = 2 \tan[\tan^{-1}x + \tan^{-1}(x^3)]$.

91. Show that $\ln[\tan(\theta + \pi/4)] =$
$$\ln[\sin \theta + \cos \theta] - \ln[\cos \theta - \sin \theta].$$

92. Solve $16^{\sin^2 x} = 2$ over the interval $[0, \pi]$.

93. Suppose that the interior angles of a triangle are α, β, and γ. Show that

$$\cos^2\alpha + \cos^2\beta + \cos^2\gamma = 1 - 2 \cos \alpha \cos \beta \cos \gamma.$$

(Hint: See Problem 44.)

94. Suppose that the interior angles of a triangle are α, β, and γ. Show that

$$\tan \alpha + \tan \beta + \tan \gamma = \tan \alpha \tan \beta \tan \gamma.$$

(Hint: See Problem 43.)

95. In Figure 28,

a) Show that $\tan \alpha = \frac{1}{2}$.

b) Show that $\tan \beta = \frac{1}{3}$.

c) Show that $\alpha + \beta = \pi/4$. *(Hint: What is the tangent of this sum?)*

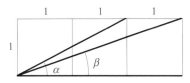

FIGURE 28

CHAPTER 7

APPLICATIONS OF TRIGONOMETRY

OBLIQUE TRIANGLES AND THE LAW OF SINES

The trigonometric functions can be used to find missing sides and angles of right triangles, as we saw in Section 5.1. Now consider the problem of solving **oblique** triangles—that is, triangles that do not contain a right angle.

In general, we will be given three parts of a triangle and will want to find three remaining parts. Finding the three remaining parts is called **solving** the triangle. There are several ways that three parts of a triangle can be given:

1. Two angles and the included side—ASA

2. Two angles and a side not included—AAS

3. Two sides and an included angle—SAS

4. Two sides and a nonincluded angle—SSA

5. Three sides—SSS

6. Three angles—AAA   (this does not determine a triangle)

We start with triangle *ABC* in Figure 1. The sides *a*, *b*, and *c* are opposite the angles α, β, and γ, respectively. Drop a perpendicular of length *h* from vertex *C* to the opposite side. This altitude divides triangle *ABC* into two right triangles. Considering each right triangle, we have

$$\sin \alpha = \frac{h}{b} \quad \text{and} \quad \sin \beta = \frac{h}{a}.$$

If we solve for *h* in each of these equations and equate the results, we obtain

$$b \sin \alpha = a \sin \beta.$$

Therefore,

$$\frac{\sin \alpha}{a} = \frac{\sin \beta}{b}.$$

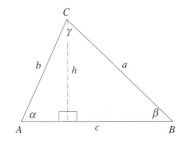

FIGURE 1

387

The last equation follows from drawing a perpendicular from vertex C to the opposite side. Suppose, instead, that we draw a perpendicular from vertex A to the opposite side (Figure 2). Using the same reasoning as before, we obtain

$$\frac{\sin \beta}{b} = \frac{\sin \gamma}{c}.$$

Combining these last two equations gives

$$\frac{\sin \alpha}{a} = \frac{\sin \beta}{b} = \frac{\sin \gamma}{c}.$$

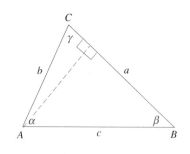

FIGURE 2

One might question how this argument might change if one of the angles is obtuse, greater than $90°$. Suppose $90° < \alpha < 180°$. In this case draw a perpendicular of length h from vertex C to an extension of the opposite side (Figure 3). From the diagram

$$\sin \alpha' = \frac{h}{b}.$$

Notice that $\alpha = 180° - \alpha'$. Therefore, $\sin \alpha = \sin \alpha'$ (why?), and we obtain the same result as in the original case,

$$\sin \alpha = \frac{h}{b}.$$

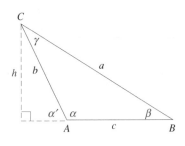

FIGURE 3

Thus, whether all three angles are acute or one angle is obtuse, we have the following relationship.

The Law of Sines

Consider any triangle with angles α, β, γ, and corresponding opposite sides a, b, c, respectively. Then,

$$\frac{\sin \alpha}{a} = \frac{\sin \beta}{b} = \frac{\sin \gamma}{c}.$$

Equivalently,

$$\frac{a}{\sin \alpha} = \frac{b}{\sin \beta} = \frac{c}{\sin \gamma}.$$

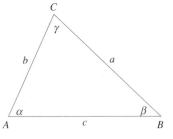

The law of sines can be useful in several situations. The next two examples illustrate how to apply it to solve triangles when we know any two angles and a side.

EXAMPLE 1   In triangle ABC suppose that $\alpha = 105°$, $\gamma = 21°$, and $c = 43$. Solve the triangle, that is, find a, b, and β.

SOLUTION Draw a picture with the given data: two angles and a nonincluded side (AAS), as in Figure 4. Since $\alpha + \beta + \gamma = 180°$,

$$\beta = 180° - (\alpha + \gamma) = 180° - (105° + 21°) = 54°.$$

Using the law of sines, we get

$$\frac{a}{\sin 105°} = \frac{43}{\sin 21°}$$

$$a = \frac{43 \sin 105°}{\sin 21°}$$

$$\approx 116.$$

Also by the law of sines we have

$$\frac{b}{\sin 54°} = \frac{43}{\sin 21°}$$

$$b = \frac{43 \sin 54°}{\sin 21°}$$

$$\approx 97.$$

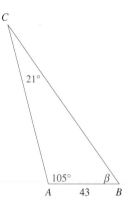

FIGURE 4

EXAMPLE 2   An island is visible from two boat launches on shore, A and B, that are 2.1 miles apart. From boat launch A, the angle of sight between the island and launch B is 56°. From boat launch B, the angle of sight between the island and launch A is 65°. What is the distance from each launch to the island?

SOLUTION We are given two angles and the included side (ASA), as in Figure 5. The remaining angle is

$$180° - (56° + 65°) = 59°.$$

Let a represent the distance from launch B to the island, and let b represent the distance from launch A to the island. Using the law of sines, we get

$$\frac{a}{\sin 56°} = \frac{2.1}{\sin 59°}$$

$$a = \frac{2.1 \sin 56°}{\sin 59°}$$

$$\approx 2.0.$$

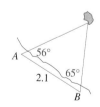

FIGURE 5

The procedure for finding b is similar:

$$\frac{b}{\sin 65°} = \frac{2.1}{\sin 59°}$$

$$b = \frac{2.1 \sin 65°}{\sin 59°}$$

$$\approx 2.2.$$

The distances from the island to launches A and B are 2.2 miles and 2.0 miles, respectively.

The law of sines can also be employed to solve a triangle when given two sides and an angle opposite one of those sides (SSA).

EXAMPLE 3   Point Q is 40.0 miles N45.0°E from point P. A car heads directly east from P. When the car is 30.0 miles from point Q, how far is it from point P?

SOLUTION Draw a picture (Figure 6). Notice that there are two points east of P that are 30 miles from point Q. To find them we consider two oblique triangles, PQR and PQR'.
    Using the law of sines, we have

$$\frac{\sin \angle PRQ}{40.0} = \frac{\sin 45.0°}{30.0}$$

$$\sin \angle PRQ = \frac{40.0 \sin 45.0°}{30.0} = \frac{2\sqrt{2}}{3}.$$

There are two angles between 0° and 180° that satisfy this equation. From our diagram we see that

$$\angle PR'Q = \sin^{-1}\left[\frac{2\sqrt{2}}{3}\right] \approx 70.5°$$

and

$$\angle PRQ = 180° - \sin^{-1}\left[\frac{2\sqrt{2}}{3}\right] \approx 109.5°.$$

Therefore, the remaining parts of triangle PQR' are

$$\angle PQR' = 180° - (45.0° + 70.5°) = 64.5°.$$

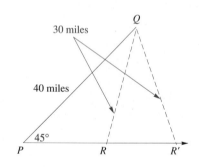

FIGURE 6

Using the law of sines again, we have

$$\frac{PR'}{\sin 64.5°} = \frac{30.0}{\sin 45.0°}$$

$$PR' = \frac{30.0 \sin 64.5°}{\sin 45.0°} \approx 38.3 \text{ miles.}$$

For triangle PQR,

$$\angle PQR = 180° - (45.0° + 109.5°) = 25.5°.$$

Therefore,

$$\frac{PR}{\sin 25.5°} = \frac{30.0}{\sin 45.0°}$$

$$PR = \frac{30.0 \sin 25.5°}{\sin 45.0°} \approx 18.3 \text{ miles.}$$

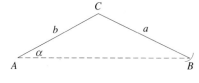

FIGURE 7

In Example 3 we were given two sides and a nonincluded angle (SSA). This case may or may not determine a unique triangle. Let's consider the possibilities.

Suppose we are given a, b, and α. To construct a triangle from these data, first draw angle α with adjacent side b (Figure 7). Several possibilities exist, depending on the length a of the side opposite angle α. Imagine trying to locate vertex B by striking a circular arc of radius a, centered at C. If $a \geq b$, this will produce a unique triangle (Figure 8).

However, if $a < b$, there are three possibilities (Figure 9).

FIGURE 8

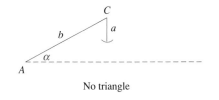

No triangle

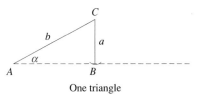

One triangle

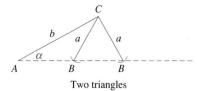

Two triangles

FIGURE 9

The last of these three cases is sometimes referred to as the **ambiguous case.** Fortunately, when you apply the law of sines, the particular situation becomes clear. You simply need to be alert when given two sides and an angle opposite one of those sides. The next three examples illustrate.

EXAMPLE 4   In triangle ABC, suppose that $\alpha = 33.0°$, $a = 14.1$, and $b = 19.8$. Solve the triangle.

SOLUTION Draw a picture (Figure 10). This case is SSA, so there may be more than one triangle with the given parts. To solve the triangle, we need to determine c, β, and γ. Starting with β, we have

$$\frac{\sin \beta}{b} = \frac{\sin \alpha}{a}.$$

Now, multiply both sides of the proportion by b and substitute the given information.

$$\sin \beta = \frac{b \sin \alpha}{a} = \frac{19.8 \sin 33°}{14.1} \approx 0.76481.$$

There are two possible values of β, $0 < \beta < 180°$, that satisfy the last equation:

$$\beta_1 = \sin^{-1}(0.76481) \approx 49.9°$$

and

$$\beta_2 = 180° - \sin^{-1}(0.76481) \approx 130.1°.$$

For each case, the corresponding third angle is

$$\gamma_1 = 180° - (\alpha + \beta_1) \approx 97.1°$$

and

$$\gamma_2 = 180° - (\alpha + \beta_2) \approx 16.9°.$$

A sketch of the two possible triangles is helpful (Figure 11).
    All that remain to be found are c_1 and c_2. In each case, we use

$$\frac{a}{\sin \alpha} = \frac{c}{\sin \gamma}$$

$$\frac{14.1}{\sin 33°} = \frac{c_1}{\sin 97.1°} \Rightarrow c_1 = \frac{14.1 \sin 97.1°}{\sin 33°} \approx 25.7$$

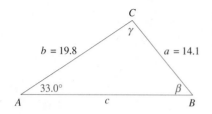

C
$b = 19.8$ $a = 14.1$
γ
33.0° β
A c B

FIGURE 10

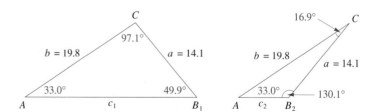

FIGURE 11

and

$$\frac{14.1}{\sin 33°} = \frac{c_2}{\sin 16.9°} \Rightarrow c_2 = \frac{14.1 \sin 16.9°}{\sin 33°} \approx 7.5.$$

EXAMPLE 5 Solve the triangle ABC for which $a = 26.5$, $c = 24.9$, and $\alpha = 38.2°$.

SOLUTION Draw a picture (Figure 12). This case is SSA, so there may be more than one triangle to solve. Using the law of sines, we have

$$\frac{\sin \gamma}{c} = \frac{\sin \alpha}{a}.$$

Substitute the given information and solve for $\sin \gamma$:

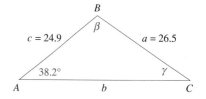

FIGURE 12

$$\frac{\sin \gamma}{24.9} = \frac{\sin 38.2°}{26.5} \Rightarrow \sin \gamma = \frac{24.9 \sin 38.2°}{26.5} \approx .58107.$$

There are two possible values of γ, $0 < \gamma < 180°$, that satisfy the last equation:

$$\gamma_1 = \sin^{-1}(0.58107) \approx 35.5° \quad \text{and} \quad \gamma_2 = 180° - \sin^{-1}(0.58107) \approx 144.5°.$$

The second value is impossible since $\alpha + \gamma_2 = 182.7°$ and the sum of any two angles of a triangle must be less than $180°$.

Thus, only one triangle is possible in this case. The remaining angle is

$$\beta = 180° - (\alpha + \gamma_1) = 180° - (38.2° + 35.5°) = 106.3°.$$

To find the remaining side, use the law of sines:

$$\frac{b}{\sin \beta} = \frac{a}{\sin \alpha}$$

$$b = \frac{26.5 \sin 106.3°}{\sin 38.2°} \approx 41.1.$$

EXAMPLE 6 Solve triangle ABC, where $b = 19$, $c = 26$, and $\beta = 62°$.

SOLUTION Draw a picture (Figure 13). Again, this case is SSA. Solve for $\sin \gamma$ in the proportion

$$\frac{\sin \gamma}{c} = \frac{\sin \beta}{b}$$

and substitute the given information:

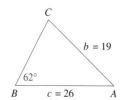

FIGURE 13

$$\sin \gamma = \frac{c \sin \beta}{b} = \frac{26 \sin 62°}{19} \approx 1.20824.$$

Because the sine can not exceed 1, there is no solution. In other words, no triangle can be formed with $b = 19$, $c = 26$, and $\beta = 62°$.

Applications

We finish this section with two applications of the law of sines.

EXAMPLE 7   To find the height h of a hill, a surveyor measures two angles of elevation from ground level. First, she measures the angle of elevation from point A to the top of the hill. Next, she moves 200 feet directly away from the hill to point B and measures the angle of elevation from there. If the measurements are $42°$ and $36°$, find the height h.

SOLUTION   Draw a picture (Figure 14). Let the top of the hill be point C, and let point D be h units directly below C.

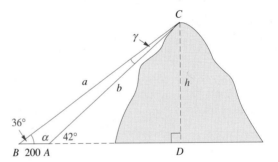

FIGURE 14

Consider triangle ABC. We know that $c = 200$ and $\beta = 36°$. Then,

$$\alpha = 180° - 42° = 138°$$

and

$$\gamma = 180° - (\alpha + \beta) = 180° - (138° + 36°) = 6°.$$

Using the law of sines, we obtain

$$\frac{b}{\sin \beta} = \frac{c}{\sin \gamma}$$

$$b = \frac{c \sin \beta}{\sin \gamma} = \frac{200 \sin 36°}{\sin 6°} \approx 1124.64.$$

Now consider right triangle ADC:

$$\sin 42° = \frac{h}{b} \Rightarrow h = b \sin 42° = 1124.64 \sin 42° \approx 753 \text{ feet.}$$

EXAMPLE 8 Two lighthouses, A and B, spot a ship at 1 P.M. From lighthouse A, which is 38.2 miles due west of B, the ship is N36.0°E. From lighthouse B the ship is N75.0°W. At 1:15 P.M., lighthouse A observes the ship's position to be N48.0°E, and lighthouse B observes the ship's position to be N75.0°W (i.e., the ship is moving straight toward B). Find the average speed of the ship between the two sightings.

SOLUTION Draw a picture (Figure 15). Let the first observed position be C and the second observed position be D. From the diagram,

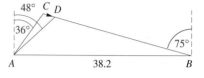

$$\angle BAD = 90° - 48° = 42°$$

$$\angle ABD = 90° - 75° = 15°$$

$$\angle BAC = 90° - 36° = 54°$$

$$\angle ACB = 180° - (54° + 15°) = 111°$$

$$\angle ADB = 180° - (42° + 15°) = 123°.$$

FIGURE 15

Consider triangle ABC. Using the law of sines, we have

$$\frac{BC}{\sin 54°} = \frac{38.2}{\sin 111°} \Rightarrow BC = \frac{38.2 \sin 54°}{\sin 111°} \approx 33.10315.$$

Now consider triangle ABD. Again, by the law of sines,

$$\frac{BD}{\sin 42°} = \frac{38.2}{\sin 123°} \Rightarrow BD = \frac{38.2 \sin 42°}{\sin 123°} \approx 30.47775.$$

The distance the ship traveled is

$$CD = BC - BD = 2.62540 \text{ miles.}$$

So the average speed of the ship is

$$\frac{2.62540 \text{ miles}}{0.25 \text{ hour}} \approx 10.5 \text{ miles per hour.}$$

EXERCISE SET 7.1

A

In Problems 1 through 12, approximate the remaining parts of triangle ABC, if possible.

1. $\alpha = 40°$, $\beta = 53°$, $a = 16.2$
2. $\beta = 78°$, $\gamma = 22°$, $b = 8.3$
3. $\beta = 104°$, $\gamma = 18°$, $a = 15$
4. $\alpha = 86°$, $\gamma = 52°$, $b = 42$
5. $a = 5.0$, $b = 3.0$, $\alpha = 70°$
6. $b = 24.4$, $c = 16.2$, $\beta = 112°$
7. $a = 8.6$, $b = 10.0$, $\alpha = 28°$

8. $a = 32, b = 37, \alpha = 25°$

9. $b = 14.0, c = 23.5, \beta = 52°$

10. $a = 47.5, c = 58.0, \alpha = 74°$

11. $\beta = 81°, \gamma = 63°, a = 20.3$

12. $\alpha = 142°, \beta = 23°, b = 127$

13. Use the law of sines to find x in Figure 16. Leave your answers in terms of radicals rather than decimals.

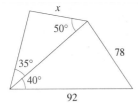

FIGURE 16

14. Find x in Figure 17. Approximate to the nearest unit.

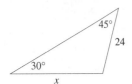

FIGURE 17

15. Find x in Figure 18. Approximate to the nearest unit.

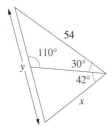

FIGURE 18

16. Find y in Problem 15 (Figure 18). Approximate to the nearest unit.

B

17. A tower stands on a hillside that makes an angle of 6° with the horizontal (Figure 19). Thirty feet down the hill from the base of the tower, the angle of elevation to the top of the tower is 54°. Find the height h of the tower.

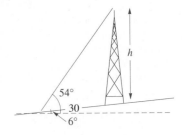

FIGURE 19

18. A bridge is to be built from A to C across a river (Figure 20). To estimate the length of the proposed bridge, Robin stands at point A facing point C. She then turns 75° and walks 60 feet to point B, where she measures $\angle ABC$ to be 42°. What is the length of the proposed bridge?

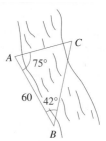

FIGURE 20

19. The orbits of earth and Venus are very nearly circular with the sun at the center (Figure 21). From earth, an angle of sight between the sun and Venus is measured to be 14°. If the radius of the orbits of Venus and earth are 68 million miles and 93 million miles, respectively, what are the possible distances from earth to Venus?

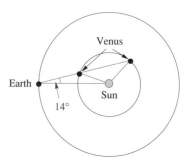

FIGURE 21

20. Two lookout stations, *A* and *B*, are 12.5 miles apart. A fire is spotted from both stations at point *C*. Station *A* measures ∠*BAC* = 14°, and station *B* finds ∠*ABC* = 82°. How far is the fire from the nearer station?

21. An airplane is headed toward an airport that is 32 miles from a mountain. The pilot makes two sightings of the mountain, 20 miles apart, as shown in Figure 22. If the airplane stays on course, how far is the airplane from the airport?

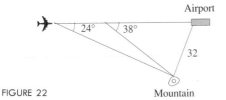

FIGURE 22

22. Two observatories, *A* and *B*, are 26 miles apart. From *A* a UFO is observed, and the angle of sight to observatory *B* is 76°. Similarly from *B* the angle of sight from the UFO to *A* is 58°. Find the distance from *A* to the UFO (see Figure 23).

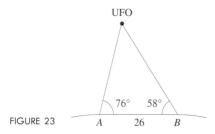

FIGURE 23

23. The crank and connecting rod of an engine are 35 and 105 cm long, respectively (see Figure 24). What angle does the crank make with the horizontal when the connecting rod makes an angle of 8°?

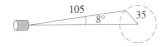

FIGURE 24

24. To find the height *h* of a hill, two angles of elevation are measured from (horizontal) ground level (Figure 25). The angle of elevation from the point *A* to the top is 63°. After the surveyor moves 140 feet directly away from the hill to point *B*, the angle of elevation to the top is measured to be 55°. Find the height *h*.

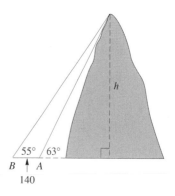

FIGURE 25

25. The angles of elevation of a balloon from points *A* and *B* on the ground are 67° and 21°, respectively (Figure 26). Points *A* and *B* are 11 miles apart, and *A*, *B*, and the balloon are in the same plane. Find the height of the balloon.

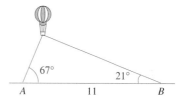

FIGURE 26

26. A boat spots a lighthouse N28°E and then proceeds east 7.5 nautical miles, where the lighthouse is again sighted N16°E. Find the distance from the boat to the lighthouse.

27. A torpedo is fired from a submarine in the direction N40°E. A ship, which is 5 miles due east of the submarine, spots the torpedo to be N64°W. One minute later, the ship observes the torpedo to be N43°W. Find the speed of the torpedo.

28. A swimmer heads N52°W from a north-south shoreline for 410 yards. She then changes direction and swims for 380 yards to the shore. What are the possible distances from her point of origin?

C

29. Show that $\dfrac{a + b}{b} = \dfrac{\sin \alpha + \sin \beta}{\sin \beta}$.

30. Using Figure 27, show that the radius of the circle circumscribed about triangle ABC is

$$r = \frac{a}{2 \sin \alpha} = \frac{b}{2 \sin \beta} = \frac{c}{2 \sin \gamma}.$$

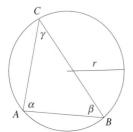

FIGURE 27

31. In Figure 28, CD bisects $\angle ACB$. Show that

$$\frac{p}{q} = \frac{r}{s}.$$

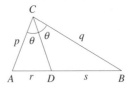

FIGURE 28

S E C T I O N 7.2 THE LAW OF COSINES

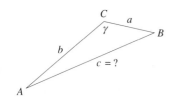

FIGURE 29

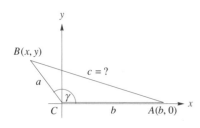

FIGURE 30

The law of sines cannot be used to solve triangles for which all three sides are given (SSS) or for which two sides and the included angle are specified (SAS).

Suppose we are given sides a and b and the included angle γ (Figure 29). Consider the problem of expressing side c in terms of a, b, and γ. One approach is to place triangle ABC on a coordinate plane with C at the origin and A at $(b, 0)$ (Figure 30).

The coordinates of B are

$$x = a \cos \gamma \quad \text{and} \quad y = a \sin \gamma.$$

Using the distance formula, we have

$$\begin{aligned}
c^2 &= (x - b)^2 + (y - 0)^2 \\
&= (a \cos \gamma - b)^2 + (a \sin \gamma - 0)^2 \\
&= a^2\cos^2\gamma - 2ab \cos \gamma + b^2 + a^2\sin^2\gamma \\
&= a^2(\cos^2 \gamma + \sin^2\gamma) + b^2 - 2ab \cos \gamma \\
&= a^2 + b^2 - 2ab \cos \gamma.
\end{aligned}$$

This is what we set out to find. If we orient triangle ABC so that A is at the origin and B is at $(c, 0)$, a similar argument leads to

$$a^2 = b^2 + c^2 - 2bc \cos \alpha.$$

In the same way, it can be shown that

$$b^2 = a^2 + c^2 - 2ac \cos \beta.$$

These three results are known as the law of cosines.

Consider any triangle with angles α, β, γ and corresponding opposite sides a, b, c, respectively. Then

$$a^2 = b^2 + c^2 - 2bc \cos \alpha$$

$$b^2 = a^2 + c^2 - 2ac \cos \beta$$

$$c^2 = a^2 + b^2 - 2ab \cos \gamma.$$

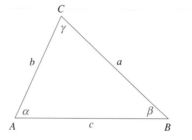

The Law of Cosines

Notice that if one of the angles, say γ, is 90°, then

$$c^2 = a^2 + b^2 - 2ab \cos 90°$$

$$= a^2 + b^2 - 2ab(0)$$

so

$$c^2 = a^2 + b^2. \qquad \text{(This should look familiar.)}$$

In this case the other two statements in the law of cosines reduce to what you would expect. For example,

$$a^2 = b^2 + c^2 - 2bc \cos \alpha$$

becomes

$$a^2 = b^2 + (a^2 + b^2) - 2bc \cos \alpha.$$

Solving for $\cos \alpha$ leads to the fundamental relationship

$$\cos \alpha = \frac{b}{c}.$$

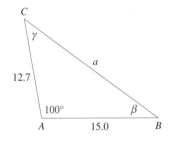

FIGURE 31

Considering the special case when γ is 90° serves as a check for the law of cosines but is not useful for solving triangles. The next two examples illustrate how the law of cosines is used to find unknown parts of an oblique triangle when we are given two sides and the included angle (SAS) or all three sides (SSS).

EXAMPLE 1 In triangle ABC (Figure 31), suppose that $\alpha = 100°$, $c = 15.0$, and $b = 12.7$. Solve the triangle, that is, find a, β, and γ.

SOLUTION By the law of cosines,

$$a^2 = b^2 + c^2 - 2bc \cos \alpha$$

$$a = \sqrt{12.7^2 + 15.0^2 - 2(12.7)(15.0) \cos 100°}$$

$$\approx 21.3.$$

The keystrokes on a calculator for this last computation (in degree mode) are as shown:

KEYSTROKES	DISPLAY
$\boxed{12.7}\ \boxed{x^2}$	$\boxed{161.29}$
$\boxed{+}\ \boxed{15}\ \boxed{x^2}$	$\boxed{225}$
$\boxed{-}\ \boxed{2}\ \boxed{\times}\ \boxed{12.7}\ \boxed{\times}$	
$\boxed{15}\ \boxed{\times}\ \boxed{100}\ \boxed{\cos}\ \boxed{=}$	$\boxed{452.449956}$
$\boxed{\sqrt{\ }}$	$\boxed{21.2708711}$

To find β, we could use the law of sines or the law of cosines. We use the law of sines because the computation is simpler:

$$\frac{\sin \beta}{b} = \frac{\sin \alpha}{a}$$

$$\sin \beta = \frac{b \sin \alpha}{a} = \frac{12.7 \sin 100°}{21.3} = 0.58718 \ldots$$

$$\beta = \sin^{-1}(0.58718 \ldots) \approx 36°.$$

Finally, $\gamma = 180° - (100° + 36°) = 44°$.

EXAMPLE 2 In triangle ABC, $a = 14$, $b = 17$, and $c = 25$. Find β and γ.

SOLUTION Using the law of cosines, $c^2 = a^2 + b^2 - 2ab \cos \gamma$, solve for $\cos \gamma$:

$$\cos \gamma = \frac{a^2 + b^2 - c^2}{2ab}$$

$$= \frac{14^2 + 17^2 - 25^2}{2(14)(17)}$$

$$= -0.2941 \ldots .$$

Therefore, $\gamma = \cos^{-1}(-0.2941\ldots) \approx 107°$.

The keystrokes on a calculator for this computation are as shown

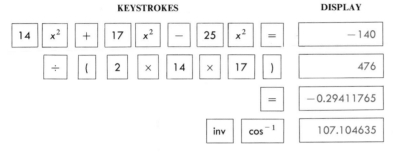

KEYSTROKES	DISPLAY
14 x^2 + 17 x^2 − 25 x^2 =	−140
÷ (2 × 14 × 17)	476
=	−0.29411765
inv cos^{-1}	107.104635

By the law of sines,

$$\sin \beta = \frac{b \sin \gamma}{c} = \frac{17 \sin 107°}{25}$$

$$\beta = \sin^{-1}\left(\frac{17 \sin 107°}{25}\right) \approx 41°.$$

Notice that we first used the law of cosines to find the largest angle (the angle opposite the longest side). When given all three sides of a triangle, it is best to use the law of cosines to find the largest angle because if the triangle has an obtuse angle, it will be identified initially. The other two angles must be acute, and we would be less likely to make an error when applying the law of sines to an acute angle.

EXAMPLE 3   In triangle ABC, $a = 11$, $b = 15$, and $c = 28$. Find the angles.

SOLUTION Using the law of cosines, $c^2 = a^2 + b^2 - 2ab \cos \gamma$, solve for $\cos \gamma$:

$$\cos \gamma = \frac{a^2 + b^2 - c^2}{2ab}$$

$$= \frac{11^2 + 15^2 - 28^2}{2(11)(15)}$$

$$= -1.3\overline{27}.$$

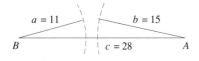

FIGURE 32

Since $-1 < \cos \gamma < 1$, there is no value for γ that will satisfy this equation. This is because no triangle can be formed when the sum of any two sides is less than the third side (Figure 32).

EXAMPLE 4 Two lighthouses, A and B, spot a ship at 1 P.M. Lighthouse A, which is 18 miles due west of B, observes the ship to be N35°E. From lighthouse B the ship is N52°W. At 1:30 P.M. lighthouse A observes the ship N60°E, and lighthouse B observes the ship N5°E. Find the average speed of the ship between sightings.

SOLUTION Don't even think about trying to solve this problem without a sketch. Let the ship's first observed position be C and the second position be D (Figure 33).

Consider triangle ABC (Figure 34). From the diagram,

$$\angle CAB = 90° - 35° = 55°$$
$$\angle CBA = 90° - 52° = 38°$$
$$\angle ACB = 180° - (55° + 38°) = 87°.$$

Using the law of sines, we have

$$\frac{AC}{\sin 38°} = \frac{18}{\sin 87°}$$

$$AC = \frac{18 \sin 38°}{\sin 87°} = 11.09711\ldots.$$

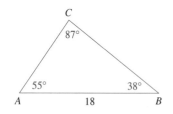

FIGURE 33

Now consider triangle ABD (Figure 35):

$$\angle ABD = 90° + 5° = 95°$$
$$\angle DAB = 90° - 60° = 30°$$
$$\angle ADB = 180° - (95° + 30°) = 55°.$$

By the law of sines,

$$\frac{AD}{\sin 95°} = \frac{18}{\sin 55°}$$

$$AD = \frac{18 \sin 95°}{\sin 55°} = 21.8903\ldots.$$

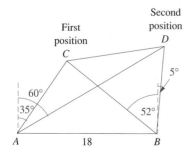

FIGURE 34

Finally, consider triangle ACD (Figure 36):

$$\angle CAD = 60° - 35° = 25°.$$

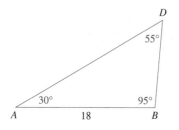

FIGURE 35

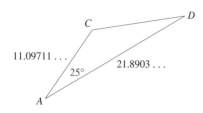

FIGURE 36

Knowing sides AC and AD and the included angle, we apply the law of cosines to find CD.

$$CD^2 = (11.09711\ldots)^2 + (21.8903\ldots)^2$$
$$-2(11.09711\ldots)(21.8903\ldots)\cos 25°$$
$$= 162.0127\ldots$$
$$CD = \sqrt{162.0127\ldots} = 12.72842\ldots.$$

Therefore the average speed of the ship was

$$\frac{\text{distance traveled}}{\text{time interval}} = \frac{12.72842\ldots \text{ miles}}{0.5 \text{ hours}} \approx 25.5 \text{ miles per hour.}$$

The law of cosines can be employed to find a mathematical function that models a particular situation.

EXAMPLE 5   Two jets A and B take off from point P at the same time in directions such that $\angle APB = 120°$. Jet A is flying at a rate of 220 mph, and jet B is traveling at 265 mph. Determine the distance between them as a function of the number of hours in flight, t.

SOLUTION After t hours, jet A will be $220t$ miles from point P, and jet B will be $265t$ miles from point P. From Figure 37, the distance between them after t hours satisfies

$$[D(t)]^2 = (220t)^2 + (265t)^2 - 2(220t)(265t)\cos 120°.$$

Solving for $D(t)$ and simplifying, we get

$$D(t) = \sqrt{(220t)^2 + (265t)^2 - 2(220t)(265t)(-\tfrac{1}{2})}$$
$$= 5t\sqrt{7077} \text{ miles.}$$

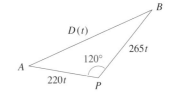

FIGURE 37

Area of a Triangle

The area of a triangle can be determined from two sides and an included angle (SAS), from all three sides (SSS), or from two angles and a side (ASA or AAS).

We start with SAS. Consider triangle ABC with altitude h in Figure 38. The area of the triangle is

$$A = \tfrac{1}{2}(\text{base})(\text{height})$$
$$= \tfrac{1}{2}ch$$

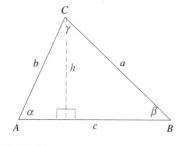

FIGURE 38

The value of h can be expressed in terms of a or b by considering the two right triangles formed by the altitude:

$$\frac{h}{b} = \sin \alpha \quad \text{or} \quad \frac{h}{a} = \sin \beta$$

$$h = b \sin \alpha \qquad h = a \sin \beta.$$

Substituting these values for h in the expression for area leads to

$$A = \tfrac{1}{2}bc \sin \alpha = \tfrac{1}{2}ac \sin \beta.$$

If an altitude is drawn from vertex A, the same line of reasoning yields a formula for the area in terms of γ or β:

$$A = \tfrac{1}{2}ab \sin \gamma = \tfrac{1}{2}ac \sin \beta.$$

> Consider any triangle with angles α, β, γ and corresponding opposite sides a, b, c, respectively. The area is given by
>
> $$A = \tfrac{1}{2}bc \sin \alpha = \tfrac{1}{2}ac \sin \beta = \tfrac{1}{2}ab \sin \gamma.$$

This formula is just three variations of the same idea: The area of any triangle is one-half the product of two sides and the sine of the included angle. This idea can be applied to find the area of a polygon with more than three sides by dividing the region into triangles.

FIGURE 39

EXAMPLE 6　Find the area of the quadrilateral in Figure 39.

　Divide the quadrilateral into two triangles with the diagonal as shown in Figure 40. The area is the sum of the areas of the triangles:

$$A = \tfrac{1}{2}(38)(63) \sin 112° + \tfrac{1}{2}(67)(40) \sin 70°$$

$$\approx 2369 \text{ square units.}$$

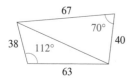

FIGURE 40

　We are now in a position to determine a formula for the area of a triangle in terms of the three sides. Starting with

$$A = \tfrac{1}{2}ab \sin \gamma$$

square both sides to get

$$A^2 = \tfrac{1}{4}a^2b^2 \sin^2\gamma = \tfrac{1}{4}a^2b^2(1 - \cos^2\gamma).$$

Using the law of cosines, replace $\cos\gamma$ with

$$\frac{a^2 + b^2 - c^2}{2ab}.$$

Then multiply both sides by 16 to obtain

$$16A^2 = 4a^2b^2\left(1 - \frac{(a^2 + b^2 - c^2)^2}{4a^2b^2}\right)$$
$$= 4a^2b^2 - (a^2 + b^2 - c^2)^2.$$

Factor this as the difference of two squares:

$$16A^2 = [2ab - (a^2 + b^2 - c^2)][2ab + (a^2 + b^2 - c^2)]$$
$$= [c^2 - (a^2 - 2ab + b^2)][(a^2 + 2ab + b^2) - c^2]$$
$$= [c^2 - (a - b)^2][(a + b)^2 - c^2].$$

Again, factor each of the two factors as the difference of two squares:

$$16A^2 = [c - (a - b)][c + (a - b)][(a + b) - c][(a + b) + c]$$
$$= [-a + b + c][a - b + c][a + b - c][a + b + c].$$

The **semiperimeter** of a triangle is $\tfrac{1}{2}(a + b + c)$. If we let

$$s = \tfrac{1}{2}(a + b + c)$$

then it follows that

$$16A^2 = [2s - 2a][2s - 2b][2s - 2c][2s]$$

or

$$16A^2 = 2[s - a] \cdot 2[s - b] \cdot 2[s - c] \cdot 2[s].$$

Dividing both sides by 16 and taking the square root leads to

$$A = \sqrt{s(s - a)(s - b)(s - c)}.$$

This is known as **Hero's** or **Heron's formula.**

Hero's Formula

> The area of any triangle with sides a, b, and c is
> $$A = \sqrt{s(s - a)(s - b)(s - c)}$$
> where $s = \frac{1}{2}(a + b + c)$.

EXAMPLE 7   Find the area of the triangle in Example 2.

SOLUTION The semiperimeter is

$$s = \tfrac{1}{2}(a + b + c) = \tfrac{1}{2}(14 + 17 + 25) = 28.$$

The area is

$$A = \sqrt{s(s - a)(s - b)(s - c)}$$
$$= \sqrt{28(28 - 14)(28 - 17)(28 - 25)}$$
$$= \sqrt{12936}$$
$$\approx 113.7.$$

The area of a triangle can also be found if two angles and a side are known.

> Consider any triangle with angles α, β, γ and corresponding opposite sides a, b, c, respectively. The area is given by
> $$A = \frac{a^2 \sin \beta \sin \gamma}{2 \sin \alpha} = \frac{b^2 \sin \alpha \sin \gamma}{2 \sin \beta} = \frac{c^2 \sin \alpha \sin \beta}{2 \sin \gamma}.$$

The proof is left as an exercise. You may have noticed that each expression requires all three angles; this isn't a problem since the third angle is easily found.

EXAMPLE 8   Determine the area of the triangle in Figure 41.

SOLUTION The third angle is $180° - (130° + 35°) = 15°$. The area is

$$A = \frac{(7.4)^2 \sin 130° \sin 15°}{2 \sin 35°} \approx 9.5 \text{ square units.}$$

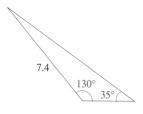

FIGURE 41

EXERCISE SET 7.2

A

In Problems 1 through 12, approximate the remaining parts of triangle ABC, if possible.

1. $a = 10.0, b = 15.0, \gamma = 24°$

2. $a = 20.0, b = 14.0, \gamma = 31°$

3. $b = 24.4, c = 16.2, \alpha = 112°$

4. $a = 47.5, c = 58.0, \beta = 74°$

5. $a = 5.0, c = 3.0, \gamma = 70°$

6. $b = 4.4, c = 6.2, \alpha = 84°$

7. $a = 8.6, b = 10.0, c = 5.2$

8. $a = 36.0, b = 11.0, c = 26.2$

9. $a = 28.2, b = 11.0, c = 13.7$

10. $a = 8.6, b = 10.0, \alpha = 24°, \beta = 55°$

11. $a = 75, c = 20, \alpha = 118°, \gamma = 21°$

12. $b = 35, c = 21, \beta = 76°, \gamma = 31°$

13. Find the measure of the largest angle of the triangle whose sides are 12.3, 14.0, and 15.7.

14. Find the measure of the smallest angle of the triangle whose sides are 18.1, 21.0, and 23.7.

15. Find the measure of the larger acute angle of the triangle whose sides are 8.1, 10.0, and 16.7.

16. Find the area of the triangle in Problem 1.

17. Find the area of the triangle in Problem 2.

18. Find the area of the triangle in Problem 3.

19. Find the area of the triangle in Problem 4.

20. Find the area of the triangle in Problem 7.

21. Find the area of the triangle in Problem 8.

B

22. Find the value of x in Figure 42.

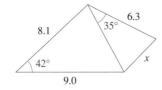

FIGURE 42

23. Find the value of θ in Figure 43.

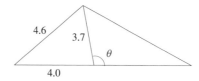

FIGURE 43

24. In Figure 44, $UX = 17.8$ and $XZ = 30.9$. Find VZ.

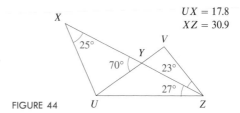

FIGURE 44

25. Three circles with radii 3, 5, and 6 are tangent to each other. Find the three angles of the triangle formed by joining their centers (Figure 45).

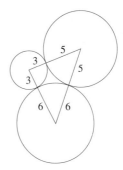

FIGURE 45

26. Two sides of a parallelogram are 5.5 and 7.2 inches. The length of the shorter diagonal is 5.1 inches. Find the area of the parallelogram.

27. A regular octagon is inscribed in a circle with radius $\frac{1}{2}$ (Figure 46).

    a) Use the law of cosines to find the length of each side.

    b) Find the perimeter of the octagon. Compare your answer with the circumference of the circle.

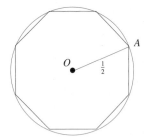

FIGURE 46

28. Repeat Problem 27 for a regular 36-sided polygon inscribed in a circle.

29. Figure 47 shows a regular octagon inscribed in a unit circle.

   a) Find the (shaded) area of triangle AOB.

   b) Find the area of the octagon. Compare your answer with the area of the circle.

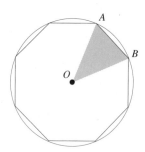

FIGURE 47

30. Repeat Problem 29 for a regular 36-sided polygon inscribed in a unit circle.

31. A regular n-sided polygon is inscribed in a circle of radius $\frac{1}{2}$. Determine the perimeter of the polygon as a function of n. What happens as $n \to \infty$?

32. A regular n-sided polygon is inscribed in a unit circle. Determine the area of the polygon as a function of n. What happens as $n \to \infty$?

33. In this section and the previous section, we have developed formulas to determine unknown parts of a triangle when we know SSS, SAS, ASA, AAS, and SSA. Does a formula exist for the case AAA (we are given only the three angles)? Why or why not?

34. A surveyor stands at a point that is 200 yards from one end of a pond and 280 yards from the other end. She notes that the bearing to the first end is N48.0°W

and the bearing of the other is N62.0°E. Find the length of the pond (Figure 48).

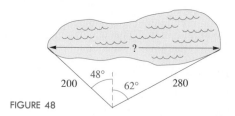

FIGURE 48

35. Two airplanes leave an airport at the same time. One flies N35°W at 160 mph, and the other flies S70°W at 170 mph.

   a) How far apart are they after 2 hours?

   b) Determine the distance between the airplanes as a function of t hours in flight.

36. A student leaves his house and walks N60.0°W for 2.0 miles to a bus stop. There he boards a bus heading N15.0°E at 32.0 mph.

   a) To the nearest 0.1 mile, how far is the student from his house after 45 minutes of riding?

   b) Determine the distance between the student and his house as a function of t hours of riding.

37. Two search helicopters, Sky King and Seahunt, leave San Diego at noon to find a boat in distress. Sky King travels due west at 100 mph, and Seahunt travels N20.0°W at 110 mph. At 1:06 P.M. Seahunt locates the boat and radios Sky King to assist. What is the bearing and distance that Sky King needs to travel?

38. A balloon for a sales promotion is tethered by two ropes 100 ft apart. The angles that the ropes make with the horizontal are 23.4° and 51.2°.

   a) Find the area of the triangle formed in Figure 49.

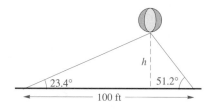

FIGURE 49

b) Find the height of the balloon above the ground (to the nearest foot). *(Hint: Equate your answer in (a) to $\frac{1}{2}$(base)(height), and solve for h.)* Compare this to Exercise 44 in Section 5.1.

39. Two ships *A* and *B* are anchored in the Gulf of Mexico. At 2:00 P.M., both ships spot Barbara kayaking to Tierra del Fuego. Ship *A*, which is 1.2 miles due south of *B*, observes Barbara to be at the direction N20.0°E. Ship *B* observes Barbara at the direction S70.0°E. At 2:10 P.M., ship *A* observes Barbara at N34.0°E, and ship *B* observes her at S55.0°E. Find Barbara's average speed.

40. Two straight roads meet at an angle of 60°. A jogger on one road is 3 miles from the intersection and running away from it at a rate of 5 miles per hour. At the same instant, another jogger is on the other road 9 miles from the intersection and moving toward it at 3 miles per hour. Assuming they run at constant speed:

a) Find the distance between them after 2 hours.

b) Determine the distance between them after *t* hours, that is, express the distance between them as a function of t, $0 \le t \le 3$.

C

41. In Problem 40, find the value of *t* for which the distance between the two joggers is a minimum. *(Hint: Refer to Section 2.4 Example 11.)*

42. Show that the area *A* of a triangle with sides *a*, *b*, *c* and corresponding angles α, β, γ is

$$A = \frac{a^2 \sin \beta \sin \gamma}{2 \sin \alpha} = \frac{b^2 \sin \alpha \sin \gamma}{2 \sin \beta} = \frac{c^2 \sin \alpha \sin \beta}{2 \sin \gamma}.$$

43. Show that the radius *r* of the inscribed circle of a triangle with sides *a*, *b*, and *c* is

$$r = \sqrt{\frac{(s-a)(s-b)(s-c)}{s}}$$

where $s = \frac{1}{2}(a + b + c)$.

44. Use the area formula

$$A = \frac{1}{2}bc \sin \alpha = \frac{1}{2}ac \sin \beta = \frac{1}{2}ab \sin \gamma$$

to derive the law of sines.

45. A point in the interior of an equilateral triangle is a distance of 10, 14, and 16 units from each of the vertices. Find the length of the side of the equilateral triangle.

VECTORS

S E C T I O N   7.3

Many physical quantities such as area, volume, energy, and time can be described by a single real number; these quantities are called **scalars.** Other important quantities such as displacement, velocity, acceleration, force, and momentum that have both a magnitude and a direction cannot be completely characterized by a single real number; these are called **vectors.**

For example, consider the change in location, or **displacement,** of a particle moving from a point *P* to a point *Q*. This can be represented by a directed line segment (arrow) with initial point *P* and terminal point *Q*, $\overrightarrow{PQ}$, as shown in Figure 50. The important idea here is the magnitude (length) and direction, not the position. If we start from another initial point *R* to another point *S* such that the displacement $\overrightarrow{RS}$ has the same magnitude and direction as $\overrightarrow{PQ}$, then the two directed line segments are called equivalent. (Figure 51). In fact, there are an infinite number of directed line segments that are equivalent to $\overrightarrow{PQ}$ since the initial point can be positioned anywhere in the plane.

FIGURE 50

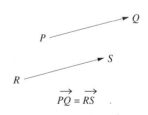

FIGURE 51

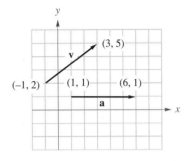

FIGURE 52

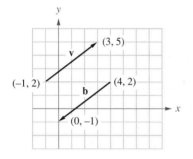

FIGURE 53

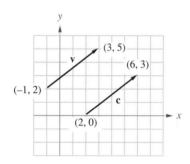

FIGURE 54

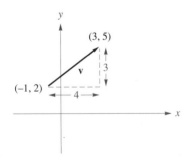

FIGURE 55

> A **vector** is a directed line segment.

Vectors are denoted by boldface letters such as **v** and **w**. Thus, the discussion above could be written as $\mathbf{v} = \overrightarrow{PQ} = \overrightarrow{RS}$.

We define the collection of all directed line segments with zero length to be the **zero vector 0.**

Problems involving vectors can often be simplified by imposing a rectangular coordinate system.

EXAMPLE 1   Let **v** be the vector from $(-1, 2)$ to $(3, 5)$ in the rectangular coordinate system. Determine whether or not each of the following vectors is equal to **v**:

   a)   **a** is the vector from $(1, 1)$ to $(6, 1)$

   b)   **b** is the vector from $(4, 2)$ to $(0, -1)$

   c)   **c** is the vector from $(2, 0)$ to $(6, 3)$

SOLUTION   If we draw a picture of **v** and **a**, it is clear that they have different directions. So $\mathbf{v} \neq \mathbf{a}$. (See Figure 52.)

Vector **b** appears to have the same magnitude and slope as **v** but is directed toward the lower left. So $\mathbf{v} \neq \mathbf{b}$. (See Figure 53.)

Vectors **c** and **v** *appear* to have the same magnitude and direction (Figure 54). To be sure, we calculate length and slope of each. Using the formulas for distance and slope, we find

$$\text{length of } \mathbf{c} = \sqrt{(6 - 2)^2 + (3 - 0)^2} = 5$$
$$\text{length of } \mathbf{v} = \sqrt{[3 - (-1)]^2 + (5 - 2)^2} = 5$$
$$\text{slope of } \mathbf{c} = \frac{3 - 0}{6 - 2} = \frac{3}{4}$$
$$\text{slope of } \mathbf{v} = \frac{5 - 2}{3 - (-1)} = \frac{3}{4}.$$

Showing that two vectors have the same slope does not imply they have the same direction. Since **v** and **c** also are directed to the upper right, $\mathbf{v} = \mathbf{c}$.

A vector can be uniquely represented by specifying the horizontal change and vertical change (from initial point to terminal point). For example, the vector **v** in Example 1 from $(-1, 2)$ to $(3, 5)$ has a horizontal change of 4 units to the right and a vertical change of 3 units up. This is denoted

$$\mathbf{v} = \langle 4, 3 \rangle$$

and is called the component form of **v**. See Figure 55.

> Suppose **v** is a vector with horizontal change v_1 (to the right is positive, to the left is negative) and vertical change v_2 (upward is positive, downward is negative). The **component form** of **v** is
>
> $$\mathbf{v} = \langle v_1, v_2 \rangle.$$
>
> The numbers v_1 and v_2 are called the **components** of **v**.

Component Form of a Vector

If a vector is positioned on a rectangular coordinate system such that its initial point is $P(x_1, y_1)$, and its terminal point is $Q(x_2, y_2)$, then the component form is (Figure 56)

$$\overrightarrow{PQ} = \langle \text{horizontal change, vertical change} \rangle$$
$$= \langle x_2 - x_1, y_2 - y_1 \rangle.$$

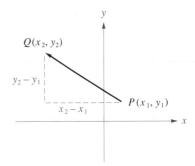

FIGURE 56

EXAMPLE 2 Find the component form of the vectors **a**, **b**, and **c** in Example 1.

SOLUTION Since the initial and terminal points of **a** are $(1, 1)$ and $(6, 1)$, respectively, the components of **a** are

$$\mathbf{a} = \langle 6 - 1, 1 - 1 \rangle = \langle 5, 0 \rangle.$$

Similarly,

$$\mathbf{b} = \langle 0 - 4, -1 - 2 \rangle = \langle -4, -3 \rangle$$
$$\mathbf{c} = \langle 6 - 2, 3 - 0 \rangle = \langle 4, 3 \rangle.$$

Given an initial point and a terminal point, we can always express the vector they determine in component form. However, if we know only the component form of a vector, it is not possible to determine the location of the initial or terminal points. We have already noted that a vector is not "changed" if it translated since the magnitude and direction remain the same. It is sometimes convenient to place the initial point at the origin of the rectangular coordinate system (Figure 57). In this case the vector is said to be in **standard position.** If the vector $\mathbf{v} = \langle v_1, v_2 \rangle$ is in standard position, then the terminal point coincides with the point $R(v_1, v_2)$ in the xy-plane. In this way the vector **v** serves to identify the point R, and **v** is called the **position vector** of R. Thus, for every vector there corresponds a point in the plane, and for every point in the plane there corresponds a vector. This correspondence implies the following.

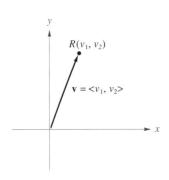

FIGURE 57

> Two vectors $\mathbf{v} = \langle v_1, v_2 \rangle$ and $\mathbf{w} = \langle w_1, w_2 \rangle$ are equal if and only if $v_1 = w_1$ and $v_2 = w_2$.

The **magnitude** (or length) of a vector **v** is also called the **norm** of the vector and is denoted $\|\mathbf{v}\|$ (Figure 58). Applying the Pythagorean theorem, we have the following:

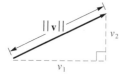

FIGURE 58

> If $\mathbf{v} = \langle v_1, v_2 \rangle$, then
> $$\|\mathbf{v}\| = \sqrt{v_1{}^2 + v_2{}^2}.$$

For example, the vector $\mathbf{v} = \langle -4, 2 \rangle$ has the norm

$$\|\mathbf{v}\| = \sqrt{(-4)^2 + 2^2} = \sqrt{20} = 2\sqrt{5}.$$

When working with vectors, always try to keep two viewpoints: geometric and algebraic (analytic). In physical applications it is often advantageous to concentrate on geometric relationships among vectors, and imposing a coordinate system may only cloud the issue. On the other hand, introducing a coordinate system can facilitate computations. We will therefore formulate operations with vectors geometrically as well as in terms of their components.

Adding Vectors

The sum $\mathbf{v} + \mathbf{w}$ of two vectors is defined as follows: Place the initial point of **w** at the terminal point of **v** (Figure 59). The vector $\mathbf{v} + \mathbf{w}$ is the vector from the initial point of **v** to the terminal point of **w**. This can be interpreted in terms of displacement as follows: The sum $\mathbf{v} + \mathbf{w}$ is the net displacement resulting from a displacement represented by **v** followed by a displacement represented by **w**. Notice that the sum $\mathbf{w} + \mathbf{v}$, obtained by placing the initial point of **v** at the terminal point of **w**, is the same vector as $\mathbf{v} + \mathbf{w}$ (Figure 60). Thus, $\mathbf{v} + \mathbf{w} = \mathbf{w} + \mathbf{v}$.

In terms of components, vector addition is straightforward (Figure 61).

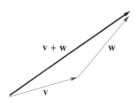

FIGURE 59

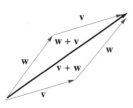

FIGURE 60

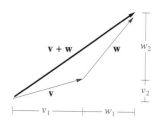

FIGURE 61

If $\mathbf{v} = \langle v_1, v_2 \rangle$ and $\mathbf{w} = \langle w_1, w_2 \rangle$ then
$$\mathbf{v} + \mathbf{w} = \langle v_1, v_2 \rangle + \langle w_1, w_2 \rangle = \langle v_1 + w_1, v_2 + w_2 \rangle.$$

Vector Addition

Consider how to define a scalar times a vector, such as $3\mathbf{v}$. The natural way to view this operation is in terms of repeated addition:

$$3\mathbf{v} = \mathbf{v} + \mathbf{v} + \mathbf{v}.$$

Geometrically this would look like Figure 62.
In terms of components, if $\mathbf{v} = \langle v_1, v_2 \rangle$, then

$$3\mathbf{v} = \mathbf{v} + \mathbf{v} + \mathbf{v}$$
$$= \langle v_1, v_2 \rangle + \langle v_1, v_2 \rangle + \langle v_1, v_2 \rangle$$
$$= \langle v_1 + v_1 + v_1, v_2 + v_2 + v_2 \rangle$$
$$= \langle 3v_1, 3v_2 \rangle.$$

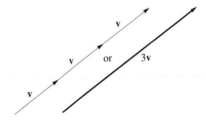

FIGURE 62

In general, if k is a scalar and $\mathbf{v}$ is a vector, the **scalar product** $k\mathbf{v}$ is defined as follows (see Figure 63):

Scalar Product

1) If $k > 0$, then $k\mathbf{v}$ is the vector whose direction is the same as that of $\mathbf{v}$ and whose length is k times the length of $\mathbf{v}$.

2) If $k < 0$, then $k\mathbf{v}$ is the vector whose direction is the opposite of $\mathbf{v}$ and whose length is $|k|$ times the length of $\mathbf{v}$.

3) If $k = 0$, $k\mathbf{v}$ is the zero vector $\mathbf{0}$.

FIGURE 63

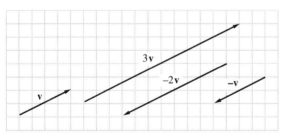

In addition, the vector $-\mathbf{v}$ is defined to be $(-1)\mathbf{v}$, as the figure illustrates.
In terms of components (see Figure 64), if $\mathbf{v} = \langle v_1, v_2 \rangle$, then

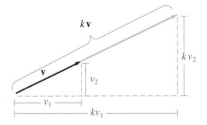

$$k\mathbf{v} = k\langle v_1, v_2 \rangle = \langle kv_1, kv_2 \rangle.$$

FIGURE 64

Subtracting Vectors

The **difference** $\mathbf{v} - \mathbf{w}$ of the two vectors $\mathbf{v}$ and $\mathbf{w}$ is defined as the vector that, when added to $\mathbf{w}$ gives the result $\mathbf{v}$ (see Figure 65).

An equivalent way to view $\mathbf{v} - \mathbf{w}$ is (see Figure 66):

$$\mathbf{v} - \mathbf{w} = \mathbf{v} + (-\mathbf{w}) = \mathbf{v} + (-1)\mathbf{w}.$$

FIGURE 65

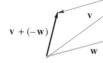

FIGURE 66

Vector Subtraction

> If $\mathbf{v} = \langle v_1, v_2 \rangle$ and $\mathbf{w} = \langle w_1, w_2 \rangle$, then
>
> $$\mathbf{v} - \mathbf{w} = \langle v_1, v_2 \rangle - \langle w_1, w_2 \rangle = \langle v_1 - w_1, v_2 - w_2 \rangle.$$

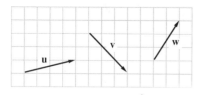

FIGURE 67

EXAMPLE 3 Using the vectors illustrated in Figure 67, graph $2\mathbf{v} + \mathbf{u}$ and $\mathbf{v} - \mathbf{u} + 3\mathbf{w}$.

SOLUTION

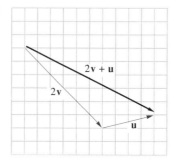

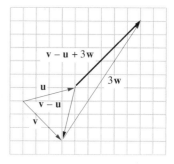

EXAMPLE 4 For the vectors $\mathbf{u}$, $\mathbf{v}$, and $\mathbf{w}$ in Example 3, find the components and the norms of $2\mathbf{v} + \mathbf{u}$ and $\mathbf{v} - \mathbf{u} + 3\mathbf{w}$.

SOLUTION First we find the components for $\mathbf{u}$, $\mathbf{v}$, and $\mathbf{w}$, based on their graphs in Example 3:

$$\mathbf{u} = \langle 4, 1 \rangle, \qquad \mathbf{v} = \langle 3, -3 \rangle, \qquad \mathbf{w} = \langle 2, 3 \rangle.$$

Thus

$$2\mathbf{v} + \mathbf{u} = 2\langle 3, -3 \rangle + \langle 4, 1 \rangle = \langle 6, -6 \rangle + \langle 4, 1 \rangle = \langle 10, -5 \rangle.$$

The norm (magnitude) of this vector is

$$\|2\mathbf{v} + \mathbf{u}\| = \sqrt{10^2 + (-5)^2} = 5\sqrt{5}.$$

Similarly,

$$v - u + 3w = \langle 3, -3 \rangle - \langle 4, 1 \rangle + 3 \langle 2, 3 \rangle$$
$$= \langle 3, -3 \rangle - \langle 4, 1 \rangle + \langle 6, 9 \rangle$$
$$= \langle 5, 5 \rangle.$$

The norm of this vector is

$$\|v - u + 3w\| = \sqrt{5^2 + 5^2} = 5\sqrt{2}.$$

Vector addition and scalar multiplication have many of the properties of addition and multiplication of real numbers. In fact, an algebraic system for vectors can be formulated based on these properties; this system turns out to be useful in engineering and science as well as in abstract mathematics.

For the vectors u, v, w and scalars c and d, the following relation-ships hold:

Properties of Vector Addition and Scalar Multiplication

1) $u + v = v + u$

2) $(u + v) + w = u + (v + w)$

3) $u + 0 = u$

4) $u + (-u) = 0$

5) $c(du) = (cd)u$

6) $c(u + v) = cu + cv$

7) $(c + d)u = cu + du$

8) $1u = u,\ 0u = 0$

We have already given a geometric proof for Property 1. To illustrate that these properties can be established geometrically or analytically, we will also prove this first property analytically.

Analytic Proof of Property 1 Let $u = \langle u_1, u_2 \rangle$ and $v = \langle v_1, v_2 \rangle$. Then

$$u + v = \langle u_1, u_2 \rangle + \langle v_1, v_2 \rangle$$
$$= \langle u_1 + v_1, u_2 + v_2 \rangle$$
$$= \langle v_1 + u_1, v_2 + u_2 \rangle$$
$$= \langle v_1, v_2 \rangle + \langle u_1, u_2 \rangle$$
$$= v + u.$$

Notice that this proof depends on the commutative law of addition for real numbers. Similarly, an analytic proof of Property 6 relies on the distributive property of real numbers.

Analytic Proof of Property 6 Let $\mathbf{u} = \langle u_1, u_2 \rangle$ and $\mathbf{v} = \langle v_1, v_2 \rangle$. Then

$$
\begin{aligned}
c(\mathbf{u} + \mathbf{v}) &= c(\langle u_1, u_2 \rangle + \langle v_1, v_2 \rangle) \\
&= c(\langle u_1 + v_1, u_2 + v_2 \rangle) \\
&= \langle c(u_1 + v_1), c(u_2 + v_2) \rangle \\
&= \langle cu_1 + cv_1, cu_2 + cv_2 \rangle \\
&= \langle cu_1, cu_2 \rangle + \langle cv_1, cv_2 \rangle \\
&= c\langle u_1, u_2 \rangle + c\langle v_1, v_2 \rangle \\
&= c\mathbf{u} + c\mathbf{v}.
\end{aligned}
$$

The proofs of the remaining properties are left as exercises.

A vector $\mathbf{u}$ such that $\|\mathbf{u}\| = 1$ is called a **unit vector.** In some applications it is useful to find a unit vector having the same direction as a given vector $\mathbf{v}$. First, note from our definition of scalar multiplication that

$$\|k\mathbf{v}\| = |k|\,\|\mathbf{v}\|.$$

To find a unit vector $\mathbf{u}$ in the same direction as $\mathbf{v}$, we need to find a scalar $k > 0$, such that $\mathbf{u} = k\mathbf{v}$, that satisfies

$$\|k\mathbf{v}\| = 1.$$

Since $k > 0$, it follows that

$$k\|\mathbf{v}\| = 1$$

$$k = \frac{1}{\|\mathbf{v}\|}.$$

Unit Vector

> A unit vector $\mathbf{u}$ having the same direction as the nonzero vector $\mathbf{v}$ is
>
> $$\mathbf{u} = \frac{1}{\|\mathbf{v}\|}\,\mathbf{v}.$$
>
> This is sometimes written as $\mathbf{u} = \mathbf{v}/\|\mathbf{v}\|$

Finding a unit vector with the same direction as $\mathbf{v}$ is called **normalizing $\mathbf{v}$.**

EXAMPLE 5   Find a unit vector in the direction of $\mathbf{v} = \langle 3, -2 \rangle$, and verify that the resulting vector has length 1.

SOLUTION First we find the length of $\mathbf{v}$:

$$\|\mathbf{v}\| = \sqrt{3^2 + (-2)^2} = \sqrt{13}.$$

Then we multiply $\mathbf{v}$ by the scalar $1/\|\mathbf{v}\|$:

$$\frac{\mathbf{v}}{\|\mathbf{v}\|} = \frac{\langle 3, -2 \rangle}{\sqrt{13}} = \left\langle \frac{3}{\sqrt{13}}, \frac{-2}{\sqrt{13}} \right\rangle.$$

Therefore, the magnitude of this vector is

$$\sqrt{\left(\frac{3}{\sqrt{13}}\right)^2 + \left(\frac{-2}{\sqrt{13}}\right)^2} = \sqrt{\frac{9}{13} + \frac{4}{13}} = 1.$$

The unit vectors $\langle 1, 0 \rangle$ and $\langle 0, 1 \rangle$, called **standard unit vectors,** are frequently used and are denoted (see Figure 68)

$$\mathbf{i} = \langle 1, 0 \rangle \quad \text{and} \quad \mathbf{j} = \langle 0, 1 \rangle.$$

Any vector $\mathbf{v}$ is expressible as a linear combination of $\mathbf{i}$ and $\mathbf{j}$:

$$\mathbf{v} = \langle v_1, v_2 \rangle = \langle v_1, 0 \rangle + \langle 0, v_2 \rangle$$
$$= v_1 \langle 1, 0 \rangle + v_2 \langle 0, 1 \rangle$$
$$= v_1 \mathbf{i} + v_2 \mathbf{j}.$$

This amounts to another notation for the component form of a vector. For example,

$$\langle 7, 3 \rangle = 7\mathbf{i} + 3\mathbf{j}$$
$$\langle \sqrt{3}, 1 \rangle = \sqrt{3}\mathbf{i} + \mathbf{j}$$
$$\langle 2, -12 \rangle = 2\mathbf{i} - 12\mathbf{j}$$
$$\langle -4, 0 \rangle = -4\mathbf{i}.$$

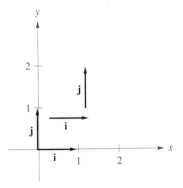

FIGURE 68

EXAMPLE 6   Find a vector with the same direction as

$$\mathbf{v} = (3\mathbf{i} + \mathbf{j}) - (2\mathbf{i} + 5\mathbf{j})$$

but with magnitude 8.

SOLUTION Simplify $\mathbf{v}$ and find its norm. We show both notations:

$$\mathbf{v} = (3\mathbf{i} + \mathbf{j}) - (2\mathbf{i} + 5\mathbf{j}) \quad \text{or} \quad \mathbf{v} = \langle 3, 1 \rangle - \langle 2, 5 \rangle$$
$$= \mathbf{i} - 4\mathbf{j} \qquad\qquad\qquad = \langle 1, -4 \rangle.$$

So

$$\|\mathbf{v}\| = \sqrt{1^2 + (-4)^2} = \sqrt{17}.$$

A unit vector in the same direction as **v** is

$$\frac{\mathbf{v}}{\|\mathbf{v}\|} = \frac{\mathbf{i} - 4\mathbf{j}}{\sqrt{17}} = \frac{1}{\sqrt{17}}\mathbf{i} - \frac{4}{\sqrt{17}}\mathbf{j} \quad \text{or} \quad \frac{\mathbf{v}}{\|\mathbf{v}\|} = \frac{\langle 1, -4 \rangle}{\sqrt{17}} = \left\langle \frac{1}{\sqrt{17}}, \frac{-4}{\sqrt{17}} \right\rangle.$$

A vector in the same direction as this unit vector (and therefore as **v**) with magnitude 8 is

$$8\left(\frac{1}{\sqrt{17}}\mathbf{i} - \frac{4}{\sqrt{17}}\mathbf{j}\right) = \frac{8}{\sqrt{17}}\mathbf{i} - \frac{32}{\sqrt{17}}\mathbf{j} \quad \text{or} \quad 8\left\langle \frac{1}{\sqrt{17}}, \frac{-4}{\sqrt{17}} \right\rangle = \left\langle \frac{8}{\sqrt{17}}, \frac{-32}{\sqrt{17}} \right\rangle.$$

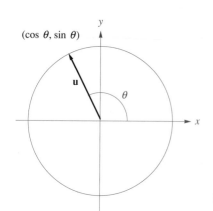

$(\cos \theta, \sin \theta)$

FIGURE 69

If **u** is a unit vector in standard position, and θ is the angle from the positive x-axis (measured counterclockwise) to **u**, then the terminal point of **u** is $(\cos \theta, \sin \theta)$ (Figure 69). Thus

$$\mathbf{u} = \langle \cos \theta, \sin \theta \rangle = (\cos \theta)\mathbf{i} + (\sin \theta)\mathbf{j}.$$

Consequently, if **v** is any vector making an angle θ with the positive x-axis (measured counterclockwise), then **v** has the same direction as **u**. Therefore,

$$\mathbf{v} = \|\mathbf{v}\|\mathbf{u} = \|\mathbf{v}\|\langle \cos \theta, \sin \theta \rangle$$
$$= \|\mathbf{v}\| \cos \theta \mathbf{i} + \|\mathbf{v}\| \sin \theta \mathbf{j}.$$

For example, the vector of length 6 making an angle of 240° with the positive x-axis is

$$\mathbf{v} = \|\mathbf{v}\|\langle \cos \theta, \sin \theta \rangle$$
$$= 6\langle \cos 240°, \sin 240° \rangle$$
$$= \langle -3, -3\sqrt{3} \rangle.$$

Vectors are used in a multitude of applications in science and engineering. One example is the representation of **velocity.** The velocity of an object is represented by a vector in the direction of movement with magnitude equal to the speed.

EXAMPLE 7   An airplane is flying in calm conditions N30°E at 350 miles per hour. Suddenly the plane encounters a wind of 60 miles per hour in the direction N70°E. Find the actual speed and direction of the airplane.

SOLUTION   We represent the velocities of the airplane and wind as vectors with respective lengths 350 and 60. The actual velocity **v** of the airplane is the sum of these two vectors; in other words, the direction is as pictured in Figure 70 and the speed is ‖**v**‖.

Geometric Approach Using the vector diagram in Figure 70, we can determine the obtuse angle to be 140°. Consider the triangle in Figure 71. Using the law of cosines gives

$$\|\mathbf{v}\| = \sqrt{350^2 + 60^2 - 2(350)(60)\cos 140°}$$
$$= 397.83648 \ldots \approx 398.$$

Using the law of sines gives

$$\sin \alpha = \frac{60 \sin 140°}{397.83648 \ldots}$$

$$\alpha = \sin^{-1} \frac{60 \sin 140°}{397.83648 \ldots} = 5.56313 \ldots \approx 6°.$$

Therefore the actual speed and direction of the airplane is N36°E at 398 miles per hour.

Analytic Approach We can represent the velocity of the airplane (Figure 72) with the vector

$$\mathbf{a} = 350\langle \cos 60°, \sin 60° \rangle = \left\langle 350\left(\frac{1}{2}\right), 350\left(\frac{\sqrt{3}}{2}\right)\right\rangle$$

$$= \langle 175, 175\sqrt{3} \rangle = \langle 175, 303.10889 \ldots \rangle$$

and the velocity of the wind (Figure 73) with the vector

$$\mathbf{w} = 60\langle \cos 20°, \sin 20° \rangle = \langle 60 \cos 20°, 60 \sin 20° \rangle$$

$$= \langle 56.38155 \ldots, 20.5212 \ldots \rangle.$$

The actual velocity of the airplane is

$$\mathbf{v} = \mathbf{a} + \mathbf{w} = \langle 231.38155 \ldots, 323.6301 \ldots \rangle.$$

The speed of the airplane is

$$\|\mathbf{v}\| = \sqrt{(231.3815 \ldots)^2 + (323.6301 \ldots)^2}$$

$$= 397.8364 \ldots \approx 398.$$

To find the direction, write **v** in the form

$$\mathbf{v} = \|\mathbf{v}\|\langle \cos \theta, \sin \theta \rangle$$

$$= 397.8364 \ldots \left\langle \frac{231.3815 \ldots}{397.8364 \ldots}, \frac{323.6301 \ldots}{397.8364 \ldots}\right\rangle.$$

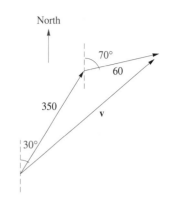

FIGURE 70

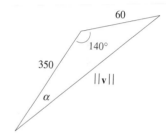

FIGURE 71

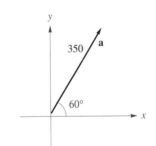

FIGURE 72

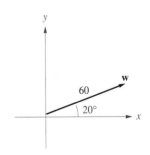

FIGURE 73

This implies that

$$\cos \theta = \frac{231.3815\ldots}{397.8364\ldots}$$

$$\theta = \cos^{-1} \frac{231.3815\ldots}{397.8364\ldots} \approx 54°.$$

Therefore, the speed of the airplane is 398 miles per hour, and the direction makes an angle of 54° with the positive x-axis. This agrees with the answer of the geometric approach.

EXERCISE SET 7.3

A

In Problems 1 through 14, refer to Figure 74.

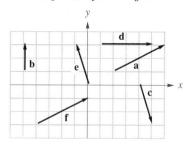

FIGURE 74

1. Name the vector that is in standard position.

2. Name the two vectors that are equal.

3. Name the two vectors whose sum is the zero vector.

4. Name the vector(s) whose second component is negative.

5. Name the two vectors whose sum is the vector **f**.

6. Graph **e** + **f**.

7. Graph **a** + **c**.

8. Graph **a** − **e**.

9. Graph **e** − **b**.

10. Graph 4**c**.

11. Graph $(-\frac{1}{2})$**a**.

12. Graph **a** + 2**e**.

13. Graph 2**c** + **d** − 3**a**.

14. Graph 4**b** − **f** + **d**.

In Problems 15 through 17, find the component form of $\mathbf{v} = \overrightarrow{PQ}$ for the given coordinates of points P and Q.

15. $P(2, 8)$, $Q(12, 14)$

16. $P(-4, 11)$, $Q(9, 4)$

17. $P(7, -8)$, $Q(0, 2)$

18. Find the terminal point of $\mathbf{v} = \langle 16, -9 \rangle$ if the initial point is (4, 5).

19. Find the terminal point of $\mathbf{v} = \langle -8, -1 \rangle$ if the initial point is (14, 3).

20. Find the initial point of $\mathbf{v} = \langle -6, 7 \rangle$ if the terminal point is (10, 2).

In Problems 21 through 26, find $\|\mathbf{v}\|$.

21. $\mathbf{v} = \langle 5, -12 \rangle$

22. $\mathbf{v} = \langle -24, -7 \rangle$

23. $\mathbf{v} = \langle 6, -6 \rangle$

24. $\mathbf{v} = \langle -9, 5 \rangle$

25. $\mathbf{v} = \langle 2, 2\sqrt{3} \rangle$

26. $\mathbf{v} = \langle -\sqrt{5}, 4 \rangle$

B

In Problems 27 through 38, let $\mathbf{u} = \langle 4, 3 \rangle$, $\mathbf{v} = \langle 5, -5 \rangle$, and $\mathbf{w} = \langle -12, 10 \rangle$. Find each quantity.

27. $2\mathbf{w} - 6\mathbf{v} + \mathbf{u}$

28. $3\mathbf{v} + 2\mathbf{u} - \mathbf{w}$

29. $\|3\mathbf{u}\|$

30. $\|-2\mathbf{v}\|$

31. $\|\mathbf{v} + \mathbf{w}\|$

32. $\|\mathbf{u} - \mathbf{w}\|$

33. $\|\mathbf{v}\| + \|\mathbf{w}\|$

34. $\|\mathbf{u}\| - \|\mathbf{w}\|$

35. $\dfrac{1}{\|\mathbf{w}\|} \|\mathbf{w}\|$

36. $\dfrac{1}{\|\mathbf{v}\|} \|\mathbf{v}\|$

37. $\mathbf{x}$, if $2\mathbf{x} + \mathbf{v} = 5\mathbf{u}$

38. $\mathbf{x}$, if $3\mathbf{x} - \mathbf{w} = 2\mathbf{v}$

39. Find a unit vector in the direction of $\mathbf{v} = \langle -4, -2\sqrt{5} \rangle$.

40. Find a unit vector in the direction of $\mathbf{v} = \langle 8, -6 \rangle$.

41. Find a vector with magnitude 20 in the direction of $\mathbf{v} = \langle 2, 2 \rangle$.

42. Find a vector with magnitude 14 in the direction of $\mathbf{v} = \langle -1, 2 \rangle$.

43. Find a vector with magnitude 5 whose first (horizontal) component is twice the second (vertical) component.

44. Find a vector with magnitude 15 whose second (vertical) component is three times the first (horizontal) component.

In Problems 45 through 52, find the component form of $\mathbf{v}$ *with the given magnitude and the angle it makes with the positive x-axis.*

45. $\|\mathbf{v}\| = 10$, $\theta = 135°$ **46.** $\|\mathbf{v}\| = 24$, $\theta = 120°$

47. $\|\mathbf{v}\| = 3$, $\theta = 180°$ **48.** $\|\mathbf{v}\| = 7$, $\theta = 270°$

49. $\|\mathbf{v}\| = \sqrt{3}$, $\theta = 210°$ **50.** $\|\mathbf{v}\| = 3\sqrt{2}$, $\theta = 315°$

51. $\|\mathbf{v}\| = 5$, $\theta = 70°$ **52.** $\|\mathbf{v}\| = 9$, $\theta = 160°$

In Problems 53 through 55, find $\|\mathbf{v}\|$ *and the angle* θ *that* $\mathbf{v}$ *makes with the positive x-axis.*

53. $\mathbf{v} = \langle 1, -\sqrt{3} \rangle$ **54.** $\mathbf{v} = \langle \sqrt{6} - \sqrt{2}, \sqrt{6} + \sqrt{2} \rangle$

55. $\mathbf{v} = \langle 1 + \sqrt{5}, \sqrt{10 - 2\sqrt{5}} \rangle$

C

56. For any triangle ABC, let D be the midpoint of side $\overline{AB}$ and let E be the midpoint of side $\overline{BC}$. Prove that $\overline{DE}$ is parallel to $\overline{AC}$ and half as long. *(Hint: Place triangle ABC on a coordinate system (Figure 75), and find the component form of relevant vectors.)*

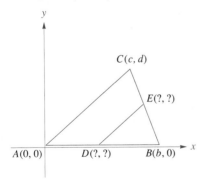

FIGURE 75

57. Show that the midpoints of the sides of any quadrilateral form a parallelogram. *(Hint: Consider a quadrilateral on a coordinate system with vertices $A(0, 0)$, $B(b, 0)$, $C(c, d)$, $D(e, f)$, and find the component form of relevant vectors.)*

58. An airplane is flying in calm conditions N20°W at 280 miles per hour. Suddenly the plane encounters a wind of 40 miles per hour in the direction N65°W. Find the actual speed and direction of the airplane.

59. A boat is traveling in calm waters S80°W at 50 miles per hour. Suddenly the vessel encounters a current of 12 miles per hour in the direction N70°E. Find the actual speed and direction of the boat.

60. An airplane pilot needs to maintain a true course and speed (relative to the ground) of N75°E at 300 miles per hour. The wind is blowing N10°E at 20 miles per hour. Approximate the required airspeed and direction (to the nearest mile per hour and degree, respectively).

61. Neglecting friction, the weight $\mathbf{w}$ of an object on an inclined plane can be considered as the sum of two perpendicular forces $\mathbf{F}_1$ and $\mathbf{F}_2$. The magnitudes of these forces $\|\mathbf{F}_1\|$ and $\|\mathbf{F}_2\|$ represent the number of pounds the object exerts down the incline and against the incline, respectively (Figure 76). If an inclined plane makes an angle of 5° with the horizontal, how many pounds of force does an object weighing 400 pounds exert down the incline?

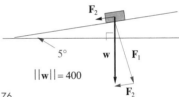

FIGURE 76

62. Repeat Problem 61 for an incline making a 10°-angle with the horizontal and an object weighing 300 pounds.

63. Prove the following properties for vectors.

    a) $(\mathbf{u} + \mathbf{v}) + \mathbf{w} = \mathbf{u} + (\mathbf{v} + \mathbf{w})$

    b) $\mathbf{u} + \mathbf{0} = \mathbf{u}$

    c) $\mathbf{u} + (-\mathbf{u}) = \mathbf{0}$

    d) $c(d\mathbf{u}) = (cd)\mathbf{u}$

    e) $(c + d)\mathbf{u} = c\mathbf{u} + d\mathbf{u}$

Consider two vectors $\mathbf{v} = \langle v_1, v_2 \rangle$ and $\mathbf{w} = \langle w_1, w_2 \rangle$. The inner product $\mathbf{v} \cdot \mathbf{w}$ is a real number defined by $\mathbf{v} \cdot \mathbf{w} = v_1 w_1 + v_2 w_2$. (This product is also called the dot product.)

In Problems 64 through 67, compute the inner product **v** · **w**.

64. **v** $= \langle 7, -1 \rangle$, **w** $= \langle 3, -4 \rangle$

65. **v** $= \langle 3, \sqrt{3} \rangle$, **w** $= \langle 4\sqrt{3}, -4 \rangle$

66. **v** $= \langle 3, 0 \rangle$, **w** $= \langle 0, -2 \rangle$

67. **v** $= \langle -6, 4 \rangle$, **w** $= \langle -6, -9 \rangle$

68. a) Suppose that θ is the smallest nonnegative angle between two vectors **v** and **w**. Use the law of cosines to show that

$$\cos \theta = \frac{\|\mathbf{v}\|^2 + \|\mathbf{w}\|^2 - \|\mathbf{v} - \mathbf{w}\|^2}{2\|\mathbf{v}\|\,\|\mathbf{w}\|}.$$

b) Use part (a) and the definition of inner product to show that

$$\cos \theta = \frac{\mathbf{v} \cdot \mathbf{w}}{\|\mathbf{v}\|\,\|\mathbf{w}\|}.$$

69. Use part (b) of Problem 68 to find the smallest nonnegative angle between the two vectors given in Problem 64.

70. Use part (b) of Problem 68 to find the smallest nonnegative angle between the two vectors given in Problem 65.

71. Use part (b) of Problem 68 to find the smallest nonnegative angle between the two vectors given in Problem 66.

72. Use part (b) of Problem 68 to find the smallest nonnegative angle between the two vectors given in Problem 67.

73. Use part (b) of Problem 68 to prove that two vectors are perpendicular if and only if **v** · **w** = 0.

S E C T I O N   7.4   COMPLEX NUMBERS IN TRIGONOMETRIC FORM

Recall that a complex number is any number that can be written in the form $a + bi$, where $i = \sqrt{-1}$ and a and b are real numbers. Because a complex number $a + bi$ is completely determined by the real numbers a and b, there is a correspondence

$$a + bi \leftrightarrow (a, b)$$

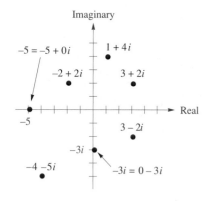

FIGURE 77

between complex numbers and ordered pairs of real numbers. Specifically, for each complex number there corresponds a unique ordered pair of real numbers; and for each ordered pair of real numbers there corresponds a unique complex number. This suggests that we can represent complex numbers geometrically in a coordinate plane, called the **complex plane** or the **Gaussian plane** (after Karl F. Gauss). The horizontal axis is called the **real axis,** which is simply the real number line in its customary orientation. The vertical axis is called the **imaginary axis.** The graphs of several complex numbers are shown in Figure 77.

The absolute value of a real number was defined in Chapter 1 as the distance between the real number and the origin. The absolute value of a complex number, called the **modulus** (plural is moduli), is defined in the same way. The distance formula leads to the following definition.

If $z = a + bi$, then the absolute value or modulus of z is defined as the distance between z and the origin:

$$|z| = \sqrt{a^2 + b^2}.$$

See Figure 78.

Absolute Value of a Complex Number

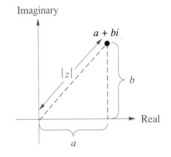

FIGURE 78

EXAMPLE 1   Find $|z|$ for each value of z:

a) $z = 12 - 5i$ **b)** $z = 2 + i$ **c)** $z = 8i$

SOLUTION

a) $|12 - 5i| = \sqrt{12^2 + (-5)^2} = \sqrt{169} = 13.$

b) $|2 + i| = \sqrt{2^2 + 1^2} = \sqrt{5}.$

c) $|8i| = \sqrt{0^2 + 8^2} = 8.$

EXAMPLE 2   Graph each of the following in the complex plane:

a) All real numbers x satisfying $|x| \le 4$

b) All complex numbers z satisfying $|z| \le 4$

SOLUTION According to the definition of absolute value, in each case we graph all numbers that are less than or equal to 4 units from the origin (Figure 79). In the graph of (a), the endpoints are included. Similarly, in (b) the boundary of the disk is solid, indicating that it is part of the graphed solution.

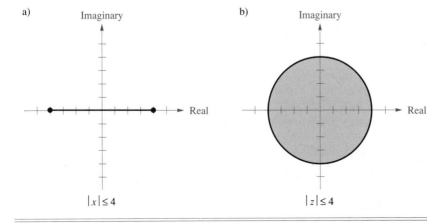

FIGURE 79

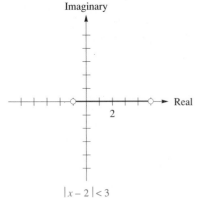

$$|x - 2| < 3$$

FIGURE 80

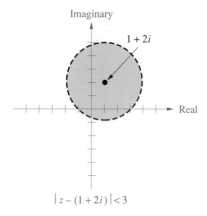

$$|z - (1 + 2i)| < 3$$

FIGURE 81

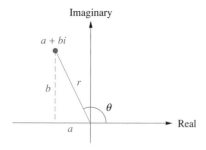

FIGURE 82

Recall from Chapter 1 that $|x_1 - x_2|$ can be interpreted as the distance between the real numbers x_1 and x_2. Similarly, $|z_1 - z_2|$ can be interpreted as the distance between the complex numbers z_1 and z_2.

EXAMPLE 3 Graph each of the following in the complex plane:

a) All real numbers x satisfying $|x - 2| < 3$

b) All complex numbers z satisfying $|z - (1 + 2i)| < 3$

SOLUTION

a) The inequality $|x - 2| < 3$ can be interpreted as the set of all real numbers x whose distance from the number 2 is less than 3 units. The graph is the open interval in Figure 80. This open interval is said to be centered at 2 with a radius of 3.

b) The inequality $|z - (1 + 2i)| < 3$ can be interpreted as the set of all complex numbers z whose distance from the number $1 + 2i$ is less than 3 units. This is a disk in the complex plane with radius 3 centered at the point corresponding to $1 + 2i$ (Figure 81). The boundary of the disk is not part of the graphed solution and is therefore dashed. This disk is said to be centered at $1 + 2i$ with a radius of 3.

The correspondence between complex numbers and points in the complex plane enables us to express any complex number in another useful form. Consider the graph of a complex number $z = a + bi$ shown in Figure 82. Let θ be the angle in standard position whose terminal side is the line segment from the origin to the point representing z. If we let r be the length of the line segment, then

$$r = |z| = \sqrt{a^2 + b^2}$$

and

$$\cos \theta = \frac{a}{r}, \qquad \sin \theta = \frac{b}{r}, \qquad \text{and} \qquad \tan \theta = \frac{b}{a}.$$

Therefore, we can substitute

$$a = r \cos \theta \qquad \text{and} \qquad b = r \sin \theta$$

in $z = a + bi$ to obtain

$$z = a + bi$$
$$= (r \cos \theta) + (r \sin \theta)i$$
$$= r(\cos \theta + i \sin \theta).$$

The **trigonometric form** of $z = a + bi$ is	*Trigonometric Form of a*

The **trigonometric form** of $z = a + bi$ is

$$z = r(\cos \theta + i \sin \theta)$$

where $r = |z| = \sqrt{a^2 + b^2}$, and $\tan \theta = b/a$. Sometimes the abbreviation

$$r(\cos \theta + i \sin \theta) = r \text{ cis } \theta$$

is used.

Trigonometric Form of a Complex Number

The angle θ is called the **argument** of z and is written

$$\theta = \arg z.$$

The trigonometric representation is not unique. In fact, there are an infinite number of trigonometric representations for a complex number because if θ is an argument of z, then so is $(\theta + 360k)°$ for any integer k. However, the representation is unique for all nonzero z if $0 \le \theta < 360°$.

The form $a + bi$ is called the **rectangular form** of the complex number. When you convert a complex number from rectangular form into trigonometric form, it is important that you know where the number is located in the complex plane.

EXAMPLE 4   Express the following complex numbers in trigonometric form with $0° \le \theta < 360°$:

a) $-2 + 2i$ **b)** $\sqrt{3} - i$ **c)** $-4i$ **d)** $-1.9 - 7.8i$

SOLUTION

a) First we consider the graph of $z = -2 + 2i$ in the complex plane (see Figure 83). Identifying $a = -2$ and $b = 2$, we have

$$r = \sqrt{(-2)^2 + 2^2} = 2\sqrt{2} \quad \text{and} \quad \tan \theta = \frac{2}{-2} = -1.$$

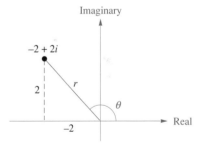

FIGURE 83

Since $-2 + 2i$ lies in the second quadrant, $\theta = 135°$. Thus,

$$-2 + 2i = 2\sqrt{2}(\cos 135° + i \sin 135°)$$

$$= 2\sqrt{2} \text{ cis } 135°.$$

b) The graph of $z = \sqrt{3} - i$ is shown in Figure 84. Using $a = \sqrt{3}$ and $b = -1$, we have

$$r = \sqrt{(\sqrt{3})^2 + (-1)^2} = 2 \quad \text{and} \quad \tan \theta = \frac{-1}{\sqrt{3}}.$$

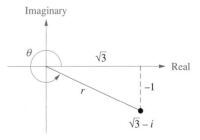

FIGURE 84

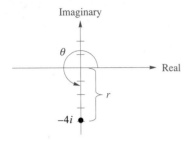

FIGURE 85

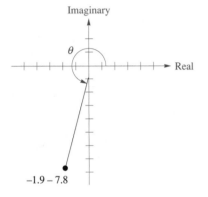

FIGURE 86

Since z lies in the fourth quadrant, $\theta = 330°$, so

$$\sqrt{3} - i = 2(\cos 330° + i \sin 330°)$$
$$= 2 \text{ cis } 330°.$$

c) Notice that for $a = 0$ and $b = -4$, $\tan \theta$ is undefined. Using the graph (Figure 85) of $z = -4i$, we see that the modulus (distance from the origin) is $r = 4$ and the argument is $\theta = 270°$:

$$-4i = 4(\cos 270° + i \sin 270°)$$
$$= 4 \text{ cis } 270°.$$

d) Since $a = -1.9$ and $b = -7.8$,

$$r = \sqrt{(-1.9)^2 + (-7.8)^2} = \sqrt{64.45} \approx 8.0$$

and

$$\tan \theta = \frac{-7.8}{-1.9}.$$

It follows that $\theta \approx 256°$ (Figure 86), and that

$$-1.9 - 7.81i \approx 8.0 \text{ cis } 256°.$$

EXAMPLE 5   Express the following complex numbers in rectangular form.

a) $5\sqrt{3}(\cos 60° + i \sin 60°)$ b)   6 cis 112°.

SOLUTION

a) $5\sqrt{3}(\cos 60° + i \sin 60°) = 5\sqrt{3} \cos 60° + 5\sqrt{3} \sin 60° \, i$

$$= 5\sqrt{3}\left(\frac{1}{2}\right) + 5\sqrt{3}\left(\frac{\sqrt{3}}{2}\right) i$$

$$= \frac{5\sqrt{3}}{2} + \frac{15}{2} i.$$

b)   Using a calculator, we obtain

$$6 \text{ cis } 112° = 6(\cos 112° + i \sin 112°)$$
$$= 6 \cos 112° + 6 \sin 112° \, i$$
$$\approx -2.2 + 5.6 \, i \quad \text{(to the nearest 0.1)}.$$

The trigonometric form is quite effective when multiplying or dividing two complex numbers. Consider the product,

$$z_1 z_2 = (r_1 \text{cis } \theta_1)(r_2 \text{cis } \theta_2)$$
$$= r_1(\cos \theta_1 + i \sin \theta_1) r_2(\cos \theta_2 + i \sin \theta_2)$$
$$= r_1 r_2 (\cos \theta_1 + i \sin \theta_1)(\cos \theta_2 + i \sin \theta_2)$$
$$= r_1 r_2 (\cos \theta_1 \cos \theta_2 + i \sin \theta_1 \cos \theta_2 + i \sin \theta_2 \cos \theta_1 - \sin \theta_1 \sin \theta_2)$$
$$= r_1 r_2 [(\cos \theta_1 \cos \theta_2 - \sin \theta_1 \sin \theta_2) + i(\sin \theta_1 \cos \theta_2 + \sin \theta_2 \cos \theta_1)].$$

Now use the addition identities for sine and cosine:

$$= r_1 r_2 [\cos(\theta_1 + \theta_2) + i \sin(\theta_1 + \theta_2)]$$
$$= r_1 r_2 \text{cis}(\theta_1 + \theta_2).$$

In other words, to multiply two complex numbers in trigonometric form, multiply the moduli and add the arguments. A similar argument, left as an exercise, reveals, that to divide two complex numbers, divide the moduli and subtract the arguments.

Let $z_1 = r_1 \text{cis } \theta_1$ and $z_2 = r_2 \text{cis } \theta_2$. The product and quotient of z_1 and z_2 are given respectively as

$$z_1 z_2 = (r_1 \text{cis } \theta_1)(r_2 \text{cis } \theta_2) = r_1 r_2 \text{cis}(\theta_1 + \theta_2)$$

and

$$\frac{z_1}{z_2} = \frac{r_1 \text{cis } \theta_1}{r_2 \text{cis } \theta_2} = \frac{r_1}{r_2} \text{cis}(\theta_1 - \theta_2).$$

Multiplication and Division in Trigonometric Form

EXAMPLE 6   Use the trigonometric form to evaluate $z_1 z_2$, for

$$z_1 = 5\sqrt{3} + 5i \quad \text{and} \quad z_2 = -2 + 2\sqrt{3}i.$$

SOLUTION For z_1,

$$r_1 = \sqrt{(5\sqrt{3})^2 + 5^2} = 10 \quad \text{and} \quad \theta_1 = \tan^{-1}\left(\frac{5}{5\sqrt{3}}\right) = 30°.$$

So $z_1 = 10 \text{ cis } 30°$.

For z_2,

$$r_2 = \sqrt{(-2)^2 + (2\sqrt{3})^2} = 4 \quad \text{and} \quad \theta_2 = 180° + \tan^{-1}\left(\frac{2\sqrt{3}}{-2}\right) = 120°.$$

So $z_2 = 4 \text{ cis } 120°$ and

$$z_1 z_2 = (10 \text{ cis } 30°)(4 \text{ cis } 120°)$$
$$= (10)(4) \text{ cis}(30° + 120°)$$
$$= 40 \text{ cis } 150°.$$

In rectangular form, this is

$$40(\cos 150° + i \sin 150°) = (40 \cos 150°) + (40 \sin 150°)i$$
$$= -20\sqrt{3} + 20i.$$

EXAMPLE 7   Find z_1/z_2 for $z_1 = -6i$ and $z_2 = (\sqrt{5} - 1) + (\sqrt{10 + 2\sqrt{5}})i$, using trigonometric form.

SOLUTION By picturing $z_1 = -6i$ in the complex plane, we obtain the trigonometric representation $z_1 = 6 \text{ cis } 270°$. The graph of z_2 lies in the first quadrant. Using a calculator gives

$$\theta_2 = \tan^{-1}\left(\frac{\sqrt{10 + 2\sqrt{5}}}{\sqrt{5} - 1}\right)$$
$$= \tan^{-1}(3.07768\ldots)$$
$$= 72°$$

and

$$r_2 = \sqrt{(\sqrt{5} - 1)^2 + (\sqrt{10 + 2\sqrt{5}})^2} = \sqrt{16} = 4.$$

Thus, $z_2 = 4 \text{ cis } 72°$.

Now, to divide, we find the quotient of the moduli and subtract the arguments:

$$\frac{z_1}{z_2} = \frac{6 \text{ cis } 270°}{4 \text{ cis } 72°} = \frac{6}{4} \text{ cis}(270° - 72°) = \frac{3}{2} \text{ cis } 198°.$$

The rectangular form of this answer is

$$\frac{3}{2} \text{ cis } 198° = \frac{3}{2}(\cos 198° + i \sin 198°)$$

$$\approx -1.43 - 0.31i.$$

(It can be shown that the exact rectangular representation for this answer is

$$\frac{3}{2} \text{ cis } 198° = -\frac{3\sqrt{10 + 2\sqrt{5}}}{8} - \frac{3\sqrt{5} - 3}{8} i).$$

In alternating-current (AC) circuits, there is opposition to current flow, called **impedance.** An AC circuit that has a resistor, an inductor, and a capacitor in series (called an *RLC* circuit) is pictured in Figure 87.

An Application: Alternating Current

If the resistance across the resistor is R, the inductive reactance across the inductor is R_L, and the capacitive reactance across the capacitor is R_C, then the **impedance** Z can be expressed as

$$Z = R + (R_L - R_C)i \qquad \text{where } R \geq 0.$$

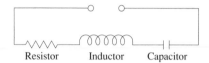

Resistor   Inductor   Capacitor

FIGURE 87

The quantities R, R_L, and R_C are measured in ohms. If I is the current (in amperes), the **voltage** V across these elements is

$$V = IZ.$$

When the impedance Z is expressed in trigonometric form, the argument θ is called the **phase angle** and has the restriction $-90° \leq \theta \leq 90°$. The **magnitude** of impedance or voltage is the absolute value of these quantities.

EXAMPLE 8   In an *RLC* alternating-current circuit, $R = 24$ ohms, $R_L = 2$ ohms, $R_C = 12$ ohms, and $I = 3$ amperes. Find the impedance, the voltage, the magnitudes for both, and the phase angle.

SOLUTION The impedance is

$$Z = R + (R_L - R_C)i$$
$$= 24 + (2 - 12)i$$
$$= 24 - 10i.$$

The magnitude of the impedance is

$$|Z| = \sqrt{24^2 + (-10)^2} = 26 \quad \text{ohms.}$$

The voltage is $V = IZ = 3(24 - 10i) = 72 - 30i$, and the magnitude is

$$|V| = \sqrt{72^2 + (-30)^2} = 78 \quad \text{volts.}$$

The phase angle is $\tan^{-1}(-10/24) \approx -23°$.

EXERCISE SET 7.4

A

Graph the complex numbers in Problems 1 through 6 in the complex plane.

1. $5 + 3i$ **2.** $-2 + 6i$ **3.** $8 - 4i$

4. $-5 - i$ **5.** -10 **6.** $7i$

In Problems 7 through 12, find the absolute values.

7. $|5 + 3i|$ **8.** $|-2 + 6i|$ **9.** $|8 - 4i|$

10. $|-5 - i|$ **11.** $|-10|$ **12.** $|7i|$

Graph each of the inequalities in Problems 13 through 20 in the complex plane.

13. a) All real numbers x satisfying $|x| \le 4$
    b) All complex numbers z satisfying $|z| \le 4$

14. a) All real numbers x satisfying $|x| < 5$
    b) All complex numbers z satisfying $|z| < 5$

15. a) All real numbers x satisfying $|x - 3| < 2$
    b) All complex numbers z satisfying $|z - 3| < 2$

16. a) All real numbers x satisfying $|x - 6| \le 1$
    b) All complex numbers z satisfying $|z - 6| \le 1$

17. a) All real numbers x satisfying $|x + 5| \le 3$
    b) All complex numbers z satisfying $|z + 5| \le 3$

18. a) All real numbers x satisfying $|x + 1| < 2$
    b) All complex numbers z satisfying $|z + 1| < 2$

19. All complex numbers z satisfying $|z - (3 + 4i)| \le 2$

20. All complex numbers z satisfying $|z - (-5 + i)| \le 3$

In Problems 21 through 32 express the complex numbers in trigonometric form, with $0° \le \theta < 360°$.

21. $1 + i$ **22** $-3 + 4i$

23. $15 - 36i$ **24.** $-7 - 24i$

25. $\sqrt{19} + 9i$ **26.** $8 + 7i$

27. $\sqrt{2} + i$ **28.** -6

29. $20i$ **30.** 14

31. $42.65 + 18.05i$ **32.** $13.2 - 9.7i$

In Problems 33 through 44, express the complex numbers in rectangular form, and graph in the complex plane.

33. $8 \operatorname{cis} 135°$ **34.** $12 \operatorname{cis} 120°$

35. $5 \operatorname{cis} 225°$ **36.** $9 \operatorname{cis} \dfrac{\pi}{6}$

37. $10 \operatorname{cis} \dfrac{5\pi}{3}$ **38.** $10\left(\cos \dfrac{7\pi}{6} + i \sin \dfrac{7\pi}{6}\right)$

39. $4\left(\cos \dfrac{\pi}{3} + i \sin \dfrac{\pi}{3}\right)$ **40.** $10 \operatorname{cis} 900°$

41. $29 \operatorname{cis} 630°$ **42.** $6 \operatorname{cis} 122°$

43. $12 \operatorname{cis} 41°$ **44.** $5 \operatorname{cis} 205°$

B

Perform the indicated operations in Problems 45 through 60. Express answers in trigonometric form with $0° \le \theta < 360°$ or $0° \le \theta < 2\pi$.

45. $(3 \operatorname{cis} 120°)(5 \operatorname{cis} 195°)$ **46.** $(6 \operatorname{cis} 135°)(8 \operatorname{cis} 15°)$

47. $(11 \operatorname{cis} 20°)(\sqrt{2} \operatorname{cis} 220°)$ **48.** $(4 \operatorname{cis} 265°)(\sqrt{3} \operatorname{cis} 155°)$

49. $6(\cos 24° + i \sin 24°) \cdot \sqrt{5}(\cos 219° + i \sin 219°)$.

50. $\sqrt{7}(\cos 74° + i \sin 74°) \cdot 12(\cos 112° + i \sin 112°)$.

51. $\dfrac{18 \operatorname{cis} 330°}{6 \operatorname{cis} 60°}$ **52.** $\dfrac{20 \operatorname{cis} 210°}{4 \operatorname{cis} 45°}$

53. $\dfrac{15 \operatorname{cis} 80°}{24 \operatorname{cis} 62°}$ **54.** $\dfrac{6 \operatorname{cis} 78°}{14 \operatorname{cis} 130°}$

55. $\dfrac{10(\cos 106° + i \sin 106°)}{22(\cos 200° + i \sin 200°)}$ **56.** $(2 \operatorname{cis} 30°)^3$

57. $\left(4 \operatorname{cis} \dfrac{\pi}{3}\right)^3$ **58.** $\left(\operatorname{cis} \dfrac{\pi}{8}\right)^4$

59. $\left(\operatorname{cis} \dfrac{3\pi}{8}\right)^4$ **60.** $(2 \operatorname{cis} 18°)^8$

In Problems 61 through 67, perform the indicated operations by first expressing the given complex numbers in trigonometric form.

61. $(2 - 2\sqrt{3}i)(1 + i)$ **62.** $(\sqrt{3} - i)(1 + \sqrt{3}i)$

63. $\dfrac{-4 + 4i}{\sqrt{3} + i}$ **64.** $\dfrac{4i}{2 - 3i}$

65. $(1 - \sqrt{3}i)^6$ **66.** $(5\sqrt{3} + 5i)^9$

67. $(4 - 4i)^8$

68. In an RLC alternating-current circuit, $R = 6$ ohms, $R_L = 10$ ohms, $R_C = 2$ ohms, and $I = 2$ amperes. Find the impedance, the voltage, the magnitudes for both, and the phase angle.

69. In an RLC alternating-current circuit, $R = 16$ ohms, $R_L = 8$ ohms, $R_C = 12$ ohms, and $I = 4$ amperes. Find the impedance, the voltage, the magnitudes for both, and the phase angle.

C

70. Prove that

$$\frac{r_1(\cos \theta_1 + i \sin \theta_1)}{r_2(\cos \theta_2 + i \sin \theta_2)} = \frac{r_1}{r_2}[\cos(\theta_1 - \theta_2) + i \sin(\theta_1 - \theta_2)].$$

71. Prove the triangle inequality: $|z_1 + z_2| \le |z_1| + |z_2|$.

72. Prove that $||z_1| - |z_2|| \le |z_1 - z_2|$. (Hint: Use the triangle inequality in Problem 7.1.)

DEMOIVRE'S THEOREM

The ease of multiplication of complex numbers in trigonometric form leads to a particularly useful way to find powers and roots. Starting with

$$z = r \text{ cis } \theta = r(\cos \theta + i \sin \theta)$$

we then have

$$z^2 = (r \text{ cis } \theta)^2$$
$$= (r \text{ cis } \theta)(r \text{ cis } \theta)$$
$$= rr \text{ cis}(\theta + \theta)$$
$$= r^2 \text{ cis } 2\theta.$$

Thus, $(r \text{ cis } \theta)^2 = r^2 \text{ cis } 2\theta$. If we multiply this result by z we get

$$z^3 = (r \text{ cis } \theta)^3$$
$$= (r \text{ cis } \theta)^2(r \text{ cis } \theta)$$
$$= (r^2 \text{cis } 2\theta)(r \text{ cis } \theta)$$
$$= r^2r \text{ cis}(2\theta + \theta)$$
$$= r^3 \text{cis } 3\theta.$$

Therefore, $(r \text{ cis } \theta)^3 = r^3 \text{cis } 3\theta$. Similarly, if we multiply this last result again by $z = r \text{ cis } \theta$, we discover that $(r \text{ cis } \theta)^4 = r^4 \text{cis } 4\theta$. Continuing this pattern leads to the following remarkable theorem, named after the French-born English mathematician, Abraham DeMoivre.

DeMoivre's Theorem

For every natural number n,

$$(r \text{ cis } \theta)^n = r^n \text{cis } n\theta.$$

In other words,

$$[r(\cos \theta + i \sin \theta)]^n = r^n(\cos n\theta + i \sin n\theta).$$

This is equivalent to saying that if $z = r \text{ cis } \theta$, then

$$|z^n| = |z|^n \quad \text{and} \quad \arg(z^n) = n \arg(z).$$

EXAMPLE 1   Find $(-1 + i)^{10}$. Express the answer in both trigonometric and rectangular form.

SOLUTION First convert $-1 + i$ into trigonometric form (Figure 88):

$$r = \sqrt{(-1)^2 + 1^2} = \sqrt{2}$$

$$\tan \theta = \frac{1}{-1} \Rightarrow \theta = \frac{3\pi}{4}.$$

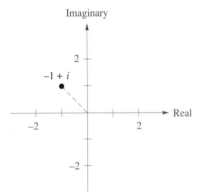

FIGURE 88

So

$$-1 + i = \sqrt{2} \text{ cis } \frac{3\pi}{4}.$$

Applying DeMoivre's theorem gives

$$(-1 + i)^{10} = \left(\sqrt{2} \text{ cis } \frac{3\pi}{4} \right)^{10}$$

$$= (\sqrt{2})^{10}\text{cis} \left(10 \, \frac{3\pi}{4} \right)$$

$$= 32 \text{ cis} \left(\frac{15\pi}{2} \right)$$

$$= 32 \text{ cis} \left(\frac{3\pi}{2} \right) \qquad \text{(trigonometric form)}$$

$$= -32i. \qquad \text{(rectangular form)}$$

Recall from our study of polynomials in Chapter 3 that a polynomial equation of degree n has n roots in the complex number system. For example, the polynomial equation

$$z^3 = 8$$

has three roots. One way to find them is to subtract 8 from both sides and then factor as the difference of two cubes:

$$(z - 2)(z^2 + 2z + 4) = 0.$$

From this we obtain the roots

$$z = 2, \; -1 + \sqrt{3}i, \quad \text{and} \; -1 - \sqrt{3}i.$$

These three solutions to $z^3 = 8$ are called the **cube roots of 8.**
    An alternate approach to solving the original equation is to first express the equation in trigonometric form. Letting $z = r \text{ cis } \theta$,

$$z^3 = 8 \Rightarrow (r \text{ cis } \theta)^3 = 8 \text{ cis } 0°.$$

Using DeMoivre's theorem, we have

$$r^3 \text{cis } 3\theta = 8 \text{ cis } 0°.$$

Two complex numbers are equal if and only if they have the same modulus and their arguments are coterminal. Therefore,

$$r^3 = 8 \Rightarrow r = 2$$

and

$$3\theta = (0 + 360k)° \Rightarrow \theta = 120k°$$

where k is an integer. Thus,

$$z = 2 \text{ cis } 120k°.$$

Substituting the values $k = 0$, $k = 1$, $k = 2$, $k = 3$, $k = 4$, and so on, gives

$$z = 2 \text{ cis } 120(0)° = 2 \text{ cis } 0° \quad \longleftarrow \quad 2$$
$$z = 2 \text{ cis } 120(1)° = 2 \text{ cis } 120° \quad \longleftarrow \quad -1 + \sqrt{3}i$$
$$z = 2 \text{ cis } 120(2)° = 2 \text{ cis } 240° \quad \longleftarrow \quad -1 - \sqrt{3}i$$
$$z = 2 \text{ cis } 120(3)° = 2 \text{ cis } 360° \quad \longleftarrow \quad 2$$
$$z = 2 \text{ cis } 120(4)° = 2 \text{ cis } 480° \quad \longleftarrow \quad -1 + \sqrt{3}i$$
$$\vdots$$

Knowing that the original equation $z^3 = 8$ has three solutions, we did not need to substitute values for $k \geq 3$. Doing so only shows that the roots begin to repeat after $k = 2$.

The method of using trigonometric forms to find the three roots of $z^3 = 8$ can be generalized:

nth Roots of a Complex Number

For any positive integer n, the solutions to the equation

$$z^n = r \text{ cis } \theta$$

are given by

$$\sqrt[n]{r} \text{ cis}\left(\frac{\theta^\circ + 360^\circ k}{n}\right) \quad \text{or} \quad \sqrt[n]{r} \text{ cis}\left(\frac{\theta + 2\pi k}{n}\right)$$

for $k = 0, 1, 2, 3, \ldots, n - 1$.

These solutions are called the **nth roots of r cis θ.**

EXAMPLE 2   Find the fourth roots of $8\sqrt{3} + 8i$. Graph the roots in the complex plane.

SOLUTION   First express $8\sqrt{3} + 8i$ in trigonometric form:

$$r = \sqrt{(8\sqrt{3})^2 + 8^2} = 16$$

and

$$\theta = \tan^{-1}\left(\frac{8}{8\sqrt{3}}\right) = \tan^{-1}\left(\frac{1}{\sqrt{3}}\right) = 30^\circ.$$

So $8\sqrt{3} + 8i = 16 \text{ cis } 30^\circ$, and the four roots have the form

$$\sqrt[4]{16} \text{ cis}\left(\frac{30^\circ + 360^\circ k}{4}\right) = 2 \text{ cis}(7.5^\circ + 90^\circ k).$$

Substituting $k = 0, 1, 2,$ and 3, we obtain the fourth roots:

$$k = 0: 2 \text{ cis } 7.5^\circ \approx 1.983 + 0.261i$$

$$k = 1: 2 \text{ cis } 97.5^\circ \approx -0.261 + 1.983i$$

$$k = 2: 2 \text{ cis } 187.5^\circ \approx -1.983 - 0.261i$$

$$k = 3: 2 \text{ cis } 277.5^\circ \approx 0.261 - 1.983i.$$

Other integral values for k just repeat the roots.

The graphs of these are shown in Figure 89. They are equally spaced along a circle of radius 2. The angle between successive roots is $\frac{360^\circ}{4} = 90^\circ$.

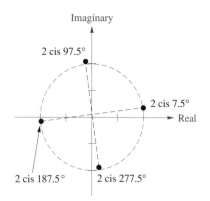

FIGURE 89

Knowing how to find the *n*th roots of a complex number can be useful when solving polynomial equations.

EXAMPLE 3   Use the fact that $x^5 - 1 = (x - 1)(x^4 + x^3 + x^2 + x + 1)$ to find all complex solutions to $x^4 + x^3 + x^2 + x + 1 = 0$.

SOLUTION The zeros for $x^5 - 1$ are the fifth roots of 1. Since a trigonometric form of 1 is 1 cis 0°, the fifth roots of 1 are given by

$$\sqrt[5]{1} \operatorname{cis}\left(\frac{0° + 360°k}{5}\right) = \operatorname{cis} 72°k.$$

With $k = 0, 1, 2, 3$, and 4, the solutions to $x^5 - 1 = 0$ are

$$k = 0: \operatorname{cis} 0° \approx 1$$
$$k = 1: \operatorname{cis} 72° \approx 0.309 + 0.951i$$
$$k = 2: \operatorname{cis} 144° \approx -0.809 + 0.588i$$
$$k = 3: \operatorname{cis} 216° \approx -0.809 - 0.588i$$
$$k = 4: \operatorname{cis} 288° \approx 0.309 - 0.951i.$$

These are also the solutions to

$$(x - 1)(x^4 + x^3 + x^2 + x + 1) = 0$$

and since $x = 1 = \operatorname{cis} 0°$ can be obtained from the first factor, the roots of $x^4 + x^3 + x^2 + x + 1 = 0$ must therefore be the remaining four:

$$\operatorname{cis} 72° \approx 0.309 + 0.951i$$
$$\operatorname{cis} 144° \approx -0.809 + 0.588i$$
$$\operatorname{cis} 216° \approx -0.809 - 0.588i$$
$$\operatorname{cis} 288° \approx 0.309 - 0.951i.$$

The fifth roots of 1 (in the preceding example) are sometimes called the fifth roots of unity. In general, for any positive integer n, the nth roots of 1 are called the nth roots of unity. They are given by

$$\operatorname{cis}\left(\frac{360°k}{n}\right) \quad \text{or} \quad \operatorname{cis}\left(\frac{2\pi k}{n}\right), \qquad k = 0, 1, 2, \ldots, n - 1.$$

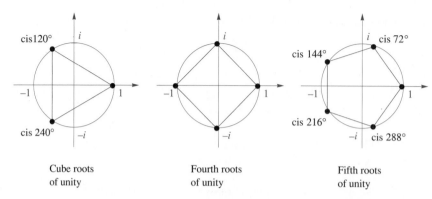

FIGURE 90

| Cube roots of unity | Fourth roots of unity | Fifth roots of unity |

If $n \geq 3$, the graph of the nth roots of unity are vertices forming a regular n-sided polygon inscribed in a unit circle. For example, the cube, fourth, and fifth roots of unity are shown in Figure 90.

We know how to raise a real number to a rational power. The process of raising a complex number to a rational power is defined in a similar way.

Rational Power of a Complex Number

> The nth roots of a nonreal complex number z is denoted $z^{1/n}$. Furthermore, if m and n are positive integers,
>
> $$z^{m/n} = (z^{1/n})^m.$$
>
> In other words, $z^{m/n}$ are the nth roots of z raised to the power m.

It can be shown that $(z^{1/n})^m = (z^m)^{1/n}$. This simply means that $z^{m/n}$ can also be calculated by first evaluating z^m and then finding the nth roots of the result.

With this new notation, if $z = r \operatorname{cis} \theta$, then

$$z^{1/n} = \sqrt[n]{r} \operatorname{cis}\left(\frac{\theta° + 360°k}{n}\right).$$

Raising both sides to the power m gives

$$z^{m/n} = \left[\sqrt[n]{r} \operatorname{cis}\left(\frac{\theta° + 360°k}{n}\right)\right]^m$$

$$= r^{m/n}\operatorname{cis}\left[\frac{m}{n} \cdot (\theta° + 360°k)\right], \qquad k = 0, 1, 2, \ldots, n - 1.$$

If we define

$$z^{-m} = \frac{1}{z^m}$$

then it can be shown that

$$z^{-m/n} = \left[\sqrt[n]{r} \, \text{cis} \left(\frac{\theta° + 360°k}{n} \right) \right]^{-m}$$

$$= r^{-m/n} \text{cis} \left[\frac{-m}{n} \cdot (\theta° + 360°k) \right], \qquad k = 0, 1, 2, \ldots, n - 1.$$

The following is a summary of these results.

Let q be the rational number $q = m/n$ or $q = -m/n$, where m and n are positive integers. If the trigonometric form of a complex number is $z = r \, \text{cis} \, \theta$, then

$$z^q = r^q \text{cis}[q \cdot (\theta + 360°k)], \qquad k = 0, 1, 2, \ldots, n - 1$$

or

$$z^q = r^q \text{cis}[q \cdot (\theta + 2\pi k)], \qquad k = 0, 1, 2, \ldots, n - 1.$$

EXAMPLE 4   Find all possible values of $(-2 + 2i)^{2/5}$.

SOLUTION The base in trigonometric form is

$$-2 + 2i = 2\sqrt{2} \, \text{cis} \, 135°.$$

Therefore,

$$(-2 + 2i)^{2/5} = [2\sqrt{2} \, \text{cis} \, 135°]^{2/5}$$

$$= (2\sqrt{2})^{2/5} \text{cis}[\tfrac{2}{5} \cdot (135° + 360°k)]$$

$$= (2^{3/2})^{2/5} \text{cis}[54° + 144°k]$$

$$= 2^{3/5} \text{cis}[54° + 144°k], \qquad k = 0, 1, 2, 3, 4.$$

The five values for $(-2 + 2i)^{2/5}$ are

$$k = 0: 2^{3/5} \text{cis} \, 54° \approx 0.891 + 1.226i$$

$$k = 1: 2^{3/5} \text{cis} \, 198° \approx -1.442 - 0.468i$$

$$k = 2: 2^{3/5} \text{cis} \, 342° \approx 1.442 - 0.468i$$

$$k = 3: 2^{3/5} \text{cis} \, 486° \approx -0.891 + 1.226i$$

$$k = 4: 2^{3/5} \text{cis} \, 630° \approx -1.516i.$$

One of the most beautiful (nonetheless useful) mathematical results relates the exponential function e^x to the trigonometric functions. First discovered by the great mathematician Leonard Euler, its verification requires calculus; however, it is a gem that you would want to see anyway.

Euler's Identity

If θ is a real number, then

$$e^{i\theta} = \text{cis } \theta.$$

Since θ is a real number, be sure that all trigonometric calculations are done in radians. For example,

$$e^i = \text{cis}(1) = \cos(1) + i\sin(1) \approx 0.540 + 0.841i$$

$$e^{i[\pi/4]} = \text{cis}\left(\frac{\pi}{4}\right) = \cos\left(\frac{\pi}{4}\right) + i\sin\left(\frac{\pi}{4}\right) = \frac{\sqrt{2}}{2} + \frac{\sqrt{2}}{2}i$$

$$e^{i\pi} = \text{cis}(\pi) = -1.$$

EXAMPLE 5   Express the complex numbers in the form $re^{i\theta}$.

a) $\sqrt{3} + i$ b) $3 - 2i$

SOLUTION

a) First write $\sqrt{3} + i$ in trigonometric form using radian measure

$$\sqrt{3} + i = 2 \text{ cis } \frac{\pi}{6}.$$

Using Euler's identity gives

$$\sqrt{3} + i = 2 \text{ cis } \frac{\pi}{6} = 2e^{i[\pi/6]} = 2e^{i\pi/6}.$$

b) Again, we first write $3 - 2i$ in trigonometric form (using radian measure), and then apply Euler's identity:

$$3 - 2i \approx \sqrt{13} \text{ cis}(5.695) = \sqrt{13}e^{5.695i}.$$

EXERCISE SET 7.5

A

33. $(-\sqrt{3} + i)^{5/4}$ **34.** $(2 + 2\sqrt{3}i)^{2/5}$

In Problems 1 through 12, use DeMoivre's theorem to find the following complex numbers. Express your answer in both trigonometric and rectangular form.

1. $(4 \text{ cis } 20°)^6$ **2.** $(4 \text{ cis } 110°)^8$

3. $(2 \text{ cis } 30°)^7$ **4.** $(2 + 2i)^8$

5. $(-1 + i)^{14}$ **6.** $(3\sqrt{2} - 3\sqrt{2}i)^5$

7. $(2\sqrt{3} + 2i)^8$ **8.** $(1 - \sqrt{3}i)^{11}$

9. $(-3\sqrt{3} + 2i)^5$ **10.** $(2 + 5i)^6$

11. $(-3 + i)^6$ **12.** $(3 - 4i)^4$

Find the indicated roots of the complex numbers in Problems 13 through 24. Express your answers in trigonometric form.

13. Fourth roots of 81 cis 132°

14. Fifth roots of 32 cis 70°

15. Cube roots of 64 cis 78°

16. Square roots of 36 cis 208°

17. Fifth roots of i

18. Fourth roots of $1 + i$

19. Sixth roots of $-\sqrt{3} + i$

20. Sixth roots of $-64i$

21. Ninth roots of $\sqrt{2} - \sqrt{2}i$

22. Eighth roots of -256

23. Fifth roots of $-12 - 5i$

24. Sixth roots of $-7 + 24i$

In Problems 25 through 28, find the indicated roots and represent them geometrically in the complex plane.

25. Sixth roots of unity   **26.** Eighth roots of unity

27. Cube roots of $125i$ **28.** Fifth roots of -32

Determine all possible values in each of Problems 29 through 34. Express your answer in trigonometric form.

29. $(8 \text{ cis } 51°)^{2/3}$ **30.** $(9 \text{ cis } 80°)^{3/2}$

31. $(-32i)^{3/5}$ **32.** $(27 - 27i)^{4/3}$

B

Find all the complex solutions to the equations in Problems 35 through 40.

35. $x^4 + 16 = 0$ **36.** $x^5 - 243 = 0$

37. $x^6 - 64i = 0$ **38.** $x^4 - 16i = 0$

39. $x^3 + 2 = 2i$ **40.** $x^5 - i = 1$

$x^n - 1 = (x - 1)(x^{n-1} + x^{n-2} + x^{n-3} + \cdots + x^2 + x + 1)$.
Use that fact to find all complex solutions to the equations in Problems 41 through 44.

41. $x^3 + x^2 + x + 1 = 0$

42. $x^5 + x^4 + x^3 + x^2 + x + 1 = 0$

43. $x^6 + x^5 + x^4 + x^3 + x^2 + x + 1 = 0$

44. $x^4 - x^3 + x^2 - x + 1 = 0$

In Problems 45 through 53, express the complex numbers in the form $re^{i\theta}$, with $0 \le \theta < 2\pi$.

45. $1 + i$ **46.** $-3 + 4i$ **47.** $15 - 36i$

48. $-7 - 24i$ **49.** $\sqrt{19} + 9i$ **50.** $8 + 7i$

51. $\sqrt{2} + i$ **52.** -6 **53.** $20i$

C

54. Use Euler's identity to find a real ith root of i; in other words find a real solution to $x^i = i$. You may assume that $(e^{z_1})^{z_2} = e^{z_1 z_2}$ for the complex numbers z_1 and z_2.

55. a) $(A + B)^3 = A^3 + 3A^2B + 3AB^2 + B^3$. Use that fact to expand $(\cos\theta + i\sin\theta)^3$ in powers of $\cos\theta$ and $\sin\theta$. Express in the form $a + bi$; that is, separate the real and imaginary parts of the expansion.

b) Use DeMoivre's theorem and the result of **(a)** to express $\cos 3\theta$ and $\sin 3\theta$ in terms of powers of $\cos\theta$ and $\sin\theta$.

Karl F. Gauss

Karl Gauss is in the company of Archimedes and Newton as one of the greatest mathematicians of all time. Dubbed the "Prince of Mathematicians," he extended practically every mathematical frontier and created new branches as well. It is not possible to overstate his genius and contribution to mathematics. He was one of the first to represent complex numbers as points in a plane and developed the algebra of complex numbers and the arithmetic of complex numbers. For example, a new theory of prime numbers emerged in which numbers such as 3, 7, and 11 are prime, but $5 = (1 + 2i)(1 - 2i)$ and $13 = (2 + 3i)(2 - 3i)$ are not. His doctoral dissertation, which he completed at age twenty-two, is a landmark, the first rigorous proof of the fundamental theorem of algebra (see Section 3.3). Generally, Gauss imposed a rigor that dramatically changed the face of all mathematics.

Born in Brunswick, Germany in 1777, Gauss was the only child of poor parents. As such, Gauss could expect only a minimal education. However, he was a child prodigy whose exceptional ability with numbers was apparent at a very early age; he detected an error in his father's payroll ledger at the age of three. Most fortunately for mathematics and science, the boy's precocity was brought to the attention of the Duke of Brunswick who graciously financed his education. Gauss attended the University of Gottingen from 1795 to 1800, where he began solving problems that had been unsolved from antiquity. He was interested in physics and astronomy and also earned lasting fame for his research in magnetism and electricity; one of many examples of his contributions is his invention of the telegraph.

Using the methods of least squares, which he invented, he discovered the planetoids Ceres in 1801 and Pallus in 1802; the mathematics he used also eventually led to the discovery of Neptune. In 1804, he was appointed professor of astronomy and director of the observatory at the University of Gottingen, where he remained until his death in 1855.

Gauss was often flooded with so many ideas and discoveries that he would be overwhelmed if he attempted to pursue them in detail. Consequently, he recorded them briefly in a diary. He was a perfectionist and would not publish his works until they had been polished to his highest standard of elegance and precision. The resulting papers were above criticism and devoid of any hint of the analysis by which the results were reached. His motto was *Pauca sed matura* (few but ripe). In a letter he once communicated that ". . . I write slowly. This is chiefly because I am never satisfied until I have said as much as possible in a few words, and writing briefly takes far more time than writing at length."

Many of his investigations were discovered in his diary after his death. He was described as a demanding professor who loved learning. Although he avoided teaching as much as he could, his presentations were superb when he did teach. In addition, he was quite open-minded, supporting a woman Ph.D. during a time when such an attitude was not common.

MISCELLANEOUS EXERCISES

In Problems 1 through 12, approximate the remaining parts of triangle ABC, if possible.

1. $\alpha = 105°$, $b = 12$, $c = 15$

2. $\beta = 66°$, $\gamma = 28°$, $b = 6.5$

3. $a = 4.6$, $b = 2.5$, $c = 5.3$

4. $\gamma = 52°$, $a = 21$, $b = 32$

5. $\alpha = 35°$, $b = 13.0$, $\gamma = 70°$

6. $a = 20.7$, $b = 26.4$, $c = 15.2$

7. $\alpha = 64°$, $\beta = 38°$, $b = 14.0$

8. $\alpha = 25°$, $a = 14$, $b = 12$

9. $\beta = 52°$, $a = 21$, $b = 16$

10. $a = 5.8$, $b = 2.3$, $c = 3.3$

11. $\alpha = 27°$, $a = 15.1$, $b = 27.3$

12. $\gamma = 32°$, $a = 25$, $c = 21$

For Problems 13 through 17, refer to Figure 91.

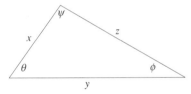

FIGURE 91

13. Express z in terms of x, θ, and y.

14. Express θ in terms of x, y, and z.

15. Express ϕ in terms of x, y, and ψ.

16. Express z in terms of ψ, θ, and y.

17. Express z in terms of x, θ, and ψ.

18. A landmark is spotted from two points A and B that are on the same side of a canyon. Points A and B are 2.6 miles apart. From A, the angle of sight between the landmark and point B is 48°. From B, the angle of sight between the landmark and point A is 65°. What is the distance from each of the points A and B to the landmark?

19. To find the height of a tree across a river, two angles of elevation are measured from ground level. The angle of elevation from a point A to the top of the tree is 52°. After the surveyor moves 30 feet directly away from the tree to a point B, the angle of elevation is measured to be 47°. Find the height of the tree.

20. A fishing boat heads in the direction N76°E from a north-south shoreline for 12 miles. Then it changes direction and travels 10 miles to the shore. What are the possible distances from the point of origin?

21. An airplane flies in the direction N56°W for 320 miles from A to B and then flies in the direction N20°E for 550 miles from B to C. What is the distance from A to C?

22. A diagonal of a parallelogram is 19 cm and forms angles of 18° and 35° with the two sides. Find the lengths of the sides of the parallelogram.

23. In Figure 92, $XY = 27.5$. Find XT.

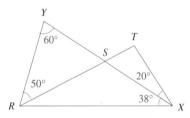

FIGURE 92

24. For the quadrilateral in Figure 93, $NQ = 6$ and $NP = 14$. Find the perimeter of the quadrilateral.

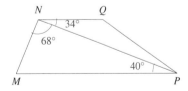

FIGURE 93

25. Two jet planes leave an airport at the same time. One heads N15°W at a rate of 460 miles per hour, and the other flies at 520 miles per hour N52°E. How far apart are the two jets after $1\frac{1}{2}$ hours?

26. Determine the distance between the jets in Problem 25 as a function of t hours of flight.

27. From an island, a triathlete swims N52°E for 10 miles to the coast. There she bicycles due north at 36 miles per hour. How far is she from the island after 40 minutes of riding?

28. In Problem 27, determine the distance between the triathlete and the island as a function of t hours of riding.

29. A baseball diamond is a square 90.0 feet on a side. The pitcher's mound is 60.5 feet from home plate. How far is it from the pitcher's mound to first base (Figure 94)?

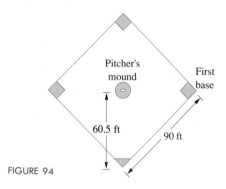

FIGURE 94

30. Find the area of triangle ABC described in Problem 1.

31. Find the area of triangle ABC described in Problem 2.

32. Find the area of triangle ABC described in Problem 3.

33. Find the area of triangle ABC described in Problem 5.

34. Find the area of the quadrilateral in Problem 24.

35. Express the area of the triangle in Figure 95 as a function of t.

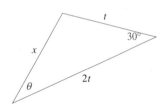

FIGURE 95

36. For the triangle in Problem 35, express x as a function of t.

37. For the triangle in Problem 35, find $\tan \theta$.

38. The base angles of an isosceles triangle have equal measure. Show that the area A of the isosceles triangle with base angles θ and included side x is $(\frac{1}{4})x^2\tan \theta$.

39. Assume that h satisfies $0 < h < \pi/2$ and construct angle h in standard position. The terminal side intersects the unit circle at $P(\cos h, \sin h)$. Construct a vertical line through $A(1, 0)$ (Figure 96.)

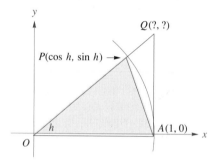

FIGURE 96

a) Determine the coordinates of the intersection of the vertical line with the terminal side of the angle h.

b) Determine the areas of triangle OAP, sector OAP, and triangle OAQ.

c) Based on the results of (b), show that

$$\cos h < \frac{\sin h}{h} < 1.$$

In Problems 40 through 42, the initial point P and terminal point Q of vector $\mathbf{v}$ are given. Find the component form of $\mathbf{v}$ and the norm of $\mathbf{v}$, $\|\mathbf{v}\|$.

40. $P(-5, 8), Q(11, 2)$ **41.** $P(4, 2), Q(-9, 6)$

42. $P(t, 3t), Q(5t, 7)$

In Problems 43 through 51, let $\mathbf{u} = \langle -12, 5 \rangle$, $\mathbf{v} = \langle 8, 6 \rangle$, and $\mathbf{w} = \langle 10, 10 \rangle$.

43. Find $2\mathbf{u} + 3\mathbf{v} - \mathbf{w}$ **44.** Find $4\mathbf{u} - 2\mathbf{v} + \mathbf{w}$

45. Find $\|2\mathbf{u}\|$ and $2\|\mathbf{u}\|$ **46.** Find $\|3\mathbf{v}\|$ and $3\|\mathbf{v}\|$

47. Find $\|\mathbf{u} + \mathbf{v}\|$ and $\|\mathbf{u}\| + \|\mathbf{v}\|$

48. Find $\|\mathbf{w} + \mathbf{v}\|$ and $\|\mathbf{w}\| + \|\mathbf{v}\|$

49. Find a unit vector in the same direction as $\mathbf{u}$.

50. Find a unit vector in the same direction as $\mathbf{v}$.

51. Find a unit vector in the same direction as $\mathbf{w}$.

In Problems 52 through 54, find the component form of $\mathbf{v}$ with the given magnitude and the angle it makes with the positive x-axis.

52. $\|\mathbf{v}\| = 12$, $\theta = 60°$ **53.** $\|\mathbf{v}\| = 18$, $\theta = 225°$

54. $\|\mathbf{v}\| = 4$, $\theta = 75°$

In Problems 55 through 57, find $\|\mathbf{v}\|$ and the smallest positive angle θ that vector $\mathbf{v}$ makes with the positive x-axis.

55. $\mathbf{v} = \langle -4\sqrt{3}, 4 \rangle$

56. $\mathbf{v} = \langle 2\sqrt{10 + 2\sqrt{5}}, 2\sqrt{5} - 2 \rangle$

57. $\mathbf{v} = \langle \sqrt{6} - \sqrt{2}, \sqrt{6} + \sqrt{2} \rangle$

58. An airplane is flying in calm conditions in the direction N32°E at 260 miles per hour. If the plane encounters a wind of 30 miles per hour in the direction N60°W, approximate the resulting speed and direction of the plane (to the nearest mile per hour and degree, respectively).

59. An airplane pilot needs to maintain a true course and speed (relative to the ground) of N55°W at 280 miles per hour. The wind is blowing N10°E at 20 miles per hour. Approximate the required airspeed and direction (to the nearest mile per hour and degree respectively).

In Problems 60 through 65, express each of the complex numbers in rectangular form.

60. $2\left(\cos \dfrac{2\pi}{3} + i \sin \dfrac{2\pi}{3} \right)$

61. $6\left(\cos \dfrac{5\pi}{4} + i \sin \dfrac{5\pi}{4} \right)$

62. $\sqrt{2}(\cos 135° + i \sin 135°)$

63. $\sqrt{3}(\cos 330° + i \sin 330°)$

64. $12 \text{ cis } 105°$

65. $10 \text{ cis } 165°$

In Problems 66 through 71, express each of the complex numbers in trigonometric form. Approximate angles to the nearest 0.01°.

66. $2 + 2i$ **67.** $10\sqrt{3} - 10i$ **68.** $9i$

69. -20 **70.** $-4 - 3i$ **71.** $24 + 7i$

In Problems 72 through 74, find $z_1 z_2$ and z_1/z_2. Express answers in trigonometric form with a positive argument.

72. $z_1 = 12\left(\cos \dfrac{5\pi}{6} + i \sin \dfrac{5\pi}{6} \right)$, $z_2 = 4\left(\cos \dfrac{\pi}{3} + i \sin \dfrac{\pi}{3} \right)$

73. $z_1 = 9 \text{ cis } 105°$, $z_2 = 3 \text{ cis } 60°$

74. $z_1 = \sqrt{6} \text{ cis } 26°$, $z_2 = \sqrt{3} \text{ cis } 140°$

In Problems 75 through 80, use DeMoivre's theorem to find the following. Express answers in trigonometric form and rectangular form.

75. $\left[2\left(\cos \dfrac{\pi}{6} + i \sin \dfrac{\pi}{6} \right) \right]^4$

76. $\left[3\left(\cos \dfrac{\pi}{4} + i \sin \dfrac{\pi}{4} \right) \right]^6$

77. $[4 \text{ cis } 36°]^5$ **78.** $[4 \text{ cis } 15°]^9$

79. $(-\sqrt{3} + i)^3$ **80.** $(3\sqrt{2} - 3\sqrt{2}i)^4$

Find the indicated roots of the complex numbers in Problems 81 through 84. Express your answers in trigonometric form.

81. Fourth roots of $16 \text{ cis } 120°$

82. Fifth roots of $243 \text{ cis } 300°$

83. Cube roots of $-64i$

84. Fourth roots of $\text{cis } 208°$

Find all the complex solutions to the equations in Problems 85 through 87. Graph the solutions in the complex plane.

85. $z^3 + 64 = 0$ **86.** $z^5 - 32i = 0$ **87.** $z^4 = 8 + 8i$

88. Graph all complex solutions to $|z - 4| < 3$.

89. Graph all complex solutions to $|z + 3| < 5$.

90. Graph all complex solutions to $|z - (4 + 2i)| \leq 3$.

CHAPTER 8

PLANE ANALYTIC GEOMETRY

LINES

In Section 2.4 we saw that the graph of the linear equation $y = mx + b$ is a nonvertical line, where m is the slope and b is the y-intercept. This equation can be written in another form that is congruous with the topics in this chapter.

> The **general form** for the equation of a line is
> $$Dx + Ey + F = 0$$
> where D and E are not both zero.

This form is appropriately named because the equation of any line, including vertical lines, can be written in general form.

EXAMPLE 1   Write the linear equation $y = \frac{4}{5}x + 3$ in general form.

SOLUTION   Subtract $\left[\frac{4}{5}x + 3\right]$ from both sides of the given equation:

$$-\tfrac{4}{5}x + y - 3 = 0.$$

Even though this is in general form, to clear fractions we multiply each side by -5:

$$4x - 5y + 15 = 0.$$

The general form of a linear equation is not unique. For example,

$$-4x + 5y - 15 = 0$$

is also a general form for the linear equation.

445

Inclination of a Line

The angle of inclination α of a line is the smallest nonnegative angle from the x-axis, or any horizontal line, to the line. Notice that, for both cases in Figure 1,

$$\tan \alpha = \frac{y_2 - y_1}{x_2 - x_1} = m.$$

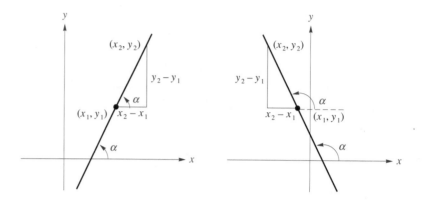

FIGURE 1

EXAMPLE 2   Find the exact angle of inclination of the line $2x + 3y - 4 = 0$. In addition, give an approximation to the nearest $0.01°$.

SOLUTION First write the equation in slope-intercept form:

$$3y = -2x + 4$$

or

$$y = -\tfrac{2}{3}x + \tfrac{4}{3},$$

so the slope is $-\tfrac{2}{3}$.

Using the formula $\tan \alpha = m$ gives $\tan \alpha = -\tfrac{2}{3}$. Since $0 \le \alpha < 180°$, the exact value is

$$\alpha = 180° + \tan^{-1}(-\tfrac{2}{3}).$$

A calculator can be used to approximate the value of α. The keystrokes and the display are shown:

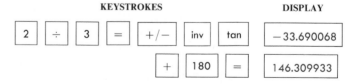

So $\alpha \approx 146.31°$.

EXAMPLE 3   Find an equation of the line having an angle of inclination of $\pi/3$, passing through $(1, 4)$. Express your answer in general form.

SOLUTION The line has slope $\tan \pi/3 = \sqrt{3}$. Using the point-slope form:

$$y - y_1 = m(x - x_1) \Rightarrow y - 4 = \sqrt{3}(x - 1).$$

In general form this equation is

$$\sqrt{3}x - y + (4 - \sqrt{3}) = 0.$$

Recall that two lines are parallel if and only if they have equal slopes, and that two lines are perpendicular if and only if their slopes are negative reciprocals of each other. In any other case, two lines intersect forming an angle θ, where $0 < \theta < \pi/2$ (Figure 2). Let the angles of inclination of the two lines be α and β. Consider the triangle with interior angles α, θ, and $180° - \beta$. The sum of these three angles must be $180°$:

$$\alpha + \theta + (180° - \beta) = 180°.$$

If we solve for θ, we get

$$\theta = \beta - \alpha.$$

This assumes $\alpha < \beta$. It may be that $\alpha > \beta$. In either case,

$$\tan \theta = \left| \tan(\beta - \alpha) \right| = \left| \frac{\tan \beta - \tan \alpha}{1 + \tan \alpha \tan \beta} \right|.$$

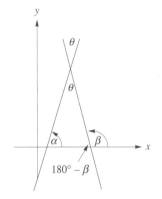

FIGURE 2

Suppose the slopes of the two lines are m_1 and m_2, that is, let $m_1 = \tan \alpha$ and $m_2 = \tan \beta$. Substituting into the last equation, we have the following result:

The smallest positive angle θ between two lines is given by

$$\tan \theta = \left| \frac{m_2 - m_1}{1 + m_1 m_2} \right|.$$

If $m_1 m_2 = -1$, the lines are perpendicular.

Angle between Two Lines

EXAMPLE 4   Find the smallest positive angle formed by the lines $3x - y + 2 = 0$ and $2x + 4y = 5$. Also give an approximation to the nearest $0.01°$.

SOLUTION First write both equations in slope-intercept form to find their slopes:

$$3x - y + 2 = 0 \Rightarrow y = 3x + 2$$

$$2x + 4y = 5 \Rightarrow y = (-\tfrac{1}{2})x + \tfrac{5}{4}.$$

Hence,

$$\tan \theta = \left| \frac{m_2 - m_1}{1 + m_1 m_2} \right| = \left| \frac{-\tfrac{1}{2} - 3}{1 + (-\tfrac{1}{2})(3)} \right| = 7 \quad \text{or} \quad \theta = \tan^{-1} 7.$$

With a calculator, an approximation to the nearest hundredth of a degree is $\theta \approx 81.87°$.

Distance from a Point to a Line

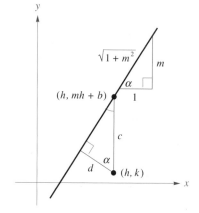

y

$\sqrt{1 + m^2}$

m

α

$(h, mh + b)$ 1

c

α

d (h, k)

x

FIGURE 3

Consider the problem of finding the distance d from a line $y = mx + b$ to a point (h, k) not on the line. What we want is the shortest distance from the point to the line, which is represented by a perpendicular line segment from the point (h, k) to the line $y = mx + b$. Draw a vertical segment from (h, k) to the line, and call the length of this segment c. Because the other endpoint is on the given line, it has coordinates $(h, mh + b)$. In general then, $c = |mh + b - k|$.

Since the slope of the line $y = mx + b$ is m, we can form a right triangle with hypotenuse on the line and legs 1 and m.

Consider the two right triangles in Figure 3. Since they have the same angles, they are similar triangles. This means that corresponding sides are in proportion, in particular,

$$\frac{d}{1} = \frac{c}{\sqrt{1 + m^2}} = \frac{|mh + b - k|}{\sqrt{1 + m^2}}.$$

It is left as an exercise to show that if $y = mx + b$ is written in the general form $Dx + Ey + F = 0$, then

$$d = \frac{|mh + b - k|}{\sqrt{1 + m^2}} = \frac{|Dh + Ek + F|}{\sqrt{D^2 + E^2}}.$$

Summarizing, we have the following:

The distance d from the point (h, k) to the line $y = mx + b$ is

$$d = \frac{|mh + b - k|}{\sqrt{1 + m^2}}.$$

If the line is in the form $Dx + Ey + F = 0$, then

$$d = \frac{|Dh + Ek + F|}{\sqrt{D^2 + E^2}}.$$

EXAMPLE 5   Find the distance from the point $(5, -3)$ to the line $x - 4y + 2 = 0$.

SOLUTION The line is in general form with $D = 1$, $E = -4$, and $F = 2$, so

$$d = \frac{|1(5) + (-4)(-3) + 2|}{\sqrt{1^2 + (-4)^2}} = \frac{19}{\sqrt{17}}.$$

EXAMPLE 6   Find the distance between the two parallel lines $y = 2x - 1$ and $y = 2x + 7$.

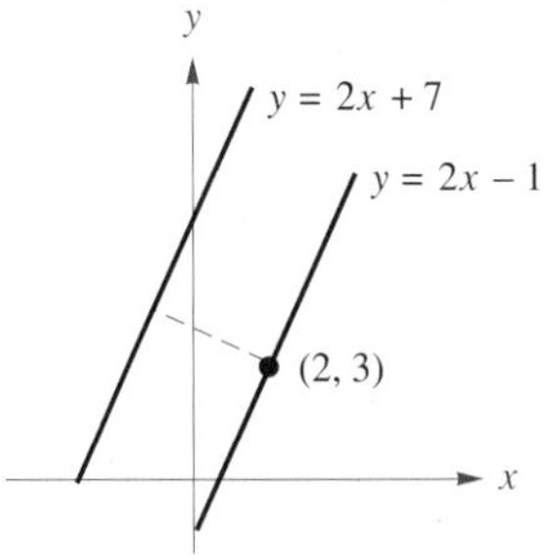

SOLUTION Our plan of attack is to pick a point on one of the lines and find the distance from this point to the other line. Pick any point (h, k) on the first line by letting h be, say 2 (Figure 4). Then

$$k = 2h - 1 = 2(2) - 1 = 3.$$

The distance from the point $(2, 3)$ to the line $y = 2x + 7$ is

$$d = \frac{|mh + b - k|}{\sqrt{1 + m^2}} = \frac{|2(2) + 7 - 3|}{\sqrt{1 + 2^2}} = \frac{8}{\sqrt{5}}.$$

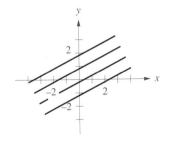

FIGURE 4

Families of Lines

The equation $y = mx + 3$ represents a line with slope m and y-intercept 3. If various values for m are substituted into the equation, we get a collection of equations. Members of this collection are

$$y = 5x + 3,$$
$$y = -4x + 3,$$
$$y = -\tfrac{1}{2}x + 3$$

and so on. The resulting graphs will be a collection, or **family,** of lines passing through $(0, 3)$. Since the slope m varies for this family, m is called a **parameter** of the family.

EXAMPLE 7   Describe the family of lines $y = \tfrac{1}{2}x + b$, where the parameter b takes on the values -1, 0, 1, and 2.

SOLUTION The family consists of all the lines with slope $\tfrac{1}{2}$ and with y-intercepts -1, 0, 1, and 2, respectively. The lines are shown in Figure 5.

FIGURE 5

EXAMPLE 8   Describe the family of all lines described by $y - 2 = m(x - 3)$, where the parameter m takes on all positive real values.

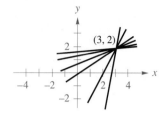

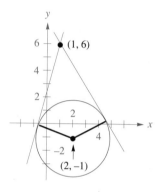

FIGURE 6

SOLUTION This is a family of all lines with positive slope passing through the point $(3, 2)$. A few representative lines are shown in Figure 6.

The next example incorporates families of lines, the distance from a line to a point, as well as tangent lines.

EXAMPLE 9 Find the slope of each of the lines passing through $(1, 6)$ that is tangent to the circle $(x - 2)^2 + (y + 1)^2 = 10$.

SOLUTION First, draw a picture (Figure 7). The graph of the circle is centered at $(2, -1)$ with radius $\sqrt{10}$. There are two lines tangent to the circle passing through $(1, 6)$. The equations for either of these two lines can be written in point-slope form:

$$y - k = m(x - h)$$
$$y - 6 = m(x - 1).$$

Now we want to use the fact that the tangent lines are $\sqrt{10}$ units from the center $(2, -1)$ because the radius of the circle is $\sqrt{10}$. To utilize the formula for the distance from a point to a line, express the tangent lines in slope-intercept form.

$$y - 6 = m(x - 1)$$
$$y = mx + (-m + 6).$$

So the slope of each line is m. The y-intercept of each line is $(-m + 6)$. The distance from the tangent lines to the center $(2, -1)$ is $\sqrt{10}$:

$$d = \frac{|mh + b - k|}{\sqrt{1 + m^2}}$$
$$\sqrt{10} = \frac{|m(2) + (-m + 6) - (-1)|}{\sqrt{1 + m^2}}.$$

Squaring both sides and simplifying, we get

$$10(1 + m^2) = (m + 7)^2.$$

Now solve for m:

$$10 + 10m^2 = m^2 + 14m + 49$$
$$9m^2 - 14m - 39 = 0$$
$$(m - 3)(9m + 13) = 0$$

which gives $m = 3$ or $m = -\frac{13}{9}$.

FIGURE 7

A

In Problems 1 through 8, find an equation of each line. Express your answer in (a) slope-intercept form, if possible, and (b) general form.

1. Passing through $(-2, 5)$ with slope $\frac{3}{2}$
2. Passing through $(1, -7)$ with slope $-\frac{1}{4}$
3. Passing through $(2, 6)$ with no slope
4. Passing through $(-3, 8)$ with no slope
5. Passing through $(1, 4)$ with angle of inclination $\frac{\pi}{4}$
6. Passing through $(-2, 3)$ with angle of inclination $\frac{3\pi}{4}$
7. Passing through $(2, -2)$ with angle of inclination $\frac{2\pi}{3}$
8. Passing through $(-3, 4)$ with angle of inclination $\frac{\pi}{6}$

In Problems 9 through 18, find the exact angle of inclination for each given line. In addition, give an approximation to the nearest 0.01°.

9. $y = 1 + 4x$
10. $y = 2 + x$
11. $y = 1 - 3x$
12. $y = 3 - 2x$
13. $2x - y + 5 = 0$
14. $6x - 11y + 2 = 0$
15. $x + y = 0$
16. $x - y = 0$
17. Passing through $(4, 4)$ and $(-1, 3)$.
18. Passing through $(0, -6)$ and $(-1, 2)$.

In Problems 19 through 28, find the positive angle between the given lines. In addition, give an approximation to the nearest 0.01°.

19. $y = x + 13$ and $y = \frac{9}{5}x$
20. $y = x - 6$ and $y = -\frac{1}{4}x$
21. $2x - y + 3 = 0$ and $x + 4y - 7 = 0$
22. $2x - y + 5 = 0$ and $x - y = 0$
23. $x + y = 0$ and $2x - y = 4$
24. $3x + y = 0$ and $x + 7y = -3$
25. $x - 3y = 6$ and $x = 5$
26. $5x - y = 6$ and $x = -2$
27. Any two lines with angles of inclination 12° and 130°, respectively.
28. Any two lines with angles of inclination 5° and 58°, respectively.

In Problems 29 through 34, find the distance from the given point to the graph of the linear equation.

29. $(5, -3)$; $y = \frac{3}{4}x - 1$
30. $(7, -2)$; $y = \frac{5}{12}x - 1$
31. $(-1, 3)$; $7x - 2y = 0$
32. $(4, 0)$; $2x + 2y - 1 = 0$
33. $(0, 0)$; $3x + 2y = 6$
34. $(2, 0)$; $3x + 2y = 12$

In Problems 35 through 46, describe the collection of all lines that satisfy the given conditions. Sketch at least four lines for each family.

35. $y = mx - 1$,   where m takes on the values -2, -1, $-\frac{1}{2}$, and 0.
36. $y = mx$,   where m takes on the values $\frac{1}{2}$, 1, 2, and $\frac{5}{2}$.
37. $y = x + b$,   where b takes on the values -1, $-\frac{1}{2}$, 0, $\frac{1}{2}$.
38. $y = -\frac{2}{3}x + b$,   where b takes on the values 3, $\frac{7}{2}$, 4, $\frac{9}{2}$.
39. $y = mx + 3$,   where $0 < m < 1$.
40. $y = mx - 2$,   where $-2 < m < -1$.
41. $y - 2 = m(x + 2)$,   where $m < 0$.
42. $y + 3 = m(x + 4)$,   where $m > 0$.
43. $2x - 4y = c$,   where c is any real number.
44. $x + 3y = c$,   where c is any real number.
45. $y = cx + c$,   where $1 \le c \le 3$.
46. $cy = x + c^2$,   where $1 \le c \le 4$.
47. Find the distance between the lines $8x - 2y = 1$ and $8x - 2y = -5$.
48. Find the distance between the lines $x + 4y = 11$ and $x + 4y = 3$.

B

49. Find the slope(s) of the line(s) that make an angle of 60° with $2x - 2y + 5 = 0$.
50. Find the slope(s) of the line(s) that make an angle of 30° with $4x - 2y - 3 = 0$.
51. Find the measure of angle A to the nearest 0.01° of the triangle with vertices $A(2, -1)$, $B(3, 1)$, and $C(-2, 4)$.
52. Find the measure of angle A to the nearest 0.01° of the triangle with vertices $A(2, -4)$, $B(4, 2)$, and $C(-5, 5)$.

53. Find the slope of all the lines passing through $(4, 13)$ that are tangent to the circle $(x + 3)^2 + (y - 9)^2 = 20$.

54. Find the slope of all the lines passing through $(7, -10)$ that are tangent to the circle $(x - 5)^2 + (y + 4)^2 = 8$.

55. Find equations of the lines described in Problem 53.

56. Find equations of the lines described in Problem 54.

57. Find an equation(s) of the line(s) containing $(2, 6)$ and at a distance of 3 from $(4, -4)$.

58. Find an equation(s) of the line(s) containing $(4, 3)$ and at a distance of 2 from $(7, -3)$.

59. Find an equation(s) of the line(s) parallel to and a distance of 2 units from $4x + 3y = 0$.

60. Find an equation(s) of the line(s) parallel to and a distance of 4 units from $x - 3y = 0$.

C

61. Show that if $y = mx + b$ is written in the general form $Dx + Ey + F = 0$, then

$$\frac{|mh + b - k|}{\sqrt{1 + m^2}} = \frac{|Dh + Ek + F|}{\sqrt{D^2 + E^2}}.$$

S E C T I O N   8.2   THE PARABOLA

In Section 2.4 we graphed the quadratic function $f(x) = ax^2 + bx + c$. In this section we revisit this curve and consider its geometric properties.

Parabola

> A **parabola** is the set of all points (in the plane) whose distances from a fixed point equal their distances from a fixed line. The fixed point is called the **focus,** and the fixed line is called the **directrix.**

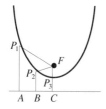

FIGURE 8

Figure 8 illustrates points P_1, P_2, and P_3, on a parabola. The points A, B, and C are on the directrix, and point F is the focus. By the definition,

$$d(F, P_1) = d(A, P_1)$$
$$d(F, P_2) = d(B, P_2)$$
$$d(F, P_3) = d(C, P_3).$$

The point P_3, that is the midpoint of the perpendicular line segment from F to C is the vertex. The line containing the vertex and the focus is called the **axis of symmetry.**

There is an easy and interesting way to generate some points on a parabola geometrically. Suppose the distance from the focus to the directrix is $2c$, so that the distance from the focus to the vertex is c. Figure 9 consists of a family of lines parallel to the directrix and spaced c units apart, along with a family of circles with their centers at the focus whose radii are c, $2c$,

3c, and so on. The intersection of the line c units to the right of the directrix and the circle with radius c is a point, the vertex, on the parabola (Why?). The intersection of the line 2c units to the right of the directrix and the circle with radius 2c are two points on the parabola. Continuing in this way generates a set of points on the parabola.

Up to this point our discussion of parabolas has been entirely geometric and independent of a coordinate system. To find an equation for a parabola, we set the focus and directrix on the Cartesian plane. For convenience let the focus be at $(0, c)$ and the directrix be the line $y = -c$, as shown in Figures 10 and 11. If (x, y) is a point on the parabola, the distance from (x, y) to $(0, c)$ must equal the distance from (x, y) to the line $y = -c$. We have

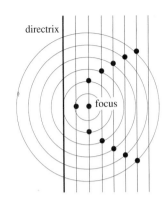

FIGURE 9

distance from (x, y) to $(0, c)$ = distance from (x, y) to $y = -c$

$$\sqrt{(x-0)^2 + (y-c)^2} = |y\text{-coordinate of } (x, y)$$
$$- y\text{-coordinate of a point on } y = -c|$$
$$\sqrt{x^2 + (y-c)^2} = |y - (-c)| = |y + c|.$$

Squaring both sides and expanding leads to

$$x^2 + y^2 - 2cy + c^2 = y^2 + 2cy + c^2.$$

If we add $2cy$ to both sides and simplify, we get

$$x^2 = 4cy.$$

The results are summarized below.

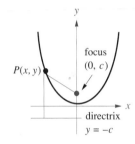

FIGURE 10

The equation for a parabola with focus at $(0, c)$, directrix $y = -c$, and vertex at the origin is

$$x^2 = 4cy.$$

The graph opens upward if $c > 0$; the graph opens downward if $c < 0$.

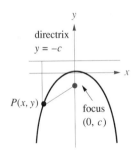

FIGURE 11

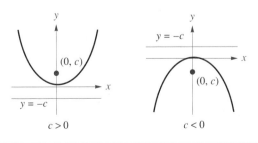

EXAMPLE 1   Graph $x^2 = -8y$. Identify the coordinates of the focus and the equation of the directrix.

SOLUTION Comparing the given equation to $x^2 = 4cy$, we have

$$4c = -8 \Rightarrow c = -2.$$

The parabola has its focus at $(0, c) = (0, -2)$. The directrix is the line

$$y = -c \Rightarrow y = -(-2) \Rightarrow y = 2.$$

The graph is shown in Figure 12.

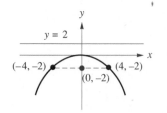

FIGURE 12

In the last example consider the points $(4, -2)$ and $(-4, -2)$ on the parabola. These two points are 4 units (horizontally) from the focus and 4 units (vertically) from the directrix. The line segment with endpoints $(4, -2)$ and $(-4, -2)$ is called a **focal chord** because it passes through the focus.

For our discussion, we will define a focal chord of a parabola as the line segment parallel to the directrix containing the focus with endpoints on the parabola. For the parabola determined by $x^2 = 4cy$, the focal chord always has a length of $|4c|$. This fact can be helpful in sketching the graph of a parabola (Figure 13).

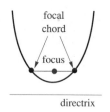

FIGURE 13

EXAMPLE 2   Sketch $x^2 = 10y$ using the focal chord.

SOLUTION Comparing the given equation to $x^2 = 4cy$, we find that the length of the (horizontal) focal chord is $4c = 10$. The focus is determined by solving $4c = 10$ to get $c = \frac{5}{2}$. Hence the focus is at $(0, \frac{5}{2})$.

Now draw a horizontal line with length 10, having the focus as a midpoint. The parabola passes through the endpoints. The graph is shown in Figure 14.

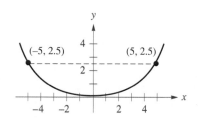

FIGURE 14

A parabola with focus at $(0, c)$ and directrix $y = -c$ is said to be in **standard position.** We have seen that the graph opens upward or downward, depending on the sign of c. There are two other possibilities to consider. A parabola with focus at $(c, 0)$ and directrix $x = -c$ is also a parabola in standard position. It opens to the right or to the left, depending on the sign of c. Furthermore, the equation for such a parabola is $y^2 = 4cx$. The results are summarized next.

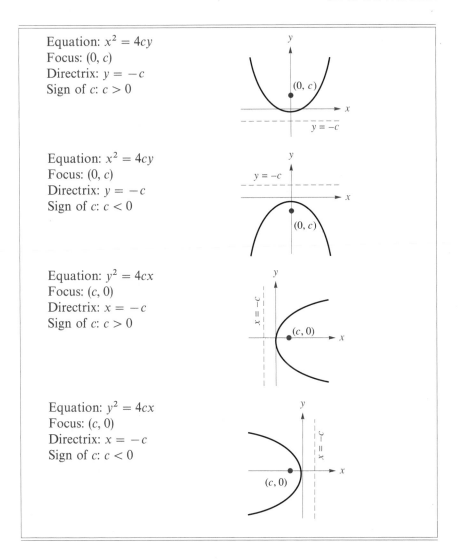

Equation: $x^2 = 4cy$
Focus: $(0, c)$
Directrix: $y = -c$
Sign of c: $c > 0$

Equation: $x^2 = 4cy$
Focus: $(0, c)$
Directrix: $y = -c$
Sign of c: $c < 0$

Equation: $y^2 = 4cx$
Focus: $(c, 0)$
Directrix: $x = -c$
Sign of c: $c > 0$

Equation: $y^2 = 4cx$
Focus: $(c, 0)$
Directrix: $x = -c$
Sign of c: $c < 0$

EXAMPLE 3   Sketch $y^2 = -12x$ using the focal chord. Include the directrix in your sketch.

SOLUTION Comparing the given equation to $y^2 = 4cx$, we find that the graph opens to the left and that the length of the (vertical) focal chord is $|4c| = |-12| = 12$. The focus is determined by solving $4c = -12$ to get $c = -3$. Hence the focus is at $(-3, 0)$, and the directrix is $x = 3$. Now draw a vertical line segment with length 12, having the focus as a midpoint. The parabola passes through the endpoints $(-3, 6)$ and $(-3, -6)$. The graph is shown in Figure 15.

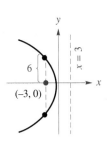

FIGURE 15

Shifting a Parabola

Recall from Section 2.5 that the graph of $y - k = f(x - h)$ is the same as the graph of $y = f(x)$ shifted h units horizontally and k units vertically. This idea of translating a graph applies as well to graphs of equations that are not functions. For example, the circle $x^2 + y^2 = 9$ can be shifted h units horizontally by substituting $x - h$ for x. Substituting $y - k$ for y translates the circle k units vertically.

Similarly, a parabola can be translated h units horizontally and k units vertically by substituting $x - h$ for x and $y - k$ for y.

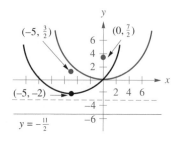

$$(x + 5)^2 = 14(y + 2)$$

FIGURE 16

EXAMPLE 4   Graph $(x + 5)^2 = 14(y + 2)$. Identify the vertex, focus, and directrix.

SOLUTION The first step is to recognize the form $x^2 = 14y$. The graph of $x^2 = 14y$ is a parabola that opens upward with vertex at the origin, focus at $(0, \frac{7}{2})$, and directrix $y = -\frac{7}{2}$. The graph of $(x + 5)^2 = 14(y + 2)$ is the same parabola shifted 5 units to the left and 2 units downward (Figure 16). The vertex is at $(-5, -2)$, the focus (recall $c = \frac{7}{2}$) is at $(-5, \frac{3}{2})$, and the directrix is $y = -\frac{11}{2}$.

The previous example illustrates a convenient form for graphing parabolas. It will be to our advantage to focus our attention on these forms in general.

Convenient Forms For
Graphing

The graph of $(x - h)^2 = 4c(y - k)$ is the parabola $x^2 = 4cy$ shifted h units horizontally and k units vertically.

The graph of $(y - k)^2 = 4c(x - h)$ is the parabola $y^2 = 4cx$ shifted h units horizontally and k units vertically.

At this point, two questions may have occurred to you:

1.   How do we recognize that an equation describes a parabola?

2.   After we recognize that the graph of an equation is a parabola, how do we get the equation into a convenient form?

A simple answer to the first question is: If an equation is quadratic in one variable and linear in the other variable, the graph is a parabola.

The **general form** for a parabola is

$$Ax^2 + Cy^2 + Dx + Ey + F = 0$$

where either $A = 0$ or $C = 0$, but not both.

An answer to the second question, how to transform any equation of a parabola into a form convenient for graphing, is: Complete the square in the quadratic variable.

EXAMPLE 5   Sketch the graph of $2y^2 - 3x + 28y + 110 = 0$.

SOLUTION This equation is quadratic in y, so we complete the square with the terms involving y. The first step is to isolate all such terms and then factor out the coefficient of y^2:

$$2y^2 + 28y = 3x - 110$$

$$2(y^2 + 14y) = 3x - 110.$$

The square of half the y-coefficient is 49. Be careful! By inserting the term 49 inside the parentheses on the left side of the equation, we are actually adding $2(49) = 98$ to the left side. Consequently, we must add 98 to the right side:

$$2(y^2 + 14y + \mathbf{49}) = 3x - 110 + \mathbf{98}$$

$$2(y + 7)^2 = 3x - 12$$

$$2(y + 7)^2 = 3(x - 4)$$

$$(y + 7)^2 = \tfrac{3}{2}(x - 4).$$

The graph of this equation is the same as that of $y^2 = \tfrac{3}{2}x$ shifted 4 units horizontally and -7 units vertically. Setting $4c = \tfrac{3}{2}$, we get $c = \tfrac{3}{8}$, so the focus for $y^2 = \tfrac{3}{2}x$ is at $(\tfrac{3}{8}, 0)$. The focus for

$$(y + 7)^2 = \tfrac{3}{2}(x - 4)$$

must then be at

$$(\tfrac{3}{8} + 4, 0 - 7) = (\tfrac{35}{8}, -7).$$

The focal chord has a length of $|4c| = \tfrac{3}{2}$. Putting all this information together, we get the graph in Figure 17.

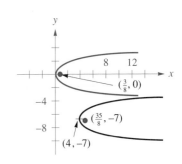

FIGURE 17

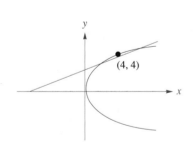

FIGURE 18

EXAMPLE 6   Find the focus and the equation of the parabola passing through the point $(8, -3)$ with vertex $(3, 2)$ and directrix parallel to the x-axis.

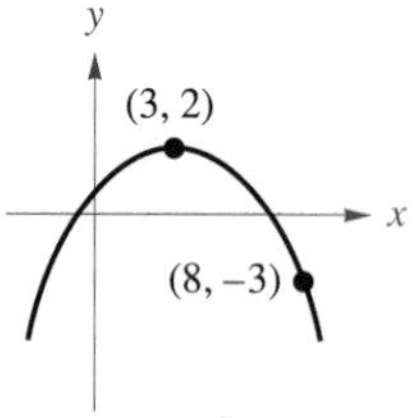

SOLUTION   Draw a picture of the information to get a sense of the way the parabola opens (Figure 18). Since the parabola opens down, it must have the form $(x - h)^2 = 4c(y - k)$, where $c < 0$. Substituting the coordinates of the vertex yields

$$(x - 3)^2 = 4c(y - 2).$$

We are given that $x = 8$ when $y = -3$, so

$$(8 - 3)^2 = 4c(-3 - 2)$$
$$25 = 4c(-5).$$

Solving for $4c$ gives

$$-5 = 4c.$$

Substituting this value for $4c$ into $(x - 3)^2 = 4c(y - 2)$ gives $(x - 3)^2 = -5(y - 2)$. Since $c = -\frac{5}{4}$, the focus is $\frac{5}{4}$ units below the vertex. The coordinates of the focus are $(3, \frac{3}{4})$.

One of the most useful properties of a parabola relate to the tangent line. A line tangent to a parabola intersects the parabola at exactly one point and is not perpendicular to the directrix. The following example shows a way to find the line tangent to a parabola at a given point.

EXAMPLE 7   Find an equation of the line tangent to $y^2 = 4x$ at the point $(4, 4)$.

SOLUTION   A sketch of the situation is shown in Figure 19. Using the point-slope form of a linear equation, we know that the tangent line belongs to the family of lines $y - 4 = m(x - 4)$. Solving for y, we get

$$y = mx - 4m + 4.$$

We need to determine the value of m so that the line has only one point $(4, 4)$ in common with the parabola $y^2 = 4x$. Also, since the directrix is parallel to the y-axis, $m \neq 0$. Substituting $mx - 4m + 4$ for y in $y^2 = 4x$, we have $(mx - 4m + 4)^2 = 4x$.

Expand the left side and collect x^2 and x terms:

$$m^2x^2 - 8m^2x + 16m^2 + 8mx - 32m + 16 = 4x$$
$$m^2x^2 + (-8m^2 + 8m - 4)x + (16m^2 - 32m + 16) = 0.$$

FIGURE 19

This equation is quadratic in x. Recall that $ax^2 + bx + c = 0$ has only one solution if $b^2 - 4ac = 0$. Consequently, the quadratic equation has only one solution when

$$(-8m^2 + 8m - 4)^2 - 4(m^2)(16m^2 - 32m + 16) = 0.$$

If the left side is expanded and simplified (with paper and pencil), the equation reduces to a quadratic equation in m that can be solved fairly easily:

$$64m^2 - 64m + 16 = 0$$

$$4m^2 - 4m + 1 = 0$$

$$(2m - 1)^2 = 0 \Rightarrow m = \tfrac{1}{2}.$$

The equation of the tangent line is $y - 4 = \tfrac{1}{2}(x - 4)$, or in slope-intercept form it is $y = \tfrac{1}{2}x + 2$.

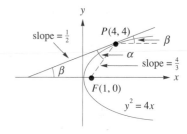

FIGURE 20

The algebra in the previous example was tedious. In calculus you will learn an easier way to find the value of m. Now that we have the equation of the tangent line, consider its angle of inclination β (Figure 20). Let α be the angle that the tangent line makes with the focal radius at the point of tangency $P(4, 4)$. The point $F(1, 0)$ is the focus for $y^2 = 4x$. The slope of the line through $F(1, 0)$ and $P(4, 4)$ is $\tfrac{4}{3}$.

Recall from Section 8.1 that the angle α between the tangent line (with slope $= \tfrac{1}{2}$) and the line through F and P (slope $= \tfrac{4}{3}$) is given by

$$\tan \alpha = \left| \frac{m_1 - m_2}{1 + m_1 m_2} \right| = \left| \frac{\tfrac{1}{2} - \tfrac{4}{3}}{1 + (\tfrac{1}{2})(\tfrac{4}{3})} \right| = \frac{1}{2}.$$

Notice that the angle of inclination β of the tangent line is also given by $\tan \beta = \tfrac{1}{2}$. From these two equations (and Figure 20) we conclude that

$$\alpha = \beta.$$

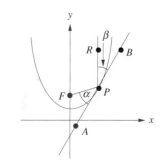

FIGURE 21

This is a specific case of an important property of parabolas in general. Figure 21 shows a tangent line at the point P and a focal radius $\overline{FP}$. The line $\overline{RP}$ is perpendicular to the directrix. The angles $\alpha = \angle APF$ and $\beta = \angle BPR$ are equal. This property is called the **reflective property** of a parabola.

There are many applications of the reflective property. Parabolic reflectors are used for telescopes, dish antennas, directional microphones, headlights, and solar collectors (Figure 22). For example, radio waves reflect off a surface at an angle equal to the angle of incidence. Because of this, satellite dishes are made such that their cross sections are parabolic. The dish collects incoming transmissions that are parallel to the axis of symmetry. The radio waves reflect from the dish to the focus, which is where the receiver is positioned.

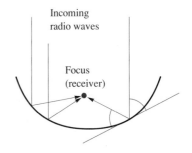

FIGURE 22

We have investigated parabolas whose directrices are either vertical or horizontal. Using the definition of a parabola, we can also derive an equation describing a parabola whose directrix is neither horizontal nor vertical.

EXAMPLE 8　Find the equation of the parabola with directrix $y = -x$ and focus at $(2, 2)$.

SOLUTION See Figure 23. The point $P(x, y)$ is on the parabola if and only if the distance from $P(x, y)$ to $F(2, 2)$ equals the distance from $P(x, y)$ to the line $y = -x$. Recall from Section 8.1 that the distance from $P(x, y)$ to the line $y = mx + b$ is

$$d = \frac{|mh + b - k|}{\sqrt{1 + m^2}} = \frac{|(-1)x + 0 - y|}{\sqrt{1 + (-1)^2}} = \frac{|-x - y|}{\sqrt{2}}.$$

The distance from $P(x, y)$ to $F(2, 2)$ is

$$d = \sqrt{(x - 2)^2 + (y - 2)^2}.$$

Equating these and squaring both sides gives

$$\sqrt{(x - 2)^2 + (y - 2)^2} = \frac{|-x - y|}{\sqrt{2}}$$

$$(x - 2)^2 + (y - 2)^2 = \frac{x^2 + 2xy + y^2}{2}.$$

Expanding and collecting like terms leads to

$$x^2 - 2xy + y^2 - 8x - 8y + 16 = 0.$$

Note the presence of the xy-term. In general, the equation of a parabola has an xy-term if and only if the directrix is neither horizontal nor vertical.

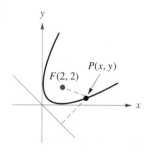

FIGURE 23

EXERCISE SET 8.2

A

In Problems 1 through 9, the vertex V and the focus F of a parabola are given. Sketch the parabola showing the vertex, focus, and directrix, and give an equation of the parabola.

1. $V(0, 0)$; $F(2, 0)$

2. $V(0, 0)$; $F(-1, 0)$

3. $V(0, 0)$; $F(0, -3)$

4. $V(0, 0)$; $F(0, 2)$

5. $V(4, 1)$; $F(4, 3)$

6. $V(-2, -2)$; $F(-2, -3)$

7. $V(0, 3)$; $F(1, 3)$

8. $V(5, 1)$; $F(2, 1)$

9. $V(-2, -1)$; $F(0, -1)$

In Problems 10 through 15, the focus F and the equation of the directrix for a parabola are given. Sketch the parabola showing the vertex, focus, and directrix, and give an equation of the parabola.

10. $F(0, 1); y = 0$

11. $F(0, -2); y = 0$

12. $F(1, 2); x = 3$

13. $F(-4, 4); x = -5$

14. $F(1, -2); y = 1$

15. $F(3, 5); y = 7$

In Problems 16 through 30, find the vertex, focus and directrix of the parabola. Sketch the parabola showing the vertex, focus, and directrix.

16. $x^2 = -4y$

17. $x^2 = 10y$

18. $(x - 3)^2 = 8(y + 2)$

19. $(x + 1)^2 = 4(y - 5)$

20. $x^2 - 4x - 4y + 8 = 0$

21. $x^2 - 10x - 8y + 49 = 0$

22. $y^2 - 2y + 8x + 33 = 0$

23. $y^2 + 6y + 8x - 7 = 0$

24. $x^2 + 12x + 12y + 36 = 0$

25. $y^2 + 8y - 4x + 16 = 0$

26. $x^2 - 10x - 2y + 23 = 0$

27. $x^2 + 2x + 6y - 11 = 0$

28. $x^2 - 6x + 2y + 10 = 0$

29. $3x^2 - 12x - 8y = 4$

30. $4y^2 - 8y + 3x = 2$

B

In Problems 31 through 40, find the equation of the parabola satisfying the given conditions.

31. The focus is $(2, 4)$, and the vertex is $(2, 1)$.

32. The focus is $(-1, 3)$, and the vertex is $(2, 3)$.

33. The focus is $(1, 2)$, the length of the focal chord is 6, and the directrix is vertical.

34. The focus is $(5, -2)$, the length of the focal chord is 10, and the directrix is horizontal.

35. The focal chord has endpoints $(3, 2)$ and $(-5, 2)$, and the parabola opens down.

36. The focal chord has endpoints $(5, 2)$ and $(5, -6)$, and the parabola opens to the right.

37. The vertex is $(2, 0)$, the parabola contains the point $(1, 2\sqrt{2})$, and the directrix is vertical.

38. The vertex is $(0, 4)$, the parabola contains the point $(\sqrt{6}, 5)$, and the directrix is horizontal.

39. The focus is $(2, 0)$, the parabola contains the point $(1, 2\sqrt{2})$, the directrix is vertical, and the parabola opens to the right.

40. The focus is $(0, \frac{11}{2})$, the parabola contains the point $(\sqrt{30}, 6)$, the directrix is horizontal, and the parabola opens upward.

41. An equation of the tangent line to $18x = y^2$ at the point $P(2, 6)$ is $y = \frac{3}{2}x + 3$. Find the acute angle between this tangent line and the line from the focus of the parabola to $P(2, 6)$. (Hint: How is the angle related to the angle of inclination of the tangent line?)

42. An equation of the tangent line to $3x = y^2$ at the point $P(3, 3)$ is $y = \frac{1}{2}x + \frac{3}{2}$. Find the acute angle between this tangent line and the line from the focus of the parabola to $P(3, 3)$. (Hint: How is the angle related to the angle of inclination of the tangent line?)

43. Find an equation of the tangent line to $x^2 = 2y$ at the point $(4, 8)$.

44. Find an equation of the tangent line to $y^2 = \frac{1}{2}x$ at the point $(2, 1)$.

For Problems 45 through 47, a cross section of parabolic mirror for a flashlight has a width of w units and a height of h units (Figure 24). How many units above the vertex should the bulb be placed? (The bulb should be at the focus.)

45. $w = 6, h = 4$

46. $w = 4, h = 6$

47. $w = 10, h = 3$

FIGURE 24

If the weight of a suspension bridge is distributed uniformly along the main cables, the shape of each cable is parabolic (Figure 25). Use this fact to answer Problems 48 and 49.

FIGURE 25

48. The two towers of a suspension bridge are 450 ft apart and extend 120 ft above the (horizontal) road surface. If the vertex of the cable is on the center of the road surface, find the height of the cable at a point 60 ft from the center of the bridge.

49. The two towers of a suspension bridge are 270 ft apart, and the vertex of the cable is on the center of the (horizontal) road surface. At a point 40 ft from the center of the bridge, the cables are 30 ft above the road surface. Find the height of the towers.

C

50. Find an equation of the parabola with focus at (1, 3) and directrix given by $y = 2x$.

51. Find an equation of the parabola with focus at $(-2, 1)$ and directrix given by $y = x$.

SECTION 8.3 ELLIPSES AND HYPERBOLAS

Take a piece of string and fix the ends (with thumbtacks or tape) so that the string is slack. Now holding the string taut with a pencil trace a curve completely around (Figure 26).

This curve is called an **ellipse.** Notice that the distance from one thumbtack to the pencil point plus the distance from the other thumbtack to the pencil point remains constant—the length of the string.

Ellipse

> Let F_1 and F_2 be two fixed points in a plane. An **ellipse** is the set of all points P such that the sum of the distances $d(P, F_1) + d(P, F_2)$ is a constant.

The fixed points F_1 and F_2 are called **foci** (foci is plural for focus). The line segment containing the foci that spans the ellipse is called the **major axis.** Similarly, the perpendicular bisector of the major axis that spans the ellipse is called the **minor axis** (Figure 27). The four endpoints of these axes are the **vertices,** and the intersection of these axes is the **center.** If we imagine a right triangle formed with the center, a focus, and a vertex on the minor axis, then we discover a useful relationship. Let

c = distance from the center to a focus

b = distance from the center to vertex on the minor axis

a = hypotenuse (distance from focus to vertex).

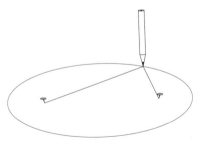

FIGURE 26

By the Pythagorean theorem (see Figure 28),

$$a^2 - c^2 = b^2.$$

This particular case shows, from the symmetry of Figure 29, that the constant sum of the distances from any point to the foci is $2a$.

To find an equation for an ellipse, we impose a coordinate system (Figure 30). For convenience, we place the center at the origin and the foci at $F_1(-c, 0)$ and $F_2(c, 0)$. Suppose $P(x, y)$ represents any point on the curve. According to our observation,

$$d(P, F_1) + d(P, F_2) = d(P_1, F_1) + d(P_1, F_2) = 2a.$$

Using the distance formula, we have

$$\sqrt{(x + c)^2 + y^2} + \sqrt{(x - c)^2 + y^2} = 2a.$$

If we isolate the first radical, square both sides, and simplify, we get

$$x^2 + 2cx + y^2 + c^2 = 4a^2 - 4a\sqrt{x^2 - 2cx + c^2 + y^2}$$
$$+ x^2 - 2cx + c^2 + y^2.$$

Isolating the term with the radical and dividing by -4, we have

$$a^2 - cx = a\sqrt{x^2 - 2cx + c^2 + y^2}.$$

Squaring both sides and simplifying leads to

$$\left(\frac{a^2 - c^2}{a^2}\right)x^2 + y^2 = a^2 - c^2.$$

Replacing $a^2 - c^2$ with b^2, and then dividing both sides by b^2, we obtain the following result.

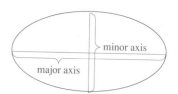

FIGURE 27

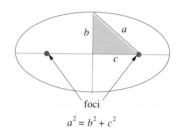

$$a^2 = b^2 + c^2$$

FIGURE 28

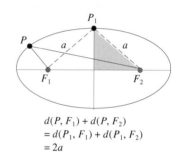

$$d(P, F_1) + d(P, F_2)$$
$$= d(P_1, F_1) + d(P_1, F_2)$$
$$= 2a$$

FIGURE 29

An equation for an ellipse with foci at $(c, 0)$ and $(-c, 0)$ has the form

$$\frac{x^2}{a^2} + \frac{y^2}{b^2} = 1.$$

where $b^2 = a^2 - c^2$.

Equation of an Ellipse

From this equation, we can see that the curve is symmetric with respect to both axes and the origin. Setting $y = 0$, the x-intercepts are $(\pm a, 0)$. The line segment with endpoints $(a, 0)$ and $(-a, 0)$ is the major axis. Similarly, the minor axis is the segment between $(0, b)$ and $(0, -b)$. From Figure 31 it should be clear that the length of the major axis is $2a$. This should make sense to you in terms of the thumbtack-and-string definition. Since we have essentially labeled the length of the string $2a$, when the pencil reaches the point

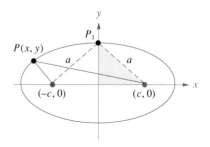

FIGURE 30

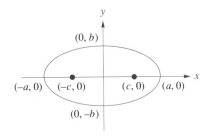

FIGURE 31

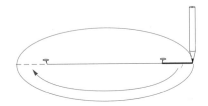

FIGURE 32

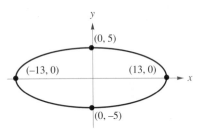

FIGURE 33

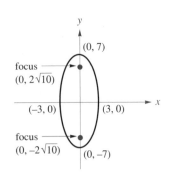

FIGURE 34

shown in Figure 32, there is a "doubling" of string between one focus and vertex. We could remove one of the equal pieces to the other side, and it would complete the major axis.

EXAMPLE 1   Graph

$$\frac{x^2}{169} + \frac{y^2}{25} = 1.$$

Determine the coordinates of the foci, and find the constant sum of the distances from any point on the ellipse to the two foci.

SOLUTION Set $y = 0$ and solve for x to find the x-intercepts:

$$\frac{x^2}{169} + \frac{0^2}{25} = 1 \Rightarrow x = \pm 13.$$

Set $x = 0$ and solve for y to find the y-intercepts:

$$\frac{0^2}{169} + \frac{y^2}{25} = 1 \Rightarrow y = \pm 5.$$

This is enough to get a sketch of the ellipse (Figure 33). In this case $a = 13$, $b = 5$, and

$$c = \sqrt{a^2 - b^2} = \sqrt{169 - 25} = 12.$$

This means that the foci are 12 units from the center on the major axis, so their coordinates are $(12, 0)$ and $(-12, 0)$. The sum of the distances from any point on the ellipse to the foci is $2a = 26$.

EXAMPLE 2   Graph

$$\frac{x^2}{9} + \frac{y^2}{49} = 1.$$

Determine the coordinates of the foci, and find the constant sum of the distances from any point on the ellipse to the two foci.

SOLUTION Set $y = 0$ and solve for x to find the x-intercepts $(\pm 3, 0)$. Set $x = 0$ and solve for y to find the y-intercepts $(0, \pm 7)$. Using these four intercepts, we can sketch the ellipse (Figure 34).

This ellipse has a vertical major axis, so $a = 7$, $b = 3$, and $c = \sqrt{49 - 9} = 2\sqrt{10}$. The foci are $2\sqrt{10}$ units from the center along the major axis. The sum of the distances from any point on the ellipse to the foci $(0, \pm 2\sqrt{10})$ is $2a = 14$.

In summary, when graphing an ellipse we first find the intercepts on the axes of symmetry. The value of a is half the length of the major axis, and the value of b is half the length of the minor axis. From the relationship

$$c = \sqrt{a^2 - b^2}$$

the foci are always on the major axis, c units from the center.

Shifting an Ellipse

We can translate an ellipse h units horizontally or k units vertically by substituting $(x - h)$ for x or $(y - k)$ for y in the equation. This puts the center of the ellipse at the point (h, k).

EXAMPLE 3   Sketch the graph of the ellipse $x^2/13 + y^2/4 = 1$ that is translated 5 units down and 3 units right. Determine the coordinates of the foci and the equation of the shifted graph.

SOLUTION The graph of $x^2/13 + y^2/4 = 1$ has x-intercepts $(\sqrt{13}, 0)$ and $(-\sqrt{13}, 0)$; the y-intercepts are $(0, 2)$ and $(0, -2)$. Since $c = \sqrt{13 - 4} = 3$, the foci are at $(3, 0)$ and $(-3, 0)$. If this graph is shifted 5 units down and 3 units to the right, the center will be at $(3, -5)$, and the foci will have coordinates $(6, -5)$ and $(0, -5)$ (Figure 35). The equation for this shifted ellipse is obtained by substituting $(x - 3)$ for x and $(y + 5)$ for y, which yields

Original equation: $\dfrac{x^2}{13} + \dfrac{y^2}{4} = 1$

Translated equation: $\dfrac{(x - 3)^2}{13} + \dfrac{(y + 5)^2}{4} = 1.$

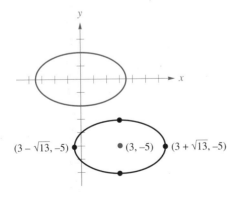

FIGURE 35

The last equation in the previous example illustrates a form that is easily graphed and should be highlighted:

Convenient Form for Graphing

The graph of

$$\frac{(x-h)^2}{a^2} + \frac{(y-k)^2}{b^2} = 1$$

is the ellipse

$$\frac{x^2}{a^2} + \frac{y^2}{b^2} = 1$$

shifted h units horizontally and k units vertically.

The graph of

$$\frac{(x-h)^2}{b^2} + \frac{(y-k)^2}{a^2} = 1$$

is the ellipse

$$\frac{x^2}{b^2} + \frac{y^2}{a^2} = 1$$

shifted h units horizontally and k units vertically.

A way to transform an equation that is quadratic in both variables into one of the preceding forms is by completing the square.

EXAMPLE 4   Graph $9x^2 + 100y^2 - 18x - 400y + 184 = 0$.

SOLUTION Our plan is to complete the square in each variable. First we group the x-terms and the y-terms and factor the leading coefficients:

$$(9x^2 - 18x) + (100y^2 - 400y) = -184$$
$$9(x^2 - 2x) + 100(y^2 - 4y) = -184.$$

To complete the square, we must be careful to add the same number to each side of the equation:

$$9(x^2 - 2x + \mathbf{1}) + 100(y^2 - 4y + \mathbf{4}) = -184 + 9(\mathbf{1}) + 100(\mathbf{4})$$
$$9(x-1)^2 + 100(y-2)^2 = 225.$$

Now, dividing both sides by 225, we obtain

$$\frac{(x-1)^2}{25} + \frac{4(y-2)^2}{9} = 1$$

which can also be written as

$$\frac{(x-1)^2}{25} + \frac{(y-2)^2}{9/4} = 1.$$

The graph of this equation is the same as the graph of

$$\frac{x^2}{25} + \frac{y^2}{9/4} = 1$$

shifted 1 unit right and 2 units up. In this case, $a = 5$ and $b = \frac{3}{2}$; so from the center $(1, 2)$, the major axis extends 5 units to the left and right. The minor axis extends $\frac{3}{2}$ units above and below the center (Figure 36).

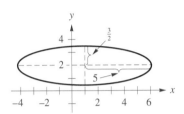

FIGURE 36

EXAMPLE 5   Graph $9x^2 + y^2 - 36x + 8y + 43 = 0$. Identify the coordinates of the foci.

SOLUTION Group the x terms and the y terms, factor the leading coefficients, and complete the square:

$$(9x^2 - 36x) + (y^2 + 8y) = -43$$

$$9(x^2 - 4x) + (y^2 + 8y) = -43$$

$$9(x^2 - 4x + \mathbf{4}) + (y^2 + 8y + \mathbf{16}) = -43 + 9(\mathbf{4}) + \mathbf{16}$$

$$9(x - 2)^2 + (y + 4)^2 = 9$$

$$\frac{(x-2)^2}{1} + \frac{(y+4)^2}{9} = 1.$$

In this case, $a = 3$ and $b = 1$. From the center $(2, -4)$, the major axis extends 3 units above and below, and the minor axis extends 1 unit to the left and right. The graph is sketched in Figure 37. Since the foci are always on the major axis, they are

$$c = \sqrt{a^2 - b^2} = \sqrt{9 - 1} = 2\sqrt{2}$$

units above and below the center; so their coordinates are $(2, -4 + 2\sqrt{2})$ and $(2, -4 - 2\sqrt{2})$.

If an equation of the form

$$\frac{(x-h)^2}{a^2} + \frac{(y-k)^2}{b^2} = 1$$

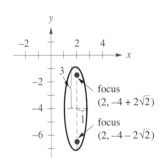

FIGURE 37

is expanded and simplified, it can always be written in the form $Ax^2 + Cy^2 + Dx + Ey + F = 0$, where A and C have the same sign. The last two examples suggest that this process can be reversed, but this is not the case. For example, if we are given the equation $2x^2 + y^2 + 8x - 2y + 15 = 0$ and apply the methods in Examples 4 and 5, we get

$$\frac{(x + 2)^2}{-3} + \frac{(y - 1)^2}{-6} = 1.$$

This equation has no graph since the left side is nonpositive.

It is also possible to end up with a form like

$$\frac{(x + 2)^2}{3} + \frac{(y - 1)^2}{6} = 0.$$

In this case, the graph of the equation has only one point, $(-2, 1)$. These situations are called **degenerate cases.**

Except for degenerate cases, you should be able to convince yourself that equations of the form $Ax^2 + Cy^2 + Dx + Ey + F = 0$, where $AC > 0$ (this means A and C have the same sign) can be expressed in the form

$$\frac{(x - h)^2}{a^2} + \frac{(y - k)^2}{b^2} = 1.$$

The **general form** for the equation of an ellipse is

$$Ax^2 + Cy^2 + Dx + Ey + F = 0$$

where $AC > 0$.

Eccentricity of an Ellipse

Consider the relationship $b^2 = a^2 - c^2$ for an ellipse. If a is held fixed and c varies over the interval $0 < c < a$, the ellipse becomes more circular as c approaches 0 and becomes more elongated as c approaches a (see Figure 38).

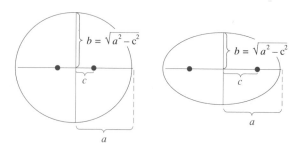

FIGURE 38

A way to quantify the elongation is by the ratio

$$e = \frac{c}{a}$$

where e is called the **eccentricity.** Consequently, for an ellipse, we get

$$0 < e < 1.$$

A value of e that is near 0 is relatively circular, and a value of e that is near 1 is relatively elongated (Figure 39).

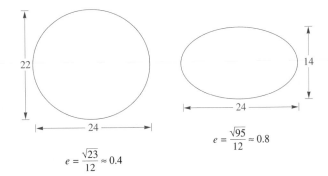

$$e = \frac{\sqrt{23}}{12} \approx 0.4$$

$$e = \frac{\sqrt{95}}{12} \approx 0.8$$

FIGURE 39

EXAMPLE 6   Find an equation of the ellipse with foci at $(-2, 5)$ and $(6, 5)$ and with eccentricity $\frac{2}{3}$.

SOLUTION Sketch the foci. The major axis runs parallel to the x-axis, and the center is at $(2, 5)$. Therefore the equation must be expressible in the form

$$\frac{(x - 2)^2}{a^2} + \frac{(y - 5)^2}{b^2} = 1.$$

All that remains is to find a^2 and b^2. The distance from the center to each focus is $c = 4$. Since the eccentricity is given as $\frac{2}{3}$,

$$e = \frac{c}{a} = \frac{2}{3}.$$

Substitute $c = 4$ and solve for a:

$$\frac{4}{a} = \frac{2}{3} \Rightarrow a = 6.$$

Using $b^2 = a^2 - c^2$ gives $b^2 = 6^2 - 4^2 = 20$. So an equation for the ellipse is

$$\frac{(x - 2)^2}{36} + \frac{(y - 5)^2}{20} = 1.$$

Hyperbolas

We have defined an ellipse as the set of all points P such that the sum of the distances from P to two fixed points is a positive constant. A **hyperbola** is the set of all points P such that the *differences of* the distances from P to two fixed points is a positive constant.

Hyperbola

> Let F_1 and F_2 be two fixed points (in a plane). A **hyperbola** is the set of all points P such that $|d(P, F_1) - d(P, F_2)|$ is a constant.

The absolute value is used because we want to include both possible differences:

$$d(P, F_1) - d(P, F_2) \quad \text{or} \quad d(P, F_2) - d(P, F_1)$$

depending on which is positive. This results in two separate curves or "branches" (Figure 40). The two fixed points F_1 and F_2 are the **foci,** and the midpoint of the segment connecting them is the **center.**

To find an equation of a hyperbola, as with the ellipse, we impose a coordinate system, placing the two foci at $F_1(-c, 0)$ and $F_2(c, 0)$, where $c > 0$. Let the constant difference be $2a$. If $P(x, y)$ is any point on the hyperbola, then

$$|d(F_1, P) - d(F_2, P)| = 2a.$$

Using the distance formula yields

$$\left| \sqrt{(x + c)^2 + y^2} - \sqrt{(x - c)^2 + y^2} \right| = 2a.$$

There are two possibilities:

$$\sqrt{(x + c)^2 + y^2} - \sqrt{(x - c)^2 + y^2} = \pm 2a.$$

The steps in simplifying this equation are very similar to the steps in our derivation of the equation for the ellipse. It is left as an exercise to show that this equation simplifies to

$$\frac{x^2}{a^2} + \frac{y^2}{a^2 - c^2} = 1.$$

Recall that for an ellipse we substituted b^2 for $a^2 - c^2$, but this is not valid for a hyperbola. In this case $c > a$ because in triangle $F_1 F_2 P$ (Figure 41),

$$d(F_1, F_2) + d(P, F_2) \geq d(P, F_1)$$

and since $d(F_1, F_2) = 2c$, it follows that $2c > 2a$. Therefore $c^2 - a^2$ is positive, and we make the substitution $b^2 = c^2 - a^2$.

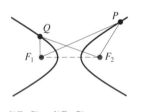

$d(F_1, P) - d(F_2, P)$
$= d(F_2, Q) - d(F_1, Q) = \text{constant}$

FIGURE 40

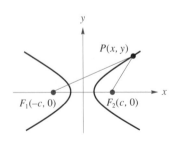

FIGURE 41

An equation for a hyperbola with foci at $(c, 0)$ and $(-c, 0)$ is

$$\frac{x^2}{a^2} - \frac{y^2}{b^2} = 1$$

where $b^2 = c^2 - a^2$.

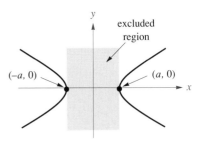

From this equation we can see that the curve is symmetric with respect to both axes and the origin. Setting $y = 0$ gives the x-intercepts $(\pm a, 0)$. Setting $x = 0$ we see there are no y-intercepts. In fact, $x^2 \geq a^2$ because, for values of x between $-a$ and a, there are no real values of y that satisfy the equation of the hyperbola. This means that no portion of the hyperbola lies between the lines $x = a$ and $x = -a$ (Figure 42).

FIGURE 42

In the equation

$$\frac{x^2}{a^2} - \frac{y^2}{b^2} = 1,$$

Asymptotes

if we solve for y, we obtain

$$y = \pm \frac{b}{a} \sqrt{x^2 - a^2} = \pm \frac{bx}{a} \sqrt{1 - \frac{a^2}{x^2}}.$$

As $|x|$ grows, the term a^2/x^2 approaches zero, so for large values of $|x|$,

$$y \approx \pm \frac{b}{a} x.$$

Thus as $x \to \pm \infty$,

$$y \to \pm \frac{b}{a} x.$$

This means the two lines

$$y = \frac{b}{a} x \quad \text{and} \quad y = -\frac{b}{a} x$$

are slant asymptotes for the hyperbola (Figure 43)

$$\frac{x^2}{a^2} - \frac{y^2}{b^2} = 1.$$

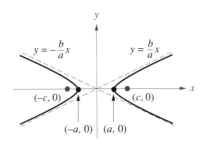

FIGURE 43

If we interchange the variables x and y, the equation of the hyperbola becomes

$$\frac{y^2}{a^2} - \frac{x^2}{b^2} = 1$$

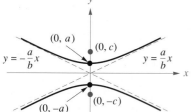

FIGURE 44

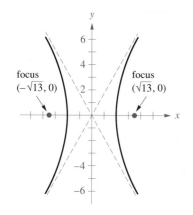

FIGURE 45

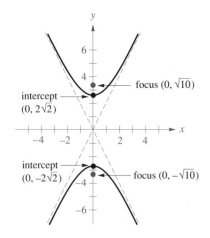

FIGURE 46

and the asymptotes are

$$x = \pm \frac{b}{a} y,$$

which can be written

$$y = \pm \frac{a}{b} x.$$

This hyperbola (Figure 44) has no x-intercepts (why?); the y-intercepts are $(0, a)$ and $(0, -a)$. The foci are located at $(0, c)$ and $(0, -c)$.

EXAMPLE 7   Graph $9x^2 - 4y^2 = 36$. Identify the coordinates of the foci and the equations of the asymptotes.

SOLUTION Divide both sides by 36 to get

$$\frac{x^2}{4} - \frac{y^2}{9} = 1.$$

Setting $y = 0$ gives the x-intercepts $(\pm 2, 0)$, which means the hyperbola opens along the x-axis. We observe that $a = 2$, $b = 3$, and

$$c = \sqrt{a^2 + b^2} = \sqrt{4 + 9} = \sqrt{13}.$$

The values of a and b give the equations of the two asymptotes:

$$y = \pm \frac{b}{a} x = \pm \frac{3}{2} x.$$

The foci are $c = \sqrt{13}$ units from the center (origin) along the x-axis. After drawing the two asymptotes and the two x-intercepts, we can sketch the hyperbola (Figure 45).

EXAMPLE 8   Graph $4x^2 - y^2 = -8$. Identify the coordinates of the foci and the equations of the asymptotes.

SOLUTION To get 1 on the right side of the equation, divide both sides by -8:

$$\frac{4x^2}{-8} - \frac{y^2}{-8} = 1, \quad \text{or} \quad \frac{y^2}{8} - \frac{x^2}{2} = 1.$$

Setting $x = 0$ and solving for y gives the y-intercepts $(0, \pm 2\sqrt{2})$, so this hyperbola opens along the y-axis. Identifying $a = 2\sqrt{2}$ and $b = \sqrt{2}$, the equations of the asymptotes are

$$y = \pm \frac{a}{b} x = \pm \frac{2\sqrt{2}}{\sqrt{2}} x = \pm 2x.$$

Figure 46 shows the intercepts and asymptotes.

Since the intercepts are on the y-axis, the foci are located on the y-axis,

$$c = \sqrt{a^2 + b^2} = \sqrt{8 + 2} = \sqrt{10}$$

units from the center of the hyperbola. After drawing the two asymptotes and the two y-intercepts, we can sketch the hyperbola (Figure 46).

For a hyperbola, it is not necessary that $a > b$, as for the ellipse. The values of a and b, as well as the direction the hyperbola opens, depend on the sign rather than the size of the denominators of the x^2 and y^2 terms:

$$\frac{x^2}{\underset{\underset{a^2}{\uparrow}}{169}} - \frac{y^2}{\underset{\underset{b^2}{\uparrow}}{25}} = 1 \qquad \frac{x^2}{\underset{\underset{a^2}{\uparrow}}{9}} - \frac{y^2}{\underset{\underset{b^2}{\uparrow}}{49}} = 1$$

$$\frac{y^2}{\underset{\underset{a^2}{\uparrow}}{169}} - \frac{x^2}{\underset{\underset{b^2}{\uparrow}}{25}} = 1 \qquad \frac{y^2}{\underset{\underset{a^2}{\uparrow}}{9}} - \frac{x^2}{\underset{\underset{b^2}{\uparrow}}{49}} = 1$$

You may have anticipated shifting any hyperbola that is centered at the origin to one that is centered at the point (h, k). As with the ellipse, there is a convenient form for graphing hyperbolas.

Convenient Form for Graphing

The graph of

$$\frac{(x - h)^2}{a^2} - \frac{(y - k)^2}{b^2} = 1$$

is the hyperbola

$$\frac{x^2}{a^2} - \frac{y^2}{b^2} = 1$$

shifted h units horizontally and k units vertically.

The graph of

$$\frac{(y - k)^2}{a^2} - \frac{(x - h)^2}{b^2} = 1$$

is the hyperbola

$$\frac{y^2}{a^2} - \frac{x^2}{b^2} = 1$$

shifted h units horizontally and k units vertically.

Recall that the graph of an equation is an ellipse if it is quadratic in both variables and the x^2 term and y^2 term have the same sign. The essential algebraic property of a hyperbola is that it also is quadratic in both variables and the x^2 term and y^2 term have opposite signs.

The **general form** for a hyperbola is

$$Ax^2 + Cy^2 + Dx + Ey + F = 0$$

where $AC < 0$.

EXAMPLE 9   Graph the equation $4x^2 - y^2 + 8x + 2y + 7 = 0$. Indicate asymptotes and foci.

SOLUTION   The equation describes a hyperbola since it is quadratic in both variables and the quadratic terms have opposite signs ($A = 4$, $C = -1$). We group, factor, and complete the square to transform the given equation into a convenient form:

$$4x^2 + 8x - y^2 + 2y = -7$$
$$4(x^2 + 2x) - (y^2 - 2y) = -7$$
$$4(x^2 + 2x + \mathbf{1}) - (y^2 - 2y + \mathbf{1}) = -7 + 4(\mathbf{1}) - (\mathbf{1})$$
$$4(x + 1)^2 - (y - 1)^2 = -4.$$

To get 1 on the right side of the equation, we divide both sides by -4:

$$\frac{(y - 1)^2}{4} - \frac{(x + 1)^2}{1} = 1.$$

This has the form,

$$\frac{(y - k)^2}{a^2} - \frac{(x - h)^2}{b^2} = 1$$

so we find that $a = 2$, $b = 1$, and $c = \sqrt{4 + 1} = \sqrt{5}$. This hyperbola opens along the y-axis. The center is at $(-1, 1)$. The asymptotes contain the center and have slopes $\pm a/b = \pm 2$.

The hyperbola crosses the vertical axis of symmetry $a = 2$ units above and below the center, that is, at $(-1, -1)$ and $(-1, 3)$. The graph is sketched in Figure 47.

The foci are $c = \sqrt{5}$ units from the center, so the coordinates of the foci are $(-1, 1 + \sqrt{5})$ and $(-1, 1 - \sqrt{5})$.

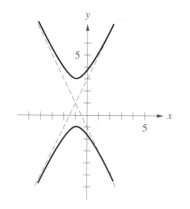

FIGURE 47

As with the ellipse, the **eccentricity** of a hyperbola is defined as

$$e = \frac{c}{a}.$$

In this case, $e > 1$ since $c = \sqrt{a^2 + b^2} > a$. The larger the value for e, the wider the hyperbola appears (Figure 48).

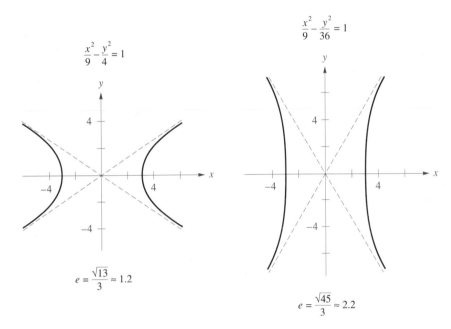

$$\frac{x^2}{9} - \frac{y^2}{4} = 1$$

$$e = \frac{\sqrt{13}}{3} \approx 1.2$$

$$\frac{x^2}{9} - \frac{y^2}{36} = 1$$

$$e = \frac{\sqrt{45}}{3} \approx 2.2$$

FIGURE 48

A summary of how to identify the graph of a second-degree equation follows. It furnishes a common bond for the graphs that we have discussed in this chapter.

$Ax^2 + Cy^2 + Dx + Ey + F = 0$ is the equation of a(n):
parabola if $A = 0$ or $C = 0$ but not both,
ellipse if $AC > 0$ (and a circle if $A = C$),
hyperbola if $AC < 0$.

The Graph of a Second Degree Equation

If A and C are both 0, then the equation is an equation for a line in general form.

There are many instances in which ellipses can be found in the real world. Many stone bridges and arches in buildings are elliptic, for aesthetic as well as structural reasons. Also, according to Newton's laws of mechanics, planets or satellites orbit in elliptic paths.

Hyperbolas have applications also. For example, the LORAN (Long Range Navigation) system uses hyperbolic grids to track ships and airplanes. Another instance is that when two particles of like charges are shot toward each other, they are repelled along a hyperbolic path.

EXERCISE SET 8.3

A

Sketch the graphs in Problems 1 through 18. Indicate the coordinates of the foci.

1. $\dfrac{x^2}{49} + \dfrac{y^2}{9} = 1$ **2.** $\dfrac{x^2}{25} + \dfrac{y^2}{4} = 1$

3. $\dfrac{x^2}{9} + \dfrac{y^2}{35} = 1$ **4.** $\dfrac{x^2}{10} + \dfrac{y^2}{36} = 1$

5. $9x^2 + y^2 = 9$ **6.** $x^2 + 16y^2 = 1$

7. $4x^2 + 25y^2 = 25$ **8.** $4x^2 + 9y^2 = 81$

9. $\dfrac{(x+3)^2}{25} + \dfrac{(y-5)^2}{16} = 1$

10. $\dfrac{(x-6)^2}{36} + \dfrac{(y-1)^2}{100} = 1$

11. $13(x-2)^2 + 4(y-5)^2 = 52$

12. $2(x+4)^2 + (y+1)^2 = 18$

13. $2x^2 + 5y^2 = -20$ **14.** $7x^2 + y^2 = -14$

15. $5x^2 + 8y^2 = 0$ **16.** $4x^2 + 3y^2 = 0$

17. $y = \sqrt{1 - 4x^2}$ **18.** $y = \sqrt{1 - 9x^2}$

Sketch the graph in Problems 19 through 24.

19. $x^2 + 4y^2 + 6x - 16y - 11 = 0$

20. $25x^2 + 4y^2 - 200x - 16y + 316 = 0$

21. $9x^2 + 4y^2 - 54x - 8y + 49 = 0$

22. $9x^2 + 16y^2 - 54x + 64y + 1 = 0$

23. $3x^2 + 4y^2 - 6x + 16y + 7 = 0$

24. $x^2 + 14y^2 + 14x - 56y + 91 = 0$

For each of the equations given in Problems 25 through 30, let P be any point on the ellipse, and let F_1 and F_2 be the foci. Find $d(P, F_1) + d(P, F_2)$.

25. $\dfrac{x^2}{49} + \dfrac{y^2}{9} = 1$ **26.** $\dfrac{x^2}{25} + \dfrac{y^2}{4} = 1$

27. $\dfrac{x^2}{9} + \dfrac{y^2}{35} = 1$ **28.** $\dfrac{x^2}{10} + \dfrac{y^2}{6} = 1$

29. $\dfrac{(x+3)^2}{64} + \dfrac{(y-5)^2}{25} = 1$

30. $\dfrac{(x-6)^2}{9} + \dfrac{(y-1)^2}{49} = 1$

In Problems 31 through 36, find an equation of the ellipse with the given properties.

31. Foci at $(-1, 1)$ and $(5, 1)$; passing through $(6, 1)$

32. Foci at $(2, 8)$ and $(2, 0)$; passing through $(2, -2)$

33. Endpoints of major axis at $(3, 2)$ and $(7, 2)$; eccentricity $\frac{1}{2}$

34. Endpoints of major axis at $(1, 1)$ and $(1, -9)$; eccentricity $\frac{2}{5}$

35. Focus at $(4, -6)$; center at $(1, -6)$; eccentricity $\frac{3}{5}$

36. Focus at $(3, 0)$; center at $(-2, 0)$; eccentricity $\frac{5}{36}$

Sketch the graphs in Problems 37 through 51. Include the asymptotes and indicate the coordinates of the foci.

37. $\dfrac{x^2}{16} - \dfrac{y^2}{9} = 1$

38. $\dfrac{x^2}{64} - \dfrac{y^2}{36} = 1$

39. $\dfrac{y^2}{9} - \dfrac{x^2}{64} = 1$

40. $y^2 - \dfrac{x^2}{16} = 1$

41. $x^2 - y^2 = 2$

42. $x^2 - y^2 = -18$

43. $4(x-2)^2 - 25(y+1)^2 = 25$

44. $(x-3)^2 - 16(y-4)^2 = 4$

45. $x^2 - (y+1)^2 = -9$

46. $144(x - 5)^2 - 16(y + 2)^2 = -144$

47. $3x^2 - 7y^2 = 0$

48. $4x^2 - 3y^2 = 0$

49. $y = \sqrt{1 + 16x^2}$

50. $y = \sqrt{4 + 36x^2}$

51. $x = \sqrt{y^2 + 1}$

Sketch the graphs in Problems 52 through 57. Include the asymptotes.

52. $x^2 - 9y^2 + 2x - 80 = 0$

53. $4x^2 - 25y^2 + 200y - 500 = 0$

54. $16x^2 - y^2 + 32x + 4y + 76 = 0$

55. $x^2 - 4y^2 + 10x - 32y - 3 = 0$

56. $x^2 - 9y^2 - 12x + 36y - 27 = 0$

57. $4x^2 - y^2 - 8x - 6y - 3 = 0$

For each of the equations in Problems 58 through 60, let P be any point on the hyperbola, and let F_1 and F_2 be the foci. Find $d(P, F_1) - d(P, F_2)$.

58. $\dfrac{x^2}{4} - \dfrac{y^2}{10} = 1$

59. $12x^2 - y^2 = 3$

60. $2x^2 - 16y^2 = -9$

In Problems 61 through 66, find an equation of the hyperbola with the given properties.

61. Foci at $(3, 2)$ and $(3, -6)$; passing through $(3, 1)$

62. Foci at $(-5, -1)$ and $(3, -1)$; passing through $(2, -1)$

63. Foci at $(0, 0)$ and $(5, 0)$; eccentricity $\frac{5}{4}$

64. Foci at $(3, 0)$ and $(3, 7)$; eccentricity $\frac{7}{4}$

65. Asymptotes $y = \frac{2}{3}x + 3$ and $y = -\frac{2}{3}x - 1$; passing through $(3, 1)$

66. Asymptotes $y = \frac{1}{2}x - 1$ and $y = -\frac{1}{2}x + 5$; passing through $(6, -3)$

B

Graph the equations in Problems 67 through 72. Indicate the coordinates of the foci and eccentricity.

67. $25x^2 + 9y^2 - 100x + 18y - 116 = 0$

68. $4x^2 + 9y^2 + 8x - 36y + 4 = 0$

69. $9y^2 - 4x^2 - 8x - 18y - 4 = 0$

70. $x^2 - y^2 - 4x + 2y - 6 = 0$

71. $x^2 - 4y^2 - 2x + 32y - 184 = 0$

72. $2x^2 + 5y^2 + 8x - 10y + 13 = 0$

73. Graph $x^2 + 27y^2 = k$, for $k = 0, 3$, and 9.

74. Graph $32x^2 + y^2 = k$, for $k = 0, 2$, and 8.

75. Graph $24x^2 - y^2 = k$, for $k = -6, 0$, and 6.

76. Graph $2x^2 - 18y^2 = k$, for $k = -2, 0$, and 2.

C

77. Line segment AB, with a length of 7 units, is placed so that endpoint A is on the y-axis and endpoint B is on the x-axis. Suppose $P(x, y)$ is a point on AB that is 2 units from A and 5 units from B. Find an equation describing the path of $P(x, y)$ as point A moves along the y-axis (from 0 to 7) and point B moves along the x-axis (Figure 49).

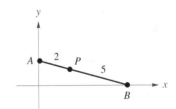

FIGURE 49

THE GEOMETRY OF CONIC SECTIONS   SECTION 8.4

We have seen that parabolas, ellipses, and hyperbolas are related algebraically by a second-degree equation. A common geometric bond for these curves is the rather surprising fact that they can be obtained by taking a **conic section,** that is, the intersection of a plane and a cone.

Start with a right circular cone that has two portions, or nappes, separated by the vertex. These nappes extend indefinitely and have no base. If a plane cuts completely across one nappe of the cone and is not perpendicular to the axis of the cone, the curve of intersection is an ellipse (if the plane cuts at right angles to the axis, the curve is a circle). If a plane cuts only one nappe, but does not cut completely through, the curve of intersection is a parabola. If a plane cuts both nappes (but not through the vertex), the curve of intersection is a hyperbola (Figure 50).

FIGURE 50

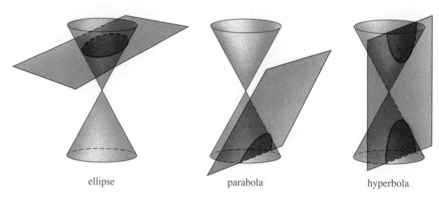

ellipse   parabola   hyperbola

The fact that each of these conic sections looks like an ellipse, parabola, or hyperbola is not enough to justify that this is indeed the case. For example, does the conic section in Figure 50 that looks like an ellipse satisfy the focal definition given in Section 8.3? To prove this, we need to show that if P is any point on the curve of intersection, then $d(P, F_1) + d(P, F_2)$ is a constant, where F_1 and F_2 are two fixed points in the plane. The proof we will present is a beautiful argument given in 1822 by the Belgian mathematician G. P. Dandelin. We use the following fact from geometry (see Figure 51):

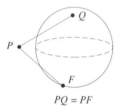

$PQ = PF$

FIGURE 51

> If PQ and PF are line segments tangent to a sphere where Q and F are the points of tangency, then $PQ = PF$.

The proof of this is left as an exercise.

Consider two spheres inscribed in the cone that are tangent to the plane at F_1 and F_2, touching the cone along two parallel circles (Figure 52). Let P be any point on the curve. Draw a line through P and the vertex of the cone. This line intersects the parallel circles at Q_1 and Q_2. Since PQ_1 and PF_1 are both tangent to the larger sphere, it follows that $PQ_1 = PF_1$. Similarly, we also get $PQ_2 = PF_2$. Combining these last two equations gives us

$$PQ_1 + PQ_2 = PF_1 + PF_2.$$

FIGURE 52

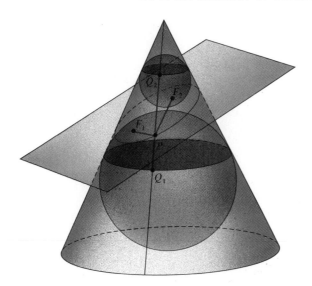

But $PQ_1 + PQ_2 = Q_1Q_2$, the constant distance along the surface of the cone between the two parallel circles. Regardless of the choice for P, $PF_1 + PF_2$ is constant. This means that the conic section is an ellipse.

The arguments that the parabola and hyperbola are also conic sections are similar. They are left as exercises.

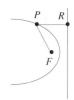

FIGURE 53

Recall that a parabola has a focus and directrix. Suppose F is the focus and P is any point on a parabola. If we draw a perpendicular from P to the directrix intersecting at R (Figure 53), then by definition,

$$PF = k(PR), \qquad \text{where } k = 1.$$

Each ellipse has two directrices, one for each focus. Suppose one focus is F_1 (Figure 54). We will show that for an ellipse,

$$PF_1 = k(PR), \qquad \text{where } k < 1.$$

Similarly, each hyperbola has two directrices, one for each focus (Figure 55). In this case we get

$$PF_1 = k(PR), \qquad \text{where } k > 1.$$

We will prove the case for an ellipse ($k < 1$) using a figure similar to the one used in Dandelin's proof. Figure 56 consists of a circular cone intersecting a plane to form an ellipse and a sphere inscribed in the cone that is tangent to the plane at one of the foci F_1. The sphere touches the cone along a circle. The plane containing this circle intersects the plane of the ellipse at line L.

Focus-Directrix Properties

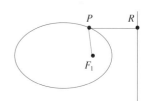

FIGURE 54

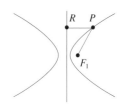

FIGURE 55

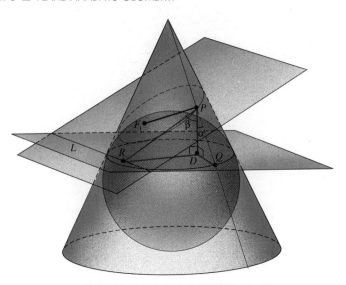

FIGURE 56

Let P be any point on the ellipse. From P draw a perpendicular line to the plane containing the circle, intersecting it at D. From P draw a perpendicular to the line L, intersecting at R. Let $\alpha = \angle QPD$ and $\beta = \angle RPD$. Then we have

$$PQ = \frac{PD}{\cos \alpha} \quad \text{and} \quad PR = \frac{PD}{\cos \beta}.$$

But $PQ = PF_1$ since they are tangents to the same sphere. Therefore,

$$\frac{PF_1}{PR} = \frac{PQ}{PR} = \frac{\cos \beta}{\cos \alpha}.$$

Notice that α and β are determined by the aperture of the cone and the intersecting planes, respectively, so these angles are independent of the choice of P. Hence, $\cos \beta / \cos \alpha = k$ is a constant, and $PF_1 = k(PR)$. The directrix is line L.

If $0 < \alpha < \beta$, as pictured, then $k < 1$, and the conic section is an ellipse. If $\alpha = \beta$, then $k = 1$, and the conic section is a parabola. If $\alpha > \beta$, then $k > 1$, and the conic section is a hyperbola.

To determine the relationship between the ellipse and one of its directrices, we place it on the Cartesian plane with foci at $(-c, 0)$ and $(c, 0)$. Let the equation of the directrix in question be $x = d$, where d is to be determined (Figure 57). We know that the distance from $(c, 0)$ to P equals a constant k times the distance from P to the directrix, where $k < 1$. Using the distance formula gives

$$\sqrt{(x - c)^2 + y^2} = k|d - x|.$$

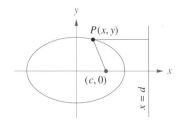

FIGURE 57

If we square both sides and collect like terms, we arrive at

$$\frac{x^2}{\left(\dfrac{k^2d^2 - c^2}{1 - k^2}\right)} + \frac{(2dk^2 - 2c)x}{k^2d^2 - c^2} + \frac{y^2}{k^2d^2 - c^2} = 1.$$

Compare this equation with the equation $x^2/a^2 + y^2/b^2 = 1$. Since there is no x-term, $2dk^2 - 2c = 0$. Also,

$$\frac{k^2d^2 - c^2}{1 - k^2} = a^2 \quad \text{and} \quad k^2d^2 - c^2 = b^2.$$

These last two equations imply $a^2(1 - k^2) = b^2$. Therefore $c^2 = a^2 - b^2 = a^2k^2$. Solving for k yields the eccentricity:

$$k = \frac{c}{a} = e.$$

From the equation $2dk^2 - 2c = 0$, we conclude that

$$d = \frac{a}{e}.$$

A similar analysis for a hyperbola reveals that $k > 1$ is also the eccentricity. The equation of one of the directrices is $x = a/e$. The relationships are pictured in Figure 58.

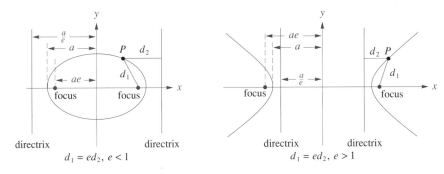

$d_1 = ed_2,\ e < 1$

$d_1 = ed_2,\ e > 1$

FIGURE 58

EXAMPLE 1 Find an equation of the ellipse with foci at $(-3, 0)$ and $(3, 0)$, with corresponding directrices $x = -6$ and $x = 6$.

SOLUTION The ellipse is centered at the origin with $c = 3$, so the equation can be written in the form

$$\frac{x^2}{a^2} + \frac{y^2}{b^2} = 1.$$

We need to determine a^2 and b^2. The directrix is $x = a/e = 6$. Solving for a, we have

$$a = 6e = 6\left(\frac{c}{a}\right) \Rightarrow a^2 = 6c = 18.$$

Using $b^2 = a^2 - c^2$ gives $b^2 = 18 - 9 = 9$. An equation for the ellipse is

$$\frac{x^2}{18} + \frac{y^2}{9} = 1.$$

EXAMPLE 2   Find an equation that describes the set of all points whose distance from the point $(9, 0)$ is $\frac{3}{2}$ times their distance from the line $x = 4$.

SOLUTION Since the eccentricity is $e = \frac{3}{2} > 1$, the graph is a hyperbola. Let (x, y) be any point on the hyperbola (see Figure 59). Using the distance formula, we have

$$\sqrt{(x - 9)^2 + y^2} = \tfrac{3}{2}|x - 4|.$$

Squaring both sides of this equation and simplifying leads to

$$\frac{x^2}{36} - \frac{y^2}{45} = 1.$$

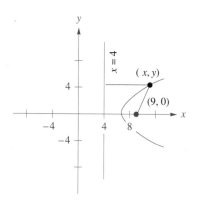

FIGURE 59

Reflective Properties with Applications

In Section 8.2 we discovered a useful reflective property of the parabola analytically. The ellipse and hyperbola also enjoy similar geometric properties that deal with tangent lines. A line tangent to an ellipse or hyperbola intersects the curve exactly once. As a way to look into the reflective properties, consider the following problem (one often found in calculus texts).

PROBLEM   Given two points F_1 and F_2 and a line not containing either point, locate point P on the line so that $PF_1 + PF_2$ is minimized.

SOLUTION Reflect F_1 through the line, obtaining the reflected point F, so the given line is the perpendicular bisector of F_1F. Now draw the straight line from F_2 to F, intersecting the given line at point P (Figure 60). The minimum distance is $PF_1 + PF_2$ because for any other point P' on the line

$$P'F_1 + P'F_2 = P'F + P'F_2 > PF + PF_2 = PF_1 + PF_2.$$

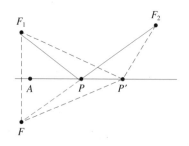

FIGURE 60

Notice that $\angle APF_1 = \angle APF$ and that $\angle APF = \angle P'PF_2$. Therefore, $\angle APF_1 = \angle P'PF_2$; in other words, P is the point such that PF_1 and PF_2 make equal angles with the given line.

Now consider a line tangent to an ellipse (with foci F_1 and F_2) at any point T. Because T is on the ellipse, $TF_1 + TF_2$ is a constant, which we have designated in Section 8.3 as $2a$. Let T' be any other point on the tangent line (Figure 61). Since T' must lie outside the ellipse,

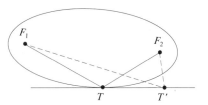

FIGURE 61

$$T'F_1 + T'F_2 > 2a = TF_1 + TF_2.$$

Thus T is the point that minimizes the sum of the distances from the foci to the tangent line. As noted earlier, this means that a line tangent to an ellipse forms equal angles with the lines joining the foci to the point of tangency (see Figure 62).

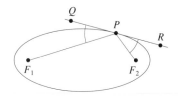

$\angle F_1PQ = \angle F_2PR$

FIGURE 62

As an application, if a source of sound or light is placed at one focus, each path reflects off the curve toward the other focus. One of the most recent applications of this reflective property is to break up stones in the kidney or bladder using a **lithotriptor.** As an alternative to surgery, the procedure essentially uses one focus as a source of either x-ray or ultrasonic waves. These waves reflect to converge on the other focus, where the doomed stone is positioned. The resulting fragments are then easily passed through the urinary tract.

The reflective property of the parabola can be verified by considering the parabola as a limiting case of ellipses whose eccentricity approaches 1. Suppose one focus F_1 remains fixed along with the nearest vertex V_1, and the other focus F_2 moves away along the major axis. The ellipse becomes more elongated. In each case the focal radii make equal angles with the tangent line. The eccentricity approaches 1 (why?), and so the limiting curve is a parabola (Figure 63).

The Greeks poetically described the parabola as "an ellipse in search of its lost focus."

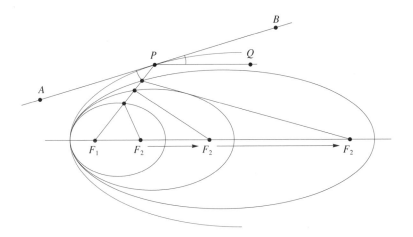

FIGURE 63

In the limiting parabola, the focal radius PF_2 is parallel to the axis of symmetry, and so

$$\angle F_1 PA = \angle QPB.$$

Using an argument very similar to the one establishing the reflective property for an ellipse, we can show that a tangent to a hyperbola bisects the angle made by the focal radii to the point of tangency. In Figure 64, $\angle F_1 PA = \angle F_2 PA$.

Suppose that a ray of light is directed toward one focus of a hyperbolic mirror. It reflects toward the other focus. This reflective property of hyperbolas is utilized in certain telescopes, which employ hyperbolic as well as elliptic and parabolic mirrors. Figure 65 illustrates.

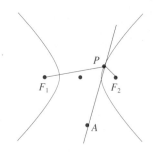

FIGURE 64

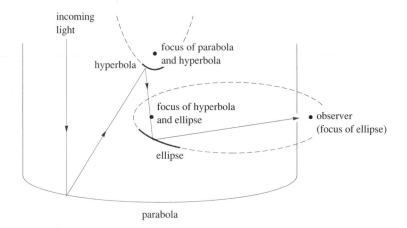

FIGURE 65

EXERCISE SET 8.4

A

In Problems 1 through 9, find the equations of directrices and the eccentricity.

1. $\dfrac{x^2}{49} + \dfrac{y^2}{9} = 1$

2. $\dfrac{x^2}{25} + \dfrac{y^2}{4} = 1$

3. $\dfrac{x^2}{16} - \dfrac{y^2}{4} = 1$

4. $\dfrac{x^2}{9} - \dfrac{y^2}{4} = 1$

5. $x^2 + 4y^2 + 6x - 16y - 11 = 0$

6. $x^2 - 9y^2 + 2x - 80 = 0$

7. $x^2 = -4y$

8. $y^2 = 10x$

9. $x^2 - 4x - 4y + 8 = 0$

10. Find an equation of the ellipse with foci at $(-6, 0)$ and $(6, 0)$; with directrices $x = -8$ and $x = 8$.

11. Find an equation of the ellipse with foci at $(0, -5)$ and $(0, 5)$; with directrices $y = -7$ and $y = 7$.

12. Find an equation of the hyperbola with foci at $(-8, 0)$ and $(8, 0)$; with directrices $x = -6$ and $x = 6$.

13. Find an equation of the hyperbola with foci at $(-10, 0)$ and $(10, 0)$; with directrices $x = -9$ and $x = 9$.

14. Find an equation that describes the set of all points whose distance from the point $(4, 0)$ is $\frac{3}{2}$ times their distance from the line $x = 2$.

15. Find an equation that describes the set of all points whose distance from the point $(3, 0)$ is $\frac{5}{2}$ times their distance from the line $x = 1$.

16. Find an equation that describes the set of all points whose distance from the point $(1, 0)$ is $\frac{1}{2}$ times their distance from the line $x = 4$.

17. Find an equation that describes the set of all points whose distance from the point $(2, 0)$ is $\frac{1}{4}$ times their distance from the line $x = 5$.

18. The earth orbits the sun in an elliptic path with the sun located at one of the foci. If the eccentricity is 0.02, and the length of the major axis is 184 million miles, how close does the earth get to the sun?

B

19. The graph of $xy = 2$ is a hyperbola whose asymptotes are the coordinate axes. Find the equations of the directrices.

20. The line $x + 2y = 4$ is tangent to the ellipse $x^2 + 4y^2 = 8$ at the point $(2, 1)$. Draw a line segment from $(2, 1)$ to each foci. What is the angle between each of these segments and the tangent line (to the nearest $0.01°$)?

21. The line $x + 3y = 12$ is tangent to the ellipse $x^2 + 3y^2 = 36$ at the point $(3, 3)$. Draw a line segment from $(3, 3)$ to each focus. What is the angle between each of these segments and the tangent line (to the nearest $0.01°$)?

C

22. The intersection of a right circular cylinder and a plane is either a circle, two parallel lines, or a curve that appears to be an ellipse. Use an argument similar to Dandelin's to show that the curve satisfies the definition of an ellipse.

23. Figure 66 shows a conic section with one sphere inscribed in each nappe of a cone, each sphere tangent to the intersecting plane at points F_1 and F_2. Show that the conic section is a hyperbola; that is, show that if Q is a point on the conic section shown, then $QF_1 - QF_2$ is a constant.

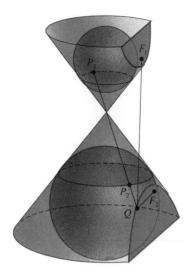

FIGURE 66

24. **a)** Show that if PQ and PF are tangent lines to a circle, where Q and F are the points of tangency, $PF = PQ$ (see Figure 67). *(Hint: The radius drawn from the center to any point of tangency forms a right angle with the tangent line.)*

   b) Show that if PQ and PF are tangent lines to a sphere, where Q and F are the points of tangency, $PF = PQ$. *(Hint: The plane containing points P, Q, and F intersects the sphere to form a circle.)*

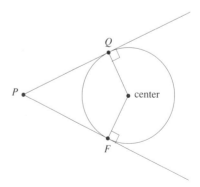

FIGURE 67

SECTION 8.5 ROTATIONS

With only a few exceptions, we have developed equations for conic sections with axes of symmetry parallel to the x-axis or y-axis. One exception was the parabola in Example 8 of Section 8.2. In that case the focus was located at $(2, 2)$, and the directrix was the line $y = -x$. The equation of this "rotated" parabola turned out to be

$$x^2 - 2xy + y^2 - 8x - 8y + 16 = 0.$$

The axis of symmetry for this parabola makes an angle of $45°$ with the x-axis. This parabola is said to be rotated $45°$. The next example develops an equation of a rotated ellipse.

EXAMPLE 1   If $Q(2, 4)$ is a point on the ellipse with foci $F_1(-10, -1)$ and $F_2(6, 1)$, find an equation for the ellipse.

SOLUTION If $P(x, y)$ is any point on the ellipse, then by definition

$$d(P, F_1) + d(P, F_2) = d(Q, F_1) + d(Q, F_2).$$

Using the distance formula (Figure 68), we have

$$\sqrt{(x + 10)^2 + (y + 1)^2} + \sqrt{(x - 6)^2 + (y - 1)^2} = 13 + 5 = 18.$$

If we isolate the first radical, square both sides, and simplify, we obtain

$$8x + y - 65 = -9\sqrt{(x - 6)^2 + (y - 1)^2}.$$

Squaring both sides again and simplifying yields

$$17x^2 - 16xy + 80y^2 + 68x - 32y - 1228 = 0.$$

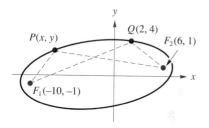

FIGURE 68

Notice that in both cases the rotated conic section has the form

$$Ax^2 + Bxy + Cy^2 + Dx + Ey + F = 0.$$

This form is called the **general second-degree equation,** and one of the goals of this section is to graph equations of this form. You may be wondering if it would be worthwhile to rotate the coordinate axes to obtain a new set of axes, x' and y', parallel or perpendicular to an axis of symmetry. Reconsider, for example, the rotated parabola (Example 8, Section 8.2) given by

$$x^2 - 2xy + y^2 - 8x - 8y + 16 = 0.$$

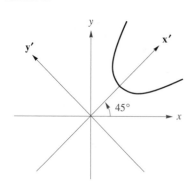

FIGURE 69

Since the axis of symmetry has an angle of inclination of 45°, we could graph this equation relative to a set of axes that is rotated 45° (Figure 69).

Similarly, the ellipse in Example 1,

$$17x^2 - 16xy + 80y^2 + 68x - 32y - 1228 = 0$$

has a major axis containing the points $(-10, -1)$ and $(6, 1)$. Therefore, the slope of the major axis is $\frac{1}{8}$, and the angle of inclination is $\tan^{-1}(\frac{1}{8}) \approx 7°$. We could graph this equation relative to a set of axes that is rotated $\tan^{-1}(\frac{1}{8})$. In other words, the x'-axis should pass through $(0, 0)$ and $(8, 1)$, as illustrated in Figure 70.

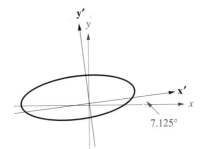

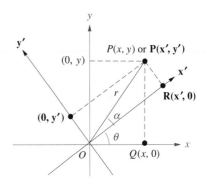

FIGURE 70

Consider the general case, in which the axes are rotated through some positive angle θ (Figure 71). If P is any point in the plane, it has two representations. Relative to the original coordinate system the coordinates for P are (x, y), and relative to the rotated system the coordinates are (x', y'). Let $r = d(O, P)$. Looking at right triangle OPR, we have

$$x' = r \cos \alpha \quad \text{and} \quad y' = r \sin \alpha.$$

By considering right triangle OPQ, we get

$$x = r \cos(\theta + \alpha) \quad \text{and} \quad y = r \sin(\theta + \alpha).$$

Expanding these last two equations using the addition identities (see Section 6.2) leads to

FIGURE 71

$$x = r(\cos \theta \cos \alpha - \sin \theta \sin \alpha)$$

and

$$y = r(\sin \theta \cos \alpha + \cos \theta \sin \alpha).$$

Finally, we distribute r and replace $r \cos \alpha$ with x' and $r \sin \alpha$ with y':

$$x = r \cos \theta \cos \alpha - r \sin \theta \sin \alpha = x' \cos \theta - y' \sin \theta$$
$$y = r \sin \theta \cos \alpha + r \cos \theta \sin \alpha = x' \sin \theta + y' \cos \theta.$$

This is the relationship between the two coordinate systems that we need.

$$x = x' \cos \theta - y' \sin \theta$$

and

$$y = x' \sin \theta + y' \cos \theta.$$

Rotation of Axes Formulas

The next example illustrates how to use these equations, given an appropriate value of θ.

EXAMPLE 2   Sketch $244x^2 - 360xy + 601y^2 = 676$ by rotating the coordinate axes an angle of $\theta = \tan^{-1}(\frac{5}{12})$.

SOLUTION The right triangle in Figure 72 shows that for $\theta = \tan^{-1}(\frac{5}{12})$, $\cos\theta = \frac{12}{13}$ and $\sin\theta = \frac{5}{13}$. The equations for rotation of axes becomes

$$x = x'(\tfrac{12}{13}) - y'(\tfrac{5}{13}) \quad \text{and} \quad y = x'(\tfrac{5}{13}) + y'(\tfrac{12}{13}).$$

Substituting these into the equation

$$244x^2 - 360xy + 601y^2 = 676$$

we get

$$244(\tfrac{12}{13}x' - \tfrac{5}{13}y')^2 - 360(\tfrac{12}{13}x' - \tfrac{5}{13}y')(\tfrac{5}{13}x' + \tfrac{12}{13}y') + 601(\tfrac{5}{13}x' + \tfrac{12}{13}y')^2 = 676.$$

Expanding and simplifying leads (eventually) to

$$(x')^2 + 4(y')^2 = 4.$$

Since $\theta = \tan^{-1}(\frac{5}{12})$, sketch the x'-axis so that it passes through $(0, 0)$ and $(12, 5)$. Now sketch the ellipse $(x')^2 + 4(y')^2 = 4$ relative to the $x'y'$-axes. The graph is sketched in Figure 73.

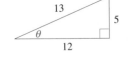

FIGURE 72

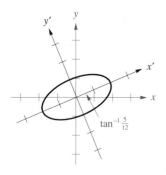

FIGURE 73

In the last example, we transformed an equation containing an xy term into an equation without an $x'y'$ term. The given angle of rotation was an important part of the process. If we repeat this procedure for the general situation, our goal being to eliminate the xy term, we will find a way to select an appropriate angle of rotation.
     Given

$$Ax^2 + Bxy + Cy^2 + Dx + Ey + F = 0$$

substitute $(x' \cos\theta - y' \sin\theta)$ for x and $(x' \sin\theta + y' \cos\theta)$ for y. After a Herculean effort of expansion and simplification, we get an equation of the form

$$A'(x')^2 + B'x'y' + C'(y')^2 + D'x' + E'y' + F' = 0$$

where the coefficient of the $x'y'$ term is

$$B' = 2(C - A) \sin\theta \cos\theta + B(\cos^2\theta - \sin^2\theta).$$

To eliminate the $x'y'$ term, we set this equal to zero and solve for θ. Using the double-angle formulas, it follows that

$$B \cos 2\theta = (A - C) \sin 2\theta.$$

This is equivalent to the following:

$$\cot 2\theta = \frac{A - C}{B}.$$

To Determine the Angle of Rotation

EXAMPLE 3   Use rotation of axes to transform

$$3x^2 - 10xy + 3y^2 = -32$$

into an equation without an $x'y'$ term. Sketch the graph.

SOLUTION Comparing the equation with

$$Ax^2 + Bxy + Cy^2 + Dx + Ey + F = 0$$

we identify $A = 3$, $B = -10$, and $C = 3$. To find an angle of rotation that eliminates the $x'y'$ term, we solve

$$\cot 2\theta = \frac{A - C}{B} = 0.$$

Of course, there are an infinite number of solutions; we pick θ such that $0 < 2\theta < 180°$. We have

$$2\theta = 90° \Rightarrow \theta = 45°.$$

Thus $\cos \theta = \sqrt{2}/2$ and $\sin \theta = \sqrt{2}/2$, so the rotation of axes formulas are

$$x = \frac{\sqrt{2}}{2} x' - \frac{\sqrt{2}}{2} y' \quad \text{and} \quad y = \frac{\sqrt{2}}{2} x' + \frac{\sqrt{2}}{2} y'.$$

Next, we substitute these values into $3x^2 - 10xy + 3y^2 = -32$:

$$3\left(\frac{\sqrt{2}}{2} x' - \frac{\sqrt{2}}{2} y' \right)^2 - 10\left(\frac{\sqrt{2}}{2} x' - \frac{\sqrt{2}}{2} y' \right)\left(\frac{\sqrt{2}}{2} x' + \frac{\sqrt{2}}{2} y' \right)$$

$$+ 3\left(\frac{\sqrt{2}}{2} x' + \frac{\sqrt{2}}{2} y' \right)^2 = -32.$$

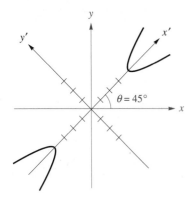

FIGURE 74

After expanding and simplifying, we reduce the equation to

$$-2(x')^2 + 8(y')^2 = -32$$

or

$$\frac{(x')^2}{16} - \frac{(y')^2}{4} = 1.$$

Consequently, the graph is a hyperbola crossing the x'-axis at $(4, 0)$ and $(-4, 0)$. The asymptotes have slopes $\frac{1}{2}$ and $-\frac{1}{2}$ (relative to the rotated axes). The graph is shown in Figure 74.

After rotation of its axes, a graph may not be centered at the origin; a translation with respect to the rotated axes may be necessary.

EXAMPLE 4   Transform $9x^2 + 24xy + 16y^2 + 130x - 160y + 25 = 0$ into an equation without an $x'y'$ term. Sketch the graph.

SOLUTION Comparing the equation with the general second-degree equation, $A = 9$, $B = 24$, and $C = 16$. The angle of rotation is given by

$$\cot 2\theta = \frac{A - C}{B} = -\frac{7}{24}.$$

Instead of solving for θ, we are really interested only in $\cos \theta$ and $\sin \theta$. Assuming $0 < 2\theta < 180°$, since $\cot 2\theta$ is negative, 2θ must be in Quadrant II. Figure 75 reveals that $\cos 2\theta = -\frac{7}{25}$. Using the half-angle identities gives us

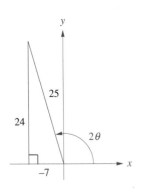

FIGURE 75

$$\cos \theta = \sqrt{\frac{1 + \cos 2\theta}{2}} = \sqrt{\frac{1 + (-\frac{7}{25})}{2}} = \frac{3}{5}$$

$$\sin \theta = \sqrt{\frac{1 - \cos 2\theta}{2}} = \sqrt{\frac{1 - (-\frac{7}{25})}{2}} = \frac{4}{5}.$$

Hence the rotation of axes formulas are

$$x = \tfrac{3}{5}x' - \tfrac{4}{5}y' \quad \text{and} \quad y = \tfrac{4}{5}x' + \tfrac{3}{5}y'.$$

If we substitute these expressions into

$$9x^2 + 24xy + 16y^2 + 130x - 160y + 25 = 0$$

we get

$$9(\tfrac{3}{5}x' - \tfrac{4}{5}y')^2 + 24(\tfrac{3}{5}x' - \tfrac{4}{5}y')(\tfrac{4}{5}x' + \tfrac{3}{5}y') + 16(\tfrac{4}{5}x' + \tfrac{3}{5}y')^2 + 130(\tfrac{3}{5}x' - \tfrac{4}{5}y')$$
$$- 160(\tfrac{4}{5}x' + \tfrac{3}{5}y') + 25 = 0.$$

Simplification leads to

$$(x')^2 - 2x' - 8y' + 1 = 0.$$

Finally, completing the square, we get

$$(x' - 1)^2 = 8y'.$$

This is a parabola with vertex at $(1, 0)$ on the x'-axis and whose axis of symmetry is parallel to the y'-axis. The rotated axes can be sketched using

$$\tan \theta = \frac{\sin \theta}{\cos \theta} = \frac{\tfrac{4}{5}}{\tfrac{3}{5}} = \frac{4}{3}.$$

So the x'-axis has a slope of $\tfrac{4}{3}$ (relative to the x-axis). The graph is shown in Figure 76.

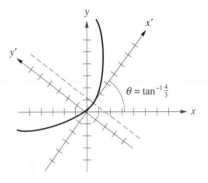

FIGURE 76

Substituting the rotation of axes formulas into a second-degree equation and simplifying is a tedious process. There is a way to identify the type of curve (ellipse, parabola, or hyperbola) without going through this process.

The Invariant $B^2 - 4AC$

It can be shown that, when

$$Ax^2 + Bxy + Cy^2 + Dx + Ey + F = 0$$

is transformed into

$$A'x^2 + B'xy + C'y^2 + D'x + E'y + F' = 0$$

using any angle of rotation, then

$$B^2 - 4AC = (B')^2 - 4A'C'$$

(see Problem 29). When θ is selected to eliminate the $x'y'$ term, B' will be zero; so

$$B^2 - 4AC = -4A'C'.$$

We know that except in the case of degenerates,

$$A'x^2 + C'y^2 + D'x + E'y + F' = 0$$

is

a parabola if $A'C' = 0$,

an ellipse if $A'C' > 0$, or

a hyperbola if $A'C' < 0$.

Therefore we have the following:

$Ax^2 + Bxy + Cy^2 + Dx + Ey + F = 0$ represents $a(n)$:
parabola if $B^2 - 4AC = 0$,
ellipse if $B^2 - 4AC < 0$, or
hyperbola if $B^2 - 4AC > 0$.

Knowing what type of a conic we have, it is sometimes easier to sketch the graph by using other techniques.

EXAMPLE 5   Sketch the graph of $xy - 2x - 3y = 0$.

SOLUTION Comparing the equation with

$$Ax^2 + Bxy + Cy^2 + Dx + Ey + F = 0$$

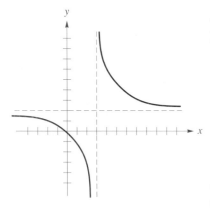

we calculate $B^2 - 4AC = 1^2 - 4(0)(0) > 0$, so the graph is a hyperbola. Rather than rotate the axes, we isolate y:

$$xy - 3y = 2x$$

$$y(x - 3) = 2x$$

$$y = \frac{2x}{x - 3}.$$

This is a rational function and can be sketched by using the techniques discussed in Section 3.4. There is a vertical asymptote at $x = 3$, a horizontal asymptote at $y = 2$, and the only intercept is $(0, 0)$. By calculating a few ordered pairs, the graph is easily sketched (Figure 77).

FIGURE 77

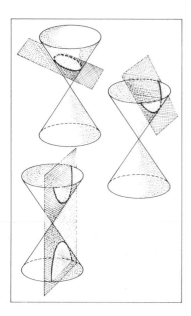

The History of Conic Sections

The discovery of the ellipse, hyperbola, and parabola as cross sections of the cone was first made by Menaechmus in about 350 B.C. Menaechmus was a student of Eudoxus (the outstanding mathematician of the third century B.C.), who in turn was a student of Plato. Menaechmus used three different cones—one with an acute vertex angle (oxytome), one with a right vertex angle (orthotome), and one with an obtuse vertex angle (amblytome)—for each of the curves. The discovery of the parabola and hyperbola was a means to Menaechmus' solution of a construction problem in geometry. His discovery of the ellipse was merely the third case, and at the time there was no apparent application for this by-product. This is an example of something that has occurred frequently in mathematics. Many mathematical discoveries have been made merely for the sake of knowing. Such pure mathematical discoveries are often found to be profoundly useful, sometimes centuries later (as with the ellipse).

The names *ellipse, parabola,* and *hyperbola* were first introduced by Apollonius of Perga (about 250 B.C.) to convey "deficiency," "equal in comparison," and "excess," respectively. Apollonius, known as The Great Geometer, was one of the outstanding mathematicians of the Hellenistic Age, second only to Archimedes. He wrote the definitive treatise on the conic sections, *Conics.* His books on the subject were original, comprehensive, and well written. Unlike Menaechmus, Apollonius obtained the three sections from the same cone. He was also the first to use nappes of a cone, which of course means that he was the first to recognize that a hyperbola has two branches. Apollonius did not know of equations for these curves since analytic geometry came much later. He did, however, work with a kind of coordinate system, with the diameter and the perpendicular from the vertex serving as the axes. The *Conics* was highly regarded. Kepler and Halley both used this treatise in their work in astronomy. Descartes drew upon it to develop his analytic geometry. The *Conics* consisted of eight books, the last of which has been lost. Little is known of Apollonius' life. It is likely that he knew Eratosthenes. He taught mathematics to Alexander the Great. We can get an idea of his attitude about mathematics from a statement he made regarding some of his theorems: "They are worthy of acceptance for the sake of the demonstration themselves, in the same way as we accept many other things in mathematics for this and no other reason."

EXAMPLE 6   Sketch the graph of $x^2 - 4xy + 4y^2 - 4x = 0$.

SOLUTION Since $B^2 - 4AC = (-4)^2 - 4(1)(4) = 0$, the graph is a parabola. Instead of rotating axes, we treat the equation as a quadratic equation in y and use the quadratic formula to isolate y:

$$4y^2 - 4xy + (x^2 - 4x) = 0$$

$$y = \frac{-b \pm \sqrt{b^2 - 4ac}}{2a}$$

$$= \frac{-(-4x) \pm \sqrt{(-4x)^2 - 4(4)(x^2 - 4x)}}{2(4)}$$

$$= \tfrac{1}{2}x \pm \sqrt{x}.$$

We can sketch this last equation as follows:

a) Graph $y = \tfrac{1}{2}x$ and $y = \sqrt{x}$, then add y-coordinates.

b) Similarly, we graph $y = \tfrac{1}{2}x$ and $y = -\sqrt{x}$ and then add y-coordinates. The parabola is the union of the graphs in (a) and (b). In Figure 78, the line $y = \tfrac{1}{2}x$ is not the axis of symmetry.

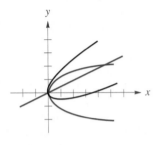

FIGURE 78

EXERCISE SET 8.5

A

In Problems 1 through 6, identify each of the following as a parabola, an ellipse, or a hyperbola. Do not sketch.

1. $x^2 - 2xy + y^2 + 5x + 2y - 45 = 0$
2. $4x^2 - 2xy - 3y^2 + 2y - 20 = 0$
3. $4x^2 + 10xy + 4y^2 + 10x + 2y - 36 = 0$
4. $3x^2 - 3xy + y^2 + 10x + 2y - 36 = 0$
5. $x^2 + 4y^2 + 2y + 10xy - 16 = 0$
6. $12x^2 + 3y^2 + 10xy + 6y - 22 = 0$

In Problems 7 through 12, find θ, the angle of rotation of axes that will remove the $x'y'$ term.

7. $x^2 - 2xy + y^2 + 5x + 2y - 45 = 0$
8. $2x^2 - xy + 12y^2 + 6x + 2y - 20 = 0$
9. $4x^2 + \sqrt{3}xy + 3y^2 + 11x + 2y - 35 = 0$
10. $3\sqrt{3}x^2 - 2xy + \sqrt{3}y^2 + 10x + 2y - 36 = 0$
11. $x^2 + 6y^2 + 2y + 12xy - 16 = 0$
12. $3x^2 + 7y^2 + 3xy + 6y - 22 = 0$

B

In Problems 13 through 20, rotate the axes to eliminate the $x'y'$ term and sketch the graph, showing both pairs of axes.

13. $29x^2 + 42xy + 29y^2 - 200 = 0$
14. $x^2 + 2xy + y^2 + 2\sqrt{2}x - 2\sqrt{2}y = 0$
15. $3x^2 + 2\sqrt{3}xy + y^2 + 8\sqrt{3}y = 8x$
16. $-23x^2 + 3y^2 + 26\sqrt{3}xy = 144$
17. $11x^2 + 10\sqrt{3}xy + y^2 = 4$
18. $25x^2 - 36xy + 40y^2 - 12\sqrt{13}x - 8\sqrt{13}y = 0$
19. $16x^2 - 24xy + 9y^2 + 65x - 80y + 25 = 0$
20. $6x^2 + 24xy - y^2 - 12x + 26y + 11 = 0$

In Problems 21 through 25, sketch the graph using the method described in Examples 5 and 6.

21. $xy + 2x - 4y = 0$
22. $8x^2 + 4xy + 5y^2 - 16 = 0$
23. $x^2 - xy - y - 4 = 0$

24. $2xy - y^2 - 4 = 0$

25. $5x^2 - 4xy + y^2 - 4 = 0$

C

26. Show that the rational function $y = 1/x$ is a hyperbola by rotating the axes.

27. Show that the rational function

$$y = \frac{2x^2 + 1}{x}$$

is a hyperbola by rotating the axes.

28. Find the equation for the axis of symmetry for the parabola in Example 6.

29. Show that the when

$$Ax^2 + Bxy + Cy^2 + Dx + Ey + F = 0$$

is transformed into

$$A'x^2 + B'xy + C'y^2 + D'x + E'y + F' = 0,$$

using any angle of rotation, then

$$B^2 - 4AC = (B')^2 - 4A'C'.$$

MISCELLANEOUS EXERCISES

In Problems 1 through 4, find an equation of each line. Express your answer in (a) slope-intercept form, if possible, and (b) general form.

1. Passing through $(-5, 7)$ with slope $\frac{4}{3}$

2. Passing through $(3, -6)$ and $(4, -5)$

3. Passing through $(2, -4)$ with angle of inclination $\pi/3$

4. Passing through $(2\sqrt{3}, 5)$, having positive slope, and forming an angle of $\pi/6$ with the line $x - \sqrt{3}y = 0$

In Problems 5 through 7, find the smallest angle of inclination for each given line. In addition, give an approximation to the nearest 0.01°.

5. $y = \dfrac{\sqrt{3}}{3}x - 8$ **6.** $5x - 2y = 6$

7. Passing through $(1, 2)$ and $(-3, 6)$.

8. Find the smallest positive angle between the lines $6x + 5y = 1$ and $y = x - 2$. Give an exact answer and give an approximation to the nearest 0.01°.

9. Find the smallest positive angle between two lines whose angles of inclination are $22°$ and $83°$.

10. Find the distance from the point $(6, -4)$ to the line $y = x - 7$.

11. Find the distance between the lines $3x + 6y = 10$ and $2x + 4y = -3$.

12. Find the distance between the lines $2x - y = 6$ and $3x + 4y = -1$.

13. Describe the collection of all lines satisfying $5x - 2y = c$, where c is any real number. Sketch at least four lines for this family.

14. Describe the collection of all lines satisfying $y = cx + 3$, where c is any real number. Sketch at least four lines for this family.

15. Find the slope(s) of the line(s) that make an angle of $30°$ with $3x - 9y - 2 = 0$.

16. Find the slope of each of the lines passing through $(7, 2)$ that is tangent to the circle $(x + 2)^2 + (y - 1)^2 = 10$.

17. Find an equation(s) of the line(s) parallel to and a distance of 5 units from $5x - 12y = 1$.

In Problems 18 through 21, sketch the graph of each parabola. Label the vertex, focus, and directrix.

18. $x^2 = -6y$

19. $(y - 3)^2 = 4(x + 1)$

20. $y^2 - 4y + 8x = 36$

21. $x^2 + 8x - 10y + 6 = 0$

In Problems 22 through 26, find an equation of the parabola that satisfies the given conditions, and sketch the graph.

22. Vertex at $(3, -4)$ and focus at $(3, 2)$.

23. The focal chord has endpoints $(3, 4)$ and $(-1, 4)$, and the parabola opens down.

24. The focus is $(1, 0)$, the parabola contains the point $(2, 2\sqrt{6})$, the directrix is vertical, and the parabola opens to the left.

25. Focus at $(2, 3)$ and directrix $x = 4$.

26. Focus at $(-2, -5)$ and directrix $y = -2$.

27. Find an equation of the tangent line to $x^2 = 8y$ at the point $(12, 18)$.

28. An equation of the tangent line to $2x = 3y^2$ at the point $P(6, 2)$ is $y = \frac{1}{6}x + 1$. Find the acute angle between this tangent line and the line from the focus of the parabola to $P(6, 2)$.

In Problems 29 through 38, graph the equation.

29. $x^2 + 9y^2 = 9$

30. $36x^2 - y^2 = -9$

31. $4x^2 - 9y^2 - 32x + 36y + 27 = 0$

32. $9x^2 + 25y^2 + 18x + 100y = 116$

33. $4x^2 + 3y^2 - 16x + 30y + 79 = 0$

34. $6x^2 - 9y^2 + 90y = 207$

35. $x^2 - 4x - 2y = 0$

36. $5x^2 + 16y^2 = -4$

37. $36x^2 - 9y^2 = 0$

38. $3y = \sqrt{12x - x^2 - 27}$

39. Find the eccentricity and coordinates of the foci for the graph of Problem 29.

40. Find the eccentricity and coordinates of the foci for the graph of Problem 30.

41. Let P be any point on the ellipse $2x^2 + 25y^2 = 8$. If F_1 and F_2 are the foci, find $d(P, F_1) + d(P, F_2)$.

42. Let P be any point on the hyperbola $25x^2 - 4y^2 = 100$. If F_1 and F_2 are the foci, find $|d(P, F_1) - d(P, F_2)|$.

43. Find an equation of the ellipse with vertices $(-7, -2)$, $(9, -2)$, $(1, 1)$, and $(1, -5)$.

44. Find an equation of the ellipse with foci $(-2, -1)$, and $(4, -1)$, passing through $(5, 0)$.

45. Find an equation of the ellipse such that the endpoints of the major axis are $(-2, 6)$ and $(-2, 0)$, and the eccentricity is $\frac{2}{3}$.

46. Find an equation of the hyperbola with foci $(-3, 2)$ and $(5, 2)$, passing through $(6, 5)$.

47. Find an equation of the hyperbola with asymptotes $y = 3x + 12$ and $y = -3x - 12$, containing the point $(-2, 3\sqrt{5})$.

48. Graph $4x^2 - y^2 = k$, for $k = -4, 0, 4$, and 16.

49. Find the eccentricity and equations of the directrices for the graph of Problem 29.

50. Find the eccentricity and equations of the directrices for the graph of Problem 30.

51. Find an equation of the hyperbola with foci at $(-12, 0)$, and $(12, 0)$; with directrices $x = -9$ and $x = 9$.

52. Find an equation of the ellipse with foci at $(0, -10)$ and $(0, 10)$; with directrices $y = -14$ and $y = 14$.

In Problems 53 through 56, identify the graph of each as a parabola, ellipse, or hyperbola. Do not sketch the graph.

53. $7x^2 - 2xy - 5y^2 + 12x = 56$

54. $50x^2 - 20xy + 2y^2 + 18x + 4y = 214$

55. $5x^2 + 2xy + y^2 - 8x + 4y = 80$

56. $5x^2 - 3xy + 2y^2 + 10x - 6y = 108$

In Problems 57 through 60, find the smallest positive angle of rotation of axes θ, that will eliminate the $x'y'$ term. Do not sketch the graph.

57. $5x^2 + 2xy + 5y^2 - 2x + 14y = 60$

58. $7x^2 + 4xy + 3y^2 - 8x + 4y = 80$

59. $5x^2 - 3xy + 2y^2 + 8x - y = 45$

60. $x^2 + 2\sqrt{3}xy + 3y^2 + 6y = 54$

In Problems 61 through 64, rotate the axes to eliminate the $x'y'$ term and sketch the graph, showing both pairs of axes.

61. $5x^2 - 4xy + 8y^2 - 36 = 0$

62. $x^2 + 4xy - 2y^2 - 48 = 0$

63. $3x^2 + 4\sqrt{3}xy - y^2 - 15 = 0$

64. $x^2 - 2\sqrt{3}xy + 3y^2 - 16\sqrt{3}x - 16y = 0$

65. Find an equation of the ellipse with foci $(-6, 0)$ and $(1, 1)$, passing through $(0, 3)$.

66. Find the distance from the center of the ellipse $3x^2 + 2xy + 3y^2 - 8x - 8y = 0$ to the line $2x + 3y - 8 = 0$.

67. Find the smallest positive angle between the lines $2x + 3y - 4 = 0$ and the major axis of the ellipse $3x^2 + 2xy + 3y^2 - 8x - 8y = 0$.

68. Identify the graph of $x^2 - xy + y + 4 = 0$. Sketch without rotating axes (i.e., isolate y).

69. Graph $(4x^2 + y^2 - 16)(x - 2y + 5) = 0$.

70. Graph $x^3 - x - xy^2 + y^2 - x + 1 = 0$. *[Hint:* $x^3 - x - xy^2 + y^2 - x + 1 = x^2(x-1) - y^2(x-1) - (x-1)].$

CHAPTER 9

POLAR COORDINATES AND PARAMETRIC EQUATIONS

POLAR COORDINATES

Suppose that you want to give instructions to a bug on how to get from the origin to a point A with rectangular coordinates (x, y) on the coordinate plane (this is a bug with exceptional intelligence). The instructions for our bug might be to travel x units horizontally and then y units vertically. This is the concept behind the rectangular coordinates; the coordinate pair (x, y) can be thought of as instructions on how to locate the point A (Figure 1).

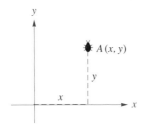

Rectangular coordinates

FIGURE 1

There are, however, other ways to instruct our bug to find the same point. Imagine telling the bug to turn counterclockwise through an angle θ so that he is directly facing the point A; now, have him walk a distance r (along the terminal side of θ) to get to the point A. Alternatively, the bug could turn clockwise and then travel a distance r to A. Or, perhaps we tell the bug to turn clockwise through an angle to face *directly away* from the point A and *back up* r units to get to A. You can probably think of other scenarios. These sets of "instructions" $[r, \theta]$ represent the concept behind **polar coordinates** (Figure 2).

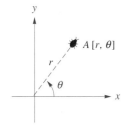 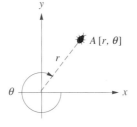 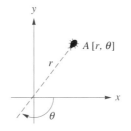

FIGURE 2

Polar Coordinates

The point P that is associated with the polar coordinates $[r, \theta]$ is the point on the terminal side of the angle θ that is r units from the origin, if $r \geq 0$. If $r < 0$, then P is the point that is a distance $|r|$ from the origin on the ray opposite the terminal side of θ.

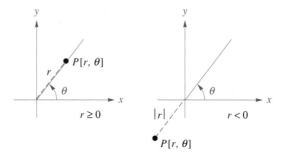

Coordinates with brackets indicate polar coordinates. Coordinates with parentheses will still be used for rectangular coordinates.

The coordinates $[0, \theta]$ for any θ represent the origin.

In the context of polar coordinates, the origin is called the **pole,** and the positive x-axis is called the **polar axis.** For a point $P[r\ \theta]$, r is the **radial distance** of P and θ is the **polar angle** of P.

As we discuss this new coordinate system, keep in mind that this is merely another method of addressing specific points on the coordinate plane.

EXAMPLE 1 Plot and label the points on the plane with the given polar coordinates.

a) $P[2, \pi/3]$ **b)** $S[3, 3\pi/2]$ **c)** $T[4, 2]$ **d)** $R[-3, \pi/2]$

SOLUTION

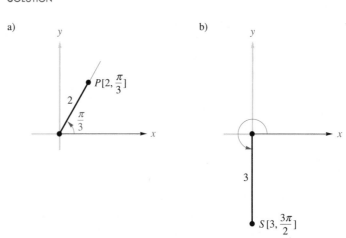

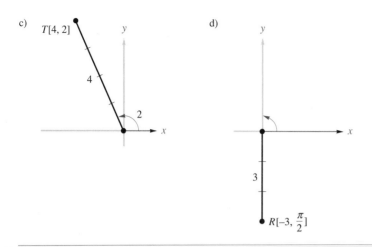

c) T[4, 2]

d)

R[-3, $\frac{\pi}{2}$]

In part (c), note that the polar angle is 2 radians.

A couple of points must be made with respect to Example 1. First, compare (a) and (c); notice that the respective signs of the radial distance and the polar angle of a point do not necessarily determine the quadrant, as in rectangular coordinates.

Second, two different ordered pairs can describe the same point, as they do in (b) and (d). In other words, the polar coordinate representation of a point is not unique (the rectangular coordinate representation of a point on the coordinate plane is unique). Indeed, any ordered pair with radial distance 3 and polar angle coterminal to $3\pi/2$ (namely, $[3, 3\pi/2 + 2n\pi]$) and any ordered pair with radial distance -3 and polar angle coterminal to $-\pi/2$ (namely, $[3, -\pi/2 + 2n\pi]$) describes the same point.

Given a point P with a polar representation $[r, \theta]$, the ordered pairs

$$[r, \theta + 2n\pi] \quad \text{and} \quad [-r, \theta + (2n + 1)\pi] \qquad (n \text{ any integer})$$

are also polar representations of P.

Representations of a Point in Polar Coordinates

EXAMPLE 2   Find all the polar representations $[r, \theta]$ of $A[2, 2\pi/3]$ such that $\pi \leq \theta \leq 3\pi$.

SOLUTION   First, we plot the point on the coordinate plane. The polar angles of the other representations of A either have terminal sides that coincide with, or are opposite to, the terminal side of $2\pi/3$. In the interval $\pi \leq \theta \leq 3\pi$, this

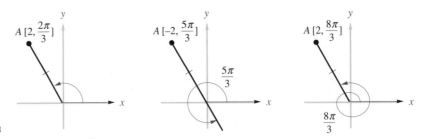

FIGURE 3

means $\theta = 5\pi/3$ or $8\pi/3$. Thus, the representations we seek are $[-2, 5\pi/3]$ and $[2, 8\pi/3]$ (Figure 3).

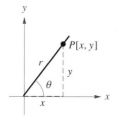

FIGURE 4

Recall (from Chapter 5) that for an angle θ in standard position with a point $P(x, y)$ on its terminal side that is $r = \sqrt{x^2 + y^2}$ units from the origin, the following relations hold (Figure 4):

$$x = r \cos \theta \quad \text{and} \quad y = r \sin \theta.$$

This suggests the following relationship between polar coordinates and rectangular coordinates.

Changing Polar Coordinates to Rectangular Coordinates

A point with a polar representation $[r, \theta]$ has a rectangular representation (x, y), where

$$x = r \cos \theta \quad \text{and} \quad y = r \sin \theta.$$

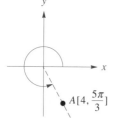

FIGURE 5

EXAMPLE 3   Plot the point and determine its rectangular coordinates:

a) $P[4, 5\pi/3]$ **b)** $Q[-6, 5\pi/4]$

SOLUTION

a)   As Figure 5 shows, the point P is in the 4th quadrant. Using the relations $x = r \cos \theta$ and $y = r \sin \theta$, we get

$$x = 4 \cos(5\pi/3) = 4(\tfrac{1}{2}) = 2$$
$$y = 4 \sin(5\pi/3) = 4(-\sqrt{3}/2) = -2\sqrt{3}.$$

Thus the rectangular coordinates for the point P are $(2, -2\sqrt{3})$.

b) In a similar fashion (Figure 6),

$$x = -6\cos(5\pi/4) = -6(-\sqrt{2}/2) = 3\sqrt{2}$$
$$y = -6\sin(5\pi/4) = -6(-\sqrt{2}/2) = 3\sqrt{2}.$$

Thus the rectangular coordinates for the point Q are $(3\sqrt{2}, 3\sqrt{2})$.

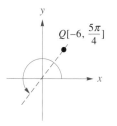

FIGURE 6

It will also be to our advantage to be able to determine a polar representation for a point whose rectangular coordinates are known. The radial distance r (assume for a moment that $r > 0$) of a point P is, by the distance formula, $\sqrt{x^2 + y^2}$ (Figure 7). By the definition of tangent function, the polar angle θ is such that

$$\tan\theta = \frac{y}{x}.$$

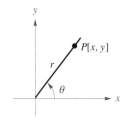

This is not to say that $\theta = \tan^{-1}(y/x)$. We must choose θ such that the terminal side of θ contains the point P (of course, this is for $r > 0$, if $r < 0$, then $r = \sqrt{x^2 + y^2}$, and we choose a θ such that P is contained in the ray that is directly opposite to the terminal side of θ).

FIGURE 7

Suppose that the point P has rectangular coordinates (x, y). Then the point P has the polar representation $[r, \theta]$, where

$$r = \sqrt{x^2 + y^2} \quad \text{and} \quad \tan\theta = \frac{y}{x}.$$

The polar angle θ must be such that its terminal side is in the same quadrant as the point P.

Changing Rectangular Coordinates to Polar Coordinates

EXAMPLE 4   Plot the point and determine its polar coordinates, subject to the restrictions given:

a) $A(4, -4)$ $r > 0$ and $0 \le \theta < 2\pi$

b) $B(0, 3)$ $r < 0$ and $\pi \le \theta < 3\pi$

SOLUTION

a) First, we plot the point; it lies in the 4th quadrant (see Figure 8). To compute r:

$$r = \sqrt{4^2 + (-4)^2} = 4\sqrt{2}.$$

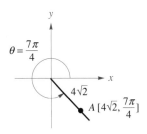

FIGURE 8

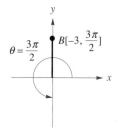

$\theta = \dfrac{3\pi}{2}$　　$B[-3, \dfrac{3\pi}{2}]$

FIGURE 9

Since $r > 0$, we need a θ such that the terminal side of θ passes through the point A and $0 \le \theta < 2\pi$. It follows then that $\theta = 7\pi/4$. Thus the polar representation of the point A that satisfies the given conditions is $[4\sqrt{2}, 7\pi/4]$.

b) Again the first step is to plot the point (Figure 9). In this particular case, it should be obvious without computation that the point B is 3 units from the origin. Since we seek a radial distance r such that $r < 0$, we get $r = -3$. The polar angle must have a terminal side that lies along the negative y-axis, and it must satisfy the condition that $\pi \le \theta < 3\pi$. The value of θ is $3\pi/2$. The polar representation of the point B that satisfies the given conditions is $[-3, 3\pi/2]$.

Just as an equation in x and y can be represented by a graph in the rectangular coordinate system, an equation in r and θ can be represented by a set of points in a polar coordinate system.

Polar Graphs of Equations

> Given an equation in r and θ, the **polar graph** of the equation is the set of points, each with a polar representation $[r, \theta]$ such that this ordered pair makes the equation true.

The techniques involved in sketching such graphs are the topics of the next section. In many cases, however, it is possible to tell much about a polar graph by examining its connection with a rectangular graph.

EXAMPLE 5 Find an equation in polar coordinates that corresponds to the equation $2x + 3y = 12$ in rectangular coordinates.

SOLUTION The plan of attack here is to use the relations

$$x = r \cos \theta \quad \text{and} \quad y = r \sin \theta$$

to transform the given equation into a corresponding polar equation. So,

$$2x + 3y = 12$$
$$2(r \cos \theta) + 3(r \sin \theta) = 12$$
$$r(2 \cos \theta + 3 \sin \theta) = 12$$

or

$$r = \frac{12}{2 \cos \theta + 3 \sin \theta}.$$

The next two examples deal with the problem of finding a corresponding rectangular equation for a given polar equation.

EXAMPLE 6   Find an equation in rectangular coordinates that corresponds to the polar equation $r = 4 \sin \theta$. Sketch the graph.

SOLUTION Since $r = \sqrt{x^2 + y^2}$ and $y = r \sin \theta$,

$$\sin \theta = \frac{y}{r} = \frac{y}{\sqrt{x^2 + y^2}}.$$

So $r = 4 \sin \theta$ becomes

$$\sqrt{x^2 + y^2} = 4 \frac{y}{\sqrt{x^2 + y^2}}.$$

Multiplying each side by $\sqrt{x^2 + y^2}$, we get

$$x^2 + y^2 = 4y.$$

By completing the square, we get

$$x^2 + (y - 2)^2 = 4,$$

a circle with center $(0, 2)$ and radius 2 (you should verify this). Notice that the circle is tangent to the x-axis at the origin (Figure 10).

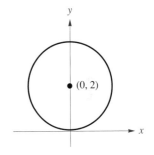

FIGURE 10 *Graph of $r = 4 \sin \theta$*

EXAMPLE 7   Find an equation in rectangular coordinates that corresponds to the polar equation $r^2 = 4 \cos 2\theta$.

SOLUTION First, using the double-angle identity, we can rewrite $r^2 = 4 \cos 2\theta$ as

$$r^2 = 4(\cos^2\theta - \sin^2\theta).$$

Since $r = \sqrt{x^2 + y^2}$, $x = r \cos \theta$, and $y = r \sin \theta$. Thus,

$$\cos^2\theta = \left(\frac{x}{r}\right)^2 = \frac{x^2}{x^2 + y^2} \quad \text{and} \quad \sin^2\theta = \left(\frac{y}{r}\right)^2 = \frac{y^2}{x^2 + y^2}$$

so $r^2 = 4(\cos^2\theta - \sin^2\theta)$ becomes

$$x^2 + y^2 = 4\left(\frac{x^2}{x^2 + y^2} - \frac{y^2}{x^2 + y^2}\right)$$

or

$$(x^2 + y^2)^2 = 4(x^2 - y^2).$$

If you don't recognize this as an equation with a familiar graph, don't despair. The graph of this equation is discussed in the next section.

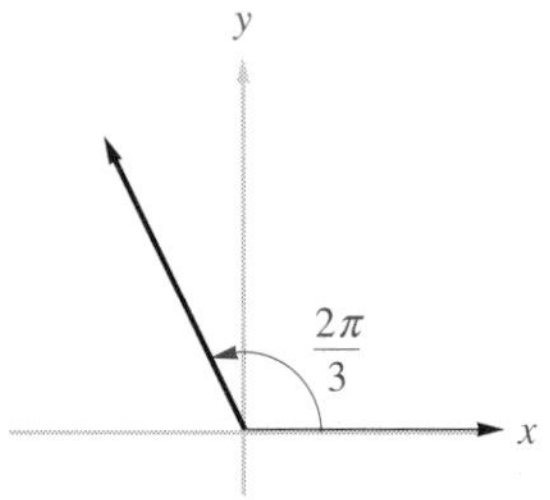

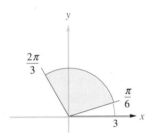

FIGURE 11

The next example illustrates one of the skills needed for calculus. In many applications, it is necessary to sketch a set of points described by equations or inequalities.

EXAMPLE 8   Sketch the points $[r, \theta]$ such that

a) $\theta = 2\pi/3, \qquad r \geq 0$

b) $\pi/6 \leq \theta \leq 2\pi/3, \qquad 0 \leq r \leq 3$

SOLUTION

a)   This set is precisely the points on the terminal side of the angle $2\pi/3$ in standard position (Figure 11).

b)   Think of a line segment of length 3 sweeping out a sector from $\theta = \pi/6$ to $\theta = 2\pi/3$. This gives us the region shown in Figure 12.

FIGURE 12

EXERCISE SET 9.1

A

In Problems 1 through 6, plot both points in polar coordinates on the same coordinate plane.

1. $A[2, \pi/4]$ and $B[-2, \pi/4]$
2. $A[-1, 5\pi/6]$ and $B[1, 5\pi/6]$
3. $A[2, \pi/2]$ and $B[-2, 3\pi/2]$
4. $A[3, -5\pi/3]$ and $B[3, 5\pi/3]$

5. $A[1, 3]$ and $B[3, 1]$
6. $A[0, 2]$ and $B[2, 0]$

In Problems 7 through 12, determine the rectangular coordinates of the points whose polar coordinates are given.

7. $[2, \pi/4]$ 8. $[5, 3\pi/4]$
9. $[-2, \pi/6]$ 10. $[-5, 3\pi]$
11. $[-12, -5\pi/3]$ 12. $[\frac{1}{2}, -3\pi/4]$

In Problems 13 through 24, determine the polar representation(s) of the point described, subject to the restrictions given.

13. Point has polar coordinates $[2, \pi/4]$; $2\pi \leq \theta \leq 3\pi$.

14. Point has polar coordinates $[2, \pi/4]$; $-\pi \leq \theta \leq 0$.

15. Point has polar coordinates $[-4, 3\pi/4]$; $0 \leq \theta \leq 4\pi$ and $r > 0$.

16. Point has polar coordinates $[-4, 3\pi/4]$; $0 \leq \theta \leq 4\pi$ and $r < 0$.

17. Point has polar coordinates $[2, \pi]$; $0 \leq \theta \leq 6\pi$ and $r > 0$.

18. Point has polar coordinates $[2, \pi]$; $-6\pi \leq \theta \leq 0$ and $r > 0$.

19. Point has rectangular coordinates $(2, -2)$; $-2\pi \leq \theta \leq 0$ and $r > 0$.

20. Point has rectangular coordinates $(2, -2)$; $0 \leq \theta \leq 2\pi$ and $r < 0$.

21. Point has rectangular coordinates $(0, -6)$; $-2\pi \leq \theta \leq 2\pi$ and $r > 0$.

22. Point has rectangular coordinates $(0, -6)$; $-2\pi \leq \theta \leq 0$ and $r > 0$.

23. Point has rectangular coordinates $(1, -\sqrt{3})$; $-4\pi \leq \theta \leq 4\pi$ and $r > 0$.

24. Point has rectangular coordinates $(-\sqrt{3}, 1)$; $-4\pi \leq \theta \leq 4\pi$ and $r < 0$.

B

In Problems 25 through 30, express the equation in polar coordinates.

25. $x^2 + y^2 = 6x$

26. $x^2 + y^2 = 4y$

27. $xy = \frac{1}{4}$

28. $y = -\sqrt{3}x$

29. $x^2 + 4xy + 4y^2 = 0$

30. $x^2 - y^2 = 4$

In Problems 31 through 36, express the equation in rectangular coordinates and sketch the graph.

31. $r = 4 \csc \theta$

32. $r = -3 \sec \theta$

33. $r = \dfrac{4}{1 - \cos \theta}$

34. $r = \dfrac{4}{4 - \cos \theta}$

35. $r^2 = 4 \csc 2\theta$

36. $r = -6 \sin \theta$

37. Approximate the rectangular coordinates of the point $P[2, 5]$ (to the nearest 0.01).

38. Approximate the rectangular coordinates of the point $P[-3, -1]$ (to the nearest 0.01).

39. Approximate the polar coordinates with $r > 0$ and $0 \leq \theta \leq 2\pi$ of the point $P(2, 8)$ (r to the nearest 0.01 and θ to the nearest 0.1 radian).

40. Approximate the polar coordinates with $r < 0$ and $0 \leq \theta \leq 2\pi$ of the point $P(8, 2)$ (r to the nearest 0.01 and θ to the nearest 0.1 radian).

In Problems 41 through 52, sketch the set of points described by the inequalities.

41. $\theta = \pi/3$, $r \leq 0$

42. $r = 2$, $0 \leq \theta \leq \pi$

43. $0 \leq r \leq 4$, $0 \leq \theta \leq 2\pi/3$

44. $0 \leq r \leq 3$, $-\pi/2 \leq \theta \leq \pi/2$

45. $\theta = 5\pi/6$, $r \geq 3$

46. $r = -3$, $\pi \leq \theta \leq 11\pi/6$

47. $2 \leq r \leq 6$, $\pi/4 \leq \theta \leq 3\pi/4$

48. $1 \leq r \leq 4$, $\pi/3 \leq \theta \leq 5\pi/4$

49. $-4 \leq r \leq -2$, $0 \leq \theta \leq \pi$

50. $-3 \leq r \leq -1$, $\pi/2 \leq \theta \leq 3\pi/4$

51. $-1 \leq r \leq 3$, $0 \leq \theta \leq \pi/2$

52. $-2 \leq r \leq 4$, $-\pi/4 \leq \theta \leq \pi/4$

53. Show that the distance between points $[a, \alpha]$ and $[b, \beta]$ is given by $d(A, B) = \sqrt{a^2 + b^2 - 2ab \cos(\alpha - \beta)}$.

C

54. Show that the equation in polar coordinates $r = a \sin \theta + b \cos \theta$ represents a circle for any real numbers a and b. Find the center and radius in terms of a and b.

55. **a)** Show that the equation in polar coordinates

$$r = \frac{c}{a \cos \theta + b \sin \theta}$$

represents a straight line for any real numbers a, b, and c (a and b not both zero).

b) Show that the equation in (a) can also be written as $r = k \csc(\theta - \alpha)$, where α is the angle of inclination of the line, and k is a real number.

S E C T I O N   9.2   **POLAR GRAPHS**

Just as with a rectangular graph of an equation, a **polar graph** tells a great deal about an equation. Our present goal is to develop techniques and skills to sketch these functions quickly just as we did with rectangular graphs.

Our concern is mostly with relations of the form $r = f(\theta)$, where f is a function. In this context, the variable θ plays the role of the independent variable, and the variable r plays the role of the dependent variable. Also, many of these functions are periodic with period 2π.

In sketching rectangular graphs, we marked the plane with a rectangular grid: horizontal lines of the form $y = k$ and vertical lines of the form $x = h$. To sketch polar graphs, it will be to our advantage to mark the plane with a **polar grid:** radial lines of the form $\theta = \alpha$ and concentric circles of the form $r = a$ (Figure 13).

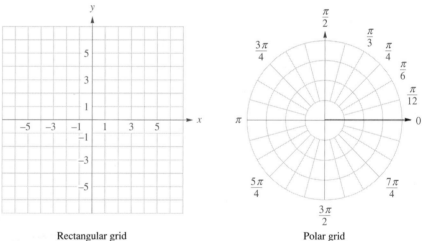

FIGURE 13   Rectangular grid   Polar grid

Keep in mind that, regardless of the grid used, it is the same coordinate plane. The choice of grid is only to facilitate the sketching of a graph.

EXAMPLE 1   Sketch the graph of $r = 4 \sin \theta$.

SOLUTION   We can generate ordered pairs $[r, \theta]$ that make the equation true. The table shows representative values for θ in the interval $0 \le \theta \le 2\pi$ and the corresponding values of r. The related points for $0 \le \theta \le \pi$ are plotted and a smooth curve is sketched (Figure 14). The arrows indicate the direction of increasing θ.

θ	r (approx.)
0	0
$\pi/6$	2
$\pi/3$	$2\sqrt{3}$ (3.46)
$\pi/2$	4
$2\pi/3$	$2\sqrt{3}$ (3.46)
$3\pi/4$	$2\sqrt{2}$ (2.83)
π	0
$7\pi/6$	-2
$4\pi/3$	$-2\sqrt{3}$ (-3.46)
$3\pi/2$	-4
2π	0

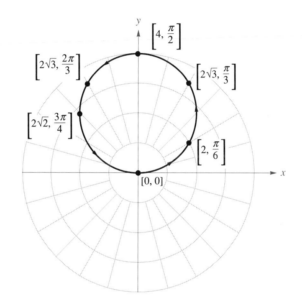

FIGURE 14

It is no surprise that the graph appears to be a circle; in Example 6 of Section 9.1 we found that the equation in rectangular coordinates that corresponds to the polar equation $r = 4 \sin \theta$ is $x^2 + (y - 2)^2 = 4$.

In Figure 14, think of $P[r, \theta]$ as a point that moves around the circle as θ increases. Notice that P travels completely around the circle counterclockwise as θ increases from 0 to π. Now, as θ increases from π to 2π, the point P simply traces out the circle again in the same manner. This is because the values of r are negative for these values of θ from π to 2π. In this interval of θ, we get different polar representations for the same points as we did for values of θ from 0 to π (for example, note that $[2, \pi/6]$ and $[-2, 7\pi/6]$ address the same point on the plane). This "retracing" of the graph is common for many polar equations.

The circle of Example 1 is an example of the general set of circles described below. We leave it to you to verify that the graphs of these equations are the circles shown.

Circles

Assume that $a > 0$. Then the polar graphs of these equations are circles:

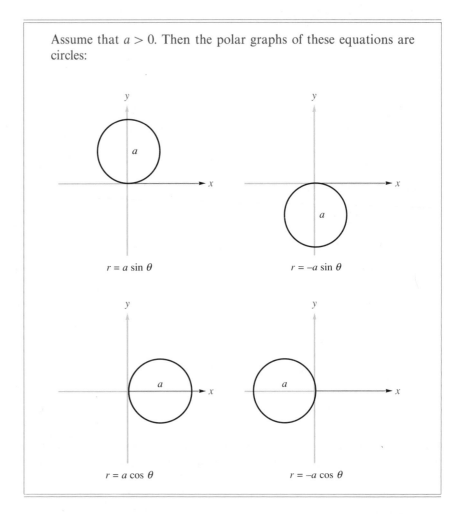

$r = a \sin \theta$

$r = -a \sin \theta$

$r = a \cos \theta$

$r = -a \cos \theta$

EXAMPLE 2   Sketch the graph of $r = 2 \sec \theta$.

SOLUTION First, we generate a table of ordered pairs $[r, \theta]$ for $0 \leq \theta \leq \pi$ and plot the points.

θ	r (approx.)
0	2
$\pi/6$	$4\sqrt{3}/3$ (2.31)
$\pi/4$	$2\sqrt{2}$ (2.83)
$\pi/3$	4
$\pi/2$	undefined
$2\pi/3$	-4
$3\pi/4$	$-2\sqrt{2}$ (-2.83)
$7\pi/6$	$-4\sqrt{3}/3$ (-2.31)
π	-2

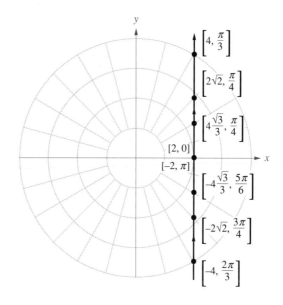

FIGURE 15

This graph seems to coincide with the vertical line with rectangular equation $x = 2$ (Figure 15). This conjecture can be verified since $r = 2 \sec \theta$ implies that

$$r = 2\frac{1}{\cos \theta}$$

or

$$r \cos \theta = 2$$

so

$$x = 2 \qquad \text{(since } x = r \cos \theta\text{).}$$

We leave it to you to show that the point $P[r, \theta]$ retraces the line for $\pi \le \theta \le 2\pi$.

In general, the polar graph of the equation of the form $r = h \sec \theta$ is a vertical line with the corresponding rectangular graph $x = h$, and the polar graph of the equation $r = k \csc \theta$ is a horizontal line with the corresponding rectangular graph $y = k$. We leave it to you to verify these results. These are generalized as follows:

Lines

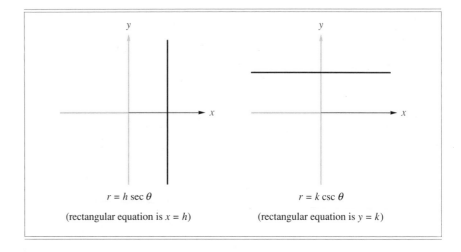

$r = h \sec \theta$

(rectangular equation is $x = h$)

$r = k \csc \theta$

(rectangular equation is $y = k$)

Just as with a rectangular graph, the ability to recognize the symmetry (if any) of a polar graph reduces the amount of labor involved in discovering its graph. The following are the tests for symmetry of polar graphs; they are analogous to the symmetry tests of Section 2.3.

Symmetry of Polar Graphs

The polar graph of an equation in r and θ is

a) symmetric with respect to the x-axis if an equivalent equation results when $[r, \theta]$ is replaced by $[r, -\theta]$ in the equation,

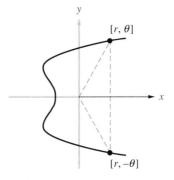

b) symmetric with respect to the y-axis if an equivalent equation results when $[r, \theta]$ is replaced by $[r, \pi - \theta]$ in the equation, and

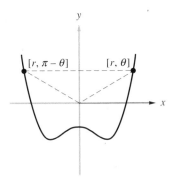

c) symmetric with respect to the origin if an equivalent equation results when $[r, \theta]$ is replaced by $[-r, \theta]$ in the equation.

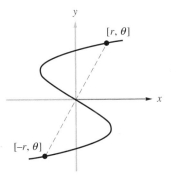

An important point must be made here: If a particular equation passes one of these symmetry tests, then the graph has the appropriate symmetry. If, however, the equation fails the test, then no information about that symmetry can be determined; in other words, the graph may or may not have that symmetry. This is different from the tests of symmetry for rectangular graphs discussed in Section 2.3. If an equation failed one of those tests, then it was assured that the rectangular graph did not have the symmetry.

EXAMPLE 3 Test the graph of $r = 2 + 4 \sin \theta$ for symmetry.

SOLUTION Apply the tests.

a) We replace $[r, \theta]$ with $[r, -\theta]$ and simplify:

$$r = 2 + 4 \sin(-\theta)$$

$$r = 2 - 4 \sin \theta \qquad (\text{since } \sin(-\theta) = -\sin \theta).$$

This is not equivalent to the original equation; the test gives no information.

b) We replace $[r, \theta]$ with $[r, \pi - \theta]$ and simplify:

$$r = 2 + 4 \sin(\pi - \theta)$$

$$r = 2 + 4 \sin \theta \quad \text{(since } \sin(\pi - \theta) = \sin \theta\text{).}$$

This is equivalent to the original equation. The graph is symmetric to the y-axis.

c) We replace $[r, \theta]$ with $[-r, \theta]$ and simplify:

$$-r = 2 + 4 \sin \theta$$

$$r = -2 - 4 \sin \theta.$$

This is not equivalent to the original equation; the test gives no information.

EXAMPLE 4   Test the graph of $r = 4 \cos 2\theta$ for symmetry.

SOLUTION

a)
$$r = 4 \cos 2(-\theta) \Rightarrow r = 4 \cos(-2\theta)$$
$$\Rightarrow r = 4 \cos 2\theta.$$

The graph is symmetric to the x-axis.

b)
$$r = 4 \cos 2(\pi - \theta) \Rightarrow r = 4 \cos(2\pi - 2\theta)$$
$$\Rightarrow r = 4 \cos(-2\theta)$$
$$\Rightarrow r = 4 \cos 2\theta.$$

The graph is symmetric to the y-axis.

c)
$$-r = 4 \cos 2\theta \Rightarrow r = -4 \cos 2\theta.$$

The test gives no information.

We can exploit the techniques that we have acquired for sketching rectangular graphs to discover the graphs of related polar graphs. This method is especially useful in sketching the polar graphs of those equations involving $\sin \theta$ and $\cos \theta$. The next two examples show how this is done.

EXAMPLE 5   Consider the equation from Example 3, $r = 2 + 4 \sin \theta$. The rectangular graph $y = 2 + 4 \sin x$ is a sinusoid with period 2π, amplitude 4, and a vertical translation of 2 (review Section 5.4 if necessary). Use this rectangular graph to sketch the polar graph of $r = 2 + 4 \sin \theta$.

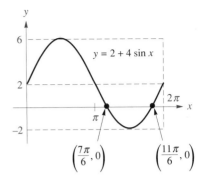

FIGURE 16

SOLUTION Examine the rectangular graph in Figure 16. For a given value of x, say $\pi/6$, the value of y, 4, is equal to the value of r for $\theta = \pi/6$. This value 4 is represented by the vertical line segment of length 4 at $x = \pi/6$ on the rectangular graph and by a radial length of 4 at $\theta = \pi/6$ on the polar graph. A few other line segments are transferred from the rectangular graph to the polar graph, and the graph over the interval $0 \le \theta \le \pi/2$ is sketched in Figure 17.

Next, we use the rectangular graph to sketch the polar graph over $\pi/2 \le \theta \le \pi$. Since the rectangular graph decreases over $\pi/2 \le \theta \le \pi$ in the same manner that it increased over the interval $0 \le \theta \le \pi/2$, the polar graph is similar in shape over both of these intervals (Figure 18).

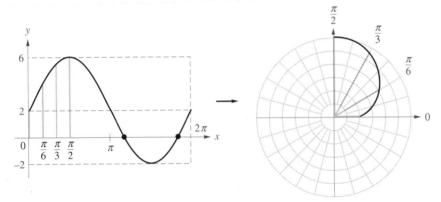

FIGURE 17

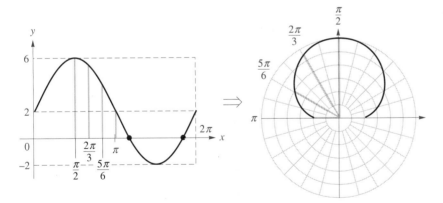

FIGURE 18

Over the interval $\pi \le \theta \le 3\pi/2$, the value of r assumes both positive and negative values. This causes the polar graph to pass through the pole at $\theta = 7\pi/6$ and to turn into the first quadrant, even though the polar angles have terminal sides in the third quadrant (Figure 19).

FIGURE 19

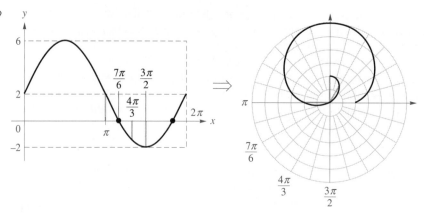

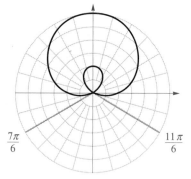

FIGURE 20

We leave it to you to complete the graph for the remaining interval $[3\pi/2, 2\pi]$ of θ (Figure 20). Notice that the graph has symmetry with respect to the y-axis, as we found in Example 3.

The graph of Example 5 is a specific member of a set of polar graphs called **limaçons** (pronounced LEE-ma-sonns). The name is the French word for slug, a snail-like creature (you may need to exercise your imagination to see this). The graph of the general equation $r = a + b \sin \theta$ is a limaçon. Limaçons can be classified by the relative magnitudes of a and b.

Limaçons

Assume that a and b are positive real numbers. The polar graph of $r = a + b \sin \theta$ is a limaçon. There are four basic shapes, depending on the relative values of a and b (these are shown in the figure). If $a < b$, there is a loop inside the limaçon. If $a = b$, a heart-shaped graph called a **cardioid** occurs. If $b < a < 2b$, there is a dimple on the graph. If $a \geq 2b$, there is no dimple; the graph is convex.

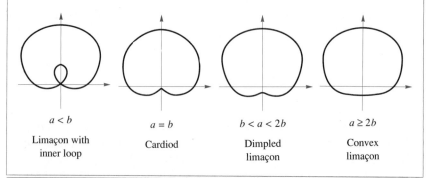

EXAMPLE 6 Sketch the graph of $r = 4 \cos 2\theta$. Use the information derived from the rectangular graph of $y = 4 \cos 2x$.

SOLUTION First, the rectangular graph of $y = 4 \cos 2x$ is a sinusoid with amplitude 4 and period π (Figure 21).

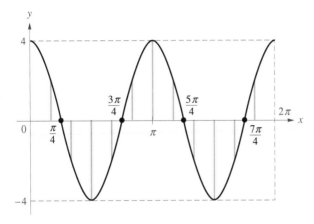

FIGURE 21

As we did in Example 5, we transfer the lengths of the vertical lines in the rectangular graph to the lengths of the corresponding radial lines in the polar graph (Figure 22).

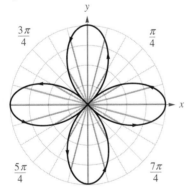

FIGURE 22

Notice that there is symmetry with respect to the x-axis and y-axis, as our symmetry tests in Example 4 predicted. However, there is also symmetry with respect to the origin, even though the equation did not pass the test for that symmetry. This is why we say that failing the test gives no information; it still may have the symmetry in question.

The graph of Example 6 is a **rose.** Roses are of the form $r = a \cos n\theta$ and $r = a \sin n\theta$. If n is odd, the rose has n leaves (the graph is traced twice as θ ranges from 0 to 2π). If n is even, the rose has $2n$ leaves (it is traced once as θ ranges from θ to 2π). Specific cases are shown in the following box.

Roses

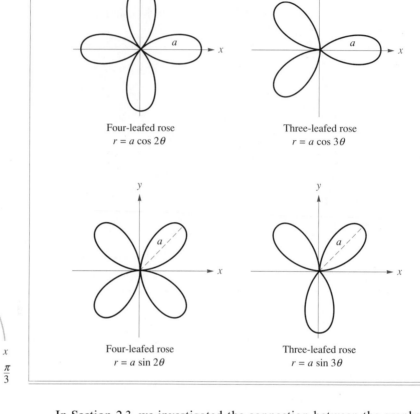

In Section 2.3, we investigated the connection between the graphs of a function $y = f(x)$ and $y = f(x - h)$. There is an analogous connection between the polar graphs of $r = f(\theta)$ and $r = f(\theta - \alpha)$.

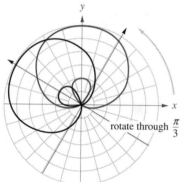

FIGURE 23

Rotation of a Polar Graph

> In general, the polar graph of $r = f(\theta - \alpha)$ is identical to the polar graph of $r = f(\theta)$ rotated through an angle of α. If $\alpha > 0$, then the rotation is counterclockwise; if $\alpha < 0$, then the rotation is clockwise.

EXAMPLE 7   Use the polar graph of the limaçon of $r = 2 + 4 \sin \theta$ to sketch the graph of $r = 2 + 4 \sin(\theta - \pi/3)$.

SOLUTION The graph in question is the same as the graph we discovered in Example 5, rotated through an angle of $\pi/3$ (Figure 23).

EXAMPLE 8 Use the polar graph of the limaçon $r = 2 + 4 \sin \theta$ in Figure 24 to sketch the polar graphs of

a) $r = 2 + 4 \cos \theta$ **b)** $r = 2 - 4 \cos \theta$ **c)** $r = 2 - 4 \sin \theta$

SOLUTION The plan of attack here is to rewrite each of these equations as $r = 2 + 4 \sin(\theta - \alpha)$ for some value of α. To do this, we use some of the reduction identities of Section 6.3.

a) Since $\cos \theta = \sin(\theta + \pi/2)$, the equation is equivalent to $r = 2 + 4 \sin(\theta + \pi/2)$, whose graph is the same as $r = 2 + 4 \sin \theta$ rotated $-\pi/2$ radians (Figure 25).

b) Since $-\cos \theta = \sin(\theta - \pi/2)$, the equation is equivalent to $r = 2 + 4 \sin(\theta - \pi/2)$, whose graph is the same as $r = 2 + 4 \sin \theta$ rotated $\pi/2$ radians (Figure 26).

c) Since $-\sin \theta = \sin(\theta - \pi)$, the equation is equivalent to $r = 2 + 4 \sin(\theta - \pi)$, whose graph is the same as $r = 2 + 4 \sin \theta$ rotated π radians (Figure 27).

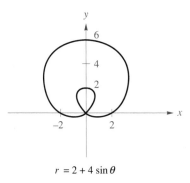

$r = 2 + 4 \sin \theta$

FIGURE 24

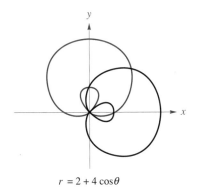

$r = 2 + 4 \cos\theta$

FIGURE 25

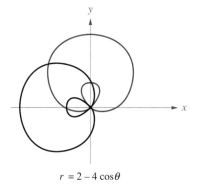

$r = 2 - 4 \cos\theta$

FIGURE 26

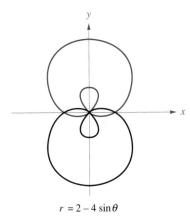

$r = 2 - 4 \sin\theta$

FIGURE 27

EXAMPLE 9 Sketch the polar graph of $r^2 = 16 \cos 2\theta$.

SOLUTION This equation is somewhat different from the other examples since it is not of the form $r = f(\theta)$. Because r^2 must be nonnegative, the only allowable values of θ are those that make the $\cos 2\theta$ nonnegative. In the interval $0 \le \theta \le 2\pi$, this means that $0 \le \theta \le \pi/4$ or $3\pi/4 \le \theta \le 5\pi/4$ or $7\pi/4 \le \theta \le 2\pi$ (to see this, examine the graph of $y = \cos 2x$ over $x \le \theta \le 2\pi$ in Figure 28).

Next, we check for symmetry.

$$r^2 = 16 \cos 2(-\theta) \Rightarrow r^2 = 16 \cos(-2\theta) \Rightarrow r^2 = 16 \cos 2\theta.$$

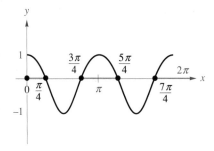

FIGURE 28

There is symmetry with respect to the x-axis.

$$r^2 = 16 \cos 2(\pi - \theta) \Rightarrow r^2 = 16 \cos(2\pi - 2\theta) \Rightarrow r^2 = 16 \cos 2\theta.$$

There is symmetry with respect to the y-axis.

$$(-r)^2 = 16 \cos 2\theta \Rightarrow r^2 = 16 \cos 2\theta.$$

There is symmetry with respect to the origin.

Marking off the excluded regions and plotting points in the interval $[0, \pi/4]$ gives us the picture in Figure 29. Using the symmetry of the graph, we get the entire graph in Figure 30. This particular polar graph is called a **lemniscate** (this name comes from the Greek *lemnisikos* which means *ribbon*).

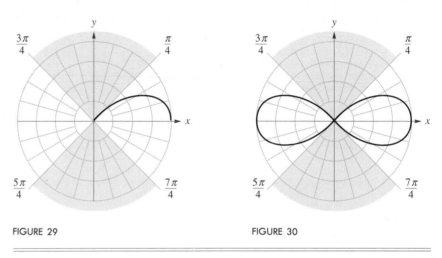

FIGURE 29   FIGURE 30

One final note on polar graphs. You may think that these graphs warrant study for esthetic reasons alone but they come up in the study of orbital bodies, acoustics, broadcasting, and internal combustion engines. In calculus and physics, you will find that they often transform a seemingly difficult problem into a much easier one.

EXERCISE SET 9.2

A

In Problems 1 through 12, the polar graph of the equation is either a circle or a line. Sketch the graph.

1. $r \cos \theta = 4$

2. $r \sin \theta = 4$

3. $r = 6 \cos \theta$

4. $r = 4 \sin \theta$

5. $r = -6 \sin \theta$

6. $r = -2 \cos \theta$

7. $r = 6 \sec \theta$

8. $r = -4 \csc \theta$

9. $r \sec \theta = 4$

10. $r \csc \theta = -2$

11. $r = -6 \sec(-\theta)$

12. $r = -2 \csc(-\theta)$

In Problems 13 through 18, test each equation for symmetry. Sketch the polar equation and note any symmetry not indicated by the tests.

13. $r = 2(1 - \sin \theta)$ **14.** $r = 4 + \cos \theta$

15. $r = 4 \sin 3\theta$ **16.** $r = 3 \cos 4\theta$

17. $r = \dfrac{4}{1 - \cos \theta}$ **18.** $r = \dfrac{8}{2 - \cos \theta}$

In Problems 19 through 27, both $y = f(x)$ and $r = f(\theta)$ are given. Sketch the rectangular graph of $y = f(x)$ and use it to sketch the polar graph of $r = f(\theta)$. Identify the graph as a limaçon or rose if applicable.

19. $y = 6 \cos 3x$, $r = 6 \cos 3\theta$

20. $y = -4 \sin 2x$, $r = -4 \sin 2\theta$

21. $y = 2 - 4 \sin x$, $r = 2 - 4 \sin \theta$

22. $y = 4 - 2 \sin x$, $r = 4 - 2 \sin \theta$

23. $y = 3 + 3 \cos x$, $r = 3 + 3 \cos \theta$

24. $y = 4 + 3 \sin x$, $r = 4 + 3 \sin \theta$

25. $y = 4 \cos 4x$, $r = 4 \cos 4\theta$

26. $y = 4 \sin 4x$, $r = 4 \sin 4\theta$

27. $y = 3 + \cos x$, $r = 3 + \cos \theta$

In Problems 28 through 33, sketch the graph of the lemniscate given.

28. $r^2 = 4 \cos 2\theta$ **29.** $r^2 = 16 \sin 2\theta$

30. $r^2 = -9 \cos 2\theta$ **31.** $r^2 = -25 \sin 2\theta$

32. $r^2 = 16(\cos^2\theta - \sin^2\theta)$ **33.** $r^2 = 18 \sin \theta \cos \theta$

B

In Problems 34 through 39, sketch the polar graph of both of the equations given. Notice that the second graph is a rotation of the first. Identify the angle of rotation.

34. $r = 2 \cos \theta$, $r = 2 \cos(\theta - \pi/4)$

35. $r = 3 \sec \theta$, $r = 3 \sec(\theta + \pi/4)$

36. $r = 2 + \cos \theta$, $r = 2 + \cos(\theta + \pi/3)$

37. $r = 4 \cos 3\theta$, $r = 4 \cos[3(\theta - \pi)]$

38. $r = 2 \cos 2\theta$, $r = 2 \cos(2\theta - \pi)$ *(Hint: Be careful with the angle of rotation!)*

39. $r = 4 \sin \theta$, $r = 2\sqrt{3} \sin \theta - 2 \cos \theta$ *[Hint: Rewrite as $r = A \sin B(\theta - \alpha)$.]*

In Problems 40 through 46, sketch the polar graph of the equation given. Notice that it is not periodic.

40. $r = \theta, \theta \geq 0$ (spiral of Archimedes)

41. $r = \theta, \theta \leq 0$ (spiral of Archimedes)

42. $r = 1/\theta, \theta > 0$ (hyperbolic spiral)

43. $r = 1/\theta, \theta < 0$ (hyperbolic spiral)

44. $r = e^\theta$ (logarithmic spiral)

45. $r^2 = \theta$ (Fermat's spiral)

46. $r^2 = 1/\theta$ (lituus spiral)

C

In Problems 47 through 52, match the equations from I through VI to the polar graphs shown.

  I. $r(\sin^3\theta + \cos^3\theta) = 6 \sin \theta \cos \theta$
      Folium of Descartes (Descartes—1638)

  II. $r^2\cos^2\theta = 9 \cot \theta - 1$
      Serpentine (Newton—1701)

  III. $r = 4 + 2 \sec \theta$
      Conchoid of Nichomedes (Nichomedes—225 B.C.)

  IV. $r = 2 \sin \theta \tan \theta$
      Cissoid of Diocles (Diocles—CA. 200 B.C.)

  V. $r^2 = 12(4 - 3 \sin^2\theta)$
      Hippopede (Proclus—CA. 75 B.C.)

  VI. $r \sin \theta(r^2\cos^2\theta + 4) = 8$
      Witch of Agnesi (Agnesi—1748)

47.

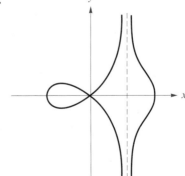

48.

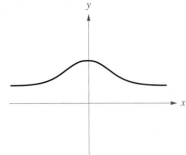

49.

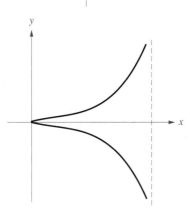

50.

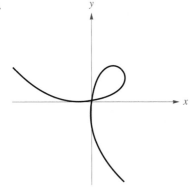

51.

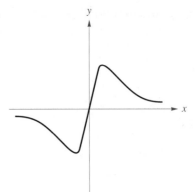

52.

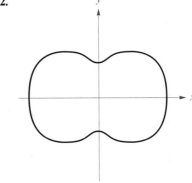

53. The equation $r^n = \cos n\theta$ has a polar graph that is a hyperbola, parabola, lemniscate, circle, straight line, or cardioid, depending on whether n equals -2, -1, $\frac{1}{2}$, 1, or 2 (but not necessarily in the same order). Determine the type of graph listed that corresponds to each of these values of n.

54. Consider two points $F_1(-a, 0)$ and $F_2(a, 0)$. Let the point $P(x, y)$ be such that $d(P, F_1) \cdot d(P, F_2) = a^2$. Show that the set of all such points lies on the polar graph of the lemniscate $r^2 = 2a^2\cos 2\theta$

SECTION   9.3   **PARAMETRIC EQUATIONS AND PLANE CURVES**

In calculus, many curves arise that are not the graphs of functions. For example, the graph of the equation $4x^2 + 16y^2 = 25$ is not that of a function (recall that this is an ellipse; it fails the vertical line test). In this section we investigate a method used to describe these more general curves. A third variable, called a parameter, is used to describe the x and y coordinates of the points on the curve. Example 1 illustrates this method.

EXAMPLE 1   Sketch the graph of the curve composed of points of the form $(\frac{1}{2}t^2, t - 2)$ for values of t in the interval $[-2, 4]$.

SOLUTION We start by compiling a table of selected values of t and corresponding values of $x = \frac{1}{2}t^2$ and $y = t - 2$. Next we plot these points on the coordinate plane and sketch a curve through successive points (Figure 31). The arrows on the curve indicate the direction of increasing t.

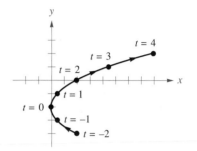

FIGURE 31

t	x	y
-2	2	-4
-1	$\frac{1}{2}$	-3
0	0	-2
1	$\frac{1}{2}$	-1
2	2	0
3	$\frac{9}{2}$	1
4	8	2

This type of curve on the coordinate plane is called, appropriately enough, a **plane curve.**

A plane curve C is a set of ordered pairs of the form $(f(t), g(t))$ for a specified domain of t. The equations

$$\begin{cases} x = f(t) \\ y = g(t) \end{cases}$$

are called **parametric equations** for the curve C. The variable t is called the **parameter.** The direction on the curve of increasing values of t is called the **orientation** of the curve.

Plane Curve

It is helpful to think of the parameter t as a variable that represents some measure of time and a plane curve as the path of a moving point P with coordinates $(f(t), g(t))$ as t increases through its domain.

The curve sketched in Example 1 looks suspiciously like a portion of a parabola. In fact, we can eliminate the parameter t from the set of parametric equations

$$\begin{cases} x = \tfrac{1}{2}t^2 \\ y = t - 2 \end{cases}$$

by solving the second equation for t and substituting the equivalent expression in y into the first equation in place of t:

$$y = t - 2 \Rightarrow t = y + 2$$
$$x = \tfrac{1}{2}t^2 \Rightarrow x = \tfrac{1}{2}(y + 2)^2.$$

This is a parabola opening to the right with vertex at $(0, -2)$. The graph of this rectangular equation in Figure 32 does seem to coincide with our sketch in Example 1.

This process of finding a corresponding rectangular equation often helps in discovering the graph of a plane curve from its parameteric representation.

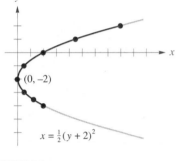

$(0, -2)$

$x = \tfrac{1}{2}(y + 2)^2$

FIGURE 32

EXAMPLE 2   Sketch the curve represented by the given equations by eliminating the parameters.

a) $x = \sqrt{t} + 2$ and $y = 4 - 2\sqrt{t}$

b) $x = -\tfrac{1}{2}\log_2 t$ and $y = 8 + \log_2 t$

SOLUTION

a)   Since there is no restriction on t, we assume the domain to be the largest interval over which the equations make sense. In this case, this means that $t \geq 0$. Also, notice that $x \geq 2$ and $y \leq 4$.

Now consider the first equation. We can solve for t:

$$x = \sqrt{t} + 2$$
$$\sqrt{t} = x - 2$$
$$t = (x - 2)^2.$$

Next, we substitute this expression for t into the equation for y:

$$y = 4 - 2\sqrt{(x - 2)^2}$$
$$y = 4 - 2(x - 2) \qquad \text{(since } x \geq 2)$$
$$y = 8 - 2x, \qquad x \geq 2.$$

We sketch the line and determine a few convenient points to show the orientation of the curve (Figure 33). Notice that the curve has an endpoint at the point (2, 4).

t	x	y
0	2	4
1	3	2
4	4	0
9	5	−2
16	6	−4

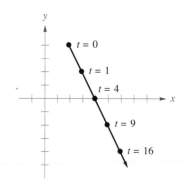

FIGURE 33

b) Recall that the domain of $\log_2 t$ is the set of all positive real numbers and the range is the set of all real numbers. This implies that $t > 0$ but that x and y can assume any real value. Solving the first equation for t and substituting, we get

$$x = -\tfrac{1}{2}\log_2 t \Rightarrow \log_2 t = -2x$$
$$\Rightarrow t = 2^{-2x}$$
$$y = 8 + \log_2 t \Rightarrow y = 8 + \log_2(2^{-2x})$$
$$\Rightarrow y = 8 + (-2x)$$
$$\Rightarrow y = 8 - 2x \qquad \text{for all real } x.$$

We sketch the line and determine a few convenient points to show the orientation of the curve (Figure 34).

t	x	y
4	−1	10
1	0	8
$\frac{1}{4}$	1	6
$\frac{1}{16}$	2	4
$\frac{1}{64}$	3	2

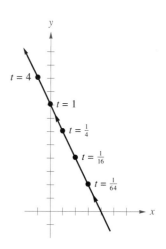

FIGURE 34

Notice that the rectangular equation for each of these sets of parametric equations is the same, $y = 8 - 2x$. Notice, however, that the orientation and the domain of the two are different.

The method we used in Example 2 to find a rectangular equation for a parametric representation of a curve depends on being able to solve one or the other of the equations for the parameter t. Unfortunately, this is sometimes extremely difficult, if not impossible. Some other tricks, however, do allow us to find a rectangular equation for a curve from its parametric representation. Example 3 shows such a trick.

EXAMPLE 3   Sketch the curve represented by the given equations by eliminating the parameter:

$$x = 3 \cos t \quad \text{and} \quad y = 3 \sin t, \qquad 0 \le t \le 2\pi.$$

SOLUTION These equations do not allow us to eliminate the variable t easily by solving one of them for t and substituting. However, solving these equations for $\cos t$ and $\sin t$, we get

$$\cos t = x/3 \quad \text{and} \quad \sin t = y/3$$

and substituting into the identity $\cos^2 t + \sin^2 t = 1$ yields

$$\left(\frac{x}{3}\right)^2 + \left(\frac{y}{3}\right)^2 = 1$$

$$\frac{x^2}{9} + \frac{y^2}{9} = 1$$

$$x^2 + y^2 = 9.$$

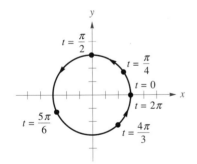

FIGURE 35

The graph of this curve is a circle centered at the origin with radius 3. Plotting a few points shows us that both of the curve's endpoints are (1, 0) and that the curve is oriented counterclockwise (Figure 35).

One skill that is important for calculus is to determine a parametric representation for a curve whose rectangular equation, orientation, and extent are given. The following table shows common representations for some of the curves encountered in calculus.

Plane Curve	Parametric Representation	Curve and Orientation
Function $y = f(x)$	$\begin{cases} x = t \\ y = f(t) \end{cases}$	
Circle $x^2 + y^2 = a^2$	$\begin{cases} x = a \cos t \\ y = a \sin t \end{cases}$	
Ellipse $\dfrac{x^2}{a^2} + \dfrac{y^2}{b^2} = 1$	$\begin{cases} x = a \cos t \\ y = b \sin t \end{cases}$	
Line segment from $P(a, b)$ to $Q(c, d)$	$\begin{cases} x = a + (c - a)t \\ y = b + (d - b)t \end{cases}$ $0 \le t \le 1$	
Graph of polar equation $r = f(\theta)$	$\begin{cases} x = f(t) \cos t \\ y = f(t) \sin t \end{cases}$	

Parametric Representations of Common Curves

The next two examples show how the general parametrizations shown in the table can be applied.

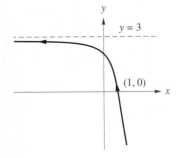

FIGURE 36

EXAMPLE 4   Determine a parametric representation of the curve with rectangular equation $y = 3 - 3^x$ and with orientation shown in the graph in Figure 36.

SOLUTION   This is an equation of a function, and if we apply the parametrization in the table, we get $x = t$ and $y = 3 - 3^t$. However, this gives us an orientation in the direction of increasing x, and the graph shows an orientation in the direction of decreasing x. We can reverse the orientation by replacing t with $-t$. This gives us the parametric representation

$$\begin{cases} x = -t \\ y = 3 - 3^{-t}. \end{cases}$$

You may wish to verify this by plotting a few points on the graph.

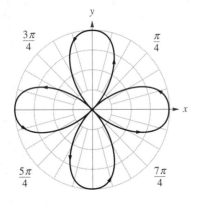

FIGURE 37

EXAMPLE 5   Determine a parametric representation of the curve with polar equation $r = 4 \cos 2\theta$ and with orientation shown in the graph in Figure 37 (see Example 6 of Section 9.2).

SOLUTION   According to the table, we can parametrize a polar equation of the form $r = f(\theta)$ by $x = f(\theta) \cos \theta$ and $y = f(\theta) \sin \theta$. This should seem reasonable. Recall in our discussion of rectangular and polar coordinates in Section 9.1 that

$$x = r \cos \theta \quad \text{and} \quad y = r \sin \theta.$$

Substituting $f(\theta)$ for r gives us this result.

In this particular case our parametric representation is

$$\begin{cases} x = 4 \cos 2\theta \cos \theta \\ y = 4 \cos 2\theta \sin \theta. \end{cases}$$

Notice that the parameter is θ instead of the usual t. We leave it to you to verify that this gives the proper orientation.

The Cycloid and the Brachistochrone Problem

Imagine a wheel rolling across a table top without slipping, with point P fixed to the edge of the wheel, as shown in Figure 38. The curve traced out by the point P as the wheel rolls along the table is called a **cycloid.** There is no easy way to write this in a rectangular form, but there is a way to determine a parametric representation of the curve.

FIGURE 38

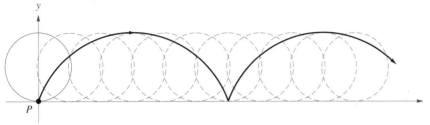

Look at Figure 39 carefully; it should be clear that the coordinates (x, y) of the point P are given by

$$\begin{cases} x = \overline{OR} - \overline{PQ} \\ y = \overline{CR} - \overline{CQ}. \end{cases}$$

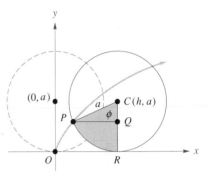

FIGURE 39

The line segment $\overline{CR}$ is a radius of the wheel; it follows that $CR = a$. The length OR is the same as the length of the arc intercepted by the angle ϕ. The length of this arc is $a\phi$, so $OR = a\phi$. An examination of the triangle CQP should convince you that $PQ = a \sin \phi$ and $CQ = a \cos \phi$. Thus a parametric representation of this cycloid is

$$\begin{cases} x = a\phi - a \sin \phi \\ y = a - a \cos \phi. \end{cases}$$

Note that ϕ is used as the parameter instead of t.

One of the classic problems of calculus, the **brachistochrone problem** (from the Greek *brachys* for "short" and *chronos* for "time"), involved this curve. Consider a rigid wire between two fixed points A and B (as shown in Figure 40), and a frictionless bead sliding down the wire under the influence of gravity from A to B. What is the shape that enables the bead to traverse the span in the shortest possible time? The obvious answer, a straight line, is incorrect; the shortest distance is not the fastest (why?). The solution is an inverted cycloid curve. The inverted cycloid possesses another interesting property: regardless of where the bead starts on this curve, the amount of time required to slide to point B is the same.

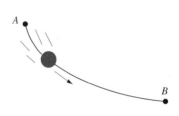

FIGURE 40

EXERCISE SET 9.3

A

In Problems 1 through 12, sketch the curve represented by the parametric equations (be sure to indicate the orientation) and write a corresponding rectangular equation by eliminating the parameter.

1. $x = t - 1, \quad y = 4 - 2t, \qquad 0 \le t \le 5$

2. $x = 5 - 2t, \quad y = 1 + 2t, \qquad -1 \le t \le 4$

3. $x = 2t^2 + 2, \quad y = \tfrac{1}{2}t, \qquad -2 \le t \le 2$

4. $x = 2t - 1, \quad y = t^2 + 2, \qquad -1 \le t \le 3$

5. $x = 2t - 2, \quad y = \sqrt{t}, \qquad 0 \le t \le 4$

6. $x = \sqrt{t} - 1, \quad y = 4 - 2t, \qquad 0 \le t \le 5$

7. $x = t - 1, \quad y = \dfrac{t}{t - 1}, \qquad t \ge 2$

8. $x = \dfrac{2t}{t + 1}, \quad y = 2 - t, \qquad t \ge 0$

9. $x = \cos t$,　$y = \cos^2 t$,　　$0 \le t \le \pi$

10. $x = \sin t$,　$y = \sin^3 t$,　　$-\pi \le t \le \pi$

11. $x = 2 \log_2 t$,　$y = 4 - \log_2 t$,　　$0 < t \le 8$

12. $x = e^t$,　$y = e^{2t}$,　　$t \ge 0$

In Problems 13 through 24, sketch the curve described and determine a parametric representation for it. Be sure to give the domain of the parameter.

13. The line segment from $(1, 2)$ to $(9, -4)$

14. The line segment from $(0, 0)$ to $(-3, 12)$

15. The parabola $y = x^2 - 4$ from $(-2, 0)$ to $(2, 0)$

16. The parabola $y = x^2 - 4$ from $(2, 0)$ to $(-2, 0)$

17. The arc of the circle $x^2 + y^2 = 9$ in the first quadrant from $A(3, 0)$ to $B(0, 3)$

18. The arc of the circle $x^2 + y^2 = 16$ in the first and second quadrants from $A(4, 0)$ to $B(-4, 0)$

19. The ellipse $9x^2 + 16y^2 = 144$, oriented counterclockwise

20. The ellipse $9x^2 + 4y^2 = 36$, oriented counterclockwise

21. The graph of polar equation $r = 2 \cos \theta$, oriented counterclockwise

22. The graph of $y = \sqrt[3]{x}$ from $(-8, -2)$ to $(8, 2)$

23. The graph of $y = \cos x$ from $(\pi, -1)$ to $(-\pi, -1)$

24. The entire graph of $y = \tan^{-1} x$, oriented from right to left

B

In Problems 25 through 30, sketch the curve represented by the parametric equations (be sure to indicate the orientation), and write a corresponding rectangular equation by eliminating the parameter. You will need to use an appropriate trigonometric identity in each case.

25. $x = \sec t$,　$y = \tan t$,　　$0 \le t < \pi/2$

26. $x = \cot t$,　$y = \csc t$,　　$-\pi/2 < t < \pi/2$

27. $x = \cos t$,　$y = \cos 2t$,　　$0 \le t \le \pi$

28. $x = \sin t$,　$y = \cos 2t$,　　$-\pi/2 \le t \le \pi/2$

29. $x = \arcsin t$,　$y = \arccos t$,　　$-1 \le t \le 1$

30. $x = \cos^2 t$,　$y = \sin^2 t$,　　$0 < t < \pi/2$

31. Find the value of k such that the graph of the rectangular equation $x^2 - y^2 = k$ in the first and fourth quadrants has the parametric representation
$$x = 3^t + 3^{-t}, \qquad y = 3^t - 3^{-t}.$$

Describe the orientation of the parametrization (see Figure 41).

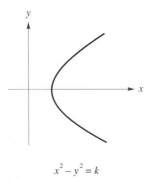

$x^2 - y^2 = k$

FIGURE 41

32. Find the value of k such that the graph of the rectangular equation $x^2 - y^2 = k$ in the second and third quadrant has the parametric representation
$$x = -2^t - \left(\frac{1}{2}\right)^t \qquad y = 2^t - \left(\frac{1}{2}\right)^t.$$

Describe the orientation of the parametrization (see Figure 42).

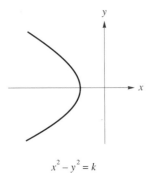

$x^2 - y^2 = k$

FIGURE 42

33. A particle starts at the point $(0, 1)$ at $t = 0$ seconds on the graph of $y = 2x + 1$ and moves in the direction of increasing x such that its x-coordinate changes $\frac{1}{2}$ unit/second. Find a set of parametric equations that gives the coordinates of the particle at time t seconds.

34. A particle starts at the point $(0, 0)$ at $t = 0$ seconds on the graph of $y = x^2$ and moves in the direction of increasing x such that its x-coordinate changes 2 units/second. Find a set of parametric equations that gives the coordinates of the particle at time t seconds.

35. The graphs of $x = f(t)$ and $y = g(t)$ are shown in Figure 43. Sketch the plane curve $(f(t), g(t))$.

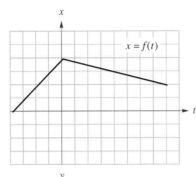

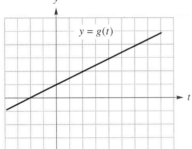

FIGURE 43

36. The graphs of $x = f(t)$ and $y = g(t)$ are shown in Figure 44. Sketch the plane curve $(f(t), g(t))$.

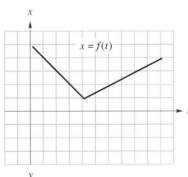

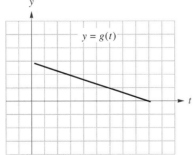

FIGURE 44

C

In Problems 37 through 42, match parametric equations I through VI to the curves shown.

I. $x = 6(t^2 - 3)$, $y = 2t(t^2 - 3)$
  Trisectrix of Catalin

II. $x = 2 \sec t$, $y = 2 \tan t \sec t$
  Kampyle of Eudoxus

III. $x = 2(1 + \sin t)$, $y = 4 \cos t(1 + \sin t)$
  Piriform

IV. $x = 6 \cos^3 t$, $y = 6 \sin^3 t$, $0 \le t \le 2\pi$
  Asteroid

V. $x = 6 \cos t - 3 \cos 2t$,
  $y = 6 \sin t - 3 \sin 2t$, $-\pi \le t \le \pi$
  Epitrochoid

VI. $x = 9 \sin \frac{1}{2}t$, $y = 8 \sin t$, $0 \le t \le 2\pi$
  Bowditch Curve

37.

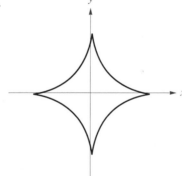

38.

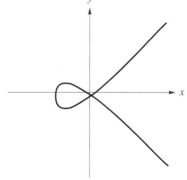

39.

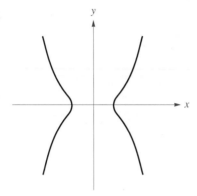

40.

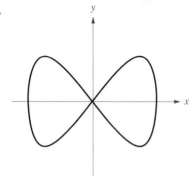

41.

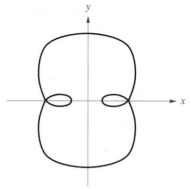

42.

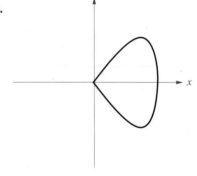

43. a) Suppose that a wheel is rolled along a table (as we did in the development of the cycloid). The curve traced by a point P that is b units from the center on the wheel ($b < a$) is a **curtate cycloid** (see Figure 45). Sketch this curve and determine a parametric representation for it.

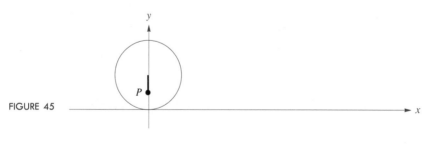

FIGURE 45

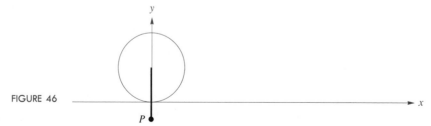

FIGURE 46

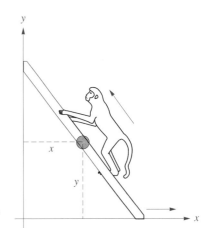

b) Suppose that we change the situation in (a) so that the point P is outside the circle but fixed to it (that is, $b > a$). The curve traced by the point P is a **prolate cycloid** (see Figure 46). Sketch this curve and determine a parametric representation for it.

44. At $t = 0$ sec, a monkey is at the midpoint of a ladder of length 20 ft (Figure 47). The base of the ladder is 8 ft from the wall. The monkey starts to climb the ladder (at a rate of 2 ft/sec), and the base of the ladder starts to slip away from the wall at a rate of 2 ft/sec. Find a set of equations $x = f(t)$ and $y = g(t)$ that describe the location of the monkey in terms of t.

FIGURE 47

POLAR REPRESENTATIONS OF THE CONICS   S E C T I O N   9.4

Polar coordinates often give us a way to examine a graph from a different perspective. This is certainly true of the graphs of the conic sections.

Back in Section 8.4, we showed that a conic section is a set of points P such that the distance from P to a fixed line (the **directrix**) and the distance from P to a fixed point (the **focus**) were in constant ratio. The value of this constant ratio (the **eccentricity**) determined the type and general shape of the conic section.

A conic section is completely determined by a fixed line ℓ, a point F (as shown in the figure), and a positive real number e such that

$$ePR = PF.$$

The value of e determines the type and the shape of the section. Specifically,

$0 < e < 1 \Rightarrow$ section is an ellipse,

$e = 1 \Rightarrow$ section is a parabola, and

$e > 1 \Rightarrow$ section is a hyperbola.

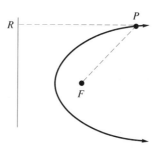

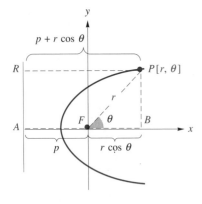

FIGURE 48

Suppose that we impose a polar coordinate system such that focus falls on the pole, and the directrix is p units to the left of the pole (this implies that the polar equation of the directrix is $r = -p \sec \theta$). Study Figure 48; you should be able to convince yourself that

$$PR = p + r \cos \theta \quad \text{and} \quad PF = r.$$

So, $ePR = PF$ becomes

$$e(p + r \cos \theta) = r.$$

Solving this equation for r, we get

$$r = \frac{ep}{1 - e \cos \theta}.$$

We summarize:

Polar Equation of a Conic Section

A polar equation of a conic section with eccentricity e, focus at the pole, and directrix of $r = -p \sec \theta$ is

$$r = \frac{ep}{1 - e \cos \theta}.$$

Keep in mind that the value of e determines whether the conic section is an ellipse ($0 < e < 1$), parabola ($e = 1$), or hyperbola ($e > 1$). The next three examples involve each of these cases.

EXAMPLE 1 Determine the type of conic section represented by the polar equation

$$r = \frac{2}{1 - \cos \theta}.$$

Sketch its graph.

SOLUTION Comparing this with the general form

$$r = \frac{ep}{1 - e \cos \theta}$$

we can surmise that $e = 1$ and $p = 2$. This implies that the section is a parabola. The directrix is 2 units to the left of the pole; its equation is $r = -2 \sec \theta$. Its graph is sketched in Figure 49.

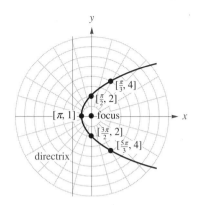

FIGURE 49

EXAMPLE 2 Determine the type of conic section represented by the polar equation

$$r = \frac{6}{4 - 2 \cos \theta}.$$

Sketch its graph.

SOLUTION We multiply the numerator and the denominator each by $\frac{1}{4}$ to put the equation in the form

$$r = \frac{ep}{1 - e \cos \theta}.$$

$$r = \frac{6}{4 - 2 \cos \theta} \cdot \frac{\frac{1}{4}}{\frac{1}{4}}$$

$$r = \frac{\frac{3}{2}}{1 - \frac{1}{2} \cos \theta}.$$

The eccentricity of this conic section is $\frac{1}{2}$; this qualifies it as an ellipse. The distance from the pole to the directrix is 3 units. A few points are plotted to complete the picture (Figure 50).

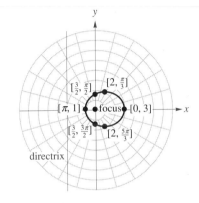

FIGURE 50

EXAMPLE 3 Determine the type of conic section represented by the polar equation

$$r = \frac{3}{2 - 3 \cos \theta}.$$

Sketch its graph.

SOLUTION This is equivalent to

$$r = \frac{\frac{3}{2}}{1 - \frac{3}{2} \cos \theta}.$$

The eccentricity is $e = \frac{3}{2}$. The distance p from the focus to the directrix is 1 (verify this). Since $\frac{3}{2} > 1$, this section is a hyperbola. Its directrix is $r = -\sec \theta$. The sketch is shown in Figure 51.

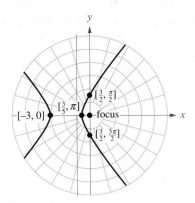

FIGURE 51

One advantage of polar representations of graphs is that rotations can be handled with relatively little difficulty. Recall that if the graph of $r = f(\theta)$ is known, then the graph of $r = f(\theta - \alpha)$ can be found by rotating the graph of $r = f(\theta)$ through an angle of α (the rotation is counterclockwise if $\alpha > 0$ and clockwise if $\alpha < 0$).

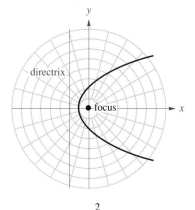

$$r = \frac{2}{1 - \cos \theta}$$

FIGURE 52

EXAMPLE 4   Given the graph (Figure 52) of $r = 2/(1 - \cos \theta)$ from Example 1, sketch the graph of

a) $\quad r = \dfrac{2}{1 - \cos\left(\theta - \dfrac{\pi}{3}\right)}$
b) $\quad r = \dfrac{2}{1 + \sin \theta}$

SOLUTION

a)   The graph of

$$r = \frac{2}{1 - \cos\left(\theta - \dfrac{\pi}{3}\right)}$$

is a rotation through an angle of $\pi/3$ of the graph of $r = 2/(1 - \cos \theta)$. Notice that the directrix also rotates. Since the equation of the original directrix was $r = -2 \sec \theta$, the equation of the new directrix is

$$r = -2 \sec\left(\theta - \frac{\pi}{3}\right).$$

The graph is shown in Figure 53.

b) Since

$$-\sin \theta = \cos\left(\theta + \frac{\pi}{2}\right),$$

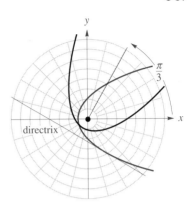

FIGURE 53

the equation

$$r = \frac{2}{1 + \sin \theta}$$

can be rewritten as

$$r = \frac{2}{1 - \cos\left(\theta + \frac{\pi}{2}\right)}.$$

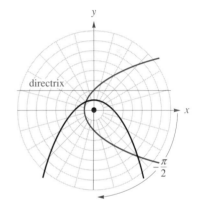

The angle of rotation then is $-\pi/2$. The directrix is $r = 2 \csc \theta$. The graph is shown in Figure 54.

FIGURE 54

Part (b) of Example 4 suggests that a polar equation of form

$$r = \frac{ep}{1 \pm e \cos \theta} \quad \text{or} \quad r = \frac{ep}{1 \pm e \sin \theta}$$

is a conic section. Each of these four forms is a rotation of the standard form

$$r = \frac{ep}{1 - e \cos \theta}$$

through an angle of $\pi/2$, π, or $3\pi/2$ (the proof of this is left as an exercise).

The orientation of the conic section determines which of the four forms is used, as follows:

Polar Equations of Conic
Sections

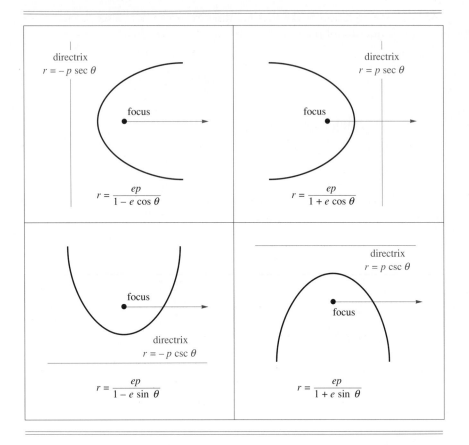

The form to use is determined by the relative positions of the directrix and the focus. The general rule is that the conic section always opens toward the focus, away from the directrix.

EXAMPLE 5   Determine a polar equation of an ellipse with focus at the origin, with directrix at $y = -3$, and with eccentricity of $\frac{1}{3}$.

SOLUTION Since the directrix is horizontal and below the focus, the form of the equation that we seek is

$$r = \frac{ep}{1 - e \sin \theta}$$

where $e = \frac{1}{3}$ and $p = 3$. Thus,

$$r = \frac{1}{1 - \frac{1}{3} \sin \theta}.$$

Multiplying numerator and denominator by 3, we get the more palatable equation,

$$r = \frac{3}{3 - \sin \theta}.$$

EXAMPLE 6 Determine a polar equation of a parabola with focus at the origin such that the closest point to the focus on the parabola is $V[2, 0]$.

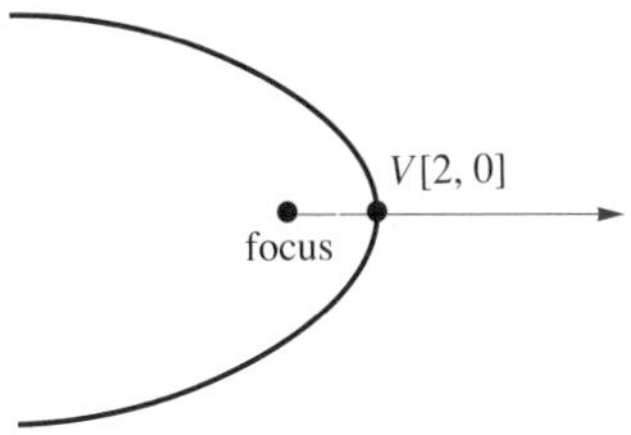

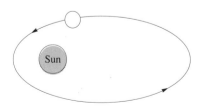

$V[2, 0]$

focus

SOLUTION The closest point to the focus on a parabola is the vertex (convince yourself of this before going on; see Figure 55). This implies that the equation is of the form

$$r = \frac{ep}{1 + e \cos \theta}.$$

FIGURE 55

Now, since the conic section is a parabola, $e = 1$. Also, since the point $V[2, 0]$ is on the graph, we can let $r = 2$ and $\theta = 0$ and solve for p:

$$2 = \frac{(1)p}{1 + (1)\cos 0}$$

$$= \frac{p}{2}$$

or $p = 4$. Our equation is

$$r = \frac{4}{1 + \cos \theta}.$$

Polar Equations and Planetary Orbits

In the early 1600s, soon after the invention of the telescope, the German astronomer Johannes Kepler discovered three properties, called Kepler's laws, that completely describe the motion of the planets in the heavens. One of Kepler's laws states that planets and other orbiting bodies of our solar system travel around the sun in an elliptical path, with the center of the sun at the focus (Figure 56). The closest point to the sun on the orbit is called the **perihelion;** the farthest is the **aphelion.**

Of course, these ellipses are of very different shapes. The variations of the shapes can be measured by the eccentricities of the ellipses. For example, the ellipse of the earth's orbit is very close to circular; its eccentricity e is about 0.02 (imagine how your life might be different if the eccentricity were

Sun

FIGURE 56

0.9). On the other hand, the orbit of Halley's comet is very eccentric ($e \approx 0.97$). A chart of bodies of our solar system and information about their elliptical orbits follows.

An astronomical unit (AU) is a standard measurement of distance used in describing distances in our solar system. By definition, 1 AU is the length of the semimajor axis of the elliptical orbit of the earth.

Body	Semimajor Axis of Orbit (in AU)	Eccentricity of Orbit
Mercury	0.3871	0.20563
Venus	0.7233	0.00679
Earth	1	0.01673
Mars	1.5237	0.09337
Jupiter	5.2028	0.07650
Saturn	9.5388	0.04844
Uranus	19.182	0.04721
Neptune	30.058	0.00858
Pluto	39.439	0.25024
Comet Halley	36.178	0.97214
Comet Kahotek	3466.7	0.99993
Asteroid Icarus	2.165	0.82712

Notice that the comets in this chart have very eccentric orbits. The paths near the sun are very close to that of a parabola (recall that a parabola has an eccentricity of one). Many comets have hyperbolic paths, but these comets were rarely named since they obviously do not make return appearances.

EXAMPLE 7   Use the solar system chart to find the distance that the planet Pluto is from the sun at perihelion and at aphelion. Determine an equation of the form

$$r = \frac{ep}{1 - e \cos \theta}$$

that describes its orbit.

SOLUTION The closest point to the focus and the farthest point from the focus are the vertices of the ellipse on the major axis (Figure 57). In terms of the equation

$$r = \frac{ep}{1 - e \cos \theta},$$

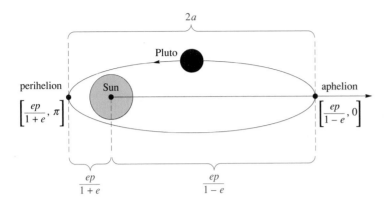

FIGURE 57

the closest point corresponds to $\theta = \pi$, and the farthest point corresponds to $\theta = 0$. Thus, the perihelion has polar coordinates

$$\left[\frac{ep}{1 + e}, \pi \right]$$

and the aphelion has polar coordinates

$$\left[\frac{ep}{1 - e}, 0 \right].$$

As Figure 57 suggests,

$$2a = \frac{ep}{1 + e} + \frac{ep}{1 - e} = \frac{2ep}{1 - e^2}$$

or $ep = a(1 - e^2)$. This means that our equation can be written

$$r = \frac{a(1 - e^2)}{1 - e \cos \theta}.$$

Using the values in the table, we obtain

$$r = \frac{39.439(1 - 0.25024^2)}{1 - 0.25024 \cos \theta}$$

or

$$r = \frac{36.693}{1 - 0.25024 \cos \theta}.$$

Using this to find the extreme values gives

$$\text{distance at perihelion} = \frac{36.693}{1 - 0.25024 \cos \pi}$$

$$= \frac{36.693}{1 + 0.25024}$$

$$= 29.570 \text{ AU}$$

$$\text{distance at aphelion} = \frac{36.693}{1 - 0.25024 \cos \theta}$$

$$= \frac{36.693}{1 - 0.25024}$$

$$= 49.308 \text{ AU}$$

EXERCISE SET 9.4

A

In Problems 1 through 12, sketch the conic section represented by the polar equation. Identify the type of conic section (parabola, ellipse, or hyperbola) and determine the polar equation of the directrix.

1. $r = \dfrac{4}{1 + \cos \theta}$

2. $r = \dfrac{3}{1 - \sin \theta}$

3. $r = \dfrac{3}{1 + 2 \sin \theta}$

4. $r = \dfrac{2}{1 + \frac{5}{2} \cos \theta}$

5. $r = \dfrac{2}{1 + \frac{2}{5} \cos \theta}$

6. $r = \dfrac{16}{4 - \cos \theta}$

7. $2r(1 + \cos \theta) = 16$

8. $\frac{2}{3}r(1 - \sin \theta) = 8$

9. $r = \dfrac{12 \sec \theta}{2 \sec \theta + 6}$

10. $r = \dfrac{9 \csc \theta}{4 \csc \theta + 3}$

11. $r = \dfrac{12}{-4 - \sin \theta}$

12. $r = \dfrac{24}{-4 + 6 \sin \theta}$

In Problems 13 through 18, the graph of the second equation is a rotation of the graph of the first. Sketch both on the same coordinate plane. Identify the angle of rotation.

13. $r = \dfrac{2}{1 + \cos \theta}$, $r = \dfrac{2}{1 + \cos\left(\theta - \dfrac{\pi}{4}\right)}$

14. $r = \dfrac{4}{1 - \sin \theta}$, $r = \dfrac{4}{1 - \sin\left(\theta + \dfrac{\pi}{4}\right)}$

15. $r = \dfrac{12}{4 + 6 \sin \theta}$, $r = \dfrac{12}{4 + 6 \sin\left(\theta + \dfrac{5\pi}{4}\right)}$

16. $r = \dfrac{3}{1 - \frac{1}{2} \cos \theta}$, $r = \dfrac{3}{1 - \dfrac{1}{2} \cos\left(\theta - \dfrac{2\pi}{3}\right)}$

17. $r = \dfrac{2}{1 + \sin \theta}$, $r = \dfrac{2}{1 + \dfrac{1}{2} \sin \theta - \dfrac{\sqrt{3}}{2} \cos \theta}$

18. $r = \dfrac{2}{1 - \cos \theta}$, $r = \dfrac{2}{1 - \dfrac{1}{2} \cos \theta - \dfrac{\sqrt{3}}{2} \sin \theta}$

B

In Problems 19 through 30, sketch the conic section described and determine a polar equation of the form

$$r = \frac{ep}{1 \pm e \cos \theta} \quad or \quad r = \frac{ep}{1 \pm e \sin \theta}.$$

Assume that one focus is at the pole.

19. A parabola with directrix at $x = 2$

Certain families leave their imprint in fields of human endeavor: the Bachs in 17th and 18th century music, the Kennedys in 20th century American politics, the Hustons of contemporary film. So it is with the Bernoulli family of Switzerland in mathematics. Over the span of 200 years, this family produced no less than a dozen important mathematicians and scientists.

Two of them, James (1654–1705) and his brother John (1667–1748) were mathematicians of the first order. They were among the first to use calculus to solve problems that had not yielded to the great minds of the past. They were able to solve the brachistochrone problem (see Section 9.3) and the problem of determining the equation of a hanging cable. James (also known as Jacques and Jakob) is given credit for the development of polar coordinates and published the first significant book on mathematical probability. John (also known as Jean and Johann) made significant contributions to trigonometry and essentially wrote the first calculus textbook (the book was published under the name of Marquis de l'Hôpital, a wealthy patron of John, for financial considerations). Among John's students was Leonhard Euler (see the box about Euler near the end of Chapter 4).

The Bernoulli brothers were rather unpleasant individuals. They were often described by their peers as petty, quarrelsome, and crabby. They often pursued the same investigations, but they were bitter adversaries who did not share the fruits of their research with each other. Indeed, when one made an important discovery, he often publically challenged the other to do the same and gloated when the other was unable to match his achievement. John, who had solved an important problem before his brother James, had this to say in a letter to a friend:

The efforts of my brother were without success; for my part, I found the skill . . . to solve it in full. It is true that it robbed me of rest for one full night . . . but the next morning, I ran to my brother who was still struggling miserably with this Gordian Knot . . . .

After John had apparently solved the brachistochrone problem, he challenged James to solve it. Later, evidence arose that John's solution had actually been incorrect and that he tried to publish his brother's subsequent solution as his own. Daniel, John's son, also became a prominent mathematician; he showed promise at an early age. Once Daniel and his father both vied for a prize from the French Academy of Science. Daniel won; John's response was to throw his son out of the house.

Bernoulli Brothers

---

20. A parabola with directrix at $y = -7$

21. A conic section with eccentricity $e = \frac{1}{2}$ and directrix at $y = -2$

22. A conic section with eccentricity $e = \frac{5}{4}$ and directrix at $y = 9$

23. A conic section with vertex at $(0,2)$ and eccentricity $e = 1$

24. A conic section with focus at $(0, 2)$ and eccentricity $e = \frac{1}{2}$

25. A conic section with rectangular equation $x^2 = 4 - 4y$

26. A conic section with rectangular equation $y^2 = 16 + 8x$

27. A conic section with rectangular equation
$5x^2 - 4y^2 - 36x + 36 = 0$

28. A conic section with rectangular equation
$7x^2 + 16y^2 + 6x - 1 = 0$

29. A conic section with eccentricity $e = \frac{1}{2}$ and a focus at $[2, \pi/2]$

30. A conic section with eccentricity $e = \frac{4}{3}$ and a focus at $[-4, \pi]$

31. Sketch the graphs of the equation

$$r = \frac{3e}{1 - e \cos \theta}$$

for $e = \frac{1}{2}$, 1, and 2, all on the same coordinate plane.

32. Sketch the graphs of the equation

$$r = \frac{4e}{1 + e \sin \theta}$$

for $e = \frac{1}{3}$, 1, and 3, all on the same coordinate plane.

33. Show that the graph of

$$r = \frac{ep}{1 - e \sin \theta}$$

is a rotation of the graph of

$$r = \frac{ep}{1 - e \cos \theta}.$$

34. Show that the graph of

$$r = \frac{ep}{1 + e \cos \theta}$$

is a rotation of the graph of

$$r = \frac{ep}{1 - e \cos \theta}$$

35. Use the solar system chart in this section to find the distance that the planet Mercury is from the sun at perihelion and at aphelion (both to the nearest 0.001 AU).

36. Use the solar system chart to find the distance that the comet Halley is from the sun at perihelion and at aphelion (both to the nearest 0.001 AU).

37. Use the solar system chart to find the distance that the asteroid Icarus is from the sun at perihelion and at aphelion (both to the nearest 0.001 AU).

38. Use the solar system chart to find the distance that the comet Kahotek is from the sun at perihelion and at aphelion (both to the nearest 0.001 AU).

39. The graph of

$$r = \frac{6}{4 - 2 \cos \theta}$$

is an ellipse (its graph is shown in Example 2 of this section) with one focus at the origin. Determine the polar coordinates of the other focus.

40. The graph of

$$r = \frac{3}{2 - 3 \cos \theta}$$

is a hyperbola (its graph is shown in Example 3 of this section) with one focus at the origin. Determine the polar coordinates of the other focus.

C

41. Kepler's third law states that the period of a planet (the time to make one complete revolution) is proportional to the length of the semimajor axis raised to the three-halves power. Use the solar system chart to determine the period (to the nearest 0.01 earth years) of all the bodies listed. Compare this with the actual values found in any astronomy reference book.

MISCELLANEOUS EXERCISES

In Problems 1 through 6, determine the rectangular coordinates of the points whose polar coordinates are given.

1. $[8, 2\pi/3]$　　**2.** $[2\sqrt{2}, \pi/4]$

3. $[5, 3\pi/2]$　　**4.** $[2, -\pi/2]$

5. $[-3, 5\pi/4]$　　**6.** $[-4, -\pi/3]$

In Problems 7 through 12, determine the polar representation of the point described, subject to the restrictions given.

7. Point has rectangular coordinates $(-3, 3)$, $0 \le \theta \le \pi$.

8. Point has rectangular coordinates $(3, -3)$,
   $\pi \leq \theta \leq 2\pi$.

9. Point has rectangular coordinates $(0, -8)$,
   $0 \leq \theta \leq 2\pi$ and $r < 0$.

10. Point has rectangular coordinates $(2\sqrt{3}, 2)$,
    $-\pi \leq \theta \leq \pi$ and $r < 0$.

11. Point has rectangular coordinates $(3, -\sqrt{3})$,
    $0 \leq \theta \leq 2\pi$ and $r < 0$.

12. Point has rectangular coordinates $(4, 2\sqrt{2})$,
    $-2\pi \leq \theta \leq 0$ and $r > 0$.

In Problems 13 through 18, sketch the set of points described by the inequalities.

13. $\theta = \pi/2$, $\qquad r \geq 0$

14. $r = 4$, $\qquad -\pi/2 \leq \theta \leq \pi$

15. $0 \leq r \leq 2$, $\qquad -\pi/3 \leq \theta \leq \pi/3$

16. $1 \leq r \leq 3$, $\qquad -\pi \leq \theta \leq \pi/2$

17. $\theta = 5\pi/3$, $\qquad r \geq -3$

18. $r = -3$, $\qquad -\pi \leq \theta \leq \pi$

In Problems 19 through 36, sketch the polar graph of the equation given.

19. $r = 4 \cos \theta$

20. $r = 3 \sin \theta$

21. $r \sin \theta = 6$

22. $r \cos \theta = -2$

23. $r = 3 \sin 2\theta$

24. $r = 4 \cos 3\theta$

25. $r = 2 + \sin \theta$

26. $r = 4 - 2 \cos \theta$

27. $r = 3 + 6 \sin \theta$

28. $r = 2 - 4 \cos \theta$

29. $r = -4 \cos 3\theta$

30. $r = 6 \sin 4\theta$

31. $r = 4 \cos 3(\theta - \pi/3)$

32. $r = -2 \sin(\theta + \pi/4)$

33. $r = 3 + 6 \sin(\theta - \pi/4)$

34. $r = 2 - 4 \cos(\theta + \pi/3)$

35. $r^2 = 4 \sin 2\theta$

36. $r^2 = -16 \cos 2\theta$

In Problems 37 through 54, sketch the curve represented by the parametric equations (be sure to indicate the orientation) and write a corresponding rectangular equation by eliminating the parameter.

37. $x = t + 2$, $\quad y = \frac{1}{2}t - 3$, $\qquad -4 \leq t \leq 4$

38. $x = 6 + \frac{1}{3}t$, $\quad y = \frac{1}{6}t$, $\qquad -12 \leq t \leq 9$

39. $x = t^2 + 4$, $\quad y = 2 - t$, $\qquad -\infty < t < \infty$

40. $x = t + 3$, $\quad y = t^2 + 6t$, $\qquad -\infty < t < \infty$

41. $x = t^2$, $\quad y = \sqrt{t} - 2$, $\qquad 0 \leq t \leq 2$

42. $x = t^2$, $\quad y = 9 - t^4$, $\qquad -\sqrt{3} \leq t \leq \sqrt{3}$

43. $x = \cos t$, $\quad y = \sin t$, $\qquad -\pi/2 \leq t \leq \pi/2$

44. $x = \sin t$, $\quad y = \cos t$, $\qquad -\pi/2 \leq t \leq \pi/2$

45. $x = \cos 2t$, $\quad y = \sin 2t$, $\qquad -\pi/2 \leq t \leq \pi/2$

46. $x = \cos \frac{1}{2}t$, $\quad y = \sin \frac{1}{2}t$, $\qquad -\pi \leq t \leq \pi$

47. $x = \sin t$, $\quad y = \sin^2 t$, $\qquad -\dfrac{\pi}{2} \leq t \leq \dfrac{\pi}{2}$

48. $x = \tan t$, $\quad y = \cot t$, $\qquad 0 < t < \pi/2$

49. $x = \cos 2t$, $\quad y = \sin t$, $\qquad -\pi/2 \leq t \leq \pi/2$

50. $x = \tan^2 t$, $\quad y = \sec^2 t$, $\qquad 0 \leq t < \pi/2$

51. $x = e^{2t}$, $\quad y = e^t$, $\qquad \ln 2 \leq t \leq \ln 5$

52. $x = 2^t$, $\quad y = 2^{-t}$, $\qquad 0 \leq t \leq 3$

53. $x = \ln t$, $\quad y = \ln(t^2)$, $\qquad 1 \leq t \leq e^2$

54. $x = \ln\sqrt{t}$, $\quad y = \ln t$, $\qquad 1 \leq t \leq e^4$

In Problems 55 through 60, sketch the curve described and determine a parametric representation for it. Be sure to give the domain of the parameter.

55. The line segment from $(0, 0)$ to $(5, 5)$

56. The line segment from $(3, 2)$ to $(1, 6)$

57. The graph of $y = 2x^2 - 8$ from $(-2, 0)$ to $(2, 0)$

58. The graph of $y = \sqrt{x} + 5$ from $(0, 5)$ to $(9, 8)$

59. The circle $x^2 + y^2 = 25$ oriented counterclockwise

60. The ellipse $4x^2 + 25y^2 = 100$ oriented counterclockwise

In Problems 61 through 69, sketch the conic section represented by the polar equation. Identify the type of conic section (parabola, ellipse, or hyperbola) and determine the polar equation of the directrix.

61. $r = \dfrac{4}{1 - \cos \theta}$

62. $r = \dfrac{6}{1 + \sin \theta}$

63. $r = \dfrac{4}{1 + 2 \sin \theta}$

64. $r = \dfrac{10}{1 + \frac{2}{5} \sin \theta}$

65. $r(1 + 2 \cos \theta) = 16$

66. $r(2 - \cos \theta) = 8$

67. $r = \dfrac{4}{1 - \cos\left(\theta + \dfrac{\pi}{4}\right)}$

68. $r = \dfrac{6}{1 + \sin\left(\theta - \dfrac{\pi}{4}\right)}$

69. $r = \dfrac{4}{1 + 2 \sin\left(\theta - \dfrac{3}{2}\pi\right)}$

In Problems 70 through 75, use the graphs in Figure 58 to sketch the graph of the given parametric equations.

70. $x = f(t), \quad y = g(t)$

71. $x = -f(t), \quad y = g(t)$

72. $x = f(t), \quad y = -g(t)$

73. $x = -f(t), \quad y = -g(t)$

74. $x = f(t) + 2, \quad y = g(t) + 3$

75. $x = f(t) - 1, \quad y = g(t) - 4$

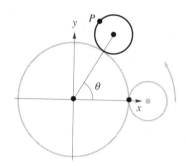

FIGURE 59

76. Make a rough sketch of an epicycloid for the following:

a) $R = 4$ **b)** $R = 2$ **c)** $R = 1$

77. A parametric representation for the epicycloid described is

$$\begin{cases} x = (1 + R)\cos\theta - \cos[(1 + R)\theta] \\ y = (1 + R)\sin\theta - \sin[(1 + R)\theta]. \end{cases}$$

The epicycloid for which $R = 2$ is called a **nephroid.** Show that this parametric representation for a nephroid can be written as

$$\begin{cases} x = 6\cos\theta - 4\cos^3\theta \\ y = 4\sin^3\theta. \end{cases}$$

78. Consider the nephroid described in Problem 77.

a) Use Problem 77 to show that
$x^2 + y^2 = 16 - 12\cos^2\theta.$

b) Use part (a) to show that $(x^2 + y^2 - 4)^3 = 108y^2$ is a nonparametric representation of a nephroid.

79. Consider the epicycloid for which $R = 1$.

a) Show that this epicycloid has the parametric representation

$$\begin{cases} x = 2\cos\theta - \cos 2\theta \\ y = 2\sin\theta - \sin 2\theta. \end{cases}$$

b) Show that $(x - 1)^2 + y^2 = 4(1 - \cos\theta)^2$

c) Show that this epicycloid is a cardioid.

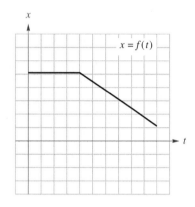

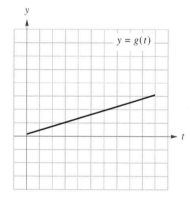

FIGURE 58

In Problems 76 through 79, consider a disk with radius 1 that rolls without slipping along the outside of a fixed disk centered at the origin with radius R. The curve traced out by a fixed point P on the circumference of the rolling disk is called an **epicycloid** *(Figure 59).*

In Problems 80 through 83, consider a disk with radius 1 that rolls without slipping along the inside of a fixed disk centered at the origin with radius R. The curve traced out by a fixed point P on the circumference of the rolling disk is called an **hypocycloid** *(Figure 60).*

80. Make a rough sketch of the hypocycloid for the following:

    a) $R = 4$ **b)** $R = 3$ **c)** $R = 2$

81. A parametric representation for the hypocycloid described is

$$\begin{cases} x = (R - 1)\cos\theta + \cos[(R - 1)\theta] \\ y = (R - 1)\sin\theta - \sin[(R - 1)\theta]. \end{cases}$$

The hypocycloid for which $R = 4$ is called an **asteroid.** Show that this parametric representation for an asteroid can be written as

$$\begin{cases} x = 4\cos^3\theta \\ y = 4\sin^3\theta. \end{cases}$$

82. Show that the asteroid described in Problem 81 is the graph of the equation $x^{2/3} + y^{2/3} = 4^{2/3}$.

83. Show that if $R = 2$, the "hypocycloid" is actually the line segment with end points $(-2, 0)$ and $(2, 0)$.

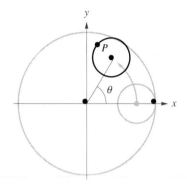

FIGURE 60

CHAPTER 10

SYSTEMS OF EQUATIONS AND INEQUALITIES

LINEAR SYSTEMS OF EQUATIONS

Quite often, the solution of a particular problem involves not just one equation with one unknown but at least two equations in at least two unknowns. Such a set of equations is called a **system of equations.** If all the equations of the system are linear, then the system is a **linear system of equations.**

In the first two sections of this chapter, we investigate methods of solving these systems. Even though we limit ourselves to no more than three equations and three unknowns, these methods also apply to larger systems of equations.

Consider this linear system of equations in two unknowns:

$$\begin{cases} y = 2x - 4 \\ y = -\frac{1}{3}x + 3. \end{cases}$$

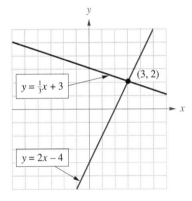

FIGURE 1

A **solution** of a system such as this is an ordered pair of numbers (x, y) that makes both equations true. We can get a good idea of what is going on by sketching the graphs of both equations (Figure 1). Since the solutions of an equation in two unknowns are represented by the points on its graph, it should seem reasonable that the solution of this system of equations is represented by the point that is the intersection of the two lines. From the sketch of this system, this point of intersection appears to be $(3, 2)$. We can check this directly by substituting $x = 3$ and $y = 2$ into each of the equations and verifying that these values make the equations true:

$$y = 2x - 4$$
$$(2) \overset{?}{=} 2(3) - 4$$
$$2 = 6 - 4. \quad \text{True}$$

and

$$y = -\frac{1}{3}x + 3$$
$$(2) \overset{?}{=} -\frac{1}{3}(3) + 3$$
$$2 = -1 + 3. \quad \text{True}$$

549

In general, the graphs of two lines on the same coordinate plane fall into one of three cases.

Systems of Two Equations of Two Unknowns

Consider a system of two equations in two unknowns:

Case 1 If the graphs of the two equations intersect at exactly one point, then the system is **consistent.** In this case, the system has exactly one solution.

Case 2 If the graphs of the equations are parallel and have no points of intersection, then the system is **inconsistent.** In this case, the system has no solutions.

Case 3 If the graphs of the equations coincide (this means they are actually the same line), then the system is **dependent.** In this case, the system has infinitely many solutions.

Example 1 shows each of these types of systems.

EXAMPLE 1 Sketch the graph of the system and determine if the system is consistent, inconsistent, or dependent.

a) $\begin{cases} 2x + 3y = 6 \\ x - 3y = 8 \end{cases}$ b) $\begin{cases} y = \frac{1}{2}x + 4 \\ y = \frac{1}{2}x - 2 \end{cases}$ c) $\begin{cases} x + y = 4 \\ -2x - 2y = -8 \end{cases}$

SOLUTION

a) By writing each of the equations in slope-intercept form

$$\begin{cases} y = -\frac{2}{3}x + 2 \\ y = \frac{1}{3}x - \frac{8}{3} \end{cases}$$

The point of intersection appears to be $(5, -1)$. However, this ordered pair fails to check in the first equation of the system

we can sketch the graph of this system. Since the slopes of the lines are not the same, the lines are not parallel; they intersect at exactly one point. This system is consistent (Figure 2).

b) The equations of this system are already written in slope-intercept form. Since they have the same slope but different y-intercepts, they are distinct parallel lines with no intersection. This system is inconsistent (Figure 3).

c) Writing each of the equations in slope-intercept form makes it apparent that these equations are equivalent. They represent the same line, namely, $y = -x + 4$. Any point on this line represents a solution to this system. This system is dependent (Figure 4).

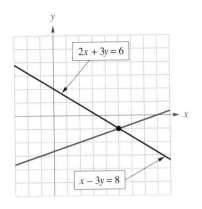

FIGURE 2

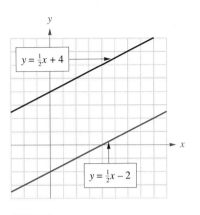

FIGURE 3

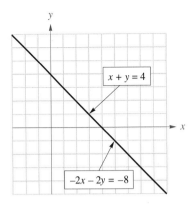

FIGURE 4

In general, it is not always easy or even possible to discover the solution of a system of two equations in two unknowns by examining its graph. To determine an exact solution, we must use algebraic methods. We introduce two such methods: the **substitution method** and the **elimination method.**

Substitution Method

The substitution method begins with using one of the equations to express one of the unknowns as an algebraic expression in terms of the other unknown. This expression is then substituted into the other equation to get an equation in one unknown. This equation, in turn, can be solved with the usual techniques of solving equations.

EXAMPLE 2   Use the substitution method to solve the system:

$$\begin{cases} 2x + 3y = 6 \\ x - 3y = 8. \end{cases}$$

(This is the system discussed in part (a) of Example 1.)

SOLUTION Solving the second equation for x yields an expression in y:

$$x = 8 + 3y.$$

Next, we substitute this expression into the first equation for x and solve the resulting equation for y:

$$2x + 3y = 6$$

$$2(8 + 3y) + 3y = 6$$

$$16 + 9y = 6$$

$$y = -\tfrac{10}{9}.$$

When selecting a variable for which to solve, try to find one with a coefficient of 1; this way, you may avoid messy fractions.

To find the corresponding value of x, we can substitute this value of y into $x = 8 + 3y$:

$$x = 8 + 3(-\tfrac{10}{9}) = \tfrac{14}{3}.$$

Our solution to the system therefore is $x = \tfrac{14}{3}$ and $y = -\tfrac{10}{9}$. Expressed as an ordered pair, the solution is $(\tfrac{14}{3}, -\tfrac{10}{9})$.

The sketch of the graph of this system in part (a) of Example 1 suggests that this answer is reasonable. Notice, however, that one would have to be very lucky to guess this solution from the graph alone.

Just as a system of two equations in two unknowns has an ordered pair of real numbers (x, y) as a solution, a system of three equations in three unknowns such as

$$\begin{cases} x + y + z = 6 \\ 2x - y + 3z = 5 \\ 3x + 4y - 7z = 1 \end{cases}$$

has an ordered triple (x, y, z) as a solution. For an ordered triple (x, y, z) to be a solution of such a system, it must be a solution to each of the three equations.

We can use the technique of substitution to solve these systems, as illustrated in the next example.

EXAMPLE 3   Use the substitution method to solve the system:

$$\begin{cases} x + y + z = 6 \\ 2x - y + 3z = 5 \\ 3x + 4y - 7z = 1. \end{cases}$$

SOLUTION We start by solving the first equation for x:

$$x = 6 - y - z.$$

Substituting $6 - y - z$ for x in the second and third equations gives us a system in two equations in two unknowns.

$$\begin{cases} 2x - y + 3z = 5 \\ 3x + 4y - 7z = 1 \end{cases}$$

$$\begin{cases} 2(6 - y - z) - y + 3z = 5 \\ 3(6 - y - z) + 4y - 7z = 1. \end{cases}$$

Simplifying and collecting terms gives

$$\begin{cases} -3y + z = -7 \\ y - 10z = -17. \end{cases}$$

Pause here and notice what we have accomplished. We have reduced this problem to solving a system of two equations in two unknowns. We can now proceed as we did in Example 2.

Solving the first of these two equations for z, we get

$$z = -7 + 3y.$$

Substituting into the second equation gives

$$y - 10(-7 + 3y) = -17$$
$$-29y + 70 = -17$$
$$-29y = -87$$
$$y = 3.$$

It follows that

$$z = -7 + 3y = -7 + 3(3) = 2$$

and

$$x = 6 - y - z = 6 - (3) - (2) = 1.$$

The solution is $(1, 3, 2)$.

The elimination method is often used for solving large systems of equations. Before we discuss this method, we investigate how to solve a special set of linear systems.

Elimination Method

EXAMPLE 4   Solve these systems.

a) $\begin{cases} 3x + 4y = 16 \\ \quad\quad 2y = 5 \end{cases}$

b) $\begin{cases} 2x + 5y - 3z = -1 \\ \quad\quad 3y - z = 2 \\ \quad\quad\quad\quad 2z = 8 \end{cases}$

SOLUTION

a) From the second equation of the system, we can deduce that $y = \frac{5}{2}$. Substituting this back into the first equation gives

$$3x + 4(\tfrac{5}{2}) = 16$$
$$3x + 10 = 16$$
$$3x = 6$$
$$x = 2.$$

The solution to the system is $(2, \frac{5}{2})$. We leave it to you to verify this solution.

b) Starting with the third equation of the system, we discover that $z = 4$. Substituting this into the second equation in the same manner as part (a) of this example gives

$$3y - (4) = 2 \Rightarrow y = 2.$$

Substituting into the first equation yields the value of x:

$$2x + 5(2) - 3(4) = -1 \Rightarrow x = \tfrac{1}{2}.$$

The solution of the system is $(\frac{1}{2}, 2, 4)$. Again, you should pause and verify this solution.

In general, solving a system of equations is not as easy as in Example 4. These two systems are examples of **upper triangular systems.** The name suggests the position of the nonzero terms on the left side of the equations:

$$\begin{cases} 3x + 4y = 16 \\ 2y = 5 \end{cases} \qquad \begin{cases} 2x + 5y - 3z = -1 \\ 3y - z = 2 \\ 2z = 8 \end{cases}.$$

In each case, the bottom equation can be solved easily for one of the variables. That value, in turn, can be substituted back into the preceding equation to determine the value of another variable. This process can be continued until the solution is known. This process is called **back-substitution.**

Now, here is the goal of the elimination method. Given a system of equations, we seek an equivalent system of equations that is upper triangular (two systems are **equivalent** if they have the same solution). Once such an upper-triangular system is found, the solution can be determined by back-substitution. To do this, we use three basic operations on a system, each of which yields an equivalent system.

Equivalent Systems

Each of these operations on a system of equations yields an equivalent system of equations:

1) Interchanging any two equations of the system.

2) Replacing an equation with a nonzero multiple of itself (a **multiple** of an equation is the result of multiplying each side of the equation by the same real number).

3) Replacing an equation with the sum of that equation and a multiple of another equation of the system.

The next three examples will help you see how these operations on systems work. You should convince yourself that they do indeed give systems with the same solution as the original.

EXAMPLE 5   Use the elimination method to solve this system:

$$\begin{cases} 2x + 5y = -2 \\ 3x - 4y = 20. \end{cases}$$

SOLUTION To make this system upper triangular, we need to come up with an equivalent system in which the $3x$ term in the second equation is eliminated. We do this by multiplying the first equation by 3 and the second equation by -2 (using operation 2 in the preceding box).

$$\begin{cases} 6x + 15y = -6 \\ -6x + 8y = -40. \end{cases}$$

Then by adding the two equations

$$+\begin{cases} 6x + 15y = -6 \\ -6x + 8y = -40 \\ \hline 23y = -46 \end{cases}$$

and then replacing the second equation in the original system by this new equation (using operation 3 in the box), we obtain

$$\begin{cases} 2x + 5y = -2 \\ -23y = 46. \end{cases}$$

The reason we choose 3 and -2 by which to multiply the first and second equations, respectively, is that this forces the coefficients of x in these equations to have opposite signs. When the equations are added together, the sum has no x term.

Now that we have arrived at an equivalent system that is upper triangular, we solve for y in the second equation and back-substitute to find x, as we did in Example 4.

$$-23y = 46 \ \Rightarrow y = -2$$

$$2x + 5(-2) = -2 \Rightarrow x = 4$$

Our solution is $(4, -2)$.

In the case of three equations in three unknowns, it is helpful to adopt a notation to indicate which of the operations is being used. In this notation, the first, second, and third equations are represented by E_1, E_2, and E_3, respectively. Then, for example, multiplying the second equation by four is denoted as $4E_2 \rightarrow E_2$. Adding three times the first equation to the third equation is represented by $3E_1 + E_3 \rightarrow E_3$.

EXAMPLE 6   Use the elimination method to solve this system:

$$\begin{cases} 2x - 3y + 5z = 29 \\ x - 2y + z = 7 \\ 3x - 7y + 2z = 16. \end{cases}$$

SOLUTION   It is to our advantage to have a 1 as the x-coefficient in the first equation. We do so by interchanging the first and second equations (we denote this by $E_1 \rightarrow E_2$ and $E_2 \rightarrow E_1$):

$$\begin{matrix} E_2 \rightarrow E_1 \\ E_1 \rightarrow E_2 \end{matrix} \begin{cases} x - 2y + z = 7 \\ 2x - 3y + 5z = 29 \\ 3x - 7y + 2z = 16. \end{cases}$$

To make this system upper triangular, we need to eliminate the x term in the second equation, the x term in the third equation, and the y term in the third equation. We start by adding -2 times the first equation to the second equation ($-2E_1 + E_2 \rightarrow E_2$) and by adding -3 times the first equation to the third equation ($-3E_1 + E_3 \rightarrow E_3$):

$$\begin{matrix} -2E_1 + E_2 \rightarrow E_2 \\ -3E_1 + E_3 \rightarrow E_3 \end{matrix} \begin{cases} x - 2y + z = 7 \\ y + 3z = 15 \\ -y - z = -5. \end{cases}$$

Now we eliminate the y term in the third equation. To do this, we add the second equation to the third equation:

$$\begin{matrix} \\ \\ -E_2 + E_3 \rightarrow E_3 \end{matrix} \begin{cases} x - 2y + z = 7 \\ y + 3z = 15 \\ 2z = 10. \end{cases}$$

This upper triangular system is equivalent to the original system. Using back-substitution as we did in Example 4, you should be able to determine (and then verify) that the solution is $(2, 0, 5)$.

EXAMPLE 7   Use the elimination method to solve this system:

$$\begin{cases} \frac{1}{3}x - y + \frac{2}{3}z = -1 \\ 2x + y + 3z = 16 \\ -\frac{3}{2}x + y = \frac{17}{2}. \end{cases}$$

SOLUTION   Before starting in earnest to transform this system to an upper-triangular system, it will help to multiply the first equation by 3 and the third equation by 2 to avoid fractional coefficients.

$$3E_1 \to E_1 \begin{cases} x - 3y + 2z = -3 \\ 2x + y + 3z = 16 \\ 2E_3 \to E_3 \end{cases} \begin{aligned} x - 3y + 2z &= -3 \\ 2x + y + 3z &= 16 \\ -3x + 2y \quad\;\; &= 17. \end{aligned}$$

Eliminating the x terms in the second and third equations yields

$$\begin{matrix} & \begin{cases} x - 3y + 2z = -3 \\ -2E_1 + E_2 \to E_2 \\ 3E_1 + E_3 \to E_3 \end{cases} & \begin{aligned} x - 3y + 2z &= -3 \\ 7y - z &= 22 \\ -7y + 6z &= 8. \end{aligned} \end{matrix}$$

And finally,

$$E_2 + E_3 \to E_3 \begin{cases} x - 3y + 2z = -3 \\ 7y - z = 22 \\ 5z = 30. \end{cases}$$

By back-substituting, we obtain

$$5z = 30 \Rightarrow z = 6$$
$$7y - (6) = 22 \Rightarrow y = 4$$
$$x - 3(4) + 2(6) = -3 \Rightarrow x = -3.$$

The solution to the system is $(-3, 4, 6)$.

Applications of Linear Systems of Equations

The importance of linear systems lies in their application in solving problems in mathematics and science, such as in the next two examples.

EXAMPLE 8 Find an equation of the form $y = ax^2 + bx + c$ whose graph is a parabola passing through $(1, -5)$, $(-3, 11)$, and $(0, -4)$.

SOLUTION To find this equation we need to determine the values of a, b, and c. If the graph passes through the point $(1, -5)$, then

$$-5 = a(1)^2 + b(1) + c \quad \text{or} \quad a + b + c = -5.$$

Likewise, the point $(-3, 11)$ gives the equation

$$11 = a(-3)^2 + b(-3) + c \quad \text{or} \quad 9a - 3b + c = 11$$

and the point $(0, -4)$ gives the equation

$$-4 = a(0)^2 + b(0) + c \quad \text{or} \quad c = -4.$$

Writing this as a system, we have

$$\begin{cases} a + b + c = -5 \\ 9a - 3b + c = 11 \\ \phantom{9a - 3b + {}} c = -4. \end{cases}$$

We employ the substitution method to solve the system:

$$\begin{cases} a + b + (-4) = -5 \\ 9a - 3b + (-4) = 11 \end{cases} \quad \text{or} \quad \begin{cases} a + b = -1 \\ 9a - 3b = 15. \end{cases}$$

From the first equation, $a = -b - 1$. Substituting this into the second equation gives

$$9(-b - 1) - 3b = 15 \Rightarrow b = -2$$

so

$$a = -(-2) - 1 = 1.$$

The solution is $a = 1$, $b = -2$, and $c = -4$. The equation of the parabola is $y = x^2 - 2x - 4$ (Figure 5).

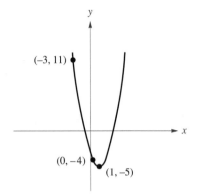

(−3, 11)

(0, −4)

(1, −5)

FIGURE 5

═══

EXAMPLE 9   Garden fertilizer has two important components, nitrogen and phosphate. Fertilizer P is 20% nitrogen and 2% phosphate, fertilizer Q is 15% nitrogen and 6% phosphate, and fertilizer R is 30% nitrogen and 0% phosphate. How many pounds of each should be mixed to make 20 pounds of fertilizer that is 19% nitrogen and 4% phosphate?

SOLUTION   Since the total amount of fertilizer is 20 pounds, it should seem clear that

$$\binom{\text{pounds}}{\text{of } P} + \binom{\text{pounds}}{\text{of } Q} + \binom{\text{pounds}}{\text{of } R} = \binom{\text{pounds}}{\text{of mix}}$$

or, using P, Q, and R to represent these quantities, we have

$$P + Q + R = 20.$$

Now, consider the nitrogen in each of the fertilizers:

$$\begin{pmatrix} \text{pounds} \\ \text{of nitrogen} \\ \text{in } P \end{pmatrix} + \begin{pmatrix} \text{pounds} \\ \text{of nitrogen} \\ \text{in } Q \end{pmatrix} + \begin{pmatrix} \text{pounds} \\ \text{of nitrogen} \\ \text{in } R \end{pmatrix} = \begin{pmatrix} \text{pounds} \\ \text{of nitrogen} \\ \text{in mix} \end{pmatrix}$$

or

$$0.20\binom{\text{pounds}}{\text{of } P} + 0.15\binom{\text{pounds}}{\text{of } Q} + 0.30\binom{\text{pounds}}{\text{of } R} = 0.19\binom{\text{pounds}}{\text{of mix}}$$

or

$$0.20P + 0.15Q + 0.30R = 3.8.$$

In a similar fashion, the phosphate amounts determine the equation

$$0.02P + 0.06Q = 0.8.$$

Putting these three equations together gives a system to solve:

$$\begin{cases} P + Q + R = 20 \\ 0.20P + 0.15Q + 0.30R = 3.8 \\ 0.02P + 0.06Q = 0.8. \end{cases}$$

Multiplying the second and third equations by 100 eliminates the decimals:

$$\begin{matrix} & \\ 100E_2 \rightarrow E_2 \\ 100E_3 \rightarrow E_3 \end{matrix} \begin{cases} P + Q + R = 20 \\ 20P + 15Q + 30R = 380 \\ 2P + 6Q = 80 \end{cases}$$

Using the elimination method yields

$$\begin{matrix} & \\ -20E_1 + E_2 \rightarrow E_2 \\ -2E_1 + E_3 \rightarrow E_3 \end{matrix} \begin{cases} P + Q + R = 20 \\ -5Q + 10R = -20 \\ 4Q - 2R = 40 \end{cases}$$

$$-\tfrac{1}{5}E_2 \rightarrow E_2 \begin{cases} P + Q + R = 20 \\ Q - 2R = 4 \\ 4Q - 2R = 40 \end{cases}$$

$$-4E_2 + E_3 \rightarrow E_3 \begin{cases} P + Q + R = 20 \\ Q - 2R = 4 \\ 6R = 24. \end{cases}$$

By back-substitution, the solution is $P = 4$, $Q = 12$, and $R = 4$. Thus, 4 lb of fertilizer P, 12 lb of fertilizer Q, and 4 lb of fertilizer R are required for the mix.

EXERCISE SET 10.1

A

In Problems 1 through 12, sketch the graph of the system and state whether the system is consistent, inconsistent, or dependent. If the system is consistent, determine the solution by using the substitution method and label the point of intersection.

1. $\begin{cases} x + y = 2 \\ x - y = 4 \end{cases}$

2. $\begin{cases} y = -2x - 3 \\ y = -2x + 5 \end{cases}$

3. $\begin{cases} y = 2x + 8 \\ x - \tfrac{1}{2}y = -4 \end{cases}$

4. $\begin{cases} x = 3 \\ x + y = 7 \end{cases}$

5. $\begin{cases} y = -2x - 5 \\ y = \tfrac{1}{2}x + 5 \end{cases}$

6. $\begin{cases} 3x - 5y = 3 \\ 15x + 5y = 21 \end{cases}$

7. $\begin{cases} \tfrac{1}{2}x + \tfrac{1}{6}y = 1 \\ \tfrac{1}{3}x + \tfrac{1}{3}y = 1 \end{cases}$

8. $\begin{cases} 2x - 3y = -1 \\ 6x + 6y = 3 \end{cases}$

9. $\begin{cases} 12x + 8y = 5 \\ -9x - 6y = 5 \end{cases}$

10. $\begin{cases} \tfrac{1}{9}x + \tfrac{1}{6}y = 1 \\ y = -\tfrac{2}{3}x + 6 \end{cases}$

11. $\begin{cases} y = -2x - 3 \\ y = \tfrac{1}{2}x + 5 \end{cases}$

12. $\begin{cases} 2x = 4y - 9 \\ x - 2y = 4 \end{cases}$

In Problems 13 through 18, solve the system by the substitution method.

13. $\begin{cases} 2x + y \quad\ = 10 \\ x - y \quad\ = 2 \\ y + z = 3 \end{cases}$

14. $\begin{cases} x \quad\ + 2z = -2 \\ 2x \quad\ - 4z = -4 \\ 3x + y \quad\ = 0 \end{cases}$

15. $\begin{cases} x + 2y - z = 2 \\ 2x - y + z = 5 \\ 3x - 2y + 2z = 5 \end{cases}$

16. $\begin{cases} x + y - 3z = -9 \\ 2x \quad\ + z = -2 \\ 3x - 2y + 2z = 0 \end{cases}$

17. $\begin{cases} 2x - y + z = 4 \\ 2x + 2y + 3z = 3 \\ 6x - 9y - 2z = 17 \end{cases}$

18. $\begin{cases} 3x + y - 2z = 4 \\ -5x \quad\ + 2z = 5 \\ -7x - y + 3z = -2 \end{cases}$

In Problems 19 through 24, note that the system given is upper-triangular. Solve the system by back-substitution.

19. $\begin{cases} -2x + 3y = 80 \\ 2y = 32 \end{cases}$

20. $\begin{cases} 3x - 9y = 6 \\ 7y = 0 \end{cases}$

21. $\begin{cases} 7x - 2y = 2 \\ 3y = 4 \end{cases}$

22. $\begin{cases} 3x - 4y + 2z = 5 \\ 2y \quad\ = 6 \\ -2z = 5 \end{cases}$

23. $\begin{cases} x - 3y + 4z = 6 \\ 2y + z = 7 \\ 3z = 15 \end{cases}$

24. $\begin{cases} -4x - 4y + 4z = 4 \\ -28y + 24z = 80 \\ 31z = 31 \end{cases}$

In Problems 25 through 30, solve the system by finding an equivalent upper-triangular system and then back-substituting.

25. $\begin{cases} x + 3y = 5 \\ -4x + y = -7 \end{cases}$

26. $\begin{cases} 2x - 6y = -8 \\ x + 3y = 2 \end{cases}$

27. $\begin{cases} 2x + 3y = 6 \\ -4x + y = -2 \end{cases}$

28. $\begin{cases} -7x - 4y = -8 \\ 12x + 7y = 2 \end{cases}$

29. $\begin{cases} 2x - 4y = -6 \\ 3x - 5y = 3 \end{cases}$

30. $\begin{cases} 9x + 8y = -6 \\ 3x + 2y = 3 \end{cases}$

B

In Problems 31 through 42, solve the system by finding an equivalent upper-triangular system and then back-substituting. Use the notation introduced in the section to show your steps (for example, $2E_1 + E_2 \to E_2$).

31. $\begin{cases} x + y + z = 3 \\ y + 3z = 16 \\ 4y + 5z = 29 \end{cases}$

32. $\begin{cases} x + 2y - z = 11 \\ 2x - y + 5z = 18 \\ 3z = 15 \end{cases}$

33. $\begin{cases} x - 2y + 4z = -2 \\ -y + z = -3 \\ 3x - 3y + z = 3 \end{cases}$

34. $\begin{cases} 4x + 3y - z = 3 \\ x + y = 4 \\ x + 2z = 7 \end{cases}$

35. $\begin{cases} x + 2y = 0 \\ x + 3y - z = 4 \\ x + 3y + 2z = -2 \end{cases}$

36. $\begin{cases} x - 2y + 3z = 7 \\ 2x + y + z = 4 \\ 3x - 2y + 2z = -10 \end{cases}$

37. $\begin{cases} x + 2y - z = -3 \\ 2x - 4y + z = -7 \\ -2x + 2y - 3z = 4 \end{cases}$

38. $\begin{cases} 4x + 2y - z = 0 \\ x + 3y + 2z = 0 \\ x + y + 3z = 4 \end{cases}$

39. $\begin{cases} 2x + 3y + z = 3 \\ -4x + y + 5z = 10 \\ x - y + z = 18 \end{cases}$

40. $\begin{cases} 2x + 3y + z = 6 \\ -4x + y + 5z = -2 \\ x - y + z = 0 \end{cases}$

41. $\begin{cases} \frac{1}{2}x - \frac{3}{4}y - \frac{1}{4}z = \frac{11}{4} \\ \frac{1}{2}x - y - 2z = -2 \\ \frac{3}{2}x - \frac{1}{2}y - z = 4 \end{cases}$

42. $\begin{cases} \frac{1}{2}x + \frac{1}{3}y + \frac{1}{6}z = 1 \\ x + \frac{1}{4}y - \frac{1}{4}z = 1 \\ y + 2z = 3 \end{cases}$

43. Solve the system

$$\begin{cases} \dfrac{2}{x} - \dfrac{1}{y} = \dfrac{1}{2} \\[2mm] \dfrac{1}{x} + \dfrac{2}{y} = \dfrac{2}{3}. \end{cases}$$

(Hint: Even though this is not a linear system, you can solve it using the techniques of this section. Let $u = 1/x$ and $v = 1/y$. Solve the resulting linear system in u and v and then determine the values of x and y.)

44. Solve the system

$$\begin{cases} x^2 + y^2 = 25 \\ x^2 - y^2 = 7. \end{cases}$$

(Hint: See Problem 43. Let $u = x^2$ and $v = y^2$.)

45. Solve the system

$$\begin{cases} x^2 + y^2 = 169 \\ x^2 - y^2 = 119. \end{cases}$$

(Hint: See Problem 43. Let $u = x^2$ and $v = y^2$.)

46. Solve the system

$$\begin{cases} \dfrac{1}{x} + \dfrac{1}{y} + \dfrac{1}{z} = 1 \\[2mm] \dfrac{2}{x} \quad + \dfrac{6}{z} = 2 \\[2mm] \dfrac{4}{x} - \dfrac{3}{y} \quad = 1. \end{cases}$$

(Hint: See Problem 43. Let $u = 1/x$, $v = 1/y$, and $w = 1/z$.)

47. Solve the system

$$\begin{cases} \cos^{-1}(x) + \sin^{-1}(y) + \tan^{-1}(z) = \pi/4 \\ 2\cos^{-1}(x) - \sin^{-1}(y) - \tan^{-1}(z) = 3\pi/4 \\ \cos^{-1}(x) \qquad\quad + \tan^{-1}(z) = \pi/12. \end{cases}$$

(Hint: See Problem 43. Let $u = \cos^{-1}(x)$, $v = \sin^{-1}(y)$ and $w = \tan^{-1}(z)$.)

48. Find an equation of the form $y = ax^2 + bx + c$ whose graph is a parabola passing through $(1, 2)$, $(-2, -7)$, and $(2, -3)$.

49. Find an equation of the form $x^2 + y^2 + ax + by + c = 0$ whose graph is a circle passing through $(1, 0)$, $(0, 1)$, and $(1, -2)$.

50. Find a function of the form $f(\theta) = a \sin \theta + b \cos \theta$ such that $f(\pi/4) = 4$ and $f(3\pi/4) = 2$.

51. Find a polynomial function of degree three whose graph passes through $(1, -1)$, $(3, -9)$, $(-3, 63)$, and the origin.

52. A coffee company wishes to blend three coffees: Colombian ($\$1.20$/lb), Kenyan ($\$2.00$/lb), and Peruvian ($\$1.00$/lb). The final blend is to weigh 5000 lb and cost $\$1.66$/lb. There must also be twice

as much Kenyan coffee as Colombian coffee in the blend. Find the amount of each coffee to be blended.

53. A mixture of nuts is to be made of peanuts, filberts, and cashews, which cost $\$0.80$/lb, $\$1.60$/lb, and $\$2.40$/lb. The final blend is to weigh 100 lb and cost $\$1.32$/lb. Furthermore, the combined weight of the cashews and the filberts must be as much as that of the peanuts. Find the amount of each type of nut to be mixed.

54. An individual invests $\$10,000$ in three stocks that pay dividends of 10%, 8%, and 6%. Given that the total return on the investment is $\$860$ and that the amount invested at 10% is twice the amount invested at 6%, find the amount invested in each stock.

55. A three-digit natural number is such that the sum of the digits is 6. If the digits are reversed, the resulting number is 99 larger than the original number. Also, the tens digit is equal to the sum of the ones digit and the hundreds digit. Find the original number.

56. A three-digit natural number is such that the sum of the digits is 12. If the digits are reversed, the resulting number is 198 less than the original number. Also, the hundreds digit is equal to the sum of the ones digit and the tens digit. Find the original number.

C

57. Consider the linear system

$$\begin{cases} a_{11}x + a_{12}y + a_{13}z = 0 \\ a_{21}x + a_{22}y + a_{23}z = 0 \\ a_{31}x + a_{32}y + a_{33}z = 0. \end{cases}$$

a) Show that this system always has the trivial solution $(0, 0, 0)$.

b) Show that if (x_0, y_0, z_0) is a solution, then $(2x_0, 2y_0, 2z_0)$ is also a solution.

c) Explain why there are infinitely many solutions to this system if there is one solution other than the trivial solution.

S E C T I O N 10.2 MATRIX SOLUTIONS OF LINEAR SYSTEMS

Solving a system of equations by either the substitution method or the elimination method can be tedious, especially for a system with larger numbers or fractions. As you have probably already discovered, the least bit of carelessness on the solver's part usually leads to a fatal error in the solution. As with most situations of this type, the introduction of appropriate mathematical notation can help ease this tedium, focus our attention on the important elements of the problem, and help us avoid the careless errors.

The important elements of a system of three equations in three unknowns, for example, are the coefficients of the variables and the constants. These real numbers form an array with three rows (one for each equation in the system) and four columns (one for each variable and one for the constants). For example, this system has a corresponding array:

$$\begin{cases} 2x + 3y - 7z = 11 \\ 4x + 5y - 6z = 19 \\ -8x + y - z = -3 \end{cases} \Rightarrow \left[\begin{array}{ccc|c} 2 & 3 & -7 & 11 \\ 4 & 5 & -6 & 19 \\ -8 & 1 & -1 & -3 \end{array} \right].$$

Such an array of numbers is called a **matrix.** Each of the numbers in the matrix is called an **entry.** A matrix with m rows and n columns is an **$m \times n$ matrix.** In general, the entry in the ith row and jth column is referred to as a_{ij}. For example, a 2×4 matrix can be written in general as

$$\begin{bmatrix} a_{11} & a_{12} & a_{13} & a_{14} \\ a_{21} & a_{22} & a_{23} & a_{24} \end{bmatrix}.$$

Matrices (matrices is the plural of matrix) play an important part in mathematics. Using them to solve systems of equations is only one of their many applications. In particular, a matrix that represents a system of equations, including the constants, is referred to as the **augmented matrix** for the system. The array of the coefficients only is called the **coefficient matrix.** The coefficient matrix for the system above is

An augmented matrix is usually written with a vertical line separating the coefficients from the constants.

$$\begin{bmatrix} 2 & 3 & -7 \\ 4 & 5 & -6 \\ -8 & 1 & -1 \end{bmatrix}.$$

EXAMPLE 1   Write the augmented matrix for the following:

a) $\begin{cases} x + 3y = 11 \\ 2x + 4y = 19 \end{cases}$

b) $\begin{cases} y - z = -3 \\ 2x + 2y - 3z = 8 \\ 4x - 6z = 12 \end{cases}$

SOLUTION

a) $\begin{bmatrix} 1 & 3 & | & 11 \\ 2 & 4 & | & 19 \end{bmatrix}$
b) $\begin{bmatrix} 0 & 1 & -1 & | & -3 \\ 2 & 2 & -3 & | & 8 \\ 4 & 0 & -6 & | & 12 \end{bmatrix}$

Notice that in part (b), a zero is entered if there is no corresponding term in the system of equations.

Since an augmented matrix is just another way to express a system of equations, this matrix can be manipulated in much the same way that we manipulated a system of equations in the previous section. The end result of this process, of course, is to determine the solution to the system of equations. The advantage of using a matrix to accomplish this is that we focus only on the important elements of the problem: the coefficients and the constants.

In the previous section, we discussed operations on systems. Their purpose was to find an equivalent, upper-triangular system and then to use back-substitution to determine the solution. For each of these operations on systems, there is an analogous **row operation** that can be used to manipulate an augmented matrix. We list them here. If they look vaguely familiar, it is because they are in essence the same operations we discussed earlier.

Matrix Row Operations

Given a matrix of a system of equations, each of these operations yields a matrix of an equivalent system of equations:

1) Interchanging two rows of the matrix.

2) Replacing a row with a nonzero multiple of itself.

3) Replacing a row with the sum of that row and a multiple of another row of the matrix.

Just as we previously used symbols (such as $5E_1 + E_3 \rightarrow E_3$) to denote the operations on a system of equations, we can use a similar notation to show these matrix row operations. One difference is that we use R_1, R_2, and so on, to represent the rows. So, for example, to indicate replacing the third row with the sum of 2 times the second row and the third row, we use $2R_2 + R_3 \rightarrow R_3$.

Solving a system by using matrix row operations on the augmented matrix is much the same as our solutions in the previous section, except that the process should seem a bit easier and more streamlined.

EXAMPLE 2 Solve the systems by using matrix row operations on the augmented matrix.

a) $\begin{cases} x + 3y = 11 \\ 2x + 4y = 19 \end{cases}$

b) $\begin{cases} y - z = -3 \\ 2x + 2y - 3z = 8 \\ 4x \quad\quad - 6z = 12 \end{cases}$

(These are the same systems as in Example 1.)

SOLUTION

a) The augmented matrix was determined in Example 1 to be

$$\begin{bmatrix} 1 & 3 & | & 11 \\ 2 & 4 & | & 19 \end{bmatrix}.$$

Our goal here, using matrix row operations, is to reduce this matrix to a matrix that represents an upper-triangular system of equations. One row operation is sufficient:

$$-2R_1 + R_2 \to R_2 \begin{bmatrix} 1 & 3 & | & 11 \\ 0 & -2 & | & -3 \end{bmatrix}.$$

This augmented matrix corresponds to the system

$$\begin{cases} x + 3y = 11 \\ \quad -2y = -3. \end{cases}$$

By back-substitution, we get $y = \frac{3}{2}$ from the second equation, and

$$x + 3(\tfrac{3}{2}) = 11 \Rightarrow x = \tfrac{13}{2}$$

gives us the solution $(\frac{13}{2}, \frac{3}{2})$. We leave it to you to check the correctness of this solution.

b) From Example 1, the augmented matrix is

$$\begin{bmatrix} 0 & 1 & -1 & | & -3 \\ 2 & 2 & -3 & | & 8 \\ 4 & 0 & -6 & | & 12 \end{bmatrix}.$$

The first step here is to interchange the first row and the second row so that the first entry in the first row is nonzero:

$$\begin{matrix} R_2 \to R_1 \\ R_1 \to R_2 \end{matrix} \begin{bmatrix} 2 & 2 & -3 & | & 8 \\ 0 & 1 & -1 & | & -3 \\ 4 & 0 & -6 & | & 12 \end{bmatrix}.$$

Next, we eliminate the entries below the first entry in the first row:

$$-2R_1 + R_3 \rightarrow R_3 \begin{bmatrix} 2 & 2 & -3 & | & 8 \\ 0 & 1 & -1 & | & -3 \\ 0 & -4 & 0 & | & -4 \end{bmatrix}.$$

Finally, we eliminate the second entry in the third row. This makes the matrix upper-triangular:

$$4R_2 + R_3 \rightarrow R_3 \begin{bmatrix} 2 & 2 & -3 & | & 8 \\ 0 & 1 & -1 & | & -3 \\ 0 & 0 & -4 & | & -16 \end{bmatrix}.$$

This matrix represents the linear system of equations

$$\begin{cases} 2x + 2y - 3z = 8 \\ y - z = -3 \\ -4z = -16. \end{cases}$$

By back-substitution, we obtain

$$-4z = -16 \Rightarrow z = 4$$

$$y - (4) = -3 \Rightarrow y = 1$$

$$2x + 2(1) - 3(4) = 8 \Rightarrow x = 9.$$

The solution to the system is $(9, 1, 4)$.

Suppose that instead of back-substituting in part (b) of Example 2, we continue to use the row operations in the following manner:

$$\tfrac{1}{2}R_1 \rightarrow R_1 \begin{bmatrix} 1 & 1 & -\tfrac{3}{2} & | & 4 \\ 0 & 1 & -1 & | & -3 \\ 0 & 0 & -4 & | & -16 \end{bmatrix}$$

$$-R_2 + R_1 \rightarrow R_1 \begin{bmatrix} 1 & 0 & -\tfrac{1}{2} & | & 7 \\ 0 & 1 & -1 & | & -3 \\ 0 & 0 & -4 & | & -16 \end{bmatrix}$$

$$-\tfrac{1}{4}R_3 \rightarrow R_3 \begin{bmatrix} 1 & 0 & -\tfrac{1}{2} & | & 7 \\ 0 & 1 & -1 & | & -3 \\ 0 & 0 & 1 & | & 4 \end{bmatrix}$$

$$\begin{matrix} \tfrac{1}{2}R_3 + R_1 \rightarrow R_1 \\ R_3 + R_2 \rightarrow R_2 \end{matrix} \begin{bmatrix} 1 & 0 & 0 & | & 9 \\ 0 & 1 & 0 & | & 1 \\ 0 & 0 & 1 & | & 4 \end{bmatrix}.$$

Think about the system of linear equations that this matrix represents:

$$\begin{cases} x & & = 9 \\ & y & = 1 \\ & & z = 4. \end{cases}$$

The solution to this system is obvious. This last matrix is in **row reduced form.** Reducing a matrix of a system of equations to this form usually takes a few more steps, but discovering the solution in this manner does not require back-substitution.

Row Reduced Form

A matrix is in **row reduced form** if it satisfies the following conditions:

1) The first nonzero element of each row is a 1 (this is called the leading 1 of the row).

2) Every other element in a column with a leading 1 is 0.

3) Each leading 1 must be in a column to the right of all the leading 1's above it.

4) If there are any rows with all 0's then these rows appear at the bottom of the matrix.

No matter what row operations are used to reduce a given matrix to row reduced form, this row reduced form is always the same. In other words, every matrix is equivalent to a unique row reduced form.

Consider these three matrices. They are all in row reduced form.

$$\begin{bmatrix} 1 & 0 & -3 \\ 0 & 1 & 12 \end{bmatrix} \quad \begin{bmatrix} 1 & 0 & 0 & 34 \\ 0 & 1 & 0 & 16 \\ 0 & 0 & 1 & -4 \end{bmatrix} \quad \begin{bmatrix} 1 & 0 & 0 & 9 \\ 0 & 1 & 3 & 2 \\ 0 & 0 & 0 & 0 \end{bmatrix}.$$

On the other hand, the following are not in row reduced form.

$$\begin{bmatrix} 0 & 1 & 34 \\ 1 & 0 & 16 \end{bmatrix} \quad \begin{bmatrix} 1 & 0 & 0 & 3 \\ 0 & 1 & 0 & 5 \\ 0 & 1 & 0 & 1 \end{bmatrix} \quad \begin{bmatrix} 0 & 1 & 4 & 1 \\ 1 & 2 & 0 & 5 \end{bmatrix}.$$

Pause here for a moment and decide why each of the last three matrices is not in row reduced form.

EXAMPLE 3 Solve the system of linear equations by reducing the augmented matrix to row reduced form.

$$\begin{cases} -3x - 8y - 51z = -43 \\ 4x + 10y + 66z = 54 \\ x + 2y + 12z = 10. \end{cases}$$

SOLUTION The augmented matrix for the system is

$$\left[\begin{array}{ccc|c} -3 & -8 & -51 & -43 \\ 4 & 10 & 66 & 54 \\ 1 & 2 & 12 & 10 \end{array}\right].$$

The plan of attack is to proceed column by column. First, we **pivot** about the element of the matrix in the first row and first column. This means that we use appropriate row operations to place a leading one in the first row and first column and zeros in the other entries of the first column (the pivot in this procedure is circled).

$$\begin{array}{c} R_3 \to R_1 \\ \\ R_1 \to R_3 \end{array} \left[\begin{array}{ccc|c} ① & 2 & 12 & 10 \\ 4 & 10 & 66 & 54 \\ -3 & -8 & -51 & -43 \end{array}\right]$$

$$\begin{array}{c} \\ -4R_1 + R_2 \to R_2 \\ 3R_1 + R_3 \to R_3 \end{array} \left[\begin{array}{ccc|c} 1 & 2 & 12 & 10 \\ 0 & 2 & 18 & 14 \\ 0 & -2 & -15 & -13 \end{array}\right].$$

Next, we move down a row and over a column. If possible, we want to pivot about this element:

$$\begin{array}{c} \\ \tfrac{1}{2}R_2 \to R_2 \\ \\ \end{array} \left[\begin{array}{ccc|c} 1 & 2 & 12 & 10 \\ 0 & ① & 9 & 7 \\ 0 & -2 & -15 & -13 \end{array}\right]$$

$$\begin{array}{c} -2R_2 + R_1 \to R_1 \\ \\ 2R_2 + R_3 \to R_3 \end{array} \left[\begin{array}{ccc|c} 1 & 0 & -6 & -4 \\ 0 & 1 & 9 & 7 \\ 0 & 0 & 3 & 1 \end{array}\right].$$

Notice the pattern evolving in the first two columns. We continue this in the third column; the pivot is the element in the third row and third column:

$$\begin{array}{c} \\ \\ \tfrac{1}{3}R_3 \to R_3 \end{array} \left[\begin{array}{ccc|c} 1 & 0 & -6 & -4 \\ 0 & 1 & 9 & 7 \\ 0 & 0 & ① & \tfrac{1}{3} \end{array}\right]$$

$$\begin{array}{c} 6R_3 + R_1 \to R_1 \\ -9R_3 + R_2 \to R_2 \\ \end{array} \left[\begin{array}{ccc|c} 1 & 0 & 0 & -2 \\ 0 & 1 & 0 & 4 \\ 0 & 0 & 1 & \tfrac{1}{3} \end{array}\right].$$

We interpret this row reduced matrix as a system. Our answer is $(-2, 4, \tfrac{1}{3})$. Again, you should check this in the original system.

In the previous section, we discussed the inconsistent and dependent systems from a geometric perspective. The next three examples concern solving these types of systems by using matrices.

EXAMPLE 4   Solve the system of linear equations by reducing the augmented matrix to row reduced form.

a) $\begin{cases} y = \frac{1}{2}x + 4 \\ y = \frac{1}{2}x - 2 \end{cases}$ b) $\begin{cases} x + y = 4 \\ -2x - 2y = -8 \end{cases}$

(Notice that these are the same systems as in Example 1 of Section 10.1.)

SOLUTION

a)   From part (b) of Example 1, we already know that this is an inconsistent system; it has no solutions. However, pretend for a moment that we are not aware of this and try to solve this system.

First, we rewrite each of these equations in the form $ax + by = c$:

$$\begin{cases} x - 2y = -8 \\ x - 2y = 4. \end{cases}$$

The augmented matrix is

$$\begin{bmatrix} 1 & -2 & | & -8 \\ 1 & -2 & | & 4 \end{bmatrix}.$$

Reducing to row reduced form, we have

$$-R_1 + R_2 \rightarrow R_2 \begin{bmatrix} 1 & -2 & | & -8 \\ 0 & 0 & | & 12 \end{bmatrix}$$

$$\tfrac{1}{12}R_2 \rightarrow R_2 \begin{bmatrix} 1 & -2 & | & -8 \\ 0 & 0 & | & 1 \end{bmatrix}.$$

This matrix represents the linear system

$$\begin{cases} x - 2y = -8 \\ 0y = 1. \end{cases}$$

Of course, the last equation, $0y = 1$, is not true, no matter what value we choose for y. This implies that this system has no solution (as we saw graphically in Example 1 of Section 10.1).

b)   From part (c) of Example 1, we already know that this is a dependent system; it has an infinite number of solutions. Trying to solve this by matrix methods gives

$$\begin{bmatrix} 1 & 1 & | & 4 \\ -2 & -2 & | & -8 \end{bmatrix}$$

$$2R_1 + R_2 \rightarrow R_2 \begin{bmatrix} 1 & 1 & | & 4 \\ 0 & 0 & | & 0 \end{bmatrix}.$$

The system is

$$\begin{cases} x + y = 4 \\ 0y = 0. \end{cases}$$

The last equation of the system, $0y = 0$, is true for any value we choose for y. To express the solution, we use a **parameter** t. Suppose that we let $y = t$. Then by back-substituting in the first equation, we get

$$x + t = 4 \Rightarrow x = 4 - t.$$

We can now express the solution as the ordered pair $(4 - t, t)$. For each real number t, there is a solution. For example,

$$t = 0 \Rightarrow (4, 0) \text{ is a solution.}$$

$$t = 3 \Rightarrow (1, 3) \text{ is a solution.}$$

$$t = -12 \Rightarrow (16, -12) \text{ is a solution.}$$

Pause here and generate a few solutions of your own. Check them in the original system.

Systems of three equations in three unknowns can also be inconsistent (no solutions) or dependent (an infinite number of solutions). In fact, you may have noticed the quiet conspiracy, so far, to avoid these types of systems in the examples or exercises. The conspiracy ends with the next two examples.

EXAMPLE 5 Solve the system of linear equations by reducing the augmented matrix to row reduced form.

$$\begin{cases} x + 3y - z = 4 \\ 2x + 4y - 5z = 6 \\ 2y + 3z = 7 \end{cases}$$

SOLUTION The augmented matrix is

$$\begin{bmatrix} 1 & 3 & -1 & | & 4 \\ 2 & 4 & -5 & | & 6 \\ 0 & 2 & 3 & | & 7 \end{bmatrix}.$$

Using row operations, we have

$$-2R_1 + R_2 \rightarrow R_2 \begin{bmatrix} 1 & 3 & -1 & | & 4 \\ 0 & -2 & -3 & | & -2 \\ 0 & 2 & 3 & | & 7 \end{bmatrix}$$

$$-\tfrac{1}{2}R_2 \rightarrow R_2 \begin{bmatrix} 1 & 3 & -1 & | & 4 \\ 0 & 1 & \frac{3}{2} & | & 1 \\ 0 & 2 & 3 & | & 7 \end{bmatrix}$$

$$\begin{matrix} -3R_2 + R_1 \rightarrow R_1 \\ \\ -2R_2 + R_3 \rightarrow R_3 \end{matrix} \begin{bmatrix} 1 & 0 & -\frac{11}{2} & | & 1 \\ 0 & 1 & \frac{3}{2} & | & 1 \\ 0 & 0 & 0 & | & 5 \end{bmatrix}$$

$$-\tfrac{1}{5}R_3 \rightarrow R_3 \begin{bmatrix} 1 & 0 & -\frac{11}{2} & | & 1 \\ 0 & 1 & \frac{3}{2} & | & 1 \\ 0 & 0 & 0 & | & 1 \end{bmatrix}.$$

The last system is in row reduced form. It represents the system

$$\begin{cases} x - \frac{11}{2}z = 1 \\ y + \frac{3}{2}z = 1 \\ 0z = 1. \end{cases}$$

Regardless of our choice for z, the last equation is false. This is an inconsistent system. There are no solutions for this system of equations.

EXAMPLE 6 Solve the system of linear equations by reducing the augmented matrix to row reduced form.

$$\begin{cases} 2x + 6y + 3z = 24 \\ 3y + z = 9 \\ 4x + 3y + 3z = 21 \end{cases}$$

SOLUTION The augmented matrix is

$$\begin{bmatrix} 2 & 6 & 3 & | & 24 \\ 0 & 3 & 1 & | & 9 \\ 4 & 3 & 3 & | & 21 \end{bmatrix}.$$

We proceed:

$$\frac{1}{2}R_1 \rightarrow R_1 \begin{bmatrix} 1 & 3 & \frac{3}{2} & \bigm| & 12 \\ 0 & 3 & 1 & \bigm| & 9 \\ 4 & 3 & 3 & \bigm| & 21 \end{bmatrix}$$

$$-4R_1 + R_3 \rightarrow R_3 \begin{bmatrix} 1 & 3 & \frac{3}{2} & \bigm| & 12 \\ 0 & 3 & 1 & \bigm| & 9 \\ 0 & -9 & -3 & \bigm| & -27 \end{bmatrix}$$

$$\frac{1}{3}R_2 \rightarrow R_2 \begin{bmatrix} 1 & 3 & \frac{3}{2} & \bigm| & 12 \\ 0 & 1 & \frac{1}{3} & \bigm| & 3 \\ 0 & -9 & -3 & \bigm| & -27 \end{bmatrix}$$

$$\begin{matrix} R_1 - 3R_2 \rightarrow R_1 \\ \\ 9R_2 + R_3 \rightarrow R_3 \end{matrix} \begin{bmatrix} 1 & 0 & \frac{1}{2} & \bigm| & 3 \\ 0 & 1 & \frac{1}{3} & \bigm| & 3 \\ 0 & 0 & 0 & \bigm| & 0 \end{bmatrix}.$$

This last matrix is in row reduced form. It represents the system

$$\begin{cases} x & + \frac{1}{2}z = 3 \\ y + \frac{1}{3}z = 3 \\ 0z = 0. \end{cases}$$

The last equation of the system is obviously true for all values we might choose for z. This is a dependent system. Suppose that we represent z with the parameter t. Then

$$x + \tfrac{1}{2}t = 3 \Rightarrow x = 3 - \tfrac{1}{2}t$$
$$y + \tfrac{1}{3}t = 3 \Rightarrow y = 3 - \tfrac{1}{3}t.$$

The finite set of solutions can be written as

$$(3 - \tfrac{1}{2}t, 3 - \tfrac{1}{3}t, t).$$

To get a feel for this set, consider these values of t and the solutions they generate:

$$t = 0 \quad \Rightarrow (3, 3, 0) \text{ is a solution}$$
$$t = 6 \quad \Rightarrow (0, 1, 6) \text{ is a solution}$$
$$t = -15 \Rightarrow (\tfrac{21}{2}, 8, -15) \text{ is a solution.}$$

Again, pause here to generate and check a few of your own solutions to this system.

Systems with fewer equations than unknowns generally have an infinite number of solutions. These systems are called **underdetermined** since they do not have enough equations to determine exactly one solution.

EXAMPLE 7   Find the solutions to the undetermined system of equations.

$$\begin{cases} 3x + 2y + 4z = 13 \\ x - 4y + z = -5 \end{cases}$$

SOLUTION The augmented matrix is

$$\begin{bmatrix} 3 & 2 & 4 & | & 13 \\ 1 & -4 & 1 & | & -5 \end{bmatrix}.$$

Using row operations, we have

$$\begin{matrix} R_2 \to R_1 \\ R_1 \to R_2 \end{matrix} \begin{bmatrix} 1 & -4 & 1 & | & -5 \\ 3 & 2 & 4 & | & 13 \end{bmatrix}$$

$$-3R_1 + R_2 \to R_2 \begin{bmatrix} 1 & -4 & 1 & | & -5 \\ 0 & 14 & 1 & | & 28 \end{bmatrix}$$

$$\tfrac{1}{14}R_2 \to R_2 \begin{bmatrix} 1 & -4 & 1 & | & -5 \\ 0 & 1 & \tfrac{1}{14} & | & 2 \end{bmatrix}$$

$$4R_2 + R_1 \to R_1 \begin{bmatrix} 1 & 0 & \tfrac{9}{7} & | & 3 \\ 0 & 1 & \tfrac{1}{14} & | & 2 \end{bmatrix}.$$

The last matrix is in row reduced form. The system it represents is

$$\begin{cases} x + \tfrac{9}{7}z = 3 \\ y + \tfrac{1}{14}z = 2. \end{cases}$$

Allowing the variable z to be represented by the parameter t, we get

$$x + \tfrac{9}{7}t = 3 \Rightarrow x = 3 - \tfrac{9}{7}t$$

$$y + \tfrac{1}{14}t = 2 \Rightarrow y = 2 - \tfrac{1}{14}t.$$

The solutions can be expressed as $(3 - \tfrac{9}{7}t, 2 - \tfrac{1}{14}t, t)$.

Most of what we have discussed in this and the previous sections also applies to larger systems of equations. We finish the section with the solution of a system of four equations in four unknowns.

EXAMPLE 8   Solve the system of linear equations by reducing the augmented matrix to row reduced form.

$$\begin{cases} 2x + 4y - z + w = 1 \\ x + 2y + 3w = 4 \\ x + y + 3z = 2 \\ 3y + 2z + w = -1. \end{cases}$$

SOLUTION The augmented matrix is

$$\left[\begin{array}{cccc|c} 2 & 4 & -1 & 1 & 1 \\ 1 & 2 & 0 & 3 & 4 \\ 1 & 1 & 3 & 0 & 2 \\ 0 & 3 & 2 & 1 & -1 \end{array}\right].$$

Using row operations to reduce this to row reduced form, we obtain

$$\begin{array}{c} R_3 \to R_1 \\ \\ R_1 \to R_3 \\ \\ \end{array} \left[\begin{array}{cccc|c} 1 & 1 & 3 & 0 & 2 \\ 1 & 2 & 0 & 3 & 4 \\ 2 & 4 & -1 & 1 & 1 \\ 0 & 3 & 2 & 1 & -1 \end{array}\right]$$

$$\begin{array}{c} \\ -R_1 + R_2 \to R_2 \\ -2R_1 + R_3 \to R_3 \\ \\ \end{array} \left[\begin{array}{cccc|c} 1 & 1 & 3 & 0 & 2 \\ 0 & 1 & -3 & 3 & 2 \\ 0 & 2 & -7 & 1 & -3 \\ 0 & 3 & 2 & 1 & -1 \end{array}\right]$$

$$\begin{array}{c} -R_2 + R_1 \to R_1 \\ \\ -2R_2 + R_3 \to R_3 \\ -3R_2 + R_4 \to R_4 \end{array} \left[\begin{array}{cccc|c} 1 & 0 & 6 & -3 & 0 \\ 0 & 1 & -3 & 3 & 2 \\ 0 & 0 & -1 & -5 & -7 \\ 0 & 0 & 11 & -8 & -7 \end{array}\right]$$

$$\begin{array}{c} \\ \\ -R_3 \to R_3 \\ \\ \end{array} \left[\begin{array}{cccc|c} 1 & 0 & 6 & -3 & 0 \\ 0 & 1 & -3 & 3 & 2 \\ 0 & 0 & 1 & 5 & 7 \\ 0 & 0 & 11 & -8 & -7 \end{array}\right]$$

$$\begin{array}{c} -6R_3 + R_1 \to R_1 \\ 3R_3 + R_2 \to R_2 \\ \\ -11R_3 + R_4 \to R_4 \end{array} \left[\begin{array}{cccc|c} 1 & 0 & 0 & -33 & -42 \\ 0 & 1 & 0 & 18 & 23 \\ 0 & 0 & 1 & 5 & 7 \\ 0 & 0 & 0 & -63 & -84 \end{array}\right]$$

$$\begin{array}{c} \\ \\ \\ -\frac{1}{63}R_4 \to R_4 \end{array} \left[\begin{array}{cccc|c} 1 & 0 & 0 & -33 & -42 \\ 0 & 1 & 0 & 18 & 23 \\ 0 & 0 & 1 & 5 & 7 \\ 0 & 0 & 0 & 1 & \frac{4}{3} \end{array}\right]$$

$$\begin{array}{c} 33R_4 + R_1 \to R_1 \\ -18R_4 + R_2 \to R_2 \\ -5R_4 + R_3 \to R_3 \\ \\ \end{array} \left[\begin{array}{cccc|c} 1 & 0 & 0 & 0 & 2 \\ 0 & 1 & 0 & 0 & -1 \\ 0 & 0 & 1 & 0 & \frac{1}{3} \\ 0 & 0 & 0 & 1 & \frac{4}{3} \end{array}\right].$$

From the last matrix, we deduce the solution to be $x = 2$, $y = -1$, $z = \frac{1}{3}$, and $w = \frac{4}{3}$, or $(2, -1, \frac{1}{3}, \frac{4}{3})$.

One obvious difference with solving larger systems is that there is an increased amount of work and a greater probability of human error. Fortunately, many calculators and computers are now capable of accurately solving large systems.

EXERCISE SET 10.2

A

In Problems 1 through 6, write the augmented matrix for the given system.

1. $\begin{cases} x + 2y = 7 \\ 3x - 4y = 12 \end{cases}$

2. $\begin{cases} 2x - 4y = -9 \\ 3x - y = -8 \end{cases}$

3. $\begin{cases} 2x - 4y + z = 14 \\ 3x - y = 21 \\ x + 2y - z = 14 \end{cases}$

4. $\begin{cases} \frac{2}{5}x + \frac{7}{9}y = 14 \\ 6x - y + 10z = 21 \\ 3y - 5z = 34 \end{cases}$

5. $\begin{cases} 2x + 3y - 5z + 4w = 34 \\ 6x - y + 10z = 21 \\ 3y - 5z = 3 \\ 9w = 12 \end{cases}$

6. $\begin{cases} 4x - y - 3z + 12w = 34 \\ 2x - 5z + 23w = -7 \end{cases}$

In Problems 7 through 12, solve the system by using row operations to reduce its augmented matrix to a matrix that represents an equivalent system that is upper-triangular and back-substitute.

7. $\begin{cases} x - 2y = 4 \\ -5x + 11y = 7 \end{cases}$

8. $\begin{cases} 5x + 9y = 4 \\ 2x + 4y = 10 \end{cases}$

9. $\begin{cases} 4x + 2y = 1 \\ 3x + 2y = -9 \end{cases}$

10. $\begin{cases} x + 4y + z = -4 \\ 2x + 9y + 3z = -9 \\ 4y - 5z = -17 \end{cases}$

11. $\begin{cases} x + y + z = 6 \\ x - 3y - 2z = -9 \\ 2x - 2y + 3z = 1 \end{cases}$

12. $\begin{cases} 2y + 3z = 1 \\ x - 2z = -9 \\ x + y = 6 \end{cases}$

In Problems 13 through 18, determine if the matrix given is in row reduced form. If not, state why not.

13. $\begin{bmatrix} 1 & 0 & 0 \\ 0 & 1 & 3 \end{bmatrix}$

14. $\begin{bmatrix} 1 & 3 & 0 \\ 0 & 0 & 1 \end{bmatrix}$

15. $\begin{bmatrix} 1 & 0 & 0 & -3 \\ 0 & 1 & 3 & 2 \end{bmatrix}$

16. $\begin{bmatrix} 1 & 0 & 0 & 12 \\ 0 & 1 & 2 & 23 \\ 0 & 0 & 1 & -9 \end{bmatrix}$

17. $\begin{bmatrix} 1 & 0 & 0 & 12 \\ 0 & 1 & 0 & 0 \\ 0 & 0 & 1 & -9 \end{bmatrix}$

18. $\begin{bmatrix} 1 & -2 & 0 & -7 & 1 \\ 0 & 0 & 1 & 0 & 3 \end{bmatrix}$

In Problems 19 through 30, solve the system by reducing its augmented matrix to row reduced form (note that these are the systems as in Problems 31 through 42 in Section 10.1).

19. $\begin{cases} x + y + z = 3 \\ y + 3z = 16 \\ 4y + 5z = 29 \end{cases}$

20. $\begin{cases} x + 2y - z = 11 \\ 2x - y + 5z = 18 \\ 3z = 15 \end{cases}$

21. $\begin{cases} x - 2y + 4z = -2 \\ -y + z = -3 \\ 3x - 3y + z = 3 \end{cases}$

22. $\begin{cases} 4x + 3y - z = 3 \\ x + y = 4 \\ x + 2z = 7 \end{cases}$

23. $\begin{cases} x + 2y = 0 \\ x + 3y - z = 4 \\ x + 3y + 2z = -2 \end{cases}$

24. $\begin{cases} x - 2y + 3z = 7 \\ 2x + y + z = 4 \\ 3x - 2y + 2z = -10 \end{cases}$

25. $\begin{cases} x + 2y - z = -3 \\ 2x - 4y + z = -7 \\ -2x + 2y - 3z = 4 \end{cases}$

26. $\begin{cases} 4x + 2y - z = 0 \\ x + 3y + 2z = 0 \\ x + y + 3z = 4 \end{cases}$

27. $\begin{cases} 2x + 3y + z = 3 \\ -4x + y + 5z = 10 \\ x - y + z = 18 \end{cases}$

28. $\begin{cases} 2x + 3y + z = 6 \\ -4x + y + 5z = -2 \\ x - y + z = 0 \end{cases}$

29. $\begin{cases} \frac{1}{2}x - \frac{3}{4}y - \frac{1}{4}z = \frac{11}{4} \\ \frac{1}{2}x - y - 2z = -2 \\ \frac{3}{2}x - \frac{1}{2}y - z = 4 \end{cases}$

30. $\begin{cases} \frac{1}{2}x + \frac{1}{3}y + \frac{1}{6}z = 1 \\ x + \frac{1}{4}y - \frac{1}{4}z = 1 \\ y + 2z = 3 \end{cases}$

41. $\begin{cases} 5x + y - z = 0 \\ -6x + y + z = 0 \\ 6x - y - z = 0 \end{cases}$

42. $\begin{cases} 5x + y - z = 2 \\ -6x + y + z = 7 \\ 6x - y - z = -1 \end{cases}$

In Problems 43 through 48, solve the underdetermined system by reducing its augmented matrix to row reduced form. State the solution in terms of a parameter t.

43. $\begin{cases} x - z = 2 \\ y + z = 7 \end{cases}$

44. $\begin{cases} x + y - 2z = 2 \\ 2y + z = 7 \end{cases}$

45. $\begin{cases} x + y - 2z = 2 \\ -3x + y + 6z = 7 \end{cases}$

46. $\begin{cases} x + 2y - 4z = 12 \\ 2x + 5y - 10z = -19 \end{cases}$

47. $\begin{cases} x - 2y + z - 2w = 0 \\ 2x + 2y - 3z + 3w = -3 \\ -x + 4y + 2z - 4w = 13 \end{cases}$

48. $\begin{cases} x + 2y + w = 0 \\ x + 3y - z - 2w = 4 \\ x + 2z - w = -1 \end{cases}$

B

In Problems 31 through 42, solve the system by reducing its augmented matrix to row reduced form. If the system is inconsistent, so state. If the system is dependent, then determine the solution in terms of a parameter t.

31. $\begin{cases} 3x - 3y = 9 \\ 2x - 2y = 5 \end{cases}$

32. $\begin{cases} -4x + 6y = 8 \\ 2x - 3y = -4 \end{cases}$

33. $\begin{cases} 6x + 12y = -18 \\ -7x - 14y = 21 \end{cases}$

34. $\begin{cases} 3x - 3y + 5z = 9 \\ 2x - y + 3z = 5 \\ x - 2y + 2z = 4 \end{cases}$

35. $\begin{cases} y + z = 1 \\ -x + 2y + z = 3 \\ x + y + 2z = 1 \end{cases}$

36. $\begin{cases} 1.5x + y - 0.5z = 1.5 \\ 0.6x - 1.8y + 1.2z = 5.4 \\ 3.6x + 2.7z = -6.3 \end{cases}$

37. $\begin{cases} 7x - y + 3z = 14 \\ 2x - y + z = 2 \\ 3x + y + z = 10 \end{cases}$

38. $\begin{cases} 7x + y - z = 6 \\ 2x + y + z = 6 \\ 3x - y + z = 1 \end{cases}$

39. $\begin{cases} 7x + y - z = 0 \\ 2x + y + z = 2 \\ 3x - y + z = 0 \end{cases}$

40. $\begin{cases} 5x + y - z = 3 \\ -6x + y + z = -4 \\ 6x - y - z = 4 \end{cases}$

In Problems 49 through 54, solve the system by reducing its augmented matrix to row reduced form.

49. $\begin{cases} x + y + z + w = 9 \\ x + y + 2z = 11 \\ x + 2z - w = 8 \\ y + 3z - w = 11 \end{cases}$

50. $\begin{cases} x - 2z + 2w = 1 \\ -2x + 3y + 4z = -1 \\ 3x + y - 2z - w = 3 \\ y + z - w = 0 \end{cases}$

51. $\begin{cases} x + y + z + 4w = 0 \\ x - y - 2z = 0 \\ x + 2y + 8z - w = 0 \\ y + 3z + w = 0 \end{cases}$

52. $\begin{cases} 2x + 2y + z - w = 0 \\ y - z + 2w = 0 \\ 2x + 2y + 3z - 2w = 0 \\ 2x - y - z = 0 \end{cases}$

53. $\begin{cases} x + z = 2 \\ y + w = 4 \\ z + v = 6 \\ x + w = 8 \\ y + v = 0 \end{cases}$

a)   b)   c)

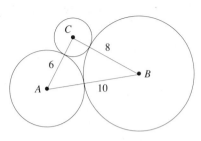

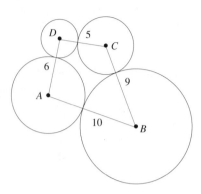

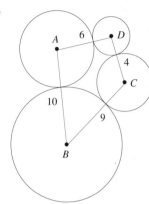

FIGURE 6

$$\begin{cases} x + y & = 6 \\ y + z & = 5 \\ z + w & = 4 \\ w + v = 3 \\ x & + v = 2 \end{cases}$$

54.

55. In each of the figures of Figure 6, assume that the circles are mutually tangent; the distances between centers are shown. Find the radius of each circle by writing a system of equations and solving the system.

56. Solve this system:

$$\begin{cases} 2x + 2y + z - w = 0 \\ \quad\quad y - z + 2w = 0. \end{cases}$$

(Hint: Reducing this to row reduced form, you will find that you need two parameters s and t to describe the solution.)

57. Solve this system:

$$\begin{cases} x + 2y + z - 3w = 0 \\ 3x + 4y - z + w = 10. \end{cases}$$

(Hint: See Problem 56.)

C

58. a) Consider the system

$$\begin{cases} a_{11}x + a_{12}y + a_{13}z = 0 \\ a_{21}x + a_{22}y + a_{23}z = 0 \\ a_{31}x + a_{32}y + a_{33}z = 0. \end{cases}$$

Suppose that (p, q, r) is a solution to this system. Show that for any real number t, (pt, qt, rt) is also a solution.

b) Consider the system with the same coefficient matrix but different constants:

$$\begin{cases} a_{11}x + a_{12}y + a_{13}z = b_1 \\ a_{21}x + a_{22}y + a_{23}z = b_2 \\ a_{31}x + a_{32}y + a_{33}z = b_3. \end{cases}$$

Suppose that a solution of this system is (x_0, y_0, z_0). Show that for any value of t, $(x_0 + pt, y_0 + qt, z_0 + rt)$ is also a solution to this system.

59. a) Solve the system

$$\begin{cases} 2x + 3y - 2z = 0 \\ x - 3y + 5z = 0 \\ 4x - 3y + 8z = 0. \end{cases}$$

b) Verify that the system

$$\begin{cases} 2x + 3y - 2z = \quad 7 \\ x - 3y + 5z = -1 \\ 4x - 3y + 8z = \quad 5 \end{cases}$$

has a solution $(2, 1, 0)$. Use part a) and the results of Problem 58 to find the complete solution of this system.

THE DETERMINANT

SECTION 10.3

Consider the following problem:

Find the area of the parallelogram in Figure 7 (in terms of the given coordinates).

One way to solve this problem is to surround the parallelogram with a rectangle, and then partition the resulting region as in Figure 8. The triangles labeled T_1 are congruent, the triangles labeled T_2 are congruent, and the rectangles labeled R are congruent (the proofs of these assertions are left as exercises). The areas of the three congruent pairs of shaded regions are

$$R = bc, \qquad T_1 = \tfrac{1}{2}ab, \qquad T_2 = \tfrac{1}{2}cd.$$

The area of the parallelogram is

$$P = (\text{area of surrounding rectangle}) - (2R + 2T_1 + 2T_2)$$
$$= (a + c)(b + d) - (2bc + ab + cd) = ad - bc.$$

Actually, this formula relies on how the two points (a, b) and (c, d) are positioned; if they traded coordinates, the area would be $bc - ad$ (Figure 9). To be safe, you should use the formula $P = |ad - bc|$.

Problems like this, as well as systems of equations and other important calculations, often lead to complicated expressions that share certain mathematical patterns. The **determinant** is a special kind of function that serves to organize many of these calculations. The domain of this function is the set of all square matrices (a square matrix is a matrix with the same number of rows as columns). The range of this function is the set of all real numbers. We begin by defining the determinant of a 2×2 matrix.

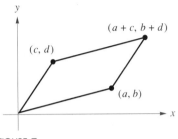

FIGURE 7

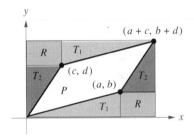

FIGURE 8

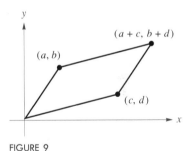

FIGURE 9

2 × 2 Determinant

Given a 2×2 matrix

$$A = \begin{bmatrix} a & b \\ c & d \end{bmatrix}$$

the determinant of A is

$$\det A = \begin{vmatrix} a & b \\ c & d \end{vmatrix} = ad - bc.$$

Even though the notation is similar to that of absolute value, there should be no confusion between the two. Absolute value is a function defined for a number; the determinant function is meaningful only for a matrix.

EXAMPLE 1   Evaluate the determinants of the following matrices.

$$A = \begin{bmatrix} 5 & 2 \\ 4 & -3 \end{bmatrix} \qquad B = \begin{bmatrix} 7 & -1 \\ 3 & 4 \end{bmatrix} \qquad C = \begin{bmatrix} 8 & -3 \\ -6 & \frac{1}{2} \end{bmatrix}$$

SOLUTION

$$\det A = \begin{vmatrix} 5 & 2 \\ 4 & -3 \end{vmatrix} = (5)(-3) - (4)(2) = -23.$$

$$\det B = \begin{vmatrix} 7 & -1 \\ 3 & 4 \end{vmatrix} = (7)(4) - (3)(-1) = 31.$$

$$\det C = \begin{vmatrix} 8 & -3 \\ -6 & \frac{1}{2} \end{vmatrix} = (8)(\tfrac{1}{2}) - (-6)(-3) = -14.$$

The next example may seem arbitrary, but problems such as these are important in the application of mathematics to science and engineering.

EXAMPLE 2   Find all values r such that

$$\begin{vmatrix} 4 - r & -2 \\ 1 & 1 - r \end{vmatrix} = 0.$$

SOLUTION Evaluating the determinant, we have

$$(4 - r)(1 - r) - (1)(-2) = 0.$$

Now multiply the two binomials and solve for r:

$$4 - 5r + r^2 + 2 = 0$$
$$r^2 - 5r + 6 = 0$$
$$(r - 2)(r - 3) = 0$$
$$r = 2 \quad \text{or} \quad r = 3.$$

To check $r = 2$, substitute into the determinant:

$$\begin{vmatrix} 4 - r & -2 \\ 1 & 1 - r \end{vmatrix} = \begin{vmatrix} 2 & -2 \\ 1 & -1 \end{vmatrix} = (2)(-1) - (1)(-2) = 0.$$

The check for $r = 3$ is left to you.

To define determinants of square matrices that are 3×3 or larger, we need the following.

Given a 3×3 matrix:

Minors and Cofactors

$$A = \begin{bmatrix} a_{11} & a_{12} & a_{13} \\ a_{21} & a_{22} & a_{23} \\ a_{31} & a_{32} & a_{33} \end{bmatrix}.$$

The **minor** of an element is the determinant of the 2×2 matrix that is obtained by deleting the row and column containing that element. In other words, the minor of the element a_{ij} is the 2×2 determinant formed by deleting the ith row and the jth column of A.

The **cofactor** of the element a_{ij} is

$$(-1)^{i+j}(\text{minor of } a_{ij}).$$

Thus, to calculate the cofactor of a_{ij}, first find the minor and then multiply by 1 or -1, depending on whether the power $(i + j)$ is even or odd.

EXAMPLE 3 Find the minor and cofactor for each of the numbers 8, -2, and 5 in the following matrix:

$$\begin{bmatrix} 1 & 4 & 8 \\ -2 & 5 & 3 \\ 3 & 7 & 6 \end{bmatrix}.$$

SOLUTION To find the minor of 8, eliminate the row and column containing 8:

$$\begin{bmatrix} \cancel{1} & 4 & \cancel{8} \\ -2 & 5 & 3 \\ 3 & 7 & 6 \end{bmatrix}.$$

The minor of 8 is the determinant

$$\begin{vmatrix} -2 & 5 \\ 3 & 7 \end{vmatrix} = (-2)(7) - (3)(5) = -29.$$

Since the entry 8 is in row 1, column 3, the cofactor of 8 is

$$(-1)^{i+j}(\text{minor of } 8) = (-1)^{1+3}(-29) = -29.$$

To find the minor of -2, eliminate the row and column containing -2:

$$\begin{bmatrix} 1 & 4 & 8 \\ -2 & 5 & 3 \\ 3 & 7 & 6 \end{bmatrix}.$$

The minor of -2 is the determinant

$$\begin{vmatrix} 4 & 8 \\ 7 & 6 \end{vmatrix} = (4)(6) - (7)(8) = -32.$$

Since -2 is in row 2, column 1, the cofactor of -2 is

$$(-1)^{i+j}(\text{minor of } -2) = (-1)^{2+1}(-32) = 32.$$

The minor of 5 in the matrix

$$\begin{bmatrix} 1 & 4 & 8 \\ -2 & 5 & 3 \\ 3 & 7 & 6 \end{bmatrix}$$

is

$$\begin{vmatrix} 1 & 8 \\ 3 & 6 \end{vmatrix} = (1)(6) - (3)(8) = -18.$$

The cofactor of 5 is

$$(-1)^{i+j}(\text{minor of } 5) = (-1)^{2+2}(-18) = -18.$$

Rather than identifying the row and column to decide on the sign of $(-1)^{i+j}$, some students prefer to use the following "checkerboard" pattern of signs:

$$\begin{bmatrix} + & - & + \\ - & + & - \\ + & - & + \end{bmatrix}.$$

For example, if we compare the matrix

$$\begin{bmatrix} 5 & 6 & 7 \\ -3 & 2 & 6 \\ 8 & 4 & 0 \end{bmatrix}$$

with the checkerboard scheme, the cofactor of the entry 4 is

$$(-1)(\text{minor of } 4) = (-1)[(5)(6) - (-3)(7)] = -51.$$

We are ready to define how to evaluate the determinant of a 3×3 matrix.

To evaluate the determinant of a 3×3 matrix:

1) Choose a row or column.

2) Find the cofactor of each entry in your chosen row or column.

3) Multiply each of the three chosen entries by their respective cofactors.

4) The sum of these three products is the determinant.

3×3 Determinant

EXAMPLE 4   Find the determinant of

$$A = \begin{bmatrix} 3 & 1 & 8 \\ -3 & 5 & 6 \\ 9 & 0 & -2 \end{bmatrix}.$$

SOLUTION First select any row or column. If we choose the second row, then

$$\det A = \begin{vmatrix} 3 & 1 & 8 \\ -3 & 5 & 6 \\ 9 & 0 & -2 \end{vmatrix}$$

$$= -3[\text{cofactor of} -3] + 5[\text{cofactor of } 5] + 6[\text{cofactor of } 6]$$

$$= -3\left[(-1)\begin{vmatrix} 1 & 8 \\ 0 & -2 \end{vmatrix}\right] + 5\left[(1)\begin{vmatrix} 3 & 8 \\ 9 & -2 \end{vmatrix}\right] + 6\left[(-1)\begin{vmatrix} 3 & 1 \\ 9 & 0 \end{vmatrix}\right]$$

$$= -3[(-1)(-2)] + 5[(1)(-78)] + 6[(-1)(-9)] = -342.$$

For an encore, we evaluate det A by using cofactors of entries in the third column:

$$\det A = \begin{vmatrix} 3 & 1 & 8 \\ -3 & 5 & 6 \\ 9 & 0 & -2 \end{vmatrix}$$

$$= 8(\text{cofactor of } 8) + 6(\text{cofactor of } 6) + -2(\text{cofactor of } -2)$$

$$= 8\left[(1)\begin{vmatrix} -3 & 5 \\ 9 & 0 \end{vmatrix}\right] + 6\left[(-1)\begin{vmatrix} 3 & 1 \\ 9 & 0 \end{vmatrix}\right] + (-2)\left[(1)\begin{vmatrix} 3 & 1 \\ -3 & 5 \end{vmatrix}\right]$$

$$= 8[-45] + 6[9] + (-2)[18] = -342.$$

The fact that this agrees with the previous answer is somewhat remarkable, yet necessary for the determinant to be a well-defined function. In other words, the determinant of a square matrix is a unique number regardless of the choice of the row or column. Showing this fact for a 3×3 matrix is straightforward but tedious. As a partial proof, we find the determinant of a general 3×3 matrix by selecting two of the six choices of row or column. Starting with

$$A = \begin{bmatrix} a_{11} & a_{12} & a_{13} \\ a_{21} & a_{22} & a_{23} \\ a_{31} & a_{32} & a_{33} \end{bmatrix}$$

we select the first row to evaluate det A:

$$\det A = a_{11}\left[(1)\begin{vmatrix} a_{22} & a_{23} \\ a_{32} & a_{33} \end{vmatrix}\right] + a_{12}\left[(-1)\begin{vmatrix} a_{21} & a_{23} \\ a_{31} & a_{33} \end{vmatrix}\right] + a_{13}\left[(1)\begin{vmatrix} a_{21} & a_{22} \\ a_{31} & a_{32} \end{vmatrix}\right]$$

$$= a_{11}[a_{22}a_{33} - a_{32}a_{23}] - a_{12}[a_{21}a_{33} - a_{31}a_{23}]$$
$$+ a_{13}[a_{21}a_{32} - a_{31}a_{22}]$$

$$= a_{11}a_{22}a_{33} - a_{11}a_{32}a_{23} - a_{12}a_{21}a_{33} + a_{12}a_{31}a_{23}$$
$$+ a_{13}a_{21}a_{32} - a_{13}a_{31}a_{22}.$$

Next, we calculate the determinant by using the entries in the second row:

$$\det A = a_{21}\left[(-1)\begin{vmatrix} a_{12} & a_{13} \\ a_{32} & a_{33} \end{vmatrix}\right] + a_{22}\left[(1)\begin{vmatrix} a_{11} & a_{13} \\ a_{31} & a_{33} \end{vmatrix}\right] + a_{23}\left[(-1)\begin{vmatrix} a_{11} & a_{12} \\ a_{31} & a_{32} \end{vmatrix}\right]$$

$$= -a_{21}[a_{12}a_{33} - a_{32}a_{13}] + a_{22}[a_{11}a_{33} - a_{31}a_{13}]$$
$$- a_{23}[a_{11}a_{32} - a_{31}a_{12}]$$

$$= -a_{12}a_{21}a_{33} + a_{13}a_{21}a_{32} + a_{11}a_{22}a_{33} - a_{13}a_{31}a_{22}$$
$$- a_{11}a_{32}a_{23} + a_{12}a_{31}a_{23}.$$

Comparing this with the first result, we have shown that the determinant has the same value whether we select the first row or the second. Similarly, it can be shown that choosing the third row or the first, second, or third columns also yields the same expression.

To evaluate the determinant of an $n \times n$ matrix, we proceed inductively, patterning after the definitions for a 3×3 matrix.

$n \times n$ Determinant

Let A be an $n \times n$ matrix. The **minor** of an entry a_{ij} is the determinant of the $(n-1) \times (n-1)$ matrix formed by deleting the ith row and jth column. The **cofactor** of a_{ij} is

$$(-1)^{i+j}(\text{minor of } a_{ij}).$$

To evaluate the determinant of an $n \times n$ matrix:

1) Choose a row or column.

2) Find the cofactor of each entry in your chosen row or column.

3) Multiply each of the n chosen entries by their respective cofactors.

4) The sum of these n products is the determinant.

As with 3×3 matrices, the determinant of an $n \times n$ matrix is the same regardless of our choice of the row or column. This fact can be useful, especially if zeros appear in any rows or columns.

EXAMPLE 5   Find the determinant of

$$A = \begin{bmatrix} 2 & 1 & -1 & 3 \\ 5 & 0 & -3 & 7 \\ -5 & 2 & 1 & 2 \\ 1 & 4 & 6 & -1 \end{bmatrix}$$

SOLUTION Since the second row has a zero as one of its entries, it is convenient to select the second row for the evaluation. By definition,

$$\det A = 5(\text{cofactor of } 5) + 0(\text{cofactor of } 0)$$

$$+ (-3)(\text{cofactor of } -3) + 7(\text{cofactor of } 7).$$

The minor of 5 is the 3×3 determinant that results from deleting the row and column where the 5 is located. Because the 5 is in the second row and the first column, $i = 2$ and $j = 1$.

$$\begin{bmatrix} 2 & 1 & -1 & 3 \\ 5 & 0 & -3 & 7 \\ -5 & 2 & 1 & 2 \\ 1 & 4 & 6 & -1 \end{bmatrix}$$

The cofactor of 5 is

$$(-1)^{2+1} \begin{vmatrix} 1 & -1 & 3 \\ 2 & 1 & 2 \\ 4 & 6 & -1 \end{vmatrix}.$$

Evaluating this 3×3 determinant using row 1 gives

$$(-1)^{2+1} \begin{vmatrix} 1 & -1 & 3 \\ 2 & 1 & 2 \\ 4 & 6 & -1 \end{vmatrix} = (-1) \left[1 \begin{vmatrix} 1 & 2 \\ 6 & -1 \end{vmatrix} + 1 \begin{vmatrix} 2 & 2 \\ 4 & -1 \end{vmatrix} + 3 \begin{vmatrix} 2 & 1 \\ 4 & 6 \end{vmatrix} \right]$$

$$= (-1)[1(-13) + 1(-10) + 3(8)] = -1.$$

Similarly, the cofactor of -3 is

$$(-1)^{2+3} \begin{vmatrix} 2 & 1 & 3 \\ -5 & 2 & 2 \\ 1 & 4 & -1 \end{vmatrix}$$

$$= (-1) \left[2 \begin{vmatrix} 2 & 2 \\ 4 & -1 \end{vmatrix} + (-1) \begin{vmatrix} -5 & 2 \\ 1 & -1 \end{vmatrix} + 3 \begin{vmatrix} -5 & 2 \\ 1 & 4 \end{vmatrix} \right]$$

$$= (-1)[(-20) + (-3) + (-66)] = 89.$$

Finally, the cofactor of 7 is

$$(-1)^{2+4} \begin{vmatrix} 2 & 1 & -1 \\ -5 & 2 & 1 \\ 1 & 4 & 6 \end{vmatrix} = (1) \left[2 \begin{vmatrix} 2 & 1 \\ 4 & 6 \end{vmatrix} + (-1) \begin{vmatrix} -5 & 1 \\ 1 & 6 \end{vmatrix} + (-1) \begin{vmatrix} -5 & 2 \\ 1 & 4 \end{vmatrix} \right]$$

$$= (1)[(16) + (31) + (22)] = 69.$$

Hence, the determinant of the original 4×4 matrix is

$$\det A = 5(\text{cofactor of } 5) + 0(\text{cofactor of } 0)$$

$$+ (-3)(\text{cofactor of } -3) + 7(\text{cofactor of } 7)$$

$$= 5(-1) + 0 + (-3)(89) + 7(69) = 211.$$

EXERCISE SET 10.3

A

In Problems 1 through 6, evaluate the determinants.

1. a) $\begin{vmatrix} 2 & 5 \\ 1 & 7 \end{vmatrix}$ **b)** $\begin{vmatrix} 3 & 1 \\ 4 & 2 \end{vmatrix}$

2. a) $\begin{vmatrix} 6 & 4 \\ 1 & 2 \end{vmatrix}$ **b)** $\begin{vmatrix} 7 & 2 \\ 2 & 5 \end{vmatrix}$

3. a) $\begin{vmatrix} -6 & 2 \\ -7 & 5 \end{vmatrix}$ **b)** $\begin{vmatrix} 8 & -1 \\ -7 & \frac{1}{2} \end{vmatrix}$

4. a) $\begin{vmatrix} -9 & -2 \\ 7 & -3 \end{vmatrix}$ **b)** $\begin{vmatrix} -5 & 6 \\ \frac{1}{3} & -7 \end{vmatrix}$

5. a) $\begin{vmatrix} \sqrt{3}+1 & -\frac{1}{2}\sqrt{2} \\ \sqrt{6} & 4 \end{vmatrix}$ **b)** $\begin{vmatrix} \cos x & \sin x \\ \sin y & \cos y \end{vmatrix}$

6. a) $\begin{vmatrix} \sqrt{5}-1 & \sqrt{2} \\ -\sqrt{10} & 3 \end{vmatrix}$ **b)** $\begin{vmatrix} \cos A & \cos B \\ \sin A & \sin B \end{vmatrix}$

In Problems 7 through 12, refer to the matrix

$$\begin{bmatrix} 4 & 1 & -7 \\ -2 & -3 & -5 \\ 3 & 6 & 2 \end{bmatrix}$$

to evaluate the following.

7. The minor of the entry -2.

8. The minor of the entry -3.

9. The cofactor of the entry -7.

10. The cofactor of the entry 1.

11. The cofactor of the entry -5.

12. The cofactor of the entry 2.

Evaluate the determinants in Problems 13 through 21.

13. $\begin{vmatrix} 4 & 1 & 2 \\ -1 & -3 & -5 \\ 3 & 2 & -1 \end{vmatrix}$ **14.** $\begin{vmatrix} 2 & 1 & 2 \\ 3 & -2 & -5 \\ -5 & 3 & -1 \end{vmatrix}$

15. $\begin{vmatrix} 5 & 1 & -2 \\ 3 & -2 & 0 \\ 5 & 6 & 1 \end{vmatrix}$ **16.** $\begin{vmatrix} 0 & 9 & -3 \\ 3 & 4 & 2 \\ 1 & 0 & 8 \end{vmatrix}$

17. $\det \begin{bmatrix} 5 & 9 & -3 \\ 0 & 4 & -6 \\ 1 & 0 & 2 \end{bmatrix}$

18. $\det \begin{bmatrix} 4 & 0 & -6 \\ 3 & 1 & 8 \\ 11 & 0 & 12 \end{bmatrix}$

19. $\begin{vmatrix} 2 & -3 & 2 & 4 \\ 0 & 3 & -4 & 1 \\ 5 & 2 & -7 & -3 \\ 1 & 0 & -1 & 4 \end{vmatrix}$

20. $\begin{vmatrix} 3 & -1 & 1 & 8 \\ 3 & 0 & -3 & 1 \\ -1 & 2 & -9 & 5 \\ 1 & 2 & 0 & 4 \end{vmatrix}$

21. $\det \begin{bmatrix} 0 & -8 & 1 & 8 \\ 3 & 5 & -3 & 1 \\ 1 & 2 & -6 & 0 \\ -1 & 6 & 5 & 4 \end{bmatrix}$

B

In Problems 22 through 26, find the values of r that satisfy each equation.

22. $\begin{vmatrix} 1-r & 6 \\ 5 & 2-r \end{vmatrix} = 0$

23. $\begin{vmatrix} 5-r & 4 \\ 1 & 2-r \end{vmatrix} = 0$

24. $\begin{vmatrix} -5-r & 4 \\ -8 & 7-r \end{vmatrix} = 0$

25. $\begin{vmatrix} 2-r & 4 & 2 \\ 0 & -3-r & -1 \\ 0 & 0 & -r \end{vmatrix} = 0$

26. $\begin{vmatrix} -r & 0 & 2 \\ 6 & 1-r & 4 \\ -1 & 0 & 3-r \end{vmatrix} = 0$

For Problems 27 through 29, refer to Figure 10.

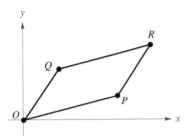

FIGURE 10

27. If the coordinates of P and Q are $P(5, 2)$ and $Q(2, 3)$, find the coordinates for R and the area of the parallelogram.

28. If the coordinates of P and Q are $P(8, 4)$ and $Q(3, 5)$, find the coordinates for R and the area of the parallelogram.

29. If the coordinates of P and R are $P(6, 2)$ and $R(8, 7)$, find the coordinates for Q and the area of the parallelogram.

For Problems 30 through 32, refer to Figure 11.

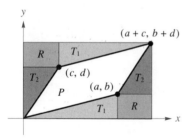

FIGURE 11

30. Verify that the quadrilateral with vertices $(0, 0)$, (a, b), (c, d), and $(a + c, b + d)$ is a parallelogram *(Hint: See Problem 51 of Section 2.1)*

31. Explain why the triangles marked T_1 are congruent.

32. Explain why the triangles marked T_2 are congruent.

To evaluate the determinant of 3×3 matrix A, the following alternate method is sometimes used. First, copy the first two columns, placing them to the right of the matrix:

$$\begin{bmatrix} a_{11} & a_{12} & a_{13} \\ a_{21} & a_{22} & a_{23} \\ a_{31} & a_{32} & a_{33} \end{bmatrix} \begin{matrix} a_{11} & a_{12} \\ a_{21} & a_{22} \\ a_{31} & a_{32} \end{matrix}$$

Next, form the products obtained by multiplying the entries on the diagonals:

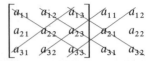

The three products formed on the downward-sloping (left to right) diagonals are assigned positive values; the three products formed on the upward-sloping (left to right) diagonals are assigned negative values:

$$\det A = \underbrace{(a_{11}a_{22}a_{33} + a_{12}a_{23}a_{31} + a_{13}a_{21}a_{32})}$$

These terms are the products of the
downward-sloping diagonals

$$-\underbrace{(a_{13}a_{22}a_{31} + a_{11}a_{23}a_{31} + a_{12}a_{21}a_{33})}$$

These terms are the products of the
upward-sloping diagonals

(This method is not valid for determinants of 4×4 or larger matrices.)

33. Use this method to evaluate the determinant in Problem 13.

34. Use this method to evaluate the determinant in Problem 14.

35. Use this method to evaluate the determinant in Problem 15.

36. Let

$$A = \begin{bmatrix} 2 & 4 & 1 \\ 2 & -5 & 3 \\ 7 & 8 & -6 \end{bmatrix}.$$

Form matrix B by replacing the second row of A with the sum of that row and 3 times the first row:

$$B = \begin{bmatrix} 2 & 4 & 1 \\ 8 & 7 & 6 \\ 7 & 8 & -6 \end{bmatrix}.$$

Compare $\det A$ and $\det B$.

37. Let

$$A = \begin{bmatrix} 5 & 4 & -2 \\ 1 & 5 & 6 \\ 3 & -7 & -1 \end{bmatrix}.$$

Form matrix B by replacing the third row of A with the sum of that row and 2 times the second row:

$$B = \begin{bmatrix} 5 & 4 & -2 \\ 1 & 5 & 6 \\ 5 & 3 & 11 \end{bmatrix}.$$

Compare det A and det B.

38. Let

$$A = \begin{bmatrix} 6 & 4 & -2 \\ 3 & 7 & 1 \\ 9 & -1 & 0 \end{bmatrix}.$$

Form matrix **B** by replacing the first row of A with $(\frac{1}{2})$ times that row:

$$B = \begin{bmatrix} 3 & 2 & -1 \\ 3 & 7 & 1 \\ 9 & -1 & 0 \end{bmatrix}.$$

Compare det A and det B.

39. Let

$$A = \begin{bmatrix} -1 & 7 & 6 \\ 4 & 3 & 2 \\ -6 & 9 & 1 \end{bmatrix}.$$

Form matrix **B** by replacing the second row of A with three times that row:

$$B = \begin{bmatrix} -1 & 7 & 6 \\ 12 & 9 & 6 \\ -6 & 9 & 1 \end{bmatrix}.$$

Compare det A and det B.

Let $A(x_1, y_1)$, $B(x_2, y_2)$, and $C(x_3, y_3)$ be given points. It can be shown that the area A of the triangle formed is

$$A = \frac{1}{2} \left| \det \begin{bmatrix} x_1 & y_1 & 1 \\ x_2 & y_2 & 1 \\ x_3 & y_3 & 1 \end{bmatrix} \right|$$

Use this for Problems 40 through 44.

40. Find the area of the triangle whose vertices are $(1, 2)$, $(2, -5)$, and $(-3, 1)$.

41. Find the area of the triangle whose vertices are $(3, 0)$, $(1, 1)$, and $(-2, 4)$.

42. Find the area of the triangle whose vertices are $(-3, 1)$, $(4, -1)$, and $(5, 9)$.

43. Use a 3×3 determinant to find the equation of the line through $(-1, 6)$ and $(4, 2)$. *(Hint: The point (x, y) lies on the line through $(-1, 6)$ and $(4, 2)$ if and only if the area of the triangle formed by these 3 points is zero.)*

44. Use a 3×3 determinant to find the equation of the line through $(-3, -7)$ and $(1, -1)$. *(Hint: The point (x, y) lies on the line through $(-3, -7)$ and $(1, -1)$ if and only if the area of the triangle formed by these 3 points is zero.)*

C

Let $A(x_1, y_1)$, $B(x_2, y_2)$, and $C(x_3, y_3)$ be given points. It can be shown that an equation of the circle through A, B, and C is given by

$$\begin{vmatrix} x^2 + y^2 & x & y & 1 \\ x_1^2 + y_1^2 & x_1 & y_1 & 1 \\ x_2^2 + y_2^2 & x_2 & y_2 & 1 \\ x_3^2 + y_3^2 & x_3 & y_3 & 1 \end{vmatrix} = 0$$

Use this for Problems 45 and 46.

45. Find an equation of the circle passing through $(-3, 0)$, $(-1, 4)$, and $(2, -5)$.

46. Find an equation of the circle passing through $(-2, 1)$, $(0, 5)$, and $(6, 5)$.

47. Computing the determinant of a 3×3 matrix involved evaluating three 2×2 determinants. How many 2×2 determinants must be evaluated to compute a 4×4 determinant? 5×5? $n \times n$?

PROPERTIES OF DETERMINANTS AND CRAMER'S RULE

S E C T I O N 10.4

The process of evaluating determinants can be simplified by using the elementary row operations discussed in Section 10.2. The next set of statements shows how these operations affect the determinant.

*Row (or Column) Operations
with Determinants*

When evaluating the determinant of an $n \times n$ matrix, each elementary row operation has the following effect:

1) The interchange of two rows produces a change of signs of the determinant.

2) Replacing a row with a nonzero multiple of itself multiplies the determinant by the same amount.

3) Replacing a row with the sum of that row and a multiple of another row does not change the determinant.

Furthermore, each of these three statements remains valid if the word *row* is replaced with the word *column*.

EXAMPLE 1   Given that

$$\begin{vmatrix} a_1 & a_2 & a_3 \\ b_1 & b_2 & b_3 \\ c_1 & c_2 & c_3 \end{vmatrix} = 12,$$

evaluate:

a) $\begin{vmatrix} a_1 & a_3 & a_2 \\ b_1 & b_3 & b_2 \\ c_1 & c_3 & c_2 \end{vmatrix}$
 b) $\begin{vmatrix} a_1 & a_2 & a_3 \\ b_1 & b_2 & b_3 \\ 3c_1 & 3c_2 & 3c_3 \end{vmatrix}$

c) $\begin{vmatrix} a_1 + 2b_1 & a_2 + 2b_2 & a_3 + 2b_3 \\ b_1 & b_2 & b_3 \\ c_1 & c_2 & c_3 \end{vmatrix}$

SOLUTION

a) Comparing this determinant with the given determinant, we see that the second and third columns have been interchanged. Therefore,

$$\begin{vmatrix} a_1 & a_3 & a_2 \\ b_1 & b_3 & b_2 \\ c_1 & c_3 & c_2 \end{vmatrix} = -\begin{vmatrix} a_1 & a_2 & a_3 \\ b_1 & b_2 & b_3 \\ c_1 & c_2 & c_3 \end{vmatrix} = -12.$$

b) Comparing this determinant with the given determinant, we see that the third row has been multiplied by 3. This means that

$$\begin{vmatrix} a_1 & a_2 & a_3 \\ b_1 & b_2 & b_3 \\ 3c_1 & 3c_2 & 3c_3 \end{vmatrix} = 3\begin{vmatrix} a_1 & a_2 & a_3 \\ b_1 & b_2 & b_3 \\ c_1 & c_2 & c_3 \end{vmatrix} = 36.$$

c) This determinant is obtained by replacing the first row with the sum of the first row and 2 times the second row. Thus, the value of the determinant is not changed:

$$\begin{vmatrix} a_1 + 2b_1 & a_2 + 2b_2 & a_3 + 2b_3 \\ b_1 & b_2 & b_3 \\ c_1 & c_2 & c_3 \end{vmatrix} = \begin{vmatrix} a_1 & a_2 & a_3 \\ b_1 & b_2 & b_3 \\ c_1 & c_2 & c_3 \end{vmatrix} = 12.$$

The third property is especially useful for simplifying the evaluation of determinants. The basic idea is to obtain a column or row with a lot of zeros before calculating the determinant. The next two examples demonstrate.

EXAMPLE 2   Evaluate

$$\begin{vmatrix} 7 & -3 & 7 \\ -2 & 2 & 5 \\ -4 & 1 & -2 \end{vmatrix}$$

by using elementary row (or column) operations.

SOLUTION There are many ways to approach this problem. Because there is an entry of 1, it would be easy to obtain two zeros in the second column. To accomplish this, add 3 times the third row to the first row. Using the notation introduced in Section 10.2, we have:

$$\begin{vmatrix} 7 & -3 & 7 \\ -2 & 2 & 5 \\ -4 & 1 & -2 \end{vmatrix} = \begin{vmatrix} -5 & 0 & 1 \\ -2 & 2 & 5 \\ -4 & 1 & -2 \end{vmatrix} \qquad (3R_3 + R_1 \rightarrow R_1).$$

Now add -2 times the third row to the second row:

$$\begin{vmatrix} -5 & 0 & 1 \\ -2 & 2 & 5 \\ -4 & 1 & -2 \end{vmatrix} = \begin{vmatrix} -5 & 0 & 1 \\ 6 & 0 & 9 \\ -4 & 1 & -2 \end{vmatrix} \qquad (-2R_3 + R_2 \rightarrow R_2).$$

The determinant is easy to calculate if we use the second column:

$$\begin{vmatrix} -5 & 0 & 1 \\ 6 & 0 & 9 \\ -4 & 1 & -2 \end{vmatrix} = 0 + 0 + 1(\text{cofactor of } 1)$$

$$= 1 \left[(-1) \begin{vmatrix} -5 & 1 \\ 6 & 9 \end{vmatrix} \right] = 51.$$

EXAMPLE 3   Evaluate

$$\begin{vmatrix} 3 & 1 & 8 \\ -3 & 5 & 6 \\ 9 & 0 & -2 \end{vmatrix}$$

by using elementary row (or column) operations.

SOLUTION   Since there is already a zero in the third row, we create another zero in the third row. Add $\frac{9}{2}$ times the third column to the first column:

$$\begin{vmatrix} 3 & 1 & 8 \\ -3 & 5 & 6 \\ 9 & 0 & -2 \end{vmatrix} = \begin{vmatrix} 39 & 1 & 8 \\ 24 & 5 & 6 \\ 0 & 0 & -2 \end{vmatrix}.$$

Now evaluate the determinant using the third row:

$$\begin{vmatrix} 39 & 1 & 8 \\ 24 & 5 & 6 \\ 0 & 0 & -2 \end{vmatrix} = -2(\text{cofactor of } -2)$$

$$= -2\left[(1) \begin{vmatrix} 39 & 1 \\ 24 & 5 \end{vmatrix} \right] = (-2)(195 - 24)$$

$$= -342.$$

In Sections 10.1 and 10.2 we developed methods for finding the solutions to systems of linear equations. For the purpose of deriving a formula for the solution of linear systems, consider the general system of two equations in two unknowns:

$$\begin{cases} a_{11}x + a_{12}y = k_1 \\ a_{21}x + a_{22}y = k_2. \end{cases}$$

Solve this system for x by multiplying the first equation by a_{22}, multiplying the second equation by $-a_{12}$, and adding the resulting equations.

$$\begin{aligned} a_{11}a_{22}x + a_{12}a_{22}y &= k_1a_{22} \\ -a_{21}a_{12}x - a_{22}a_{12}y &= -k_2a_{12} \\ \hline (a_{11}a_{22} - a_{21}a_{12})x &= k_1a_{22} - k_2a_{12}. \end{aligned}$$

If $a_{11}a_{22} - a_{21}a_{12} \neq 0$, we can solve for x:

$$x = \frac{k_1a_{22} - k_2a_{12}}{a_{11}a_{22} - a_{21}a_{12}}.$$

This formula for x is much easier to remember in a form using determinants:

$$x = \frac{\begin{vmatrix} k_1 & a_{12} \\ k_2 & a_{22} \end{vmatrix}}{\begin{vmatrix} a_{11} & a_{12} \\ a_{21} & a_{22} \end{vmatrix}}.$$

Similarly, if we solve the original system for y, we get

$$y = \frac{\begin{vmatrix} a_{11} & k_1 \\ a_{21} & k_2 \end{vmatrix}}{\begin{vmatrix} a_{11} & a_{12} \\ a_{21} & a_{22} \end{vmatrix}}.$$

Notice that in both cases the denominator is the determinant of the coefficient matrix. (Recall from Section 10.2 that the coefficient matrix is the 2×2 array of coefficients). Also notice that if the first column of the denominator is replaced with the column consisting of k_1 and k_2, the resulting 2×2 determinant is the numerator for x. Similarly, the numerator for y is the 2×2 determinant that results from replacing the second column of the denominator with k_1 and k_2. This scheme for the solution of two equatons in two unknowns is known as **Cramer's rule.**

Cramer's Rule for Two Equations in Two Unknowns

Given the system

$$\begin{cases} a_{11}x + a_{12}y = k_1 \\ a_{21}x + a_{22}y = k_2, \end{cases}$$

let D be the 2×2 determinant of the coefficient matrix:

$$D = \begin{vmatrix} a_{11} & a_{12} \\ a_{21} & a_{22} \end{vmatrix}.$$

Let D_x be the determinant formed by replacing the column of coefficients for x with k_1 and k_2, respectively:

$$D_x = \begin{vmatrix} k_1 & a_{12} \\ k_2 & a_{22} \end{vmatrix}.$$

Similarly, let D_y be the determinant formed by replacing the column of coefficients for y with k_1 and k_2, respectively:

$$D_y = \begin{vmatrix} a_{11} & k_1 \\ a_{21} & k_2 \end{vmatrix}.$$

If $D \neq 0$, then the solution to the system is unique and is given by

$$x = \frac{D_x}{D} \quad \text{and} \quad y = \frac{D_y}{D}.$$

EXAMPLE 4 Use Cramer's rule to solve

$$\begin{cases} 3x + 4y = -1 \\ 2x - 6y = 5. \end{cases}$$

SOLUTION The determinant of the coefficient matrix is

$$D = \begin{vmatrix} 3 & 4 \\ 2 & -6 \end{vmatrix} = -18 - 8 = -26.$$

Also,

$$D_x = \begin{vmatrix} -1 & 4 \\ 5 & -6 \end{vmatrix} = -14 \quad \text{and} \quad D_y = \begin{vmatrix} 3 & -1 \\ 2 & 5 \end{vmatrix} = 17.$$

So,

$$x = \frac{D_x}{D} = \frac{-14}{-26} = \frac{7}{13} \quad \text{and} \quad y = \frac{D_y}{D} = \frac{17}{-26} = -\frac{17}{26}.$$

The solution is $\left(\dfrac{7}{13}, -\dfrac{17}{26}\right)$.

EXAMPLE 5 Use Cramer's rule to solve the following:

a) $\begin{cases} -2x + y = 1 \\ 8x - 4y = 5 \end{cases}$ 　　　　 **b)** $\begin{cases} -2x + y = 3 \\ 8x - 4y = -12 \end{cases}$

SOLUTION

a) We start by computing the values of D, D_x, and D_y.

$$D = \begin{vmatrix} -2 & 1 \\ 8 & -4 \end{vmatrix} = 0 \qquad D_x = \begin{vmatrix} 1 & 1 \\ 5 & -4 \end{vmatrix} = -9$$

$$D_y = \begin{vmatrix} -1 & 1 \\ 8 & 5 \end{vmatrix} = -13.$$

So, we find that

$$x = \frac{-9}{0} \quad \text{is undefined and} \quad y = \frac{-13}{0} \quad \text{is undefined.}$$

Solving this system by other means shows that this system is inconsistent (pause and solve it).

b) Computing the values of D, D_x, and D_y, we get

$$D = \begin{vmatrix} -2 & 1 \\ 8 & -4 \end{vmatrix} = 0 \qquad D_x = \begin{vmatrix} 3 & 1 \\ -12 & -4 \end{vmatrix} = 0$$

$$D_y = \begin{vmatrix} -2 & 3 \\ 8 & -12 \end{vmatrix} = 0$$

So we find that

$$x = \frac{0}{0} \quad \text{is indeterminant} \quad \text{and} \quad y = \frac{0}{0} \quad \text{is indeterminant.}$$

Solving this system by other means shows that this system is dependent (pause and solve it).

Cramer's rule should not be used in either part (a) or part (b) since $D = 0$ in each case. We followed through with the calculations of x and y to make a point.

It can be shown that if Cramer's rule produces "solutions" in which the denominator is zero and each of the numerators is not zero, then the system is inconsistent. If any of the solutions have the form $\frac{0}{0}$, then the system is dependent.

Cramer's rule generalizes easily to systems of n equations in n unknowns. However, for systems in which n is larger than 3, the calculations become tedious, and the matrix method discussed in Section 10.2 should probably be used. For this reason, we state Cramer's rule only for 3 linear equations in 3 unknowns.

Given the system

$$\begin{cases} a_{11}x + a_{12}y + a_{13}z = k_1 \\ a_{21}x + a_{22}y + a_{23}z = k_2 \\ a_{31}x + a_{32}y + a_{33}z = k_3 \end{cases}$$

Cramer's Rule for Three Equations in Three Unknowns

let

$$D = \begin{vmatrix} a_{11} & a_{12} & a_{13} \\ a_{21} & a_{22} & a_{23} \\ a_{31} & a_{32} & a_{33} \end{vmatrix} \quad D_x = \begin{vmatrix} k_1 & a_{12} & a_{13} \\ k_2 & a_{22} & a_{23} \\ k_3 & a_{32} & a_{33} \end{vmatrix}$$

$$D_y = \begin{vmatrix} a_{11} & k_1 & a_{13} \\ a_{21} & k_2 & a_{23} \\ a_{31} & k_3 & a_{33} \end{vmatrix} \quad D_z = \begin{vmatrix} a_{11} & a_{12} & k_1 \\ a_{21} & a_{22} & k_2 \\ a_{31} & a_{32} & k_3 \end{vmatrix}.$$

If $D \neq 0$, then the unique solution to the system of equations is

$$x = \frac{D_x}{D} \qquad y = \frac{D_y}{D} \qquad z = \frac{D_z}{D}.$$

EXAMPLE 6 Use Cramer's rule to solve

$$\begin{cases} x + 2y - z = -3 \\ 2x \quad\quad + z = 1. \\ 3x - 4y + 2z = 1 \end{cases}$$

SOLUTION The determinant D is

$$D = \begin{vmatrix} 1 & 2 & -1 \\ 2 & 0 & 1 \\ 3 & -4 & 2 \end{vmatrix}.$$

To simplify the calculation, add 2 times the first row to the third row:

$$D = \begin{vmatrix} 1 & 2 & -1 \\ 2 & 0 & 1 \\ 5 & 0 & 0 \end{vmatrix}.$$

Use the third row to evaluate:

$$D = 5\left[(1)\begin{vmatrix} 2 & -1 \\ 0 & 1 \end{vmatrix}\right] = 5[2] = 10.$$

The calculation for D_x is very similar:

$$D_x = \begin{vmatrix} -3 & 2 & -1 \\ 1 & 0 & 1 \\ 1 & -4 & 2 \end{vmatrix} = \begin{vmatrix} -3 & 2 & -1 \\ 1 & 0 & 1 \\ -5 & 0 & 0 \end{vmatrix} = -5\left[(1)\begin{vmatrix} 2 & -1 \\ 0 & 1 \end{vmatrix}\right] = -10.$$

To simplify the evaluation of D_y, add 3 times the third row to the first row, and add -1 times the third row to the second row:

$$D_y = \begin{vmatrix} 1 & -3 & -1 \\ 2 & 1 & 1 \\ 3 & 1 & 2 \end{vmatrix} = \begin{vmatrix} 10 & 0 & 5 \\ -1 & 0 & -1 \\ 3 & 1 & 2 \end{vmatrix} = 1\left[(-1)\begin{vmatrix} 10 & 5 \\ -1 & -1 \end{vmatrix}\right] = 5.$$

We leave it to you to verify the steps for D_z:

$$D_z = \begin{vmatrix} 1 & 2 & -3 \\ 2 & 0 & 1 \\ 3 & -4 & 1 \end{vmatrix} = \begin{vmatrix} 1 & 2 & -3 \\ 2 & 0 & 1 \\ 5 & 0 & -5 \end{vmatrix} = 2\left[(-1)\begin{vmatrix} 2 & 1 \\ 5 & -5 \end{vmatrix}\right] = 30.$$

The solutions for the system are

$$x = \frac{D_x}{D} = \frac{-10}{10} = -1 \qquad y = \frac{D_y}{D} = \frac{5}{10} = \frac{1}{2} \qquad z = \frac{D_z}{D} = \frac{30}{10} = 3.$$

Most proofs of Cramer's rule are long and systematic. However, one very elegant proof by D. E. Whitford and M. S. Klamkin was published in the American Mathematical Monthly in 1953. Given the system

Proof of Cramer's Rule for Three Equations in Three Unknowns

$$\begin{cases} a_{11}x + a_{12}y + a_{13}z = k_1 \\ a_{21}x + a_{22}y + a_{23}z = k_2 \\ a_{31}x + a_{32}y + a_{33}z = k_3 \end{cases}$$

Whitford and Klamkin started with the expression

$$x \begin{vmatrix} a_{11} & a_{12} & a_{13} \\ a_{21} & a_{22} & a_{23} \\ a_{31} & a_{32} & a_{33} \end{vmatrix} = \begin{vmatrix} a_{11}x & a_{12} & a_{13} \\ a_{21}x & a_{22} & a_{23} \\ a_{31}x & a_{32} & a_{33} \end{vmatrix}$$

(recall that replacing a column with a multiple of itself multiplies the determinant by that amount).

Now, replacing the first column with the sum of y times the second column and the first column gives

$$= \begin{vmatrix} a_{11}x + a_{12}y & a_{12} & a_{13} \\ a_{21}x + a_{22}y & a_{22} & a_{23} \\ a_{31}x + a_{32}y & a_{32} & a_{33} \end{vmatrix}.$$

Similarly, replacing the first column with the sum of z times the third column and the first column gives

$$= \begin{vmatrix} a_{11}x + a_{12}y + a_{13}z & a_{12} & a_{13} \\ a_{21}x + a_{22}y + a_{23}z & a_{22} & a_{23} \\ a_{31}x + a_{32}y + a_{33}z & a_{32} & a_{33} \end{vmatrix}.$$

Comparing the first column to the given system leads to

$$= \begin{vmatrix} k_1 & a_{12} & a_{13} \\ k_2 & a_{22} & a_{23} \\ k_3 & a_{32} & a_{33} \end{vmatrix}.$$

Connecting the beginning and end of this string of equations, we have

$$x \begin{vmatrix} a_{11} & a_{12} & a_{13} \\ a_{21} & a_{22} & a_{23} \\ a_{31} & a_{32} & a_{33} \end{vmatrix} = \begin{vmatrix} k_1 & a_{12} & a_{13} \\ k_2 & a_{22} & a_{23} \\ k_3 & a_{32} & a_{33} \end{vmatrix}.$$

This says that $xD = D_x$; so if $D \neq 0$, $x = D_x/D$.

The proofs for solutions y and z are very similar and are left as an exercise.

If the material in Sections 10.1 and 10.2 is still fairly fresh in your mind, you now have several methods at your command for solving linear systems of equations. This kind of power deserves good judgment. To help you make a wise choice of method of solution, we summarize some of the limitations of Cramer's rule.

Cramer's rule can be applied only to systems with the same number of equations as unknowns. The rule gives a solution only for those cases in which there is a unique solution. The computations of the determinants for systems with more than 3 unknowns are so tedious that Cramer's rule becomes impractical.

EXERCISE SET 10.4

A

Solve the systems in Problems 1 through 20 using Cramer's rule, if possible.

1. $\begin{cases} 3x + 2y = 10 \\ x - 4y = 22 \end{cases}$

2. $\begin{cases} 2x + y = 1 \\ 6x - 5y = -37 \end{cases}$

3. $\begin{cases} 3x + y = 5 \\ -6x + 7y = 17 \end{cases}$

4. $\begin{cases} 3x + 4y = 2 \\ 5x - 8y = -13 \end{cases}$

5. $\begin{cases} 3x + y = 4 \\ x + 5y - 9 = 0 \end{cases}$

6. $\begin{cases} 2x + 7y - 1 = 0 \\ 2x - 8y = 5 \end{cases}$

7. $\begin{cases} 3y + 2x = x + 8 \\ x = 4y - 5 \end{cases}$

8. $\begin{cases} y - 2 = x \\ 6x - 5y = 2y - 1 \end{cases}$

9. $\begin{cases} 6x - 10y = 1 \\ 9x - 15y = -4 \end{cases}$

10. $\begin{cases} 3x + 12y = 2 \\ 2x + 8y = 5 \end{cases}$

11. $\begin{cases} 3x + 2y + z = 4 \\ x - y + z = 5 \\ 2x - y = 11 \end{cases}$

12. $\begin{cases} 6x + y - z = 7 \\ x + 2y + z = -1 \\ 3y + 2z = -4 \end{cases}$

13. $\begin{cases} 4x + y - 4z = 4 \\ x - y + 2z = -2 \\ 2x - y + 8z = 0 \end{cases}$

14. $\begin{cases} -4x - 4y - z = 0 \\ 3x + 6y + 2z = 3 \\ x - 2y + 5z = 4 \end{cases}$

15. $\begin{cases} 3x - y - 4z = 1 \\ 5x - y + z = -2 \\ 2x - y = 0 \end{cases}$

16. $\begin{cases} y - 4z = 0 \\ 2x + 2z = 3 \\ x + 3y + 5z = 4 \end{cases}$

17. $\begin{cases} 4x + y - 4z = 4 \\ x - y + 2z = -2 \end{cases}$

18. $\begin{cases} -4x - 4y - z = 0 \\ 3x + 6y + 2z = 3 \end{cases}$

19. $\begin{cases} 2x + y - z = 6 \\ x - 4y + 2z = -1 \\ 5x - 2y = 11 \end{cases}$

20. $\begin{cases} x - y + 3z = 1 \\ 3x + 3y + z = 0 \\ 7x - y + 13z = 7 \end{cases}$

B

Let

$$\begin{vmatrix} a_1 & a_2 & a_3 \\ b_1 & b_2 & b_3 \\ c_1 & c_2 & c_3 \end{vmatrix} = 18.$$

Use this to find the determinants in Problems 21 through 26.

21. $\begin{vmatrix} a_1 & a_2 & a_3 \\ c_1 & c_2 & c_3 \\ b_1 & b_2 & b_3 \end{vmatrix}$

22. $\begin{vmatrix} a_3 & a_2 & a_1 \\ b_3 & b_2 & b_1 \\ c_3 & c_2 & c_1 \end{vmatrix}$

23. $\begin{vmatrix} a_1 & a_2 & 5a_3 \\ b_1 & b_2 & 5b_3 \\ c_1 & c_2 & 5c_3 \end{vmatrix}$

24. $\begin{vmatrix} a_1 & a_2 & a_3 \\ 2c_1 & 2c_2 & 2c_3 \\ b_1 & b_2 & b_3 \end{vmatrix}$

25. $\begin{vmatrix} a_1 & a_2 & 4a_1 + a_3 \\ b_1 & b_2 & 4b_1 + b_3 \\ c_1 & c_2 & 4c_1 + c_3 \end{vmatrix}$

26. $\begin{vmatrix} c_1 & c_2 & c_3 \\ 2a_1 + b_1 & 2a_2 + b_2 & 2a_3 + b_3 \\ a_1 & a_2 & a_3 \end{vmatrix}$

In Problems 27 through 32, solve for the indicated variables using Cramer's rule.

27. Solve for a and b

$$\begin{cases} 3xa + x^2b = 4 \\ 5a + 2xb = -1 \end{cases}.$$

28. Solve for s and t

$$\begin{cases} 4s + xt = 6 \\ 5xs + 2x^2t = 8 \end{cases}.$$

29. Solve for u and v

$$\begin{cases} ue^x + ve^{2x} = 1 \\ ue^x + 2ve^{2x} = 0 \end{cases}.$$

30. Solve for u and v

$$\begin{cases} ue^{3x} + ve^{4x} = 1 \\ 3ue^{3x} + 4ve^{4x} = 0 \end{cases}.$$

31. Solve for x' and y'

$$\begin{cases} x'\cos\theta - y'\sin\theta = x \\ x'\sin\theta + y'\cos\theta = y \end{cases}.$$

32. Solve for x' and y'

$$\begin{cases} x'\sec\theta + y'\tan\theta = x \\ x'\tan\theta + y'\sec\theta = y \end{cases}.$$

33. Show by direct computation that

$$\begin{vmatrix} a_{11} & a_{12} \\ a_{21} & a_{22} \end{vmatrix} = -\begin{vmatrix} a_{12} & a_{11} \\ a_{22} & a_{21} \end{vmatrix}.$$

34. Show by direct computation that

$$k\begin{vmatrix} a_{11} & a_{12} \\ a_{21} & a_{22} \end{vmatrix} = \begin{vmatrix} a_{11} & a_{12} \\ ka_{21} & ka_{22} \end{vmatrix}.$$

35. Show by direct computation that

$$\begin{vmatrix} a_{11} & a_{12} \\ a_{21} & a_{22} \end{vmatrix} = \begin{vmatrix} a_{11} + ka_{21} & a_{12} + ka_{22} \\ a_{21} & a_{22} \end{vmatrix}.$$

In Problems 36 and 37, extend Cramer's rule to solve the system.

36. $\begin{cases} 2x + y - z + w = 3 \\ x - 4y + 2w = -2 \\ 5x - 2y + z = 0 \\ x - y + z = 1 \end{cases}.$

37. $\begin{cases} x - y + 3z - 2w = 1 \\ 3x + z + w = 6 \\ 6x - 4y + 4z = 0 \\ 5x - 2z - w = 2 \end{cases}.$

C

38. Show that the determinant of an upper triangular matrix is the product of its main diagonal; that is,

$$\begin{vmatrix} a_{11} & a_{12} & a_{13} \\ 0 & a_{22} & a_{23} \\ 0 & 0 & a_{33} \end{vmatrix} = a_{11}a_{22}a_{33}.$$

39. Use elementary row operations of determinants to transform the determinant in Example 3 of this section to an upper triangular form. Evaluate using Problem 38.

40. Use elementary row operations of determinants to transform the determinant in Example 5 of Section 10.3 to an upper triangular form. Explain why the determinant is the product of the terms on the main diagonal.

41. Show that the system of equations

$$\begin{cases} ax + by = c \\ kax + kby = d \end{cases}$$

is not consistent. Under what conditions is this system dependent?

42. Show that

$$\begin{vmatrix} \cos\theta & \sin\theta \\ -r\sin\theta & r\cos\theta \end{vmatrix} = r.$$

43. Show that

$$\begin{vmatrix} \sin\phi\cos\theta & p\cos\phi\cos\theta & -p\sin\phi\sin\theta \\ \sin\phi\sin\theta & p\cos\phi\sin\theta & p\sin\phi\cos\theta \\ \cos\phi & -p\sin\phi & 0 \end{vmatrix} = p^2\sin\phi.$$

SECTION 10.5 PARTIAL FRACTIONS

Back in Section 1.4, much of the discussion dealt with simplifying expressions such as

$$\frac{3}{x+2} + \frac{7}{x-4}$$

by finding a common denominator, building the fractions, and adding. You should pause here and verify, for example, that

$$\frac{3}{x+2} + \frac{7}{x-4} = \frac{10x+2}{(x+2)(x-4)}.$$

In calculus, it will be important to rewrite a given rational expression into a sum or difference of simpler rational expressions. In this case, simpler means a denominator of lesser degree. This process is called **partial fraction decomposition.**

One of the amazing facts about real polynomials (recall that real polynomials are polynomials with real coefficients) is that in theory they can be factored into linear factors and quadratic factors that are irreducible over the real numbers (see Problem 46 of this section for a sketch of the proof). A quadratic factor $ax^2 + bx + c$ is **irreducible over the real numbers,** or **irreducible,** if it cannot be factored over the real numbers. (You may wish to pause here and review Section 3.3.) There is a way to quickly determine if a given quadratic expression is irreducible. Applying the factor theorem of Section 3.2, the quadratic expression $ax^2 + bx + c$ can be factored into real linear factors $(x - r_1)(x - r_2)$ if and only if r_1 and r_2 are roots of the equation $ax^2 + bx + c = 0$. Under what conditions does a quadratic equation have real roots? Recall that the discriminant $b^2 - 4ac$ answers this question.

Test for Irreducible Quadratic Expressions

> The quadratic expression $ax^2 + bx + c$ is irreducible over the real numbers if and only if $b^2 - 4ac < 0$.

In practice, it is not always so easy to factor a polynomial into linear and irreducible quadratic factors, but the polynomials encountered in calculus usually yield to the common factoring techniques.

Suppose we start with the rational expression

$$\frac{x+9}{(x-3)(x+1)}$$

and try to find two simpler rational expressions for which this expression is the sum. It should seem reasonable that the denominators of the two simpler expressions could be $(x - 3)$ and $(x + 1)$. In fact, the partial fraction decomposition is

$$\frac{x + 9}{(x - 3)(x + 1)} = \frac{3}{x - 3} + \frac{-2}{x + 1}.$$

This result is easily verified by adding the two terms on the right side and comparing that sum with the left side. The difficult part obviously is determining the numerators of the partial fractions. The next example shows how this is done.

EXAMPLE 1   Find the partial fraction decomposition of

$$\frac{2x + 1}{x^2 - 4}.$$

SOLUTION We start by factoring the denominator:

$$\frac{2x + 1}{x^2 - 4} = \frac{2x + 1}{(x - 2)(x + 2)}.$$

This tells us the possible denominators for the partial fractions. Because the original rational expression is proper (the degree of the numerator is less than the degree of the denominator), each of the partial fractions is also proper. The degree of each of the denominators is 1, so this forces the numerators to be constants. Since these are yet undetermined, we call them A and B:

$$\frac{2x + 1}{(x - 2)(x + 2)} = \frac{A}{(x - 2)} + \frac{B}{(x + 2)}.$$

The problem now is reduced to finding these constants A and B. We multiply each side of this equation by $(x + 2)(x - 2)$ to clear the equation of fractions:

$$2x + 1 = A(x + 2) + B(x - 2).$$

Next, we simplify the left side and collect similar terms:

$$2x + 1 = Ax + 2A + Bx - 2B$$

$$2x + 1 = (A + B)x + (2A - 2B).$$

These two expressions are equal for all values of x. This can happen only if the coefficient of x and the constant term are the same for both sides of the equation. This gives a system of equations:

$$\begin{cases} A + B = 2 \\ 2A - 2B = 1 \end{cases}$$

$$-2E_1 + E_2 \rightarrow E_2 \begin{cases} A + B = 2 \\ -4B = -3. \end{cases}$$

By back substitution, $A = \frac{5}{4}$ and $B = \frac{3}{4}$. Thus the partial fraction decomposition of the original rational expression is

$$\frac{2x + 1}{x^2 - 4} = \frac{\frac{5}{4}}{(x - 2)} + \frac{\frac{3}{4}}{(x + 2)}$$

or

$$\frac{2x + 1}{x^2 - 4} = \frac{5}{4(x - 2)} + \frac{3}{4(x + 2)}.$$

In general, each factor of the denominator of a rational expression contributes one or more terms to the partial fraction decomposition of the rational expression. The usual approach to finding a decomposition is first to find the form of the decomposition and then to compute the undetermined coefficients of the numerators.

The table that follows shows how the factors of the denominator dictate the type of terms in the decomposition. Don't despair if the table seems a bit confusing right now; the next three examples show how to find the appropriate terms using this table.

Types of Terms in a Partial Fraction Decomposition

Factor in the Denominator	Corresponding Term in the Partial Fraction Decomposition
First power of linear factor $ax + b$	$\dfrac{A}{ax + b}$
Second power of a linear factor $(ax + b)^2$	$\dfrac{A}{ax + b} + \dfrac{B}{(ax + b)^2}$
First power of an irreducible quadratic factor $ax^2 + bx + c$	$\dfrac{Ax + B}{ax^2 + bx + c}$
Second power of an irreducible quadratic factor $(ax^2 + bx + c)^2$	$\dfrac{Ax + B}{ax^2 + bx + c} + \dfrac{Cx + D}{(ax^2 + bx + c)^2}$

The results in the table can be generalized to factors that are higher powers of linear factors and irreducible quadratic factors (see Problems 42 through 45). For now, though, we limit our investigation to those factors shown in our table.

EXAMPLE 2 Find the partial fraction decomposition of

$$\frac{4x - 5}{x(2x + 1)^2}.$$

SOLUTION The denominator is the product of a linear factor x and the square of a linear factor $(2x + 1)^2$. The form of the partial fraction decomposition is

$$\frac{4x - 5}{x(2x + 1)^2} = \frac{A}{x} + \frac{B}{2x + 1} + \frac{C}{(2x + 1)^2}.$$

Multiplying through by $x(2x + 1)^2$ to clear the fractions from the equation, we get

$$4x - 5 = A(2x + 1)^2 + Bx(2x + 1) + Cx.$$

Next, we collect terms of the right side and compare corresponding coefficients of each side:

$$0x^2 + 4x - 5 = (4A + 2B)x^2 + (4A + B + C)x + A.$$

This gives the system of linear equations:

$$\begin{cases} 4A + 2B & = & 0 \\ 4A + B + C & = & 4 \\ A & = & -5 \end{cases}$$

Substituting $A = -5$ into the first equation gives us

$$4(-5) + 2B = 0 \Rightarrow B = 10$$

so,

$$4(-5) + (10) + C = 4 \Rightarrow C = 14.$$

The partial fraction decomposition is

$$\frac{4x - 5}{x(2x + 1)^2} = \frac{-5}{x} + \frac{10}{2x + 1} + \frac{14}{(2x + 1)^2}.$$

EXAMPLE 3   Find the partial fraction decomposition of

$$\frac{2x^2 - 4x + 5}{(x + 3)(x^2 + 2x + 4)}.$$

SOLUTION The denominator is the product of a linear factor and an irreducible quadratic factor ($x^2 + 2x + 4$ is irreducible since $b^2 - 4ac = (2)^2 - 4(1)(4) = -12$). The form of the partial fraction decomposition is

$$\frac{2x^2 - 4x + 5}{(x + 3)(x^2 + 2x + 4)} = \frac{A}{x + 3} + \frac{Bx + C}{x^2 + 2x + 4}.$$

Clearing the fractions and combining similar terms yields

$$2x^2 - 4x + 5 = A(x^2 + 2x + 4) + (Bx + C)(x + 3)$$
$$= (A + B)x^2 + (2A + 3B + C)x + (4A + 3C).$$

Equating coefficients yields the system

$$\begin{cases} A + B = 2 \\ 2A + 3B + C = -4. \\ 4A + 3C = 5 \end{cases}$$

Using the matrix methods of Section 10.2, we have

$$\begin{bmatrix} 1 & 1 & 0 & | & 2 \\ 2 & 3 & 1 & | & -4 \\ 4 & 0 & 3 & | & 5 \end{bmatrix}$$

$$\begin{matrix} \\ -2R_1 + R_2 \to R_2 \\ -4R_1 + R_3 \to R_3 \end{matrix} \begin{bmatrix} 1 & 1 & 0 & | & 2 \\ 0 & 1 & 1 & | & -8 \\ 0 & -4 & 3 & | & -3 \end{bmatrix}$$

$$\begin{matrix} -R_2 + R_1 \to R_1 \\ \\ 4R_2 + R_3 \to R_3 \end{matrix} \begin{bmatrix} 1 & 0 & -1 & | & 10 \\ 0 & 1 & 1 & | & -8 \\ 0 & 0 & 7 & | & -35 \end{bmatrix}$$

$$\begin{matrix} \\ \\ \tfrac{1}{7}R_3 \to R_3 \end{matrix} \begin{bmatrix} 1 & 0 & -1 & | & 10 \\ 0 & 1 & 1 & | & -8 \\ 0 & 0 & 1 & | & -5 \end{bmatrix}$$

$$\begin{matrix} R_3 + R_1 \to R_1 \\ -R_3 + R_2 \to R_2 \\ \\ \end{matrix} \begin{bmatrix} 1 & 0 & 0 & | & 5 \\ 0 & 1 & 0 & | & -3 \\ 0 & 0 & 1 & | & -5 \end{bmatrix}.$$

The solution to the system of equations is $A = 5$, $B = -3$, and $C = -5$. The partial fraction decomposition is

$$\frac{2x^2 - 4x + 5}{(x + 3)(x^2 + 2x + 4)} = \frac{5}{x + 3} + \frac{-3x - 5}{x^2 + 2x + 4}.$$

EXAMPLE 4   Find the partial fraction decomposition of

$$\frac{2x^2 - 3x + 4}{x(x^2 + 1)^2}.$$

SOLUTION Using the table, the partial fraction decomposition has the form

$$\frac{2x^2 - 3x + 4}{x(x^2 + 1)^2} = \frac{A}{x} + \frac{Bx + C}{x^2 + 1} + \frac{Dx + E}{(x^2 + 1)^2}.$$

Proceeding, we have

$$2x^2 - 3x + 4 = A(x^2 + 1)^2 + (Bx + C)x(x^2 + 1) + (Dx + E)x$$
$$= A(x^4 + 2x^2 + 1) + B(x^4 + x^2) + C(x^3 + x) + Dx^2 + Ex$$
$$= (A + B)x^4 + Cx^3 + (2A + B + D)x^2 + (C + E)x + A.$$

Equating coefficients gives this system:

$$\begin{cases} A + B & = & 0 \\ \quad C & = & 0 \\ 2A + B \quad + D & = & 2 \\ \quad C \quad + E & = & -3 \\ A & = & 4. \end{cases}$$

Even though this is a system of five equations in five unknowns, it is relatively easy to solve by substitution. We leave it to you to determine that the solution of this system is

$$A = 4, \quad B = -4, \quad C = 0, \quad D = -2, \quad \text{and} \quad E = -3.$$

So,

$$\frac{2x^2 - 3x + 4}{x(x^2 + 1)^2} = \frac{4}{x} + \frac{-4x}{x^2 + 1} + \frac{-2x - 3}{(x^2 + 1)^2}.$$

The determination of the partial fraction decomposition can be very tedious for a rational expression with many factors in the denominator. The next example illustrates a trick to sidestep much of this tedium.

EXAMPLE 5   Find the partial fraction decomposition of

$$\frac{x + 7}{x^4 - 13x^2 + 36}.$$

SOLUTION

$$\frac{x + 7}{x^4 - 13x^2 + 36} = \frac{x + 7}{(x^2 - 9)(x^2 - 4)}$$

$$= \frac{x + 7}{(x - 3)(x + 3)(x - 2)(x + 2)}$$

$$= \frac{A}{x - 3} + \frac{B}{x + 3} + \frac{C}{x - 2} + \frac{D}{x + 2}$$

$$x + 7 = A(x + 3)(x - 2)(x + 2) + B(x - 3)(x - 2)(x + 2)$$
$$+ C(x - 3)(x + 3)(x + 2) + D(x - 3)(x + 3)(x - 2).$$

If we were to proceed as we have in the last three examples, we would be faced with the enormous task of simplifying the left side, collecting similar terms, and solving the resulting system of four equations in four unknowns. An alternative method would certainly be welcome here.

You can assume that the last equation is true for all real values of x. Specifically, it is true for $x = 3$. Substituting $x = 3$ into the equation gives

$$(3) + 7 = A(6)(1)(5) + B(0)(1)(5) + C(0)(6)(5) + D(0)(6)(1).$$

Notice that choosing $x = 3$ makes all but one term on the right side equal to zero. Simplifying gives

$$10 = 30A \Rightarrow A = \tfrac{1}{3}.$$

Other judicious choices for x are shown below, along with their results:

$$x = -3 \Rightarrow 4 = -30B \Rightarrow B = -\tfrac{2}{15}$$
$$x = 2 \quad \Rightarrow 9 = -20C \Rightarrow C = -\tfrac{9}{20}$$
$$x = -2 \Rightarrow 5 = 20D \quad \Rightarrow D = \tfrac{1}{4}.$$

The partial fraction decomposition is

$$\frac{x + 7}{x^4 - 13x^2 + 36} = \frac{\tfrac{1}{3}}{x - 3} + \frac{-\tfrac{2}{15}}{x + 3} + \frac{-\tfrac{9}{20}}{x - 2} + \frac{\tfrac{1}{4}}{x + 2}$$

$$= \frac{1}{3(x - 3)} - \frac{2}{15(x + 3)} - \frac{9}{20(x - 2)} + \frac{1}{4(x + 2)}.$$

The particular method of finding the coefficients of the partial fractions used in the last example is called the **Heaviside method** (after Oliver Heaviside [1850–1925], an English engineer and scientist). This method helps find coefficients associated with linear factors. It can be used to find some or all the partial fraction coefficients in most problems.

So, far we have applied partial fraction decomposition to proper rational expressions. When faced with an improper rational expression, we can use polynomial division to reduce it to the sum of a polynomial and a proper rational expression. The proper rational expression in turn can be decomposed into its partial fractions.

Recall that an improper rational expression is one in which the numerator is of degree greater than or equal to the degree of the denominator.

EXAMPLE 6 Find the partial fraction decomposition of

$$\frac{2x^4 - x^3 - 5x^2 + 6x - 11}{x^2 - 4}.$$

SOLUTION Since the degree of the denominator is two and the degree of the numerator is four, this is an improper rational expression. In Example 1 of Section 3.2, polynomial division was used to show that

$$\frac{2x^4 - x^3 - 5x^2 + 6x - 11}{x^2 - 4} = 2x^2 - x + 3 + \frac{2x + 1}{x^2 - 4}.$$

In Example 1 of this section, we found that

$$\frac{2x + 1}{x^2 - 4} = \frac{5}{4(x - 2)} + \frac{3}{4(x + 2)}$$

so,

$$\frac{2x^4 - x^3 - 5x^2 + 6x - 11}{x^2 - 4} = 2x^2 - x + 3 + \frac{5}{4(x - 2)} + \frac{3}{4(x + 2)}.$$

EXERCISE SET 10.5

A

In Problems 1 through 12, the rational expression has a denominator that has only linear factors. Find its partial fraction decomposition.

1. $\dfrac{12}{(x + 5)(x - 1)}$

2. $\dfrac{7}{(x - 4)(x + 3)}$

3. $\dfrac{3x - 31}{x^2 - 9x + 8}$

4. $\dfrac{5x - 2}{x^2 - 2x - 8}$

5. $\dfrac{4}{x^2 - 2x}$

6. $\dfrac{2}{x^2 + 4x}$

7. $\dfrac{3x + 5}{x^2 + 2x + 1}$

8. $\dfrac{6x}{(4x^2 - 4x + 1)}$

9. $\dfrac{1}{x^3 - x}$

10. $\dfrac{6}{x^3 + 4x^2 + 3x}$

11. $\dfrac{x - 2}{(x^2 + x - 6)(x^2 + 2x - 3)}$

12. $\dfrac{x + 4}{(x^2 - 4)(x^2 + 2x - 8)}$

In Problems 13 through 24, the rational expression has a denominator that may have linear or quadratic factors. Find its partial fraction decomposition.

13. $\dfrac{4}{x(x^2 + 1)}$

14. $\dfrac{2}{x(x^2 + 9)}$

15. $\dfrac{2x^2 - 10x + 2}{(x - 2)(x^2 + 2x + 2)}$

16. $\dfrac{3x - 2}{(x^2 - x + 3)(x + 1)}$

17. $\dfrac{x^3}{(x^2 + 1)^2}$

18. $\dfrac{3x^2}{(x^2 + x + 2)^2}$

19. $\dfrac{7}{2x^4 + 13x^2 + 15}$

20. $\dfrac{2}{x(x^2 - 9)}$

21. $\dfrac{5x^2 - 21x + 13}{(x + 2)(x^2 - 6x + 9)}$

22. $\dfrac{2}{x(x^2 + 4x + 4)}$

23. $\dfrac{1}{x^2 - 2}$

24. $\dfrac{\sqrt{5}\,x}{x^2 - 5}$

B

In Problems 25 through 33, find the partial fraction decomposition of the rational expression. If the expression is an improper rational expression, use polynomial division to rewrite it as a polynomial and a proper expression. You may find it useful to review the rational root theorem of Section 3.3.

25. $\dfrac{13x + 1}{(x^3 - 8)}$

26. $\dfrac{x^2 + 5}{x^3 + 1}$

27. $\dfrac{1}{x^4 - 16}$

28. $\dfrac{x^3}{x^2 - 1}$

29. $\dfrac{x^4}{x^2 - 9}$

30. $\dfrac{x^3 + 2x + 1}{x^2 + 2x + 1}$

31. $\dfrac{2x^3 + 9x^2 + 3x + 2}{2x^3 - x^2 + 2x - 1}$

32. $\dfrac{x^3 - 14}{2x^4 - x^3 + x^2 - 2x - 6}$

33. $\dfrac{12}{x^4 - 2x^3 - 7x^2 + 8x + 12}$

34. Find the partial fraction decomposition (in terms of $\cos \theta$) for
$$\frac{\cos \theta}{\cos^2\theta - 2 \cos \theta - 3}.$$

35. Find the partial fraction decomposition (in terms of $\sin \theta$) for
$$\frac{1}{2 \sin^2\theta - 1}.$$

36. Find the partial fraction decomposition (in terms of e^x) for
$$\frac{4}{e^{2x} - e^x}.$$

37. Find the partial fraction decomposition (in terms of $\ln x$) for
$$\frac{\ln(x^2)}{(\ln x + 2)(\ln x - 2)}.$$

38. Suppose that a student incorrectly sets up a partial fraction decomposition as
$$\frac{x^2}{x^2 - 4} = \frac{A}{x - 2} + \frac{B}{x + 2}.$$
Proceed with this incorrect solution and arrive at a system of equations. Explain what is wrong with this system and why this solution fails.

39. Suppose that a student sets up a partial fraction decomposition as
$$\frac{x - 4}{x^2 - 6x + 8} = \frac{A}{x - 4} + \frac{B}{x - 2}.$$
Proceed with this solution and arrive at a system of equations. Solve the system to determine the values of A and B. Explain how this student could have arrived at the same result without using partial fraction decomposition.

40. Consider the partial fraction decomposition:
$$\frac{2x + 3}{x(x^2 + 1)} = \frac{A}{x} + \frac{Bx + C}{x^2 + 1}.$$
This leads to $2x + 3 = A(x^2 + 1) + (Bx + C)x$. Use the Heaviside method with $x = 0$ and $x = i$ and the fact that two complex numbers $a + bi$ and $c + di$ are equal if and only if $a = c$ and $b = d$ to find the values of A, B, and C.

41. Use the Heaviside method outlined in Problem 40 to find the partial fraction decomposition of

$$\frac{2}{(x-1)(x^2+4)}.$$

42. Given a rational expression, the occurrence of a third power of a linear factor in its denominator results in the terms

$$\frac{A}{(ax+b)} + \frac{B}{(ax+b)^2} + \frac{C}{(ax+b)^3}$$

in its partial fraction decomposition. Use this to find the partial fraction decomposition of

$$\frac{3x-4}{x(2x-1)^3}.$$

43. Find the partial fraction decomposition of

$$\frac{x}{(x-1)^2(x+2)^3}.$$

(Refer to Problem 42.)

44. Given a rational expression, the occurrence of a third power of an irreducible quadratic factor in its denominator results in the terms

$$\frac{Ax+B}{(ax^2+bx+c)} + \frac{Cx+D}{(ax^2+bx+c)^2} + \frac{Ex+F}{(ax^2+bx+c)^3}$$

in its partial fraction decomposition. Use this to find the partial fraction decomposition of

$$\frac{x^5+3x^3-x^2+2x-2}{(x^2+2)^3}.$$

45. Find the partial fraction decomposition of

$$\frac{3x^5+6x^3+3x+4}{(x^2+1)^3}.$$

(Refer to Problem 44.)

C

46. a) Consider a complex number $a+bi$ and its conjugate $a-bi$. Define

$$q(x) = (x-(a+bi))(x-(a-bi)).$$

Show that $q(x)$ is a real quadratic function.

b) Suppose that $P(x)$ is a real polynomial function of degree N with a complex zero $a+bi$. Use the conjugate root theorem of Section 3.3 and the result of part (a) to show that $P(x) = q(x)R(x)$, where $q(x)$ is a real, irreducible quadratic function, and $R(x)$ is a real polynomial function of degree $N-2$.

c) Use the result of part (b) to show that the real polynomial function $P(x)$ of part (b) can be written as

$$P(x) = q_1(x)q_2(x)q_3(x)\cdots q_n(x)r_1(x)r_2(x)\cdots r_m(x)$$

where $q_i(x)$ $(i = 1, 2, \ldots, n)$ represents an irreducible quadratic factor and $r_i(x)$ $(i = 1, 2, \ldots, m)$ represents a linear factor. (It is possible that for a given real polynomial there are no irreducible quadratic factors or that there are no linear factors. It is also possible that the quadratic factors or the linear factors are not distinct.)

47. Consider a proper rational function that is the ratio of two polynomial functions $N(x)$ and $D(x)$, such that $D(x)$ is the product of distinct linear factors; that is,

$$\frac{N(x)}{D(x)} = \frac{N(x)}{(x-r_1)(x-r_2)(x-r_3)\cdots(x-r_n)}$$

such that $r_i \neq r_j$ if $i \neq j$. It follows that the partial fraction decomposition is

$$\frac{N(x)}{D(x)} = \frac{A_1}{(x-r_1)} + \frac{A_2}{(x-r_2)} + \frac{A_3}{(x-r_3)} + \cdots + \frac{A_n}{(x-r_n)}.$$

Show that

$$A_1 = \frac{N(r_1)}{(r_1-r_2)(r_1-r_3)\cdots(r_1-r_n)}$$

$$A_2 = \frac{N(r_2)}{(r_2-r_1)(r_2-r_3)\cdots(r_2-r_n)}$$

$$A_3 = \frac{N(r_3)}{(r_3-r_1)(r_3-r_2)\cdots(r_3-r_n)}$$

$$\vdots$$

$$A_n = \frac{N(r_n)}{(r_n-r_1)(r_n-r_2)\cdots(r_n-r_{n-1})}.$$

S E C T I O N 10.6 SYSTEMS OF NONLINEAR EQUATIONS

Many of the systems of equations that will stand between you and a solution to a problem in calculus will involve a system of equations in two unknowns that is not linear. The techniques for solving these **nonlinear systems** are similar to those of the linear systems that we have seen so far in this chapter.

Just as with a linear system in two unknowns, each solution of a nonlinear system is the set of coordinates of a point of intersection of the graphs of the equations. If the graphs do not intersect, then the system has no solutions.

EXAMPLE 1 Sketch the graphs of the equations of the system

$$\begin{cases} y - x^2 + 4 = 0 \\ y = \frac{1}{2}x + 1 \end{cases}$$

Determine the coordinates of the point of intersection.

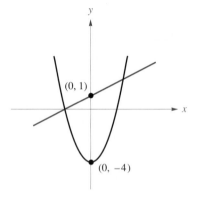

FIGURE 12

SOLUTION The graph of first equation is a parabola opening up with a vertex at $(0, -4)$. The graph of the second is a line with slope $\frac{1}{2}$ and y-intercept 1. Their graphs are shown in Figure 12.

To solve the system, we can substitute the $\frac{1}{2}x + 1$ for y in the first equation; this gives an equation in x only, which can be solved:

$$y - x^2 + 4 = 0$$
$$(\tfrac{1}{2}x + 1) - x^2 + 4 = 0$$
$$-x^2 + \tfrac{1}{2}x + 5 = 0$$
$$2x^2 - x - 10 = 0$$
$$(2x - 5)(x + 2) = 0$$
$$2x - 5 = 0 \quad \text{or} \quad x + 2 = 0$$
$$x = \tfrac{5}{2} \quad \text{or} \qquad x = -2.$$

Each of these solutions, $\frac{5}{2}$ or -2, corresponds to one of the points of intersection of the graphs. We can find the y-coordinate of each of these points by evaluating $y = \frac{1}{2}x + 1$ for $x = \frac{5}{2}$ or $x = -2$:

$$y = \tfrac{1}{2}(\tfrac{5}{2}) + 1 \qquad y = \tfrac{1}{2}(-2) + 1$$
$$y = \tfrac{9}{4} \qquad\qquad y = 0.$$

This means that our solutions are $(\frac{5}{2}, \frac{9}{4})$ and $(-2, 0)$ (Figure 13). Of course, these solutions should be checked in the original system.

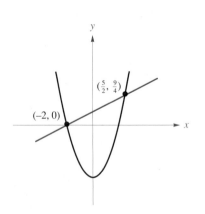

FIGURE 13

In certain cases, the method of linear combinations can be used to solve a nonlinear system. The next example demonstrates this technique.

EXAMPLE 2 Sketch the graphs of the equations of the system

$$\begin{cases} x^2 + 4y^2 = 100 \\ x^2 + y^2 = 52 \end{cases}$$

Determine the coordinates of the point of intersection.

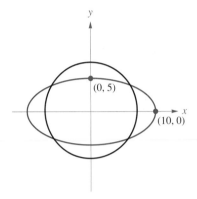

SOLUTION The graph of first equation is an ellipse centered at the origin with vertices (10, 0), (10, 0), (0, −5), and (0, 5); the graph of the second is a circle centered at the origin with radius $\sqrt{52} = 2\sqrt{13}$ (this is about 7.2) (Figure 14).

Looking at the sketch, there apparently are four solutions to the system; each one is symmetric with respect to the x-axis, the y-axis, or the origin with one of the remaining three.

FIGURE 14

While this is not a linear system of equations in x and y, it is a linear system in x^2 and y^2:

$$(-1)\begin{cases} x^2 + 4y^2 = 100 \\ x^2 + y^2 = 52 \end{cases}$$
$$+\begin{cases} x^2 + 4y^2 = 100 \\ -x^2 - y^2 = -52 \end{cases}$$
$$\overline{\phantom{+\{}3y^2 = 48}$$
$$y^2 = 16$$
$$y = 4 \quad \text{or} \quad y = -4$$

Substituting the value $y = 4$ into $x^2 + y^2 = 52$ (actually, either equation will do) yields

$$x^2 + (4)^2 = 52$$
$$x^2 = 36$$
$$x = 6 \quad \text{or} \quad x = -6.$$

This implies that (6, 4) and (−6, 4) are solutions. In a similar fashion, substituting the value $y = -4$ into $x^2 + y^2 = 52$ uncovers the solutions (6, −4) and (−6, −4). These solutions agree with the previous observations on the symmetry of the solutions (Figure 15). Again, checking these solutions is left to you.

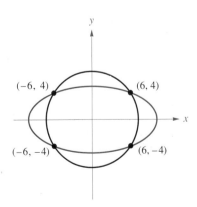

FIGURE 15

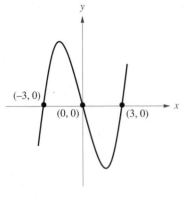

FIGURE 16

The solution of a system of equations by substitution depends upon being able to eventually solve an equation with one unknown. For a non-linear system, these equations may require more effort than for a linear system. The next two examples are typical cases.

EXAMPLE 3 Sketch the graphs of the equations of the system

$$\begin{cases} y = x^3 - 9x \\ y + 6x = 2 \end{cases}.$$

Determine the coordinates of the point of intersection.

SOLUTION The first equation is a third-degree polynomial; it can be factored without much difficulty: $y = x(x - 3)(x + 3)$. Using the techniques of Chapter 3, we arrive at the graph in Figure 16.

The second equation is a line; its graph intersects the graph of the first equation at either three points, two points, or one point. In a situation like this, it is very difficult to determine the actual case from the picture (Figure 17).

Solving by substitution:

$$y + 6x = 2$$

$$(x^3 - 9x) + 6x = 2$$

$$x^3 - 3x - 2 = 0.$$

Solving this equation is going to take a bit of effort. Recall the rational root theorem of Section 3.4, which suggests the potential rational roots for this

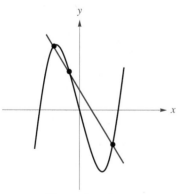

Three points of
intersection

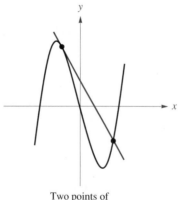

Two points of
intersection

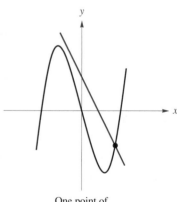

One point of
intersection

FIGURE 17

equation, namely ±1 and ±2. Using synthetic division, our efforts bear fruit when we check -1:

$$
\begin{array}{r|rrrr}
-1 & 1 & 0 & -3 & -2 \\
   &   & -1 & 1 & 2 \\
\hline
   & 1 & -1 & -2 & \,0.
\end{array}
$$

So -1 is a root of the equation. Moreover, the synthetic division tableau shows that the equation can be factored as

$$(x + 1)(x^2 - x - 2) = 0.$$

So,

$$x + 1 = \ \ 0 \quad \text{or} \quad x^2 - x - 2 = 0$$

$$x = -1 \quad \text{or} \quad (x - 2)(x + 1) = 0$$

$$x - 2 = 0 \quad \text{or} \quad x + 1 = \ \ 0$$

$$x = 2 \quad \text{or} \quad x = -1.$$

There are two distinct, real zeros, 2 and -1. Substituting each value of x into the equation $y + 6x = 2$ we get the solutions $(-1, 8)$ and $(2, -10)$ (Figure 18).

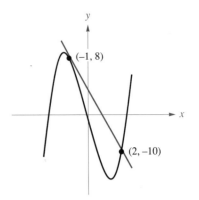

Two points of intersection

FIGURE 18

If the graphs of the equations of a system do not intersect, then the system has no real solutions. This can be determined graphically from the sketch of the system or algebraically from the solution of the system.

EXAMPLE 4   Solve the system of equations

$$\begin{cases} x + y^2 + 4 = 0 \\ (x - 2)^2 + y^2 = 24. \end{cases}$$

SOLUTION Since there is a y^2 term in both equations, it seems reasonable to solve the first equation for y^2 and substitute into the second:

$$y^2 = -x - 4$$

$$(x - 2)^2 + y^2 = 24$$

$$(x - 2)^2 + (-x - 4) = 24.$$

Solving for x yields

$$x^2 - 4x + 4 - x - 4 = 24$$

$$x^2 - 5x - 24 = 0$$

$$(x - 8)(x + 3) = 0 \Rightarrow x = 8 \quad \text{or} \quad x = -3.$$

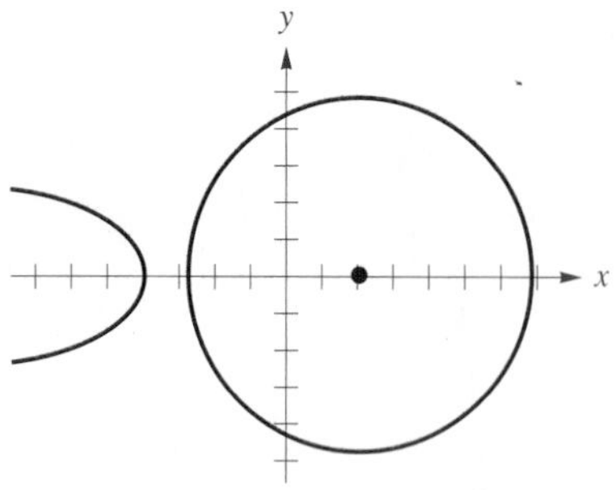

y

FIGURE 19

But notice what happens when we try to determine y for $x = 8$:

$$y^2 = -x - 4 = -(8) - 4 = -12.$$

There is no real number y such that $y^2 = -12$. In a similar fashion, letting $x = -3$ implies that $y^2 = -1$, which is also impossible for all real numbers y.

This system has no real solutions. Looking at the graphs of the equations in Figure 19, it should be clear that this is so.

The Intersection of Polar Curves

Determining the points of intersection of two polar graphs is not quite as straightforward as it is for rectangular graphs. The difficulty arises from the fact that unlike rectangular graphs, any point on the polar graph can be described in an infinite number of ways with polar coordinates. This means that the algebraic solution may not yield all the points of intersection.

EXAMPLE 5 Sketch the graphs of the polar equations and determine the coordinates of the points of intersection.

$$\begin{cases} r = 2 + 4 \sin \theta \\ r = 8 \sin \theta. \end{cases}$$

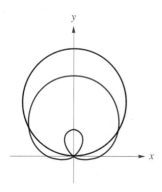

FIGURE 20

SOLUTION The graph of the first equation is a limaçon with an inner loop (see Example 8 of Section 9.2). The graph of the second equation is a circle that is tangent to and above the x-axis; its diameter is 8 (Figure 20).

By substitution, we obtain

$$8 \sin \theta = 2 + 4 \sin \theta$$

$$4 \sin \theta = 2$$

$$\sin \theta = \tfrac{1}{2}$$

$$\theta = \pi/6, \ 5\pi/6.$$

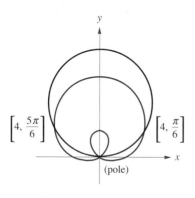

FIGURE 21

If $\theta = \pi/6$, then $r = 8 \sin \theta = 4$; $[4, \pi/6]$ is a solution. If $\theta = 5\pi/6$, then $r = 8 \sin \theta = 4$; $[4, 5\pi/6]$ is another solution. Something seems to be amiss here, though. We determined only two solutions to the system, but the graph reveals that there are three points of intersection. We did not discover the apparent intersection at the pole. The graph of the first equation passes through the pole since the ordered pair $[0, -\pi/6]$ makes the first equation true. On the other hand, the graph of the second equation passes through the pole since the ordered pair $[0, 0]$ makes the second equation true. These are different representations, but they represent the same point. This is why we didn't find this particular point when solving the system algebraically. The points of intersection are shown in Figure 21.

We summarize:

To determine the intersection of two polar graphs:

1) Sketch the graphs; identify the points of intersection.

2) Solve the system of equations to determine the coordinates of points of intersection.

3) Determine the coordinates of the points not found in solving the system by trying other polar representations in the equations.

Intersection of Polar Graphs

EXERCISE SET 10.6

A

In Problems 1 through 18, sketch the graphs of the equations of the system and determine the points of intersection by solving the system.

1. $\begin{cases} x^2 + y^2 = 25 \\ y = x - 1 \end{cases}$

2. $\begin{cases} x^2 + y^2 = 169 \\ y - x = 7 \end{cases}$

3. $\begin{cases} 4x^2 + 9y^2 = 144 \\ 2x + 3y = 12 \end{cases}$

4. $\begin{cases} 9x^2 + y^2 = 45 \\ x = \frac{1}{3}y - 1 \end{cases}$

5. $\begin{cases} x^2 - y^2 = 1 \\ y = x^2 \end{cases}$

6. $\begin{cases} y^2 - x^2 = 9 \\ y = x^2 - 11 \end{cases}$

7. $\begin{cases} x^2 + y^2 = 8 \\ xy = 4 \end{cases}$

8. $\begin{cases} x^2 - y^2 = 3 \\ xy = -2 \end{cases}$

9. $\begin{cases} (x - 2)^2 + (y + 1)^2 = 9 \\ y + 2x = 11 \end{cases}$

10. $\begin{cases} (x - 4)^2 + y^2 = 4 \\ y^2 = x - 2 \end{cases}$

11. $\begin{cases} x^2 + (y + 2)^2 = 4 \\ y = 12 - x^2 \end{cases}$

12. $\begin{cases} (x - 3)^2 + (y - 6)^2 = 18 \\ y = |x - 3| \end{cases}$

13. $\begin{cases} y = \log_2 x \\ y = 4 - \log_2 x \end{cases}$

14. $\begin{cases} y = -\log_3 x \\ y = 2 + \log_3 x \end{cases}$

15. $\begin{cases} y = 2^x \\ y = 4^x - 2 \end{cases}$

16. $\begin{cases} y = 3^x \\ 2y = 9^x \end{cases}$

17. $\begin{cases} y = \dfrac{x}{x^2 - 4} \\ y = x \end{cases}$

18. $\begin{cases} y = \dfrac{2x^2}{x^2 - 4} \\ y = x^2 \end{cases}$

B

In Problems 19 through 24, solve the system (you need not sketch the graphs of the equations).

19. $\begin{cases} \sqrt{x} - \sqrt{y} = 2 \\ x - 9y = 0 \end{cases}$

20. $\begin{cases} \sqrt{x} + \sqrt{y} = 8 \\ x - y = 0 \end{cases}$

21. $\begin{cases} \log_2(2x - y) = 3 \\ \log_3(x + y) = 2 \end{cases}$

22. $\begin{cases} 2^{x + 3y} = 16 \\ 5^{x^2 + y} = 25 \end{cases}$

23. $\begin{cases} |x + 2y| = 3 \\ x - y^2 = 0 \end{cases}$

24. $\begin{cases} |x - y| = 3 \\ y = \sqrt{x} \end{cases}$

In Problems 25 through 30, sketch the graphs of the polar equations of the system. Determine the points of intersection.

25. $\begin{cases} r = 4 \cos \theta \\ r = 2 \end{cases}$

26. $\begin{cases} r = -8 \sin \theta \\ r = 4\sqrt{2} \end{cases}$

27. $\begin{cases} r = 2 + 4 \sin \theta \\ r = 6 \sin \theta \end{cases}$

28. $\begin{cases} r = 3 - 6 \cos \theta \\ r = 9 \cos \theta \end{cases}$

29. $\begin{cases} r^2 = 16 \cos 2\theta \\ r = 2\sqrt{2} \end{cases}$

30. $\begin{cases} r^2 = 4 \cos 2\theta \\ r = 2 \cos \theta \end{cases}$

31. A rectangle has area of 24 square feet and a perimeter of 35 feet. Find the dimensions of the rectangle.

32. A rectangle has area of 60 cm² and a diagonal of 13 cm. Find the dimensions of the rectangle.

33. A box with a square base and no top has volume of 20 ft³ and a surface area of 44 ft². Find the dimensions of the rectangle.

34. A cone has a radius half as long as its height. Given that the volume of the cone is 9π ft², determine its height and its radius.

35. A right triangle has area 24 in.² and perimeter 24 in. Find the lengths of the sides of the triangle.

36. An isosceles triangle has area 60 in.² and perimeter 36 in. Find the lengths of the sides of the triangle.

37. Solve the system

$$\begin{cases} 9^x + y^2 = 90 \\ 3^x - 2y = 3 \end{cases}$$

38. Solve the system

$$\begin{cases} \dfrac{y}{x} = 4 \\ \dfrac{\log_2 x}{\log_2 y} = \dfrac{1}{-6} \log_2\left(\dfrac{x}{y}\right) \end{cases}$$

39. Solve the system

$$\begin{cases} y = x^3 - x^2 \\ y - x + 3 = 2x^2 \end{cases}$$

(Hint: you may need the rational root theorem.)

40. Sketch the graphs of the equations of the system and determine the points of intersection by solving the system:

$$\begin{cases} y = x^5 - 5x^3 + 3x + 2 \\ y = -x + 2. \end{cases}$$

(Hint: the graph of the first equation was discussed in Example 4 of Section 3.2. You may need the rational root theorem.)

41. Sketch the graphs of the equations of the system and determine the points of intersection by solving the system:

$$\begin{cases} y = \sqrt{3} \cos x & 0 \le x \le 2\pi \\ y = \sin 2x & 0 \le x \le 2\pi. \end{cases}$$

42. Sketch the graphs of the equations of the system and determine the points of intersection by solving the system.

$$\begin{cases} y = 4 \cos x + 1 & 0 \le x \le 2\pi \\ y = -4 \cos 2x & 0 \le x \le 2\pi \end{cases}$$

(Give approximate answers to the nearest 0.001.)

In Problems 43 through 48, sketch the system and determine the number of solutions (you don't need to solve the system).

43. $\begin{cases} y = 2^x \\ y = 4 - x^2 \end{cases}$ **44.** $\begin{cases} y = 3^{-x} \\ xy = -4 \end{cases}$

45. $\begin{cases} y = \sin x \\ y = \frac{1}{2}x \end{cases}$ **46.** $\begin{cases} y = \cos x \\ y = -\frac{1}{2}x \end{cases}$

47. $\begin{cases} y = \arctan x \\ x^2 - y^2 = 4 \end{cases}$ **48.** $\begin{cases} y = \ln|x| \\ y = \dfrac{4}{1 + x^2} \end{cases}$

C

49. Consider the system

$$\begin{cases} x^2 + 4xy - y^2 = 16 \\ 2x^2 - 7xy + 2y^2 = -4. \end{cases}$$

a) Make the substitution $y = mx$ to get a system in m and x.

b) Solve this system for m and x.

c) State the solutions in the form (x, mx).

50. Solve the system using the method of Problem 49.

$$\begin{cases} 3x^3 - 2x^2y + 3y^3 = 16 \\ 5x^2y + 7xy^2 = 16 \end{cases}$$

Describe the type of system for which this method of solution works.

SYSTEMS OF INEQUALITIES

S E C T I O N   10.7

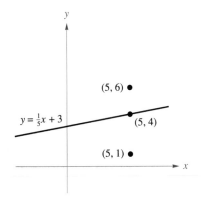

FIGURE 22

As we have seen, the graph of an equation in two unknowns is the set of all points whose coordinates are solutions to the equation. In a similar fashion, the graph of an **inequality in two unknowns** is the set of all points whose coordinates are solutions to the inequality. Some inequalities such as $y \le x^2$ and $3x - y \ge 4$ allow equality as well. Others, such as $2y < x - 4$ and $x + y > 4x$, do not allow equality; these are called **strict inequalities.**

For example, consider the strict inequality $y > \frac{1}{5}x + 3$. Suppose that we want to determine its graph. The graph of the **associated equation** $y = \frac{1}{5}x + 3$ is a straight line. The point $(5, 4)$ is on the graph of this equation since $(4) = \frac{1}{5}(5) + 3$. The point $(5, 6)$ is above the graph; this should seem reasonable since $(6) > \frac{1}{5}(5) + 3$ (Figure 22).

Any point (x, y) above the line is such that $y > \frac{1}{5}x + 3$. This set of points above the line is exactly the graph of the inequality. We use a dotted line to indicate that the points on the line are not part of the graph of the inequality (Figure 23.) Had we originally been concerned with the inequality $y \ge \frac{1}{5}x + 3$, then any point on the line would have made the inequality true; to indicate that this boundary line is part of the graph, we would use a solid line (Figure 24).

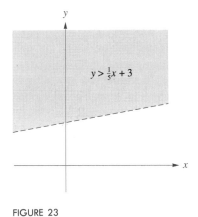

FIGURE 23

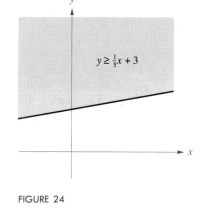

FIGURE 24

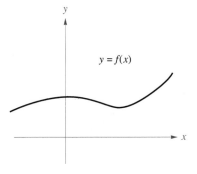

FIGURE 25

Let $f(x)$ be a function such as $\frac{1}{5}x + 3$. The graph of the equation $y = f(x)$ partitions the coordinate plane into three sets of points: those on the line, those above the line, and those below the line (Figure 25). The graphs of the inequalities $y > f(x)$, $y \ge f(x)$, $y < f(x)$, and $y \le f(x)$ are very much related to the graph of $y = f(x)$ and these three sets of points. The following table summarizes these relations.

Graphs of Inequalities in Two Unknowns

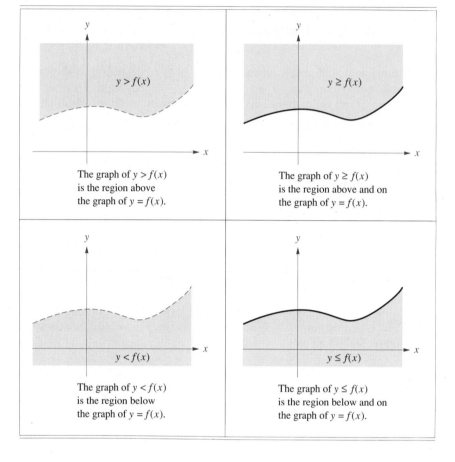

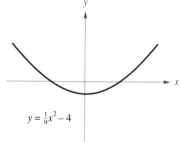

$y = \frac{1}{9}x^2 - 4$

FIGURE 26

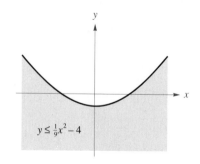

$y \leq \frac{1}{9}x^2 - 4$

FIGURE 27

To apply these generalizations the inequality must first be solved for y. This is not always desirable or possible. Part (b) of the next example shows an alternative method of determining the graph of an inequality.

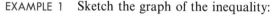

EXAMPLE 1   Sketch the graph of the inequality:

    a) $y \leq \frac{1}{9}x^2 - 4$ **b)** $2x - 3y > 12$

SOLUTION

    a) We start by sketching the graph of the associated equation $y = \frac{1}{9}x^2 - 4$. This is a parabola opening up with vertex at $(0, -4)$ (Figure 26).

        Since the inequality specifies "less than or equal to," the sketch of the graph of this inequality is the set of points on or below the graph of the equation. We use a solid line in sketching the parabola since this is not a strict inequality (Figure 27).

b) The graph of the associated equation $2x - 3y = 12$ is a straight line (Figure 28). Now, instead of solving this inequality for y and applying the appropriate result from the table, suppose that we proceed as follows. Choose a point that is not on the line, say $(0, 0)$, and determine if this point makes the inequality true or false. If these coordinates make the inequality true, then the solution must be the side that contains the point $(0, 0)$. On the other hand, if the inequality proves to be false, then the region we seek is the other side of the line. So,

$$2(0) - 3(0) \overset{?}{>} 12$$

$$0 > 12. \qquad \text{False}$$

The graph of the inequality is the region below the line (Figure 29). Notice that the line is dotted since it is not part of the solution set.

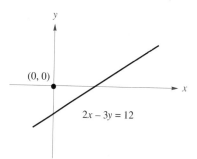

FIGURE 28

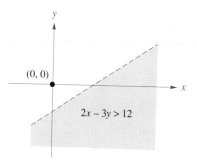

FIGURE 29

When the associated equation of an inequality cannot be written as $y = f(x)$, we can use a test point that is not on the graph of the equation. If the coordinates of this point make the inequality true, then the region that contains the point is the graph of the inequality.

EXAMPLE 2   Sketch the graph of

$$\frac{x^2}{9} + \frac{y^2}{49} \leq 1.$$

SOLUTION The graph of the associated equation of this inequality,

$$\frac{x^2}{9} + \frac{y^2}{49} = 1,$$

is discussed in Example 2 of Section 8.3. Its graph is shown in Figure 30. The graph of this equation divides the coordinate plane into two regions: the inside of the ellipse and the outside of the ellipse. One of these regions, along with the graph of the ellipse, is our solution. To determine which one, we pick a test point, say $(0, 0)$, and determine if these coordinates make the inequality true or false:

$$\frac{(0)^2}{9} + \frac{(0)^2}{49} \overset{?}{\leq} 1$$

$$0 \leq 1. \qquad \text{True}$$

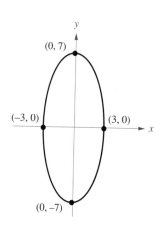

FIGURE 30

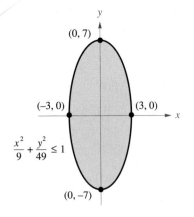

$$\frac{x^2}{9} + \frac{y^2}{49} \leq 1$$

FIGURE 31

The point is in the interior of the ellipse. The graph of the inequality is shown in Figure 31.

Just as with a system of equations, the graph of a system of inequalities is the intersection of the graphs of the individual inequalities of the system. Determining the region described by a system of inequalities plays an important role in the solution of many problems in calculus.

EXAMPLE 3 Sketch the graph of the following system:

$$\begin{cases} 2x - y < 0 \\ x + 2y < 4. \end{cases}$$

SOLUTION The first inequality has as an associated equation $2x - y = 0$. Using a point not on the line, say $(1, -1)$, as a test point, we have

$$2(1) - (-1) \overset{?}{<} 0$$

$$3 < 0. \qquad \text{False}$$

Notice that the point $(0, 0)$ is on the line. This means it cannot be used as a test point.

The region we seek is the half-plane above the line $2x - y = 0$, as shown in Figure 32.

The graph of the second inequality is bounded by the line $x + 2y = 4$. Using $(0, 0)$ as a test point, we have

$$(0) + 2(0) \overset{?}{<} 4$$

$$0 < 4. \qquad \text{True}$$

The graph, therefore, is the half-plane below the graph of the line $x + 2y = 4$ (Figure 33).

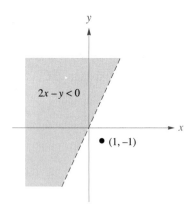

FIGURE 32

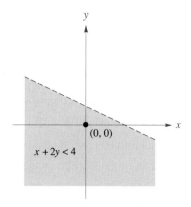

FIGURE 33

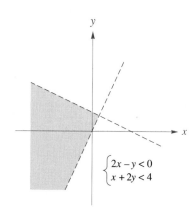

FIGURE 34

The intersection of these two sets is the solution of the system of inequalities, as shown (Figure 34).

EXAMPLE 4   Sketch the graph of the system

$$\begin{cases} \dfrac{x^2}{9} + \dfrac{y^2}{49} \leq 1 \\ x^2 + y^2 \geq 16 \end{cases}$$

FIGURE 35

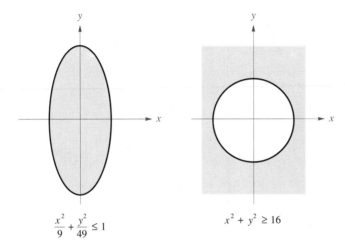

$\dfrac{x^2}{9} + \dfrac{y^2}{49} \leq 1$ $x^2 + y^2 \geq 16$

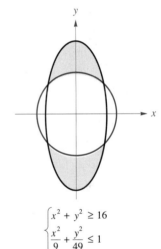

$$\begin{cases} x^2 + y^2 \geq 16 \\ \dfrac{x^2}{9} + \dfrac{y^2}{49} \leq 1 \end{cases}$$

FIGURE 36

SOLUTION The sketch of the first inequality is shown in Example 2. The associated equation of the second inequality is $x^2 + y^2 = 16$. The graph of this equation is a circle, centered at the origin, with radius 4. We leave it to you to pick a test point and verify that the region we seek is outside this circle. Both graphs are shown in Figure 35.

The intersection of these two sets is precisely those regions inside the ellipse and outside the circle, along with the boundaries of these regions (Figure 36).

EXAMPLE 5   Sketch the graph of the inequality $x^2 \leq y \leq 2x + 3$.

SOLUTION Recall from Section 1.1 that the solution to a continued inequality such as $a \leq t \leq b$ is equivalent to the intersection of the solutions of the inequalities $t \geq a$ and $t \leq b$. So this inequality is the same as the system

$$\begin{cases} y \geq x^2 \\ y \leq 2x + 3. \end{cases}$$

The graph of this inequality is the set of points in the plane that are on or above the parabola $y = x^2$ and on or below the line $y = 2x + 3$ (Figure 37).

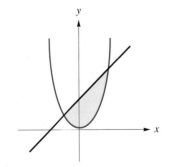

FIGURE 37

EXAMPLE 6   Sketch the graph of the system

$$\begin{cases} -2 - \tfrac{1}{3}x \le y \le \sqrt{x}. \\ 1 \le x \le 9 \end{cases}$$

SOLUTION The graph of the first inequality is the set of points that are on or above the graph of $y = -2 - \tfrac{1}{3}x$ and on or below the graph of $y = \sqrt{x}$, as shown in Figure 38.

The second continued inequality is the set of points between the vertical lines $x = 1$ and $x = 9$ (Figure 39).

The intersection of these two graphs is the solution to the inequality (Figure 40).

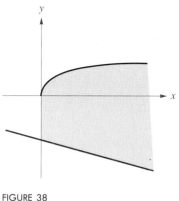

FIGURE 38

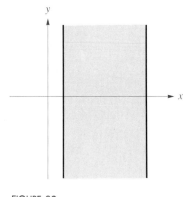

FIGURE 39

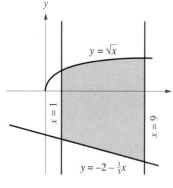

FIGURE 40

Inequalities Involving Polar Coordinates

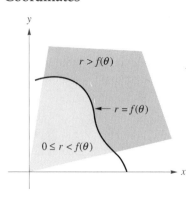

FIGURE 41

Regions on the coordinate plane can also be described by using inequalities in terms of polar coordinates.

Consider the equation $r = f(\theta)$ and its polar graph. The points inside the graph of this equation can be characterized by the polar inequality $0 \le r < f(\theta)$; those outside can be characterized by the polar inequality $r > f(\theta)$ (Figure 41).

EXAMPLE 7   Sketch the graph of $2 - 2 \sin \theta \le r \le 2$.

SOLUTION This inequality is equivalent to the system

$$\begin{cases} r \ge 2 - 2 \sin \theta. \\ r \le 2 \end{cases}$$

The first inequality is the exterior of the graph of the cardioid $r = 2 - 2\sin\theta$. The second inequality represents the interior of the circle centered at the pole, with a radius of 2 (Figure 42).

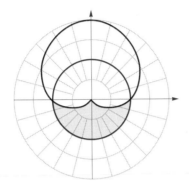

FIGURE 42

EXERCISE SET 10.7

A

In Problems 1 through 12, sketch the solution to the inequality. Use dotted lines for boundaries that are not part of the solutions, and use solid lines for boundaries that are part of the solution.

1. $y \le \frac{1}{2}x - 4$

2. $y \ge \frac{1}{4}x + 2$

3. $2x - 3y > 12$

4. $x + 5y < 10$

5. $1 + x^2 < y$

6. $4 - y^2 < x$

7. $x^2 + y^2 < 16$

8. $x^2 + y^2 \ge 25$

9. $16(x - 4)^2 + 9(y + 3)^2 \le 144$

10. $x^2 - y^2 \ge 16$

11. $xy \ge 4$

12. $x^2 y \ge 16$

In Problems 13 through 18, sketch the solution to the continued inequality.

13. $x^2 \le y \le 4$

14. $-6 \le y \le 3 - x^2$

15. $x^2 - 4 \le y \le \frac{1}{2}x$

16. $x \le y \le x^2 - 16x$

17. $\log_2 x \le y \le 3$

18. $2^x \le y \le \frac{3}{2}x + 1$

In Problems 19 through 30, sketch the solution to the system of inequalities.

19. $\begin{cases} y \le \frac{1}{2}x - 4 \\ 2x + 3y \ge 12 \end{cases}$

20. $\begin{cases} y \ge \frac{1}{4}x + 2 \\ 5x + y \le 10 \end{cases}$

21. $\begin{cases} 2x - 3y \ge 6 \\ x^2 + y^2 \le 16 \end{cases}$

22. $\begin{cases} x + 2y \le 8 \\ 9x^2 + 4y^2 \le 144 \end{cases}$

23. $\begin{cases} y - x^2 < 0 \\ y < 4x - x^2 \end{cases}$

24. $\begin{cases} 4 - y^2 < x \\ y^2 < x + 4 \end{cases}$

25. $\begin{cases} x^2 + y^2 \le 16 \\ 9x^2 + 16y^2 \ge 144 \end{cases}$

26. $\begin{cases} x^2 + y^2 \ge 9 \\ x^2 - y^2 \le 9 \end{cases}$

27. $\begin{cases} y \ge 3^{-x} \\ y \le 4 - 3^x \end{cases}$

28. $\begin{cases} y \le 4 + 2^{-x} \\ y \ge 4 - 2^x \end{cases}$

29. $\begin{cases} y \le \log_2 x \\ y \ge \log_4 x \end{cases}$

30. $\begin{cases} y \le 1 + \ln x \\ y \ge 2\ln x \end{cases}$

B

In Problems 31 through 42, sketch the solution to the system of inequalities. Use dotted lines for boundaries that are not part of the solutions, and used solid lines for boundaries that are part of the solution.

31. $\begin{cases} x \ge 0 \\ y \ge 0 \\ 5x + 2y \le 20 \end{cases}$

32. $\begin{cases} x \ge 0 \\ y \ge 0 \\ y \le 2x + 10 \end{cases}$

33. $\begin{cases} x \ge 0 \\ y \le 4 \\ y \ge \log_2 x \end{cases}$

34. $\begin{cases} y \ge 0 \\ 0 \le x \le 2 \\ y \ge 3^x \end{cases}$

John von Neumann

John von Neumann (pronounced noy'-men) is considered by many as the greatest mathematical genius of the 20th century. Born in Budapest, Hungary in 1903, von Neumann showed remarkable ability and interest in science as a child. There are numerous accounts of his fantastic capacity for mental calculations. He also possessed a photographic memory.

As an undergraduate, von Neumann first studied chemistry but soon became interested in mathematics. He attended the University of Berlin, the Technische Hochschule in Zurich, and the University of Budapest. He was appointed Privat-dozent at the University of Berlin at the age of 24 (unusually young for such an appointment). While teaching at the University of Hamburg in 1930, von Neumann accepted an invitation as a guest lecturer for one year at Princeton University. He decided to stay in the United States on the faculty of Princeton, and in 1933 he became one of the first permanent members (along with Einstein) of the Institute for Advanced Studies. His interests were always diverse, but he leaned toward abstract, theoretical fields until the late 1930s when he became interested in more applied questions. He soon became involved in the Manhattan Project and developed a technique that accelerated the production of the first atomic bomb. A computer was necessary to carry out the volume of calculations involved in his research, and von Neumann is largely responsible for the invention of the digital computer. Following this, he spent considerable ef-

fort developing automata and information theory. In 1955, he learned that he had cancer and so worked vigorously on his research of the analogies between the human brain and computers. His work was not finished when he died in 1957.

Von Neumann is considered to be the founder of the theory of games, a branch of mathematics that analyzes the strategies of two or more participants (competitors) who are trying to gain as much as possible. Employing matrices, he formulated theorems for determining optimal strategies for each participant. The theory of games has widespread application, especially in economics.

There are many accounts of von Neumann's extraordinary calculating ability. One instance was when a colleague spent the entire night doing long calculations to gain insight into a particular problem. The next day, von Neumann was introduced to the problem and suggested that doing the previously mentioned calculations would be a good approach. Unaware of the hours of work already done the night before, he proceeded to do the calculations *in his head* at such a pace that it was clear he would finish in a matter of minutes! The capable mathematician who was up working all night stood by listening to von Neumann rattle off several correct results. When von Neumann was about to arrive at the final calculation, the bleary-eyed colleague shouted the answer at him, unable to control his emotions. Seconds later the stunned and no doubt impressed von Neumann looked at him and said, "Why yes, that's correct!"

---

35. $\begin{cases} y \geq 0 \\ 0 \leq x \leq 2\pi \\ y \leq 2 + \sin x \end{cases}$

36. $\begin{cases} y \leq 0 \\ -\pi \leq x \leq \pi \\ y \leq -4 + \cos x \end{cases}$

37. $\begin{cases} \frac{1}{2}x \leq y \leq 4 - x \\ 0 \leq x \leq 2 \end{cases}$

38. $\begin{cases} x - 2 \leq y \leq x + 3 \\ -2 \leq x \leq 2 \end{cases}$

39. $\begin{cases} \frac{1}{2}x^2 - 8 \leq y \leq x^2 - 6x \\ 0 \leq x \leq 3 \end{cases}$

40. $\begin{cases} x^3 - 9x \leq y \leq x \\ -3 \leq x \leq 0 \end{cases}$

41. $\begin{cases} -2^x \leq y \leq 2^x \\ -4 \leq x \leq 4 \end{cases}$

42. $\begin{cases} \cos x \leq y \leq 1 \\ 0 \leq x \leq 2\pi \end{cases}$

43. Sketch the graph of **(a)** $y \leq 4/x$ and **(b)** $xy \leq 4$.

44. Sketch the graph of **(a)** $y \geq -2/x$ and **(b)** $xy \geq -2$.

45. Sketch the graph of **(a)** $y \leq \log_2(x^2)$ and
(b) $y \leq 2 \log_2 x$.

46. Sketch the graph of **(a)** $2 \leq \ln x + \ln y$ and
(b) $2 \leq \ln(xy)$.

47. Sketch the graph of

a) $y \leq \dfrac{x}{x^2 - 16}$ **b)** $y(x^2 - 16) \leq x$.

(Hint: The graph of the associated equation for these inequalities is discussed in Example 6 of Section 3.4.)

48. Sketch the graph of

a) $y \leq \dfrac{x^2 - 9}{x - 2}$ **b)** $y(x - 2) \leq x^2 - 9$.

(Hint: The graph of the associated equation for these inequalities is discussed in Example 9 of Section 3.4.)

In Problems 49 through 54, an expression is given in terms of x and y. Determine the ordered pairs (x, y) for which the expression is defined, and sketch the region on the coordinate plane that corresponds to these values of x and y.

49. $\sqrt{x^2 - y}$

50. $\sqrt{7 - x + y^2}$

51. $\log_2(16 - x^2 - y^2)$

52. $\ln(4x^2 + 9y^2 - 144)$

53. $\dfrac{1}{\sqrt{6 - x - y}}$

54. $\dfrac{\ln(3 - x)}{\sqrt{4 + x - y}}$

In Problems 55 through 60, sketch the solution to the polar inequality.

55. $0 \leq r \leq 4 \sin \theta$

56. $0 \leq r \leq 4 \cos \theta$

57. $3 \leq r \leq 4 + 2 \sin \theta$

58. $1 \leq r \leq 2 + 2 \cos \theta$

59. $2 \leq r \leq 4 \cos 3\theta$

60. $3 \leq r \leq 6 \sin 3\theta$

In Problems 61 through 66, sketch the solution to the system of polar inequalities. Assume in all cases that $r \geq 0$ and $0 \leq \theta \leq 2\pi$.

61. $\begin{cases} 0 \leq r \leq 4 \cos \theta \\ 0 \leq \theta \leq \pi/3 \end{cases}$

62. $\begin{cases} 0 \leq r \leq 2 \\ 0 \leq r \leq 2 + 2 \cos \theta \end{cases}$

63. $\begin{cases} 0 \leq r \leq 4 \cos \theta \\ 0 \leq r \leq 4 \sin \theta \end{cases}$

64. $\begin{cases} 0 \leq r \leq 2 \cos \theta \\ 0 \leq r \leq 4 + 2 \sin \theta \end{cases}$

65. $\begin{cases} 0 \leq r \leq \theta \\ \pi/3 \leq \theta \leq 2\pi/3 \end{cases}$

66. $\begin{cases} 0 \leq r \leq 1/\theta \\ \pi/6 \leq \theta \leq \pi/2 \end{cases}$

C

67. The graph of $r = 2 + 4 \sin \theta$ is a limaçon with an inner loop. Write a system of equations that describes the region inside the inner loop.

68. The graph of $r = 4 \sin 3\theta$ is a rose with three petals. Write a system of equations that describes the region inside the petal that is symmetric to the y-axis.

MISCELLANEOUS EXERCISES

In Problems 1 through 6, sketch the graph of the system, and state whether the system is consistent, inconsistent, or dependent. If the system is consistent, determine the solution by using the substitution method and label the point of intersection. If the system is dependent, determine the solution in terms of a parameter t.

1. $\begin{cases} 2x + y = 0 \\ 6x - 2y = 5 \end{cases}$

2. $\begin{cases} \frac{1}{2}x + y = 4 \\ x - 4y = -6 \end{cases}$

3. $\begin{cases} 4x - 6y = 12 \\ -x + \frac{3}{2}y = 6 \end{cases}$

4. $\begin{cases} \frac{3}{2}x + 6y = -6 \\ y = -\frac{1}{4}x - 1 \end{cases}$

5. $\begin{cases} 5x + 4y = 8 \\ 3y = 6 - \frac{15}{4}x \end{cases}$

6. $\begin{cases} y = \frac{2}{5}x + 4 \\ 4y = \frac{8}{5}x - 12 \end{cases}$

In Problems 7 through 15, solve the system by finding an upper-triangular system and then back-substituting. If the system is dependent, determine the solution in terms of a parameter t.

7. $\begin{cases} x - 3y = 3 \\ -2x + 7y = -2 \end{cases}$

8. $\begin{cases} 2x + 10y = -6 \\ -3x - 11y = 13 \end{cases}$

9. $\begin{cases} x + 3y + 2z = -6 \\ y + z = 3 \\ 3y + 4z = 20 \end{cases}$

10. $\begin{cases} x - y + 4z = 2 \\ -2x + 2y - z = -11 \\ y + 3z = -4 \end{cases}$

11. $\begin{cases} 2x + 4y - z = 0 \\ -4x + y = -31 \\ 3x + 6y + 5z = 13 \end{cases}$

12. $\begin{cases} x - 5y - 2z = 1 \\ 2x + 3z = 12 \\ 3x - 5y + z = 13 \end{cases}$

13. $\begin{cases} x + y - z = 1 \\ 2x - 2z = 1 \\ 8x + 2y - 8z = 5 \end{cases}$

14. $\begin{cases} x + 3y + z + w = 3 \\ x + 3y + 4z + 4w = -6 \\ y + z + w = -3 \\ 2x + 6y + 5z + 9w = 5 \end{cases}$

15. $\begin{cases} x + y - w = 4 \\ y - z + w = 3 \\ 2x + y + z - 5w = 3 \\ -y + z = -2 \end{cases}$

16. The composition of three types of steel are given in the table below. How many tons of each type should be melted and mixed together to get 20 tons of a steel that is 72.1% iron, 8.0% nickel, and 19.9% chromium?

	Iron	Nickel	Chromium
Type A	72%	8%	20%
Type B	68%	11%	21%
Type C	75%	6%	19%

17. According to the laws of physics, there are three currents I_1, I_2, and I_3 through the electrical network in Figure 43. Furthermore, these currents satisfy the system of equations:

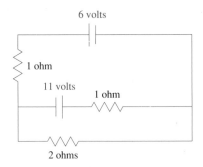

FIGURE 43

$\begin{cases} I_1 - I_2 + I_3 = 0 \\ I_1 - 2I_3 = 6. \\ I_2 + 2I_3 = 11 \end{cases}$

Solve this system to find the three currents (measured in amperes).

18. Find an equation of the form $y = ax^2 + bx + c$ whose graph is a parabola passing through $(2, 8)$, $(-2, 4)$, and $(-6, 16)$.

19. Find a function of the form $f(x) = a + b \ln(x)$ whose graph passes through $(e^2, 17)$ and $(1/e, -4)$.

In Problems 20 through 31, solve the system by reducing its augmented matrix to row echelon form. If the system is inconsistent, state so. If the system is dependent, then determine the solution in terms of a parameter t.

20. $\begin{cases} y + 1 = \frac{3}{2}x \\ -3x + 2y = 1 \end{cases}$

21. $\begin{cases} -4x + 6y = -8 \\ 6x - 9y = 12 \end{cases}$

22. $\begin{cases} 2x + 5y = -1 \\ x - 14y = 7 \end{cases}$

23. $\begin{cases} 2x + 3y - 2z = 4 \\ x - y + z = 6 \\ 3x + 7z = 10 \end{cases}$

24. $\begin{cases} x + 2y + 3z = 2 \\ y + z = 1 \\ x + y + 2z = 1 \end{cases}$

25. $\begin{cases} 4x - 2y = -4 \\ 9x + 2y + z = 3 \\ x - 2y - 2z = -2 \end{cases}$

26. $\begin{cases} 7x - 5y + z = 7 \\ 2x - y - 3z = 1 \\ -3x + 3y + z = 11 \end{cases}$

27. $\begin{cases} 2x - y - 5z = 2 \\ -x + 6y - 3z = -12 \end{cases}$

28. $\begin{cases} x - 4y - z = -8 \\ y - z = 2 \\ x + y - 6z = 2 \end{cases}$

29. $\begin{cases} x + 2y - z = 0 \\ 5x + y - 3z = -6 \end{cases}$

30. $\begin{cases} x + 2y - 2w = 7 \\ 2x - 5y + 3z - w = -1 \\ -3x + y - z + 3w = 0 \\ -4x - 6y + 3w = 12 \end{cases}$

$$31.\begin{cases} 8x - 7y - z - 2w = -31 \\ x + 2y - 3z - 2w = -8 \\ 3x + 4y + 2z + 3w = 10 \\ -x - 6y + 8z + w = 0 \end{cases}$$

In Problems 32 and 33, solve the system.

$$32.\begin{cases} \dfrac{2}{x} + \dfrac{1}{y} = 7 \\ \dfrac{4}{x} - \dfrac{3}{y} = -6 \end{cases}$$

$$33.\begin{cases} \dfrac{1}{x} + \dfrac{2}{y} - \dfrac{1}{z} = 3 \\ \dfrac{2}{x} + \dfrac{4}{y} - \dfrac{3}{z} = 5 \\ \dfrac{1}{x} \quad - \dfrac{6}{z} = -1 \end{cases}$$

In Problems 34 through 42, evaluate each of the determinants.

34. $\begin{vmatrix} 3 & -2 \\ 6 & 4 \end{vmatrix}$ **35.** $\begin{vmatrix} 13 & 23 \\ 5 & -6 \end{vmatrix}$ **36.** $\begin{vmatrix} \frac{2}{3} & 12 \\ -\frac{1}{4} & -6 \end{vmatrix}$

37. $\begin{vmatrix} \tan\theta & \sec\theta \\ \sec\theta & \tan\theta \end{vmatrix}$

38. $\begin{vmatrix} -2 & -5 & 7 \\ 4 & -6 & 3 \\ 0 & 1 & 8 \end{vmatrix}$

39. $\begin{vmatrix} -12 & -6 & 3 \\ 4 & 16 & 9 \\ 11 & -8 & 2 \end{vmatrix}$

40. $\begin{vmatrix} 5 & -1 & 6 \\ 10 & -2 & 12 \\ 4 & -4 & 3 \end{vmatrix}$

41. $\begin{vmatrix} 4 & 10 & -2 & 0 \\ -2 & -5 & 1 & 1 \\ 9 & 6 & 3 & -2 \\ -2 & -5 & 12 & 7 \end{vmatrix}$

42. $\begin{vmatrix} 0 & 2 & -5 & 1 \\ -3 & 7 & 0 & 2 \\ -1 & 1 & 3 & 0 \\ 2 & -1 & 0 & 4 \end{vmatrix}$

43. Find all values of r that satisfy

$$\begin{vmatrix} 2-r & -3 \\ 2 & 1+r \end{vmatrix} = -12.$$

44. Find all values of r that satisfy

$$\begin{vmatrix} 3-r & 2 & 1 \\ 4 & 4-r & 5 \\ 0 & 0 & -r \end{vmatrix} = -96$$

45. Use determinants to find the area of the triangle with vertices $(1, -1)$, $(3, 2)$, and $(0, 1)$. Refer to the formula given in the exercises for Section 10.3 prior to that section's Problems 40 through 44 (p. 587).

46. Find the area of the parallelogram with vertices $(1, 1)$, $(4, 2)$, $(6, 5)$, and $(3, 4)$.

For Problems 47 through 58, use Cramer's rule to solve the system of equations.

47. The system of equations in Problem 1.

48. The system of equations in Problem 2.

49. The system of equations in Problem 7.

50. The system of equations in Problem 8.

51. The system of equations in Problem 9.

52. The system of equations in Problem 10.

53. The system of equations in Problem 11.

54. The system of equations in Problem 17.

55. The system of equations in Problem 20.

56. The system of equations in Problem 22.

57. The system of equations in Problem 23.

58. The system of equations in Problem 24.

59. Solve for x and y (in terms of a):

$$\begin{cases} (1+a)x + ay = 5 \\ ax + (2+a)y = -4 \end{cases}$$

60. Solve for a and b:

$$\begin{cases} a + b = 1 \\ a + e^2 b = e \end{cases}$$

61. If the vertices of a polygon are (x_1, y_1), (x_2, y_2), (x_3, y_3), ..., (x_n, y_n), listed counterclockwise, then the area of the polygon is

$$\frac{1}{2}\left[\begin{vmatrix} x_1 & x_2 \\ y_1 & y_2 \end{vmatrix} + \begin{vmatrix} x_2 & x_3 \\ y_2 & y_3 \end{vmatrix} + \begin{vmatrix} x_3 & x_4 \\ y_3 & y_4 \end{vmatrix} + \cdots + \begin{vmatrix} x_n & x_1 \\ y_n & y_1 \end{vmatrix}\right].$$

Find the area of the six-sided figure with vertices $P_1(6, 1)$, $P_2(3, 5)$, $P_3(-4, 2)$, $P_4(-2, 1)$, $P_5(1, 1)$, and $P_6(2, -1)$.

62. Refer to Problem 61. Find the area of the five-sided figure determined by $P_1(8, 3)$, $P_2(-1, 4)$, $P_3(-4, 2)$, $P_4(2, -4)$, and $P_5(6, 2)$.

63. Determine the constants h, k, and r so that the circle given by $(x - h)^2 + (y - k)^2 = r^2$ passes through $(0, 0)$, $(2, 0)$, and $(2, 2)$.

In Problems 64 through 75, find the partial decomposition of the rational expression.

64. $\dfrac{7x - 15}{x^2 - 3x}$

65. $\dfrac{-x - 9}{(x - 1)(x + 4)}$

66. $\dfrac{x + 4}{(x - 1)(3x + 7)}$

67. $\dfrac{6x^2 - 2x - 20}{x^2(2x - 5)}$

68. $\dfrac{13}{(x^2 + 3)(2x - 1)}$

69. $\dfrac{4x^2 - x + 18}{(x^2 + x + 2)(x + 2)}$

70. $\dfrac{-x - 5}{x^3 - 1}$

71. $\dfrac{9x^3 - x^2 - 65x + 77}{(3x - 5)^2(x^2 + 1)}$

72. $\dfrac{4x^3 - 3x^2 - 52x - 35}{x^2 - x - 12}$

73. $\dfrac{8x^3 - 6x^2 - 10x + 3}{4x^2 + 3x}$

74. $\dfrac{x + 6}{(x - 1)^3}$

75. $\dfrac{6x^2 + x + 159}{(x - 3)(x + 3)^3}$

In Problems 76 through 83, sketch the graphs of the equations of the system and determine the points of intersection by solving the system.

76. $\begin{cases} x - 2y = -8 \\ y = x^2 - 1 \end{cases}$

77. $\begin{cases} x^2 + 2x = y - 1 \\ y - x = 3 \end{cases}$

78. $\begin{cases} x = y^2 + 2 \\ x^2 + 9y^2 = 18 \end{cases}$

79. $\begin{cases} x^2 + 2y^2 = 22 \\ y^2 - x^2 = 5 \end{cases}$

80. $\begin{cases} x^2 + y^2 = 25 \\ y - x^2 = -5 \end{cases}$

81. $\begin{cases} xy = 6 \\ x + 4y = 14 \end{cases}$

82. $\begin{cases} 4x - 7y = 7 \\ (x - 1)y = 1 \end{cases}$

83. $\begin{cases} x^2 + y^2 - 18x - 6y = -65 \\ x^2 + y^2 = 25 \end{cases}$

In Problems 84 through 89, solve the system of equations.

84. $\begin{cases} 3x^2 - 4y^2 = 8 \\ x^2 + 2xy = 8 \end{cases}$

85. $\begin{cases} \sqrt{x} - \sqrt{y} = 6 \\ x - 16y = 0 \end{cases}$

86. $\begin{cases} x^2 + 2x + y^2 - y = 26 \\ x^2 - 3x + y^2 + 4y = 25 \end{cases}$

87. $\begin{cases} x^2 - y^2 + 2xy = 4 \\ x^2 + y^2 = 4 \end{cases}$

88. $\begin{cases} \log(x - 3) + y = 7 \\ \log x = y - 6 \end{cases}$

89. $\begin{cases} \tan(x + y) = 1 \\ \cos(x - y) = 0 \end{cases}$

In Problems 90 through 93, sketch the graphs of the polar equations of the system. Determine the points of intersection by solving the system.

90. $\begin{cases} r = 1 + 2\cos\theta \\ r = 3\cos\theta \end{cases}$

91. $\begin{cases} r = 3 + 3\sin\theta \\ r = 6\sin\theta \end{cases}$

92. $\begin{cases} r = 4\cos\theta \\ r(2 - \cos\theta) = 3 \end{cases}$ (Hint: See Example 2, Section 9.4.)

93. $\begin{cases} r = 2 \\ r(3 + 2\sin\theta) = 4 \end{cases}$

In Problems 94 through 99, sketch the solution to the inequality.

94. $y \geq \frac{1}{4}x - 3$

95. $5x + 3y < -12$

96. $y > 6 - x^2$

97. $x^2 + 4y^2 \leq 16$

98. $y \leq \sqrt{9 - x^2}$

99. $y \leq \sqrt{x + 2}$

In Problems 100 through 108, sketch the solution to the system of inequalities.

100. $\begin{cases} x - 2y \leq -8 \\ y \geq 1 - x \end{cases}$

101. $\begin{cases} y - 1 \leq -(x + 2)^2 \\ (x + 2)^2 + (y - 1)^2 \leq 2 \end{cases}$

102. $\begin{cases} x > y^2 + 2 \\ y < -x + 4 \end{cases}$

103. $\begin{cases} x^2 + 9y^2 \leq 9 \\ x^2 - y^2 \leq 1 \end{cases}$

104. $\begin{cases} y < e^x \\ y > \ln x \end{cases}$

105. $\begin{cases} x^2 y \geq 4 \\ y \geq x^{2/3} \end{cases}$

106. $\begin{cases} x \geq 0 \\ y \geq 0 \\ 3x + 4y \leq 12 \end{cases}$

107. $\begin{cases} y \geq 0 \\ 0 \leq x \leq 2\pi \\ y \leq 2 \sin x \end{cases}$

108. $\begin{cases} -x^2 - 1 \leq y \leq x^2 + 1 \\ |x| - 3 \leq y \leq -|x| + 3 \end{cases}$

In Problems 109 through 111, an expression is given in terms of x and y. Determine the ordered pairs (x, y) for which the expression is defined, and sketch the region on the coordinate plane that corresponds to these values of x and y.

109. $\sqrt{x - y}$

110. $\ln(x^2 - y^2)$

111. $\dfrac{\sqrt{9 - x^2 - y^2}}{\log(1 + y^2 - x^2)}$

In Problems 112 through 115, sketch the solution to the system of polar inequalities. Assume in all cases that $0 \leq \theta < 2\pi$.

112. $\begin{cases} 0 \leq r \leq 4 \sin \theta \\ \pi/3 \leq \theta \leq 2\pi/3 \end{cases}$

113. $\begin{cases} 2 \leq r \leq 5 \\ \pi/6 \leq \theta \leq 2\pi/3 \end{cases}$

114. $\begin{cases} 1 \leq r \leq 3 \cos \theta \\ r \cos \theta \leq 2 \end{cases}$

115. $\begin{cases} r \leq 2 + 2 \sin \theta \\ r(2 - \cos \theta) \geq 3 \end{cases}$

CHAPTER 11

SEQUENCES AND SERIES

SEQUENCES AND SUMMATION

In Section 4.2 we examined the quantity $(1 + h)^{1/h}$ for values of h near zero. By letting $h = 1$, 0.1, 0.01, 0.001, and 0.0001, we obtained a succession of approximations to e:

First approximation: 2

Second approximation: 2.59374

Third approximation: 2.70481

Fourth approximation: 2.71692

Fifth approximation: 2.71815

Theoretically this ordered list, or **sequence,** could go on indefinitely. It is conventional to use subscript notation:

$$a_1 = 2$$
$$a_2 = 2.59374$$
$$a_3 = 2.70481$$
$$a_4 = 2.71692$$
$$a_5 = 2.71815$$
$$\vdots$$

The notation $\{a_n\}$ is used to represent the sequence a_1, a_2, a_3, a_4, a_5, ... Notice that for each natural number n there corresponds exactly one value a_n. This last statement is an idea that we have encountered before.

A sequence is a function whose domain is the set of natural numbers.	*Sequence*

The **ordered pairs** for a sequence $\{a_n\}$ are $(1, a_1)$, $(2, a_2)$, $(3, a_3)$, $(4, a_4)$, and so on. The **terms** of a sequence $\{a_n\}$ are a_1, a_2, a_3, a_4, and so on. The **general term** is a_n (also called the nth term).

Since the domain of a sequence is the set of natural numbers, the graph of a sequence differs from the graph of a function of a real variable.

EXAMPLE 1   For the sequence

$$\left\{ a_n = \frac{n}{n+1} \right\}$$

a)   list the first five terms,

b)   plot the first five ordered pairs, and

c)   graph the continuous function

$$y = \frac{x}{(x+1)}.$$

SOLUTION

a)   To find the first five terms of the sequence, we substitute the values $n = 1, 2, 3, 4,$ and 5 into the formula for a_n:

$$a_1 = \frac{(1)}{(1)+1} = \frac{1}{2} \qquad a_2 = \frac{(2)}{(2)+1} = \frac{2}{3} \qquad a_3 = \frac{(3)}{(3)+1} = \frac{3}{4}$$

$$a_4 = \frac{(4)}{(4)+1} = \frac{4}{5} \qquad a_5 = \frac{(5)}{(5)+1} = \frac{5}{6}.$$

b)   The first five ordered pairs are $(1, \frac{1}{2})$, $(2, \frac{2}{3})$, $(3, \frac{3}{4})$, $(4, \frac{4}{5})$, and $(5, \frac{5}{6})$. The graph is shown in Figure 1.

c)   The graph of $y = x/(x+1)$ can be sketched by using the techniques of graphing rational functions outlined in Section 3.4; you might want to pause here to review that section. If we divide the denominator into the numerator, the function can be rewritten as

$$y = 1 - \frac{1}{x+1}$$

or

$$y - 1 = -\frac{1}{x+1}$$

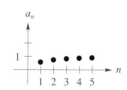

FIGURE 1

The graph can be obtained by reflecting the graph of $y = 1/x$ through the x-axis, then translating one unit left and one unit up. The graph is shown in Figure 2.

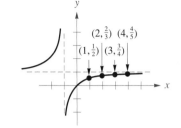

Note that the points for the sequence

$$\left\{ a_n = \frac{n}{n+1} \right\}$$

lie on the graph of $y = x/(x+1)$. In graphing a rational function such as $y = x/(x+1)$ (see Section 3.4), the line $y = 1$ is a horizontal asymptote since

$$\text{as } x \to \infty, \qquad \frac{x}{x+1} \to 1.$$

Similarly,

$$\text{as } n \to \infty, \qquad \frac{n}{n+1} \to 1$$

and we say the **sequence approaches** 1.

EXAMPLE 2   For the sequence $\{a_n = (-\frac{2}{3})^n\}$:

a)   list the first five terms

b)   graph the first five ordered pairs on the same axes with the graph of the continuous functions $f(x) = (\frac{2}{3})^x$ and $g(x) = -(\frac{2}{3})^x$.

SOLUTION

a)   Substituting $n = 1, 2, 3, 4$, and 5 into the function $(-\frac{2}{3})^n$ gives

$$a_1 = (-\tfrac{2}{3})^1 = -\tfrac{2}{3}$$
$$a_2 = (-\tfrac{2}{3})^2 = \tfrac{4}{9}$$
$$a_3 = (-\tfrac{2}{3})^3 = -\tfrac{8}{27}$$
$$a_4 = (-\tfrac{2}{3})^4 = \tfrac{16}{81}$$
$$a_5 = (-\tfrac{2}{3})^5 = -\tfrac{32}{243}$$

b)   The first five ordered pairs of the sequence are $(1, -\frac{2}{3})$, $(2, \frac{4}{9})$, $(3, -\frac{8}{27})$, $(4, \frac{16}{81})$, and $(5, -\frac{32}{243})$. These ordered pairs lie alternately on the graphs of the exponential functions $f(x)$ and $g(x)$, as illustrated in Figure 3.

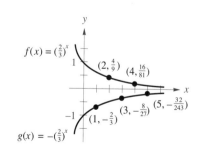

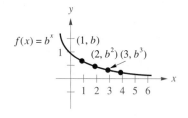

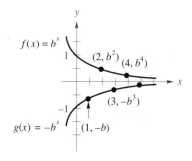

FIGURE 4

FIGURE 5

The graph of the sequence $\{a_n = (-2/3)^n\}$ in Example 2 suggests that as n increases, the ordered pairs of the sequence approach the horizontal axis. In other words, this sequence approaches 0 (Figure 4).

We can generalize this idea by using the graph of the exponential functions $f(x) = b^x$ and $g(x) = -b^x$, where $0 < b < 1$. The sequence $\{a_n = b^n\}$ has ordered pairs that lie on the graph of $f(x)$. The sequence $\{a_n = (-b)^n\}$ has points that lie alternately on the graphs of $f(x)$ and $g(x)$. The exponential functions $f(x) = b^x$ and $g(x) = -b^x$, $0 < b < 1$ approach 0 as x grows without bound (Figure 5). If we let c represent b or $-b$, then we have the following conclusion.

> As n grows without bound, the sequence $\{a_n = c^n\}$ approaches 0, provided $-1 < c < 1$.

In the previous examples, each sequence was given by a formula for the general term, such as

$$\frac{n}{n+1} \quad \text{or} \quad \left(-\frac{2}{3}\right)^n.$$

There are many important sequences for which the general term is not known or not even possible. One example is the sequence of prime positive integers: 2, 3, 5, 7, 11, 13, 17, ...

Some sequences are defined in a way that uses previous terms to find the next; this type of definition is called **recursive**. An interesting example is the Fibonacci sequence.

EXAMPLE 3 The Fibonacci sequence $\{f_n\}$ is defined as

$$f_n = \begin{cases} 1 & \text{for } n = 1 \text{ and } 2 \\ f_{n-1} + f_{n-2} & \text{for } n = 3, 4, 5, \ldots \end{cases}$$

List the first six terms.

SOLUTION The first two terms are $f_1 = 1$ and $f_2 = 1$. For $n = 3, 4, 5, 6$:

$$f_3 = f_2 + f_1 = 1 + 1 = 2$$
$$f_4 = f_3 + f_2 = 2 + 1 = 3$$
$$f_5 = f_4 + f_3 = 3 + 2 = 5$$
$$f_6 = f_5 + f_4 = 5 + 3 = 8.$$

The first six terms of the Fibonacci sequence are 1, 1, 2, 3, 5, and 8.

One of the most important ways to create a sequence is by addition. In calculus, you will often be faced with expressions of the form

$$a_n = u_1 + u_2 + u_3 + u_4 + \cdots + u_n.$$

For a more compact notation of this sum, the Greek capital letter Σ (sigma) is used:

$$\sum_{k=1}^{n} u_k = u_1 + u_2 + u_3 + \cdots + u_n.$$

The equation $k = 1$ under the Σ indicates that k takes on integer values from 1 to the value above Σ, in this case n. The variable k is called the **index.** The index does not need to run from 1 to n. For example, the sum of the first four positive odd integers could be written as

$$
\begin{array}{cccc}
k=1 & k=2 & k=3 & k=4 \\
\downarrow & \downarrow & \downarrow & \downarrow
\end{array}
$$

$$\sum_{k=1}^{4} (2k - 1) = (2(1) - 1) + (2(2) - 1) + (2(3) - 1) + (2(4) - 1)$$

$$= 1 + 3 + 5 + 7.$$

This could also be written as

$$\sum_{k=3}^{6} (2k - 5) = (2(3) - 5) + (2(4) - 5) + (2(5) - 5) + (2(6) - 5)$$

$$= 1 + 3 + 5 + 7.$$

EXAMPLE 4 Express the following in expanded form (without sigma notation). Simplify as much as possible.

a) $\displaystyle\sum_{k=2}^{5} k(3k + 2)$ **b)** $\displaystyle\sum_{k=1}^{5} \frac{(-1)^k x^{k-1}}{k^2}$

SOLUTION

a) Substitute 2, 3, 4, and 5 for k in succession into the formula $k(3k + 2)$ and add the resulting values:

$$\sum_{k=2}^{5} k(3k + 2) = 2(3(2) + 2) + 3(3(3) + 2)$$

$$+ 4(3(4) + 2) + 5(3(5) + 2)$$

$$= 16 + 33 + 56 + 85 = 190.$$

b) Substitute 1, 2, 3, 4, and 5 for k successively into the formula for each term, and add:

$$\sum_{k=1}^{5} \frac{(-1)^k x^{k-1}}{k^2} = \frac{(-1)^1 x^{1-1}}{1^2} + \frac{(-1)^2 x^{2-1}}{2^2} + \frac{(-1)^3 x^{3-1}}{3^2}$$

$$+ \frac{(-1)^4 x^{4-1}}{4^2} + \frac{(-1)^5 x^{5-1}}{5^2}$$

$$= -1 + \frac{x}{4} - \frac{x^2}{9} + \frac{x^3}{16} - \frac{x^4}{25}.$$

Summation notation can be used to define a sequence. In general, many important sequences are of the form

$$\left\{ a_n = \sum_{k=1}^{n} u_k \right\}.$$

EXAMPLE 5   Write the first five terms of the sequence

$$\left\{ a_n = \sum_{k=1}^{n} k^2 \right\}.$$

SOLUTION

$$a_1 = \sum_{k=1}^{1} k^2 = 1^2 = 1$$

$$a_2 = \sum_{k=1}^{2} k^2 = 1^2 + 2^2 = 5$$

$$a_3 = \sum_{k=1}^{3} k^2 = 1^2 + 2^2 + 3^2 = 14$$

$$a_4 = \sum_{k=1}^{4} k^2 = 1^2 + 2^2 + 3^2 + 4^2 = 30$$

$$a_5 = \sum_{k=1}^{5} k^2 = 1^2 + 2^2 + 3^2 + 4^2 + 5^2 = 55.$$

Therefore, the first five terms of the sequence are 1, 5, 14, 30, and 55.

It would be an understatement to say that someone studying calculus will be concerned with sums and summation notation. The following properties can be used to facilitate the work.

Properties of Summation

Assume that u_k and w_k are functions (of k) and that c represents a constant. For any positive integer n, the following summation properties hold:

1) $\displaystyle\sum_{k=1}^{n} c = nc$

2) $\displaystyle\sum_{k=1}^{n} cu_k = c \sum_{k=1}^{n} u_k$

3) $\displaystyle\sum_{k=1}^{n} (u_k + w_k) = \sum_{k=1}^{n} u_k + \sum_{k=1}^{n} w_k$

4) $\displaystyle\sum_{k=1}^{n} (u_k - w_k) = \sum_{k=1}^{n} u_k - \sum_{k=1}^{n} w_k$

5) $\displaystyle\sum_{k=1}^{n} k = \frac{n(n+1)}{2}$ or $\dfrac{n^2}{2} + \dfrac{n}{2}$

6) $\displaystyle\sum_{k=1}^{n} k^2 = \frac{n(n+1)(2n+1)}{6}$ or $\dfrac{n^3}{3} + \dfrac{n^2}{2} + \dfrac{n}{6}$

7) $\displaystyle\sum_{k=1}^{n} k^3 = \left[\frac{n(n+1)}{2}\right]^2$ or $\dfrac{n^4}{4} + \dfrac{n^3}{2} + \dfrac{n^2}{4}$

Properties 5, 6 and 7 will be proved in Section 11.5; however, we offer geometric verifications in Problems 64 through 66 of this section. We now prove properties 2 and 3, and the proofs of properties 1 and 4 are left for you as exercises.

To prove property 2, we use the distributive property of algebra:

$$\sum_{k=1}^{n} cu_k = cu_1 + cu_2 + cu_3 + \cdots + cu_{n-1} + cu_n$$

$$= c(u_1 + u_2 + u_3 + \cdots + u_{n-1} + u_n)$$

$$= c \sum_{k=1}^{n} u_k.$$

This completes the proof of property 2.

The proof of the property 3 uses the associative and commutative properties of algebra:

$$\sum_{k=1}^{n} (u_k + w_k) = (u_1 + w_1) + (u_2 + w_2) + (u_3 + w_3) + \cdots + (u_n + w_n)$$

$$= (u_1 + u_2 + u_3 + \cdots + u_n) + (w_1 + w_2 + w_3 + \cdots + w_n)$$

$$= \sum_{k=1}^{n} u_k + \sum_{k=1}^{n} w_k.$$

This completes the proof of property 3.

EXAMPLE 6   Use the properties of summation to evaluate

$$\sum_{k=1}^{40} (2k^3 - k^2 + 3k - 8).$$

SOLUTION First use properties 3 and 4 to break up the given problem into four summations:

$$\sum_{k=1}^{40} (2k^3 - k^2 + 3k - 8) = \sum_{k=1}^{40} 2k^3 - \sum_{k=1}^{40} k^2 + \sum_{k=1}^{40} 3k - \sum_{k=1}^{40} 8.$$

Next, factor out constant factors using property 2:

$$\sum_{k=1}^{40} 2k^3 - \sum_{k=1}^{40} k^2 + \sum_{k=1}^{40} 3k - \sum_{k=1}^{40} 8 = 2\sum_{k=1}^{40} k^3 - \sum_{k=1}^{40} k^2 + 3\sum_{k=1}^{40} k - \sum_{k=1}^{40} 8.$$

Finally, apply the formulas in properties 1, 5, 6, and 7 (with $n = 40$):

$$2\sum_{k=1}^{40} k^3 - \sum_{k=1}^{40} k^2 + 3\sum_{k=1}^{40} k - \sum_{k=1}^{40} 8$$

$$= 2\left[\frac{40(40+1)}{2}\right]^2 - \frac{40(40+1)(2(40)+1)}{6} + 3\left[\frac{40(40+1)}{2}\right] - 8(40)$$

$$= 1{,}344{,}800 - 22{,}140 + 2{,}460 - 320 = 1{,}324{,}800.$$

EXERCISE SET 11.1

A

For each sequence $\{a_n\}$ in Problems 1 through 18, (a) list the first five terms and (b) graph the first five ordered pairs.

1. $\{a_n = 3n - 1\}$

2. $\{a_n = 2n + 3\}$

3. $\{a_n = (-1)^n + n^2\}$

4. $\{a_n = (-1)^{n+1} + 2n\}$

5. $\{a_n = 1 - \cos n\pi\}$

6. $\left\{a_n = \cos^2\left(\frac{n\pi}{4}\right)\right\}$

7. $\left\{a_n = \left(\frac{n+1}{n}\right)^n\right\}$

8. $\left\{a_n = \left(\frac{2n+3}{2n}\right)^n\right\}$

9. $\left\{a_n = \frac{n}{2n+1}\right\}$

10. $\left\{a_n = \frac{n-1}{3n}\right\}$

11. $\left\{a_n = \frac{(-1)^{n+1}}{2n-1}\right\}$

12. $\left\{a_n = \frac{(-1)^{n+1}}{n^2}\right\}$

13. $\{a_n = 2^n\}$

14. $\{a_n = 5^n\}$

15. $\left\{a_n = \left(\frac{2}{3}\right)^n\right\}$

16. $\left\{a_n = \left(\frac{3}{4}\right)^n\right\}$

17. $\left\{a_n = \left(-\frac{1}{2}\right)^n\right\}$

18. $\left\{a_n = \left(-\frac{1}{4}\right)^n\right\}$

In Problems 19 through 27, (a) graph f(x) and (b) determine the value that {a_n} approaches, if any.

19. $f(x) = \dfrac{1}{x}$; $\left\{ a_n = \dfrac{1}{n} \right\}$

20. $f(x) = \dfrac{1}{x+2}$; $\left\{ a_n = \dfrac{1}{n+2} \right\}$

21. $f(x) = \dfrac{2x}{x+1}$; $\left\{ a_n = \dfrac{2n}{n+1} \right\}$

22. $f(x) = \dfrac{3x-1}{x}$; $\left\{ a_n = \dfrac{3n-1}{n} \right\}$

23. $f(x) = \dfrac{x}{x^2+2}$; $\left\{ a_n = \dfrac{n}{n^2+2} \right\}$

24. $f(x) = \dfrac{2x}{x^2+3}$; $\left\{ a_n = \dfrac{2n}{n^2+3} \right\}$

25. $f(x) = \left(\dfrac{1}{2}\right)^x$; $\left\{ a_n = \left(\dfrac{1}{2}\right)^n \right\}$

26. $f(x) = \left(\dfrac{5}{6}\right)^x$; $\left\{ a_n = \left(\dfrac{5}{6}\right)^n \right\}$

27. $f(x) = -\left(\dfrac{1}{3}\right)^x$; $\left\{ a_n = -\left(\dfrac{1}{3}\right)^n \right\}$

Find the first five terms of the sequences {a_n} in Problems 28 through 40.

28. $a_1 = 1,\ a_n = 2a_{n-1}$

29. $a_1 = 2,\ a_2 = 1,\ a_n = 3a_{n-1} + 2a_{n-2}$

30. $a_n = \begin{cases} -2, & n = 1 \\ na_{n-1}, & n \geq 2 \end{cases}$

31. $a_n = \begin{cases} -3, & n = 1 \\ 5n + a_{n-1}, & n \geq 2 \end{cases}$

32. $a_n = \begin{cases} 216, & n = 1 \\ \dfrac{a_{n-1}}{6}, & n \geq 2 \end{cases}$

33. $a_n = \sum\limits_{k=1}^{n} (3k - 5)$

34. $a_n = \sum\limits_{k=1}^{n} (4k + 1)$

35. $a_n = \sum\limits_{k=1}^{n} \dfrac{2}{k+1}$

36. $a_n = \sum\limits_{k=1}^{n} \dfrac{3}{2k-1}$

37. $a_n = \sum\limits_{k=1}^{n} \dfrac{x^{k-1}}{k}$

38. $a_n = \sum\limits_{k=1}^{n} \dfrac{x^k}{3k}$

39. $a_n = \sum\limits_{k=1}^{n} k(x-3)^{k-1}$

40. $a_n = \sum\limits_{k=1}^{n} 2^k(x-1)^{k-1}$

B

Using the properties of summation, evaluate Problems 41 through 46.

41. $\sum\limits_{k=1}^{30} (k^2 - 5k + 2)$

42. $\sum\limits_{k=1}^{50} (k^2 - 10k - 12)$

43. $\sum\limits_{k=1}^{70} (2k^3 - k + 3)$

44. $\sum\limits_{k=1}^{80} (3k^3 - 5k^2 + 6)$

45. $\sum\limits_{k=1}^{100} (k+1)(k^2 - 3)$

46. $\sum\limits_{k=1}^{100} (2k-1)(k^2 + 3)$

47. a) Use the properties of summation to evaluate

$$\frac{1}{n} \sum_{k=1}^{n} \frac{k}{n}, \quad \text{for } n = 20.$$

b) Use the properties of summation to evaluate

$$\frac{1}{n} \sum_{k=1}^{n} \frac{k}{n}, \quad \text{for } n = 100.$$

c) Use the properties of summation to determine the value approached by

$$\frac{1}{n} \sum_{k=1}^{n} \frac{k}{n} \quad \text{as } n \to \infty.$$

48. a) Use the properties of summation to evaluate

$$\frac{1}{n} \sum_{k=1}^{n} \frac{2k^2}{n^2}, \quad \text{for } n = 30.$$

b) Use the properties of summation to evaluate

$$\frac{1}{n} \sum_{k=1}^{n} \frac{2k^2}{n^2}, \quad \text{for } n = 100.$$

c) Use the properties of summation to determine the value approached by

$$\frac{1}{n} \sum_{k=1}^{n} \frac{2k^2}{n^2} \quad \text{as } n \to \infty.$$

C

Express the sums in Problems 49 through 58 in terms of summation notation (answers are not unique).

49. $3 + 6 + 9 + 12 + 15 + 18$

50. $5 + 10 + 15 + 20 + 25 + 30$

51. $3 + 7 + 11 + 15 + 19$ **52.** $1 + 7 + 13 + 19 + 25$

53. $\frac{2}{3} + \frac{4}{5} + \frac{6}{7} + \frac{8}{9}$ **54.** $\frac{3}{2} + \frac{6}{5} + \frac{9}{8} + \frac{12}{11}$

55. $1 - \frac{1}{2} + \frac{1}{4} - \frac{1}{6} + \frac{1}{8} - \frac{1}{10}$

56. $1 - \frac{1}{4} + \frac{1}{9} - \frac{1}{16} + \frac{1}{25} - \frac{1}{36}$

57. $\dfrac{1}{2 \cdot 4} + \dfrac{x}{4 \cdot 6} + \dfrac{x^2}{6 \cdot 8} + \cdots + \dfrac{x^n}{(2n + 2)(2n + 4)}$

58. $1 + \dfrac{x^2}{4} + \dfrac{x^4}{8} + \dfrac{x^6}{16} + \cdots + \dfrac{x^{2n-2}}{2^n}$

59. Prove the first summation property:

$$\sum_{k=1}^{n} c = nc.$$

60. Prove the fourth summation property:

$$\sum_{k=1}^{n} (u_k - w_k) = \sum_{k=1}^{n} u_k - \sum_{k=1}^{n} w_k.$$

61. For any positive number P, $\sqrt{P}$ can be approximated by the sequence

$$a_n = \begin{cases} \dfrac{P}{2}, & n = 1 \\ \dfrac{1}{2} a_{n-1} + \dfrac{P}{2a_{n-1}}, & n \geq 2. \end{cases}$$

Approximate $\sqrt{5}$ by letting $P = 5$ and finding the first three terms of the resulting sequence.

62. For any number P, $\sqrt[3]{P}$ can be approximated by the sequence

$$a_n = \begin{cases} \dfrac{P}{6}, & n = 1 \\ \dfrac{2}{3} a_{n-1} + \dfrac{P}{3a_{n-1}^2}, & n \geq 2. \end{cases}$$

Approximate $\sqrt[3]{12}$ by letting $P = 12$ and finding the the four terms of the resulting sequence.

63. How many squares of any size are on an 8 by 8 checker-board?

64. Explain how the summation formula

$$\sum_{k=1}^{n} k = \frac{n(n + 1)}{2}$$

can be verified by the following geometric argument (Figure 6):

$$1 + 2 + 3 + \cdots + n = \frac{n^2}{2} + \frac{n}{2} = \frac{n(n + 1)}{2}.$$

FIGURE 6

65. Explain how the summation formula

$$\sum_{k=1}^{n} k^2 = \frac{n(n + 1)(2n + 1)}{6}$$

can be verified by the following geometric argument (Figure 7):

$$1^2 + 2^2 + 3^2 + \cdots + n^2 = \frac{1}{3} n(n + 1)\left(n + \frac{1}{2}\right)$$
$$= \frac{n(n + 1)(2n + 1)}{6}.$$

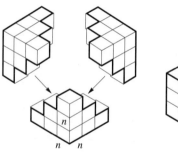
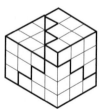
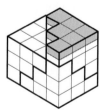
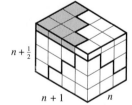

FIGURE 7

66. Explain how the summation formula

$$\sum_{k=1}^{n} k^3 = \left[\frac{n(n + 1)}{2}\right]^2$$

can be verified by the following geometric argument (Figure 8):

$$1^3 + 2^3 + 3^3 + 4^3 + \cdots + n^3$$

$$= (1 + 2 + 3 + \ldots + n)^2 = \left[\frac{n(n + 1)}{2}\right]^2.$$

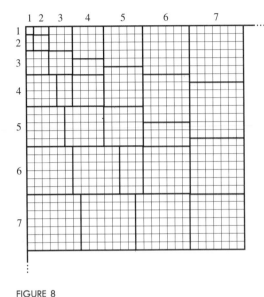

FIGURE 8

67. Let

$$r = \frac{1 + \sqrt{5}}{2} \quad \text{and} \quad s = \frac{1 - \sqrt{5}}{2}.$$

Use your calculator to evaluate Benet's formula,

$$\frac{r^n - s^n}{r - s},$$

for $n = 7$, 8, and 9. Compare your results with f_7, f_8, and f_9 in Example 3.

[Problem 64: Ian Richards, University of Minnesota, Minneapolis. Reprinted from the Mathematics Magazine 1984 with the permission of the Mathematical Association of America.

Problem 65: Man-Keung Sin, University of Hong Kong. Reprinted from the Mathematics Magazine 1984 with the permission of the Mathematical Association of America.

Problem 66: J. Barry Love, National Liberty Corp., Valley Forge, PA. Reprinted from the Mathematics Magazine 1977 with the permission of the Mathematical Association of America.]

ARITHMETIC AND GEOMETRIC SEQUENCES   S E C T I O N   11.2

For the purpose of approximating areas, volumes, probabilities, and other important quantities, a recurring initial step in calculus is to divide an interval on the x-axis into a number of subintervals. The following example is typical.

PROBLEM   *Divide the interval $[2, 5]$ into 12 subintervals of equal length. Identify the points of division.*

SOLUTION First draw a picture (Figure 9). The length of each subinterval must be

$$\frac{\text{length of } [2, 5]}{\text{number of subintervals}} = \frac{5 - 2}{12} = 0.25.$$

FIGURE 9

Starting at the left endpoint of the interval, add 0.25 successively to get each point of division:

$$x_1 = 2 + 0.25 = 2.25$$

$$x_2 = 2.25 + 0.25 = 2.50$$

$$x_3 = 2.50 + 0.25 = 2.75$$

$$\vdots$$

$$x_n = 4.75 + 0.25 = 5.0.$$

The sequence 2.25, 2.50, 2.75, . . . , 4.75, 5.0 is an example of an arithmetic sequence. An **arithmetic sequence** is typified by a constant difference between each term, in this case 0.25. A few more examples, including the difference between successive terms, are shown:

Arithmetic sequence	Difference
5, 8, 11, 14, 17, 20, 23	3
14, 12.5, 11, 9.5, 8	−1.5
2.7, 2.7, 2.7, 2.7, 2.7, 2.7, 2.7	0

Arithmetic Sequence

> A sequence in which each term differs from the succeeding term by a constant is called an **arithmetic sequence**. This constant difference between terms is called the **common difference** and is denoted by d.

Every term of an arithmetic sequence can be written in terms of the first term a_1 and the common difference d as follows:

$$a_1$$

$$a_2 = a_1 + d$$

$$a_3 = a_2 + d = (a_1 + d) + d = a_1 + 2d$$

$$a_4 = a_3 + d = (a_1 + 2d) + d = a_1 + 3d$$

$$a_5 = a_4 + d = (a_1 + 3d) + d = a_1 + 4d$$

$$\vdots$$

Notice that in each case the coefficient for d is one less than the number of the term (subscript).

> The general term a_n of an arithmetic sequence is
> $$a_n = a_1 + (n - 1)d.$$

EXAMPLE 1 Find the 23rd term for the arithmetic sequence $-11, -9.5, -8,$ $-6.5, \ldots.$

SOLUTION For this sequence, $a_1 = -11$ and $d = a_2 - a_1 = 1.5$. To find the 23rd term, substitute these values and $n = 23$ into the formula $a_n = a_1 + (n - 1)d$:

$$a_{23} = -11 + (23 - 1)1.5 = 22.$$

The general term for the sequence in Example 1 is $a_n = -11 + (n - 1)(\frac{3}{2})$. Notice that this can also be written as $a_n = \frac{3}{2}n - \frac{25}{2}$. Keeping in mind that n is the independent variable, you can consider this sequence as a linear function with slope $\frac{3}{2}$. A comparison of the graphs of the ordered pairs for a_n and for $f(x) = \frac{3}{2}x - \frac{25}{2}$ might clarify this idea (Figure 10). In general, an arithmetic sequence corresponds to a linear function whose slope is the common difference.

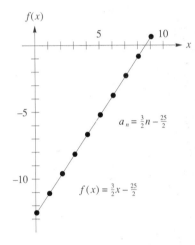

FIGURE 10

EXAMPLE 2 The 8th and 12th terms of an arithmetic sequence are 7 and $\frac{11}{2}$, respectively. Find the 36th term.

SOLUTION We are given the ordered pairs $(8, 7)$ and $(12, \frac{11}{2})$ of a linear function. The slope is

$$\frac{y_2 - y_1}{x_2 - x_1} = \frac{\frac{11}{2} - 7}{12 - 8} = -\frac{3}{8}.$$

Using the point-slope form of the equation for a line

$$y - y_1 = m(x - x_1)$$

we have

$$y - 7 = -\tfrac{3}{8}(x - 8).$$

For the sequence, a_n plays the role of y, and n plays the role of x. Therefore,

$$a_n - 7 = -\tfrac{3}{8}(n - 8)$$

or if we isolate a_n, then

$$a_n = -\tfrac{3}{8}n + 10.$$

To find the 36th term, substitute $n = 36$:

$$a_{36} = -\tfrac{3}{8}(36) + 10 = -\tfrac{7}{2}.$$

There is a traditional story about the great mathematician Karl Friedrich Gauss when he was a young boy attending primary school. The teacher assigned the class the problem of adding the numbers 1, 2, 3, 4, . . . , 50 (notice that this is an arithmetic sequence with common difference $d = 1$). Little Gauss took only a moment to hand his answer (correct, of course) to the teacher on his slate with scarcely any work on it. One can only speculate how the young genius found the sum so quickly, but many believe his line of reasoning went something like the following.

Let $S = 1 + 2 + 3 + \cdots + 50$, the required total. Reversing the order of addition does not affect the sum, so $S = 50 + 49 + 48 + \cdots + 1$. Adding these two equations gives

$$\begin{aligned}
S &= 1 + 2 + 3 + \cdots + 50 \\
S &= 50 + 49 + 48 + \cdots + 1 \\
\hline
2S &= 51 + 51 + 51 + \cdots + 51.
\end{aligned}$$

The bottom line is best evaluated by multiplication:

$$2S = 50(51)$$

so dividing by 2 gives

$$S = \frac{50(51)}{2} = 1275.$$

We can apply this idea to finding the sum of the terms of any arithmetic sequence. As we saw earlier, any arithmetic sequence can be written as

$$a_1, a_1 + d, a_1 + 2d, a_1 + 3d, \ldots, a_1 + (n - 1)d.$$

We want to find

$$S_n = a_1 + [a_1 + d] + [a_1 + 2d] + \cdots + [a_1 + (n - 1)d].$$

In the same way as in our story of young Gauss, reverse the order of addition and then add the equations vertically:

$$\begin{aligned}
S_n &= a_1 + [a_1 + d] + [a_1 + 2d] + \cdots + [a_1 + (n-1)d] \\
S_n &= [a_1 + (n-1)d] + [a_1 + (n-2)d] + [a_1 + (n-3)d] + \cdots + a_1 \\
\hline
2S_n &= [2a_1 + (n-1)d] + [2a_1 + (n-1)d] + [2a_1 + (n-1)d] + \cdots + [2a_1 + (n-1)d].
\end{aligned}$$

Since the last equation has $[2a_1 + (n-1)d]$ added repeatedly n times, we have

$$2S_n = n[2a_1 + (n-1)d].$$

Finally, dividing both sides by 2 gives

$$S_n = \frac{n}{2}[2a_1 + (n-1)d].$$

Using the fact that $a_n = a_1 + (n-1)d$, we can also write this formula as

$$S_n = \frac{n}{2}[a_1 + a_n].$$

The sum S_n of the first n terms of an arithmetic sequence $\{a_n = a_1 + (n-1)d\}$ is

$$S_n = \frac{n}{2}[2a_1 + (n-1)d].$$

An alternate form is

$$S_n = \frac{n}{2}[a_1 + a_n].$$

EXAMPLE 3   The first term of an arithmetic sequence is -4 and the common difference is $\frac{76}{11}$. If $S_n = 408$, find the number of terms n.

SOLUTION Substitute $S_n = 408$, $a_1 = -4$, and $d = \frac{76}{11}$ into the formula $S_n = n/2\,[2a_1 + (n-1)d]$, which gives

$$408 = \frac{n}{2}\left[-8 + (n-1)\left(\frac{76}{11}\right)\right].$$

Simplifying gives

$$408 = \frac{38n^2}{11} - \frac{82n}{11}.$$

Simplifying and factoring leads to two possible answers:

$$0 = 19n^2 - 41n - 2244$$

$$0 = (19n + 187)(n - 12)$$

$$n = -\tfrac{187}{19} \quad \text{or} \quad n = 12.$$

Since n must be a positive integer, the number of terms is $n = 12$.

Suppose A dollars are in an account paying 9% interest per year. After one year the amount in the account will be

$$A + 0.09(A) = A(1.09).$$

If this amount remains in the account for a second year, another 9% interest will be applied, and the account will grow to

$$A(1.09) + 0.09[A(1.09)] = A(1.09)^2.$$

Similarly, after 3 years the amount will grow to $A(1.09)^3$. To calculate the amount at the beginning of the nth year, we simply find the nth term of the sequence:

$$A, A(1.09), A(1.09)^2, A(1.09)^3, A(1.09)^4, \ldots .$$

This is an example of a **geometric sequence,** characterized by the fact that each term is some constant (in this example 1.09) times the preceeding term.

Geometric Sequence

A sequence in which each term (after the first) is obtained by multiplying the preceding term by a constant is called a **geometric sequence.** This constant multiplier between terms is called the **common ratio** and is denoted as r.

A few more examples are shown:

Geometric sequence	Ratio
5, 10, 20, 40, 80, 160	2
72, 12, 2, $\frac{1}{3}$, $\frac{1}{18}$, $\frac{1}{108}$, $\frac{1}{648}$	$\frac{1}{6}$
271, -27.1, 2.71, -0.271, 0.0271	$-\frac{1}{10}$

Every term of a geometric sequence can be written in terms of the first term a_1 and the common ratio r as follows:

$$a_1$$
$$a_2 = a_1 r$$
$$a_3 = a_2 r = a_1 r^2$$
$$a_4 = a_3 r = a_1 r^3$$
$$a_5 = a_4 r = a_1 r^4$$
$$\vdots$$

Notice that in each case the exponent for r is one less than the number of the term (subscript).

The general term a_n of a geometric sequence is

$$a_n = a_1 r^{n-1}.$$

EXAMPLE 4   Find the ninth term of the geometric sequence whose first term is $\frac{1}{4}$ and whose common ratio is $-\frac{2}{3}$.

SOLUTION Substitute $a_1 = \frac{1}{4}$ and $r = -\frac{2}{3}$ into the formula $a_n = a_1 r^{n-1}$ with $n = 9$:

$$a_9 = \frac{1}{4}\left(-\frac{2}{3}\right)^{9-1} = \frac{2^6}{3^8} = \frac{64}{6561}.$$

EXAMPLE 5   Find the common ratio of the geometric sequence whose second and fifth terms are -288 and 121.5, respectively.

SOLUTION Substitute -288 for a_2 and 121.5 for a_5 into the formula $a_n = a_1 r^{n-1}$:

$$\begin{cases} -288 = a_1 r \\ 121.5 = a_1 r^4. \end{cases}$$

Multiply the first equation in the system by r^3:

$$\begin{cases} -288 r^3 = a_1 r^4 \\ 121.5 = a_1 r^4. \end{cases}$$

The solution must then satisfy $-288 r^3 = 121.5$, which yields

$$r = \sqrt[3]{\frac{-121.5}{288}} = -\frac{3}{4}.$$

Consider the problem of adding the first n terms of a geometric sequence. The general way to write the first n terms of a geometric sequence is

$$a_1, a_1 r, a_1 r^2, a_1 r^3, \ldots, a_1 r^{n-1}.$$

We want a formula to find

$$S_n = a_1 + a_1 r + a_1 r^2 + a_1 r^3 + \cdots + a_1 r^{n-2} + a_1 r^{n-1}.$$

If we multiply both sides of this equation by $-r$, we get

$$-rS_n = -a_1r - a_1r^2 - a_1r^3 - \cdots - a_1r^{n-1} - a_1r^n.$$

Now add these two expressions for S_n and $-rS_n$ to get

$$
\begin{aligned}
S_n &= a_1 + a_1r + a_1r^2 + \cdots + a_1r^{n-1} \\
-rS_n &= \quad\ \ -a_1r - a_1r^2 - \cdots - a_1r^{n-1} - a_1r^n \\
\hline
S_n - rS_n &= a_1 \qquad\qquad\qquad\qquad\qquad\quad - a_1r^n.
\end{aligned}
$$

Now, if we factor out S_n on the left side of this equation and divide both sides by $(1 - r)$, we have the following formula.

The sum S_n of the first n terms of a geometric sequence $\{a_n = a_1r^{n-1}\}$ is

$$S_n = \frac{a_1 - a_1r^n}{1 - r}, \qquad r \neq 1.$$

EXAMPLE 6   Find the sum of the first eight terms of the geometric sequence whose first term is $\frac{5}{6}$ and whose common ratio is -3.

SOLUTION Substitute $a_1 = \frac{5}{6}$ and $r = -3$ into the formula

$$S_n = \frac{a_1 - a_1r^n}{1 - r}.$$

This yields

$$S_8 = \frac{\dfrac{5}{6} - \dfrac{5}{6}(-3)^8}{1 - (-3)} = -1366\tfrac{2}{3}.$$

EXAMPLE 7   Evaluate $\displaystyle\sum_{k=1}^{6} \left[11 + (\tfrac{1}{2})^{k-1}\right].$

SOLUTION Break up the summation into two summations:

$$\sum_{k=1}^{6} \left[11 + (\tfrac{1}{2})^{k-1}\right] = \sum_{k=1}^{6} 11 + \sum_{k=1}^{6} (\tfrac{1}{2})^{k-1}.$$

On the right side of the equation, the first summation is

$$\sum_{k=1}^{6} 11 = 6(11).$$

The second summation can be written as

$$\sum_{k=1}^{6} (\tfrac{1}{2})^{k-1} = 1 + (\tfrac{1}{2}) + (\tfrac{1}{2})^2 + (\tfrac{1}{2})^3 + (\tfrac{1}{2})^4 + (\tfrac{1}{2})^5.$$

This is the sum of the first six terms of a geometric sequence whose first term is 1 and whose common ratio is $\tfrac{1}{2}$. We have

$$\sum_{k=1}^{6} [11 + (\tfrac{1}{2})^{k-1}] = \sum_{k=1}^{6} 11 + \sum_{k=1}^{6} (\tfrac{1}{2})^{k-1}$$

$$= 6(11) + \frac{1 - (\tfrac{1}{2})^6}{1 - \tfrac{1}{2}}$$

$$= 66 + \tfrac{63}{32} = 67\tfrac{31}{32}.$$

EXERCISE SET 11.2

A

For the arithmetic sequences in Problems 1 through 9, find a_1, d, a_5, and a_n.

1. $2, 9, 16, 23, \ldots$ **2.** $57, 44, 31, 18, \ldots$

3. $6, 6\tfrac{2}{5}, 6\tfrac{4}{5}, 7\tfrac{1}{5}, \ldots$

4. $17.7, 17.3, 16.9, 16.5, \ldots$

5. $\sqrt{3} - 2, 2\sqrt{3}, 3\sqrt{3} + 2, 4\sqrt{3} + 4, \ldots$

6. $4\sqrt{2} + 9, 2\sqrt{2} + 6, 3, -2\sqrt{2}, \ldots$

7. $4y - 8, 3y - 4, 2y, y + 4, \ldots$

8. $13a + 2b, 11a + 3b, 9a + 4b, 7a + 5b, \ldots$

9. $\dfrac{x^2 - 1}{x}, x, \dfrac{x^2 + 1}{x}, \dfrac{x^2 + 2}{x}, \ldots$

For the geometric sequences in Problems 10 through 20, find a_1, r, a_5, and a_n.

10. $11, 22, 44, 88, \ldots$ **11.** $28, 14, 7, \tfrac{7}{2}, \ldots$

12. $-27, 18, -12, 8, \ldots$

13. $125, -100, 80, -64, \ldots$

14. $\tfrac{4}{3}, \tfrac{2}{9}, \tfrac{1}{27}, \tfrac{1}{162}, \ldots$

15. $3.2, 1.28, .512, .2048, \ldots$

16. $3, 3\sqrt{5}, 15, 15\sqrt{5}, \ldots$

17. $1, 1 + \sqrt{2}, 3 + 2\sqrt{2}, 7 + 5\sqrt{2}, \ldots$

18. $3x, 6x^3, 12x^5, 24x^7, \ldots$

19. $4x^6, 4x^3, 4, \dfrac{4}{x^3}, \ldots$

20. $\ln x^2, \ln x^4, \ln x^8, \ln x^{16}, \ldots$

Each sequence in Problems 21 through 28 is either arithmetic or geometric. Classify each and find the indicated term.

21. $\tfrac{1}{3}, \tfrac{1}{2}, \tfrac{2}{3}, \tfrac{5}{6}, \ldots; a_{10}$ **22.** $-8, -3, 2, 7, \ldots; a_{16}$

23. $\tfrac{16}{3}, -\tfrac{4}{3}, \tfrac{1}{3}, -\tfrac{1}{12}, \ldots; a_9$

24. $-9, -5, -1, 3, \ldots; a_{12}$

25. $3, -\tfrac{3}{2}, \tfrac{3}{4}, -\tfrac{3}{8}, \ldots; a_{10}$ **26.** $16, 12, 9, \tfrac{27}{4}, \ldots; a_9$

27. $\tfrac{11}{8}, \tfrac{3}{8}, -\tfrac{5}{8}, -\tfrac{13}{8}, \ldots; a_{11}$

28. $e^x, -e^{3x}, e^{5x}, -e^{7x}, \ldots; a_{14}$

The terms of the sums in Problems 29 through 42 are from arithmetic or geometric sequences. Compute the indicated sums.

29. $2 + 5 + 8 + \cdots + 32$

30. $43 + 41 + 39 + \cdots + 9$

31. $1 - 3 + 9 - 27 + \cdots + 6561$

32. $-192 + 96 - 48 + \cdots + \tfrac{3}{2}$

33. $\displaystyle\sum_{k=1}^{12} (4k - 19)$ **34.** $\displaystyle\sum_{k=1}^{10} 108(\tfrac{5}{6})^{k-1}$

35. $\displaystyle\sum_{k=1}^{10} 30(\tfrac{3}{2})^{k-1}$

36. $\displaystyle\sum_{k=3}^{14} \left[\frac{6+k}{2}\right]$

37. $\displaystyle\sum_{k=2}^{10} (-1)^k(2)^{k+1}$

38. $\displaystyle\sum_{k=2}^{11} (-1)^{k+1}y^{2k-2}$

39. $\displaystyle\sum_{k=1}^{20} (5k+2b)$

40. $(3a+b)+(3a+4b)+(3a+7b)+\cdots+(3a+34b)$

41. $1+x+x^2+\cdots+x^{11}$

42. $1-x+x^2-x^3+\cdots+x^{10}$

B

Evaluate each sum in Problems 43 through 48.

43. $\displaystyle\sum_{k=1}^{10} (4k+2^{k-1})$

44. $\displaystyle\sum_{k=1}^{8} [(-3)^{k-1}+(2k-5)]$

45. $\displaystyle\sum_{k=1}^{7} [(\tfrac{1}{2})^k + 12(\tfrac{1}{3})^{k-1}]$

46. $\displaystyle\sum_{k=1}^{7} [(\tfrac{2}{3})^{k-1} + 40(\tfrac{3}{2})^{k-1}]$

47. $\displaystyle\sum_{k=1}^{10} (k^2+2^k)$

48. $\displaystyle\sum_{k=1}^{8} (k^3+3^k)$

49. Find the 20th term of the arithmetic sequence whose 3rd and 12th terms are 16 and 13, respectively.

50. Find the 33rd term of the arithmetic sequence whose 7th and 13th terms are 6 and 11, respectively.

51. Find the 7th term of the geometric sequence whose 2nd and 5th terms are $-\sqrt{3}$ and $\tfrac{1}{3}$, respectively.

52. Find the 10th term of the geometric sequence whose 4th and 7th terms are 4 and 12, respectively.

53. How many integers between 10 and 200 are multiples of 3? Find their sum.

54. How many integers between 23 and 201 are multiples of 4? Find their sum.

55. Divide the interval $[3, 6]$ into 12 subintervals of equal length. Identify the points of division.

56. Divide the interval $[-1, 5]$ into 18 subintervals of equal length. Identify the points of division.

57. How much is saved after 5 years if $350 is deposited every month into an account paying 10% compounded monthly?

58. How much is saved after 8 years if $1720 is deposited every year into an account paying 11% compounded annually?

59. Neglecting air resistance, a falling object will fall 16 feet after 1 second, 48 additional feet after 2 seconds, an additional 80 feet after 3 seconds, and another 112 feet after 4 seconds. Look for a pattern to make an educated guess as to how far the object would fall after 10 seconds.

60. A concert hall has 35 rows of seats with 30 seats in the first row, 31 seats in the second row, 32 seats in the third row, and so on. What is the seating capacity?

C

61. How many multiples of 2 and multiples of 3 are between 1 and 1000? (This is not the same as the number of multiples of 6 between 1 and 1000.)

62. A set of nested squares is drawn so that the vertices of the next (inner) square are at the midpoints of the preceding (outer) square. If the first (largest) square has a side of 4, what is the perimeter of the twelfth square (Figure 11)?

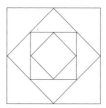

FIGURE 11

63. A trail construction crew parks its truck at the same spot every day and walks up a trail it has been building to continue its construction. They can walk 4 mph and can build trail at 0.5 mph. They spend a total of 8 hours either walking or building trail and must return to the same parking spot every day. On the first day, trail construction began at the parking lot, but each day more time is spent walking and less is spent building trail. If n is the number of days spent working on the trail, and L_n is the entire length of the trail at the end of the nth day, find a formula for L_n whose only variable is n.

INFINITE SERIES

S E C T I O N   11.3

Consider the following problem:

PROBLEM   *A ball that is dropped from a height of 16 feet always bounces back to 49% of its previous height. How much time will elapse before the ball comes to rest?*

SOLUTION We need a formula that relates the distance an object falls to the time it takes to fall. If we neglect wind resistance (or any other kind of resistance), after t seconds the height s of a falling object will be

$$s = s_0 - 16t^2$$

where s_0 is the initial height. When our ball is first dropped, $s_0 = 16$ ft. To find the time the ball takes to reach the ground, we set $s = 0$ and solve for t:

$$0 = 16 - 16t^2 \Rightarrow t_1 = 1.$$

The ball then bounces back to $(0.49)(16)$ feet and falls again. The time it takes to reach this height is the same time it takes to fall to the ground. To find this time, set $s_0 = (0.49)(16)$, $s = 0$, and again solve for t:

$$0 = (0.49)(16) - 16t^2 \Rightarrow t_2 = 0.7.$$

Hence the total time to rise and fall is $2t_2 = 2(0.7)$.

Continuing this way for successive bounces, we can find the times between each rise and fall:

$$0 = 16 - 16t^2 \qquad\quad \Rightarrow t_1 = 1$$
$$0 = (0.49)(16) - 16t^2 \Rightarrow t_2 = (0.7) \Rightarrow 2t_2 = 2(0.7)$$
$$0 = (0.49)^2(16) - 16t^2 \Rightarrow t_3 = (0.7)^2 \Rightarrow 2t_3 = 2(0.7)^2$$
$$0 = (0.49)^3(16) - 16t^2 \Rightarrow t_4 = (0.7)^3 \Rightarrow 2t_4 = 2(0.7)^3$$
$$0 = (0.49)^4(16) - 16t^2 \Rightarrow t_5 = (0.7)^4 \Rightarrow 2t_5 = 2(0.7)^4$$
$$\vdots$$

We know from experience that the ball will eventually come to rest, even though it bounces an indefinite number of times. Faithfully sticking with our model, we find that the total time our ball bounces will be

$$1 + 2(0.7) + 2(0.7)^2 + 2(0.7)^3 + \cdots.$$

The ($\cdots$) indicates that we continue the addition indefinitely. This is an example of an **infinite series.** Even though we cannot actually add an infinite number of terms, in many instances we can make some observations about such a series. The key idea is to form a sequence $\{S_n\}$ in which S_1 is the first term of the series, S_2 is the sum of the first two terms of the series, S_3 is the sum of the first three terms of the series, and so on. We will return to the series for the bouncing ball soon (Example 2) but to facilitate the discussion of infinite series in general we need the following information.

Infinite Series

Consider the sequence

$$\left\{ S_n = \sum_{k=1}^{n} u_k \right\} \qquad \text{as } n \to \infty$$

that is, $S_1 = u_1$, $S_2 = u_1 + u_2$, $S_3 = u_1 + u_2 + u_3$, and so on. If there is a number S such that $S_n \to S$ as $n \to \infty$, we write

$$S = u_1 + u_2 + u_3 + \cdots$$

or, using sigma notation,

$$S = \sum_{k=1}^{\infty} u_k.$$

If there is no number S such that $S_n \to S$ as $n \to \infty$, we write

$$\sum_{k=1}^{\infty} u_k \qquad \text{diverges.}$$

The expression $\sum_{k=1}^{\infty} u_k$ is called an **infinite series.** The sequence $\{S_n = \sum_{k=1}^{n} u_k\}$ is called the sequence of partial sums. If S exists, the series is called convergent, and S is called the sum of the series.

It is usually difficult to find an expression for S_n that lends itself to finding S (if the sum exists). However, a formula can be found if the terms of the series u_k are from a geometric sequence. In this case, the series is called a **geometric series.** Thus a series of the form

$$\sum_{k=1}^{\infty} a_1 r^{n-1}$$

is a geometric series. We know from Section 11.2 that

$$S_n = a_1 + a_1 r + a_1 r^2 + a_1 r^3 + \cdots + a_1 r^{n-2} + a_1 r^{n-1}$$

can be expressed as

$$S_n = \frac{a_1 - a_1 r^n}{1 - r}, \qquad r \neq 1.$$

Recall from Section 11.1 that if $-1 < r < 1$, r^n approaches 0 as n grows without bound. Therefore,

$$S_n \to \frac{a_1 - a_1(0)}{1 - r} = \frac{a_1}{1 - r}, \quad \text{as } n \to \infty.$$

If $r > 1$, S_n does not approach a limiting value. This is because r^n grows without bound as n grows without bound. Similarly, if $r < -1$, S_n does not approach a limiting value. This result is summarized as follows.

If $-1 < r < 1$, the infinite geometric series

$$\sum_{k=1}^{\infty} a_1 r^{k-1} = a_1 + a_1 r + a_1 r^2 + a_1 r^3 + \cdots$$

has the sum

$$\frac{a_1}{1 - r}.$$

Sum of an Infinite Geometric Series

EXAMPLE 1   Find the sum of the following series:

a) $\displaystyle\sum_{k=1}^{\infty} 3(-0.8)^{k-1}$

b) $\displaystyle\sum_{k=1}^{\infty} 10(\tfrac{5}{3})^{k-1}$.

SOLUTION Both series are geometric. For **(a)** we have $a_1 = 3$ and $r = -0.8$. Since $-1 < r < 1$, the sum is

$$\frac{a_1}{1 - r} = \frac{3}{1 - (-0.8)} = \frac{5}{3}.$$

The series in **(b)** does not have a sum, since $r = \frac{5}{3} > 1$.

Returning to the introductory problem of the bouncing ball, we are now in a position to find the total time elapsed until the ball comes to rest.

EXAMPLE 2   Find the sum of the series $1 + 2(0.7) + 2(0.7)^2 + 2(0.7)^3 + \cdots$

SOLUTION If we start with the first term, the series is not a geometric series. However, from the second term onward the series is geometric with common ratio $r = 0.7$:

$$1 + [2(0.7) + 2(0.7)^2 + 2(0.7)^3 + \cdots] = 1 + \frac{a_1}{1 - r} = 1 + \frac{2(0.7)}{1 - 0.7} = 5\frac{2}{3}.$$

Even though the ball bounces an infinite number of times according to our model, it comes to rest after $5\frac{2}{3}$ seconds.

The sum of a geometric series can be used to express any real number with an infinitely repeating decimal as a ratio of two integers.

EXAMPLE 3   Express the following repeating decimals as a ratio of two integers:

   a) $0.\overline{36}$ b) $2.68\overline{148}$

SOLUTION

   a)   This repeating decimal can be written as

   $$0.\overline{36} = 0.36 + 0.0036 + 0.000036 + \cdots$$

   which is a geometric series with $a_1 = 0.36$ and $r = 0.01$. Using the formula for the sum of a geometric series gives

   $$0.\overline{36} = \frac{a_1}{1 - r} = \frac{0.36}{1 - 0.01} = \frac{0.36}{0.99} = \frac{4}{11}.$$

   b)   The first three digits are not part of the repeating block, so we separate accordingly:

   $$2.68\overline{148} = 2.68 + 0.00\overline{148}$$

   $$= 2.68 + (0.00148 + 0.00000148 + 0.00000000148 + \cdots)$$

   The expression in the parentheses is a geometric series with $a_1 = 0.00148$ and $r = 0.001$. The sum of the series enclosed in parentheses is

   $$\frac{a_1}{1 - r} = \frac{0.00148}{1 - 0.001} = \frac{0.00148}{0.999} = \frac{1}{675}.$$

   Adding 2.68 to this result gives

   $$2.68\overline{148} = 2.68 + \frac{1}{675} = \frac{268}{100} + \frac{1}{675} = \frac{362}{135}.$$

It is sometimes possible to generate an infinite series for a given rational function of x.

EXAMPLE 4 Find an infinite geometric series for

a) $\dfrac{5}{1 - 2x}$ **b)** $\dfrac{6}{2 + x}$.

State the domain for which the sum of the series equals the rational function.

SOLUTION In both cases, we want first to write the rational function in the form $a_1/(1 - r)$. Then we can identify a_1 and r and use the relationship

$$\frac{a_1}{1 - r} = a_1 + a_1 r + a_1 r^2 + a_1 r^3 + \cdots.$$

a) Since the function is in the desired form, we recognize $a_1 = 5$ and $r = 2x$, so

$$\frac{5}{1 - 2x} = 5 + 5(2x) + 5(2x)^2 + 5(2x)^3 + \cdots$$

$$= 5 + 10x + 5(4x^2) + 5(8x^3) + \cdots$$

$$= 5 + 10x + 20x^2 + 40x^3 + \cdots.$$

Because this is a geometric series, it is valid for $-1 < r < 1$, which means $-1 < 2x < 1 \Rightarrow -\frac{1}{2} < x < \frac{1}{2}$.

b) This function is not in the form we want. Multiplying numerator and denominator by $\frac{1}{2}$ gives

This constant term needs to be 1. $\longrightarrow$ $\dfrac{6}{2 + x} = \dfrac{\frac{1}{2}(6)}{\frac{1}{2}(2 + x)} = \dfrac{3}{1 + x/2} = \dfrac{3}{1 - (-x/2)}.$

In this form we recognize that $a_1 = 3$ and $r = -x/2$, so

$$\frac{6}{2 + x} = \frac{3}{1 - (-x/2)}$$

$$= 3 + 3\left(-\frac{x}{2}\right) + 3\left(-\frac{x}{2}\right)^2 + 3\left(-\frac{x}{2}\right)^3 + \cdots$$

$$= 3 - \frac{3x}{2} + \frac{3x^2}{4} - \frac{3x^3}{8} + \cdots.$$

Since the interval for which this is valid is $-1 < r < 1$,

$$-1 < -\frac{x}{2} < 1 \Rightarrow -2 < x < 2.$$

If an infinite series is not geometric, it may be very difficult to find a simple formula for S_n, the nth partial sum. However, in some instances there are techniques that render S_n into a form that makes the sum of the series apparent. In the next example we use the technique of finding partial fractions. You might want to pause and review Section 10.5.

EXAMPLE 5   Use the nth partial sum S_n to find the sum of the infinite series

$$\sum_{k=1}^{\infty} \frac{6}{(3k+3)(3k)} = \frac{6}{6 \cdot 3} + \frac{6}{9 \cdot 6} + \frac{6}{12 \cdot 9} + \cdots$$

SOLUTION This series is not geometric. The kth term can be expressed in terms of partial fractions:

$$\frac{6}{(3k+3)(3k)} = \frac{A}{3k+3} + \frac{B}{3k}.$$

Multiplying both sides by $(3k+3)(3k)$ gives

$$6 = A(3k) + B(3k+3).$$

Letting $k = 0$ and $k = -1$ in this equation leads to

$$6 = 0 + B(0+3) \qquad \Rightarrow B = 2$$
$$6 = A(-3) + B(-3+3) \Rightarrow A = -2.$$

Thus the kth term can be expressed as

$$\frac{6}{(3k+3)(3k)} = \frac{2}{3k} - \frac{2}{3k+3}.$$

Consequently, we can write S_n as

$$S_n = \frac{6}{6 \cdot 3} + \frac{6}{9 \cdot 6} + \frac{6}{12 \cdot 9} + \cdots + \frac{6}{(3n+3)(3n)}$$

$$= \left(\frac{2}{3} - \frac{2}{6}\right) + \left(\frac{2}{6} - \frac{2}{9}\right) + \left(\frac{2}{9} - \frac{2}{12}\right) + \cdots + \left(\frac{2}{3n} - \frac{2}{3n+3}\right)$$

$$= \frac{2}{3} + \left(-\frac{2}{6} + \frac{2}{6}\right) + \left(-\frac{2}{9} + \frac{2}{9}\right) + \cdots + \left(-\frac{2}{3n} + \frac{2}{3n}\right) - \frac{2}{3n+3}$$

$$= \frac{2}{3} - \frac{2}{3n+3}.$$

As $n \to \infty$, the rational expression $2/(3n+3) \to 0$. Therefore, S_n approaches $\frac{2}{3}$; that is, the sum of the series is $\frac{2}{3}$.

The last example illustrates a method called **telescoping.** Notice that the key to finding the simple formula for S_n was to write the kth term in the series as a difference of consecutive terms. Generally speaking, the telescoping series

$$\sum_{k=1}^{\infty} (w_k - w_{k+1})$$

has the partial sum

$$S_n = (w_1 - w_2) + (w_2 - w_3) + (w_3 - w_4) + \cdots + (w_n - w_{n+1})$$

$$= w_1 + (-w_2 + w_2) + (-w_3 + w_3) + \cdots + (-w_n + w_n) - w_{n+1}$$

$$= w_1 - w_{n+1}.$$

If a series can be expressed in the form

$$\sum_{k=1}^{\infty} (w_k - w_{k+1})$$

then the series has the partial sum $S_n = w_1 - w_{n+1}$.

Telescoping Series

Looking at this expression for S_n, we see that a telescoping series converges to a sum if and only if w_{n+1} approaches a finite value as $n \to \infty$.

We emphasize that the sum of a series is the value that S_n approaches, if any, as $n \to \infty$. Do not assume that a given infinite series has a sum; many do not.

EXAMPLE 6   For the series $\sum_{k=1}^{\infty} (2k - 1)$, find a simple formula for S_n to determine whether the series converges or diverges.

SOLUTION The expression $S_n = 1 + 3 + 5 + \cdots + (2n - 1)$ is the sum of the first n terms of an arithmetic sequence with $a_1 = 1$, $d = 2$, and $a_n = 2n - 1$. Therefore,

$$S_n = \left(\frac{n}{2}\right)(a_1 + a_n) = \left(\frac{n}{2}\right)(1 + 2n - 1) = n^2.$$

As $n \to \infty$, $S_n \to \infty$; hence the series diverges.

It is sometimes possible to decide whether a series is convergent or not, even if we are unable to find a simple formula for S_n.

EXAMPLE 7 Show that the **harmonic series**

$$\sum_{k=1}^{\infty} \frac{1}{k} = 1 + \frac{1}{2} + \frac{1}{3} + \frac{1}{4} + \cdots$$

is divergent.

SOLUTION First write several terms and group them so that the denominator of the last term in each group is a power of 2:

$$1 + (\tfrac{1}{2}) + (\tfrac{1}{3} + \tfrac{1}{4}) + (\tfrac{1}{5} + \tfrac{1}{6} + \tfrac{1}{7} + \tfrac{1}{8}) + (\tfrac{1}{9} + \cdots + \tfrac{1}{16}) + \cdots.$$

Now we focus our attention on the following specific partial sums:

$$S_2 = 1 + \tfrac{1}{2}$$

$$S_4 = 1 + \tfrac{1}{2} + (\tfrac{1}{3} + \tfrac{1}{4}) > 1 + \tfrac{1}{2} + (\tfrac{1}{4} + \tfrac{1}{4}) = 1 + \tfrac{2}{2}$$

$$S_8 = 1 + \tfrac{1}{2} + (\tfrac{1}{3} + \tfrac{1}{4}) + (\tfrac{1}{5} + \tfrac{1}{6} + \tfrac{1}{7} + \tfrac{1}{8})$$

$$> 1 + \tfrac{1}{2} + (\tfrac{1}{4} + \tfrac{1}{4}) + (\tfrac{1}{8} + \tfrac{1}{8} + \tfrac{1}{8} + \tfrac{1}{8}) = 1 + \tfrac{3}{2}.$$

Continuing in this way, we have $S_{16} > 1 + \tfrac{4}{2}$, $S_{32} > 1 + \tfrac{5}{2}$, and in general,

$$S_{2^k} > 1 + \frac{k}{2}.$$

As $k \to \infty$,

$$\left(1 + \frac{k}{2}\right) \to \infty.$$

It follows that S_n could not possibly approach a number as n grows without bound, consequently the series diverges.

In calculus you will see that one of the applications of infinite series is the representation of important functions as series. Although it is beyond the scope of this text to justify the validity of such representations, we offer a few significant results:

$$e^x = 1 + x + \frac{x^2}{1 \cdot 2} + \frac{x^3}{1 \cdot 2 \cdot 3} + \frac{x^4}{1 \cdot 2 \cdot 3 \cdot 4} + \cdots$$

$$\cos x = 1 - \frac{x^2}{1 \cdot 2} + \frac{x^4}{1 \cdot 2 \cdot 3 \cdot 4} - \frac{x^6}{1 \cdot 2 \cdot 3 \cdot 4 \cdot 5 \cdot 6} + \cdots$$

$$\sin x = x - \frac{x^3}{1 \cdot 2 \cdot 3} + \frac{x^5}{1 \cdot 2 \cdot 3 \cdot 4 \cdot 5} - \frac{x^7}{1 \cdot 2 \cdot 3 \cdot 4 \cdot 5 \cdot 6 \cdot 7} + \cdots$$

For economy in notation, the factorial symbol ! is used.

Let k be a nonnegative integer. The value of $k!$, called **k factorial**, is defined as

$$k! = \begin{cases} 1, & \text{if } k = 0 \\ 1 \cdot 2 \cdot 3 \cdot \ \cdots \ \cdot k, & \text{if } k = 1, 2, 3, \ldots. \end{cases}$$

Factorial

For example,

$$0! = 1$$
$$1! = 1$$
$$2! = (1)(2) = 2$$
$$3! = (1)(2)(3) = 6$$
$$4! = (1)(2)(3)(4) = 24.$$

Using this notation the series representation for e^x, $\cos x$, and $\sin x$ can be written as

$$e^x = \sum_{k=0}^{\infty} \frac{x^k}{k!}, \qquad \cos x = \sum_{k=0}^{\infty} (-1)^k \frac{x^{2k}}{(2k)!}, \qquad \sin x = \sum_{k=0}^{\infty} (-1)^k \frac{x^{2k+1}}{(2k+1)!}.$$

EXERCISE SET 11.3

A

In Problems 1 through 16, determine whether the infinite geometric series has a sum and, if so, find that sum.

1. $60 + 30 + 15 + \cdots$

2. $180 + 60 + 20 + \cdots$

3. $-98 - 28 - 8 - \cdots$

4. $-48 - 36 - 27 - \cdots$

5. $24 + 36 + 54 + \cdots$

6. $36 + 48 + 64 + \cdots$

7. $1 - \frac{1}{2} + \frac{1}{4} - \frac{1}{8} + \cdots$

8. $3 - \frac{1}{2} + \frac{1}{12} - \frac{1}{72} + \cdots$

9. $\displaystyle\sum_{k=1}^{\infty} 10(0.1)^{k-1}$

10. $\displaystyle\sum_{k=1}^{\infty} 25(0.2)^{k-1}$

11. $\displaystyle\sum_{k=1}^{\infty} (0.4)2^{k-1}$

12. $\displaystyle\sum_{k=1}^{\infty} (0.7)5^{k-1}$

13. $\displaystyle\sum_{k=1}^{\infty} \left(-\frac{3}{5}\right)^k$

14. $\displaystyle\sum_{k=1}^{\infty} \left(-\frac{8}{15}\right)^k$

15. $3 + 3\sqrt{2} + 6 + \cdots$

16. $2\sqrt{2} + 2 + \sqrt{2} + \cdots$

In Problems 17 through 22, express each of the following repeating decimals as the ratio of two integers by considering an infinite geometric series.

17. $0.\overline{5}$

18. $0.\overline{8}$

19. $0.\overline{72}$

20. $0.0\overline{81}$

21. $6.3\overline{42}$

22. $5.2\overline{486}$

In Problems 23 through 28, find an infinite geometric series that has the given rational expression as the sum, and state the domain for which the sum exists.

23. $\dfrac{3}{1-x}$

24. $\dfrac{4}{1+x}$

25. $\dfrac{4}{1-3x}$

26. $\dfrac{3}{3-x}$

27. $\dfrac{4}{2-x}$

28. $\dfrac{10}{2+x}$

In Problems 29 through 34, the nth partial sum S_n of a series is given. By considering the behavior of S_n as $n \to \infty$, determine whether the series is convergent or divergent. When it is convergent, state the sum.

29. $S_n = \dfrac{2n - 1}{3n + 5}$

30. $S_n = \dfrac{3}{4n + 1}$

31. $S_n = \dfrac{5n - 2}{3}$

32. $S_n = \dfrac{n(n + 2)}{3n + 2}$

33. $S_n = 3 - \dfrac{1}{n}$

34. $S_n = 2 + \dfrac{n}{n + 1}$

B

In Problems 35 through 42, find a simple formula for S_n to determine whether the series converges or diverges. If the series is convergent, state the sum.

35. $\displaystyle\sum_{k=1}^{\infty} \dfrac{1}{k(k + 1)}$

36. $\displaystyle\sum_{k=1}^{\infty} \dfrac{20}{(4k - 2)(4k + 2)}$

37. $\displaystyle\sum_{k=1}^{\infty} \dfrac{2}{(2k - 1)(2k + 1)}$

38. $\displaystyle\sum_{k=1}^{\infty} \dfrac{6k + 3}{k^2(k + 1)^2}$

39. $\displaystyle\sum_{k=1}^{\infty} \dfrac{8}{(2k - 1)(2k + 3)}$

40. $\displaystyle\sum_{k=2}^{\infty} \dfrac{4}{(k - 1)(k + 1)}$

41. $\displaystyle\sum_{k=1}^{\infty} (3k + 1)$

42. $\displaystyle\sum_{k=1}^{\infty} (k + 2)$

43. At the beginning of this section we determined how long it would take a ball that bounces to 49% of its previous height to come to rest, if it was dropped from a height of 16 feet. How long would it take if the ball bounces to 64% of its previous height and if all other assumptions remain the same?

44. How long would it take if the ball (see Problem 43) bounces to 81% of its previous height and if all other assumptions remain the same?

45. What is the total distance traveled by the ball in Problem 43?

46. What is the total distance traveled by the ball in Problem 44?

47. A piece of equipment worth $80,000 depreciates 15% of its present value every year. In other words, during the first year $80,000(0.15) = $12,000 is depreciated, during the second year $68,000(0.15) = $10,200 is depreciated, and so on. If this goes on indefinitely, what is the total depreciation?

For Problems 48 through 51, refer to the following series:

$$e^x = \sum_{k=0}^{\infty} \dfrac{x^k}{k!}, \qquad \sin x = \sum_{k=0}^{\infty} (-1)^k \dfrac{x^{2k+1}}{(2k + 1)!},$$

$$\cos x = \sum_{k=0}^{\infty} (-1)^k \dfrac{x^{2k}}{(2k)!}.$$

48. Use S_3 in the series for $\sin x$ to approximate $\sin 0.1$.

49. Use S_3 in the series for $\cos x$ to approximate $\cos 0.4$.

50. Use S_4 in the series for e^x to approximate $e^{1.5}$.

51. Use S_4 in the series for e^x to approximate $e^{0.5}$.

C

52. Using the series for $\sin x$:

   a) Graph S_1; that is, graph $y = x$, for $-\pi \le x \le \pi$.

   b) Graph S_2; that is, graph $y = x - (\tfrac{1}{6})x^3$, for $-\pi \le x \le \pi$.

   c) Graph S_3; that is, graph $y = x - (\tfrac{1}{6})x^3 + (\tfrac{1}{120})x^5$, for $-\pi \le x \le \pi$.

53. Using the series for $\cos x$:

   a) Graph S_1; that is, graph $y = 1$, for $-\pi \le x \le \pi$.

   b) Graph S_2; that is, graph $y = 1 - (\tfrac{1}{2})x^2$, for $-\pi \le x \le \pi$.

   c) Graph S_3; that is, graph $y = 1 - (\tfrac{1}{2})x^2 + (\tfrac{1}{24})x^4$, for $-\pi \le x \le \pi$.

54. a) Using the series for e^x, find and simplify a series for $e^{i\theta}$, where $i = \sqrt{-1}$.

   b) Multiply each term in the series for $\sin \theta$ by i, then add to the series for $\cos \theta$.

   c) Compare the results from (a) and (b). State the identity suggested.

   d) Using the identity obtained in (c), evaluate $e^{i\pi}$.

55. Two convergent series can be multiplied to obtain another series. Starting with

$$\left[\dfrac{1}{1 - x}\right]\left[\dfrac{x}{1 - x}\right]$$

$$= [1 + x + x^2 + x^3 + \cdots][x + x^2 + x^3 + x^4 + \cdots]$$

we can expand the right side of the equation, much as we multiply two polynomials. The procedure is to first multiply $1[x + x^2 + x^3 + x^4 + \cdots]$, then $x[x + x^2 + x^3 + x^4 + \cdots]$, then

$x^2[x + x^2 + x^3 + x^4 + \cdots]$, and so on. Then we add like terms:

$$1[x + x^2 + x^3 + x^4 + \cdots] = x + x^2 + x^3 + x^4 + \cdots$$

$$x[x + x^2 + x^3 + x^4 + \cdots] = \qquad x^2 + x^3 + x^4 + \cdots$$

$$x^2[x + x^2 + x^3 + x^4 + \cdots] = \qquad\qquad x^3 + x^4 + \cdots$$

$$\vdots$$

Continue this procedure to obtain the first five terms of the series for

$$\left[\frac{1}{1-x}\right]\left[\frac{x}{1-x}\right],$$

where $|x| < 1$.

56. An **arithmetic-geometric** series has the form

$$\sum_{k=1}^{\infty} [a + (k-1)d]r^{k-1}$$
$$= a + (a+d)r + (a+2d)r^2 + (a+3d)r^3 + \cdots, \qquad |r| < 1.$$

This can be separated as follows:

$$(a + ar + ar^2 + ar^3 + \cdots) + d(r + 2r^2 + 3r^3 + 4r^4 + \cdots).$$

With the help of the result of Problem 55, find the sum.

57. Since the harmonic series diverges (see Example 7), if squares that have sides of $1, \frac{1}{2}, \frac{1}{3}, \frac{1}{4}, \ldots$ are placed on a horizontal line side-by-side, the length of the line must be unbounded (Figure 12). However, can all the squares to the right of the 1-by-1 square be packed inside the 1-by-1 square? Show that the strategy for packing shown in Figure 13 will work by calculating the total necessary width of the container.

58. On the average, the number of expected coin flips until a head appears is $\frac{1}{2}(1) + \frac{1}{4}(2) + \frac{1}{8}(3) + \frac{1}{16}(4) + \cdots$. Use the result of Problem 56 to find the sum.

59. The Cantor set is defined as follows. Starting with the closed interval $[0, 1]$, remove the (middle third) open interval $(\frac{1}{3}, \frac{2}{3})$:

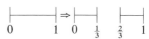

Next, from each of the remaining two intervals remove the (middle third) open intervals $(\frac{1}{9}, \frac{2}{9})$ and $(\frac{7}{9}, \frac{8}{9})$, respectively:

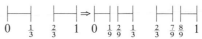

From each of the four remaining intervals remove the (middle third) open intervals $(\frac{1}{27}, \frac{2}{27})$, $(\frac{7}{27}, \frac{8}{27})$, $(\frac{19}{27}, \frac{20}{27})$, and $(\frac{25}{27}, \frac{26}{27})$. If you continue this process indefinitely, what remains is the Cantor set.

a) Which of the following numbers are in the Cantor set? $0, \frac{1}{3}, \frac{1}{2}, \frac{1}{9}, \frac{1}{6}, \frac{5}{6}, \frac{2}{9}, 1$

b) Name five other numbers that are in the Cantor set.

c) What is the total length that is removed from the original interval $[0, 1]$ to form the Cantor set?

60. Suppose we have two identical (well-shuffled) decks of distinct cards. Turn up the top card of each deck. If the cards match perfectly (a 7 of hearts and a 7 of spades is *not* a match), we stop. If not, turn up the second card in each deck to check for a perfect match. If these cards match, we stop; if not, turn up the third card in each deck to see if they match. Continuing this

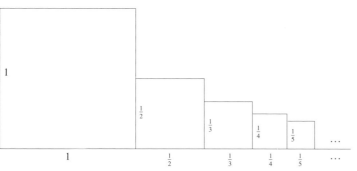

FIGURE 12

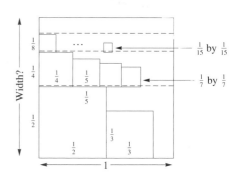

FIGURE 13

process, we might go through the decks without getting a (perfect) match. The probabilities that we *will* get a match using identical decks of 1, 2, 3, 4, and 5 cards are shown in the following table:

Number of Cards in Each Deck	Probability of a Match
1	1
2	$\dfrac{1}{2} = 1 - \dfrac{1}{2!}$
3	$\dfrac{4}{6} = 1 - \dfrac{1}{2!} + \dfrac{1}{3!}$
4	$\dfrac{15}{24} = 1 - \dfrac{1}{2!} + \dfrac{1}{3!} - \dfrac{1}{4!}$
5	$\dfrac{76}{120} = 1 - \dfrac{1}{2!} + \dfrac{1}{3!} - \dfrac{1}{4!} + \dfrac{1}{5!}$

Continue with this pattern to estimate the probability of getting a match with decks of 52 cards. (Hint: Use the infinite series for e^x with $x = -1$.)

61. Find the sum $\displaystyle\sum_{k=1}^{\infty} \dfrac{k}{(k+1)!}$.

62. A series for $\tan x$ can be obtained by dividing $\sin x$ by $\cos x$ as follows:

$$1 - \frac{x^2}{2!} + \frac{x^4}{4!} - \frac{x^6}{6!} + \cdots \overline{\left) \; x - \frac{x^3}{3!} + \frac{x^5}{5!} - \frac{x^7}{7!} + \cdots \right.}^{\displaystyle x}$$

In the preceding step, our initial guess, x, comes from dividing the leading term of the divisor, 1, into the leading term of the dividend, x. Now, multiply our guess by the divisor, and then subtract:

$$1 - \frac{x^2}{2!} + \frac{x^4}{4!} - \frac{x^6}{6!} + \cdots \overline{\left) \; x - \frac{x^3}{3!} + \frac{x^5}{5!} - \frac{x^7}{7!} + \cdots \right.}^{\displaystyle x + \frac{x^3}{3}}$$

$$-\left(x - \frac{x^3}{2!} + \frac{x^5}{4!} - \frac{x^7}{6!} + \cdots\right)$$

$$\overline{\qquad \frac{x^3}{3} - \frac{x^5}{30} + \frac{x^7}{840} - \cdots.}$$

The next guess, $x^3/3$, is shown. Continue this process for two more steps to obtain the first four terms of this series for $\tan x$.

S E C T I O N   11.4   THE BINOMIAL THEOREM

In calculus and other branches of mathematics it is sometimes necessary to multiply expressions such as $(3x + y)^6$. This can be quite a tedious exercise. In search of a formula for expanding expressions of the form $(a + b)^n$, we begin by considering specific cases and look for a pattern:

$$(a + b)^1 = a + b$$
$$(a + b)^2 = a^2 + 2ab + b^2$$
$$(a + b)^3 = a^3 + 3a^2b + 3ab^2 + b^3$$
$$(a + b)^4 = a^4 + 4a^3b + 6a^2b^2 + 4ab^3 + b^4$$
$$(a + b)^5 = a^5 + 5a^4b + 10a^3b^2 + 10a^2b^3 + 5ab^4 + b^5$$

In each case, for the expansion of $(a + b)^n$ with $n = 1, 2, 3, 4,$ or 5:

The first term is a^n, and the last term is b^n.

The second term is $na^{n-1}b$, and the next-to-last term is nab^{n-1}.

The powers of a decrease by 1 from left to right.

The powers of b increase by 1 from left to right.

Consequently, if you are a trusting sort, you might anticipate that the form for the next case will be

$$(a + b)^6 = a^6 + 6a^5b + ?a^4b^2 + ?a^3b^3 + ?a^2b^4 + 6ab^5 + b^6$$

(where $?$ denotes a missing coefficient). All that remains is to find a way to calculate these coefficients for each term. We return to the expansion for $(a + b)^5$ and list the important players:

Term Number	Coefficient	Exponent of a
1	1	5
2	5	4
3	10	3
4	10	2
5	5	1
6	1	0

Notice that for any term

$$\frac{\text{(coefficient of the term)(exponent of } a)}{\text{number of the term}} = \text{coefficient of the next term}$$

This relationship can be used to calculate the coefficients from left to right.

EXAMPLE 1   Find the expansion of $(a + b)^6$.

SOLUTION The form for the expansion is

$$(a + b)^6 = a^6 + \mathbf{6}a^5b + ?a^4b^2 + ?a^3b^3 + ?a^2b^4 + 6ab^5 + b^6.$$

The coefficient and exponent of a for the second term are 6 and 5, respectively; so the coefficient for the next term is $(5)(6)/2 = 15$. Thus we have

$$(a + b)^6 = a^6 + 6a^5b + \mathbf{15}a^4b^2 + ?a^3b^3 + ?a^2b^4 + 6ab^5 + b^6.$$

The coefficient and exponent of a for the third term are 15 and 4, respectively; so the coefficient for the next term is $(15)(4)/3 = 20$. We now have

$$(a + b)^6 = a^6 + 6a^5b + 15a^4b^2 + \mathbf{20}a^3b^3 + ?a^2b^4 + 6ab^5 + b^6$$

Continuing, the next coefficient is $(3)(20)/4 = 15$, and we have

$$(a + b)^6 = a^6 + 6a^5b + 15a^4b^2 + 20a^3b^3 + \mathbf{15}a^2b^4 + 6ab^5 + b^6.$$

We leave it to you to verify that by proceeding in this way, the completed expansion is

$$(a + b)^6 = a^6 + 6a^5b + 15a^4b^2 + 20a^3b^3 + 15a^2b^4 + 6ab^5 + b^6.$$

Consider the general case $(a + b)^n$, for any positive integer n. Starting with the first term,

$$(a + b)^n = a^n + na^{n-1}b + ?a^{n-2}b^2 + ?a^{n-3}b^3 + \cdots + nab^{n-1} + b^n.$$

Using the procedure in the previous example, we obtain the next few terms:

$$a^n + na^{n-1} + \frac{n(n-1)}{1 \cdot 2} a^{n-2}b + \frac{n(n-1)(n-2)}{1 \cdot 2 \cdot 3} a^{n-3}b^3 + ?a^{n-4}b^4 + \cdots + b^n$$

From the factorial notation introduced in Section 11.3, it follows that a general formula for expanding a binomial is the following:

Binomial Theorem

$$(a + b)^n = a^n + na^{n-1}b + \frac{n(n-1)}{2!} a^{n-2}b^2$$

$$+ \frac{n(n-1)(n-2)}{3!} a^{n-3}b^3 + \cdots$$

$$+ \frac{n(n-1)(n-2)(n-3) \cdots (n-k+1)}{k!} a^{n-k}b^k + \cdots$$

$$+ nab^{n-1} + b^n.$$

The coefficients for the terms in a binomial expansion are called **binomial coefficients.** They can be written even more economically as follows:

Term	Coefficient
a^n	$1 = \dfrac{n!}{0!n!}$
$a^{n-1}b$	$n = \dfrac{n!}{1!(n-1)!}$
$a^{n-2}b^2$	$\dfrac{n(n-1)}{2!} = \dfrac{n!}{2!(n-2)!}$
$\vdots$	$\vdots \qquad\qquad \vdots$
$a^{n-k}b^k$	$\dfrac{n(n-1)(n-2)\ldots(n-k+1)}{k!} = \dfrac{n!}{k!(n-k)!}$
$\vdots$	$\vdots \qquad\qquad \vdots$

From this observation we introduce the following notation.

$$\binom{n}{k} = \frac{n!}{k!(n-k)!}, \qquad k = 0, 1, 2, \ldots, n.$$

The symbol $\binom{n}{k}$ is usually read "n choose k." This notation is used frequently in the study of probability and statistics.

EXAMPLE 2   Evaluate $\binom{4}{0}, \binom{4}{1}, \binom{4}{2}, \binom{4}{3},$ and $\binom{4}{4}$.

SOLUTION

$$\binom{4}{0} = \frac{4!}{0!(4-0)!} = \frac{4!}{0!4!} = 1 \qquad \text{(recall that } 0! = 1\text{)}$$

$$\binom{4}{1} = \frac{4!}{1!(4-1)!} = \frac{4!}{1!3!} = \frac{4 \cdot 3!}{1!3!} = 4$$

$$\binom{4}{2} = \frac{4!}{2!(4-2)!} = \frac{4!}{2!2!} = \frac{4 \cdot 3 \cdot 2!}{2!2!} = \frac{4 \cdot 3}{2!} = 6$$

$$\binom{4}{3} = \frac{4!}{3!(4-3)!} = \frac{4!}{3!1!} = \frac{4 \cdot 3!}{1!3!} = 4$$

$$\binom{4}{4} = \frac{4!}{4!(4-4)!} = \frac{4!}{4!0!} = 1.$$

The binomial theorem can now be stated as follows.

$$(a+b)^n = \binom{n}{0}a^n + \binom{n}{1}a^{n-1}b + \binom{n}{2}a^{n-2}b^2 + \cdots$$

$$+ \binom{n}{k}a^{n-k}b^k + \cdots + \binom{n}{n-1}ab^{n-1} + \binom{n}{n}b^n.$$

In summation notation,

$$(a+b)^n = \sum_{k=0}^{n} \binom{n}{k}a^{n-k}b^k.$$

Binomial Theorem
(Alternate Form)

This form is particularly useful when only certain terms of a binomial expansion are required.

EXAMPLE 3   Find the given terms in the expansion of $(x + h)^{12}$.

a) x^4h^8 term   **b)** x^2h^{10} term

SOLUTION

a)   Since $n = 12$, the coefficient for the x^4h^8 term is

$$\binom{12}{8} = \frac{12!}{8!(12 - 8)!} = \frac{12 \cdot 11 \cdot 10 \cdot 9 \cdot 8!}{8!4!} = \frac{12 \cdot 11 \cdot 10 \cdot 9}{4 \cdot 3 \cdot 2} = 495$$

Thus the desired term is $495x^4h^8$.

b)   The coefficient for the x^2h^{10} term is

$$\binom{12}{10} = \frac{12!}{10!(12 - 10)!} = \frac{12 \cdot 11 \cdot 10!}{10!2!} = 66$$

so the x^2h^{10} term is $66x^2h^{10}$.

To apply the binomial theorem, it is important to focus on form.

EXAMPLE 4   Expand and simplify $(2x - y^2)^7$.

SOLUTION   Using the binomial theorem with $n = 7$, we have

$$(a + b)^7 = \binom{7}{0}a^7 + \binom{7}{1}a^6b + \binom{7}{2}a^5b^2 + \binom{7}{3}a^4b^3 + \binom{7}{4}a^3b^4$$

$$+ \binom{7}{5}a^2b^5 + \binom{7}{6}ab^6 + \binom{7}{7}b^7$$

$$= a^7 + 7a^6b + 21a^5b^2 + 35a^4b^3 + 35a^3b^4 + 21a^2b^5 + 7ab^6 + b^7.$$

For $(2x - y^2)^7$, replace a with $2x$ and replace b with $(-y^2)$:

$$(2x - y^2)^7 = (2x)^7 + 7(2x)^6(-y^2) + 21(2x)^5(-y^2)^2$$

$$+ 35(2x)^4(-y^2)^3 + 35(2x)^3(-y^2)^4$$

$$+ 21(2x)^2(-y^2)^5 + 7(2x)(-y^2)^6 + (-y^2)^7$$

$$= 128x^7 - 448x^6y^2 + 672x^5y^4 - 560x^4y^6 + 280x^3y^8$$

$$- 84x^2y^{10} + 14xy^{12} - y^{14}.$$

EXAMPLE 5 For $f(x) = x^5 - 1$, find and simplify

$$\frac{f(x + h) - f(x)}{h}.$$

SOLUTION

$$\frac{f(x + h) - f(x)}{h} = \frac{[(x + h)^5 - 1] - [x^5 - 1]}{h}$$

$$= \frac{x^5 + 5x^4h + 10x^3h^2 + 10x^2h^3 + 5xh^4 + h^5 - 1 - x^5 + 1}{h}$$

$$= \frac{5x^4h + 10x^3h^2 + 10x^2h^3 + 5xh^4 + h^5}{h}$$

$$= 5x^4 + 10x^3h + 10x^2h^2 + 5xh^3 + h^4$$

If the binomial coefficients are written in a triangular array, the result is known as Pascal's triangle:

			1						Row 0
		1		1					Row 1
	1		2		1				Row 2
1		3		3		1			Row 3
1		4		6		4		1	Row 4
1	5		10		10		5	1	Row 5
1	6	15	20	15	6	1			Row 6

$\vdots$

$(a + b)^0 = 1$

$(a + b)^1 = 1a + 1b$

$(a + b)^2 = 1a^2 + 2ab + 1b^2$

$(a + b)^3 = 1a^3 + 3a^2b + 3ab^2 + 1$

etc.

Each row gives the coefficients for the expansion of $(a + b)^n$ for $n = 0, 1, 2, 3, \ldots$. For example, the expansion for $(a + b)^4$ is

$$(a + b)^4 = 1a^4 + 4a^3b + 6a^2b^2 + 4ab^3 + 1b^4.$$

Notice that these coefficients can be found in the fourth row of Pascal's triangle.

Generating the rows of the triangle is based on the following observation: Except for the 1's at the beginning and end, each entry is the sum of the two numbers diagonally above. To illustrate, the seventh row is obtained by starting with a 1, then adding 1 and 6 (from row six) to get the second entry, then adding 6 and 15 to get the third entry, and so on. The following diagram illustrates.

Sixth row   1   6   15   20   15   6   1

Seventh row   1   7   21   35   35   21   7   1

Pascal's triangle can be used to determine the coefficients of the expansion for $(a + b)^n$, as long as n isn't too large. After a certain point it becomes impractical, and it is almost necessary to use the binomial theorem.

Binomial Series

There is a way to extend the binomial theorem to obtain an infinite series. Consider the expansion

$$(1 + x)^r = 1 + rx + \frac{r(r-1)}{2!}x^2 + \frac{r(r-1)(r-2)}{3!}x^3 + \cdots.$$

If r is a nonnegative integer, we can follow this pattern for the coefficients (from left to right) without regard to the last term, and we will obtain the correct expansion. To illustrate, we could expand $(1 + x)^4$ by letting $r = 4$:

$$(1+x)^4 = 1 + 4x + \frac{4(4-1)}{2!}x^2 + \frac{4(4-1)(4-2)}{3!}x^3$$
$$+ \frac{4(4-1)(4-2)(4-3)}{4!}x^4$$
$$+ \frac{4(4-1)(4-2)(4-3)(4-4)}{5!}x^5$$
$$+ \frac{4(4-1)(4-2)(4-3)(4-4)(4-5)}{6!}x^6 + \cdots$$
$$= 1 + 4x + 6x^2 + 4x^3 + 1x^4 + 0x^5 + 0x^6 + \cdots.$$

In this case each of the coefficients past x^4 is zero because of the presence of the factor $(r - 4)$ in its numerator. In general, if r is any nonnegative integer, this expansion always produces the correct finite expansion.

Now try letting $r = -1$ in the same expansion:

$$(1 + x)^{-1} = 1 + -1x + \frac{-1(-1-1)}{2!}x^2 + \frac{-1(-1-1)(-1-2)}{3!}x^3$$
$$+ \frac{-1(-1-1)(-1-2)(-1-3)}{4!}x^4$$
$$+ \frac{-1(-1-1)(-1-2)(-1-3)(-1-4)}{5!}x^5$$
$$+ \frac{-1(-1-1)(-1-2)(-1-3)(-1-4)(-1-5)}{6!}x^6 + \cdots$$
$$= 1 - x + x^2 - x^3 + x^4 - x^5 + x^6 + \cdots.$$

This is an infinite geometric series with common ratio $(-x)$ and first term 1. Using the formula for the sum of a geometric series from Section 11.3, we find that if $|x| < 1$, the expansion is correct! This rather surprising result inspires the following definition.

For any real number r, the series   *Binomial Series*

$$(1 + x)^r = 1 + rx + \frac{r(r - 1)}{2!} x^2 + \frac{r(r - 1)(r - 2)}{3!} x^3$$

$$+ \frac{r(r - 1)(r - 2)(r - 3)}{4!} x^4$$

$$+ \frac{r(r - 1)(r - 2)(r - 3)(r - 4)}{5!} x^5$$

$$+ \frac{r(r - 1)(r - 2)(r - 3)(r - 4)(r - 5)}{6!} x^6 + \cdots$$

is called the **binomial series.**

If r is a nonnegative integer the series terminates, otherwise the series is infinite.

It can be shown by using calculus that the equation in this representation is valid if and only if $|x| < 1$.

EXAMPLE 6

a) Find the first four terms of the binomial series for $\sqrt{1 + x}$.

b) Use the result of part (a) to approximate $\sqrt{1.2}$.

SOLUTION

a) Since $\sqrt{1 + x} = (1 + x)^{1/2}$, let $r = \frac{1}{2}$:

$$(1 + x)^{1/2} = 1 + \frac{1}{2} x + \frac{\frac{1}{2}(\frac{1}{2} - 1)}{2!} x^2 + \frac{\frac{1}{2}(\frac{1}{2} - 1)(\frac{1}{2} - 2)}{3!} x^3 + \cdots$$

$$= 1 + \frac{1}{2} x - \frac{1}{2^2 2!} x^2 + \frac{3}{2^3 3!} x^3 - \cdots$$

$$= 1 + \frac{1}{2} x - \frac{1}{8} x^2 + \frac{1}{16} x^3 - \cdots \qquad |x| < 1$$

b) To approximate $\sqrt{1.2}$, let $x = 0.2$ in the series we derived in part (a):

$$\sqrt{1.2} = (1 + 0.2)^{1/2}$$

$$= 1 + \tfrac{1}{2}(0.2) - \tfrac{1}{8}(0.2)^2 + \tfrac{1}{16}(0.2)^3 - \cdots$$

$$\approx 1 + 0.1 - 0.005 + 0.0005 - \cdots$$

Using the first four terms, we get $\sqrt{1.2} \approx 1.0955$.

EXAMPLE 7   Find the first four terms of the binomial series for $(4 - 2x)^{-3}$.

SOLUTION   First factor out 4, then apply the binomial series to the remaining factor.

$$(4 - 2x)^{-3} = 4^{-3}\left(1 - \frac{x}{2}\right)^{-3}$$

$$= \frac{1}{64}\left(1 - \frac{x}{2}\right)^{-3}$$

$$= \frac{1}{64}\left(1 + (-3)\left(-\frac{x}{2}\right) + \frac{(-3)(-3-1)}{2!}\left(-\frac{x}{2}\right)^2\right.$$

$$\left. + \frac{(-3)(-3-1)(-3-2)}{3!}\left(-\frac{x}{2}\right)^3 + \cdots\right)$$

$$= \frac{1}{64}\left(1 + \frac{3}{2}x + \frac{3}{2}x^2 + \frac{5}{4}x^3 + \cdots\right).$$

EXERCISE SET 11.4

A

In Problems 1 through 8, evaluate the expression.

1. $\binom{5}{2}$ **2.** $\binom{6}{4}$ **3.** $\binom{8}{1}$

4. $\binom{18}{9}$ **5.** $\binom{15}{6}$ **6.** $\binom{11}{3}$

7. $\binom{14}{0}$ **8.** $\binom{6}{6}$

In Problems 9 through 23 expand using the binomial theorem.

9. $(x + h)^4$ **10.** $(x - h)^3$ **11.** $(x + 3)^3$

12. $(x - 2)^4$ **13.** $(2x + y)^4$ **14.** $(2x - y)^5$

15. $(2x + 3y)^6$ **16.** $(x - 3y)^6$ **17.** $(2x - y)^8$

18. $(2x + y)^8$ **19.** $\left(\dfrac{x}{2} - 3y\right)^7$ **20.** $\left(2x + \dfrac{y}{2}\right)^8$

21. $(\sqrt{p} + q^2)^6$ **22.** $(m^3 - \sqrt{n})^7$

23. $(2j^4 + 3\sqrt{k})^5$

Write out the first nine rows of Pascal's triangle and use them to expand the binomials in Problems 24 through 26.

24. $(x + 2y)^6$ **25.** $(3x + 2y)^7$ **26.** $(x - 2y)^8$

B

Solve for k in Problems 27 through 32.

27. $\binom{5}{1} = \binom{5}{k}, k \neq 1$ **28.** $\binom{8}{2} = \binom{8}{k}, k \neq 2$

29. $\binom{11}{3} = \binom{11}{k}, k \neq 3$ **30.** $\binom{21}{8} = \binom{21}{k}, k \neq 8$

31. $\binom{n}{5} = \binom{n}{k}, k \neq 5, n \geq 5$

32. $\binom{n}{r} = \binom{n}{k}, k \neq r, n \geq r$

33. Use Pascal's triangle to find 11^n for $n = 1, 2, 3, 4,$ and 5.

34. Find the a^9b^2 term in the expansion of $(a + b)^{11}$.

35. Find the x^3y^8 term in the expansion of $(x + y)^{11}$.

36. Find the x^9y^6 term in the expansion of $(2x + y^2)^{12}$.

37. Find the p^3q^4 term in the expansion of $(\sqrt{p} + 2q)^{10}$.

38. Find the sixth term in the expansion of $(a - 3b)^9$.

39. Find the fifth term in the expansion of $(2m - n)^{13}$.

In Problems 40 through 45, expand the expression.

40. $(x^2y - z^3)^5$ **41.** $(x^2 + 2yz)^6$ **42.** $(x^3 + \sqrt{2})^8$

43. $(xy^4 - \sqrt{3})^4$ **44.** $\left(x^2 - \dfrac{1}{x}\right)^8$ **45.** $\left(x^3 - \dfrac{1}{x^2}\right)^6$

In Problems 46 through 49, find and simplify $\dfrac{f(x + h) - f(x)}{h}$.

46. $f(x) = x^4 - x$ **47.** $f(x) = x^5 + x$

48. $f(x) = 3x^4 - x^2 + 2$ **49.** $f(x) = 2x^4 + x^2 - 4$

Find the binomial series for each function in Problems 50 through 59. Show the first four terms.

50. $\sqrt{1 - x}$ **51.** $\sqrt[3]{1 + x}$ **52.** $\dfrac{1}{\sqrt{1 + x}}$

53. $\dfrac{1}{\sqrt{1 - x}}$ **54.** $(1 + x)^{-5}$ **55.** $(1 + x)^{-4}$

56. $(1 + x^2)^{1/2}$ **57.** $(1 - x^3)^{1/4}$ **58.** $(8 - 4x)^{2/3}$

59. $(9 + 18x)^{3/2}$

C

60. Approximate $\sqrt[3]{1.4}$, using the answer to Problem 51.

61. Approximate $\sqrt[3]{1.1}$, using the answer to Problem 51.

62. Approximate $\sqrt{0.7}$, using the answer to Problem 50.

63. Show that
$$\binom{n}{r-1} + \binom{n}{r} = \binom{n+1}{r}.$$

64. The formula for $\sum_{k=1}^{n} k^3$ given in Section 11.1 can be derived as follows.

Starting with
$$(n + 1)^4 = n^4 + 4n^3 + 6n^2 + 4n + 1$$

subtract n^4 from both sides:
$$(n + 1)^4 - n^4 = 4n^3 + 6n^2 + 4n + 1.$$

Since this is true for any value n, write the last equation for successive values $n = 1, 2, 3, 4, \ldots, n$:

$$2^4 - 1^4 = 4(1^3) + 6(1^2) + 4(1) + 1$$

$$3^4 - 2^4 = 4(2^3) + 6(2^2) + 4(2) + 1$$

$$4^4 - 3^4 = 4(3^3) + 6(3^2) + 4(3) + 1$$

$$\vdots$$

$$(n + 1)^4 - n^4 = 4n^3 + 6n^2 + 4n + 1.$$

If we add these equations, the left side has many cancellations. Adding columns on the right side, we get

$$(n + 1)^4 - 1 = 4 \sum_{k=1}^{n} k^3 + 6 \sum_{k=1}^{n} k^2 + 4 \sum_{k=1}^{n} k + \sum_{k=1}^{n} 1.$$

If we know the formulas for $\sum_{k=1}^{n} k^2$, $\sum_{k=1}^{n} k$, and $\sum_{k=1}^{n} 1$, we can solve for $\sum_{k=1}^{n} k^3$:

$$4 \sum_{k=1}^{n} k^3 = (n + 1)^4 - 1 - 6 \sum_{k=1}^{n} k^2 - 4 \sum_{k=1}^{n} k - \sum_{k=1}^{n} 1$$

$$= (n + 1)^4 - 1 - 6 \frac{n(n + 1)(2n + 1)}{6} - 4 \frac{n(n + 1)}{2} - n$$

$$= (n + 1)^4 - n(n + 1)(2n + 1) - 2n(n + 1) - (n + 1).$$

Factoring out $(n + 1)$ and simplifying leads to

$$4 \sum_{k=1}^{n} k^3 = (n + 1)[(n + 1)^3 - n(2n + 1) - 2n - 1]$$

$$= (n + 1)[n^2(n + 1)].$$

From this we get the correct result:

$$\sum_{k=1}^{n} k^3 = \left[\frac{n(n + 1)}{2} \right]^2.$$

Use this method to find a formula for $\sum_{k=1}^{n} k^4$.

S E C T I O N 11.5 **MATHEMATICAL INDUCTION**

One of the most fundamental forms of reasoning proceeds from a particular sequence of observations to a general rule for all cases of the phenomenon. This type of reasoning, called inductive reasoning, is usually convincing in most disciplines. However, in mathematics a conjecture based only on specific cases must be proven for all cases before it is accepted as fact. For example, one might make the following observation

$$11^1 = \quad 11 \quad \text{and} \quad 1 + 1 = 2^1,$$

$$11^2 = \quad 121 \quad \text{and} \quad 1 + 2 + 1 = 2^2,$$

$$11^3 = \quad 1331 \quad \text{and} \quad 1 + 3 + 3 + 1 = 2^3,$$

$$11^4 = 14641 \quad \text{and} \quad 1 + 4 + 6 + 4 + 1 = 2^4.$$

From these cases it would be natural to conjecture that for any positive integer n, the sum of the digits of 11^n is 2^n. But this is false for $n = 5$:

$$11^5 = 161051 \quad \text{and} \quad 1 + 6 + 1 + 0 + 5 + 1 \neq 2^5.$$

As another example, consider the sequence $a_n = n^2 - n + 41$:

$$a_1 = 41, \quad \text{which is a prime number}$$

$$a_2 = 43, \quad \text{which is a prime number}$$

$$a_3 = 47, \quad \text{which is a prime number}$$

$$a_4 = 53, \quad \text{which is a prime number}$$

$$a_5 = 61, \quad \text{which is a prime number}$$

Continuing to calculate successive cases reveals that $a_6, a_7, a_8, \ldots, a_{40}$ are all prime numbers. Are these enough cases to accept the conjecture that for all positive integers n, $a_n = n^2 - n + 41$ is a prime number? No! The 41st case fails:

$$a_{41} = 41^2 - 41 + 41 = 41^2$$

which is *not* prime.

The point is that verifying a general statement for any finite number of values n does not prove the statement for every value of n. A way to disprove a false conjecture is by finding a counterexample. To prove a true statement, the following principle is often used.

Suppose we have a statement involving each positive integer n. The statement is true for every $n \geq 1$ if the following two conditions can be demonstrated:

1) The statement is true for $n = 1$.

2) The assumption that the statement is true for $n = k$ implies that the statement is true for $n = k + 1$.

Principle of Mathematical Induction

The idea of mathematical induction is similar to a row of dominoes lined up so that if the first one is knocked over, all succeeding dominoes will fall (Figure 14.) Knocking over the first domino is analogous to demonstrating the first condition, that the statement is true for $n = 1$. Having the dominoes lined up so that any one (say the kth one) falling over will knock the next one (the $k + 1$st one) over corresponds to the second condition, which says that if the statement is true for $n = k$, then the statement will be true for $n = k + 1$.

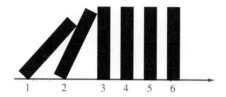

FIGURE 14

EXAMPLE 1   Prove that for every positive integer n,

$$1 + 3 + 5 + 7 + \cdots + (2n - 1) = n^2.$$

SOLUTION The proof is by mathematical induction and consists of two steps.

1) Substituting $n = 1$ into the equation gives $1 = 1^2$, which is true.

2) We start with the assumption that the equation holds for $n = k$; that is, we assume that

$$1 + 3 + 5 + 7 + \cdots + (2k - 1) = k^2.$$

We need to show that this assumption implies that the equation holds for $n = k + 1$; that is, we need to show that

$$1 + 3 + 5 + 7 + \cdots + (2(k + 1) - 1) \stackrel{?}{=} (k + 1)^2$$

which simplifies to

$$1 + 3 + 5 + 7 + \cdots + (2k - 1) + (2k + 1) \stackrel{?}{=} (k + 1)^2.$$

Knowing the task at hand, we start with

$$1 + 3 + 5 + 7 + \cdots + (2k - 1) = k^2$$

and add $(2k + 1)$ to both sides:

$$1 + 3 + 5 + 7 + \cdots + (2k - 1) + (2k + 1) = k^2 + 2k + 1$$
$$= (k + 1)^2.$$

Since both steps 1 and 2 have been established, the statement

$$1 + 3 + 5 + 7 + \cdots + (2n - 1) = n^2$$

is true for every integer n, by the principle of mathematical induction.

In Section 11.1 we stated that, for every positive integer n,

$$1^2 + 2^2 + 3^2 + \cdots + n^2 = \frac{n(n + 1)(2n + 1)}{6}$$

and that this would be proved.

EXAMPLE 2   Prove that, for every positive integer n,

$$1^2 + 2^2 + 3^2 + \cdots + n^2 = \frac{n(n + 1)(2n + 1)}{6}.$$

SOLUTION   The proof is by mathematical induction.

1)   The equation holds for $n = 1$:

$$1^2 = \frac{1(1 + 1)(2(1) + 1)}{6} = 1.$$

2) We assume the equation holds for $n = k$ and show that this implies the equation holds for $n = k + 1$. So we begin with

$$1^2 + 2^2 + 3^2 + \cdots + k^2 = \frac{k(k + 1)(2k + 1)}{6}.$$

Adding $(k + 1)^2$ to both sides gives

$$1^2 + 2^2 + 3^2 + \cdots + k^2 + (k + 1)^2 = \frac{k(k + 1)(2k + 1)}{6} + (k + 1)^2$$

$$= \frac{k(k + 1)(2k + 1)}{6} + \frac{6(k + 1)^2}{6}$$

$$= \frac{k(k + 1)(2k + 1) + 6(k + 1)^2}{6}$$

$$= \frac{(k + 1)[k(2k + 1) + 6(k + 1)]}{6}$$

$$= \frac{(k + 1)(k + 2)(2k + 3)}{6}.$$

Therefore,

$$1^2 + 2^2 + 3^2 + \cdots + (k + 1)^2 = \frac{[k + 1][(k + 1) + 1][2(k + 1) + 1]}{6}.$$

This shows that if the equation holds for $n = k$, then the equation holds for $n = k + 1$.

By the principle of mathematical induction,

$$1^2 + 2^2 + 3^2 + \cdots + n^2 = \frac{n(n + 1)(2n + 1)}{6}$$

holds for every positive integer n.

Reflecting on the two previous proofs, it is natural to question what we proved! At first glance it may seem that in step 2 we assume the very statement that we are charged to prove. Actually, step 2 only establishes the implication that if the proposition is true for $n = k$, then the proposition must be true for $n = k + 1$. Realizing this should help you appreciate the importance of step 1, which shows that the statement is true for $n = 1$. A mistake that some students make is skipping the first step.

EXAMPLE 3   Prove or disprove that $1 \cdot 1! + 2 \cdot 2! + 3 \cdot 3! + \cdots + n \cdot n! = (n + 1)!$ for all positive integers n.

SOLUTION Suppose a student carelessly skipped the first step. Then assume the equation is true for $n = k$:

$$1 \cdot 1! + 2 \cdot 2! + 3 \cdot 3! + \cdots + k \cdot k! = (k + 1)!.$$

Add $(k + 1)(k + 1)!$ to both sides:

$$1 \cdot 1! + 2 \cdot 2! + \cdots + k \cdot k! + (k + 1)(k + 1)! = (k + 1)! + (k + 1)(k + 1)!$$
$$= (k + 2)(k + 1)!$$
$$= (k + 2)!.$$

Thus, if the equation holds for $n = k$, then the equation must be true for $n = k + 1$. However, the proposition is not true for all positive integers n because it is not true for $n = 1$. In fact, there is no value for n for which it can be proved to hold.

For example, if $n = 3$, we get

$$1 \cdot 1! + 2 \cdot 2! + 3 \cdot 3! \overset{?}{=} 4!$$
$$1 + 4 + 18 \overset{?}{=} 24$$
$$23 \neq 24$$

This disproves the statement.

Sequences that are defined recursively may often have an explicit formula for a_n. Mathematical induction is perfect for proving the explicit formula.

EXAMPLE 4   For the sequence

$$a_1 = 2, \, a_{n+1} = 1 + \frac{1}{na_n} \qquad \text{for } n > 1,$$

find an explicit formula for a_n, and prove it by using mathematical induction.

SOLUTION Calculate the first several cases and look for a pattern:

$$a_1 = 2, \, a_2 = \tfrac{3}{2}, \, a_3 = \tfrac{4}{3}, \, a_4 = \tfrac{5}{4}, \, a_5 = \tfrac{6}{5}, \ldots.$$

We conjecture that

$$a_n = \frac{n + 1}{n}, \qquad \text{for every positive integer } n.$$

The proof, using mathematical induction, has two parts.

1) The conjecture is true for $n = 1$:

$$a_1 = \frac{1 + 1}{1} = 2.$$

2) Assume the conjecture is true for $n = k$:

$$a_k = \frac{k + 1}{k}.$$

By the recursive definition, we have

$$a_{k+1} = 1 + \frac{1}{ka_k}.$$

Combining this with the induction assumption gives

$$a_{k+1} = 1 + \frac{1}{ka_k} = 1 + \frac{1}{k\left(\dfrac{k + 1}{k}\right)} = 1 + \frac{1}{k + 1} = \frac{(k + 1) + 1}{(k + 1)}.$$

This shows that if the conjecture is true for $n = k$, then it is true for $n = k + 1$.

By mathematical induction, the conjecture is true for every positive integer n.

Many mathematical statements are true for all integers beyond a certain point, rather than for all positive integers. In such an instance, step 1 must be modified, as illustrated in the next example.

EXAMPLE 5 Prove $n! > 5n$ for every integer $n \geq 4$.

SOLUTION The proof is by mathematical induction; however, since the inequality is not true for $n = 1, 2$, or 3, we must modify the first step.

1) The inequality is true for $n = 4$:

$$4! = 24 > 20 = 5(4).$$

2) Assume the inequality is true for $n = k$:

$$k! > 5k.$$

Multiply both sides by the positive quantity $(k + 1)$:

$$(k + 1)k! > (k + 1)5k.$$

Now, since we are considering values of $k \geq 4$, we know that $(k + 1)5k > 5(k + 1)$. Therefore,

$$(k + 1)! = (k + 1)k! > (k + 1)5k > 5(k + 1).$$

This establishes the implication that if the inequality is true for $n = k$, then the inequality must hold for $n = k + 1$.

Therefore, by mathematical induction the inequality holds for every integer $n \geq 4$.

Recall that in Section 11.4 the binomial theorem was developed by looking at various cases and relying on a pattern. Technically, we should have named it the binomial conjecture at that point. It must be proven to realize the status of being a theorem. Mathematical induction may be used as follows to prove it.

Proof of the Binomial Theorem

Prove for every positive integer n,

$$(a + b)^n = a^n + na^{n-1}b + \frac{n(n-1)}{2!}a^{n-2}b^2 + \frac{n(n-1)(n-2)}{3!}a^{n-3}b^3 + \cdots$$

$$+ nab^{n-1} + b^n.$$

1) The theorem is true for $n = 1$, since

$$(a + b)^1 = a^1 + b^1.$$

2) Assume the theorem is true for $n = k$.

$$(a + b)^k = a^k + ka^{k-1}b + \frac{k(k-1)}{2!}a^{k-2}b^2$$

$$+ \frac{k(k-1)(k-2)}{3!}a^{k-3}b^3 + \cdots + kab^{k-1} + b^k.$$

Multiply both sides of the above assumption by $(a + b)$.

$$(a + b)(a + b)^k = (a + b)\left(a^k + ka^{k-1}b + \frac{k(k-1)}{2!} a^{k-2}b^2 \right.$$

$$+ \frac{k(k-1)(k-2)}{3!} a^{k-3}b^3 + \cdots$$

$$\left. + kab^{k-1} + b^k \right)$$

$$= a\left(a^k + ka^{k-1}b + \frac{k(k-1)}{2!} a^{k-2}b^2 \right.$$

$$+ \frac{k(k-1)(k-2)}{3!} a^{k-3}b^3 + \cdots$$

$$\left. + kab^{k-1} + b^k \right)$$

$$+ b\left(a^k + ka^{k-1}b + \frac{k(k-1)}{2!} a^{k-2}b^2 \right.$$

$$+ \frac{k(k-1)(k-2)}{3!} a^{k-3}b^3 + \cdots$$

$$\left. + kab^{k-1} + b^k \right).$$

Now distribute a and b and collect similar terms in decreasing powers of a:

$$(a + b)^{k+1} = a^{k+1} + (kb + b)a^k + \left(\frac{k(k-1)}{2!} b^2 + kb^2 \right) a^{k-1}$$

$$+ \left(\frac{k(k-1)(k-2)}{3!} b^3 + \frac{k(k-1)}{2!} b^3 \right) a^{k-2} + \cdots$$

$$+ (b^k + kb^k)a + b^{k+1}.$$

Finally, in the second term factor out b, in the third term factor out b^2, and so on. Simplifying leads to

$$(a + b)^{k+1} = a^{k+1} + (k + 1)a^kb + \left(\frac{k(k+1)}{2!} \right) a^{k-1}b^2$$

$$+ \left(\frac{(k+1)k(k-1)}{3!} \right) a^{k-2}b^3 + \cdots$$

$$+ (k + 1)ab^k + b^{k+1}.$$

This is the binomial theorem with $n = k + 1$. By mathematical induction, the theorem is true for all positive integers n.

EXERCISE SET 11.5

A

For Problems 1 through 21, use the principle of mathematical induction to prove that the statements are true for all positive integers n.

1. $1 + 2 + 3 + \cdots + n = \dfrac{n(n + 1)}{2}$

2. $2 + 4 + 6 + \cdots + 2n = n(n + 1)$

3. $3 + 7 + 11 + \cdots + (4n - 1) = n(2n + 1)$

4. $1 + 4 + 7 + \cdots + (3n - 2) = \dfrac{n(3n - 1)}{2}$

5. $2 + 2^2 + 2^3 + \cdots + 2^n = 2(2^n - 1)$

6. $\frac{1}{2} + \frac{1}{4} + \frac{1}{8} + \cdots + (\frac{1}{2})^n = 1 - (\frac{1}{2})^n$

7. $2^2 + 4^2 + 6^2 + \cdots + (2n)^2 = \dfrac{2n(n + 1)(2n + 1)}{3}$

8. $1(2) + 2(3) + 3(4) + \cdots n(n + 1) = \dfrac{n(n + 1)(n + 2)}{3}$

9. $\dfrac{1}{1(2)} + \dfrac{1}{2(3)} + \dfrac{1}{3(4)} + \cdots + \dfrac{1}{n(n + 1)} = \dfrac{n}{n + 1}$

10. $1^3 + 2^3 + 3^3 + \cdots + n^3 = \left[\dfrac{n(n + 1)}{2}\right]^2$

11. $2^3 + 4^3 + 6^3 + \cdots + (2n)^3 = 2n^2(n + 1)^2$

12. $\sin(\phi + n\pi) = (-1)^n \sin \phi$

13. $\cos(\phi + n\pi) = (-1)^n \cos \phi$

B

14. $\sin x + \sin 3x + \sin 5x + \cdots + \sin(2n - 1)x = \dfrac{\sin^2 nx}{\sin x}$

15. $\cos x + \cos 3x + \cos 5x + \ldots + \cos(2n - 1)x = \dfrac{\sin 2nx}{2 \sin x}$

16. $2^n > n$ 17. $3^n > n + 1$

18. $n^n \geq n!$ 19. $e^n > n + 1$

20. $1 + 2(2) + 3(2)^2 + 4(2)^3 + \cdots + n(2)^{n-1} = (n - 1)2^n + 1$

21. $\dfrac{1}{2!} + \dfrac{2}{3!} + \dfrac{3}{4!} + \cdots + \dfrac{n}{(n + 1)!} = 1 - \dfrac{1}{(n + 1)!}$

For Problems 22 through 25, use the princple of mathematical induction to prove that the statements are true for the indicated integers n.

22. $2^n > 3n, n \geq 4$

23. $n! > 10n, n \geq 5$

24. $(1 + x)^n > 1 + nx, x > 0, n \geq 2$

25. $n! > 2^n, n \geq 4$

The sequences in Problems 26 through 31 are defined recursively. Write the first terms, look for a pattern, and make a conjecture for an explicit formula for a_n. Then prove it using mathematical induction.

26. $a_1 = 1, a_{n+1} = a_n + 2n + 1$

27. $a_1 = 1, a_{n+1} = na_n + a_n$

28. $a_1 = 2, a_{n+1} = a_n\left(2 - \dfrac{2}{n + 1}\right)$

29. $a_1 = \frac{1}{2}, a_{n+1} = \left(a_n + \dfrac{1}{n}\right)\left(\dfrac{n}{n + 1}\right)$

30. $a_1 = 0, a_2 = 2, a_{n+2} = 3a_{n+1} - 2a_n$

31. $a_1 = 1, a_2 = 0, a_{n+2} = 2a_{n+1} - a_n$

In Problems 32 through 40, either disprove the statement by citing a counterexample or prove the statement using mathematical induction.

32. $n^2 - n + 17$ is a prime number for every positive integer n.

33. $2^n < n^3 + 2$ for every positive integer n.

34. $n^5 - n$ is divisible by 5 for every positive integer n.

35. $9^n + 7$ is divisible by 8 for every positive integer n.

36. $\dfrac{1}{1(3)} + \dfrac{1}{3(5)} + \dfrac{1}{5(7)} + \cdots + \dfrac{1}{(2n - 1)(2n + 1)} = \dfrac{n}{2n + 1}$, for all positive integers n.

37. $\dfrac{3}{1(2)^2} + \dfrac{5}{2(3)^2} + \dfrac{7}{3(4)^2} + \cdots + \dfrac{2n + 1}{n(n + 1)^2} = \dfrac{n(n + 2)}{(n + 1)^2}$, for all positive integers n.

38. $\sqrt{n^3 - 2n^2 + 11n - 6}$ is an even integer for every positive integer n.

39. $n^2 + n$ is an even integer for every positive integer n.

40. $1 + 4 + 7 + \cdots + (3n - 2) = 2n^2 - 2n + 1$ for every positive integer n.

41. Prove that the sum of the interior angles of a polygon with n sides is $180°(n - 2)$ for $n \geq 3$.

42. Prove that the number of diagonals in a polygon with n sides is $n(n-3)/2$ for $n \geq 3$.

C

43. Suppose a board has 3 pegs, and that stacked on one of the pegs are n discs (with holes in the centers) of decreasing size. Figure 15 shows $n = 4$ discs. A problem known as the Tower of Hanoi puzzle requires the player to transfer the entire stack of discs to one of the other pegs by moving only one disc at a time onto another peg. If no disc can be placed on another disc of smaller size, what is the least number of moves needed to complete the task? Prove your answer by mathematical induction.

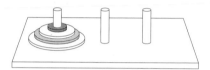

FIGURE 15

44. a) Let

$$r = \frac{1+\sqrt{5}}{2} \quad \text{and} \quad s = \frac{1-\sqrt{5}}{2}.$$

Evaluate Benet's formula

$$\frac{r^n - s^n}{r - s}$$

for $n = 1$, 2, and 3.

b) Recall the Fibonacci sequence $\{f_n\}$ defined in Example 3 of Section 11.1:

$$f_n = \begin{cases} 1 & \text{for } n = 1 \text{ or } 2 \\ f_{n-1} + f_{n-2} & \text{for } n = 3, 4, 5, \cdots \end{cases}$$

Prove by mathematical induction that

$$f_n = \frac{r^n - s^n}{r - s}.$$

Isaac Newton

One of the supreme intellects of the human race was Isaac Newton. Born on Christmas day, 1642 (the year of Galileo's death), Newton was a frail child who was raised by his grandmother in the farming village of Woolsthrope in England. During his childhood, he amused himself by making working mechanical toys, such as a water clock, windmill, and a toy carriage propelled by its rider. His uncle, who was a Cambridge graduate, noticed unusual mechanical ability in his nephew and arranged for the boy to attend Cambridge in 1661. As an undergraduate, Newton first studied chemistry and was not particularly interested in mathematics. He soon began reading the works of Euclid, Kepler, Galileo, Descartes, and Fermat and developed a fascination for mathematics. He was also fortunate to attract the attention of Isaac Barrow, a talented mathematician and professor of mathematics at Cambridge.

Soon after Newton graduated, the universities closed during 1665 and 1666 due to the outbreak of the plague. Newton went home to the family farm, site of the well-known legend of the apple falling on his head and inspiring his famous formulation of the law of gravitation. What is truly amazing is that he also made three other phenomenal discoveries during those two years on the farm: the binomial series, the groundwork for calculus, and the resolution of light into the spectrum of colors with a prism.

Newton was always a sensitive, introverted scholar; he seldom published his discoveries, and many of them had to be coaxed out of him. He regarded many of his mathematical discoveries as tools to help him investigate scientific problems. Throughout his life he was a tenacious problem solver. Once he put a needle into his eye to investigate the optics of the human eye. He was never concerned about his appearance and often skipped meals that were prepared for him while concentrating on a problem. There is one story of Newton leading a horse when he became absorbed in some problem. He later found himself standing with a bridle in his hand but with no horse!

Newton returned to Cambridge in 1667 and earned his Master's degree. In 1669 he succeeded Isaac Barrow as Lucasian chair of mathematics at Cambridge. In 1670 he built the first reflecting telescope. During the next 25 years, using his law of gravitation he formulated the motion of planets. He also developed basic theories of optics and thermodynamics. In 1687 he published what is generally considered the most important scientific achievement of the human mind—*Philosophae Naturalis Principia Mathematica* (The Mathematical Principles of Natural Philosophy).

The year 1696 marked a dark turn of events for both mathematics and science. Newton left Cambridge to take the position of Warden (and later Master) of the London Mint. During this period of his life, his attention was on theological and philosophical studies. With few exceptions, he could not be persuaded to enter discussions about science and mathematics. Although his *Optics* was published in 1704, his genius was virtually wasted during the last 25 years of his life. When a problem stimulated his curiosity, though, he occasionally reminded his contemporaries of his unrivaled ability. One problem (the so-called brachistochrone problem, posed by John Bernoulli), which had stumped all the minds of Europe for over a year, came to Newton's attention one day after working at the Mint. Newton solved the problem that night and submitted his solution to the Royal Society anonymously. When Bernoulli saw the solution he had no doubt about its author; as Bernouli put it, "I recognize the lion by his paw!"

During his long life, Newton received many honors and seemed to enjoy his fame. In 1703 he was elected president of the Royal Society, and in 1705 he was knighted by Queen Anne. Upon his death, he was buried in Westminster Abbey with the accolades and pageantry befitting his stature. Voltaire attended the funeral and said "I have seen a professor of mathematics, only because he was great in his vocation, buried like a king who had done good to his subjects."

Newton described himself as "only like a boy playing on the seashore and diverting myself in now and then finding a smoother pebble or prettier shell than ordinary, whilst the great ocean of truth lay all undiscovered before me."

MISCELLANEOUS EXERCISES

For each sequence $\{a_n\}$ in Problems 1 through 8 (a) list the first five terms and (b) graph the first five ordered pairs.

1. $a_n = 4n - 3$

2. $a_n = \dfrac{n}{3n - 2}$

3. $a_n = \dfrac{(-1)^n}{n^2}$

4. $a_n = \sin \dfrac{(n-1)\pi}{6}$

5. $a_n = 48(\frac{1}{2})^n$

6. $a_1 = 3, a_2 = 1, a_n = a_{n-1} + 2a_{n-2}$

7. $a_n = \displaystyle\sum_{k=0}^{n} \dfrac{(-1)^k}{(k+1)!}$

8. $a_n = \displaystyle\sum_{k=1}^{n} (2k - 1)$

In Problems 9 through 13 (a) graph $f(x)$ and (b) determine the value that $\{a_n\}$ approaches, if any.

9. $f(x) = (\frac{3}{4})^x; a_n = (\frac{3}{4})^n$

10. $f(x) = \dfrac{4x}{2x + 3}; a_n = \dfrac{4n}{2n + 3}$

11. $f(x) = \tan^{-1} x; a_n = \tan^{-1} n$

12. $f(x) = \dfrac{x}{\sin x}; a_n = \dfrac{n}{\sin n}$

13. $f(x) = \sqrt{x}; a_n = \sqrt{n}$

14. Determine the value that $\left\{ a_n = n \ln\left(1 + \dfrac{1}{n}\right) \right\}$ approaches, if any. (*Hint: See Section 4.2.*)

Find the first five terms of the sequences $\{a_n\}$ in Problem 15 and 16.

15. $a_n = \displaystyle\sum_{k=1}^{n} \dfrac{x^{k-1}}{k^2}$

16. $a_n = \displaystyle\sum_{k=1}^{n} \dfrac{(-1)^k(x+2)^{k-1}}{2k+1}$

17. Use the properties of summation to evaluate

$$\sum_{k=1}^{40} (k^3 - 2k + 3).$$

18. a) Use the properties of summation to evaluate

$$\frac{1}{n} \sum_{k=1}^{n} \frac{(k-1)^2}{n^2}, \qquad \text{for } n = 25.$$

b) Use the properties of summation to evaluate

$$\frac{1}{n} \sum_{k=1}^{n} \frac{(k-1)^2}{n^2}, \qquad \text{for } n = 100.$$

c) Use the properties of summation to determine the value that

$$\frac{1}{n} \sum_{k=1}^{n} \frac{(k-1)^2}{n^2}$$

approaches as $n \to \infty$.

Each sequence in Problems 19 through 27 is either arithmetic or geometric. Determine the common difference d (if arithmetic) or the common ratio r (if geometric). Also determine $a_1, a_6,$ and a_n.

19. $0, \pi, 2\pi, 3\pi, \ldots$

20. $10, 4, \frac{8}{5}, \frac{16}{25}, \ldots$

21. $-18, 24, -32, \frac{128}{3}, \ldots$

22. $12, 3, -6, -15, \ldots$

23. $4, \frac{11}{3}, \frac{10}{3}, 3, \ldots$

24. $3\sqrt{2} + 1, \sqrt{2} + 1, -\sqrt{2} + 1, -3\sqrt{2} + 1, \ldots$

25. $\sqrt{3} + 1, 3\sqrt{2} + \sqrt{6}, 6\sqrt{3} + 6, 18\sqrt{2} + 6\sqrt{6}, \ldots$

26. $\sin \theta, 1 - \cos^2\theta, \sin^3\theta, \sin^4\theta, \ldots$

27. $\ln x, \ln\left(\dfrac{1}{x^2}\right), \ln x^4, \ln\left(\dfrac{1}{x^8}\right), \ldots$

In Problems 28 through 30, the given information describes an arithmetic sequence. Determine the general term a_n and the first four terms.

28. $a_1 = 5, a_2 = -14$

29. $a_3 = 2, a_9 = 4$

30. $a_6 = 5, a_{10} = -1$

In Problems 31 through 33, the given information describes a geometric sequence. Determine the general term a_n and the first four terms.

31. $a_1 = 2, a_2 = -5$

32. $a_2 = 2, a_4 = 18, r > 0$

33. $a_3 = 3, a_5 = 1$

The terms of the sums in Problems 34 through 43 are from arithmetic or geometric sequences. Compute the indicated sums.

34. $-5 + -2 + 1 + \cdots + 25$

35. $432 - 72 + 12 - \cdots + \frac{1}{108}$

36. $-1 + 2 - 4 + 8 + \cdots + 2048$

37. $2 + 6 + 18 + \cdots + 13122$

38. $\displaystyle\sum_{k=1}^{8} 48\left(\frac{1}{4}\right)^{k-1}$

39. $\displaystyle\sum_{k=1}^{80} [7 + (k-3)2]$

40. $\displaystyle\sum_{k=1}^{50} \frac{k+3}{2}$

41. $\displaystyle\sum_{k=1}^{10} 5(-2)^{k-2}$

42. $4x + (3x + y) + (2x + 2y) + \cdots + (-5x + 9y)$

43. $1 - x^2 + x^4 - x^6 + \cdots - x^{14}$

Evaluate each sum in Problems 44 and 45.

44. $\displaystyle\sum_{k=1}^{8} [3^{k-1} + 5k]$

45. $\displaystyle\sum_{k=1}^{10} [(-2)^{k-1} + k^2]$

46. Find the 21st term of the arithmetic sequence whose 3rd and 12th terms are 20 and 14, respectively.

47. Find the 3rd term of the geometric sequence whose 4th and 7th terms are 200 and $\frac{8}{5}$, respectively.

48. Express $\ln x^2 + \ln x^4 + \ln x^6 + \cdots + \ln x^{100}$ in the form $a \ln x$, where a is a constant.

In Problems 49 through 57, determine whether the infinite geometric series has a sum and, if so, find that sum.

49. $40 + 20 + 10 + \cdots$

50. $180 - 120 + 80 + \cdots$

51. $1 + \frac{5}{6} + \frac{25}{36} + \cdots$

52. $54 - 6 + \frac{2}{3} - \cdots$

53. $1 - \frac{4}{3} + \frac{16}{9} - \cdots$

54. $8 + 10 + \frac{25}{2} + \cdots$

55. $\displaystyle\sum_{k=1}^{\infty} 300(\frac{4}{9})^{k-1}$

56. $\displaystyle\sum_{k=1}^{\infty} 28(-\frac{3}{7})^{k}$

57. $\displaystyle\sum_{k=1}^{\infty} (\sin^2 x)^k, \ 0 < x < \frac{\pi}{2}$

In Problems 58 through 60, express the repeating decimals as the ratio of two integers by considering an infinite geometric series.

58. $0.\overline{36}$

59. $0.2\overline{16}$

60. $8.2\overline{27}$

61. For what values of x does the following equation hold?

$$\frac{2}{1-x} = 2 + 2x + 2x^2 + 2x^3 + \cdots$$

62. For what values of x does the following equation hold?

$$\frac{1}{1+5x} = 1 - 5x + 25x^2 - 125x^3 + \cdots$$

63. For what values of x do the following equations hold?

a) $\dfrac{8}{4-x} = \dfrac{2}{1-(x/4)} = 2 + \dfrac{x}{2} + \dfrac{x^2}{8} + \dfrac{x^3}{32} + \cdots$

b) $\dfrac{8}{4-x} = \dfrac{8}{1-(x-3)}$

$$= 8 + 8(x-3) + 8(x-3)^2 + \cdots$$

In Problems 64 through 66, find an infinite geometric series that has the given rational expression as its sum and state the domain for which the sum exists.

64. $\dfrac{6}{1-x}$

65. $\dfrac{3}{1+2x}$

66. $\dfrac{-12}{4-x}$

In Problems 67 through 71, find a simple formula for nth partial sum S_n to determine whether the series converges or diverges; when convergent, state the sum.

67. $\displaystyle\sum_{k=1}^{\infty} \frac{5}{2k(2k+2)}$

68. $\displaystyle\sum_{k=1}^{\infty} \frac{6}{(3k-1)(3k+2)}$

69. $\displaystyle\sum_{k=1}^{\infty} \frac{10}{(5k-2)(5k+3)}$

70. $\displaystyle\sum_{k=1}^{\infty} \frac{16}{k(k+2)}$

71. $\displaystyle\sum_{k=1}^{\infty} (2k-1)$

Expand and simplify the expressions in Problems 72 through 75.

72. $(2a - 3)^4$

73. $(x^2 + 2y)^5$

74. $\left(x^2 + \dfrac{1}{x}\right)^6$

75. $(\tan x - 1)^4$

76. Find the x^9-term in the expansion of $(2x^3 + 1)^8$.

77. Find the x^6y^3-term in the expansion of $(x^2 + 3y)^6$.

78. Expand (a) $(\sqrt{5} + \sqrt{3})^4$ and (b) $(\sqrt{5} - \sqrt{3})^4$. Add these results to show that $247 < (\sqrt{5} + \sqrt{3})^4 < 248$.

Find the binomial series for each function in Problems 79 through 81. Show the first four terms.

79. $\sqrt[4]{1-x}$

80. $\sqrt{1+2x}$

81. $\dfrac{1}{\sqrt[3]{1-x}}$

82. Approximate $\sqrt[4]{0.9}$ using the result of Problem 79.

83. Prove by mathematical induction that $(xy)^n = x^n y^n$, for all positive integers n.

84. Prove by mathematical induction that
$$1^2 + 3^2 + \cdots + (2n - 1)^2 = \tfrac{1}{3}(4n^3 - n),$$
for all positive integers n.

85. Prove by mathematical induction that
$$1 + 8 + 16 + \cdots + 8(n - 1) = (2n - 1)^2$$
for all positive integers n.

86. Prove by mathematical induction that $1 + 2n \le 3^n$, for all nonnegative integers n.

87. Prove by mathematical induction that $(2n + 1)^2 - 1$ is divisible by 8, for all positive integers n.

88. Prove or disprove: $n^2 + 21n + 1$ is a prime number, for all positive integers n.

89. Prove or disprove:

$$\frac{1}{1 \cdot 2 \cdot 3} + \frac{1}{2 \cdot 3 \cdot 4} + \frac{1}{3 \cdot 4 \cdot 5} + \cdots$$

$$+ \frac{1}{n(n + 1)(n + 2)} = \frac{n(n + 3)}{4(n + 1)(n + 2)}$$

for all positive integers n.

CHAPTER 12

SOLID ANALYTIC GEOMETRY

COORDINATE SPACE

Coordinate geometry in three dimensions is merely a generalization of the familiar coordinate geometry in two dimensions. Starting with the xy-plane, we erect another axis, the **z-axis,** perpendicular to both the x-axis and the y-axis (Figure 1). One of the obvious problems with this coordinate system in three dimensions is that it is being represented in perspective by an image in two dimensions on this page. These perspectives may be drawn from any convenient point of view. One of the most important skills to acquire for calculus is the ability to see three-dimensions in your mind.

   Regardless of the point of view of the perspective, we always use the same relative orientation of the three axes. Imagine grasping the z-axis with your right hand, thumb pointing in the positive direction of the z-axis (Figure 2). The direction in which your fingers wrap is the same direction as the rotation of the x-axis to the y-axis. This coordinate system is called **right-handed** (for obvious reasons). There is a left-handed three-dimensional co-ordinate system, which is different (try as you might, it is impossible to align a right-handed system to a left-handed system). The right-handed system is the coordinate system of choice in almost all fields and disciplines.

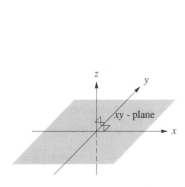

FIGURE 1

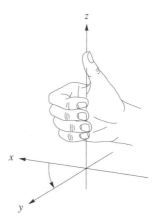

FIGURE 2

685

FIGURE 3 *Right-handed coordinate systems from different points of view*

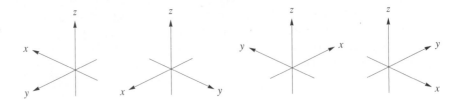

The point at which all three axes intersect is called the **origin.** The horizontal plane that contains the *x*-axis and the *y*-axis is called the **xy-plane.** In a similar fashion, the vertical plane containing the *x*-axis and the *z*-axis is called the **xz-plane,** and the vertical plane containing the *y*-axis and the *z*-axis is called the **yz-plane.** Just as the *x*-axis and the *y*-axis divide the plane into four quadrants, these three planes divide space into eight **octants** (Figure 4).

The first octant is the octant in which the values of *x*, *y*, and *z* are all positive. There is no conventional way to number the other octants.

Recall from Section 1.1 that there is a one-to-one correspondence between the set of real numbers ℝ and the set of points on the coordinate line. In Section 2.1 we saw that there was a one-to-one correspondence between the set of ordered pairs of real numbers ℝ² and the points on the coordinate plane. In coordinate space, the analogous correspondence is between the set of **ordered triples** (which we denote by ℝ³) and points in space (Figure 5).

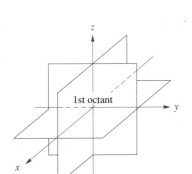

FIGURE 4

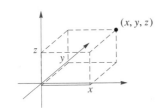

FIGURE 5

EXAMPLE 1   Plot the points in coordinate space that correspond to the ordered triples $P(2, 4, 5)$ and $Q(-5, -2, 3)$.

SOLUTION The points are shown in Figure 6.

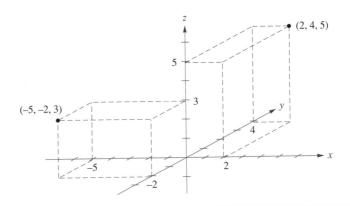

FIGURE 6

Similar to our discussion of the coordinate plane, our main concern now is the graphs of sets of ordered triples. For example, consider the set of all points corresponding to the set of ordered triples with a y-coordinate of zero (if the context is understood, this set can be described by the equation $y = 0$). These are precisely the points that are in the xz-plane (Figure 7).

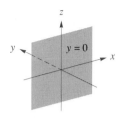

FIGURE 7 *The equation $y = 0$ describes the xz-plane*

EXAMPLE 2   Sketch the graphs in space of these equations.

a) $x = 5$ b) $z = -3$

c) $z = y$ (1st octant only)

SOLUTION

a) This is the set of ordered triples in which the x-coordinate is 5. This is a plane that is parallel to the yz-plane; it intersects the x-axis at the point $(5, 0, 0)$ (Figure 8).

b) This is the set of ordered triples in which the z-coordinate is -3 (Figure 9). This is a horizontal plane three units below the xy-plane. It intersects the z-axis at the point $(0, 0, -3)$.

c) This is the set of ordered triples in which the z-coordinate and y-coordinate are the same (Figure 10). Notice that the plane forms a 45° angle with both the xy-plane and the xz-plane.

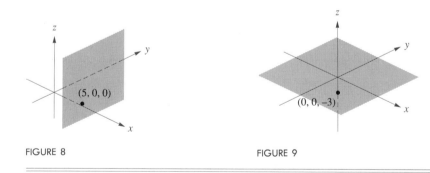

FIGURE 8   FIGURE 9

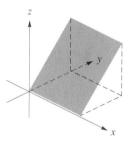

FIGURE 10

Finding the distance between two points in coordinate space is very much the same as on the plane. Suppose that we wish to find the distance from the point $P(x_1, y_1, z_1)$ and $Q(x_2, y_2, z_2)$, as shown in Figure 11. The

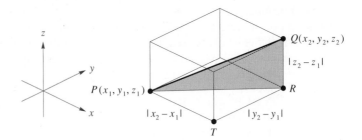

FIGURE 11

distance $d(P, Q)$ is the hypotenuse of the right triangle PRQ. Applying the Pythagorean theorem to the shaded triangle, we get

$$d(P, Q)^2 = d(P, R)^2 + d(R, Q)^2.$$

Now, from the picture,

$$d(R, Q) = |z_2 - z_1|.$$

The line segment $\overline{PR}$ in turn is the hypotenuse of the triangle PRT, so

$$d(P, R)^2 = d(R, T)^2 + d(T, P)^2$$
$$= |x_2 - x_1|^2 + |y_2 - y_1|^2.$$

Putting this all together, we get

$$d(P, Q)^2 = d(P, R)^2 + d(R, Q)^2$$
$$= |x_2 - x_1|^2 + |y_2 - y_1|^2 + |z_2 - z_1|^2$$
$$= (x_2 - x_1)^2 + (y_2 - y_1)^2 + (z_2 - z_1)^2.$$

By taking square roots, we arrive at a general distance formula for coordinate space. Notice the use of the delta notation that was introduced in Section 2.1.

Distance Formula for
Coordinate Space

> Given points $P(x_1, y_1, z_1)$ and $Q(x_2, y_2, z_2)$, the distance between P and Q is given by
>
> $$d(P, Q) = \sqrt{(\Delta x)^2 + (\Delta y)^2 + (\Delta z)^2}$$
>
> where $\Delta x = x_2 - x_1$, $\Delta y = y_2 - y_1$, and $\Delta z = z_2 - z_1$.

EXAMPLE 3   Given the points $P(2, 4, 5)$ and $Q(-5, -2, 3)$, find the distance between them (notice that these are the points plotted in Example 1).

SOLUTION By the formula,

$$d(P, Q) = \sqrt{(-5 - 2)^2 + (-2 - 4)^2 + (3 - 5)^2}$$
$$= \sqrt{49 + 36 + 4} = \sqrt{89} \approx 9.43.$$

EXAMPLE 4   Find the set of points in space that are equidistant from the points $P(2, 8, 6)$ and $Q(-1, 0, -3)$.

SOLUTION Compare this with Example 3 of Section 2.1. In that example, we found that the set of all points equidistant from two given points on the coordinate plane is a line. By analogy, the solutions of our current problem are the points in a plane that is perpendicular to the line segment $\overline{PQ}$, as Figure 12 suggests.

Suppose that the point $R(x, y, z)$ is equidistant from P and from Q. Then

$$d(P, R) = d(R, Q)$$
$$\sqrt{(x - 2)^2 + (y - 8)^2 + (z - 6)^2} = \sqrt{(x + 1)^2 + y^2 + (z + 3)^2}.$$

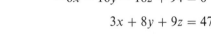

FIGURE 12

Squaring both sides and simplifying gives us

$$-6x - 16y - 18z + 94 = 0$$
$$3x + 8y + 9z = 47.$$

The next example involves finding a distance from a point to a line. In cases such as these, a sketch of the situation usually is a big help.

EXAMPLE 5   Find the distance from the point $P(8, 4, 3)$ to the x-axis.

SOLUTION First, examine Figure 13. Our goal is to find $d(P, Q)$, the hypotenuse of the triangle PQR. By the Pythagorean theorem,

$$d(P, Q) = \sqrt{4^2 + 3^2} = 5.$$

FIGURE 13

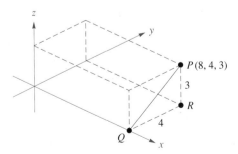

One of the direct results of the distance formula in the plane is the derivation of the equation of a circle centered at the point (h, k) with radius r. The three-dimensional analogy of this is the derivation of the equation of a sphere.

By definition, a sphere is a set of points that are equidistant from a given point. Suppose that we want to determine the equation of a sphere of radius r and center at the point $C(h, k, m)$ (Figure 14). Let $P(x, y, z)$ denote a point on the sphere. Then the distance from $C(h, k, m)$ to $P(x, y, z)$ is the radius r. By the distance formula,

$$r = \sqrt{(x - h)^2 + (y - k)^2 + (z - m)^2}.$$

Squaring both sides, we get an equation for a sphere in coordinate space.

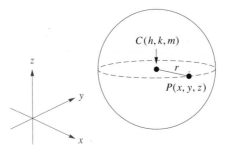

FIGURE 14

Equation of a Sphere

An equation whose graph is the sphere with radius r and center (h, k, m) is

$$(x - h)^2 + (y - k)^2 + (z - m)^2 = r^2.$$

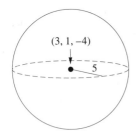

FIGURE 15

EXAMPLE 6   Identify the center and radius of the sphere whose equation is given.

$$(x - 3)^2 + (y - 1)^2 + (z + 4)^2 = 25.$$

SOLUTION From the formula, the center is at $(3, 1, -4)$, and the radius is 5 (Figure 15).

EXAMPLE 7   The graph of this equation is a sphere. Find its center and its radius.

$$x^2 + y^2 + z^2 + 2x - 8z + 5 = 0.$$

SOLUTION To determine the radius and the center, we need to rewrite the equation in the form

$$(x - h)^2 + (y - k)^2 + (z - m)^2 = r^2$$

by completing squares. Then

$$(x^2 + 2x \quad) + y^2 + (z^2 - 8z \quad) = -5$$
$$(x^2 + 2x + 1) + y^2 + (z^2 - 8z + 16) = -5 + 1 + 16$$
$$(x + 1)^2 + y^2 + (z - 4)^2 = 12.$$

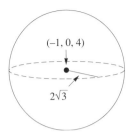

FIGURE 16

The center of the sphere is $(-1, 0, 4)$, and the radius is $2\sqrt{3}$ (Figure 16).

EXERCISE SET 12.1

A

In Problems 1 through 6, determine if the given axes represent a right-handed system.

1.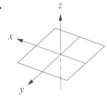

2.

3.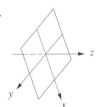

4.

5.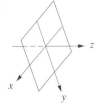

6.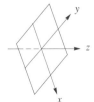

In Problems 7 through 12, sketch a three-dimensional coordinate system, plot both points given, and find the distance between them. Pick a convenient point of view.

7. $(0, 0, 4)$ and $(-6, 0, 0)$

8. $(5, 3, 0)$ and $(0, 2, 7)$

9. $(-2, 3, 0)$ and $(-6, -1, 0)$

10. $(1, 3, 4)$ and $(2, 4, 1)$

11. $(-1, -2, -5)$ and $(5, 1, 1)$

12. $(2, 2, 1)$ and $(4, 4, 2)$

13. Find the distance from the point $(2, 3, 4)$ to the xy-plane.

14. Find the distance from the point $(2, 3, 4)$ to the yz-plane.

15. Find the distance from the point $(2, 3, 4)$ to the y-axis.

16. Find the distance from the point $(2, 3, 4)$ to the z-axis.

17. Find the distance from the point $(2, 3, 4)$ to the plane $z = -3$. (Hint: see Example 2.)

18. Find the distance from the point $(2, 3, 4)$ to the plane $x = 5$. (Hint: see Example 2.)

B

In Problems 19 through 30, sketch the graph of the set of points described by the equation.

19. $x = -1$ **20.** $z = 6$ **21.** $y = -4$

22. $x = -3$ **23.** $z = 2$ **24.** $y = 2$

25. All the ordered triples (x, y, z) such that $x = 2$ and $y = 3$

26. All the ordered triples (x, y, z) such that $y = -2$ and $z = 6$

27. All the ordered triples (x, y, z) such that $x = 1$ and $z = -3$

28. The plane parallel to the z-axis and passing through the points $(2, 0, 0)$ and $(0, 4, 0)$

29. The plane parallel to the x-axis and passing through the points $(0, 0, -2)$ and $(0, -2, 0)$

30. The plane parallel to the y-axis and passing through the points $(0, 0, -2)$ and $(3, 0, 0)$

In Problems 31 through 42, determine the center and the radius of the sphere described. If no such sphere exists, explain why.

31. $(x - 2)^2 + y^2 + (z - 4)^2 = 16$

32. $x^2 + (y + 2)^2 + (z - 3)^2 = 20$

33. $(x - 2)^2 + (y - 5)^2 + (z + 1)^2 = 32$

34. $x^2 + y^2 + z^2 + 4x - 6y + 4 = 0$

35. $x^2 + y^2 + z^2 + 8x - 6y + 8z + 5 = 0$

36. $2x^2 + 2y^2 + 2z^2 + 2x - 8z - 8 = 0$

37. $x^2 + y^2 + z^2 + 2x - 2z + 2 = 0$

38. $x^2 + y^2 + z^2 + 2x - 8z + 3 = 0$

39. Tangent to the yz-plane at $(0, 2, 5)$ and the xy-plane at $(5, 2, 0)$

40. Tangent to the xy-plane at $(-3, 2, 0)$ and the xz-plane at $(-3, 0, 2)$

41. Tangent to the yz-plane at $(0, 4, 3)$ and the x-axis at $(5, 0, 0)$

42. Tangent to the xy-plane at $(5, 12, 0)$ and the z-axis at $(0, 0, 13)$

43. Determine if the points $A(-3, -2, -1)$, $B(1, 6, 11)$, and $C(-1, 2, 5)$ are collinear. *(Hint: See Exercise 43 of Section 2.1.)*

44. Determine if the points $P(-8, 1, 2)$, $Q(-3, 4, 8)$, and $R(2, 5, 14)$ are collinear. *(Hint: See Exercise 43 of Section 2.1.)*

45. Determine if the triangle with vertices $M(2, 1, 3)$, $N(7, 6, 3)$, and $P(2, 6, 8)$ is an equilateral triangle, an isosceles triangle, or a scalene triangle. *(Hint: See Exercise 45 of Section 2.1.)*

46. Determine if the triangle with vertices $A(-3, -2, -7)$, $B(12, 23, 3)$, and $C(7, 23, 8)$ is an equilateral triangle, an isosceles triangle, or a scalene triangle. *(Hint: See Exercise 45 of Section 2.1.)*

47. Compute the lengths of the sides of the triangle with vertices $H(8, 1, -2)$, $J(10, 2, 2)$, and $K(6, 1, -1)$. Determine if it is a right triangle, a obtuse triangle, or an acute triangle. *(Hint: See Exercise 47 of Section 2.1.)*

48. Compute the lengths of the sides of the triangle with vertices $A(2, 1, -4)$, $B(-1, 3, -4)$, and $C(6, 6, 2)$ and determine if it is a right triangle, a obtuse triangle, or an acute triangle. *(Hint: See Exercise 47 of Section 2.1.)*

49. Determine the measure of the nearest $0.1°$ of the interior angles of the triangle with vertices $P(2, 3, 4)$, $Q(0, 0, -2)$, and $R(3, 0, 0)$. *(Hint: You may wish to review the law of cosines in Section 7.2.)*

50. Determine the area of the triangle with vertices $F(2, 3, 4)$, $G(0, 0, -2)$, and $H(3, 0, 0)$. *(Hint: You may wish to review Hero's formula in Section 7.2.)*

C

51. Consider a point $P(3, 5, -4)$ in coordinate space. Find the point symmetric to P with respect to the following: **(a)** y-axis, **(b)** xy-plane, **(c)** origin, and **(d)** the plane $z = -3$

52. Derive the coordinates of the midpoint M of $P(x_1, y_1, z_1)$ and $Q(x_2, y_2, z_2)$. Verify that M is the midpoint of P and Q by showing that these three points are collinear.

53. Determine the median of the triangle with vertices $A(2, 3, 4)$, $B(0, 0, -2)$, and $C(3, 0, 0)$. *(Hint: See Exercise 58 of Section 2.1.)*

PLANES AND LINES

S E C T I O N 12.2

One of the most basic sets on the coordinate plane is that of a straight line. Any line on the plane can be represented by an equation of the form $ax + by = c$ for some constants a, b, and c (Figure 17). A point $P(x, y)$ is on the line if and only if (x, y) is a solution to the equation.

Furthermore, if $a \neq 0$, then this line has an x-intercept of c/a (recall that we find the x-intercept by letting $y = 0$ and solving for x). Likewise, if $b \neq 0$, then this line has an y-intercept of c/b. If $a = 0$, then the equation can also be expressed as $y = c/b$, and the corresponding line is horizontal and has no x-intercept (Figure 18). On the other hand, if $b = 0$, then the equation can also be expressed as $x = c/a$, and the corresponding line is vertical and has no y-intercept (Figure 19).

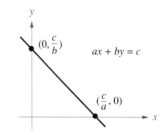

FIGURE 17

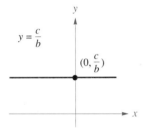

FIGURE 18

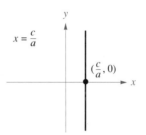

FIGURE 19

Now consider an equation of the form $ax + by + cz = d$ (for constants a, b, c, and d) and the set of points in coordinate space that it represents. The **x-intercept** is the point in the set that is on the x-axis. Suppose that $a \neq 0$. To find the x-intercept, we determine the ordered triple of the form $(x, 0, 0)$ that makes the equation true:

$$ax + b(0) + c(0) = d$$

$$ax = d$$

$$x = \frac{d}{a}.$$

The x-intercept is $(d/a, 0, 0)$. In a similar fashion, if $b \neq 0$, then the **y-intercept** is $(0, d/b, 0)$, and if $c \neq 0$, then **z-intercept** is $(0, 0, d/c)$ (Figure 20).

As you may already suspect, there is a connection between the general equation of a line on the coordinate plane and the equation $ax + by + cz = d$ in coordinate space. In Example 4 of Section 12.1, we saw that the equation $3x + 8y + 9z = 47$ represents a plane in coordinate space. This is true in general (see Problem 56).

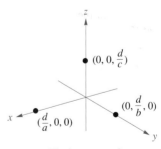

The intercepts of $ax + by + cz = d$

FIGURE 20

The Equation of a Plane

The set of points (x, y, z) in space that satisfy the equation

$$ax + by + cz = d$$

forms a plane. Furthermore:

If $a \neq 0$,   then the x-intercept is $\left(\dfrac{d}{a}, 0, 0\right)$.

If $b \neq 0$,   then the y-intercept is $\left(0, \dfrac{d}{b}, 0\right)$.

If $c \neq 0$,   then the z-intercept is $\left(0, 0, \dfrac{d}{c}\right)$.

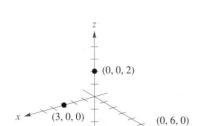

FIGURE 53 *The intercepts of*
$4x + 2y + 6z = 12$

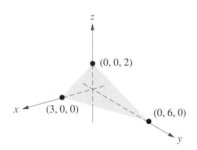

FIGURE 22

EXAMPLE 1   Sketch the set of points in the first octant that are solutions of the equation

$$4x + 2y + 6z = 12.$$

SOLUTION First we plot the intercepts. Letting $y = 0$ and $z \doteq 0$, we get

$$4x = 12$$

or

$$x = 3.$$

The x-intercept is $(3, 0, 0)$. If $x = 0$ and $z = 0$, then $y = 6$; the y-intercept is $(0, 6, 0)$. If $x = 0$ and $y = 0$ then $z = 2$; the z-intercept is $(0, 0, 2)$ (Figure 21). These three points define the plane in question (Figure 22).

In Example 1, notice the points of the plane that are in the xy-plane. They form a line passing through the points $(3, 0, 0)$ and $(0, 6, 0)$. This line is the **trace** of the plane in the xy-plane. The xy-plane is precisely those points

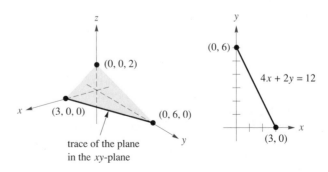

FIGURE 23

with z-coordinate equal to zero. Considering the xy-plane as a distinct co-ordinate plane, we can sketch this trace by letting $z = 0$ in the equation to get an equation in x and y only, namely, $4x + 2y = 12$ (Figure 23).

The trace in the xz-plane can be similarly determined by letting $y = 0$, and the trace in the yz-plane can be determined by letting $x = 0$ (Figure 24).

FIGURE 24

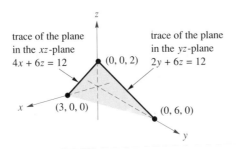

trace of the plane in the xz-plane $4x + 6z = 12$

$(0, 0, 2)$

trace of the plane in the yz-plane $2y + 6z = 12$

$(3, 0, 0)$

$(0, 6, 0)$

EXAMPLE 2 Sketch the set of points in first octant that are solutions of the equation

$$2x + 5y = 10.$$

SOLUTION Our plan of attack is this: Determine the intercepts, determine the traces, and shade appropriately. To find the intercepts:

$$2x + 5(0) = 10 \Rightarrow x\text{-intercept is } (5, 0, 0)$$

$$2(0) + 5y = 10 \Rightarrow y\text{-intercept is } (0, 2, 0)$$

$$2(0) + 5(0) = 10 \Rightarrow \text{there is no } z\text{-intercept.}$$

To find the traces (Figure 25):

$$z = 0 \Rightarrow xy\text{-plane is } 2x + 5y = 10$$

$$y = 0 \Rightarrow xz\text{-plane is } 2x = 10 \text{ or } x = 5$$

$$x = 0 \Rightarrow yz\text{-plane is } 5y = 10 \text{ or } y = 2.$$

This gives us our final sketch (Figure 26). Notice that the plane is parallel to the z-axis. This seems reasonable since there is no z-intercept.

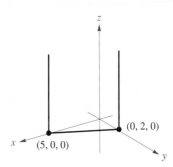

$(5, 0, 0)$ $(0, 2, 0)$

FIGURE 25

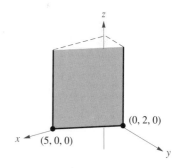

$(5, 0, 0)$ $(0, 2, 0)$

FIGURE 26

EXAMPLE 3 Sketch this plane in the first octant:

$$2x - y + 3z = 6.$$

SOLUTION First, we sketch the intercepts and the three traces of the plane (Figure 27). Notice that the y-intercept is negative, so we extend the y-axis in the negative direction to accommodate it. Shading helps to picture the plane. Two points of view are offered in Figure 28.

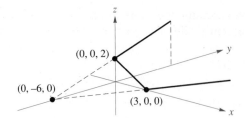

FIGURE 27

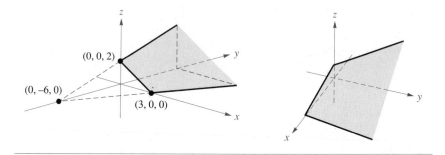

FIGURE 28

One of the important skills required by calculus is that of visualizing intersecting planes. The next three examples are practice for this skill.

EXAMPLE 4   Sketch these planes on the same axes:

$$\begin{cases} 4x + 2y + 6z = 12 \\ 2x + 5y \quad\;\; = 10. \end{cases}$$

(Note that these are the planes from Examples 1 and 2.)

SOLUTION First, we sketch the three traces of each plane (Figure 29). Next, we determine the intersection of the two planes. This is the dotted line shown. To finish, we shade appropriately to add perspective. .

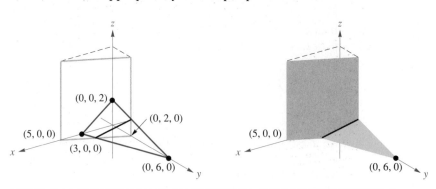

FIGURE 29

The next two examples involve systems of three equations. Notice, however, the apparently different types of intersections: One is a point, the other is a line.

EXAMPLE 5   Sketch these three planes on the same set of axes:

$$\begin{cases} x & = 1 \\ x + y + z = 5 \\ x \quad\ + z = 3. \end{cases}$$

SOLUTION First, we sketch the three traces of each of the three planes (Figure 30).

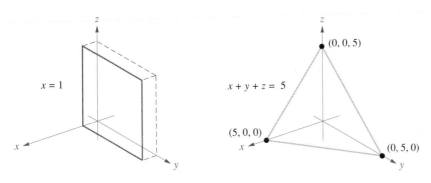
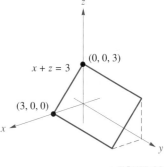

FIGURE 30

Next, we superimpose them on the same coordinate system and sketch the lines of intersection between each pair of intersecting planes. Shading appropriately finishes the job (Figure 31).

FIGURE 31

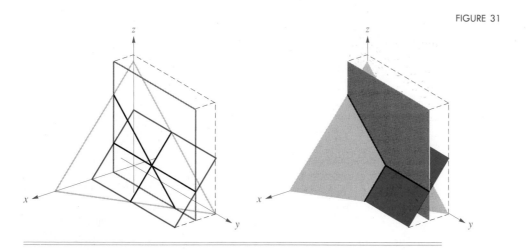

EXAMPLE 6   Sketch these three planes on the same set of axes:

$$\begin{cases} 2x + 6y + 3z = 24 \\ 3y + z = 9 \\ 4x + 3y + 3z = 21. \end{cases}$$

SOLUTION Again we start by sketching the intercepts and the traces of each of the three planes (Figure 32). Combining these pictures and shading appropriately, we get Figure 33. The planes seem to intersect along a line, but this conjecture needs further verification.

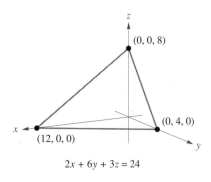

$2x + 6y + 3z = 24$

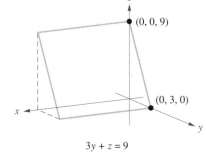

$3y + z = 9$

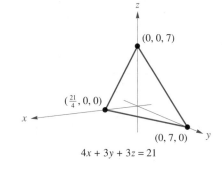

$4x + 3y + 3z = 21$

FIGURE 32

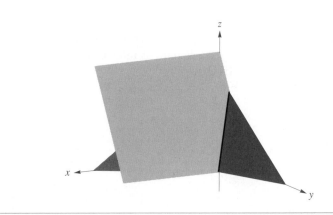

FIGURE 33

The points of intersection of intersecting planes can be determined algebraically by solving the system of linear equations in three unknowns that represents those planes (you may wish to pause here and review the techniques of solution in Chapter 10). You should recall that there are three types of possible solutions for each system: one ordered triple, no ordered triples, or an infinite number of ordered triples, which can be represented parametrically.

EXAMPLE 7   Find the intersection of the three planes sketched in Example 5.

SOLUTION The coordinates of a point in the intersection of these planes is the solution to the system of equations:

$$\begin{cases} x & = 1 \\ x + y + z = 5 \\ x \quad\quad + z = 3. \end{cases}$$

The first equation gives us $x = 1$. Using the substitution method of Section 10.1, we replace x by 1 in the third equation to determine z:

$$x + z = 3$$
$$1 + z = 3$$
$$z = 2.$$

Using these values for x and z, we determine these values of y from the second equation:

$$x + y + z = 5$$
$$1 + y + 2 = 5$$
$$y = 2.$$

The solution to this system of equations is $(1, 2, 2)$. This seems to agree with Figure 34.

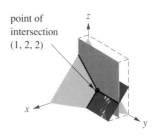

point of intersection $(1, 2, 2)$

FIGURE 34

EXAMPLE 8   Find the intersection of the three planes sketched in Figure 33.

SOLUTION In Example 6 of Section 10.2 we found that the system

$$\begin{cases} 2x + 6y + 3z = 24 \\ \quad\quad 3y + z = 9 \\ 4x + 3y + 3z = 21 \end{cases}$$

has a parametric solution:

$$(3 - \tfrac{1}{2}t, \; 3 - \tfrac{1}{3}t, \; t) \qquad \text{for all real numbers } t.$$

This certainly is an infinite set of points. The only way that the intersection of three distinct planes can be an infinite number of points is if the intersection is a straight line (Figure 35). The conjecture made earlier that the intersection is a line is indeed true.

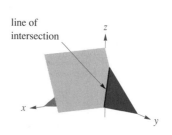

line of intersection

FIGURE 35

The fact that the points represented by the parametric solution represent a straight line in Example 8 can be generalized, as we do now.

Line in Space

> Suppose that h, k, and m and u, v, and w are constants and t is any real number. Then the points represented by
>
> $$(h + ut, k + vt, m + wt)$$
>
> form a straight line.

The next two examples deal with the intersection of a line and a plane. Just as with the intersection of two or more planes, there are three possibilities: The line may intersect the plane at exactly one point, it may be on the plane (an infinite number of points of intersection), or it may be parallel but not on the plane (no points of intersection)

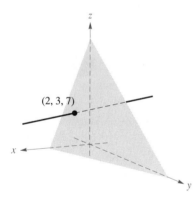

(2, 3, 7)

FIGURE 36

EXAMPLE 9   Find the intersection of the line $(6 - t, 1 + \frac{1}{2}t, -1 + 2t)$ and the plane $2x + y + z = 14$.

SOLUTION   Stated in other terms, we need to find a t such that when $6 - t$, $1 + \frac{1}{2}t$, and $-1 + 2t$ are substituted for x, y, and z, respectively, in the equation of the plane, the equation is true. So,

$$2(6 - t) + (1 + \tfrac{1}{2}t) + (-1 + 2t) = 14$$
$$12 + \tfrac{1}{2}t = 14$$
$$t = 4.$$

The point of intersection is $(6 - (4), 1 + \frac{1}{2}(4), -1 + 2(4))$, or $(2, 3, 7)$ (Figure 36).

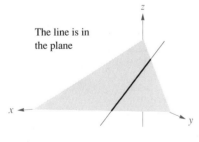

The line is in the plane

FIGURE 37

EXAMPLE 10   Find the intersection of the line $(t + 1, 3, -\frac{1}{2}t)$ and the plane $x + 3y + 2z = 10$.

SOLUTION   Proceeding as we did in Example 9, we have

$$(t + 1) + 3(3) + 2(-\tfrac{1}{2}t) = 10$$
$$10 = 10.$$

This equation is an identity; it is true for all real values of t. This, in turn, implies that every point on the line is in the plane (Figure 37).

EXERCISE SET 12.2

A

In Problems 1 through 6, match each plane with one of the equations I through VI. Copy or trace the picture and label the intercepts.

1.

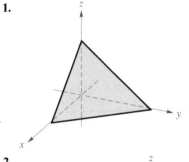

2.

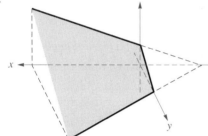

3.

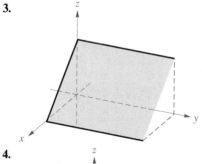

4.

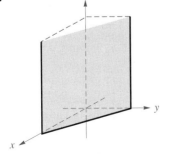

5.

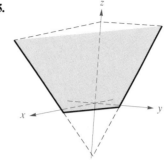

6.

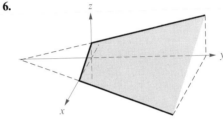

I. $2x + 7y = 12$
II. $2x - y + 4z = 4$
III. $35x + 20y + 28z = 140$
IV. $2x + 2y - 3z = 6$
V. $3x - 6y - 8z = -12$
VI. $4x + \frac{8}{3}z = 8$

In Problems 7 through 15, sketch the plane in the first octant by determining the intercepts, sketching the traces, and shading.

7. $2x + 5z = 20$ **8.** $y + 4z = 8$

9. $2x + 4y + 5z = 20$ **10.** $4x + 5y + 2z = 20$

11. $2x + 4y - 2z = 8$ **12.** $-x + 2y + 4z = 6$

13. $4x - y + 2z = 20$ **14.** $-3x - y + 2z = 0$

15. $x - y - 2z = 0$

In Problems 16 through 33, sketch the planes in the first octant. You may want to try two or more views of the axes.

16. $\begin{cases} x = 4 \\ z = 3 \end{cases}$ **17.** $\begin{cases} y = 2 \\ x = 5 \end{cases}$

18. $\begin{cases} x + y = 6 \\ z = 2 \end{cases}$ **19.** $\begin{cases} x + 2z = 8 \\ x \quad\;\; = 4 \end{cases}$

20. $\begin{cases} 2x + 3y \quad\;\; = 12 \\ \quad\;\; 3y + 2z = 12 \end{cases}$ **21.** $\begin{cases} x + \quad\;\; 3z = 6 \\ \quad 2y + \; z = 4 \end{cases}$

22. $\begin{cases} 2x + y + 2z = 4 \\ x + 2y + 3z = 6 \end{cases}$

23. $\begin{cases} 2x + 4y + z = 8 \\ x + 2y + 3z = 6 \end{cases}$

24. $\begin{cases} 2x - y + 2z = 4 \\ x + y + z = 3 \end{cases}$

25. $\begin{cases} 4x + 2y - z = 4 \\ x + y + z = 4 \end{cases}$

26. $\begin{cases} x = 3 \\ y = 6 \\ z = 2 \end{cases}$

27. $\begin{cases} y = 2 \\ x = 4 \\ z = 6 \end{cases}$

28. $\begin{cases} x = 3 \\ x + 3y = 6 \\ x + z = 2 \end{cases}$

29. $\begin{cases} y = 2 \\ x + y = 4 \\ 2x + z = 6 \end{cases}$

30. $\begin{cases} 2y + 5z = 10 \\ 5y + 2z = 10 \\ 5x + 10y + 2z = 20 \end{cases}$

31. $\begin{cases} 2x + 3z = 12 \\ 3x + 2z = 12 \\ 2x + 3y + 2z = 12 \end{cases}$

32. $\begin{cases} 2x - y + 2z = 6 \\ x + y + z = 6 \\ 3x + 3y + z = 6 \end{cases}$

33. $\begin{cases} x + y + z = 4 \\ 2x - y + 2z = 6 \\ x + y + 2z = 8 \end{cases}$

B

34. Find the intersection of the planes in Problem 22.

35. Find the intersection of the planes in Problem 23.

36. Find the intersection of the planes in Problem 24.

37. Find the intersection of the planes in Problem 28.

38. Find the intersection of the planes in Problem 29.

39. Find the intersection of the planes in Problem 30.

40. Find the intersection of the planes in Problem 31.

41. Find the intersection of the planes in Problem 32.

42. Find the intersection of the planes in Problem 33.

43. Find the intersection of the line $(t, 3t, 2t)$ and the plane $x + y + 2z = 8$.

44. Find the intersection of the line $(-2t, -t, 5t)$ and the plane $x + y + 2z = 14$.

45. Find the intersection of the line $(-4t, 2t, t)$ and the plane $x + y + 2z = 16$.

46. Find the intersection of the line $(1 + t, 2 - t, 2 + t)$ and the plane $x - 3y + 2z = 0$.

47. Find the intersection of the line $(1 + t, 2 - t, 2 + t)$ and the plane $x - 3y + 2z = -1$.

48. Find the intersection of the line $(1, 3, 2t)$ and the plane $x + y + 2z = 16$.

49. Sketch the system

$$\begin{cases} 2x + 3y + 4z = 6 \\ 2x + 3y + 4z = 12 \\ x + y = 3. \end{cases}$$

Show that this system of equations is inconsistent and explain why there is no point of intersection.

50. Sketch the system

$$\begin{cases} 2x + 4z = 8 \\ 4x + 2z = 8 \\ x + z = 3. \end{cases}$$

Show that this system of equations is inconsistent and explain why there is no point of intersection.

51. If h, k, and m are nonzero, then a line with parametric representation $(h + ut, k + vt, m + wt)$ can be written as

$$\frac{x - h}{u} = \frac{y - k}{v} = \frac{z - m}{w}.$$

These are the **symmetric equations** of the line. Find the symmetric equations of the line with the parametric representation $(1 + t, 6 + 4t, 5 - 2t)$.

C

52. The lines $(1 + t, 3 + \frac{2}{3}t, 1 - t)$ and $(-2t, 3 - t, t)$ are intersecting lines. Determine the point of intersection.

53. Determine the intersection of the lines $(1 + 4t, 8 - 6t, 2 + 4t)$ and $(7 - t, -1 + \frac{3}{2}t, -4 + t)$.

54. Solve the system of equations in Problem 32 by reducing the system to row echelon form. At each step, sketch the planes corresponding to the system at that time.

55. Solve the system of equations in Problem 33 by reducing the system to row echelon form. At each step, sketch the planes corresponding to the system at that time.

56. Consider the set of all points (x, y, z) that satisfy the equation $ax + by + cz = d$. (Assume that a, b and c are not all zero.) Given a fixed point (x_0, y_0, z_0) that is in this set, also consider two points $P(x_0 - a, y_0 - b, z_0 - c)$ and $Q(x_0 + a, y_0 + b, z_0 + c)$. Show that for any point $X(x, y, z)$ in the set,

$$d(P, X) = d(Q, X)$$

$$= \sqrt{(x_0 - x)^2 + (y_0 - y)^2 + (z_0 - z)^2 + a^2 + b^2 + c^2}.$$

Explain why this shows that this set of points forms a plane in coordinate space. (*Hint: See Example 4 of Section 12.1.*)

FUNCTIONS AND SURFACES S E C T I O N 12.3

In our discussion of functions in Section 2.2, we saw that a real function g accepts as input a real number x and produces as output another real number $g(x)$ (Figure 38). For the sake of our present discussion, we call such a function a **single-variable function,** or a **function of one variable,** since the value of the function is dependent on the value of one variable, x.

In applications, not all functions are dependent on only one variable. For example, the area of a rectangle is a function of both its width and its length. In chemistry, the pressure of a fixed mass of gas in a container is a function not only of the volume of the container but also of the temperature. In engineering, the strength of a beam is a function of its horizontal dimension and of the square of its vertical dimension. The cost of your automobile insurance depends on your age and your driving record, among other things.

These are all examples of **multivariable** functions (Figure 39). Their values are dependent on more than one variable. Much of calculus is devoted

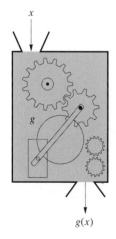

x

g

$g(x)$

FIGURE 38

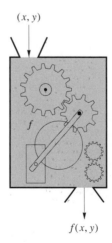

(x, y)

f

$f(x, y)$

FIGURE 39

to the study of these functions. Our present concern is with functions of two variables.

A Function of Two Variables

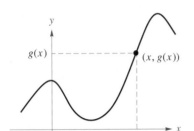

FIGURE 40 *The graph of a function g of one variable*

> A **function of two variables** is a rule f that assigns to each ordered pair (x, y) in a region of the xy-plane exactly one real number $f(x, y)$.

Just as the graph of a function g of one variable is the set of ordered pairs $(x, g(x))$ on the coordinate plane (Figure 40), the graph of a function f of two variables is the set of ordered triples $(x, y, f(x, y))$ in coordinate space. The domain of f is a region in the xy-plane (Figure 41).

Just as with functions of one variable, the domain for a function of two variables is the set of those inputs for which the function makes sense. The domains are usually represented as inequalities in two variables. These inequalities were discussed in Section 10.7 (you may wish to pause here and review this section).

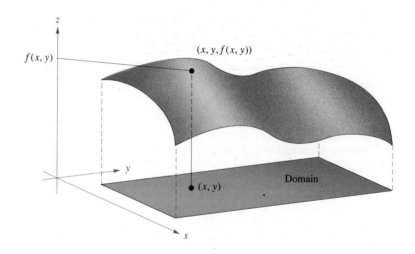

FIGURE 41 *The graph of a function f of two variables*

EXAMPLE 1   Given the function $f(x, y) = \sqrt{64 - 4x^2 - y^2}$.

a)   Determine the domain of f. Sketch the region on the xy-plane.

b)   Find $f(1, 2)$, $f(2, 1)$, and $f(5, 2)$.

SOLUTION

a)   Since the radicand in the function must be nonnegative, the domain of f is the set of all ordered pairs (x, y) such that

$$64 - 4x^2 - y^2 \geq 0$$

or

$$4x^2 + y^2 \le 64.$$

The region in the xy-plane that represents the domain of the function is the ellipse and its interior as shown in Figure 42.

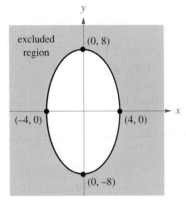

FIGURE 42

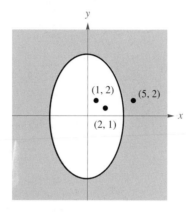

FIGURE 43

b)
$$f(1, 2) = \sqrt{64 - 4(1)^2 - (2)^2} = \sqrt{56} = 2\sqrt{14}$$
$$f(2, 1) = \sqrt{64 - 4(2)^2 - (1)^2} = \sqrt{47}.$$

The ordered pair $(5, 2)$ is not in the domain. It lies outside the ellipse sketched in part (a) (Figure 43).

In general, sketching the graphs of these functions of two variables is best left to computers and calculators capable of such a task. Figure 44 is a computer-generated graph of the function $f(x, y) = x^2 - y^2$. This picture

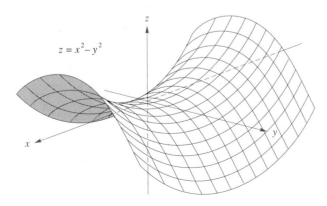

FIGURE 44

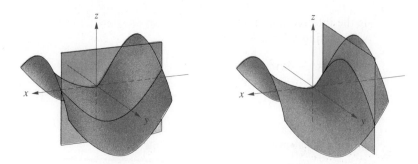

FIGURE 45

of the graph is made up of individual plane curves, each formed by the intersection of the surface and a plane parallel to the xz-plane or to the yz-plane (Figure 45). These are the **traces** of the surface in these planes. This type of perspective is called a **mesh perspective.**

EXAMPLE 2   The graph of the function $F(x, y) = x^3 - 4xy^2$ is shown in Figure 46. Find and describe the family of curves of the traces in the planes parallel to **(a)** the yz-plane, **(b)** the xz-plane

This surface is nicknamed "monkeysaddle." Imagine a monkey sitting astride this surface; there is a place for its tail and each leg.

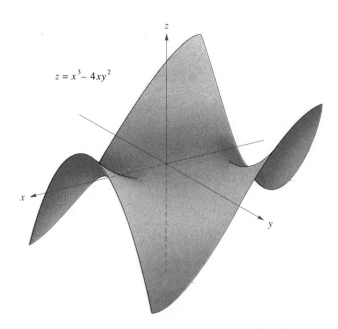

$$z = x^3 - 4xy^2$$

FIGURE 46

SOLUTION

a)   The set of planes that are parallel to the yz-plane are those of the form $x = k$, for any constant k. We find the plane curve that is the intersection of any of these planes and the surface by substituting k for x in $z = F(x, y)$:

$z = k^3 - 4ky^2 \Rightarrow$ parabola opening downward for positive values of k and opening upward for negative values of k (Figure 47).

b) The traces parallel to the xz-plane are of the form $y = k$. We proceed as above:

$z = x^3 - 4xk^2 \Rightarrow$ graph of cubic polynomial with two turning points (except for $k = 0$) (Figure 48).

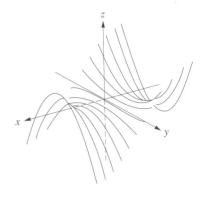

FIGURE 47

There is another way for people with marginal artistic talent to get a grasp on the shapes and extents of these graphs. Suppose that we sketch the horizontal traces (those in the planes $z = k$ for constants k) on the surface $z = f(x, y)$ and project them down into the domain on the xy-plane. These are called the **level curves** of f. These are curves in the domain of a function over which the value of the function remains constant.

This is the familiar scheme behind topographic maps used by hikers and backpackers to negotiate wilderness areas. On a topographic map, lines of constant elevation in the three-dimensional landscape are represented by curves on the two dimensional map. Each curve is labeled with its corresponding elevation (Figure 49).

The level curves of a function give a two-dimensional picture of the surface associated with the function. With a little practice, you will be able to draw a family of level curves for a given function and be able to see the surface in your mind.

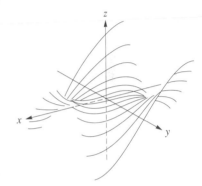

FIGURE 48

FIGURE 49

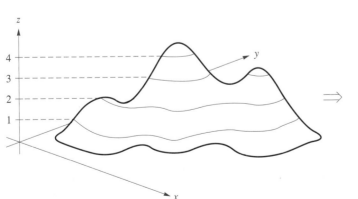

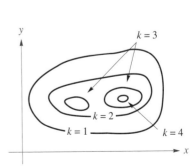

EXAMPLE 3   Consider the function $f(x, y) = x^2 - y^2$ (its graph is shown in Figure 46). Sketch the level curves $f(x, y) = k$ for $k = -4, -1, 0, 1, 4$.

SOLUTION   In the cases of $k = \pm 1$ or ± 4, the level curves are in the family of hyperbolas with $y = x$ or $y = -x$ as asymptotes. The points corresponding to $x^2 - y^2 = 0$ are exactly the asymptotes of this family since if $x^2 - y^2 = 0$, then either $y = x$ or $y = -x$ (Figure 50).

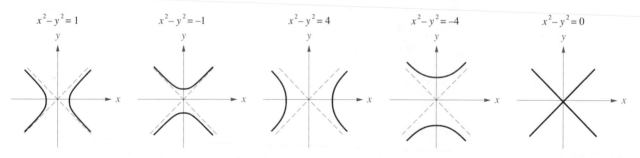

FIGURE 50

In Figure 51 these curves are superimposed on the same plane. Pause here and compare this "map" of the function with its graph in Figure 44.

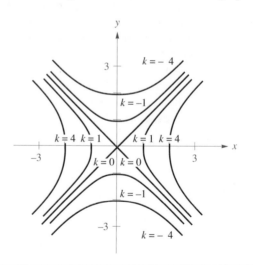

FIGURE 51   *Level curves of* $f(x, y) = x^2 - y^2$

The use of level curves is widespread in science. In meteorology, the level curves of a temperature function are called *isotherms,* and the level curves of atmospheric pressure are called *isobars* (you may have noticed these on the nightly news or in the newspaper). In physics, the level curves in an electrostatic field are called *lines of equipotential.* In geology, the level curves on the gravitational force over the surface of the earth are called *lines of geoidal height.*

EXAMPLE 4 Sketch the domain and the level curves ($k = -2, -1, 0, 1, 2,$ $3, 4$) of the function $L(x, y) = \log_2(16 - x^2 - y^2)$. (Its graph is shown in Figure 52.)

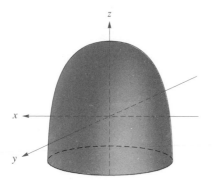

FIGURE 52
$z = \log_2(16 - x^2 - y^2)$

SOLUTION The argument of a logarithmic function must be positive, so

$$16 - x^2 - y^2 > 0$$

or

$$x^2 + y^2 < 16.$$

The domain is the interior of the circle centered at the origin with radius 4. The points on the circle are not in the domain.

The level curve for $k = 3$ is a circle of radius $2\sqrt{2}$ with center at the origin:

$$k = 3 \Rightarrow \log_2(16 - x^2 - y^2) = 3$$
$$\Rightarrow 16 - x^2 - y^2 = 2^3$$
$$\Rightarrow x^2 + y^2 = 8.$$

The other level curves, also circles with center at the origin, can be derived in a similar fashion. The results are given with approximations (to the nearest 0.01) of the radii; you should verify them:

$$k = -2 \Rightarrow x^2 + y^2 = 15\tfrac{3}{4} \Rightarrow r \approx 3.97$$
$$k = -1 \Rightarrow x^2 + y^2 = 15\tfrac{1}{2} \Rightarrow r \approx 3.94$$
$$k = 0 \quad \Rightarrow x^2 + y^2 = 15 \quad \Rightarrow r \approx 3.87$$
$$k = 1 \quad \Rightarrow x^2 + y^2 = 14 \quad \Rightarrow r \approx 3.74$$
$$k = 2 \quad \Rightarrow x^2 + y^2 = 12 \quad \Rightarrow r \approx 3.46$$
$$k = 3 \quad \Rightarrow x^2 + y^2 = 8 \quad \Rightarrow r \approx 2.83$$
$$k = 4 \quad \Rightarrow x^2 + y^2 = 0 \quad \Rightarrow r = 0$$

In general, the level curves are members of the family of curves $x^2 + y^2 = 16 - 2^k$. Notice that as $k \to -\infty$, the radius tends toward 4, the radius of the domain. Compare the sketch of these curves (Figure 53) with the graph of the function in Figure 52.

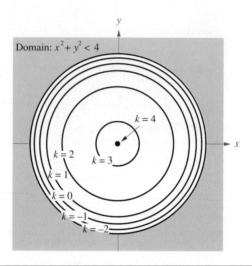

FIGURE 53

Cylinders

Another type of surface arises frequently in calculus and its applications. Consider a line parallel to a fixed line. Suppose that this line is allowed to move along a given fixed curve in a plane, all the time remaining parallel to the fixed line (Figure 54).

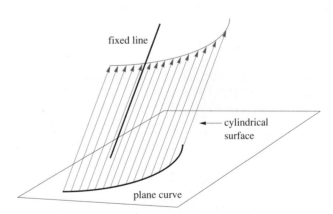

FIGURE 54

FIGURE 55 *The surface that is usually called a cylinder is a special case in which the plane curve is a circle*

The resulting surface that is swept out by the line is a **cylinder** (Figure 55). In practice, the cylinders of interest to us are those in which the role of the fixed line is played by one of the coordinate axes. These cylinders all share a characteristic: In their equations, one of the variables x, y, or z is conspicuous by its absence.

Equations of Cylinders

An equation in x and y only has a cylindrical graph that is parallel to the z-axis.

An equation in x and z only has a cylindrical graph that is parallel to the y-axis.

An equation in y and z only has a cylindrical graph that is parallel to the x-axis.

EXAMPLE 5 Sketch the graph of $z = 4 - x^2$.

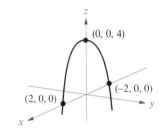

FIGURE 56

SOLUTION The variable y is missing in the equation. The surface is parallel to the y-axis. The trace in the xz-plane is a parabola (Figure 56). If we allow a line parallel to the y-axis to move along this parabola, it sweeps out the surface, a **parabolic cylinder,** shown in Figure 57.

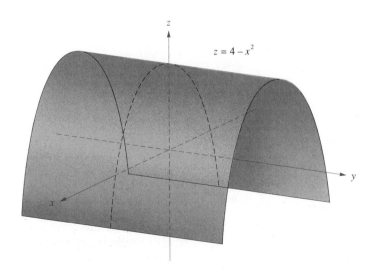

FIGURE 57

EXAMPLE 6   Sketch the graph of

$$\frac{(x-4)^2}{16} + \frac{(y-2)^2}{4} = 1.$$

SOLUTION   The surface is parallel to the z-axis. The trace in the xy-plane is an ellipse with center (4, 2) (Figures 58 and 59).

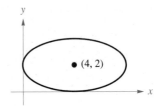

FIGURE 58

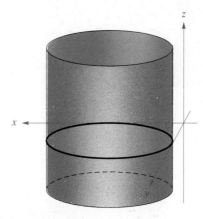

FIGURE 59

Quadric Surfaces

In the discussion of conic sections in Chapter 8, we saw that a curve represented on the coordinate plane by a second-degree equation of the form

$$Ax^2 + Bxy + Cy^2 + Dx + Ey + F = 0$$

is a circle, ellipse, parabola, hyperbola, or a degenerate conic section such as a point or a line. Without the Bxy term, it represents a conic section with axes of symmetry that are parallel to the coordinate axes.

Catalog of Quadric Surfaces

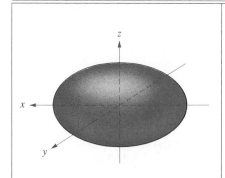

Ellipsoid

$$\frac{x^2}{a^2} + \frac{y^2}{b^2} + \frac{z^2}{c^2} = 1$$

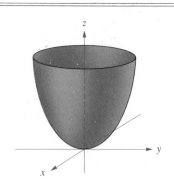

Paraboloid

$$z = \frac{x^2}{a^2} + \frac{y^2}{b^2}$$

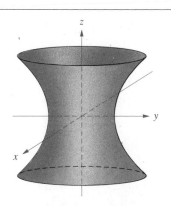

Hyperboloid of one sheet

$$\frac{x^2}{a^2} + \frac{y^2}{b^2} - \frac{z^2}{c^2} = 1$$

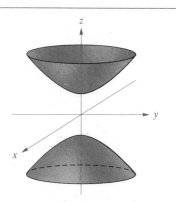

Hyperboloid of two sheets

$$\frac{x^2}{a^2} + \frac{y^2}{b^2} - \frac{z^2}{c^2} = -1$$

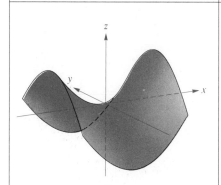

Hyperbolic Paraboloid

$$z = \frac{x^2}{a^2} - \frac{y^2}{b^2}$$

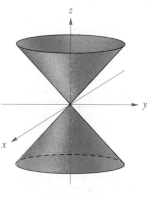

Elliptic Cone

$$z^2 = \frac{x^2}{a^2} + \frac{y^2}{b^2}$$

In three dimensions, the general second-degree equation takes the form

$$Ax^2 + By^2 + Cz^2 + Dxy + Exz + Fyz + Gx + Hy + Jz + K = 0.$$

The family of surfaces that these equations represent are called **quadric surfaces.** In the box is a catalog of six quadric surfaces. What distinguishes these surfaces is that their traces are all conic sections. Their names are derived from the conic sections of their traces. For example, the surface in Figure 45 is a hyperbolic paraboloid, since the horizontal traces (the level curves) are hyperbolas, and both sets of vertical traces are parabolas. The next example shows how quadric surfaces can be identified and sketched from the general equation.

EXAMPLE 7   Identify and sketch the quadric surface $x^2 + y^2 - 4z = 0$.

SOLUTION First, we consider the traces in the planes $x = 0$, $y = 0$, and $z = 0$:

$$x = 0 \Rightarrow \text{trace is } y^2 - 4z = 0 \text{ or } z = \tfrac{1}{4}y^2.$$

$$y = 0 \Rightarrow \text{trace is } x^2 - 4z = 0 \text{ or } z = \tfrac{1}{4}x^2.$$

$$z = 0 \Rightarrow \text{trace is } x^2 + y^2 = 0; \text{ this a degenerate circle.}$$
$$\text{Its graph is the point } (0, 0).$$

To determine the other horizontal traces, we let $z = k$:

$$z = k \Rightarrow \text{trace is } x^2 + y^2 - 4k = 0 \text{ or } x^2 + y^2 = 4k.$$

These are circles that increase in radius as k increases. The vertical traces in the planes parallel to both the xz-plane and the yz-plane are parabolas (you can verify this). This is a paraboloid (Figure 60).

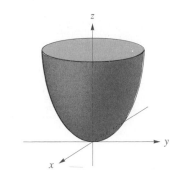

FIGURE 60

EXERCISE SET 12.3

A

In Problems 1 through 12, refer to f, g, or h and evaluate the expression. If the expression is meaningless, state why.

$f(x, y) = 2x - y^3 \quad g(x, y) = y \cos x \quad h(x, y) = \log_3(x^2 + y)$

1. **a)** $f(4, 1)$
    b) $f(1, 4)$

2. **a)** $g(0, 3)$
    b) $g(3, 0)$

3. **a)** $h(0, 3)$
    b) $h(3, 0)$

4. **a)** $f(\tfrac{1}{2}, 3)$
    b) $f(3, \tfrac{1}{2})$

5. **a)** $g(0, 0)$
    b) $h(0, 0)$

6. **a)** $f(-2, -9)$
    b) $h(-2, -9)$

7. **a)** $f(0, 2a)$
    b) $2f(0, a)$

8. **a)** $g(4t, \pi/4)$
    b) $4g(t, \pi/4)$

9. a) $h(3n, 18n^2)$

  b) $3[h(n, 2n^2)]$

11. a) $g(t^2, \pi/3)$

  b) $[g(t, \pi/3)]^2$

10. a) $f(a^3, a)$

  b) $f(a, a^3)$

12. a) $h(n, n^2)$

  b) $2[h(\sqrt{n}, n)]$

In Problems 13 through 18, determine the domain of the function. Sketch the region on the xy-plane. Use solid lines for portions of the boundary included in the domain and dashed lines for portions not included in the boundary.

13. $F(x, y) = \sqrt{3x - y}$

14. $H(x, y) = \dfrac{12}{\sqrt{y - x^2}}$

15. $\phi(x, y) = \sin^{-1}(x^2 + y^2)$

16. $\alpha(x, y) = \ln(x^2 - y^2)$

17. $R(x, y) = \sqrt{\dfrac{x^2 + y^2}{x^2 - y^2}}$

18. $D(x, y) = \sqrt{\dfrac{x^2 - y^2}{x^2 + y^2}}$

In Problems 19 through 24, the graph of one of the functions I through VI is sketched. Match each graph to the corresponding equation.

  I. $z = (\sin x)(\sin y)$

  II. $z = 4 - x^2 - y^2$

  III. $z = \sin(x^2 + y^2)$

  IV. $z = y^3 \cos x$

  V. $z = \dfrac{-1}{x^2 + y^2}$

  VI. $z = e^{-(x^2 + 2y^2)}$

19.

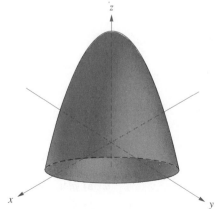

20.

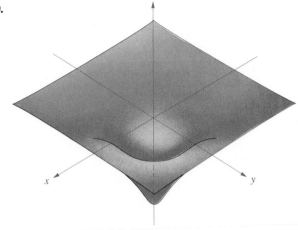

21.

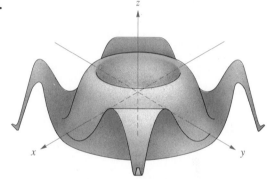

22.

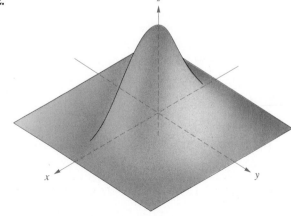

23.

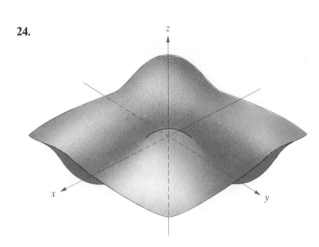

24.

B

In Problems 25 through 30, sketch the level curves $f(x, y) = k$ for the given values of k.

25. $f(x, y) = 2x + 3y, k = -2, -1, 0, 1, 2$

26. $f(x, y) = x - y^2, k = -4, -1, 0, 1, 4$

27. $f(x, y) = \dfrac{x}{y}, k = -8, -1 - \frac{1}{8}, 0, \frac{1}{8}, 1, 8$

28. $f(x, y) = \dfrac{y^2}{x}, k = -4, -1, -\frac{1}{4}, 0, \frac{1}{4}, 1, 2$

29. $f(x, y) = 2 \sin xy, k = -2, -1, 0, 1, 2$

30. $f(x, y) = 4 \cos^2(x + y), k = -4, -1, 0, 1, 4$

In Problems 31 through 36, sketch the cylindrical graphs of the equation given.

31. $z^2 = y$ **32.** $x = y^2$

33. $y = \sqrt{16 - x^2}$ **34.** $y = \sqrt{2 - z}$

35. $z = \sin x$ **36.** $y = \cos x$

In Problems 37 through 45, identify and sketch the quadratic surface represented by the equation given.

37. $x^2 + 9y^2 + 9z^2 = 36$ **38.** $9x^2 + 4y^2 + 4z^2 = 36$

39. $x - z^2 - y^2 = 0$ **40.** $z - x^2 - 9y^2 = 0$

41. $x^2 - y^2 + z^2 = 1$ **42.** $x^2 - y^2 - 4z^2 = 4$

43. $x^2 = 4z^2 + y^2$ **44.** $y^2 = x^2 + z^2$

45. $x - z^2 - y^2 = 1$ (Hint: See Problem 39.)

46. Given the function $f(x, y) = x^2 + 2xy - y^2$, simplify the expressions

$$\frac{f(x + \Delta x, y) - f(x, y)}{\Delta x} \quad \text{and} \quad \frac{f(x, y + \Delta y) - f(x, y)}{\Delta y}.$$

47. Given the function $g(x, y) = 2x/y$, simplify the expressions

$$\frac{g(x + \Delta x, y) - g(x, y)}{\Delta x} \quad \text{and} \quad \frac{g(x, y + \Delta y) - g(x, y)}{\Delta y}.$$

C

48. The graph of $4x^2 + 16y^2 + z^2 = 64$ is an ellipsoid. Show that the horizontal traces are ellipses with the same eccentricity. What is the eccentricity?

49. Sketch the family of surfaces $x^2 + y^2 - z^2 = k$ for $k = -4, -1, 0, 1, 4$. What is the special role being played by the surface for which $k = 0$?

50. Sketch the paraboloid $z = 4x^2 + 16y^2$. The horizontal traces are ellipses with foci on a curve in the xz-plane. Find an equation that describes this curve in the xz-plane in terms of x and z.

51. Sketch the paraboloid $z = 4x^2 + 16y^2$. The traces in the planes $y = k$ are parabolas with foci on a curve in the yz-plane. Find an equation that describes this curve in the yz-plane in terms of y and z.

Mathematics and Computers

Ever since the invention of the first mechanical adding machine by Blaise Pascal in 1642, there has been a unique symbiotic relation between mathematics and computing machinery. The development of each in some way has depended on the development of the other.

Throughout the 17th and 18th centuries, advancements in manufacturing technology allowed the construction of many machines like Pascal's that accomplished certain computational tasks. In the 1840s, Charles Babbage, an English mathematician, designed the first general-purpose calculating machine. What made this machine significant is that it could store a set of instructions for a specific task. Augusta Ada Lovelace (the daughter of the English poet Lord Byron) was instrumental in the development of this machine, and she wrote the first set of instructions for it. Because of this, she is considered by many to be the first true computer programmer (in recognition of her contributions, a programming language developed for the United States government in the late seventies was named Ada). Unfortunately, the technology of the day was unable to meet Babbage's exacting standards; the machine was never completed.

It was not until World War II that the art and science of computing came into its own. The Germans used a sophisticated code, the Enigma Code, for their secret communications. The success of the Allied forces depended upon breaking this code. A brilliant young English mathematician, Alan Turing, designed an ingenious machine that accomplished this and turned the tide of the war. His work during this period set the foundation for modern computing. Even though Turing was a genuine war hero, he was hounded and ostracized because of his unconventional lifestyle. He committed suicide at the age of 48.

Nowadays, computers are becoming cheaper, faster, and more powerful. They have become an integral part of modern life. Their impact on classroom education will prove to be as significant as that of pencil and paper or printed material. Computing machinery has lifted the burden of sketching graphs (such as those in this chapter) and performing tedious calculations from the shoulders of mathematicians, allowing them to pursue more important investigations.

The use of these new tools in mathematical research is not without controversy, however. This has been most evident in the recent investigations of the four-color conjecture.

The four-color conjecture is simply stated: Any map can be colored with only four colors in such a way that no two bordering states share the same color. Now, it is obvious that more than three colors are required (examine Kentucky and its neighboring states on a map of the United States to see why). Also, no one has yet devised a map that cannot be colored with four colors. This property of maps was recognized as early as 1852, but its proof has tantalized and defeated great mathematicians for more than one hundred years. In their investigations, however, these mathematicians developed a body of knowledge that has been used to solve applied problems in fields such as communications, transportation, and the design of integrated chips.

In 1976, Kenneth Appel and Wolfgang Haken used a computer to investigate the four-color problem. They were able to reduce the problem to a large set of possibilities. Using thousands of hours of computer time, they were able to verify each of them. This work, in turn, has been verified by other computers, being far too long to be verified by humans. However, it is now generally accepted that the conjecture is true.

Is this a proof? Many argue that the body of mathematical knowledge has not been advanced by this work. If we agree that a proof is an argument that convinces the human reader of the truth of the conjecture, then the answer is no. The conjecture still waits to be proved in the traditional way by a future mathematician (perhaps you!).

The purpose of computing is insight, not numbers.

R. W. HAMMING.

MISCELLANEOUS EXERCISES

In Problems 1 through 3, find the distance between the points given.

1. $(2, 2, 1)$ and $(5, 4, 2)$

2. $(4, -3, 3)$ and $(0, 1, 6)$

3. $(8, 3, -1)$ and $(5, 3, 3)$

4. Find the distance from the point $(5, 4, 2)$ to the plane $y = -6$.

In Problems 5 through 10, sketch the graph of the set of all points (x, y, z) satisfying the equation(s).

5. $x = 4$

6. $z = -2$

7. $y = 1$

8. $x = 3$ and $y = 4$

9. $z = 0$ and $y = x$

10. $x = -5$ and $z = 2$

In Problems 11 through 16, determine the center and the radius of the sphere described. If no such sphere exists, explain why.

11. $(x - 5)^2 + (y + 1)^2 + (z - 4)^2 = 49$

12. $(x - 2)^2 + y^2 + z^2 = 10$

13. $x^2 + y^2 - 8y + z^2 = 20$

14. $x^2 + y^2 + z^2 - 8x + 2y - 6z = -17$

15. $x^2 + y^2 + 4y + z^2 - 2z + 8 = 0$

16. $x^2 + y^2 + z^2 - 12x + 52 = 0$

17. Find the equation of the sphere tangent to the xz-plane at $(4, 0, 7)$ and the yz-plane at $(0, 4, 7)$.

18. Determine if the triangle with vertices $(9, 5, 2)$, $(4, 0, 2)$, and $(4, 5, 7)$ is an equilateral, isosceles, or scalene triangle. *(Hint: See Exercise 45 of Section 2.1.)*

In Problems 19 through 24, sketch the plane in the first octant by determining the intercepts, sketching the traces, and shading.

19. $3y + 2z = 12$

20. $x + 6y = 6$

21. $2x + y + 3z = 18$

22. $x + 2y + 2z = 5$

23. $2x - 4y + z = 8$

24. $-x + y - 3z = 0$

In Problems 25 through 32, sketch the planes in the first octant. You may want to try two or more views of the axes.

25. $\begin{cases} y = 5 \\ z = 3 \end{cases}$

26. $\begin{cases} z = 5 \\ x = 5 \end{cases}$

27. $\begin{cases} 2x + y = 8 \\ y = 4 \end{cases}$

28. $\begin{cases} x + 3z = 9 \\ x = 5 \end{cases}$

29. $\begin{cases} x + 2y = 10 \\ 3x + 4z = 12 \end{cases}$

30. $\begin{cases} 3x + y = 12 \\ x + 2z = 8 \end{cases}$

31. $\begin{cases} z = 3 \\ x + 3y = 9 \\ x + z = 5 \end{cases}$

32. $\begin{cases} y = 3 \\ x + y = 7 \\ 2x + z = 11 \end{cases}$

33. Find the intersection of the planes in Problem 27.

34. Find the intersection of the planes in Problem 29.

35. Find the intersection of the line $(-2t, t, 5t)$ and the plane $2x + 3y + z = 12$.

36. Find the intersection of the line $(3 + t, -5t, 1 - t)$ and the plane $3x + 2y - z = 2$.

37. Determine the point of intersection of the line $(1 + 2t, 1 - t, 6 - t)$ and the line $(10 + t, 3t, -1 - 3t)$; if the lines do not intersect, so state.

38. Determine the point of intersection of the line $(5 + 3t, 2 - t, 2 + t)$ and the line $(1 + t, 3t, 1 - 2t)$; if the lines do not intersect, so state.

In Problems 39 through 42, determine the domain of the function. Sketch the region on the xy-plane. Use solid lines for portions of the boundary included in the domain and dashed lines for portions not included in the boundary.

39. $f(x, y) = \sqrt{x - 2y + 3}$

40. $G(x, y) = \ln(4 - x^2 - 4y^2)$

41. $\phi(x, y) = \cos^{-1}(x + y)$

42. $h(x, y) = \sqrt{\dfrac{x^2}{y - x^2}}$

In Problems 43 through 45, sketch the level curves $f(x, y) = k$ for the given values of k.

43. $f(x, y) = y - |x|$, $\quad k = -1, 0, 1, 2$

44. $f(x, y) = \dfrac{y + 1}{x}$, $\quad k = -2, -1, 0, 1$

45. $f(x, y) = x - \sin^{-1} y$, $\quad k = 0, \dfrac{\pi}{2}, \pi, \dfrac{3\pi}{2}$

In Problems 46 through 49, identify and sketch the quadric surface represented by the given equation.

46. $x^2 + 4y^2 - z = 0$

47. $9x^2 + 36y^2 + 16z^2 = 144$

48. $x^2 - 4y^2 - 4z^2 = 16$

49. $x^2 + y^2 - z^2 = 9$

ANSWERS TO ODD-NUMBERED PROBLEMS

Section 1.1

1. integers: $-\frac{12}{4}$; rational numbers: $-\frac{12}{4}$ and $\frac{5}{3}$; irrational numbers: $\sqrt{7}$ and $\sqrt{19}$; increasing order: $-\frac{12}{4}$, $\frac{5}{3}$, $\sqrt{7}$, and $\sqrt{19}$
3. integers: 0; rational numbers: 0, 3.14, and $\sqrt{\frac{4}{9}}$; irrational numbers: π; increasing order: 0, $\sqrt{\frac{4}{9}}$, 3.14, π **5.** integers: $\sqrt{4}$, $\sqrt[3]{-8}$; rational numbers: $-\frac{3}{2}$, $\sqrt{4}$, $\sqrt[3]{-8}$; irrational numbers: $\pi/2$; increasing order: $\sqrt[3]{-8}$, $-\frac{3}{2}$, $\pi/2$, $\sqrt{4}$ **7.** $(4, \infty)$
9. $(-2, 4)$ **11.** $(-4, 2]$
13. $(-\infty, -2]$ or $(2, \infty)$ **15.** $(-5, 2]$
17. $(-3, 1]$ or $(4, \infty)$ **19.** $x < -2$
21. $4 \le x \le 9$ **23.** $-4 \le x < 0$
25. $x < -2$ or $x > 2$
27. $-1 \le x \le 6$ **29.** $-5 < x < -2$ or $x > 0$ **31.** $4 - \sqrt{10}$
33. $6 - \sqrt{13}$ **35.** $\sqrt{10} - \pi$
37. x^2 **39.** $n^3 - 5$ **41.** 1
43.

$\begin{array}{ccccccccccc} -5 & -4 & -3 & -2 & -1 & 0 & 1 & 2 & 3 & 4 & 5 \end{array}$

45.

$\begin{array}{cccccccccccc} -6 & -5 & -4 & -3 & -2 & -1 & 0 & 1 & 2 & 3 & 4 & 5 & 6 \end{array}$

47.

$\begin{array}{ccccccccccc} -3 & -2 & -1 & 0 & 1 & 2 & 3 & 4 & 5 & 6 & 7 \end{array}$

49.

$\begin{array}{cccccc} -1 & 0 & 1 & 2 & 3 & 4 \end{array}$

51.

$\begin{array}{cccc} 0 & 3 & 2\pi \end{array}$

53.

$\begin{array}{ccc} -1.5 & -1 & -0.5 \end{array}$

55. $2 - \sqrt{7} \approx -0.6458$,
$\sqrt{5} + \sqrt{3} \approx 3.9681$, $2 + \sqrt{7} \approx 4.6458$

---

57. $(\frac{1}{2})(\sqrt{2} + \sqrt{6}) = \sqrt{2 + \sqrt{3}} \approx 1.9319$
59. $\dfrac{\sqrt{5} + \sqrt{3}}{\sqrt{5} - \sqrt{3}} = 4 + \sqrt{15} = \dfrac{1}{4 - \sqrt{15}}$
≈ 7.8730 **61.** $|x - 2| < 5$
63. a) Answers vary. Possible answers are: 2.31, 2.32, 2.33, 2.34, 2.35, 2.36
b) Answers vary.
67. a) Answers vary. Possible answers are: $3\pi/4$, $\frac{1}{2}(\sqrt{5} + \sqrt{6})$, $\frac{1}{2}\sqrt{22}$, $\frac{5}{3}\sqrt{2}$, $\sqrt{11} - 1$, $\sqrt[3]{13}$ **b)** Answers vary.

Section 1.2

1. a) $\dfrac{1}{x^{2/3}}$ **b)** $2/y$ **c)** $\dfrac{1}{2y}$
d) $\dfrac{x^5}{y^2}$
3. a) $x^{2/5}$ **b)** $x^{-3/2}$ **c)** $x^{-4/7}$
d) $x^{5/2}$
5. a) $\sqrt[5]{x^2}$ **b)** $\sqrt[7]{x^{-3}}$ **c)** $\sqrt{x^{-3}}$
d) $\sqrt[9]{x^2}$
7. $2x^4$ **9.** $\dfrac{25x^5}{4}$ **11.** $3x^6y^4$
13. $\dfrac{1}{5x^2}$ **15.** $\dfrac{x^9}{y^6}$ **17.** $\dfrac{x^{10}y^5}{z^{15}}$
19. $xy\sqrt[3]{x^2y^2}$ **21.** $2x^4$ **23.** $\sqrt[12]{x}$
25. $20x - 7$
27. $3x^3y^2 - 5x^3y - x^2y^3 + 2xy^2$
29. $6x^2 - x - 1$
31. $8x^3 + 12x^2y + 6xy^2 + y^3$
33. $x^3 - x$ **35.** $x^3 + 125$
37. $2x^2 + 10x + 8$
39. $48x^5 + 48x^3 + 12x$
41. $x^4 - 16$
43. $112x^6 - 48x^5 + 5x^4$
45. $(3x + 1)(x - 1)$
47. $x(x + 2)(x - 5)$
49. $(5x + 6)(5x - 6)$
51. $2(x - 2)(x^2 + 2x + 4)$
53. $x(x - 3)(x^2 + 3x + 9)$
55. $(x - 1)(x^2 + 4)$
57. $x + 4\sqrt{x} + 4$ **59.** y
61. $x^2 + 2x - 1$ **63.** $x^3 + 2x^2 + x$

---

65. $x^6 + y^6$ **67.** $12x^2 + 16$
69. $x^{4/3} + 3x^{2/3}$
71. $x^{8/3} - 2x^{5/3} + x^{2/3}$ **73.** $x^7 - 1$
75. $x^2(x + 3y)^2(x - 3y)^2$
77. $(2x + 1)(x^2 + 4)$
79. $(x^2 - 3)(x^2 + 1)$
81. $(x^2 + 4y^4)(x + 2y^2)(x - 2y^2)$
83. $(x - 4)^3$ **85.** $(x + 6)(x + 1)$
87. $x^{2/3}(2x^2 - 5x + 2)$
89. $x^{1/2} - 2x^{2/3} + 4x^{-1} - 2x^{-2}$
91. a) $(x + \sqrt{6})(x - \sqrt{6})$
b) $(\sqrt{2}x + 3)(\sqrt{2}x - 3)$
c) $3(x + \sqrt{2})(x - \sqrt{2})$
d) $x^2 + 7$
95. a) $(x^2 + y^2)^2 - x^2y^2$
b) $(x^2 + xy + y^2)(x^2 - xy + y^2)$
97. $(x^2 + \sqrt{2}xy + 2y^2)(x^2 - \sqrt{2}xy + 2y^2)$

Section 1.3

1. x^2 **3.** $\dfrac{x - 3}{x + 4}$ **5.** $\dfrac{2x(x + 2)}{x^2 + 2}$
7. $\dfrac{-x - 3}{6x + 7}$ **9.** $\dfrac{x + y}{x - y}$ **11.** $\dfrac{x + 7}{x - 7}$
13. $\dfrac{x + 1}{x - 1}$ **15.** $\dfrac{x + 1}{x + 2}$
17. $\dfrac{7x - 2}{(x + 1)(x - 2)}$ **19.** $\dfrac{-x - 25}{(x + 5)(x - 5)}$
21. $\dfrac{5x}{(x + 4)(x - 1)(x - 6)}$
23. $\dfrac{10x^2 + 7x - 27}{3(x + 1)(2x + 5)}$ **25.** $\dfrac{-1}{x(x + 2)}$
27. $\dfrac{-1}{x + 1}$ **29.** $\dfrac{2}{(x + 1)(x + h + 1)}$
31. $\dfrac{2x - h}{x^2(x - h)^2}$ **33.** $\dfrac{1 - x}{x}$
35. $\dfrac{x^2y^2}{x^4 - y^4}$ **37.** $x^3 - 12x - 16$
39. $x - x^3$ **41.** $\dfrac{x - 4}{(x + 2)(x - 6)}$

43. $\dfrac{\sqrt{x}(x^2 + 1)}{x}$ **45.** $\dfrac{\sqrt{1 - x^2}}{1 - x^2}$

47. $\dfrac{x^2 - 1}{x^2 + 1}$

49. $\dfrac{\sqrt{3x + 5}\sqrt{x - 4}(6x - 7)}{2(x - 4)(3x + 5)}$

51. $\dfrac{2\sqrt{x^2 + 2}}{(x^2 + 2)^2}$ **53.** $\dfrac{9x^2 - 7x^4}{3(1 - x^2)^{4/3}}$

55. $\dfrac{\sqrt{x}(2 - 3x^2)}{2x(2 + x^2)^2}$ **57.** $\dfrac{3}{x^4} + 2$

59. $\sqrt{1 + x^2}$ **61.** $\dfrac{\sqrt{x} - 1}{x - 1}$

63. $\sqrt{x} + \sqrt{6}$ **65.** $\dfrac{2x\sqrt{1 - x^2} + 1}{2x^2 - 1}$

67. $1 - \dfrac{6}{x + 2}$ **69.** $x - 1 + \dfrac{11}{x + 4}$

71. $3x^3 - 3x^2 + x - 1 - \dfrac{4}{x + 1}$

73. $1 - \dfrac{6x + 1}{x^2 - 2}$ **75.** $2x + \dfrac{4}{2x^2 + 1}$

77. $2x^2 + x + 3 + \dfrac{3}{x + 2}$

79. a) $\frac{11}{5}, \frac{21}{11}, \frac{43}{21}, \frac{85}{43}$ **b)** 2   **81.** π

Section 1.4

1. $x = 24$ **3.** $x = \frac{5}{2}$ **5.** $x = -3$

7. $b = \dfrac{2A - ha}{h}$ **9.** $m = \dfrac{FR}{v^2}$

11. $R = \dfrac{R_1 R_2}{R_1 + R_2}$ **13.** $x = -\frac{4}{3}$ or $\frac{3}{4}$

15. $x = -\frac{3}{2}$ or $\frac{1}{5}$ **17.** $x = -\frac{1}{2}, 0$, or 1

19. $x = \pm\sqrt{5}$ **21.** $x = 4 \pm 2\sqrt{3}$

23. $x = -4 \pm \sqrt{2}/2$

25. $x = \dfrac{5 \pm \sqrt{361}}{4} = 6$ or $-\frac{7}{2}$

27. $x = \dfrac{5 \pm \sqrt{89}}{4}$

29. no real solutions

31. $(x - 3)^2 = 9 \Rightarrow x = 0$ or 6

33. $(x - 1)^2 = -3 \Rightarrow$ no real solutions

35. $(x + \frac{3}{4})^2 = \frac{45}{16} \Rightarrow x = \dfrac{-3 \pm 3\sqrt{5}}{4}$

37. $x = 4$ or -3 **39.** $x = 4$ or 20

41. $x = -1$ or -3 **43.** $x = 3$

45. $-2 + 2\sqrt[3]{3}$ **47.** $x = 2$

49. $x = \dfrac{1 \pm \sqrt{5}}{2}$ **51.** no solution

53. $x = -8, 0$, or 8   **55.** $x = \frac{7}{3}$ or $\frac{5}{3}$

57. $x = \pm\sqrt{3}$ **59.** $x = 1$ or 262144

61. $x = \frac{8}{3}$ or $\frac{13}{4}$ **63.** $x = 1$

65. $x = 4$ **67.** $x = 4$

69. no real roots   **71.** $x = 0$

75. a) $\dfrac{1 + \sqrt{5}}{2}$ **b)** $\dfrac{1 + \sqrt{5}}{2}$

77. 1.7693   **79.** 2.3307

Section 1.5

1. $x > -17$ **3.** $x \geq 13$ **5.** $x \leq \frac{25}{7}$

7. $2 < x < 5$ **9.** $14 \leq x \leq 64$

11. $\frac{1}{5} < x < 2$ **13.** $-4 \leq x \leq 2$

15. $-5 \leq x \leq 2$ **17.** $-2 \leq x \leq 3$

19. $x < -1$ or $x > 3$ **21.** $x \leq -\frac{5}{3}$ or $x \geq \frac{1}{3}$ **23.** $-\frac{4}{3} \leq x \leq 4$

25. $x < -7$ or $-4 < x < 2$

27. $x = 0$ or $3 \leq x \leq 16$

29. $-\frac{5}{2} \leq x \leq 2$ or $x \geq 3$

31. $x < -5$ or $x > 4$

33. $-7 \leq x < 5$ **35.** $x < -6$ or $2 < x < 12$ **37.** $-5 < x \leq 0$ or $x > 5$

39. $-\frac{29}{2} \leq x < -10$

41. $x < -1$ or $2 < x < \frac{7}{2}$

43. $-25 \leq x < -5$ or $-5 < x < 5$

45. $x > 1$ **47.** $\frac{5}{3} < x \leq \frac{5}{2}$

49. a) $2.4 < x < 3.6$ **b)** $\delta = 0.6$

51. a) $5.995 < x < 6.005$

b) $\delta = 0.005$

53. a) $x < -3$ or $0 < x < 3$

b) $-3 < x < 3$

55. a) $-2 < x < 4$

b) $x < -2$ or $0 < x < 4$

61. $(-\infty, 1]$ or $[3, \infty)$

Section 1.6

1. The textbook costs $27.25, the calculator costs $8.75   **3.** $3\frac{1}{2}$ by 8 inches

5. Mike's rate is 67.5 mph, Sherry's rate is 62.5 mph   **7.** $1\frac{1}{2}$ feet

9. 25 gallons   **11.** 20 ml

13. $3\frac{1}{3}$ lbs of peanuts, $6\frac{2}{3}$ lbs of cashews

15. $4300 at 13%, $8600 at 9%, total $12,900

17. The side of the square base cannot exceed 8 cm

19. 5 Farmalls and 7 Blackwelders

21. $\frac{6}{7}$ day, or about 20 hours, 34 minutes

23. 12 desks   **25.** 24 km

27. 4 mph   **29.** 17 in. by 17 in.

31. Between $0.70 and $0.80

Section 1.7

1. $0 + 5i$ or $0 - 5i$

3. $0 + \sqrt{11}i$ or $0 - \sqrt{11}i$

5. $3 + 2i$ or $3 - 2i$

7. $-1 + 2\sqrt{3}i$ or $-1 - 2\sqrt{3}i$

9. $0 + \dfrac{\sqrt{7}}{2}i$ or $0 - \dfrac{\sqrt{7}}{2}i$

11. $5 + \frac{4}{3}i$ or $5 - \frac{4}{3}i$ **13.** $6 - 6i$

15. $-8 + 12i$ **17.** $-4 + 8i$

19. $5 - 7i$ **21.** $15 - 8i$

23. $\frac{1}{5} + \frac{7}{5}i$ **25.** $13 + 11i$

27. $1 + 3i$ **29.** $-16 + 2i$

31. $-414 - 154i$ **33.** $-\frac{1}{4}$

35. 58   **37.** $-14 + 6i$

39. $19 - 33i$ **41.** $\frac{2}{125} + \frac{11}{125}i$

43. $14 + 6i$ **45.** 4

47. $-2 \pm 2\sqrt{2}i$ **49.** $1 \pm \sqrt{2}i$

51. $-\dfrac{1}{6} \pm \dfrac{\sqrt{11}}{6}i$ **53.** 1 or -5

55. $0, -2 + 2i$, or $-2 - 2i$

57. $0 + 3i, 0 - 3i, 0 + i$, or $0 - i$

59. $0 + \sqrt{3}i$ or $0 - \sqrt{3}i$

61. $-1, -5, \dfrac{1}{2} + \dfrac{\sqrt{3}}{2}i, \dfrac{1}{2} - \dfrac{\sqrt{3}}{2}i$

63. $3, -\dfrac{3}{2} + \dfrac{3\sqrt{3}}{2}i, -\dfrac{3}{2} - \dfrac{3\sqrt{3}}{2}i$

65. $\dfrac{4}{3}, -\dfrac{2}{3} + \dfrac{2\sqrt{3}}{3}i$, or $-\dfrac{2}{3} - \dfrac{2\sqrt{3}}{3}i$

67. $1, -1, \dfrac{1}{2} + \dfrac{\sqrt{3}}{2}i, \dfrac{1}{2} - \dfrac{\sqrt{3}}{2}i,$

$-\dfrac{1}{2} + \dfrac{\sqrt{3}}{2}i$, or $-\dfrac{1}{2} - \dfrac{\sqrt{3}}{2}i$

71. a) -1 **b)** $-i$ **c)** i

d) 1

73. $x^2 - 6x + 10 = 0$

Miscellaneous Exercises for Chapter 1

1. integers: $\sqrt[3]{8}$; rational numbers: $\sqrt[3]{8}$, $\frac{3}{5}$; irrational numbers: $-\pi/4, \sqrt{3}$; increasing order: $-\pi/4, \frac{3}{5}, \sqrt{3}$, and $\sqrt[3]{8}$

3. integers: $-2, \sqrt[3]{64}, \sqrt{9}$; rational numbers: $-2, \sqrt[3]{64}, \sqrt{9}, 3.14$; irrational numbers: $\sqrt{4.9}$; increasing order: $-2, \sqrt{4.9}, \sqrt{9}, 3.14, \sqrt[3]{64}$ **5.** $[-6, 1)$

7. $[-8, 1]$ **9.** $(-1, 5]$ or $(-\infty, -2)$

11. $-2 < x \leq 5$ **13.** $x < -3$ or $-1 \leq x < 0$ **15.** $0 \leq x < 4$

17. $x^2 + 1$

19.

21.

23. Answers vary. Possible answers are
6.09, 6.10, 6.11, . . . , 6.28

25. a) $\dfrac{4}{x^3}$ **b)** $\dfrac{8}{x^{1/3}}$ **c)** $\dfrac{1}{2x^{1/3}}$

d) $\dfrac{q^2}{p^5}$

27. a) $\sqrt{x^3}$ **b)** $\sqrt[4]{x^{-1}}$ **c)** $\sqrt[5]{x^{-2}}$

d) $\sqrt[3]{x^8}$ **29.** $-5x^{11}$ **31.** $\dfrac{27x^{11}}{y^{10}}$

33. $\dfrac{a^{13/2}b^{3/2}}{c^{13/2}}$ **35.** $\sqrt[6]{x^5}$

37. $-6x^3 + 11x^2 - 34x + 4$

39. $-42x^5 + 336x^2 - 672x$
41. $x^{1/2} - 125$ **43.** $x^3(3x + 2)(x - 1)$
45. $3xy(2 - x)(4 + 2x + x^2)$
47. $(x^2 + 9)(x^4 + 1)(x + 3)(x - 3)$
49. $x^{2/5}(3x^2 - x + 6)$ **51.** $-3x$
53. $2x + h + 5$ **55.** $\dfrac{(2x + 1)(x - 7)}{5x^2(x + 1)}$

57. $\dfrac{7x + 5}{(2x + 1)(x + 2)}$

59. $\dfrac{3(2x^2 + 13x + 1)}{(x + 5)(x + 2)(x - 1)}$

61. $\dfrac{x(3x + 1)}{3(x + 1)}$ **63.** $\dfrac{2x - y}{2x^{3/2}}$

65. $\dfrac{-50x^2 + 30x - 2}{\sqrt{3 - 4x}}$

67. $5x^2 + x + 2 + \dfrac{3}{x + 4}$

69. $3 + \dfrac{2x + 6}{x^2 + x + 1}$ **71.** $x = \frac{14}{19}$

73. $x = 2$ or $x = -5$
75. $x = 6$ or $x = -6$ **77.** $x = \frac{7}{4}$
79. $x = 1$ or $x = -1$

81. $x = \dfrac{3 + \sqrt{29}}{2}$ or $x = \dfrac{3 - \sqrt{29}}{2}$

83. $x = \sqrt{3}/3$

85. $x = \dfrac{2 + \sqrt{5}}{2}$ or $\dfrac{2 \pm \sqrt{3}}{2}$

87. $14 < x \le 64$ **89.** $x \le -2$ or $x \ge \frac{1}{4}$
91. $3 < x \le 9$ **93.** $4\frac{4}{9}$ minutes
95. 48 mph   **97.** 33 mph and 38 mph
99. $1 + 6i$ **101.** $-38 + 16i$

103. $8i$ **105.** $\dfrac{2}{41} - \dfrac{23}{41}i$

107. $-i$ **109.** $1 + i,\ 1 - i$

111. $\dfrac{5 \pm \sqrt{47}\,i}{4}$

C H A P T E R   2

Section 2.1

1. $d(A, B) = 2\sqrt{41},\ M(-2, -1)$
3. $d(A, B) = 7,\ M(-\frac{1}{2}, 8)$
5. $d(A, B) = 5\sqrt{3}/3,\ M(5\sqrt{3}/6, 0)$

7.

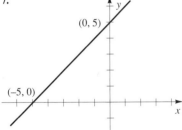

9.

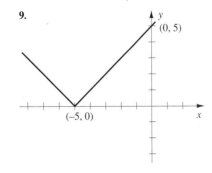

11.

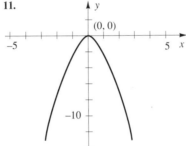

13.

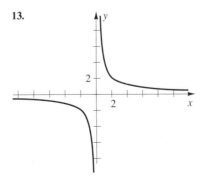

15.

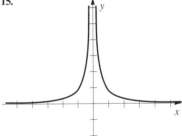

17.

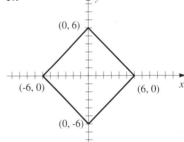

19. Center: $(0, 0)$; Radius: 2
21. Center: $(0, -6)$; Radius: 6
23. Center: $(3, -3)$; Radius: 5

25.

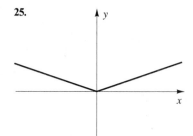

27.

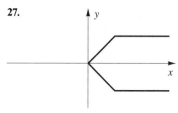

29.

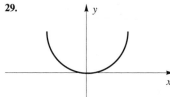

31. $(x - 2)^2 + (y - 3)^2 = 16$
33. $(x + 1)^2 + (y - \frac{3}{2})^2 = \frac{9}{4}$
35. $(x - 2)^2 + (y + 4)^2 = 0$ (the graph is the point $(2, -4)$)
37. $(x + 3)^2 + (y - 3)^2 = 9$
39. $x^2 + y^2 = 64$
41. $x^2 + (y + 2)^2 = 36$ **43.** Yes
45. Isosceles   **47.** Obtuse
51. Yes   **53.** No   **55.** Yes

Section 2.2

1. Domain: $\mathbb{R}$, Range: $[-5, +\infty)$
3. Domain: $\mathbb{R}$, Range: $[5, +\infty)$
5. Domain: $\mathbb{R}$, Range: $[0, 1)$
7. Domain: $x \neq 2$
9. Domain: $(-\infty, \frac{5}{11}]$
11. Domain $(3, +\infty)$
13. a) -6 **b)** 2
15. a) $\frac{2}{13}$ **b)** 8
17. a) 2 **b)** -4
19. a) $2a - 4|a|$ **b)** $2a - 4|a|$
21. a) $a^2 - 3a - 1$ **b)** $-a^2 - 3a + 1$
23. a) $\dfrac{6x + 37}{x + 6}$ **b)** $\dfrac{1}{x + 12}$
25. $12 + 2h$ **27.** $\dfrac{-2}{x(x + h)}$

29. $2(x + a)$
31. $h(-9) = 9$, $h(9) = \frac{9}{2}$, $h(\frac{9}{4}) = \frac{3}{2}$,
Domain: $\mathbb{R}$ **33.** $(-\infty, 2)$ or $(2, \infty)$
35. 6   **37.** $A(p) = \dfrac{p^2}{16}$, $p > 0$
39. $L(w) = 48/w$, $w > 0$
41. $A(s) = \dfrac{\sqrt{3}}{4} s^2$, $s > 0$
43. $d(t) = 25t$, $t > 0$
45. $d(x) = \sqrt{225 + x^2} + \sqrt{100 + (30 - x)^2}$,
$0 \le x \le 30$
47. $A(L) = \dfrac{L^2}{16} + \dfrac{(24 - L)^2}{4\pi}$, $0 \le L \le 24$
49. $A(x) = 40x - 2x^2$, $0 < x < 20$
51. $D(x) = \sqrt{5x^2 - 20x + 25}$,
$-\infty < x < \infty$
53. $C(x) = 2x + 5\sqrt{400 + (30 - x)^2}$,
$0 \le x \le 30$
55. $S(r) = 2\pi r^2 + 116/r$, $r > 0$
57. $d(t) = 3\sqrt[3]{t/2\pi}$, $t \ge 0$

59. $\dfrac{-2}{\sqrt{x + \Delta x}\sqrt{x}(\sqrt{x + \Delta x} + \sqrt{x})}$

61. $\dfrac{1}{\sqrt[3]{x^2} + \sqrt[3]{x^2 + x\Delta x} + \sqrt[3]{(x + \Delta x)^2}}$

Section 2.3

1. No   **3.** No   **5.** Yes
7. intercepts: $(\pm 2, 0)$, $(0, -8)$; pos: $(-\infty, -2)$, $(2, \infty)$; neg: $(-2, 2)$; incr: $(-1, 0)$, $(1, \infty)$; decr: $(-\infty, -1)$, $(0, 1)$; turning points: $(-1, -9)$, $(0, -8)$, $(1, -9)$
9. intercepts: $(\pm 3, 0)$, $(0, 3)$; pos: $[0, 3)$; neg: $(-3, 0)$; incr: nowhere; decr: $(-3, 0)$, $(0, 3)$; turning points: none
11. intercepts: none; pos: $(0, +\infty)$; neg: $(-\infty, 0)$; incr: nowhere; decr: $(-\infty, 0)$, $(0, +\infty)$; turning points: none
13. $y + 2 = |x + 3|$
15. $y + 4 = \sqrt{25 - (x - 3)^2}$
17. $y - 1 = (x - 2)^3$
19.

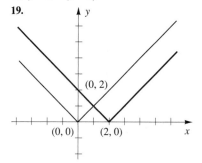

21.

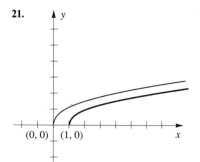

23.

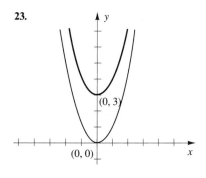

25.

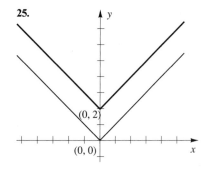

27.

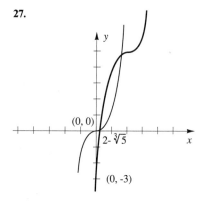

29.

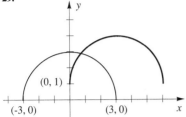

31.

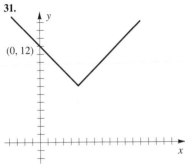

33.

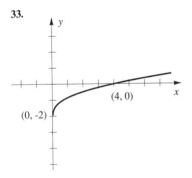

35.

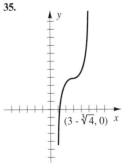

37.

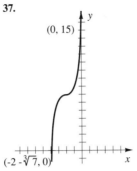

39.

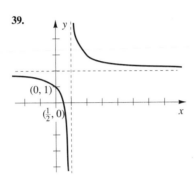

41.

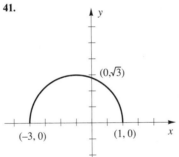

43.

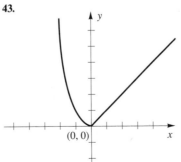

45.

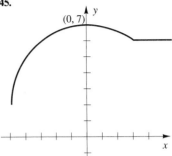

47.

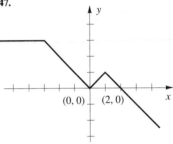

49. odd

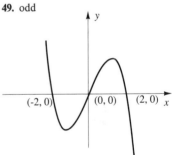

51. even

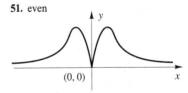

53. even

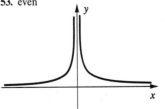

55. odd   **57.** neither   **59.** even
61. n is even $\Rightarrow f$ is an even function;
n is odd $\Rightarrow f$ is an odd function
63. Q is not necessarily even or odd
65.

Section 2.4

1.

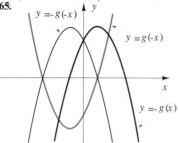

3.

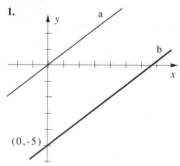

5.

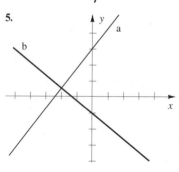

7. $y = 3x - 5$ **9.** $y = \frac{3}{7}x + \frac{1}{7}$
11. $y = \frac{7}{12}x - 15$ **13.** $y = -\frac{1}{3}x + 5$
15. $f(x) = \frac{3}{2}x + \frac{1}{2}$ **17.** $h(x) = -3$
19.

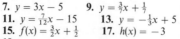

21.

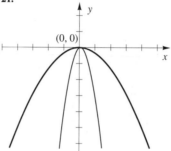

23.

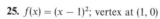

25. $f(x) = (x - 1)^2$; vertex at $(1, 0)$

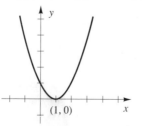

27. $f(x) = (x + 1)^2 + 2$; vertex at $(-1, 2)$

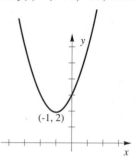

29. $f(x) = -(x - 2)^2 + 4$; vertex at $(2, 4)$

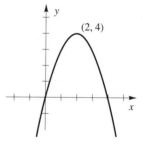

31. $f(x) = -\frac{1}{2}(x + 2)^2 + 2$; vertex at $(-2, 2)$

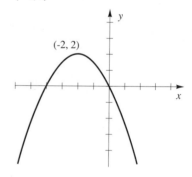

33. $f(x) = -\frac{1}{4}(x - 2)^2 + 4$; vertex at $(2, 4)$

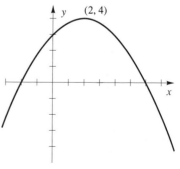

35. $f(x) = 2(x + \sqrt{3})^2 - 3$; vertex at $(-\sqrt{3}, -3)$

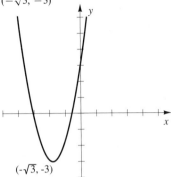

$(-\sqrt{3}, -3)$

37. Maximum value of 6 at $x = -2$
39. Maximum value of 4 at $x = 2$
41. Maximum value of $-\frac{15}{8}$ at $x = \frac{5}{4}$
43. 6 and 6   **45.** 22,500 ft
47. 630 units   **49.** 5625 m^2
51. 150 ft by 300 ft
53. a) Domain: $[0, 6]$ **b)** 18
55. $1158.33
57. $(-\infty, -4)$ or $(-\frac{1}{2}, \infty)$
59. $[-\frac{1}{4}, 3]$ **61.** no real solutions
63. Maximum value of 29 at $x = \sqrt[3]{5}$
65. Minimum value of $2\sqrt{3}$ at $x = 2$
67. Maximum value of $\frac{1}{2}$ at $x = -1$
71. $\frac{1}{2}(y_1 + y_2) \geq \sqrt{y_1 y_2}$

Section 2.5

1.

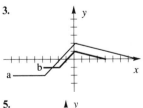

3.

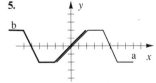

5.

7.

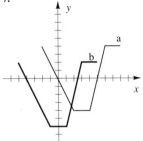

9.

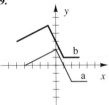

11.

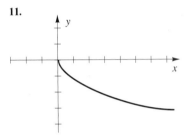

13.

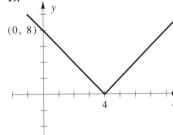

15.

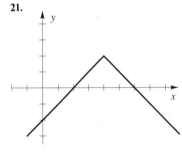

17.

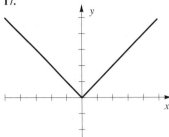

19.

$(0, 8)$

21.

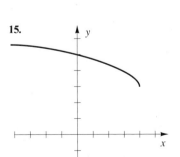

23.

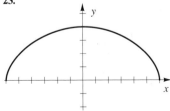

25.

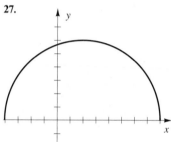

27.

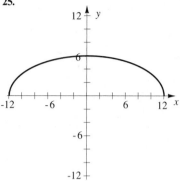

29.

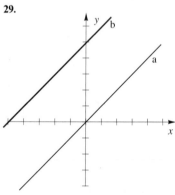

31.

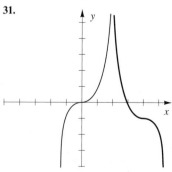

33.

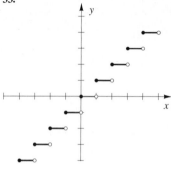

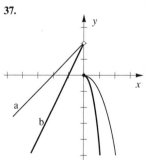

35.

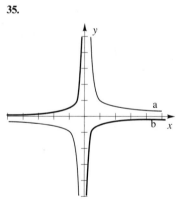

37.

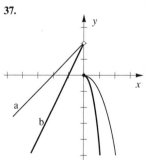

39.

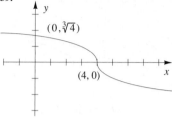

41.

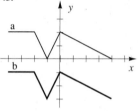

43.

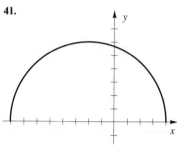

45.

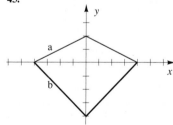

47.

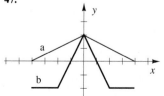

49.

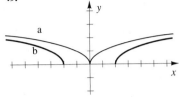

51. $[-\frac{4}{3}, -1)$ or $[-\frac{2}{3}, \frac{2}{3})$ or $[1, \frac{4}{3})$
53. $(2, 8)$
55.

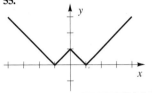

57.

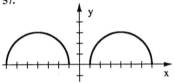

Section 2.6

1. $(f + g)(x) = 3x^4 + 5x^2 + 9$,
Domain: $\mathbb{R}$; $(f - g)(x) = x^4 - 7x^2 - 9$,
Domain: $\mathbb{R}$;
$(fg)(x) = 2x^8 + 11x^6 + 12x^4 - 9x^2$,
Domain: $\mathbb{R}$;
$(f/g)(x) = (2x^4 - x^2)/(x^2 + 3)^2$,
Domain: $\mathbb{R}$
3. $(f + g)(x) = (4x + 2)/x^2$, Domain:
$x \neq 0$; $(f - g)(x) = (2x - 2)/x^2$,
Domain: $x \neq 0$; $(fg)(x) = (3x + 6)/x^3$,
Domain: $x \neq 0$; $(f/g)(x) = 3x/(x + 2)$,
Domain: $x \neq 0, -2$
5. $(f + g)(x) = (2x^2 + 6x + 8)/(x^2 + 4x)$,
Domain: $x \neq 0, -4$;
$(f - g)(x) = (-6x - 8)/(x^2 + 4x)$,
Domain: $x \neq 0, -4$;
$(fg)(x) = (x + 2)/(x + 4)$, Domain:
$x \neq 0, -4$; $(f/g)(x) = x^2/(x + 2)(x + 4)$,
Domain: $x \neq 0, -4, -2$
7. $(f \circ g)(x) = 5/x$; $(g \circ f)(x) = 1/5x$

9. $(f \circ g)(x) = x - 1 - \dfrac{2}{\sqrt{x - 1}}$;

$(g \circ f)(x) = \dfrac{\sqrt{x^4 - x^2 - 2x}}{x}$

11. $(f \circ g)(x) = x^2 + \dfrac{1}{x^2}$

$(g \circ f)(x) = x^2 - 2 + \dfrac{1}{x^2 - 2}$

13. $(f \circ g)(x) = x$, $(g \circ f)(x) = x$
15. $(f \circ g)(x) = x - \frac{3}{2}$, $(g \circ f)(x) = x - 3$
17. $(f \circ g)(x) = 4$, $(g \circ f)(x) = \frac{1}{16}$
19. $(f \circ f)(x) = 4x - 3$
21. $(f \circ f)(x) = -|x|$
23. $(f \circ f)(x) = -1/x$
25.

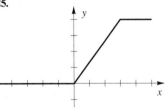

27.

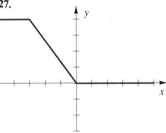

29.

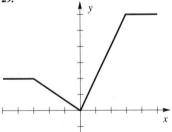

31.

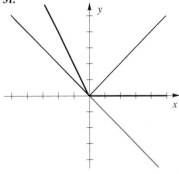

33.

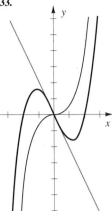

35.

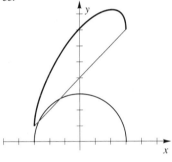

37.

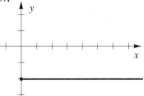

39.

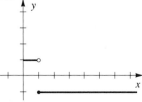

41.

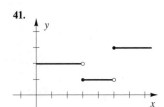

43. $f(x) = x^2 + x + 6$, $g(x) = 3x - 1$
45. $f(x) = 1/x$, $g(x) = (x - 1)^2$
47. $f(x) = \sqrt{x}$, $g(x) = 8 - x^3$
49. $f(x) = \sqrt{25 - x}$, $g(x) = (2x + 1)^2$
51. $f(x) = 2/(5 - x)$, $g(x) = (1 - x)^3$
53. $f(x) = |x| + \dfrac{3}{x^3}$, $g(x) = 2 - x$

55. a) $r(t) = 3t + 6$
   b) $V(t) = 36\pi(t + 2)^3$
57. $f(x) = 2u(x) - 3u(x - 3)$
59. $f(x) = -u(x) + 2u(x - 2) - 2u(x - 4)$
        $+ 2u(x - 6) - 2u(x - 8)$

17.

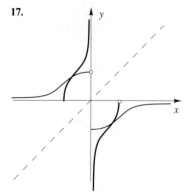

19. $h^{-1}(x) = x/5 + 3$
21. $G^{-1}(x) = 3 - x/2$
23. $d^{-1}(x) = \sqrt[3]{x - 4}$
25. $p^{-1}(x) = x^3 - 4$
27. $r^{-1}(x) = 2/x + 3$
29. $S^{-1}(x) = 4x/(1 - x)$

Section 2.7

1. Yes   **3.** No   **5.** No   **7.** No
9. Yes   **11.** No
13.

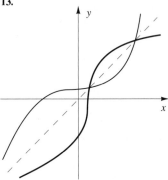

15.

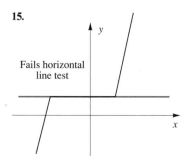

Fails horizontal line test

31.

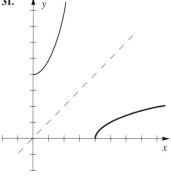

33.

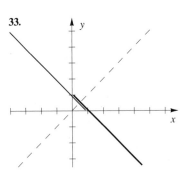

35.

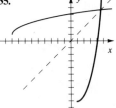

37.

39.

41.

45. $F^{-1}(x) = \frac{5}{9}(x - 32)$

47.

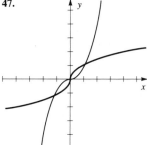

51. $f^{-1}(x) = \dfrac{5x + 2}{x - 1}$,

    $f(x^{-1}) = \dfrac{2x + 1}{1 - 5x}$,

    $[f(x)]^{-1} = \dfrac{x - 5}{x + 2}$

Miscellaneous Exercises for Chapter 2

1. $M = (\frac{5}{2}, 0)$, distance is 25

3. $\left(\frac{3\sqrt{2}}{2}, \sqrt{2}\right)$, distance is 2

5. $(x - 2)^2 + (y - 6)^2 = 4$

7. $(x - 4)^2 + (y + \frac{15}{2})^2 = \frac{289}{4}$

9. $(x - 6)^2 + (y - 5)^2 = 25$

11. $y = -\frac{7}{4}x + 14$

13. $y = \frac{4}{3}x - 2$ **15.** $y = -\frac{3}{4}x + \frac{37}{4}$

17. origin   **19.** origin   **21.** x-axis

23. $\mathbb{R}$ **25.** $x \neq 3$

27. $[-1, 0)$ or $[1, \infty)$ **29.** $[3, \infty)$

31. 12   **33.** $-2x - h$

35. $3x^2 + 3xh + h^2$ **37.** $\dfrac{1}{(x + h)x}$

39. $\dfrac{-1}{(x + h)x}$ **41.** $\dfrac{2}{\sqrt{2x + 2h} + \sqrt{2x}}$

43.
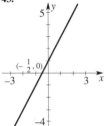

45.
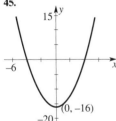

47.
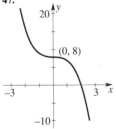

49.
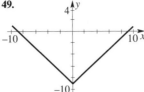

51.
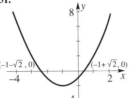

53.
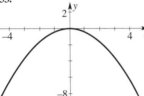

55.
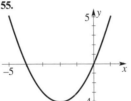

57.
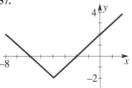

59.
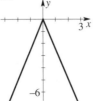

61.
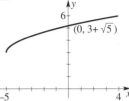

63.
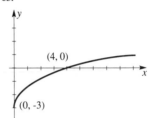

65.
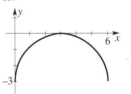

67.
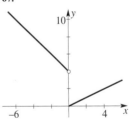

69.
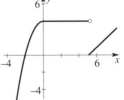

71.

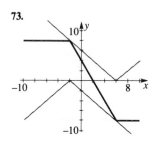

73.

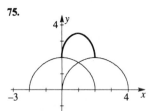

75.

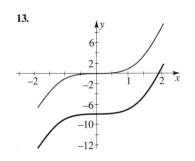

77. neither   **79.** odd   **81.** odd
83. $(u \circ p)(x)$ **85.** $(u \circ r)(x)$
87. $(r \circ q)(x)$ **89.** $(q \circ r)(x)$
91. $(q \circ q)(x)$ **93.** $(u \circ u)(x)$
95. a) $(q \circ r \circ p)(x)$ **b)** $(r \circ q \circ p)(x)$
97. Domain: $[-6, 5]$; Range: $[-1, 3]$
99. g is odd
101. a) -1 **b)** 1 **c)** -1

103.

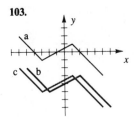

105.

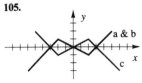

107.

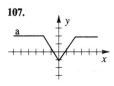

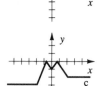

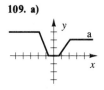

109. a)   **b)**

113. $f^{-1}(x) = \frac{1}{2}x + 4$
115. $f^{-1}(x) = \frac{x + 4}{x}$
117. No, $x = 4, 6$
119. a) $f^{-1}(x) = \sqrt{2x} + 4$

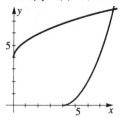

b) $f^{-1}(x) = -\sqrt{2x} + 4$

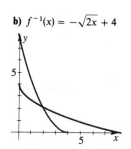

121. Minimum value of -2 at $x = 0$
123. Minimum value of -7 at $x = 2$
125. Maximum value of $\frac{2}{3}$ at $x = 0$
127. a) 600   **b)** $P(x) = 160x - 0.4x^2$
c) 200   **131.** $(a, b), (-a, b), (a, -b)$
133. $A(x) = (4 + x)(8 - \frac{1}{2}x^2), 0 \le x \le 4$
135. $h(t) = \dfrac{\sqrt{256 - (4 + 1.5t)^2}}{16}$

C H A P T E R   3

Section 3.1

1. No, y does not grow without bound as $x \to \pm\infty$.
3. No, there is a discontinuity.
5. No, the graph is not smooth.   7. IV
9. I   11. III

13.

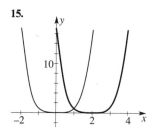

15.

17.

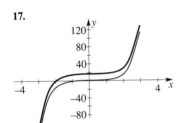

27.

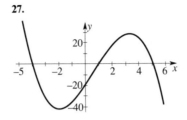

37.

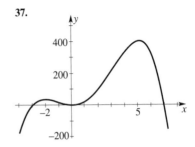

19.

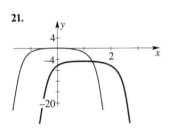

29.

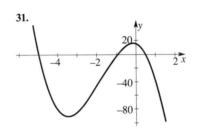

39.

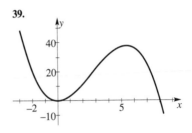

21.

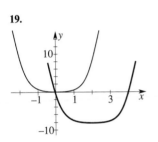

31.

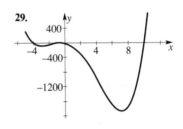

41.

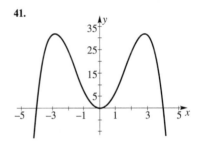

23.

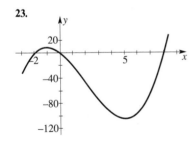

33.

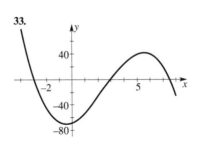

43.

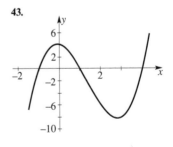

25.

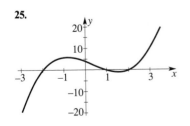

35.

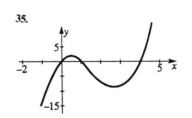

45.

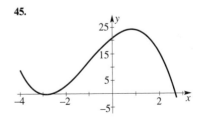

47.

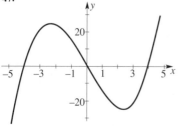

49.

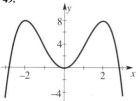

51.

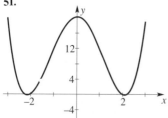

53. $(-\infty, -5]$ or $[-1, \frac{1}{2}]$
55. $(-4, 0)$ or $(0, 4)$
57. $(-\infty, -\sqrt{5}]$ or $[2, 2]$ or $[\sqrt{5}, \infty)$
59. odd function

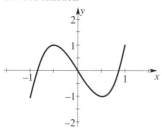

61.

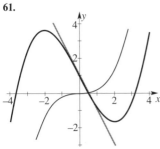

63.

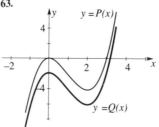

69. $(-2, 16)$ or $(2, -16)$

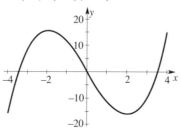

71. $(-1, 121)$ or $(4, -4)$

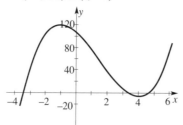

73. Maximum value is 18,000 cm³

Section 3.2

1. $Q(x) = x^2 - 5$; $R(x) = 7$;
$P(x) = (2x + 1)(x^2 - 5) + 7$
3. $Q(x) = x^2 - 3x - 1$; $R(x) = 5x - 5$;
$P(x) = (x^2 + 2x - 1)(x^2 - 3x - 1)$
     $+ 5x - 5$
5. $Q(x) = x^3 + 2x - 1$; $R(x) = 14x + 1$;
$P(x) = (x^2 - 3)(x^3 + 2x - 1) + 14x + 1$
7. $P(2) = 11$;
$P(x) = (x - 2)(3x^3 + 4x + 9) + 11$
9. $P(6) = -178$;
$P(x) = (x - 6)(x^3 + 2x^2 - 5x - 30) - 178$
11. $P(\frac{2}{3}) = 3$;
$P(x) = (x - \frac{2}{3})(12x^3 + 6x^2 - 3x - 3) + 3$
13. $P(4) = 2181$; $P(x) =$
$(x - 4)(\frac{1}{2}x^4 + 9x^3 + 34x^2 + 137x + 543)$
                 $+ 2181$
15. $P(2) = 1481$;
$P(x) = (x - 2)(6x^7 + 12x^6 + 24x^5 + 48x^4$
     $+ 96x^3 + 186x^2 + 372x + 745)$
     $+ 1481$

17. $P(x) = (x - 3)(2x^2 - 5x + 2)$
19. $P(x) = (x + 2)(3x^3 + x^2 - 6)$
21. $P(x) = (x + \frac{2}{3})(6x^3 - 18x + 21)$
23. $P(x) =$
$(x + \sqrt{5})(2x^2 + (7 - 2\sqrt{5})x - 7\sqrt{5})$
25. $x = 1$ and $x = 2$
27. $x = -4$ and $x = -1$
29. $x = -\frac{1}{2}$, $x = \sqrt{7}$ and $x = 4$
31.

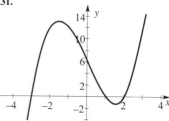

33.

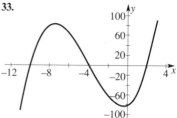

35.

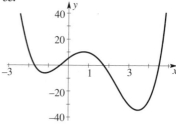

37. $(-\infty, -3)$ or $(1, 2)$
39. $[-10, -4]$ or $[2, +\infty)$
41. $(-\sqrt{3}, 2 - \sqrt{6})$ or $(\sqrt{3}, 2 + \sqrt{6})$
43. Let $f(x) = x^{76} - 4x^{54} + x^{14} + 2$.
Then $f(1) = 0$, so by the factor theorem,
$(x - 1)$ is a factor of $f(x)$.
45. Let $f(x) = x^n - a^n$. Then $f(a) = 0$, so
by the factor theorem, $(x - a)$ is a factor of
$f(x)$.   **49.** $f(-4) = 10, f(-3) = -4$
51. $f(-2) = 28, f(-1) = -4$
53. 1.82   **55.** $-\frac{11}{19}$
57. $-2 \pm \sqrt{2}$

Section 3.3

1. Rat'l zeros: 3, 4; Real zeros: 3, 4,
$2 + 2\sqrt{3}, 2 - 2\sqrt{3}$; Complex zeros: 3, 4,
$2 + 2\sqrt{3}, 2 - 2\sqrt{3}$;

$P(x) = (x - 3)(x - 4)(x - (2 + 2\sqrt{3}))$
$\times (x - (2 - 2\sqrt{3}))$

3. Rat'l zeros: $-2, 2$; Real zeros: $-2, 2$;
Complex zeros: $-2, 2, -1 + i\sqrt{3}$,
$-1 - i\sqrt{3}$; $P(x) = (x - 2)(x + 2)^2$
$\times (x - (-1 - i\sqrt{3}))(x - (-1 + i\sqrt{3}))$

5. Rat'l zeros: $-2, 2$; Real zeros: $-2, 2$,
$\sqrt{3}, -\sqrt{3}$; Complex zeros: $-2, 2, \sqrt{3}$,
$-\sqrt{3}$; $P(x) = (x + 2)^2(x - 2)^2(x - \sqrt{3})$
$\times (x + \sqrt{3})$

7. $P(x) = -2x^3 - 10x^2 - 6x + 18$

9. $P(x) = x^4 + 6x^3 + 4x^2 + 64$

11. $P(x) = x^3 + 5x^2 + 3x - 9$

13. Rat'l roots: $4, \frac{8}{3}$; Real roots: 4,
$\frac{8}{3}, \sqrt{6}/2, -\sqrt{6}/2$; Complex roots: 4,
$\frac{8}{3}, \sqrt{6}/2, -\sqrt{6}/2$

15. Rat'l roots: none; Real roots: none;
Complex roots: $4\sqrt{2}i, -4\sqrt{2}i$,
$(2 + \sqrt{6}i)/2, (2 - \sqrt{6}i)/2$

17. Rat'l roots: $-8, -2, -10$; Real roots:
$-8, -2, -10, -1 + \sqrt{7}, -1 - \sqrt{7}$;
Complex roots: $-8, -2, -10, -1 + \sqrt{7}$,
$-1 - \sqrt{7}$

19. Potential rat'l zeros: $\pm 1, \pm 5$,
$\pm\frac{1}{2}, \pm\frac{5}{2}$; actual rat'l zeros: $\frac{1}{2}$

21. Potential rat'l zeros: $\pm 1, \pm 2, \pm 3$,
$\pm\frac{1}{2}, \pm 6, \pm\frac{3}{2}$; actual rat'l zeros: 2

23. Potential rat'l zeros: $\pm 1, \pm 3, \pm\frac{1}{2}$,
$\pm\frac{3}{2}$; actual rat'l zeros: none

25. Zeros: $-5, -3, 2$;
$P(x) = (x + 3)(x - 2)(x + 5)$

27. Zeros: $-5, -\sqrt{3}, \sqrt{3}, 7$;
$g(x) = (x + 5)(x - 7)(x + \sqrt{3})(x - \sqrt{3})$

29. Zeros: $2, -3, \frac{11}{2}$;
$R(x) = 2(x - 2)^2(x + 3)^2(x - \frac{11}{2})$

31.

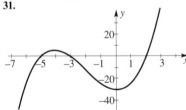

33.

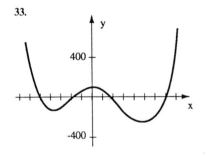

35.

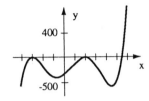

37. $-2 - 2i, -2 + 2i, 2 + 2\sqrt{3}, 2 - 2\sqrt{3}$

39. $-2, \sqrt{5}, -\sqrt{5}, 2 + 2i, 2 - 2i$

41. $-i, i, -1, 1, 4$

43. $2, -1 + \sqrt{3}i, -1 - \sqrt{3}i$

45. $-3, -\frac{3}{2} \pm \frac{3\sqrt{3}}{2}i$

49. $-8, -\sqrt{3}, \sqrt{3}$

Section 3.4

1. Domain: $x \neq 3, -6, -1$; zeros: $-3, 8$

3. Domain: $x \neq 4, 5$; zero: -8

5. Domain: $x \neq 4, -2$; zero: 1

7.

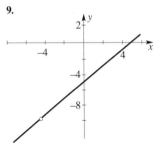

9.

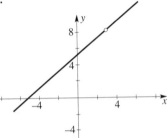

11.

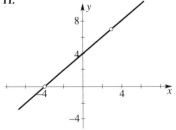

13.

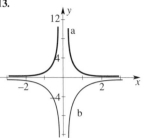

15.

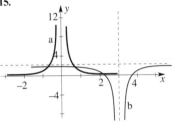

17.

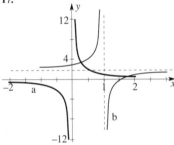

19. IV   **21.** V   **23.** III

25.

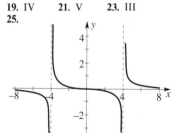

27.

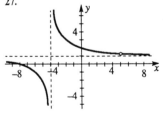

29.

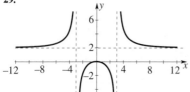

31.

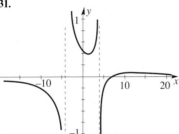

33.

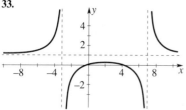

35.

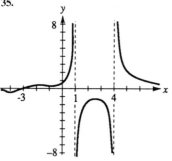

37.

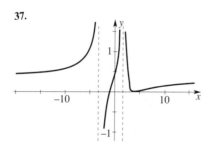

39.

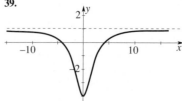

41.

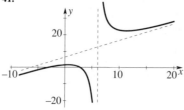

43.

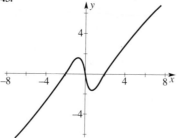

45.

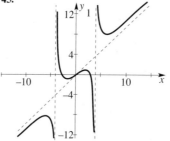

47.

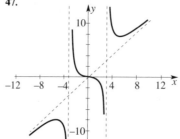

49. $(-\infty, -4)$ or $(0, 4)$ **51.** $(-3, 1)$ or $(4, +\infty)$ **53.** $[-3, 3]$ or $(6, +\infty)$
55.

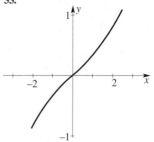

57. $f(x) = \dfrac{1}{2}x^2 + \dfrac{1}{2x^2}$. At the extremes, the graph behaves as $y = \frac{1}{2}x^2$. Near $x = 0$, the graph behaves as $y = \dfrac{1}{2x^2}$.

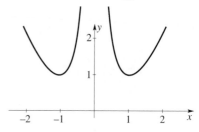

59. $(\frac{2}{3}, 2)$

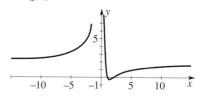

Section 3.5

1. Domain: $(-\infty, -1)$ or $(1, +\infty)$; zeros: none

3. Domain: $[2, +\infty)$; zeros: none

5. Domain: $(-2\sqrt{2}, 0)$ or $(0, 2\sqrt{2})$; zero: $\frac{5}{2}$

7.
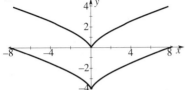

9.
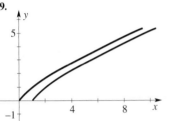

11.
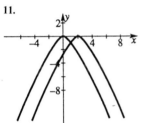

13.
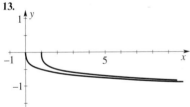

15.
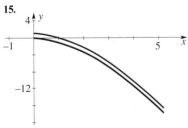

17.
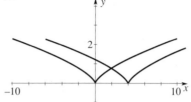

19.
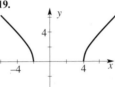

21.
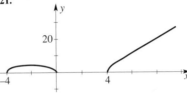

23.
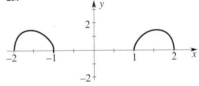

25.
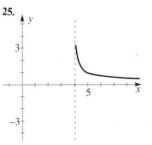

27.
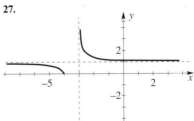

29.

31.
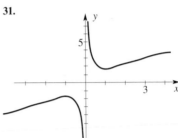

33.
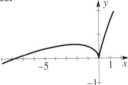

35.

37.
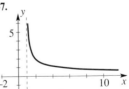

39.
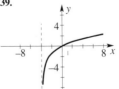

41.

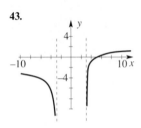

Miscellaneous Exercises for Chapter 3

13.

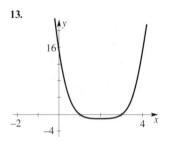

1.

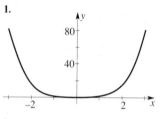

43.

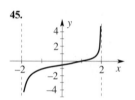

3.

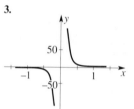

15.

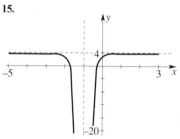

45.

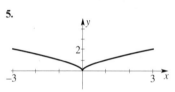

5.

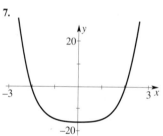

17.

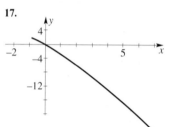

47.

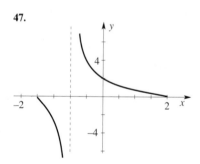

7.

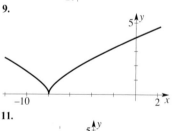

19.

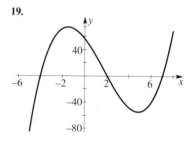

49. $(-\infty, 0)$ or $(1, +\infty)$

51. Domain of f_1: $[2, +\infty)$
    Domain of f_2: $(-\infty, -3)$ or $[2, +\infty)$

53.

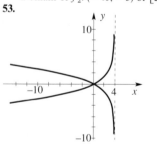

9.

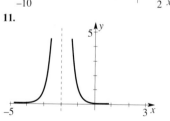

11.

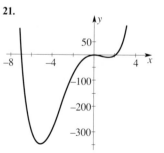

21.

23.

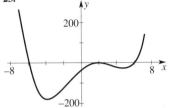

25.

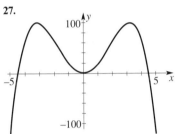

27.

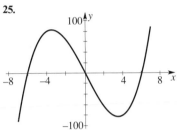

29.

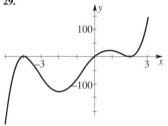

31.

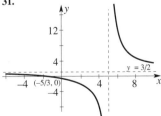

33.

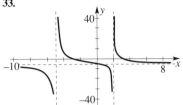

35.

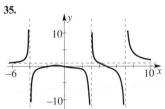

37.

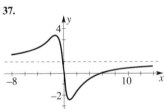

39.

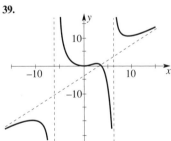

41.

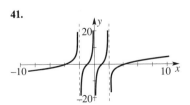

43.

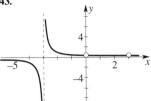

45.

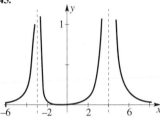

47.

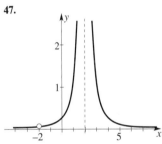

49.

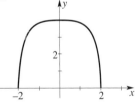

51.

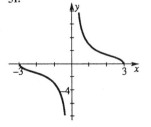

53.

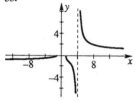

55.

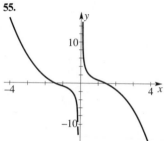

57.

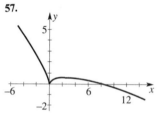

59.

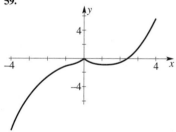

61. $P(4) = 80$;
$P(x) = (x - 4)$
$\qquad \times (2x^2 + 5x + 21) + 80$
63. $P(6) = 434$;
$P(x) = (x - 6)$
$\qquad \times (x^3 + 4x^2 + 12x + 72) + 434$
65. $P(\frac{2}{3}) = 0$
$P(x) = (x - \frac{2}{3})$
$\qquad \times (12x^3 + 6x^2 - 3x - 3)$
67. $2, -1 + \sqrt{2}, -1 - \sqrt{2}$ **69.** $-5, \frac{1}{2}$
71. $-\frac{7}{3}, \frac{5}{2}, -3\sqrt{3}, 3\sqrt{3}$

73. $(-\infty, -2)$ or $(0, 2)$

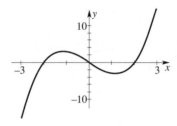

75. $(-\infty, -\sqrt{13})$ or $(\sqrt{13}, \infty)$

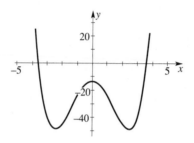

77. $(-\infty, -2\sqrt{5}], 0,$ or $[2\sqrt{5}, \infty)$

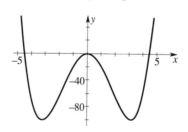

79. $(-\infty, 0)$ or $(3, \infty)$

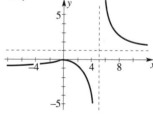

81. $(0, 2]$ or $(5, \infty)$

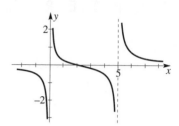

83. $(0, 2)$

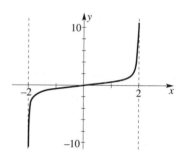

85.a)

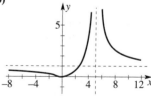

b)

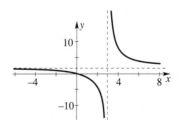

87. $P(x) = -4x^3 + 20x^2 + 52x + 28$
89. $P(x) = 2x^4 - 6x^2 - 8$

CHAPTER 4

Section 4.1

1. 31.5443   **3.** 23.1332   **5.** 0.2034
7. a) $x = 4$ **b)** $x = \frac{1}{2}$
9. a) $x = -1$ **b)** $x = 0$
11. a) no solution   **b)** no solution
13. $x = -\frac{1}{4}$
15. Horizontal asymptote at $y = 0$.

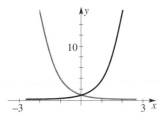

17. Horizontal asymptote at $y = 0$.

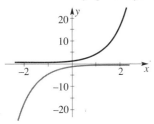

19. Horizontal asymptote at $y = 0$.

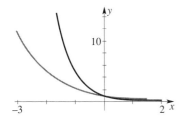

21. a) Horizontal asymptote at $y = 0$.
   b) Horizontal asymptote at $y = 3$.

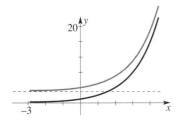

23. Horizontal asymptote at $y = 0$.

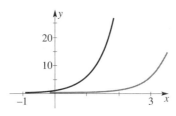

25. a) Horizontal asymptote at $y = 0$.
   b) Horizontal asymptote at $y = 1$.

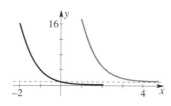

27. Answers vary. One possibility is
$g(x) = 3x - 1$ and $f(x) = 10^x$.
29. Answers vary. One possibility is
$g(x) = |x|$ and $f(x) = (\frac{1}{2})^x$.
31. Answers vary. One possibility is
$g(x) = 2^x$ and $f(x) = \sqrt{x}$.
33. Horizontal asymptote at $y = 0$.

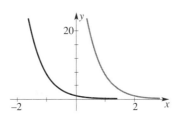

35. Horizontal asymptote at $y = 0$.

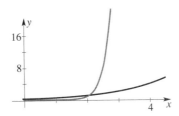

37. a) Horizontal asymptote at $y = 0$.
   b) Horizontal asymptote at $y = 0$.

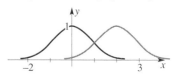

39.

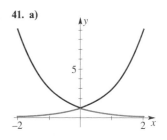

41. a)

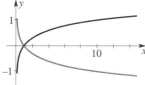

b) Horizontal asymptote at $y = 0$.
43. a) Vertical asymptote at $x = 0$.

b) Vertical asymptote at $x = 0$.

45. $8358.37 after 3 years, $9133.41 after
$4\frac{1}{2}$ yrs.   **47.** $6695.51
49. 7.50% compounded annually.
51. 1.6 lbs in 2 days, less than 0.2 pounds
after 20 days.   **53.** 6400, 4 hrs
55. a) $2.61   **b)** $2.69   **c)** $2.71
b) $2.72
57. a) About 29 mg after 3 hrs
b) 7 hrs   **59.** $218.61
61. a) False   **b)** True

63. Three solutions

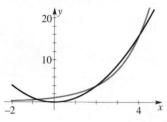

65. $[-1, 1]$

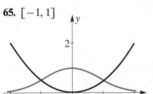

67. a) Answers vary. One possibility is
$f(x) = 10^x$.   **b)** $f(x) = 3^x$

Section 4.2

1. 2.6533   **3.** 2.7188   **5.** 4.7115
7. -56.9264 **9.** 69.1355
11. Horizontal asymptote at $y = 0$.

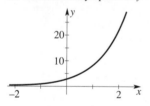

13. Horizontal asymptote at $y = 2$.

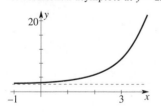

15. Horizontal asymptote at $y = -4$.

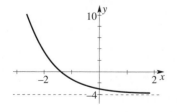

17. Answers vary. One possibility is
$f(x) = e^x$, $g(x) = 3 - x$.
19. $f(x) = e^x$, $g(x) = |x|$
21. Answers vary. One possibility is
$f(x) = e^x$, $g(x) = (x - 5)/4$
23. $x = \frac{2}{3}$ **25.** $e^3 \approx 20.1$
27. $e^{10} \approx 22{,}026$ **29.** e
31. Horizontal asymptote at $y = 0$.

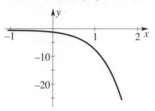

33.

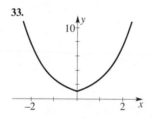

35. Horizontal asymptote at $y = 0$.

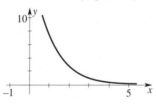

37.

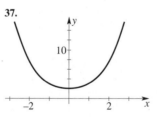

39. \$24,241.12
41. 8% compounded quarterly
43. a) 25,000   **b)** 369,119,539
45. a) 95,694,000   **b)** 72,530,000 (an
error of less than 4%)
47. $P(t) = 10000(\sqrt{2.5})^t$
or $P(t) = 10000(2.5)^{t/2}$
49. a) $T(t) = 75 + 275(\frac{27}{55})^{t/20}$ **b)** $170°$
51. 5068 million m^3 **53.** About 16
55. $P(0.75) = 2.1015625$, $e^{0.75} \approx 2.117000$
57. e^{-5x} or $\dfrac{1}{e^{5x}}$

Section 4.3

1. $\log_5 125 = 3$ **3.** $\log_{36} 6 = \frac{1}{2}$
5. $\log_4 0.0625 = -2$ **7.** $\log_5 12 = x$
9. $\log_{2.7} 5 = k$ **11.** $9^{1/2} = 3$
13. $10^3 = 1000$ **15.** $2^{5t} = 1.4$
17. $(2.7)^{3t^2} = 2$ **19.** $2 < \log_5 80 < 3$
21. $3 < \log_{10} 3250 < 4$
23. $-1 < \log_3(\frac{1}{2}) < 0$
25. $f^{-1}(x) = \log_{10} x$ **27.** $f^{-1}(x) = 4^x$
29. $f^{-1}(x) = 3^x - 5$ **31.** $x > -\frac{1}{2}$
33. $x \neq 0$ **35.** $x > 2$
37. Vertical asymptote at $x = 0$.

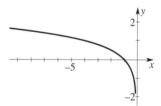

39. Vertical asymptote at $x = -3$.

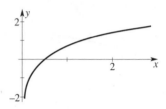

41. Vertical asymptote at $x = 0$.

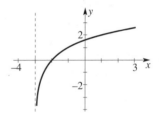

43. Vertical asymptote at $x = 2$.

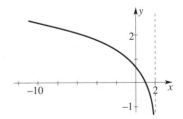

45. Vertical asymptote at $x = 0$.

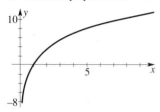

47. Vertical asymptote at $x = -3$.

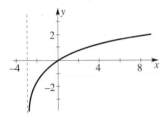

49. Vertical asymptote at $x = 0$.

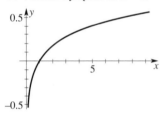

51. Vertical asymptote at $x = 0$.

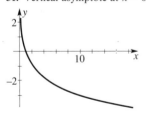

53. Vertical asymptote at $x = 0$.

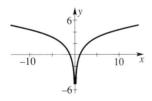

55. Vertical asymptote at $x = -2$.

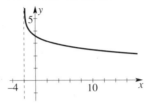

57. $\log_b \dfrac{5a^3}{x}$ **59.** $\log_b \dfrac{u+2}{(u-3)^2}$

61. $\log_b 8x^2$ **63.** $\log_b \dfrac{\sqrt{3x}}{x^4}$

65. $2\log_b x - \log_b y$
67. $\log_b A + \frac{1}{3}\log_b B$
69. $\log_b 5 + 7\log_b x + \frac{1}{3}\log_b y - \frac{1}{2}\log_b z$
71. 0.301   **73.** -0.176 **75.** 0.778
77. 1.477   **79.** 0.699
81. Vertical asymptote at $x = -1$.

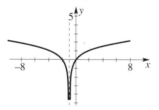

83. $\ln\left(\dfrac{B}{A^k}\right) = \ln C$ **85.** $e^{-0.26} \approx 77\%$

87. -2 **93.** $\sqrt{x}$

Section 4.4

1. 1.6350   **3.** -2.6260 **5.** 45.3692
7. -235.3711 **9.** 455.5051
11. $x = 250$ **13.** $x = \dfrac{e^3}{12} \approx 1.6738$
15. $x = 2$
17. $x = 10^{-1/3} \approx 0.4642$ or $x = 10$
19. $x = \log 15 \approx 1.1761$
21. $x = \ln 19 \approx 2.9444$
23. $t = \log_9 112 = \dfrac{\log 112}{\log 9} \approx 2.1475$
25. $t = \dfrac{1}{3}\log_{2.5} 24 = \dfrac{1}{3}\dfrac{\log 24}{\log 2.5} \approx 1.1561$
27. $t = \log_{12} 30 = \dfrac{\log 30}{\log 12} \approx 1.3687$
29. $x = \log_{6/5} 8 = \dfrac{\log 8}{\log \frac{6}{5}} \approx 11.4054$
31. $x = \dfrac{\log 25}{\log \frac{5}{3}} \approx 6.3013$

33. $x = \log_7(2 \pm \sqrt{3}) \approx 0.6768$
or -0.6768
35. $x = \log_3(4 \pm \sqrt{15}) \approx 1.8782$
or -1.8782
37. $x = \log_2 6 = \dfrac{\log 6}{\log 2} \approx 2.5850$
39. $x = 3$ or -3
41. $x = 10^{(\log 3 \log 6)/(\log 144)} \approx 1.4860$
43. $\dfrac{\log \frac{25}{18}}{4 \log 1.025} \approx 3.3$ years
45. $-\dfrac{11 \log 0.4}{\log 2} \approx 14.5$ years
47. $-\dfrac{5 \log 2}{\log 0.9125} \approx 37.8$ hours
49. $\log 1500 \approx 3.2$
51. $100 \ln 16 \approx 277$ minutes
53. **a)** $10^{-5.8} \approx 0.00000158$ or
1.58×10^{-6} **b)** 3.4   **c)** lower
55. $10^{1.8} \approx 63$ watts
57. 10^{12} times louder   **59.** $268,337
61. $\dfrac{\ln \frac{4}{15}}{-200} \approx 0.0066$ seconds
63. **a)** 69,458,000   **b)** 2023
65. 96 days   **67.** 10 or 100

Miscellaneous Exercises for Chapter 4

1. 4.0302   **3.** 2.3524   **5.** 0.9698
7. Horizontal asymptote at $y = 0$.

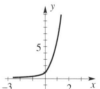

9. Horizontal asymptote at $y = 0$.

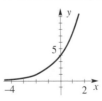

11. Horizontal asymptote at $y = 0$.

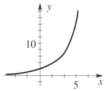

13. Horizontal asymptote at $y = 0$.

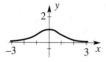

15. Horizontal asymptote at $y = 0$.

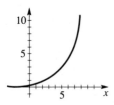

17. Horizontal asymptote at $y = 0$.

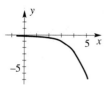

19. Vertical asymptote at $x = 0$.

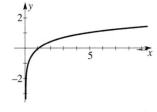

21. Vertical asymptote at $x = -5$.

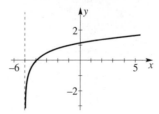

23. Vertical asymptote at $x = 0$.

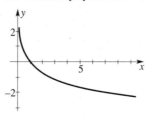

25. Vertical asymptote at $x = 0$.

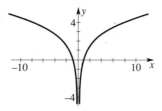

27. Vertical asymptote at $x = -3$.

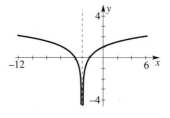

29.

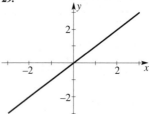

31. One possibility is $f(x) = e^x$ and $g(x) = \sqrt{x + 3}$.   **33.** One possibility is $f(x) = \sqrt[4]{x}$ and $g(x) = \log x$.
35. $3 < \log 2323 < 4$ **37.** e^3
39. $\frac{1}{2}(7^x + 3)$ **41.** $\log_b(2t^3)$
43. $\ln\left(\dfrac{3x}{y^3}\right)$
45. $\frac{1}{2}\ln x - \frac{3}{2}\ln y + 4\ln z + \ln 3$
47. $\frac{1}{12} \approx 0.0833$ **49.** -6 or $\frac{11}{2}$
51. no solution
53. $5 + e^2 \approx 12.3891$, $5 - e^2 \approx -2.3891$
55. $e^{0.7506} \approx 2.1183$ **57.** $\frac{3}{4}$
59. 0.9871 **61.** 7.3795
63. -2.5335 **65.** 1.1948
67. $c = 4$ and $f(x) = 5e^x$ **69.** 4.7 mg
71. a) About 23.5 feet
b) $t = e^{(L-13)/6.5} - 1$; about 9 years old
73. $\$15,\!959.69$
75. $n(t) = 1000(1.00916)^t$; $n(60) \approx 1728$
77. About 75.26 minutes (75 min. 16 sec.)
79. About 0.09 secs. (90 milliseconds)

CHAPTER 5

Section 5.1

1. $\sin\theta = \frac{4}{5}$, $\cos\theta = \frac{3}{5}$, $\tan\theta = \frac{4}{3}$, $\cot\theta = \frac{3}{4}$, $\sec\theta = \frac{5}{3}$, $\csc\theta = \frac{5}{4}$
3. $\sin\theta = 2\sqrt{14}/9$, $\cos\theta = \frac{5}{9}$, $\tan\theta = 2\sqrt{14}/5$, $\cot\theta = 5/2\sqrt{14}$, $\sec\theta = \frac{9}{5}$, $\csc\theta = 9/2\sqrt{14}$

5. $\sin\theta = \frac{3}{7}$, $\cos\theta = 2\sqrt{10}/7$, $\tan\theta = 3/2\sqrt{10}$, $\cot\theta = 2\sqrt{10}/3$, $\sec\theta = 7/2\sqrt{10}$, $\csc\theta = \frac{7}{3}$
7. a) 0.4848 **b)** 0.8746 **c)** 0.8746
d) 0.4848
9. a) 1.8871 **b)** 1.8871 **c)** 0.8693
d) 0.8693

11. a) $6.89°$ **b)** $83.11°$ **c)** $60.46°$
d) $51.78°$
13. 8.5 **15.** 15.7 **17.** 12.8
19. $\sin\theta = \frac{2}{3}$, $\cos\theta = \sqrt{5}/3$, $\tan\theta = 2/\sqrt{5}$, $\cot\theta = \sqrt{5}/2$, $\sec\theta = 3/\sqrt{5}$, $\csc\theta = \frac{3}{2}$
21. $\sin\theta = 7/\sqrt{53}$, $\cos\theta = 2/\sqrt{53}$, $\tan\theta = \frac{7}{2}$, $\cot\theta = \frac{2}{7}$, $\sec\theta = \sqrt{53}/2$, $\csc\theta = \sqrt{53}/7$

23. $\sin\theta = 1/\sqrt{5}$, $\cos\theta = 2/\sqrt{5}$, $\tan\theta = \frac{1}{2}$, $\cot\theta = 2$, $\sec\theta = \sqrt{5}/2$, $\csc\theta = \sqrt{5}$

29. $\sin\theta = x$, $\cos\theta = \sqrt{1-x^2}$, $\tan\theta = \dfrac{x}{\sqrt{1-x^2}}$, $\cot\theta = \dfrac{\sqrt{1-x^2}}{x}$, $\sec\theta = \dfrac{1}{\sqrt{1-x^2}}$, $\csc\theta = 1/x$

31. $\sin\theta = 4/x$, $\cos\theta = \dfrac{\sqrt{x^2-16}}{x}$, $\tan\theta = \dfrac{4}{\sqrt{x^2-16}}$, $\cot\theta = \dfrac{\sqrt{x^2-16}}{4}$, $\sec\theta = \dfrac{x}{\sqrt{x^2-16}}$, $\csc\theta = x/4$

33. $\sin\theta = \dfrac{3}{\sqrt{4x^2+9}}$, $\cos\theta = \dfrac{2x}{\sqrt{4x^2+9}}$, $\tan\theta = 3/2x$, $\cot\theta = 2x/3$, $\sec\theta = \dfrac{\sqrt{4x^2+9}}{2x}$, $\csc\theta = \dfrac{\sqrt{4x^2+9}}{3}$

35. 2.3 mi
37. S35°E   **39.** 118 ft   **41.** 306 m
43. 193 m; the plane is flying at an allowable altitude   **45.** $5\sqrt{2}$

47. $A(\theta) = 72\tan\theta$ **49.** $\theta(t) = \tan^{-1}\left(\dfrac{t^2}{15}\right)$

51. $\theta(x) = \tan^{-1}(6/x) - \tan^{-1}(3/x)$
53. $\angle EGB \approx 42°$, $\angle DGB \approx 56°$, $\angle HGB \approx 68°$

Section 5.2

1. 140°   **3.** 120°   **5.** 300°
7. $3\pi/2$ (4.7)   **9.** $9.8 - 2\pi$ (3.5)
11. $\pi^2 - 2\pi$ (3.6)   **13.** $2\pi/3$ (2.09)

15. $\dfrac{-10\pi}{9}$ (-3.49)   **17.** $5\pi^2/9$ (5.48)

19. 630°   **21.** $540/\pi$ (172°)
23. $180\sqrt{2}/\pi$ (81°)   **25.** 2.9
27. 7.1   **29.** 98°   **31.** $\pi - \theta$
33. $\pi + \theta$ **35.** $\pi/2 + \theta$
37. 10°, 16.0 million miles
39. 4377 mi, 27500 mi,   **41.** 6082
43. 128°   **45.** $A = \frac{1}{2}sr$
47. Quadrant I   **49.** Quadrant I
51. $9\sqrt{3} + 6\pi$

Section 5.3

1. $\sin\theta = \frac{4}{5}$, $\cos\theta = \frac{3}{5}$, $\tan\theta = \frac{4}{3}$

3. $\sin\theta = \frac{4}{5}$, $\cos\theta = -\frac{3}{5}$, $\tan\theta = -\frac{4}{3}$
5. $\sin\theta = -\frac{4}{5}$, $\cos\theta = -\frac{3}{5}$, $\tan\theta = \frac{4}{3}$
7. $\sin\theta = 2/\sqrt{13}$, $\cos\theta = -3/\sqrt{13}$, $\tan\theta = -\frac{2}{3}$
9. $\sin\theta = -3/\sqrt{13}$, $\cos\theta = 2/\sqrt{13}$, $\tan\theta = -\frac{3}{2}$
11. $\sin\theta = -1/2$, $\cos\theta = \sqrt{3}/2$, $\tan\theta = -1/\sqrt{3}$
13. reference angle is 60°; $\sin(-300°) = \sqrt{3}/2$; $\cos(-300°) = \frac{1}{2}$; $\tan(-300°) = \sqrt{3}$
15. reference angle is 30°; $\sin 570° = -\frac{1}{2}$; $\cos 570° = -\sqrt{3}/2$; $\tan 570° = 1/\sqrt{3}$
17. terminal side falls on positive x-axis; $\sin 720° = 0$; $\cos 720° = 1$; $\tan 720° = 0$
19. reference angle is $\pi/3$; $\sin(-2\pi/3) = -\sqrt{3}/2$; $\cos(-2\pi/3) = -\frac{1}{2}$; $\tan(-2\pi/3) = \sqrt{3}$
21. reference angle is $\pi/4$; $\sin(11\pi/4) = \sqrt{2}/2$; $\cos(11\pi/4) = -\sqrt{2}/2$; $\tan(11\pi/4) = -1$
23. terminal side falls on negative x-axis; $\sin 13\pi = 0$; $\cos 13\pi = -1$; $\tan 13\pi = 0$
25. 0.9397   **27.** -0.0374
29. -0.9511 **31.** 0.9074
33. 0.9744   **35.** 0.8534
37. $\sin\theta = -\frac{4}{5}$, $\tan\theta = -\frac{4}{3}$, $\cot\theta = -\frac{3}{4}$, $\sec\theta = \frac{5}{3}$, $\csc\theta = -\frac{5}{4}$
39. $\sin\theta = 2/\sqrt{53}$, $\cos\theta = -7/\sqrt{53}$, $\cot\theta = -\frac{7}{2}$, $\sec\theta = -\sqrt{53}/7$, $\csc\theta = \sqrt{53}/2$,
41. $\sin\theta = -4\sqrt{2}/7$, $\cos\theta = \sqrt{17}/7$, $\tan\theta = -4\sqrt{34}/17$, $\cot\theta = -\sqrt{34}/8$, $\csc\theta = -7/4\sqrt{2}$
45. a) k **b)** $-k$ **c)** $-k$ **d)** k
47. approx. is 0.19866933; error approx. is 2.54×10^{-9} **49.** approximation is $\frac{131}{240}$ (0.545833); value by calculator is 0.546302

Section 5.4

1. a) $(-1, 0)$ **b)** $(-1, 0)$
3. a) $(-1/\sqrt{2}, -1/\sqrt{2})$
b) $(-1/\sqrt{2}, 1/\sqrt{2})$
5. a) $(-1, 0)$ **b)** $(-1, 0)$
7. a) $(-\frac{1}{2}, \sqrt{3}/2)$ **b)** $(-\frac{1}{2}, -\sqrt{3}/2)$
9. a) $(1, 0)$ **b)** $(-1, 0)$
11. a) $(0, 1)$ **b)** $(0, 1)$
13. a) $(-a, -b)$ **b)** (a, b)
c) $(-a, -b)$ **d)** $(-a, b)$
15. a) $(-a, -b)$ **b)** (a, b)
c) $(-a, -b)$ **d)** $(-a, b)$
17. a) $(-a, -b)$ **b)** (a, b)
c) $(-a, -b)$ **d)** $(-a, b)$

19.
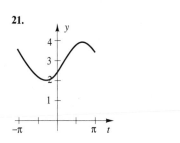

21.
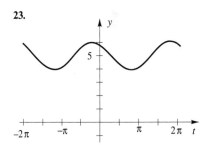

23.
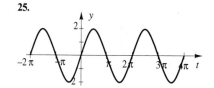

25.
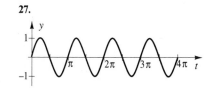

27.
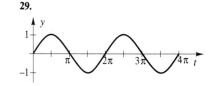

29.

31.

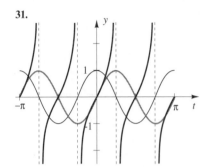

33.

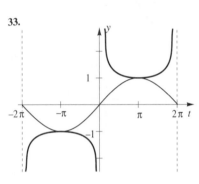

35.

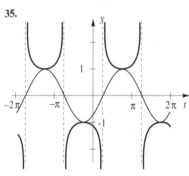

37.

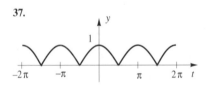

39.

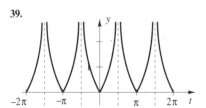

41.

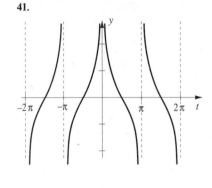

43. $y = -\sin t$　　**45.** $y = -\cos t$
47. $y = \sin t$　　**49.** $y = \cos t$
51. $y = \sin t \cdot$　　**53.** $y = \cot t$
55. $y = \cot t$　　**57.** $y = -\cot t$
59. $(0.2837, 0.9589)$
61. $(0.4121, 0.9111)$　　**67.** $(-\sqrt{2}, -\sqrt{2})$

Section 5.5

1. (h, k): $(0, 0)$
amp: 2
per: $2\pi/3$

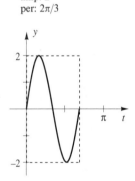

3. (h, k): $(0, 0)$
amp: 1/2
per: π

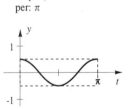

5. (h, k): $(\pi/3, 0)$
amp: 2
per: $2\pi/3$

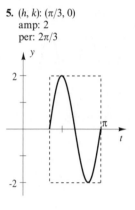

7. (h, k): $(0, 0)$
amp: 5/2
per: 2

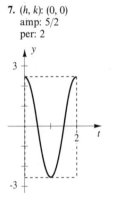

9. (h, k): $(\pi/4, 0)$
amp: 1
per: π

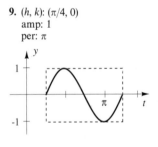

11. (h, k): $(3\pi/2, 3)$
amp: 3
per: 2π

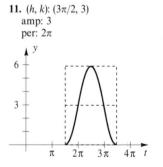

13.

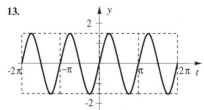

(*h*, *k*): (0, 0); amp: 3/2; per: π

15.

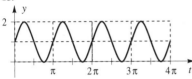

(*h*, *k*): (0, 1); amp: 1; per: π

17.

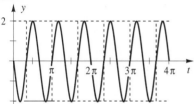

(*h*, *k*): (π/3, 0); amp: 2; per: 2π/3

19.

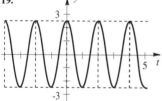

(*h*, *k*): (0, 0); amp: 5/2; per: 2

21.

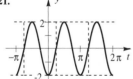

(*h*, *k*): (π/4, 0); amp: 2; per: π

23.

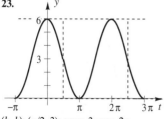

(*h*, *k*): (π/2, 3); amp: 3; per: 2π

25. XI, VI   **27.** V, IX   **29.** I, VIII

31.

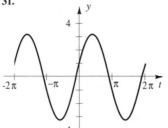

33.

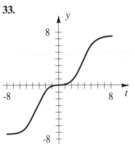

35.

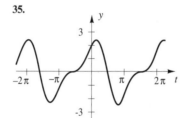

37.

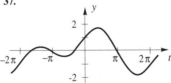

39.

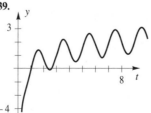

41.

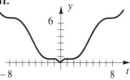

43.

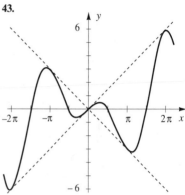

45.

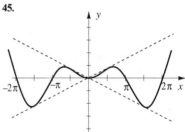

47.

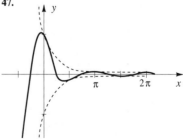

49.

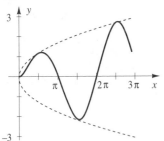

Miscellaneous Exercises for Chapter 5

1. $\sin \theta = \frac{1}{3}$, $\cos \theta = 2\sqrt{2}/3$,
$\tan \theta = 1/2\sqrt{2}$, $\cot \theta = 2\sqrt{2}$,
$\sec \theta = 3/2\sqrt{2}$, $\csc \theta = 3$
3. $\sin \theta = 2/\sqrt{5}$, $\cos \theta = 1/\sqrt{5}$,
$\tan \theta = 2$, $\cot \theta = \frac{1}{2}$, $\sec \theta = \sqrt{5}$,
$\csc \theta = \sqrt{5}/2$
5. $\sin \theta = 3/\sqrt{13}$, $\cos \theta = 2/\sqrt{13}$,
$\tan \theta = \frac{3}{2}$, $\cot \theta = \frac{2}{3}$, $\sec \theta = \sqrt{13}/2$,
$\csc \theta = \sqrt{13}/3$ **7.** $19°$ **9.** $63°$
11. a) $-\sqrt{3}$ **b)** $\sqrt{3}$ **c)** 1
d) undefined
13. a) $-1/\sqrt{2}$ **b)** $\frac{1}{2}$ **c)** $-\frac{1}{2}$
d) 0
15. a) 1 **b)** $-2/\sqrt{3}$ **c)** $2/\sqrt{3}$
d) -2
17. a) $(1/\sqrt{2}, -1/\sqrt{2})$ **b)** $(\frac{1}{2}, -\sqrt{3}/2)$
c) $(\sqrt{3}/2, -\frac{1}{2})$ **d)** $(\frac{1}{2}, \sqrt{3}/2)$
19.

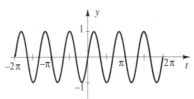

(h, k): $(0, 0)$; amp: 1; per: $2\pi/3$

51.

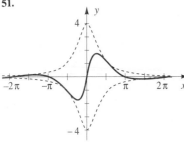

21.

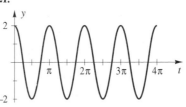

(h, k): $(0, 0)$; amp: 2; per: π

23.

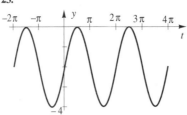

(h, k): $(0, -2)$; amp: 2; per: 2π

53.

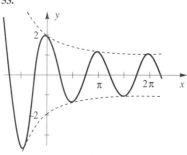

55.

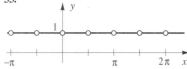

57. $\pi/2$

59. $y = 8 \sin\left[\dfrac{\pi}{8}(x - 3)\right] + 12$

61. Warmest day is 192 (July 12); coldest day is 9 (Jan. 10).   **63.** 10.3 hours

25.

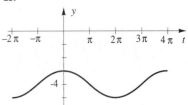

(h, k): $(0, -4)$; amp: 1; per: 4π

27.

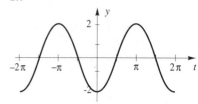

(h, k): $(\pi/2, 0)$; amp: 2; per: 2π

29.

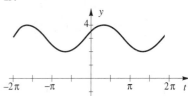

(h, k): $(\pi/3, 3)$; amp: 1; per: 2π

31.

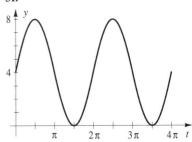

(h, k): $(-\pi, 4)$; amp: 4; per: 2π

33.

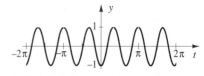

(h, k): $(-\pi/3, 0)$; amp: 1; per: $2\pi/3$

35.

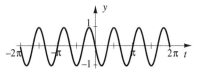

(h, k): $(-\pi, 0)$; amp: 1; per: $2\pi/3$

37.

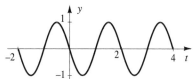

(h, k): $(-3, 0)$; amp: 2; per: 2

39.

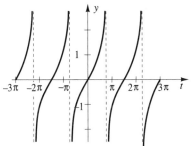

(h, k): $(0, 0)$; per: $3\pi/2$

41.

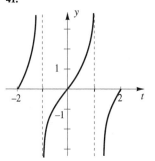

(h, k): $(0, 0)$; per: 2

43.

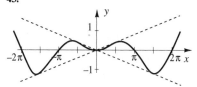

45.

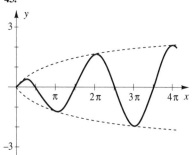

47.

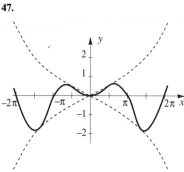

49.

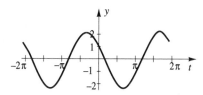

51.

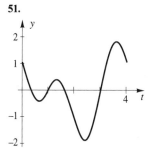

53.

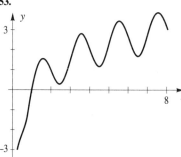

55.

57.

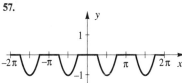

59.

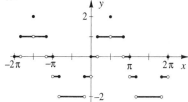

61. 22.6 in. **63.** 101° **65.** 44 ft
67. 12.4 mi. **69.** $\sin \theta = -1/\sqrt{10}$,
$\tan \theta = -\frac{1}{3}$, $\cot \theta = -3$, $\sec \theta = \sqrt{10}/3$,
$\csc \theta = -\sqrt{10}$ **71.** Quadrant IV
73. a) $12\sqrt{2}$
b) $84\pi - 72 - 72\sqrt{3}$ (67.2)
75. $14 + 2 \tan \theta + 2 \sec \theta$ **77.** 92π

C H A P T E R   6

Section 6.1

1. $\sin \theta$ **3.** -1 **5.** 1 **7.** $\sec^2\theta$
9. $-\cos^2 x \sin x$ **11.** $\tan \alpha + \sin \alpha$

13. $\sec^2\alpha - \tan \alpha$ **15.** $\dfrac{2\cos x}{\sin^2 x}$

17. $4|\sec x|$
49. Try $\theta = \pi/6 : \sqrt{3}/2 \neq \frac{3}{2}$
51. Try $\theta = 0 : 1 \neq \pi/4$
53. Try $\theta = \pi/4 : 2 \neq \sqrt{2}$
55. Let $\theta = \pi/6 : 1 \neq \sqrt{3}/2$
57. Try $\theta = 0 : 0 \neq 1$
59. Let $\theta = 3\pi/2 : 1 \neq -1$

73. $\sin \theta = \sqrt{1 - x^2}$, $\tan \theta = \dfrac{\sqrt{1 - x^2}}{x}$,

$\sec \theta = \dfrac{1}{x}$, $\csc \theta = \dfrac{1}{\sqrt{1 - x^2}}$,

$\cot \theta = \dfrac{x}{\sqrt{1 - x^2}}$

75. $\sin \theta = \dfrac{1}{\sqrt{1 + x^2}}$, $\cos \theta = \dfrac{x}{\sqrt{1 + x^2}}$,

$\csc \theta = \sqrt{1 + x^2}$, $\sec \theta = \dfrac{\sqrt{1 + x^2}}{x}$,

$\tan \theta = \dfrac{1}{x}$ **77.** $\sin \theta = \dfrac{-\sqrt{x^2 - 1}}{x}$,

$\tan \theta = -\sqrt{x^2 - 1}$, $\csc \theta = \dfrac{-x}{\sqrt{x^2 - 1}}$,

$\sec \theta = x$, $\cot \theta = \dfrac{-1}{\sqrt{x^2 - 1}}$

79. $\dfrac{\sec \theta}{a}$ **81.** $\cot^4\theta$

83. $\left(\dfrac{\sec \theta}{a}\right)^3$ **85.** $a \sec \theta$

87. $\sin^2\theta$ **89.** $a^5\sec^5\theta$ **91.** $a \tan \theta$

93. $\dfrac{\cos \theta \cot \theta}{a^2}$ **95.** $(a \tan \theta)^3$

Section 6.2

1. a) $\dfrac{\sqrt{6} - \sqrt{2}}{4}$ **b)** $\sqrt{6} + \sqrt{2}$

3. a) $\dfrac{\sqrt{6} - \sqrt{2}}{4}$ **b)** $\sqrt{6} + \sqrt{2}$

5. a) $-2 - \sqrt{3}$ **b)** $-2 + \sqrt{3}$
7. a) $\dfrac{-\sqrt{6} + \sqrt{2}}{4}$ **b)** $-\dfrac{\sqrt{3} + \sqrt{2}}{2}$

9. a) $\frac{1}{2}$ **b)** $\sqrt{3}/2$
11. a) $\sqrt{2}/2$ **b)** 1 **19.** $\sqrt{3} \sin \theta$
21. $-\frac{3}{2}\cos \theta + (\sqrt{3}/2) \sin \theta$
23. $\sqrt{2}\cos \theta + (1 - \sqrt{2}) \sin \theta$

37. a) $\dfrac{10 + 2\sqrt{30}}{21}$ **b)** $\dfrac{10 - 2\sqrt{30}}{21}$

c) $\dfrac{5\sqrt{5} - 4\sqrt{6}}{21}$ **d)** $\dfrac{5\sqrt{5} + 4\sqrt{6}}{21}$

39. a) $\frac{3}{5}$ **b)** $\frac{44}{125}$ **c)** $\frac{117}{44}$
d) $-\frac{3}{4}$

41. a) $-\frac{4}{3}$ **b)** $\dfrac{\sqrt{5} - 2\sqrt{15}}{10}$

c) $\dfrac{3\sqrt{10}}{10}$ **d)** $\frac{1}{3}$

47. $\sqrt{2}\cos(t - \pi/4)$
49. $2\cos(t - \pi/6)$
51. $\sqrt{2}\cos(5t - 3\pi/4)$

Section 6.3

1. a) $\frac{1}{2}\sqrt{2 - \sqrt{3}}$ **b)** $\dfrac{2}{\sqrt{2 - \sqrt{3}}}$

3. a) $-\frac{1}{2}\sqrt{2 + \sqrt{2}}$ **b)** $\dfrac{-2}{\sqrt{2 + \sqrt{2}}}$

5. a) $1 + \sqrt{2}$ **b)** $\sqrt{2} - 1$
7. a) $\frac{1}{2}\sqrt{2 - \sqrt{2}}$ **b)** $-\sqrt{2}/4$
9. a) $-\sqrt{2} - 1$ **b)** $\frac{1}{2}$
11. $\frac{1}{2}(\cos 2t + \cos 10t)$
13. $\frac{1}{2}[\sin \frac{2}{3}\alpha + \sin 2\alpha]$
15. $\frac{1}{2}(\cos 2x + \cos(4x + \pi/2)$, or
$\frac{1}{2}(\cos 2x - \sin 4x)$
17. $2\cos(5x/2)\cos(x/2)$
19. $-2\sin(3\theta/2)\sin(\theta/2)$ **21.** 0
35. a) $\frac{4}{5}$ **b)** $\frac{3}{5}$ **c)** $\frac{4}{3}$
37. a) $-\frac{1}{10}\sqrt{50 - 20\sqrt{5}}$ **b)** $\frac{3}{5}$
c) $\dfrac{-10}{\sqrt{50 - 20\sqrt{5}}}$
39. a) $\frac{4}{5}$ **b)** $-\frac{3}{5}$ **c)** $-\frac{3}{4}$

41. $\frac{1}{2}(\cos t + \cos 3t)$

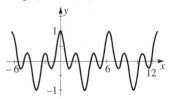

43. $\frac{1}{2}[\cos(t/2) - \cos(3t/2)]$

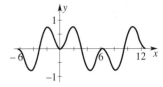

45. $\frac{1}{2}[\sin t - \sin(t/2)]$

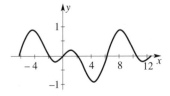

47. $\frac{5}{16} - \frac{1}{32}\cos 6\theta + \frac{3}{16}\cos 4\theta$
$- \frac{15}{32}\cos 2\theta$
49. $\frac{1}{4}\cos 3\theta + \frac{3}{4}\cos \theta$
51. $\frac{1}{16}\cos 5\theta + \frac{5}{16}\cos 3\theta + \frac{5}{8}\cos \theta$

57. $\dfrac{u^2 + 1}{(u - 1)^2}$ **59.** $\dfrac{1 - 2u - u^2}{2u}$

61. $\dfrac{2u(1 + u^2)}{u^4 + 4u^2 - 1}$

63.

a)

b)

Section 6.4

1. a) $\theta = \pi/6 \pm 2\pi n$ or $\theta = 5\pi/6 \pm 2\pi n$
b) $\pi/6$

3. a) $\theta = \pi/6 \pm \pi n$ **b)** $\pi/6$
5. a) $\theta = 3\pi/4 \pm 2\pi n$ or $\theta = 5\pi/4 \pm 2\pi n$
b) $3\pi/4$ **7.** 0   **9.** $5\pi/6$
11. $\pi/3$ **13.** $\pi/4$ **15.** meaningless,
$\sqrt{3}$ is not in the domain   **17.** $-\pi/2$
19. 0.24   **21.** 2.17   **23.** 1.13
25. 1.57   **27.** meaningless
29. 0.88
31. a) $\pi/5$ **b)** $\pi/5$
33. a) $2\pi/9$ **b)** $\pi - 2$
35. a) $\frac{4}{5}$ **b)** $\frac{12}{13}$
37. a) $-\frac{7}{2}$ **b)** $\frac{14}{9}$
39. a) $\frac{12}{5}$ **b)** $\frac{15}{8}$
41. $\dfrac{2\sqrt{2} + 3\sqrt{10}}{14}$ **43.** $\frac{336}{625}$
45. $\sqrt{70}/14$
47.

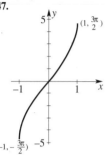

49.

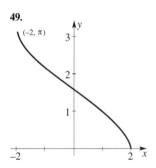

51.

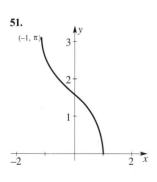

53.

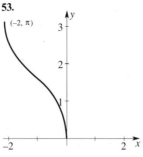

55.

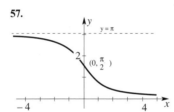

57.

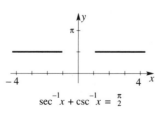

59. $\sqrt{1 - u}$ **61.** $3/2n$
63. $\dfrac{8t}{\sqrt{1 - 64t^2}}$
65.

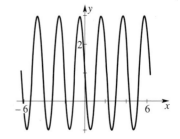

$$\sec^{-1}x + \csc^{-1}x = \frac{\pi}{2}$$

69. a) $\pi/2$ **b)** $-\pi/2$

71.

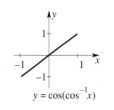

$$y = \cos(\cos^{-1}x)$$

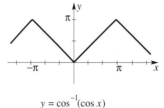

$$y = \cos^{-1}(\cos x)$$

Section 6.5

1. a) $\pi/3 \pm 2n\pi$, $2\pi/3 \pm 2n\pi$
b) $\pi/3$, $2\pi/3$
3. a) $\pi/3$, $5\pi/3$ **b)** $2\pi/3$, $4\pi/3$
5. a) $3\pi/4$, $5\pi/4$ **b)** $\pi/4$, $7\pi/4$
7. $\pi/6$, $11\pi/6$ **9.** $\pi/4$, $3\pi/4$, $5\pi/4$, $7\pi/4$
11. 0, $\pi/4$, π, $5\pi/4$ **13.** 0, $\pi/6$, $5\pi/6$, π
15. $\pi/3$, $2\pi/3$, $4\pi/3$, $5\pi/3$
17. no solution   **19.** 0, $2\pi/3$
21. $2\pi/3$ **23.** $-11\pi/12$, $\pi/12$
25. $\pi/12$, $5\pi/12$, $13\pi/12$, $17\pi/12$
27. $2\pi/9$, $5\pi/9$, $8\pi/9$
29. $-7\pi/8$, $-\pi/8$, $\pi/8$, $7\pi/8$
31. $\pi/6$, $5\pi/6$ **33.** 0, π
35. $\pi/4$, $3\pi/4$, $5\pi/4$, $7\pi/4$ **37.** 1.9, -1.9
39. -1.3, 1.3, -2.3, 2.3
41. $\pi/12$, $5\pi/12$ **43.** 0, 0.5
45. $\pi/2$, $7\pi/6$ **47.** $-\pi/2$, $-3\pi/8$,
$-\pi/8$, $\pi/8$, $3\pi/8$, $\pi/2$
49. $\pi/4$ **51.** $-35\pi/18$, $-31\pi/18$,
$-23\pi/18$, $-19\pi/18$, $-11\pi/18$,
$-7\pi/18$, $\pi/18$, $5\pi/18$, $13\pi/18$,
$17\pi/18$, $25\pi/18$, $29\pi/18$

53.

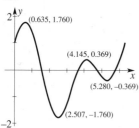

55.

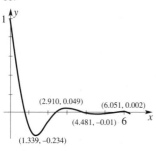

Miscellaneous Exercises for Chapter 6

1. $\sec\theta$ **3.** $\cot\theta$ **5.** $\cos\theta$
7. $\sin\theta$ **9.** $\tan\theta$ **11.** $\cos\theta$

49. a) $\dfrac{-5\sqrt{21} + 2\sqrt{39}}{40}$

b) $-\dfrac{5\sqrt{21} + 2\sqrt{39}}{40}$

c) $\dfrac{-10 - 3\sqrt{91}}{40}$ **d)** $\dfrac{-10 + 3\sqrt{91}}{40}$

51. a) $\sqrt{\tfrac{7}{10}}$ **b)** $-\sqrt{\tfrac{7}{10}}$ **c)** $-\tfrac{17}{25}$
d) $-\tfrac{17}{25}$

53. a) $\sqrt{1-x^2}$ **b)** $\sqrt{1-x^2}$

c) $\dfrac{x}{\sqrt{1-x^2}}$ **d)** $-x$

55. $f(t) = 5\cos(t - \tan^{-1}(-\tfrac{4}{3}))$

57. a) $\tfrac{4}{5}$ **b)** $\tfrac{3}{5}$ **c)** $\dfrac{3\sqrt{3} + 4}{5}$
d) $\sqrt{2}/5$

59. a) $-\pi/2$ **b)** π **c)** $\pi/4$
d) $-\pi/4$ **61.** $\tfrac{3}{5}$ **63.** $-\sqrt{3}/2$
65. $\sqrt{5}/5$ **67.** $\pi/3$ **69.** $\pi/3, 2\pi/3$
71. $0, \pi/3, \pi$ **73.** $-\pi/2, \pi/2$
75. $-\pi/3, \pi/3, -1.9, 1.9$ **77.** $0, 2\pi/3$
79. $-5\pi/6, -\pi/6, \pi/6, 5\pi/6$ **81.** 5.4
83. $-2\pi/3, \pi/2, 2\pi/3$ **89.** $\pi/4$

C H A P T E R 7

Section 7.1

1. $\gamma = 87°$, $b = 20.1$, $c = 25.2$
3. $\alpha = 58°$, $b = 17.2$, $c = 5.5$
5. $\beta = 34°$, $\gamma = 76°$, $c = 5.2$
7. Two cases: $c = 16.0$, $\gamma = 119°$, $\beta = 33°$; or $c' = 1.6$, $\gamma' = 5°$, $\beta' = 147°$
9. Triangle not possible
11. Triangle not possible
13. $24\sqrt{2}$ **15.** 37.4 **17.** 38 feet
19. 154 million miles or 26 million miles
21. 51 miles   **23.** $25°$ or $155°$
25. 3.6 miles   **27.** 1.42 miles per minute (85 miles per hour)

Section 7.2

1. $c = 7.1$, $\alpha = 35°$, $\beta = 121°$
3. $a = 34.0$, $\beta = 42°$, $\gamma = 26°$
5. Triangle not possible   **7.** $\alpha = 59°$, $\beta = 89°$, $\gamma = 32°$
9. Triangle not possible
11. Triangle not possible   **13.** $73°$
15. $25°$ **17.** 72.1 square units

19. 1324.1 square units
21. 76.5 square units   **23.** $107°$
25. $46°, 54°$, and $80°$

27. a) $\tfrac{1}{2}\sqrt{2 - \sqrt{2}} \approx 0.38$

b. $4\sqrt{2 - \sqrt{2}} \approx 3.06$, circumference is $\pi \approx 3.14$

29. a) $\sqrt{2}/4$

b) $2\sqrt{2}$, area of circle is π.

31. $\dfrac{n}{2}\sqrt{2 - 2\cos(360°/n)}$;

As $n \to \infty$, $\dfrac{n}{2}\sqrt{2 - 2\cos(360°/n)} \to \pi$

33. No. Three angles do not determine a triangle.   **35. a)** 402 miles

b) $d(t) = \sqrt{545 - 544\cos 75°}$
37. N31°E, 133 miles   **39.** 1.8 mph
41. $\tfrac{30}{49}$ hour (36 minutes 44 seconds)
45. $2\sqrt{129}$

Section 7.3

1. e   **3.** c and e
5. b and d

7.

9.

11.

13.

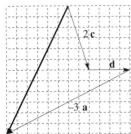

15. $\langle 10, 6 \rangle$ **17.** $\langle -7, 10 \rangle$
19. $\langle 6, 2 \rangle$ **21.** 13   **23.** $6\sqrt{2}$
25. 4   **27.** $\langle -50, 53 \rangle$ **29.** 15
31. $\sqrt{74}$ **33.** $5\sqrt{2} + 2\sqrt{61}$
35. 1   **37.** $\langle \frac{15}{2}, 10 \rangle$
39. $\langle -\frac{2}{3}, -\sqrt{5}/3 \rangle$
41. $\langle 10\sqrt{2}, 10\sqrt{2} \rangle$
43. $\langle 2\sqrt{5}, \sqrt{5} \rangle$ or $\langle -2\sqrt{5}, -\sqrt{5} \rangle$
45. $\langle -5\sqrt{2}, 5\sqrt{2} \rangle$ **47.** $\langle -3, 0 \rangle$
49. $\langle -\frac{3}{2}, -\sqrt{3}/2 \rangle$
51. $\langle 5\cos 70°, 5\sin 70° \rangle \approx \langle 1.7, 4.7 \rangle$
53. $\|\mathbf{v}\| = 2$, $\theta = 300°$ **55.** $\|\mathbf{v}\| = 4$,
$\theta = 36°$ **59.** 38 mph, S83°W
61. 35 pounds   **65.** $8\sqrt{3}$ **67.** 0
69. 45°   **71.** 90°

Section 7.4

1.

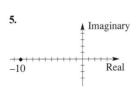

Imaginary

$5 + 3i$

Real

3.

Imaginary

Real

$8 - 4i$

5.

Imaginary

-10 Real

7. $\sqrt{34}$ **9.** $4\sqrt{5}$ **11.** 10

13. a)

Imaginary

Real

b)

Imaginary

Real

15. a)

Imaginary

Real

b)

Imaginary

Real

17. a)

Imaginary

Real

b)

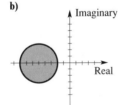

Imaginary

Real

19.

Imaginary

Real

21. $\sqrt{2}(\cos 45° + i \sin 45°) = \sqrt{2}$ cis 45°
23. $39(\cos 292.62° + i \sin 292.62°)$
    $= 39$ cis 292.62°
25. $10(\cos 64.16° + i \sin 64.16°)$
    $= 10$ cis 64.16°
27. $\sqrt{3}(\cos 35.26° + i \sin 35.26°)$
    $= \sqrt{3}$ cis 35.26°
29. $20(\cos 90° + i \sin 90°) = 20$ cis 90°

31. $46.31(\cos 22.94° + i \sin 22.94°)$
    $= 46.31$ cis 22.94°
33. $-4\sqrt{2} + 4\sqrt{2}i$

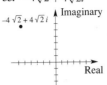

$-4\sqrt{2} + 4\sqrt{2}i$ Imaginary

Real

35. $\dfrac{-5\sqrt{2}}{2} - \dfrac{5\sqrt{2}}{2}i$

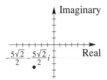

Imaginary

$-\frac{5\sqrt{2}}{2} - \frac{5\sqrt{2}}{2}i$ Real

37. $5 - 5\sqrt{3}i$

Imaginary

Real

$5 - 5\sqrt{3}i$

39. $2 + 2\sqrt{3}i$

Imaginary

$2 + 2\sqrt{3}i$

Real

41. $-29i$

Imaginary

$10i$

-10 10   Real

$-30i$ $-29i$

43. $9.06 + 7.87i$

Imaginary

$9.06 + 7.87i$

Real

45. 15 cis 315° = $15(\cos 315° + i \sin 315°)$
47. $11\sqrt{2}$ cis 240°
    $= 11\sqrt{2}(\cos 240° + i \sin 240°)$

49. $6\sqrt{5}$ cis 243°
$= 6\sqrt{5}(\cos 243° + i \sin 243°)$
51. 3 cis 270° = 3(cos 270° + i sin 270°)
53. $\frac{5}{8}$ cis 18° = $\frac{5}{8}$(cos 18° + i sin 18°)
55. $\frac{5}{11}$ cis 266° = $\frac{5}{11}$(cos 266° + i sin 266°)
57. 64 cis π = 64(cos π + i sin π)
59. cis$3\pi/2$
61. $4\sqrt{2}$ cis 345°
$= (2 + 2\sqrt{3}) + (2 - 2\sqrt{3})i$
63. $2\sqrt{2}$ cis 105°
$= (1 - \sqrt{3}) + (1 + \sqrt{3})i$
65. 64 cis 1800° = 64
67. 1048576 cis 2520° = 1048576
69. Impedance: $z = 16 - 4i$, magnitude is $4\sqrt{17}$. Voltage: $V = 64 - 16i$, magnitude is $16\sqrt{17}$. Phase angle is $\tan^{-1}(-\frac{1}{4}) \approx -14.04°$.

Section 7.5

1. 4096 cis 120° = $-2048 + 2048\sqrt{3}i$
3. 128 cis 210° = $-64\sqrt{3} - 64i$
5. 128 cis 90° = 128i
7. 65536 cis 240° = $-32768 - 32768\sqrt{3}i$
9. $961\sqrt{31}$ cis 74.74° $\approx 1408.16 + 5162i$
11. 1000 cis 249.39° = $-352 - 936i$
13. 3 cis 33°, 3 cis 123°, 3 cis 213°, 3 cis 303°   **15.** 4 cis 26°, 4 cis 146°, 4 cis 266°   **17.** cis 18°, cis 90°, cis 162°, cis 234°, cis 306°   **19.** $\sqrt[6]{2}$ cis 25°, $\sqrt[6]{2}$ cis 85°, $\sqrt[6]{2}$ cis 145°, $\sqrt[6]{2}$ cis 205°, $\sqrt[6]{2}$ cis 265°, $\sqrt[6]{2}$ cis 325°
21. $\sqrt[9]{2}$ cis 35°, $\sqrt[9]{2}$ cis 75°, $\sqrt[9]{2}$ cis 115°, $\sqrt[9]{2}$ cis 155°, $\sqrt[9]{2}$ cis 195°, $\sqrt[9]{2}$ cis 235°, $\sqrt[9]{2}$ cis 275°, $\sqrt[9]{2}$ cis 315°, $\sqrt[9]{2}$ cis 355°
23. $\sqrt[5]{13}$ cis 40.52°, $\sqrt[5]{13}$ cis 112.52°, $\sqrt[5]{13}$ cis 184.52°, $\sqrt[5]{13}$ cis 256.52°, $\sqrt[5]{13}$ cis 328.52°,
25.

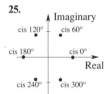

27.

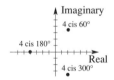

29. 4 cis 34°, 4 cis 154°, 4 cis 274°
31. 8 cis 18°, 8 cis 90°, 8 cis 162°, 8 cis 234°, 8 cis 306°
33. $2\sqrt[4]{2}$ cis 7.5°, $2\sqrt[4]{2}$ cis 97.5°, $2\sqrt[4]{2}$ cis 187.5°, $2\sqrt[4]{2}$ cis 277.5°
35. 2 cis 45°, 2 cis 135°, 2 cis 225°, 2 cis 315°   **37.** 2 cis 15°, 2 cis 75°, 2 cis 135°, 2 cis 195°, 2 cis 255°, 2 cis 315°
39. $\sqrt{2}$ cis 45°, $\sqrt{2}$ cis 165°, $\sqrt{2}$ cis 285°   **41.** cis 90°, cis 180°, cis 270°   **43.** cis$(\frac{360}{7})$°, cis$(\frac{720}{7})$°, cis$(\frac{1080}{7})$°, cis$(\frac{1440}{7})$°, cis$(\frac{1800}{7})$°, cis $(\frac{2160}{7})$°
45. $\sqrt{2}e^{i\pi/4}$ **47.** $39e^{i\theta}$, where $\theta = 2\pi - \tan^{-1}(12/5)$. An approximation is $39e^{i(5.107)}$ **49.** $10e^{i\theta}$, where $\theta = \tan^{-1}(9/\sqrt{19})$. An approximation is $\sqrt{19} + 9i \approx 10e^{i(1.120)}$
51. $\sqrt{3}e^{i\theta}$, where $\theta = \tan^{-1}(1/\sqrt{2})$. An approximation is $\sqrt{2} + i \approx \sqrt{3}e^{i(0.615)}$.
53. $20e^{i\pi/2}$
55. a) $\cos^3\theta - 3\cos\theta\sin^2\theta + i(3\cos^2\theta\sin\theta - \sin^3\theta)$
b) $\cos 3\theta = \cos^3\theta - 3\cos\theta\sin^2\theta$, $\sin 3\theta = 3\cos^2\theta\sin\theta - \sin^3\theta$

Miscellaneous Exercises for Chapter 7

1. $a = 21.5$, $\beta = 33°$, $\gamma = 42°$
3. $\alpha = 60°$, $\beta = 28°$, $\gamma = 92°$
5. $a = 7.7$, $\beta = 75°$, $c = 12.6$
7. $a = 20.4$, $\gamma = 78°$, $c = 22.2$
9. Triangle not possible
11. Ambiguous case: $c = 33.0$, $\beta = 55°$, $\gamma = 98°$; $c' = 15.7$, $\beta' = 125°$, $\gamma' = 28°$
13. $z = \sqrt{x^2 + y^2 - 2xy\cos\theta}$
15. $\phi = \sin^{-1}\left(\dfrac{x\sin\psi}{y}\right)$
17. $z = \dfrac{x\sin\theta}{\sin(\theta + \psi)}$ **19.** 198 feet
21. 700 miles   **23.** 12.7 miles
25. 815 miles   **27.** 31 miles
29. 63.7 feet   **31.** 10.8 square units
33. 47.2 square units   **35.** $\frac{1}{2}t^2$

37. $\tan\theta = \dfrac{4 + \sqrt{3}}{13}$
39. a) (1, tan h)
b) triangle $AOP = \frac{1}{2}\sin h$, sector $AOP = \frac{1}{2}h$, triangle $OAQ = \frac{1}{2}\tan h$
41. $\langle -13, 4\rangle$, $\|\mathbf{v}\| = \sqrt{185}$
43. $\langle -10, 18\rangle$ **45.** $\|2\mathbf{u}\| = 26 = 2\|\mathbf{u}\|$
47. $\|\mathbf{u} + \mathbf{v}\| = \sqrt{137}$, $\|\mathbf{u}\| + \|\mathbf{v}\| = 23$
49. $\langle -\frac{12}{13}, \frac{5}{13}\rangle$ **51.** $\left\langle \dfrac{1}{\sqrt{2}}, \dfrac{1}{\sqrt{2}}\right\rangle$
53. $\langle -9\sqrt{2}, -9\sqrt{2}\rangle$
55. $\|\mathbf{v}\| = 8$, $\theta = 150°$
57. $\|\mathbf{v}\| = 4$, $\theta = 75°$
59. N59°W at 272 mph
61. $-3\sqrt{2} - 3\sqrt{2}i$ **63.** $3/2 - \dfrac{\sqrt{3}}{2}i$
65. $-\dfrac{5\sqrt{6} + 5\sqrt{2}}{2} + \dfrac{5\sqrt{6} - 5\sqrt{2}}{2}i$
67. 20(cos 330° + i sin 330°) = 20 cis 330°
69. 20(cos 180° + i sin 180°) = 20 cis 180°
71. 25(cos 16.26° + i sin 16.26°) = 25 cis 16.26°
73. $z_1z_2 = 27$ cis 165° and $z_1/z_2 = 3$ cis 45°
75. $16(\cos 2\pi/3 + i \sin 2\pi/3) = -8 + 8\sqrt{3}i$
77. 1024 cis 180° = -1024
79. 8 cis 90° = 8i **81.** 2 cis 30°, 2 cis 120°, 2 cis 210°, 2 cis 300°
83. 4 cis 90°, 4 cis 210°, 4 cis 330°
85. 4 cis 60°, 4 cis 180°, 4 cis 300°

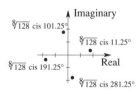

87. $\sqrt[8]{128}$ cis 11.25°, $\sqrt[8]{128}$ cis 101.25°, $\sqrt[8]{128}$ cis 191.25°, $\sqrt[8]{128}$ cis 281.25°

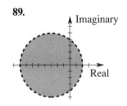

89.

Imaginary

Real

CHAPTER 8

Section 8.1

1. a) $y = \frac{3}{2}x + 8$ **b)** $3x - 2y + 16 = 0$
3. a) not expressible in slope-intercept form   **b)** $x - 2 = 0$
5. a) $y = x + 3$ **b.** $x - y + 3 = 0$
7. a) $y = -\sqrt{3}x + (2\sqrt{3} - 2)$
b) $\sqrt{3}x + y + (2 - 2\sqrt{3}) = 0$
9. $\tan^{-1}(4) \approx 75.96°$
11. $180° + \tan^{-1}(-3) \approx 108.43°$
13. $\tan^{-1}(2) \approx 63.43°$ **15.** $135°$
17. $\tan^{-1}(\frac{1}{5}) \approx 11.31°$
19. $\tan^{-1}(\frac{2}{7}) \approx 15.94°$
21. $\tan^{-1}(\frac{9}{2}) \approx 77.47°$
23. $\tan^{-1}(3) = 71.57°$
25. $90° - \tan^{-1}(\frac{1}{3}) \approx 71.57°$
27. $62°$ **29.** $\frac{23}{5}$ **31.** $13/\sqrt{53}$
33. $6/\sqrt{13}$

35.

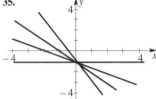

37.

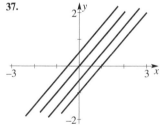

39.

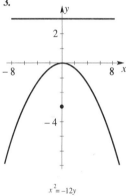

The family of all lines crossing the y-axis at $(0, 3)$ with slopes between 0 and 1.

41.

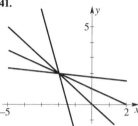

The family of all lines passing through $(-2, 2)$ with negative slope.

43.

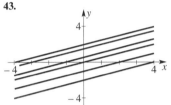

The family of all lines with slope $\frac{1}{2}$.

45.

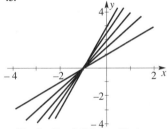

The family of all lines with slopes between 1 and 3, inclusive, that cross the x-axis at $(-1, 0)$.

47. $3/\sqrt{17}$ **49.** $-2 + \sqrt{3}, -2 - \sqrt{3}$
51. $65.22°$ **53.** $m = 2$ or $m = -\frac{2}{29}$
55. $y - 13 = 2(x - 4)$ or $y - 13 = -\frac{2}{29}(x - 4)$

57. $y - 6 = \left(\dfrac{20 + 3\sqrt{95}}{5}\right)(x - 2)$ or

$y - 6 = \left(\dfrac{20 - 3\sqrt{95}}{5}\right)(x - 2)$

59. $4x + 3y = 10$ or $4x + 3y = -10$

Section 8.2

1.

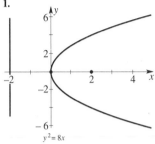

$y^2 = 8x$

3.

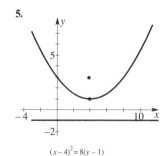

$x^2 = -12y$

5.

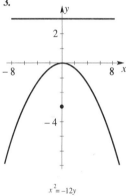

$(x - 4)^2 = 8(y - 1)$

7.

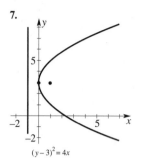

$(y-3)^2 = 4x$

9.

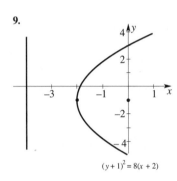

$(y+1)^2 = 8(x+2)$

11.

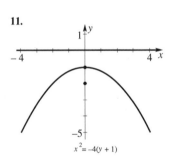

$x^2 = -4(y+1)$

13.

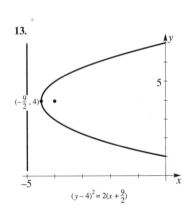

$(y-4)^2 = 2(x+\frac{9}{2})$

15.

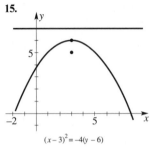

$(x-3)^2 = -4(y-6)$

17. vertex $(0, 0)$; focus $(0, \frac{5}{2})$; directrix $y = -\frac{5}{2}$

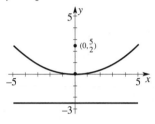

19. vertex $(-1, 5)$; focus $(-1, 6)$; directrix $y = 4$

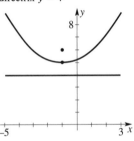

21. vertex $(5, 3)$; focus $(5, 5)$; directrix $y = 1$

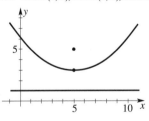

23. vertex $(2, -3)$; focus $(0, -3)$; directrix $x = 4$

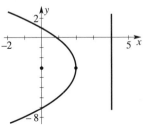

25. vertex $(0, -4)$; focus $(1, -4)$; directrix $x = -1$

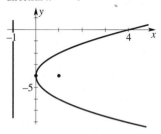

27. vertex $(-1, 2)$; focus $(-1, \frac{1}{2})$; directrix $y = \frac{7}{2}$

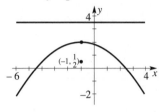

29. vertex $(2, -2)$; focus $(2, -\frac{4}{3})$; directrix $y = -\frac{8}{3}$

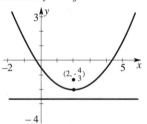

31. $(x-2)^2 = 12(y-1)$
33. $(y-2)^2 = 6(x+\frac{1}{2})$ or $(y-2)^2 = -6(x-\frac{5}{2})$
35. $(x+1)^2 = -8(y-4)$
37. $y^2 = -8(x-2)$ **39.** $y^2 = 8x$
41. $\tan^{-1}(\frac{3}{2}) \approx 56.31°$
43. $y = 4x - 8$ **45.** $\frac{9}{16}$ units
47. $\frac{25}{12}$ units   **49.** $\frac{10935}{32} \approx 342$ feet
51. $x^2 + 2xy + y^2 + 8x - 4y + 10 = 0$

Section 8.3

1.

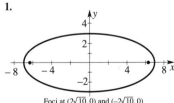

Foci at $(2\sqrt{10}, 0)$ and $(-2\sqrt{10}, 0)$

3.

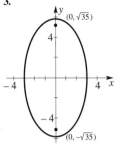

Foci at $(0, \sqrt{26})$ and $(0, -\sqrt{26})$

11.

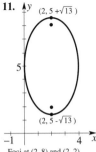

Foci at $(2, 8)$ and $(2, 2)$

23. $\dfrac{(x-1)^2}{4} + \dfrac{(y+2)^2}{3} = 1$

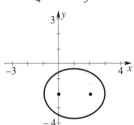

Foci at $(2, -2)$ and $(0, -2)$

13. No real ordered pairs (x, y).

5.

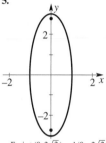

Foci at $(0, 2\sqrt{2})$ and $(0, -2\sqrt{2})$

15.

25. 14   **27.** $2\sqrt{35}$ **29.** 16

31. $\dfrac{(x-2)^2}{16} + \dfrac{(y-1)^2}{7} = 1$

33. $\dfrac{(x-5)^2}{4} + \dfrac{(y-2)^2}{3} = 1$

35. $\dfrac{(x-1)^2}{25} + \dfrac{(y+6)^2}{16} = 1$

17.

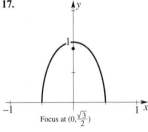

Focus at $(0, \frac{\sqrt{3}}{2})$

7.

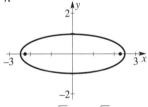

Foci at $(\frac{\sqrt{21}}{2}, 0)$ and $(-\frac{\sqrt{21}}{2}, 0)$

19. $\dfrac{(x+3)^2}{36} + \dfrac{(y-2)^2}{9} = 1$

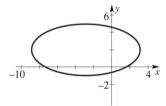

37.

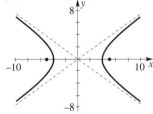

Foci at $(5, 0)$ and $(-5, 0)$

9.

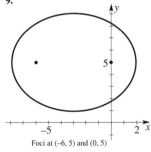

Foci at $(-6, 5)$ and $(0, 5)$

21. $\dfrac{(x-3)^2}{4} + \dfrac{(y-1)^2}{9} = 1$

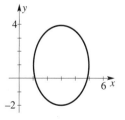

39.

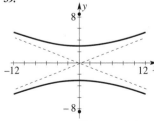

Foci at $(0, \sqrt{73})$ and $(0, -\sqrt{73})$

41.

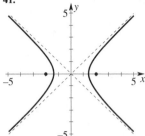

Foci at (2, 0) and (−2, 0)

43.

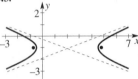

Foci at $(2 + \frac{\sqrt{29}}{2}, -1)$ and $(2 - \frac{\sqrt{29}}{2}, -1)$

45.

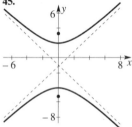

Foci at $(0, -1 + 3\sqrt{2})$ and $(0, -1 - 3\sqrt{2})$

47.

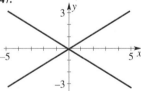

49.

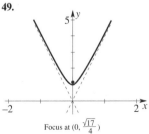

Focus at $(0, \frac{\sqrt{17}}{4})$

51.

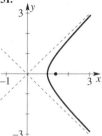

Focus at $(\sqrt{2}, 0)$

53. $\dfrac{x^2}{25} - \dfrac{(y-4)^2}{4} = 1$

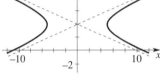

55. $\dfrac{(y+4)^2}{9} - \dfrac{(x+5)^2}{36} = 1$

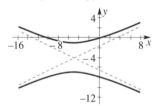

57. $\dfrac{(y+3)^2}{2} - \dfrac{(x-1)^2}{1/2} = 1$

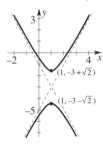

59. ± 1 **61.** $\dfrac{(y+2)^2}{9} - \dfrac{(x-3)^2}{7} = 1$

63. $\dfrac{(x - \frac{5}{2})^2}{4} - \dfrac{4y^2}{9} = 1$

65. $\dfrac{(x+3)^2}{36} - \dfrac{(y-1)^2}{16} = 1$

67.

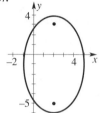

Foci at (2, 3) and (2, −5), eccentricity is $\frac{4}{5}$.

69.

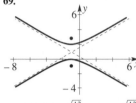

Foci at $(-1, 1 + \frac{\sqrt{13}}{2})$ and $(-1, 1 - \frac{\sqrt{13}}{2})$
eccentricity is $\frac{\sqrt{13}}{2}$.

71.

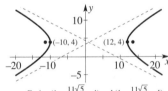

Foci at $(1 + \frac{11\sqrt{5}}{2}, 4)$ and $(1 - \frac{11\sqrt{5}}{2}, 4)$
eccentricity is $\frac{\sqrt{5}}{2}$.

73.

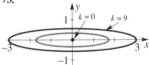

75.

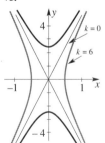

77. $\dfrac{x^2}{4} + \dfrac{y^2}{25} = 1, x \geq 0, y \geq 0$

Section 8.4

1. $x = 49\sqrt{10}/20$, $x = -49\sqrt{10}/20$, eccentricity $= 2\sqrt{10}/7$

3. $x = 8\sqrt{5}/5$, $x = -8\sqrt{5}/5$, eccentricity $= \sqrt{5}/2$

5. $x = -3 + 4\sqrt{3}$, $x = -3 - 4\sqrt{3}$, eccentricity $= \sqrt{3}/2$

7. $y = 1$, eccentricity $= 1$

9. $y = 0$, eccentricity $= 1$

11. $\dfrac{x^2}{10} + \dfrac{y^2}{35} = 1$

13. $\dfrac{x^2}{90} - \dfrac{y^2}{10} = 1$

15. $21x^2 - 4y^2 - 26x - 11 = 0$

17. $15x^2 + 16y^2 - 54x + 39 = 0$

19. $x + y = 2$ and $x + y = -2$

21. $39.23°$

Section 8.5

1. parabola   **3.** hyperbola

5. hyperbola   **7.** $45°$ **9.** $30°$

11. $90° + \frac{1}{2}\tan^{-1}(-\frac{12}{5}) \approx 56°$

13. rotation $= 45°$ $\dfrac{x'^2}{4} + \dfrac{y'^2}{25} = 1$

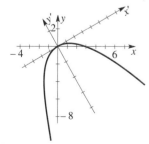

15. rotation $= 30°$ $x'^2 = -4y'$

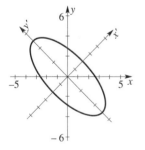

17. rotation $= 30°$ $4x'^2 - y'^2 = 1$

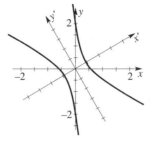

19. rotation $= 53.13°$ $(y' - 2)^2 = x' + 3$

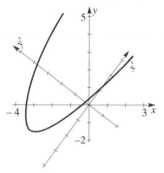

21. $y = \dfrac{-2x}{x - 4}$

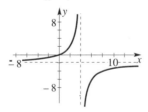

23. $y = \dfrac{(x + 2)(x - 2)}{x + 1}$

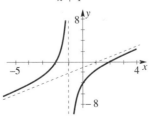

25. $y = 2x \pm \sqrt{4 - x^2}$

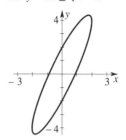

Miscellaneous Exercises for Chapter 8

1. a) $y = \frac{4}{3}x + \frac{41}{3}$

  b) $4x - 3y + 41 = 0$

3. a) $y = \sqrt{3}x - (2\sqrt{3} + 4)$

  b) $\sqrt{3}x - y - (2\sqrt{3} + 4) = 0$ **5.** $30°$

7. $135°$ **9.** $61°$ **11.** $29\sqrt{5}/30$

13. The family of all lines with slope $\frac{5}{2}$

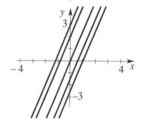

15. $\dfrac{\sqrt{3} + 3}{3\sqrt{3} - 1}$ or $\dfrac{\sqrt{3} - 3}{3\sqrt{3} + 1}$

17. $5x - 12y + 64 = 0$ or $5x - 12y - 66 = 0$

19. vertex $(-1, 3)$; focus $(0, 3)$; directrix $x = -2$

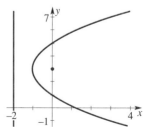

21. vertex $(-4, -1)$; focus $(-4, \frac{3}{2})$; directrix $y = -\frac{7}{2}$

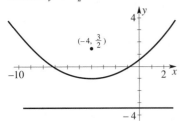

23. $(x - 1)^2 = -4(y - 5)$

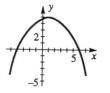

25. $(y - 3)^2 = -4(x - 3)$

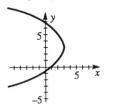

27. $y = 3x - 18$

29.

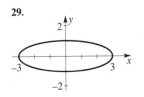

31.

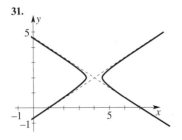

33.

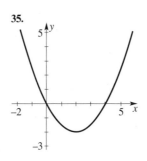

35.

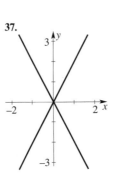

37.

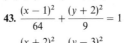

39. eccentricity $= 2\sqrt{2}/3$; foci at $(2\sqrt{2}, 0)$ and $(-2\sqrt{2}, 0)$ **41.** 4

43. $\dfrac{(x - 1)^2}{64} + \dfrac{(y + 2)^2}{9} = 1$

45. $\dfrac{(x + 2)^2}{5} + \dfrac{(y - 3)^2}{9} = 1$

47. $\dfrac{y^2}{9} - \dfrac{(x + 4)^2}{1} = 1$

49. eccentricity $= 2\sqrt{2}/3$; directrices: $x = 9\sqrt{2}/4$ and $x = -9\sqrt{2}/4$

51. $\dfrac{x^2}{108} - \dfrac{y^2}{36} = 1$ **53.** hyperbola

55. ellipse   **57.** 45°   **59.** 67.5°

61.

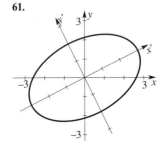

63.

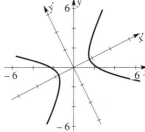

65. $31x^2 - 14xy + 79y^2 + 162x - 114y - 369 = 0$

67. $\tan^{-1} 5 \approx 78.7°$

69.

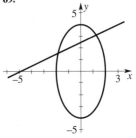

CHAPTER 9

Section 9.1

1.

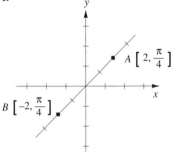

3. *A* and *B* are the same point.

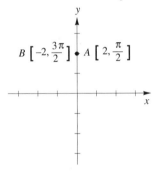

5.

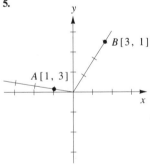

7. $(\sqrt{2}, \sqrt{2})$ **9.** $(-\sqrt{3}, -1)$
11. $(-6, -6\sqrt{3})$ **13.** $[2, 9\pi/4]$
15. $[4, 7\pi/4]$, $[4, 15\pi/4]$
17. $[2, \pi]$, $[2, 3\pi]$, $[2, 5\pi]$
19. $[2\sqrt{2}, -\pi/4]$
21. $[6, -\pi/2]$, $[6, 3\pi/2]$

23. $[2, -7\pi/3]$, $[2, -\pi/3]$,
$[2, 5\pi/3]$, $[2, 11\pi/3]$
25. $r = 6 \cos \theta$
27. $r^2 = \frac{1}{2} \csc 2\theta$
29. $\tan \theta = -1/2$

31.

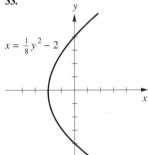

33.

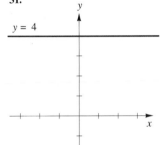

35.

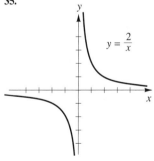

37. $(0.57, -1.92)$ **39.** $[8.25, 1.3]$

41.

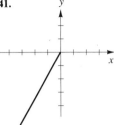

43.

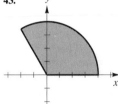

45.

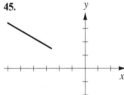

47.

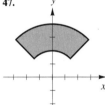

49.

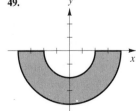

51.

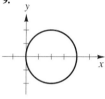

Section 9.2

1.

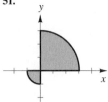

3.

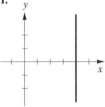

5.

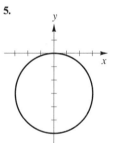

7.

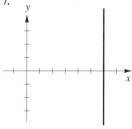

9.

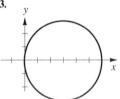

11.

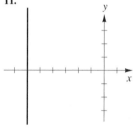

13.

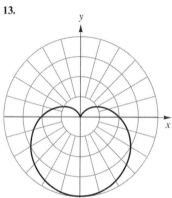

15.

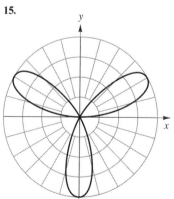

17.

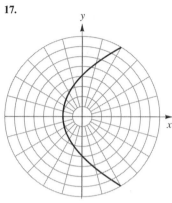

19.

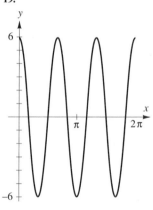

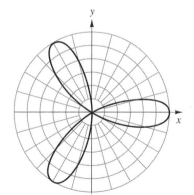

21.

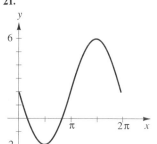 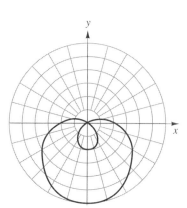

23.

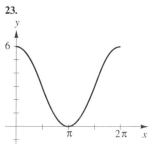 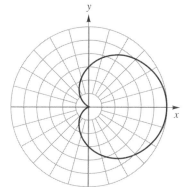

25.

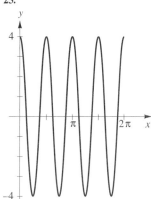 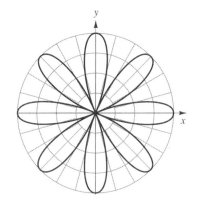

27.

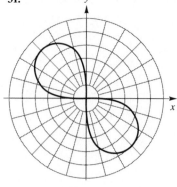

31.

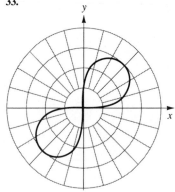

37.

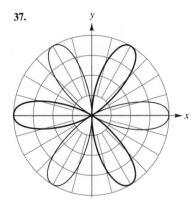

33.

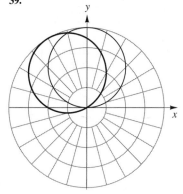

39.

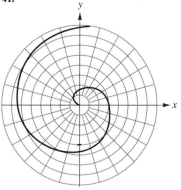

29.

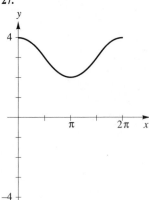

35.

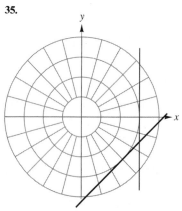

41.

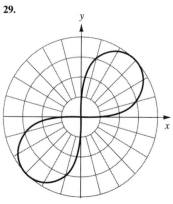

43.

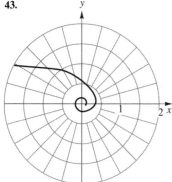

45.

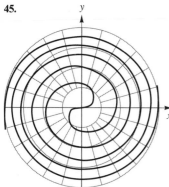

47. III   **49.** IV   **51.** II

Section 9.3

1. $y = 2 - 2x$

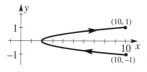

3. $x - 2 = 8y^2$

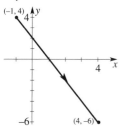

5. $x + 2 = 2y^2$

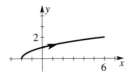

7. $y = \dfrac{x+1}{x}$

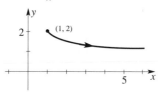

9. $y = x^2$

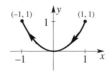

11. $y = 4 - \frac{1}{2}x$

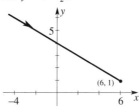

13. $x = 8t + 1$, $y = 2 - 6t$; $0 \le t \le 1$

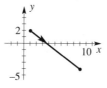

15. $x = t$, $y = t^2 - 4$; $-2 \le t \le 2$

17. $x = 3 \cos t$, $y = 3 \sin t$; $0 \le t \le \pi/2$

19. $x = 4 \cos t$, $y = 3 \sin t$; $0 \le t \le 2\pi$

21. $x = 2 \cos^2 t$, $y = 2 \cos t \sin t$; $0 \le t \le \pi$

23. $x = -t$, $y = \cos t$; $-\pi \le t \le \pi$

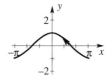

25. $x^2 - y^2 = 1$

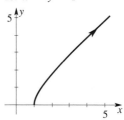

27. $y = 2x^2 - 1$

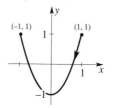

29. $x + y = \pi/2$

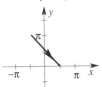

31. $k = 4$, orientation is in the upward direction

33. $x = \frac{1}{2}t$, $y = t + 1$

35.

37. IV   **39.** VI   **41.** V

Section 9.4

1.

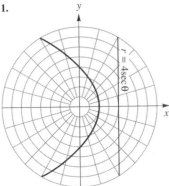

parabola; directrix is $r = 4 \sec \theta$

3.

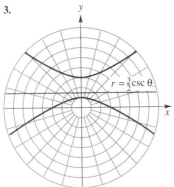

hyperbola; directrix is $r = \frac{3}{2} \csc \theta$

5.

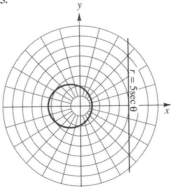

ellipse; directrix is $r = 5 \sec \theta$

7.

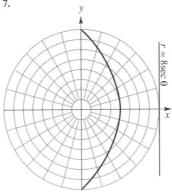

parabola; directrix is $r = 8 \sec \theta$

9.

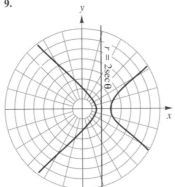

hyperbola; directrix is $r = 2 \sec \theta$

11.

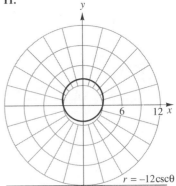

ellipse; directrix is $r = -12 \csc \theta$

13. $\alpha = \pi/4$

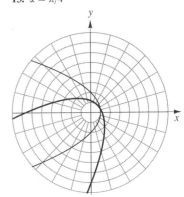

15. $\alpha = -5\pi/4$

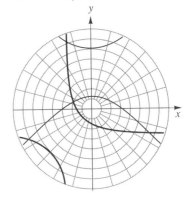

17. $\alpha = \pi/3$

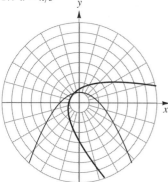

19. $r = \dfrac{2}{1 + \cos \theta}$

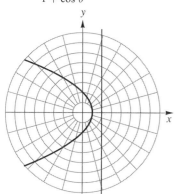

21. $r = \dfrac{1}{1 - \frac{1}{2} \sin \theta}$

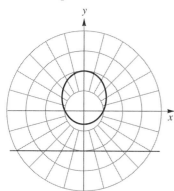

23. $r = \dfrac{4}{1 + \sin \theta}$

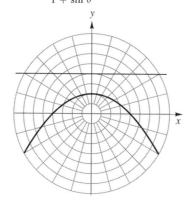

25. $r = \dfrac{2}{1 + \sin \theta}$

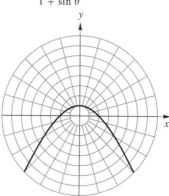

27. $r = \dfrac{3}{1 + \frac{3}{2} \cos \theta}$

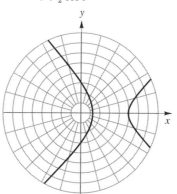

29. $r = \dfrac{\frac{3}{2}}{1 - \frac{1}{2} \sin \theta}$

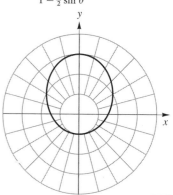

31.

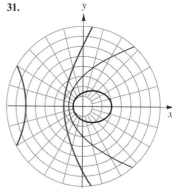

35. perihelion: 0.3075, aphelion: 0.4667
37. perihelion: 0.3743, aphelion: 3.9557
39. $[2, 0]$

Miscellaneous Exercises for Chapter 9

1. $(-4, 4\sqrt{3})$ **3.** $(0, -5)$
5. $(3/\sqrt{2}, 3/\sqrt{2})$ **7.** $[3\sqrt{2}, 3\pi/4]$
9. $[-8, \pi/2]$ **11.** $[-2\sqrt{3}, 5\pi/6]$
13.

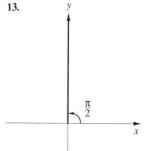

15.

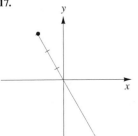

17.

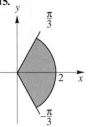

19.

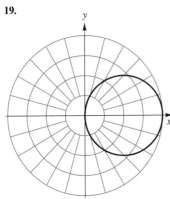

21.

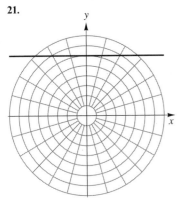

23.

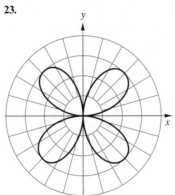

25.

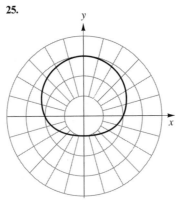

27.

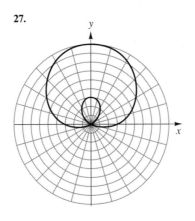

29.

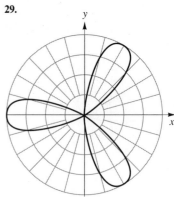

31.

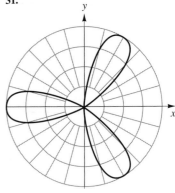

33.

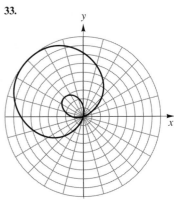

35.

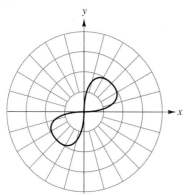

37. $y = \frac{1}{2}x - 4$

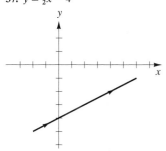

39. $x = y^2 - 4y + 8$

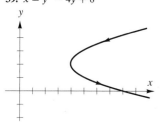

41. $x = (y + 2)^4$

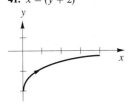

43. $x = \sqrt{1 - y^2}$

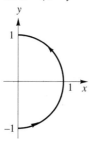

45. $x^2 + y^2 = 1$

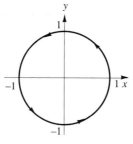

47. $y = x^2$

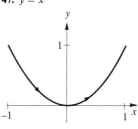

49. $x = 1 - 2y^2$

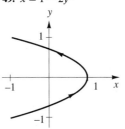

51. $x = y^2$

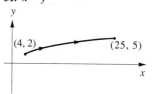

53. $y = 2x$

55. $x = 5t,\ y = 5t \qquad 0 \le t \le 1$
57. $x = t,\ y = 2t^2 - 8 \qquad -2 \le t \le 2$
59. $x = 5\cos t,\ y = 5\sin t \qquad 0 \le t \le 2\pi$

61. parabola; directrix is $r = -4\sec\theta$

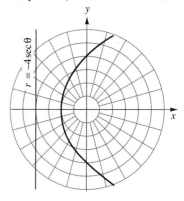

63. hyperbola; directrix is $r = 2\csc\theta$

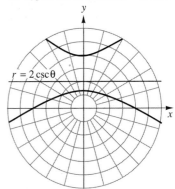

65. hyperbola; directrix is $r = 8 \sec \theta$

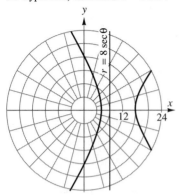

69. hyperbola; directrix is $r = 2 \sec \theta$

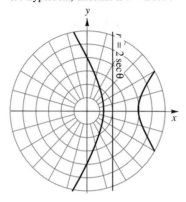

73.

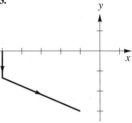

67. parabola; directrix is
$r = -4 \sec(\theta + \pi/4)$

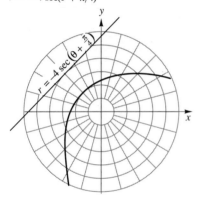

71.

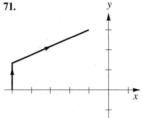

75.

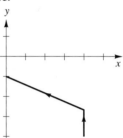

C H A P T E R 10

Section 10.1

1.

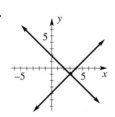

Consistent, solution is $(3, -1)$

3.

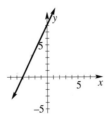

Dependent

5.

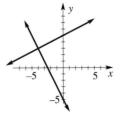

Consistent, solution is $(-4, 3)$

7.

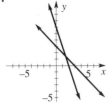

Consistent, solution is $(\frac{3}{2}, \frac{3}{2})$

9.

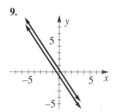

Inconsistent

11.

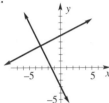

Consistent, solution is $(-\frac{16}{5}, \frac{17}{5})$

13. $x = 4, y = 2, z = 1$
15. $x = 5, y = -8, z = -13$
17. $x = \frac{13}{3}, y = \frac{5}{3}, z = -3$
19. $x = -16, y = 16$
21. $x = \frac{2}{3}, y = \frac{4}{3}$
23. $x = -11, y = 1, z = 5$
25. $(2, 1)$
27. $(\frac{6}{7}, \frac{10}{7})$ **29.** $(21, 12)$
31. $(-3, 1, 5)$ **33.** $(4, 3, 0)$
35. $(-4, 2, -2)$ **37.** $(-3, \frac{1}{2}, 1)$
39. $(\frac{115}{21}, -\frac{215}{42}, \frac{311}{42})$
41. $(4, -2, 3)$ **43.** $(3, 6)$
45. $(12, 5), (-12, 5), (12, -5),$ and
$(-12, -5)$ **47.** $(\frac{1}{2}, \frac{1}{2}, -1)$
49. $x^2 + y^2 + 2x + 2y - 3 = 0$
51. $p(x) = -x^3 + 3x^2 - 3x$
53. 50 pounds of peanuts, 35 pounds of filberts, and 15 pounds of cashews
55. 132

Section 10.2

1. $\begin{bmatrix} 1 & 2 & | & 7 \\ 3 & -4 & | & 12 \end{bmatrix}$

3. $\begin{bmatrix} 2 & -4 & 1 & | & 14 \\ 3 & -1 & & | & 21 \\ 1 & 2 & -1 & | & 14 \end{bmatrix}$

5. $\begin{bmatrix} 2 & 3 & -5 & 4 & | & 34 \\ 6 & -1 & 10 & 0 & | & 21 \\ 0 & 3 & -5 & 0 & | & 3 \\ 0 & 0 & 0 & 9 & | & 12 \end{bmatrix}$

7. $(58, 27)$ **9.** $(10, -\frac{39}{2})$
11. $(2, 3, 1)$ **13.** Row reduced form
15. Row reduced form
17. Row reduced form
19. $(-3, 1, 5)$ **21.** $(4, 3, 0)$
23. $(-4, 2, -2)$ **25.** $(-3, \frac{1}{2}, 1)$
27. $(\frac{115}{21}, -\frac{215}{42}, \frac{311}{42})$ **29.** $(4, -2, 3)$
31. Inconsistent   **33.** $(-3 - 2t, t)$
35. Inconsistent
37. Dependent: $(\frac{12}{5} - \frac{2}{5}t, \frac{14}{5} + \frac{1}{5}t, t)$
39. $(0, 1, 1)$ **41.** $(\frac{2}{11}t, \frac{1}{11}t, t)$
43. $(2 + t, 7 - t, t)$
45. $(-\frac{5}{4} + 2t, \frac{13}{4}, t)$
47. $(1 + \frac{6}{7}t, 2 + \frac{9}{28}t, 3 + \frac{25}{14}t, t)$
49. $(2, 1, 4, 2)$ **51.** $(0, 0, 0, 0)$
53. $(0, -4, 2, 8, 4)$
55. a) The radii of A, B, and C are 4, 6, and 2, respectively   **b)** The radii of A, B, C, and D are $6 - t$, $4 + t$, $5 - t$, and t, respectively, where $t < 5$
c) Not possible with the given distances between centers.
57. $(10 - 7t + 3s, -5 + 5t - 2s, s, t)$

Section 10.3

1. a) 9   **b)** 2
3. a) -16 **b)** -3
5. a) $5\sqrt{3} + 4$
b) $\cos x \cos y - \sin x \sin y = \cos(x + y)$
7. 44   **9.** -3 **11.** -21
13. 50   **15.** -69 **17.** -2
19. 6   **21.** -1428 **23.** $r = 1$ or 6
25. $-3, 0, 2$ **27.** $R(7, 5)$; area is 11
29. $R(2, 5)$; area is 26
31. They are both triangles with legs a and b.
33. 50   **35.** -69
37. They are both 305.
39. $\det A = 227, \det B = 3 \det A = 681$
41. $\frac{3}{2}$

43. $\begin{vmatrix} x & y & 1 \\ -1 & 6 & 1 \\ 4 & 2 & 1 \end{vmatrix} = 0 \Rightarrow 4x + 5y - 26 = 0$

45. $-30(x^2 + y^2) + 120x + 630 = 0 \Rightarrow$
$x^2 + y^2 - 4x - 21 = 0$
47. A 4×4 involves 12 2×2 determinants, a 5×5 involves 60 2×2 determinants. An $n \times n$ involves

$n(n - 1)(n - 2) \ldots (6)(5)(4)(3)$ 2×2 determinants.

Section 10.4

1. $(6, -4)$ **3.** $(\frac{2}{3}, 3)$ **5.** $(\frac{11}{14}, \frac{23}{14})$
7. $(\frac{17}{7}, \frac{13}{7})$
9. inconsistent $(D_x \neq 0$ and $D = 0)$
11. $(4, -3, -2)$ **13.** $(\frac{1}{2}, 3, \frac{1}{4})$
15. $(-\frac{7}{13}, -\frac{14}{13}, -\frac{5}{13})$
17. dependent $(D_x = 0$ and $D = 0)$
19. dependent $(D_x = 0$ and $D = 0)$
21. -18 **23.** 90   **25.** 18

27. $a = \dfrac{x + 8}{x}, b = \dfrac{-3x - 20}{x^2}$

29. $u = \dfrac{2}{e^x}, v = \dfrac{-1}{e^{2x}}$

31. $x' = x \cos \theta + y \sin \theta,$
$y' = -x \sin \theta + y \cos \theta$
37. $(\frac{6}{5}, \frac{17}{5}, \frac{8}{5}, \frac{4}{5})$ **39.** -342
41. Dependent if and only if $d = kc$.

Section 10.5

1. $\dfrac{-2}{x + 5} + \dfrac{2}{x - 1}$ **3.** $\dfrac{4}{x - 1} + \dfrac{-1}{x - 8}$

5. $\dfrac{-2}{x} + \dfrac{2}{x - 2}$ **7.** $\dfrac{3}{x + 1} + \dfrac{2}{(x + 1)^2}$

9. $-\dfrac{1}{x} + \dfrac{\frac{1}{2}}{x - 1} + \dfrac{\frac{1}{2}}{x + 1}$

11. $\dfrac{\frac{1}{16}}{x - 1} + \dfrac{-\frac{1}{16}}{x + 3} + \dfrac{-\frac{1}{4}}{(x + 3)^2}$

13. $\dfrac{4}{x} + \dfrac{-4x}{x^2 + 1}$

15. $\dfrac{-1}{x - 2} + \dfrac{3x - 2}{x^2 + 2x + 2}$

17. $\dfrac{x}{x^2 + 1} + \dfrac{-x}{(x^2 + 1)^2}$ **19.** $\dfrac{2}{2x^2 + 3} + \dfrac{-1}{x^2 + 5}$

21. $\dfrac{3}{x + 2} + \dfrac{2}{x - 3} + \dfrac{-1}{(x - 3)^2}$

23. $\dfrac{\left(-\frac{\sqrt{2}}{4}\right)}{x + \sqrt{2}} + \dfrac{\left(\frac{\sqrt{2}}{4}\right)}{x - \sqrt{2}}$

25. $\dfrac{\frac{9}{4}}{x - 2} + \dfrac{-\frac{9}{4}x + 4}{x^2 + 2x + 4}$

27. $\dfrac{-\frac{1}{32}}{x+2} + \dfrac{\frac{1}{32}}{x-2} + \dfrac{-\frac{1}{8}}{x^2+4}$

29. $x^2 + 9 + \dfrac{-\frac{27}{2}}{x+3} + \dfrac{\frac{27}{2}}{x-3}$

31. $1 + \dfrac{\frac{24}{5}}{2x-1} + \dfrac{\frac{13}{5}x + \frac{9}{5}}{x^2+1}$

33. $\dfrac{-\frac{3}{5}}{x+2} + \dfrac{\frac{3}{5}}{x-3} + \dfrac{-1}{x-2} + \dfrac{1}{x+1}$

35. $\dfrac{-\frac{1}{2}}{\sqrt{2}\sin\theta+1} + \dfrac{\frac{1}{2}}{\sqrt{2}\sin\theta-1}$

37. $\dfrac{1}{\ln x + 2} + \dfrac{1}{\ln x - 2}$

41. $\dfrac{\frac{2}{5}}{x-1} + \dfrac{-\frac{2}{5}x - \frac{2}{5}}{x^2+4}$

43. $\dfrac{\frac{1}{27}}{(x-1)^2} + \dfrac{-\frac{1}{27}}{(x+2)^2} + \dfrac{-\frac{2}{9}}{(x+2)^3}$

45. $\dfrac{3x}{x^2+1} + \dfrac{4}{(x^2+1)^3}$

Section 10.6

1.

(4, 3) and (−3, −4)

3.

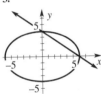

(6, 0) and (0, 4)

5.

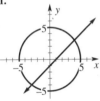

no points of intersection

7.

(2, 2) and (−2, −2)

9.

no points of intersection

11.

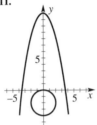

no points of intersection

13.

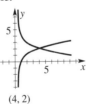

(4, 2)

15.

(1, 2)

17.

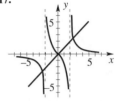

$(\sqrt{5}, \sqrt{5}), (-\sqrt{5}, -\sqrt{5}), (0, 0)$

19. (9, 1)　　**21.** $(\frac{17}{3}, \frac{10}{3})$
23. (1, 1), (9, −3)

25.

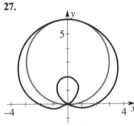

(2, 60°) and (2, 300°)

27.

(6, 90°) and the pole

29.

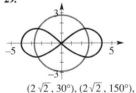

$(2\sqrt{2}, 30°), (2\sqrt{2}, 150°),$
$(2\sqrt{2}, 210°), (2\sqrt{2}, 330°)$

31. $\frac{3}{2}$ by 16 feet

33. Two solutions: 2 by 2 by 5 feet, or
$-1 + \sqrt{41}$ by $-1 + \sqrt{41}$ by
$\dfrac{21 + \sqrt{41}}{40}$ feet

35. 6 in. and 8 in.　　**37.** (2, 3)
39. (3, 18), (1, 0), (−1, −2)

41.

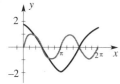

$(\frac{\pi}{3}, \frac{\sqrt{3}}{2}), (\frac{\pi}{2}, 0), (\frac{2\pi}{3}, -\frac{\sqrt{3}}{2}), (\frac{3\pi}{2}, 0)$

43.

Two solutions

45.

Three solutions

47.

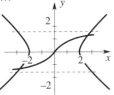

Two solutions

49. a) $\begin{cases} x^2 + 4mx^2 - m^2x^2 = 16 \\ 2x^2 - 7mx^2 + 2m^2x^2 = -4 \end{cases}$

   b) $m = 3$ and $x = \pm 2$, or $m = \frac{3}{7}$ and

     $x = \pm\dfrac{14\sqrt{31}}{31}$

   c) $(2, 6), (-2, -6), (14\sqrt{31}/31, 6\sqrt{31}/31),$
     and $(-14\sqrt{31}/31, -6\sqrt{31}/31)$

Section 10.7

1.

3.

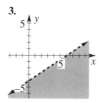

5.

7.

9.

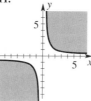

11.

13.

15.

17.

19.

21.

23.

25.

27.

29.

31.

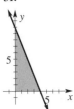

33.

35.

37.

39.

41.

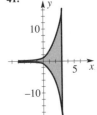

43. a)

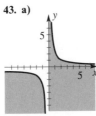

b)

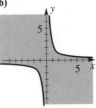

45. a)

b)

47. a)

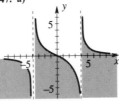

b)

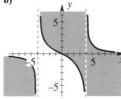

49. Defined for $x^2 - y \geq 0$.

51. Defined for $16 - x^2 - y^2 > 0$

53.

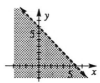

55.

57.

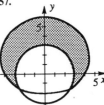

59.

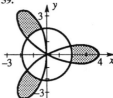

61.

63.

65.

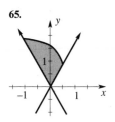

67. $\begin{cases} 0 \le r \le 2 + 4\sin\theta \\ 7\pi/6 \le \theta \le 11\pi/6 \end{cases}$

Miscellaneous Exercises for Chapter 10

1.

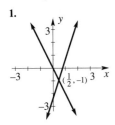

3. inconsistent

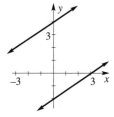

5. dependent: $x = \frac{8}{5} - \frac{4}{5}t$, $y = t$

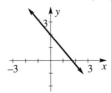

7. $(15, 4)$ **9.** $(-4, -8, 11)$
11. $(7, -3, 2)$ **13.** $(\frac{1}{2} + t, \frac{1}{2}, t)$
15. $(3 - t, 2 + t, t, 1)$
17. I_1, I_2, I_3 are 8, 9, and 1 amperes
19. $f(x) = 3 + 7\ln x$ **21.** $(2 + \frac{3}{2}t, t)$
23. $(\frac{9}{2}, -2, -\frac{1}{2})$ **25.** $(0, 2, -1)$
27. $(3t, t - 2, t)$ **29.** $(-\frac{4}{3} + \frac{5}{9}t, \frac{2}{3} + \frac{2}{9}t, t)$
31. $(-2, 1, 0, 4)$ **33.** $(\frac{1}{5}, -2, 1)$
35. -193 **37.** -1 **39.** -2418
41. -726 **43.** $-4, 5$ **45.** $\frac{7}{2}$
47. $(\frac{1}{2}, -1)$ **49.** $(15, 4)$
51. $(-4, -8, 11)$ **53.** $(7, -3, 2)$
55. inconsistent   **57.** $(\frac{9}{2}, -2, -\frac{1}{2})$
59. $\left(\dfrac{9a + 10}{3a + 2}, \dfrac{-9a - 4}{3a + 2} \right)$
61. $x = \frac{55}{2}$ **63.** $h = 1, k = 1, r = \sqrt{2}$
65. $\dfrac{1}{x + 4} + \dfrac{-2}{x - 1}$
67. $\dfrac{2}{x} + \dfrac{4}{x^2} + \dfrac{2}{2x - 5}$
69. $\dfrac{9}{x + 2} + \dfrac{-5x}{x^2 + x + 2}$
71. $\dfrac{2}{(3x - 5)^2} + \dfrac{x + 3}{x^2 + 1}$
73. $2x - 3 + \dfrac{1}{x} + \dfrac{-5}{4x + 3}$
75. $\dfrac{1}{x - 3} + \dfrac{-1}{x + 3} + \dfrac{-35}{(x + 3)^3}$
77.

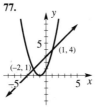

79.

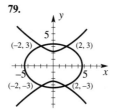

81.

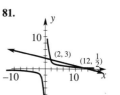

83.

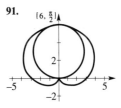

85. $x = 64$ and $y = 4$

87. $(2, 0), (-2, 0), (\sqrt{2}, \sqrt{2}), (-\sqrt{2}, -\sqrt{2})$

89. $\left(\dfrac{7\pi}{8} + k\pi, \dfrac{3\pi}{8} + n\pi \right), \left(\dfrac{3\pi}{8} + k\pi, \dfrac{7\pi}{8} + n\pi \right)$
where k and n are integers.

91.

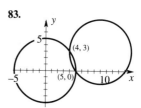

93.

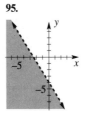

95.

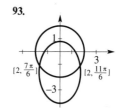

97.

99.

101.

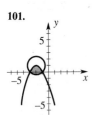

103.

105.

107.

109.

111.

113.

115.

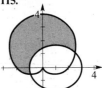

C H A P T E R 11

Section 11.1

1. a) 2, 5, 8, 11, 14

b)

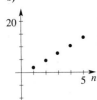

3. a) 0, 5, 8, 17, 24

b)

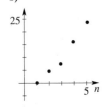

5. a) 2, 0, 2, 0, 2

b)

7. a) 2, $\frac{9}{4}$, $\frac{64}{27}$, $\frac{625}{256}$, $\frac{7776}{3125}$

b)

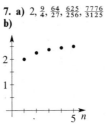

9. a) $\frac{1}{3}$, $\frac{2}{5}$, $\frac{3}{7}$, $\frac{4}{9}$, $\frac{5}{11}$

b)

11. a) 1, $-\frac{1}{3}$, $\frac{1}{5}$, $-\frac{1}{7}$, $\frac{1}{9}$

b)

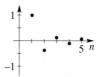

13. a) 2, 4, 8, 16, 32
b)

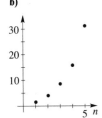

15. a) $\frac{2}{3}, \frac{4}{9}, \frac{8}{27}, \frac{16}{81}, \frac{32}{243}$
b)

17. a) $-\frac{1}{2}, \frac{1}{4}, -\frac{1}{8}, \frac{1}{16}, -\frac{1}{32}$
b)

19. a)

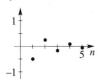

b) The sequence approaches 0.

21. a)

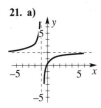

b) The sequence approaches 2.

23. a)

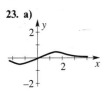

b) The sequence approaches 0.

25. a)

b) The sequence approaches 0.

27. a)

b) The sequence approaches 0.

29. 2, 1, 7, 23, 83   **31.** $-3, 7, 22, 42, 67$
33. $-2, -1, 3, 10, 20$
35. $1, \frac{5}{3}, \frac{13}{6}, \frac{77}{30}, \frac{29}{10}$
37. $1, 1+\frac{x}{2}, 1+\frac{x}{2}+\frac{x^2}{3},$

$$1+\frac{x}{2}+\frac{x^2}{3}+\frac{x^3}{4}, 1+\frac{x}{2}+\frac{x^2}{3}+\frac{x^3}{4}+\frac{x^4}{5}$$

39. $1, 1+2(x-3),$
$1+2(x-3)+3(x-3)^2,$
$1+2(x-3)+3(x-3)^2+4(x-3)^3,$
$1+2(x-3)+3(x-3)^2+4(x-3)^3+5(x-3)^4$
41. 7190   **43.** 12,348,175
45. 25,825,400
47. a) $\frac{21}{40}$ **b)** $\frac{101}{200}$ **c)** $\frac{1}{2}$

49. $\sum_{k=1}^{6} 3k$ **51.** $\sum_{k=1}^{5} (4k-1)$

53. $\sum_{k=1}^{4} \frac{2k}{2k+1}$ **55.** $1+\sum_{k=1}^{5} \frac{(-1)^k}{2k}$

57. $\sum_{k=0}^{n} \frac{x^k}{(2k+2)(2k+4)}$ **61.** $2.236\overline{1}$

63. $\sum_{k=1}^{8} k^2 = 204$
67. $12.969\ldots, 21,$ and $33.988\ldots$

Section 11.2

	a_1	d	a_5	a_n
1.	2	7	30	$7n-5$
3.	6	$\frac{2}{5}$	$7\frac{3}{5}$	$\frac{2n+28}{5}$
5.	$\sqrt{3}-2$	$\sqrt{3}+2$	$5\sqrt{3}+6$	$n(\sqrt{3}+2)-4$
7.	$4y-8$	$-y+4$	8	$n(4-y)+5y-12$
9.	$\frac{x^2-1}{x}$	$\frac{1}{x}$	$\frac{x^2+3}{x}$	$\frac{n+x^2-2}{x}$

	a_1	r	a_5	a_n
11.	28	$\frac{1}{2}$	$\frac{7}{4}$	$28(\frac{1}{2})^{n-1}$
13.	125	$-\frac{4}{5}$	$\frac{256}{5}$	$125(-\frac{4}{5})^{n-1}$
15.	3.2	0.4	0.08192	$3.2(0.4)^{n-1}$
17.	1	$1+\sqrt{2}$	$17+12\sqrt{2}$	$(1+\sqrt{2})^{n-1}$
19.	$4x^6$	$1/x^3$	$4/x^6$	$4x^{9-3n}$

21. arithmetic; $a_{10} = \frac{11}{6}$
23. geometric; $a_9 = \frac{1}{12288}$
25. geometric; $a_{10} = -\frac{3}{512}$
27. arithmetic; $a_{11} = -\frac{69}{8}$ **29.** 187
31. 4921   **33.** 84   **35.** $\frac{870375}{256}$
37. 1368   **39.** $40b+1050$
41. $\dfrac{1-x^{12}}{1-x}$ **43.** 1243   **45.** $\frac{590477}{31104}$
47. 2431   **49.** $\frac{31}{3}$ **51.** $\frac{1}{9}$
53. There are 63 multiples of 3 between 10 and 200; their sum is 6615.
55.

3   3.5   4   4.5   5   5.5   6 x

57. \$27,102.98   **59.** 1600 feet
61. 666
63. $L_n = 16(1-(\frac{7}{9})^n)$

Section 11.3

1. 120   **3.** $-\frac{686}{5}$ **5.** diverges
7. $\frac{2}{3}$ **9.** $\frac{100}{9}$ **11.** diverges
13. $-\frac{3}{8}$ **15.** diverges   **17.** $\frac{5}{9}$
19. $\frac{8}{11}$ **21.** $\frac{2093}{330}$
23. $3+3x+3x^2+3x^3+\ldots; |x|<1$
25. $4+12x+36x^2+108x^3+\ldots;$
$|x|<\frac{1}{3}$
27. $2+x+\dfrac{x^2}{2}+\dfrac{x^3}{4}+\ldots; |x|<2$
29. $S_n \to \frac{2}{3}$ **31.** diverges

33. $S_n \to 3$ **35.** $S_n = 1-\dfrac{1}{n+1}; S_n \to 1$

37. $S_n = 1-\dfrac{1}{2n+1}; S_n \to 1$

39. $S_n = \dfrac{8}{3}-\dfrac{2}{2n+1}-\dfrac{2}{2n+3}; S_n \to \frac{8}{3}$

41. $S_n = \dfrac{3}{2}n^2+\dfrac{5}{2}n;$ diverges.

43. 9 seconds   **53. a)**
45. $72\frac{8}{9}$ feet
47. \$80,000
49. 0.92107
51. 1.64583

b)

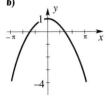

c)

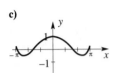

55. $x + 2x^2 + 3x^3 + 4x^4 + 5x^5 + \dots$

57. $\frac{1}{2} + \frac{1}{4} + \frac{1}{8} + \frac{1}{16} + \dots = 1$

59. a) $0, \frac{1}{3}, \frac{1}{9}, \frac{2}{9}, 1$ **b)** answers vary
c) 1   **61.** 1

Section 11.4

1. 10   **3.** 8   **5.** 5005   **7.** 1
9. $x^4 + 4x^3h + 6x^2h^2 + 4xh^3 + h^4$
11. $x^3 + 9x^2 + 27x + 27$
13. $16x^4 + 32x^3y + 24x^2y^2 + 8xy^3 + y^4$
15. $64x^6 + 576x^5y + 2160x^4y^2$
$\qquad\qquad + 4320x^3y^3 + 4860x^2y^4$
$\qquad\qquad + 2916xy^5 + 729y^6$
17. $256x^8 - 1024x^7y + 1792x^6y^2$
$\qquad\quad - 1792x^5y^3 + 1120x^4y^4 - 448x^3y^5$
$\qquad\quad + 112x^2y^6 - 16xy^7 + y^8$
19. $\dfrac{x^7}{128} - \dfrac{21x^6y}{64} + \dfrac{189x^5y^2}{32} - \dfrac{945x^4y^3}{16}$
$\qquad + \dfrac{2835x^3y^4}{8} - \dfrac{5103x^2y^5}{4}$
$\qquad + \dfrac{5103xy^6}{2} - 2187y^7$
21. $p^3 + 6p^2q^2\sqrt{p} + 15p^2q^4 + 20pq^6\sqrt{p}$
$\qquad + 15pq^8 + 6q^{10}\sqrt{p} + q^{12}$
23. $32j^{20} + 240j^{16}\sqrt{k} + 720j^{12}k$
$\qquad + 1080j^8k\sqrt{k} + 810j^4k^2 + 243k^2k$
25. $2187x^7 + 10206x^6y + 20412x^5y^2$
$\qquad + 22680x^4y^3 + 15120x^3y^4$
$\qquad + 6048x^2y^5 + 1344xy^6 + 128y^7$
27. $k = 4$ **29.** $k = 8$
31. $k = n - 5$ **33.** $11^n = (10 + 1)^n$:
11, 121, 1331, 14641, and 161051
35. $165x^3y^8$ **37.** $3360p^3q^4$
39. $366080m^9n^4$

41. $x^{12} + 12x^{10}yz + 60x^8y^2z^2 + 160x^6y^3z^3$
$\qquad + 240x^4y^4z^4 + 192x^2y^5z^5 + 64y^6z^6$
43. $x^4y^{16} - 4\sqrt{3}x^3y^{12} + 18x^2y^8$
$\qquad - 12\sqrt{3}xy^4 + 9$
45. $x^{18} - 6x^{13} + 15x^8 - 20x^3$
$\qquad + \dfrac{15}{x^2} - \dfrac{6}{x^7} + \dfrac{1}{x^{12}}$
47. $5x^4 + 10x^3h + 10x^2h^2 + 5xh^3 + h^4 + 1$
49. $8x^3 + 12x^2h + 8xh^2 + 2h^3 + 2x + h$
51. $1 + \frac{1}{3}x - \frac{1}{9}x^2 + \frac{5}{81}x^3 - \dots$
53. $1 + \frac{1}{2}x + \frac{3}{8}x^2 + \frac{5}{16}x^3 + \dots$
55. $1 - 4x + 10x^2 - 20x^3 + \dots$
57. $1 - \frac{1}{4}x^3 - \frac{3}{32}x^6 - \frac{7}{128}x^9 - \dots$
59. $27(1 + 3x + \frac{3}{2}x^2 - \frac{1}{2}x^3 - \dots)$
61. $(1 + 0.1)^{1/3} \approx 1.032284$

Miscellaneous Exercises for Chapter 11

1. a) 1, 5, 9, 13, 17
b)

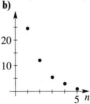

3. a) $-1, \frac{1}{4}, -\frac{1}{9}, \frac{1}{16}, -\frac{1}{25}$
b)

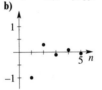

5. a) $24, 12, 6, 3, \frac{3}{2}$
b)

7. a) $1, \frac{1}{2}, \frac{2}{3}, \frac{5}{8}, \frac{19}{30}$
b)

9. a)

b) a_n approaches 0

11. a)

b) a_n approaches $\pi/2$

13. a)

b) a_n does not approach a value as $n \to \infty$.

15. $1, 1 + \dfrac{x}{4}, 1 + \dfrac{x}{4} + \dfrac{x^2}{9},$

$1 + \dfrac{x}{4} + \dfrac{x^2}{9} + \dfrac{x^3}{16}, 1 + \dfrac{x}{4} + \dfrac{x^2}{9} + \dfrac{x^3}{16} + \dfrac{x^4}{25}$

17. 670,880   **19.** Arithmetic: $d = \pi$,
$a_1 = 0, a_6 = 5\pi$, and $a_n = (n - 1)\pi$
21. Geometric: $r = -\frac{4}{3}, a_1 = -18$,
$a_6 = \frac{2048}{27}$, and $a_n = -18(-\frac{4}{3})^{n-1}$
23. Arithmetic: $d = -\frac{1}{3}, a_1 = 4, a_6 = \frac{7}{3}$,
and $a_n = 4 + (n - 1)(-\frac{1}{3})$
25. Geometric: $r = \sqrt{6}, a_1 = \sqrt{3} + 1$,
$a_6 = 108\sqrt{2} + 36\sqrt{6}$, and
$a_n = (\sqrt{3} + 1)(\sqrt{6})^{n-1}$
27. Geometric: $r = -2, a_1 = \ln x$,
$a_6 = \ln(1/x^{32}) = -32\ln x$, and
$a_n = (-2)^{n-1}\ln x$
29. $a_n = n/3 + 1$; first four terms are
$\frac{4}{3}, \frac{5}{3}, 2, \frac{7}{3}$
31. $a_n = 2(-\frac{5}{2})^{n-1}$; first four terms are 2,
$-5, \frac{25}{2}, -\frac{125}{4}$
33. $a_n = 9(\sqrt{3}/3)^{n-1}$ or $9(-\sqrt{3}/3)^{n-1}$;
first four terms are $9, 3\sqrt{3}, 3, \sqrt{3}$ or the
first four terms are $9, -3\sqrt{3}, 3, -\sqrt{3}$.
35. $\frac{39991}{108}$
37. 19,682   **39.** 6,560   **41.** $\frac{1705}{2}$

43. $\dfrac{1 - x^{16}}{1 + x^2}$ **45.** 44   **47.** 1000

49. 80   **51.** 6   **53.** diverges

55. 540   **57.** $\tan^2 x$ **59.** $\frac{107}{495}$

61. $-1 < x < 1$

63. a) $-4 < x < 4$ **b)** $2 < x < 4$

65. $3 - 6x + 12x^2 - 24x^3 + \ldots$;
$-\frac{1}{2} < x < \frac{1}{2}$

67. $S_n = \dfrac{5}{2}\left[\dfrac{1}{2} - \dfrac{1}{2n+2}\right]$ sum is $\frac{5}{4}$.

69. $S_n = \dfrac{2}{3} - \dfrac{2}{5n+3}$, sum is $\frac{2}{3}$.

71. $S_n = n^2$, diverges.

73. $x^{10} + 10x^8 y + 40x^6 y^2 + 80x^4 y^3$
$+ 80x^2 y^4 + 32y^5$

75. $\tan^4 x - 4\tan^3 x + 6\tan^2 x$
$- 4\tan x + 1$

77. $540x^6 y^3$

79. $1 - \frac{1}{4}x - \frac{3}{32}x^2 - \frac{7}{128}x^3 + \ldots$

81. $1 + \frac{1}{3}x + \frac{2}{9}x^2 + \frac{14}{81}x^3 + \ldots$

C H A P T E R   12

Section 12.1

1. Yes   **3.** No   **5.** Yes

7. $2\sqrt{13}$

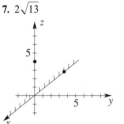

9. $4\sqrt{2}$

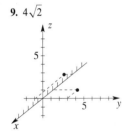

11. 9

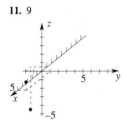

13. 4   **15.** $2\sqrt{5}$ **17.** 7

19.

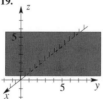

21.

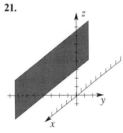

23.

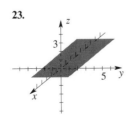

25.

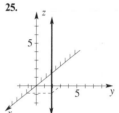

27.

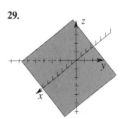

29.

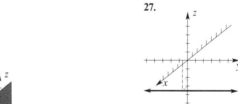

31. Center is $(2, 0, 4)$; radius is 4

33. Center is $(2, 5, -1)$; radius is $4\sqrt{2}$

35. Center is $(-4, 3, -4)$; radius is 6

37. Not a sphere; graph is the point
$(-1, 0, 1)$

39. Center is $(5, 2, 5)$; radius is 5

41. Center is $(5, 4, 3)$; radius is 5

43. Yes   **45.** Equilateral triangle

47. Right triangle

49. $\angle RPQ = 29.7°$; $\angle PQR = 44.5°$;
$\angle QRP = 105.8°$.

Section 12.2

1. III　　　**3.** VI　　　**5.** IV

7.

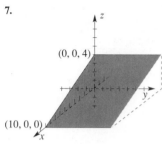

9.

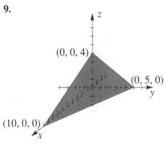

11.

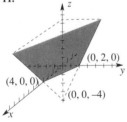

13

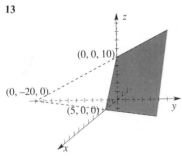

15.

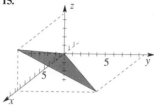

17.

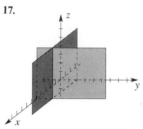

19.

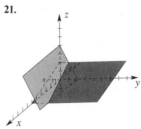

21.

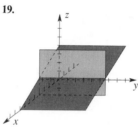

23.

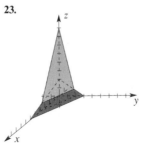

25.

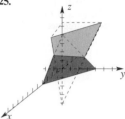

27.

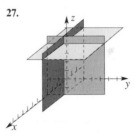

29.

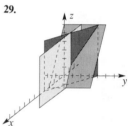

31.

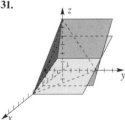

33.

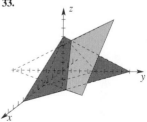

35. $(\frac{18}{5} - 2t, t, \frac{4}{5})$ **37.** $(3, 1, -1)$

39. $(\frac{4}{7}, \frac{10}{7}, \frac{10}{7})$ **41.** $(-2, 2, 6)$

43. $(1, 3, 2)$ **45.** no intersection

47. $(1, 2, 2)$

49. parallel planes

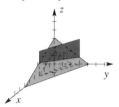

51. $x - 1 = \frac{1}{4}(y - 6) = -\frac{1}{2}(z - 5)$

Section 12.3

1. a) 7   **b)** -62

3. a) 1   **b)** 2

5. a) 0   **b)** undefined

7. a) $-8a^3$ **b)** $-2a^3$

9. a) $3 + 2\log_3 n$ **b)** $3 + 6\log_3 n$

11. a) $\dfrac{\pi}{3}\cos t^2$ **b)** $\dfrac{\pi^2}{9}\cos^2 t$

13.

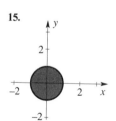

15.

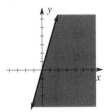

17.

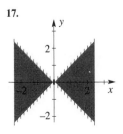

19. II   **21.** III   **23.** IV

25.

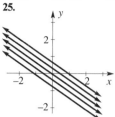

27.

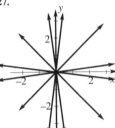

29.

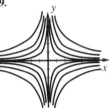

31.

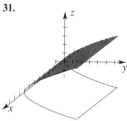

33.

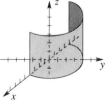

35.

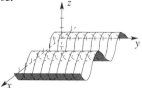

37.

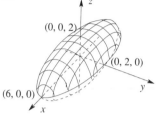

ellipsoid

39.

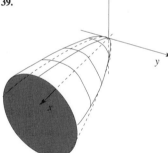

paraboloid

41.

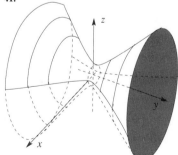

hyperboloid of one sheet

43.

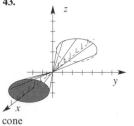

cone

45.

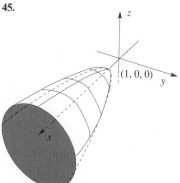

paraboloid

47. $2/y; \dfrac{-2x}{y(y + \Delta y)}$

51. $z = 16y^2 + 1$

Miscellaneous Exercises for Chapter 12

1. $\sqrt{14}$ **3.** 5

5.

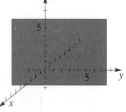

7.

9.

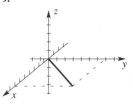

11. center: $(5, -1, 4)$; radius is 7
13. center: $(0, 4, 0)$; radius is 6
15. no sphere possible
17. $(x - 4)^2 + (y - 4)^2 + (z - 7)^2 = 16$
19.

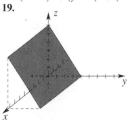

21.

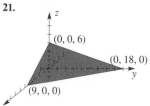

(0, 0, 6)
(0, 18, 0)
(9, 0, 0)

23.

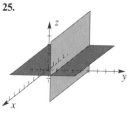

25.

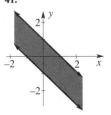

27.

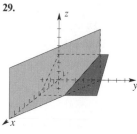

29.

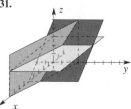

31.

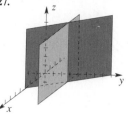

33. $(2, 4, t)$ **35.** $(-6, 3, 15)$
37. $(9, -3, 2)$
39.

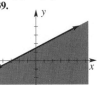

41.

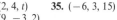

43.

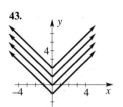

45.

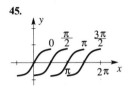

47.

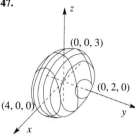

ellipsoid

49.

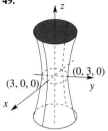

hyperboloid of one sheet

INDEX

GRAPHS

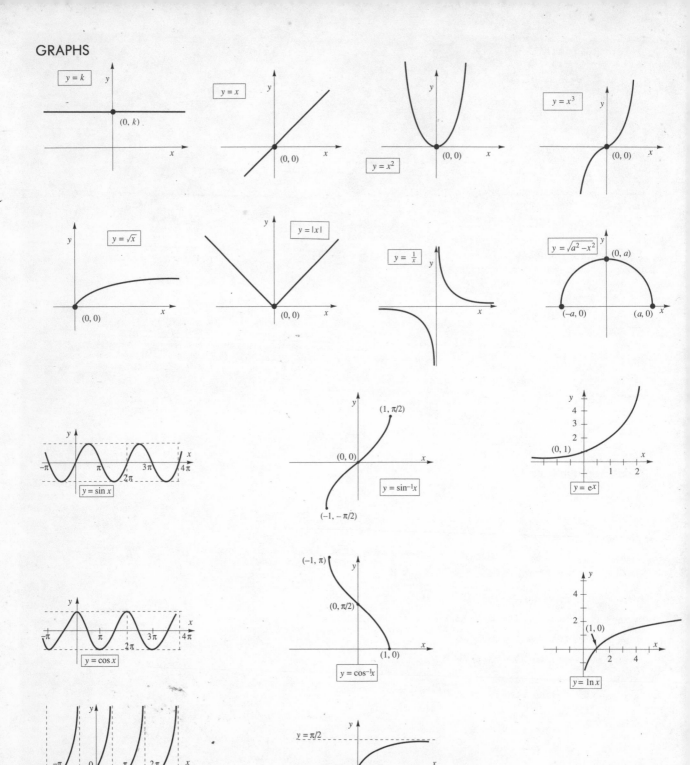